W9-CIE-861

Methods in Enzymology

Volume 97
BIOMEMBRANES
Part K
Membrane Biogenesis:
Assembly and Targeting
(Prokaryotes, Mitochondria,
and Chloroplasts)

METHODS IN ENZYMOLOGY

EDITORS-IN-CHIEF

Sidney P. Colowick   Nathan O. Kaplan

Methods in Enzymology

Volume 97

Biomembranes

Part K

*Membrane Biogenesis: Assembly and Targeting
(Prokaryotes, Mitochondria, and Chloroplasts)*

EDITED BY

*Sidney Fleischer
Becca Fleischer*

DEPARTMENT OF MOLECULAR BIOLOGY
VANDERBILT UNIVERSITY
NASHVILLE, TENNESSEE

Editorial Advisory Board

1983

ACADEMIC PRESS

A Subsidiary of Harcourt Brace Jovanovich, Publishers

New York London
Paris San Diego San Francisco São Paulo Sydney Tokyo Toronto

ACADEMIC PRESS, INC.
111 Fifth Avenue, New York, New York 10003

United Kingdom Edition published by
ACADEMIC PRESS, INC. (LONDON) LTD.
24/28 Oval Road, London NW1   7DX

Library of Congress Cataloging in Publication Data
Main entry under title:

Membrane biogenesis, assembly and targeting
    (Prokaryotes, Mitochondria, and Chloroplasts)

    (Methods in enzymology; v. 97, pt. K)
    Bibliography: p.
    Includes index.
    1. Membranes (Biology) 2. Membrane proteins.
I. Fleischer, Sidney. II. Fleischer, Becca.
III. Series. [DNLM: 1. Cell membrane. 2. Membranes--
Enzymology. W1 ME9615K v. 31, etc. / [QH 601 B6192
1974]]
QP601.M49 vol. 97, pt. K [QH601] 574.19'25s 83-2776
ISBN 0-12-181997-3   [574.87'5]

PRINTED IN THE UNITED STATES OF AMERICA

83 84 85 86   9 8 7 6 5 4 3 2 1

Table of Contents

Section I. Prokaryotic Membranes

A. General Methods

B. Outer Membrane

C. Inner Membrane

Section II. Mitochondria

Section III. Chloroplasts

Section IV. Summary of Membrane Proteins

Addendum

Contributors to Volume 97

Article numbers are in parentheses following the names of contributors.
Affiliations listed are current.

CAROLE ARGAN (38), *Department of Biochemistry, McGill University, Montreal, Quebec H3G 1Y6, Canada*

MONIQUE ARPIN (39), *Unité de Biologie des Membranes, Département de Biologie Moleculaire, Institut Pasteur, 75724 Paris Cedex 15, France*

GIUSEPPE ATTARDI (41), *Division of Biology, California Institute of Technology, Pasadena, California 91125*

JONATHAN BECKWITH (1, 2), *Department of Microbiology and Molecular Genetics, Harvard Medical School, Boston, Massachusetts 02115*

JOHN BENNETT (44), *Department of Biology, Brookhaven National Laboratory, Upton, New York 11973*

ROLAND BENZ (27), *Fakultät für Biologie der Universität Konstanz, University of Konstanz, D-7750 Konstanz, Federal Republic of Germany*

LAWRENCE BOGORAD (47), *The Biological Laboratories, Harvard University, Cambridge, Massachusetts 02138*

PETER C. BÖHNI (30), *Department of Biochemistry, Biocenter, University of Basel, CH-4056 Basel, Switzerland*

WILLIAM S. A. BRUSILOW (19), *Department of Biological Sciences, Stanford University, Stanford, California 94305*

MICHAEL P. CAULFIELD (6), *Bacterial Physiology Unit, Harvard Medical School, Boston, Massachusetts 02115*

NAM-HAI CHUA (45), *Laboratory of Plant Molecular Biology, The Rockefeller University, New York, New York 10021*

DAVID A. CLAYTON (40), *Department of Pathology, Stanford University School of Medicine, Stanford, California 94305*

JACK COLEMAN (12), *Department of Biochemistry, State University of New York at Stony Brook, Stony Brook, New York 11794*

IAN CROWLESMITH (11), *Excerpta Medica, 1016 ED Amsterdam, The Netherlands*

CHARLES J. DANIELS (15), *Department of Biochemistry, Dalhousie University, Halifax, Nova Scotia B3H-4H7, Canada*

TAKAYASU DATE (4, 5), *Department of Biological Chemistry, Kanazawa Medical University, Uchinada-cho, Ishikawa-ken 920-02, Japan*

GÜNTHER DAUM (30), *Institut für Biochemie und Lebensmittelchemie, Technische Universität Graz, A-8010 Graz, Austria*

BERNARD D. DAVIS (6), *Bacterial Physiology Unit, Harvard Medical School, Boston, Massachusetts 02115*

GUIDO DE VOS (47), *The Biological Laboratories, Harvard University, Cambridge, Massachusetts 02138*

CAROL L. DIECKMANN (35, 36), *Department of Biological Sciences, Columbia University, New York, New York 10027*

MICHAEL G. DOUGLAS (33, 34), *Department of Biochemistry, University of Texas Health Science Center, San Antonio, Texas 78284*

PHILIP L. FELGNER (42), *Syntex Research, Palo Alto, California 94304*

B. G. FORDE (43), *Department of Biochemistry, Rothamsted Experimental Station, Harpenden, Herts AL5 2JQ, England*

HELMUT FREITAG (27), *Institut für Biochemie der Universität Göttingen, D-3400 Göttingen, Federal Republic of Germany*

KONRAD GAMON (11), *Département de Biochimie Médicale, Centre Médicale Universitaire, University of Geneva, CH-1211 Geneva 4, Switzerland*

SUSAN M. GASSER (23, 32), *Laboratoire de Différenciation Cellulaire, University of Geneva, CH-1211 Geneva 4, Switzerland*

NICHOLAS J. GAY (20), *Laboratory of Molecular Biology, Medical Research Council Centre, Cambridge CB2 2QH, England*

FRANK GIBSON (18), *Department of Biochemistry, John Curtin School of Medical Research, Australian National University, Canberra, A.C.T. 2601, Australia*

JOEL M. GOODMAN (13), *University of Texas Health Science Center, Dallas, Texas 75235*

EARL J. GUBBINS (47), *The Biological Laboratories, Harvard University, Cambridge, Massachusetts 02138*

ROBERT P. GUNSALUS (19), *Department of Microbiology and the Molecular Biology Institute, University of California, Los Angeles, California 90024*

E. HACK (43), *Department of Botany, University of Edinburgh, Edinburgh EH9 3JH, Scotland*

SIMON J. S. HARDY (7), *Department of Biology, University of York, Heslington, York YO1 5DD, England*

RICK HAY (23), *Department of Biochemistry, Biocenter, University of Basel, CH-4056 Basel, Switzerland*

BERND HENNIG (25), *Institut für Biochemie der Universität Göttingen, D-3400 Göttingen, Federal Republic of Germany*

MASAYORI INOUYE (12), *Department of Biochemistry, State University of New York at Stony Brook, Stony Brook, New York 11794*

SUMIKO INOUYE (12), *Department of Biochemistry, State University of New York at Stony Brook, Stony Brook, New York 11794*

KOREAKI ITO (5), *Institute for Virus Research, Kyoto University, Sakyo-ku, Kyoto, Japan*

DAVID A. JANS (18), *Department of Biochemistry, John Curtin School of Medical Research, Australian National University, Canberra, A.C.T. 2601, Australia*

LARS-GÖRAN JOSEFSSON (8), *Department of Microbiology, University of Dublin, Dublin 2, Ireland*

ENNO KREBBERS (47), *Abteilung Saedler, Max-Planck-Institut für Züchtungsforschung, D-3400 Göttingen, Federal Republic of Germany*

ROBERT C. LANDICK (15), *Department of Biological Sciences, Stanford University, Stanford, California 94305*

IGNACIO M. LARRINUA (47), *The Biological Laboratories, Harvard University, Cambridge, Massachusetts 02138*

C. J. LEAVER (43), *Department of Botany, University of Edinburgh, Edinburgh EH9 3JH, Scotland*

ANTHONY W. LINNANE (28), *Department of Biochemistry, Monash University, Clayton, Victoria 3168, Australia*

CAROL J. LUSTY (38), *The Public Health Research Institute of the City of New York, Inc., New York, New York 10016*

HENRY R. MAHLER (37), *Department of Chemistry, Indiana University, Bloomington, Indiana 47405*

SANGKOT MARZUKI (28), *Department of Biochemistry, Monash University, Clayton, Victoria 3168, Australia*

SHIRO MATSUURA (39), *Department of Physiology, Kansai Medical University, Moriguchi, Osaka, Japan*

PHYLLIS C. MCADA (33), *Department of Biochemistry, University of Texas Health Science Center, San Antonio, Texas 78284*

JANICE L. MESSER (42), *Department of Biochemistry, Michigan State University, East Lansing, Michigan 48824*

PETER MODEL (14), *The Rockefeller University, New York, New York 10021*

JULIO MONTOYA (41), *Departamento de Bioquimica, Facultad de Veterinaria, Universidad de Zaragoza, Zaragoza, Spain*

TAKASHI MORIMOTO (39), *Department of Cell Biology, New York University School of Medicine, New York, New York 10016*

JOHN E. MULLET (45), *Department of Biochemistry and Biophysics, Texas A & M University, College Station, Texas 77843*

BERNARD J. MULLIGAN (47), *The Biological Laboratories, Harvard University, Cambridge, Massachusetts 02138*

KAREN M. T. MUSKAVITCH (47), *The Biological Laboratories, Harvard University, Cambridge, Massachusetts 02138*

DAVID R. NELSON (49), *Department of Biochemistry, University of Texas Health Science Center, San Antonio, Texas 78284*

NATHAN NELSON (46), *Department of Biology, Technion, Haifa, Israel*

DOROTHEA-CH. NEUGEBAUER (21), *Zoologisches Institut, Lehrstuhl für Neurophysiologie, Universität Münster, D-4400 Münster, Federal Republic of Germany*

WALTER NEUPERT (25, 26, 27), *Institut für Biochemie der Universität Göttingen, D-3400 Göttingen, Federal Republic of Germany*

JENNIFER B. K. NIELSEN (16), *Merck Sharp & Dohme Research Laboratories, Rahway, New Jersey 07065*

HIROSHI NIKAIDO (9), *Department of Microbiology and Immunology, University of California, Berkeley, California 94720*

WOLFGANG OERTEL (22), *Department of Membrane Biochemistry, Max-Planck-Institut für Biochemie, D-8033 Martinsried, Munich, Federal Republic of Germany*

DIETER OESTERHELT (21, 22), *Department of Membrane Biochemistry, Max-Planck-Institut für Biochemie, D-8033 Martinsried, Munich, Federal Republic of Germany*

I. OHAD (48), *Department of Biological Chemistry, The Hebrew University, 91906 Jerusalem, Israel*

KAREN O'MALLEY (34), *Department of Medicine, Stanford University Medical Center, Palo Alto, California 94303*

ELIZABETH A. ORR (47), *The Biological Laboratories, Harvard University, Cambridge, Massachusetts 02138*

P. OVERATH (17), *Max-Planck-Institut für Biologie, D-7400 Tübingen, Federal Republic of Germany*

J. C. OWENS (48), *The Biological Laboratories, Harvard University, Cambridge, Massachusetts 02138*

DALE L. OXENDER (15), *Department of Biological Chemistry, University of Michigan Medical School, Ann Arbor, Michigan 48109*

PHILIP S. PERLMAN (37), *Department of Genetics, The Ohio State University, Columbus, Ohio 43210*

ANGELA PERZ (24), *Department of Cytogenetics, GBF—Gesellschaft für Biotechnologische Forschung mbH, D-3300 Braunschweig, Federal Republic of Germany*

MARCEL POUCHELET (38), *Institut du Cancer de Montreal, Montreal, Quebec H2L 4M1, Canada*

RICHARD A. RACHUBINSKI (38), *Department of Biochemistry, McGill University, Montreal, Quebec H3G 1Y6, Canada*

LINDA L. RANDALL (7, 8), *Biochemistry/Biophysics Program, Washington State University, Pullman, Washington 99164-4660*

YVES RAYMOND (38), *Institut du Cancer de Montreal, Montreal, Quebec H2L 4M1, Canada*

GRAEME A. REID (29, 31), *Department of Biochemistry, Biocenter, University of Basel, CH-4056 Basel, Switzerland*

NEAL C. ROBINSON (49), *Department of Biochemistry, University of Texas Health Science Center, San Antonio, Texas 78284*

STEVEN R. RODERMEL (47), *The Biological Laboratories, Harvard University, Cambridge, Massachusetts 02138*

RIMA ROZEN (38), *Department of Biochemistry, McGill University, Montreal, Quebec H3G 1Y6, Canada*

MARJORIE RUSSEL (14), *The Rockefeller University, New York, New York 10021*

RUDI SCHANTZ (47), *The Biological Laboratories, Harvard University, Cambridge, Massachusetts 02138*

MAXIME SCHWARTZ (10), *Unité de Génétique Moléculaire, Institut Pasteur, 75724 Paris Cedex 15, France*

WALTER SEBALD (24), *Department of Cytogenetics, GBF—Gesellschaft für Biotechnologische Forschung mbH, D-3300 Braunschweig, Federal Republic of Germany*

GORDON C. SHORE (38), *Department of Biochemistry, McGill University, Montreal, Quebec H3G 1Y6, Canada*

THOMAS J. SILHAVY (1, 2), *Laboratory of Genetics and Recombinant DNA, National Cancer Institute, Frederick Cancer Research Facility, Frederick, Maryland 21701*

PAMELA SILVER (4, 13), *Department of Biological Chemistry and Molecular Biology, Harvard University, Cambridge, Massachusetts 02138*

ROBERT D. SIMONI (19), *Department of Biological Sciences, Stanford University, Stanford, California 94305*

ANDRE A. STEINMETZ (47), *The Biological Laboratories, Harvard University, Cambridge, Massachusetts 02138*

MORDECHAI SUISSA (29), *Department of Biochemistry, Biocenter, University of Basel, CH-4056 Basel, Switzerland*

PHANG C. TAI (6), *Bacterial Physiology Unit, Harvard Medical School, Boston, Massachusetts 02115*

DOUGLAS P. TAPPER (40), *Department of Pathology, Stanford University School of Medicine, Stanford, California 94305*

R. M. TEATHER (17), *Agriculture Canada, Research Branch, Animal Research Institute, Ottawa, Ontario K1A 0C6, Canada*

ALEXANDER TZAGOLOFF (35, 36), *Department of Biological Sciences, Columbia University, New York, New York 10027*

RICHARD A. VAN ETTEN (40), *Department of Pathology, Stanford University School of Medicine, Stanford, California 94305*

ADELHEID VIEBROCK (24), *Friedrich Miescher-Institute, Postfach 273, CH-4002 Basel, Switzerland*

HEIKE VOGELSANG (22), *Department of Membrane Biochemistry, Max-Planck-Institut für Biochemie, D-8033 Martinsreid, Munich, Federal Republic of Germany*

JOHN E. WALKER (20), *Laboratory of Molecular Biology, Medical Research Council Centre, Cambridge CB2 2QH, England*

COLIN WATTS (13), *MRC Laboratory of Molecular Biology, Cambridge CB2 2QH, England*

M. WETTERN (48), *Institute of Botany, University of Braunschweig, Braunschweig, Federal Republic of Germany*

WILLIAM WICKNER (3, 4, 5, 13), *Department of Biological Chemistry, University of California School of Medicine, Los Angeles, California 90024*

RICHARD S. WILLIAMS (44), *Department of Biological Sciences, University of Warwick, Coventry CV4 7AL, England*

JOHN E. WILSON (42), *Department of Biochemistry, Michigan State University, East Lansing, Michigan 48824*

P. B. WOLFE (3), *Department of Biological Chemistry, University of California, Los Angeles, California 90024*

J. K. WRIGHT (17), *Max-Planck-Institut für Biologie, D-7400 Tübingen, Federal Republic of Germany*

YUKUN K. YE (47), *The Biological Laboratories, Harvard University, Cambridge, Massachusetts 02138*

RICHARD ZIMMERMANN (5, 26), *Institut für Biochemie der Universität Göttingen, D-3400 Göttingen, Federal Republic of Germany*

HORST-PETER ZINGSHEIM (21), *Department of Physiology, University of Otago Medical School, Dunedin, New Zealand*

C. ZWIZINSKI (3), *Physiologish-Chemisches Institut der Georg August Universität, D-3400 Göttingen, Federal Republic of Germany*

Preface

Volumes 96 to 98, Parts J, K, and L of the Biomembranes series, focus on methodology to study membrane biogenesis, assembly, targeting, and recycling. This field is one of the very exciting and active areas of research. Future volumes will deal with transport and other aspects of membrane function.

We were fortunate to have the advice and good counsel of our Advisory Board. Additional valuable input to this volume was obtained from Drs. William T. Wickner, Gottfried Schatz, Alex Tzagoloff, and R. J. Ellis. We were gratified by the enthusiasm and cooperation of the participants in the field whose contributions and suggestions have enriched and made possible these volumes. The friendly cooperation of the staff of Academic Press is gratefully acknowledged.

SIDNEY FLEISCHER
BECCA FLEISCHER

METHODS IN ENZYMOLOGY

EDITED BY

Sidney P. Colowick and Nathan O. Kaplan

VANDERBILT UNIVERSITY
SCHOOL OF MEDICINE
NASHVILLE, TENNESSEE

DEPARTMENT OF CHEMISTRY
UNIVERSITY OF CALIFORNIA
AT SAN DIEGO
LA JOLLA, CALIFORNIA

METHODS IN ENZYMOLOGY

EDITORS-IN-CHIEF

Sidney P. Colowick   Nathan O. Kaplan

VOLUME XXXV. Lipids (Part B)
Edited by JOHN M. LOWENSTEIN

VOLUME XXXVI. Hormone Action (Part A: Steroid Hormones)
Edited by BERT W. O'MALLEY AND JOEL G. HARDMAN

VOLUME XXXVII. Hormone Action (Part B: Peptide Hormones)
Edited by BERT W. O'MALLEY AND JOEL G. HARDMAN

VOLUME XXXVIII. Hormone Action (Part C: Cyclic Nucleotides)
Edited by JOEL G. HARDMAN AND BERT W. O'MALLEY

VOLUME XXXIX. Hormone Action (Part D: Isolated Cells, Tissues, and Organ Systems)
Edited by JOEL G. HARDMAN AND BERT W. O'MALLEY

VOLUME XL. Hormone Action (Part E: Nuclear Structure and Function)
Edited by BERT W. O'MALLEY AND JOEL G. HARDMAN

VOLUME XLI. Carbohydrate Metabolism (Part B)
Edited by W. A. WOOD

VOLUME XLII. Carbohydrate Metabolism (Part C)
Edited by W. A. WOOD

VOLUME XLIII. Antibiotics
Edited by JOHN H. HASH

VOLUME XLIV. Immobilized Enzymes
Edited by KLAUS MOSBACH

VOLUME XLV. Proteolytic Enzymes (Part B)
Edited by LASZLO LORAND

VOLUME XLVI. Affinity Labeling
Edited by WILLIAM B. JAKOBY AND MEIR WILCHEK

VOLUME XLVII. Enzyme Structure (Part E)
Edited by C. H. W. HIRS AND SERGE N. TIMASHEFF

VOLUME XLVIII. Enzyme Structure (Part F)
Edited by C. H. W. HIRS AND SERGE N. TIMASHEFF

VOLUME XLIX. Enzyme Structure (Part G)
Edited by C. H. W. HIRS AND SERGE N. TIMASHEFF

VOLUME 80. Proteolytic Enzymes (Part C)
Edited by LASZLO LORAND

VOLUME 81. Biomembranes (Part H: Visual Pigments and Purple Membranes, I)
Edited by LESTER PACKER

VOLUME 82. Structural and Contractile Proteins (Part A: Extracellular Matrix)
Edited by LEON W. CUNNINGHAM AND DIXIE W. FREDERIKSEN

VOLUME 83. Complex Carbohydrates (Part D)
Edited by VICTOR GINSBURG

VOLUME 84. Immunochemical Techniques (Part D: Selected Immunoassays)
Edited by JOHN J. LANGONE AND HELEN VAN VUNAKIS

VOLUME 85. Structural and Contractile Proteins (Part B: The Contractile Apparatus and the Cytoskeleton)
Edited by DIXIE W. FREDERIKSEN AND LEON W. CUNNINGHAM

VOLUME 86. Prostaglandins and Arachidonate Metabolites
Edited by WILLIAM E. M. LANDS AND WILLIAM L. SMITH

VOLUME 87. Enzyme Kinetics and Mechanism (Part C: Intermediates, Stereochemistry, and Rate Studies)
Edited by DANIEL L. PURICH

VOLUME 88. Biomembranes (Part I: Visual Pigments and Purple Membranes, II)
Edited by LESTER PACKER

VOLUME 89. Carbohydrate Metabolism (Part D)
Edited by WILLIS A. WOOD

VOLUME 90. Carbohydrate Metabolism (Part E)
Edited by WILLIS A. WOOD

VOLUME 91. Enzyme Structure (Part I)
Edited by C. H. W. HIRS AND SERGE N. TIMASHEFF

Methods in Enzymology

Volume 97
BIOMEMBRANES
Part K
Membrane Biogenesis:
Assembly and Targeting
(Prokaryotes, Mitochondria,
and Chloroplasts)

Section I

Prokaryotic Membranes

A. General Methods
Articles 1 through 8

B. Outer Membrane
Articles 9 through 12

C. Inner Membrane
Articles 13 through 22

[1] Genetic Analysis of Protein Export in *Escherichia coli*

By JONATHAN BECKWITH and THOMAS J. SILHAVY

One of the areas of biological research with the most productive interactions between those working in prokaryotic systems and those in eukaryotic ones is the study of protein secretion. Advances in the last few years have demonstrated a striking similarity in the steps involved in initiating the passage of proteins through biological membranes in widely varying organisms. The gram-negative bacterium *Escherichia coli* has been a major focus of study among prokaryotic organisms. In this bacterium, three major noncytoplasmic compartments exist to which proteins must be localized. These are (*a*) the cytoplasmic membrane, (*b*) the outer membrane, and (*c*) the periplasmic space between the two membranes. It appears that all outer membrane and periplasmic proteins examined may be exported by a common mechanism analogous to that seen for secretion of proteins into the lumen of the rough endoplasmic reticulum (RER) in eukaryotic cells. Furthermore, some, *but not all* (e.g., the galactoside permease), proteins located in the cytoplasmic membrane also appear to be exported by this same mechanism. In particular, those proteins where it is thought that nearly the entire length except for a short carboxy-terminal fragment is protruding into the periplasmic space exhibit features common to the periplasmic and outer membrane proteins.

These exported proteins in *E. coli* and another gram-negative bacterium, *Salmonella typhimurium,* are all synthesized initially as longer precursors with amino-terminal extensions of from 18 to 26 amino acids.[1,2] These signal peptides exhibit sequences quite comparable to those seen for eukaryotic secretory proteins. They include one to three positively charged amino acids at their amino terminus followed by sequences of at least 15 amino acids uncharged and rich in hydrophobic amino acids. The prokaryotic sequences do not, for the most part, show any significant homologies among themselves and are not distinguishable from eukaryotic sequences. That is, if we were to show a signal sequence from a prokaryotic protein and that from a eukaryotic protein, the reader would be unable to tell which was which.

The most startling evidence for a common mechanism for protein

---

[1] In 18 cases, the amino acid sequence of this signal peptide has been determined. In 4 others, only a longer precursor of the appropriate molecular weight has been shown.

[2] S. Michaelis and J. Beckwith, *Annu. Rev. Microbiol.* **36**, 435 (1982).

METHODS IN ENZYMOLOGY, VOL. 97

secretion in the two systems comes from studies in which the gene for a eukaryotic secretory polypeptide is expressed in a prokaryotic system. Specifically, it is possible by the appropriate construction to introduce the gene for rat preproinsulin, including the eukaryotic signal sequence, into *E. coli* so that the gene is transcribed and translated. The *eukaryotic* signal sequence of preproinsulin serves to promote transfer of the polypeptide across the *prokaryotic* cytoplasmic membrane, and, further, the signal peptide is cleaved at the *correct* position to yield proinsulin.[3,4] Less extensive experiments on the converse situation have suggested that when the gene for a prokaryotic β-lactamase is introduced into the yeast *Saccharomyces cerevisiae* the β-lactamase is correctly processed.[5]

All this evidence leads us to suppose that in the initial steps of secretion portions of the cytoplasmic membrane of bacteria are analogous to the rough endoplasmic reticulum in eukaryotic cells. This leads us further to suggest that an understanding of these aspects of secretion in gram-negative bacteria should shed light on what may well be a universal mechanism. *Escherichia coli* offers the enormous advantage of ready genetic analysis. In the following discussion, we describe the kinds of genetic studies that are being done or can be done in *E. coli,* what problems can be studied, and recent progress.

Questions to Which the Genetic Approach Applies

The Function and Structure – Function Analysis of the Signal Sequence

There was initially strong circumstantial evidence that the signal sequence was important in initiating secretion. However, more direct evidence was derived from genetic analysis. In many cases, mutations that alter the signal sequence of a specific protein inhibit the secretion of that protein.

The first signal sequence mutation described was that of Lin *et al.,*[6] which was isolated by a complex approach. Since then, more generalizable techniques have been developed. These include the use of gene fusions between genes for cytoplasmic and exported proteins for isolating signal sequence mutants and *in vitro* mutagenesis techniques.

[3] K. Talmadge, J. Kaufman, and W. Gilbert, *Proc. Natl. Acad. Sci. U.S.A.* **77,** 3988 (1980).
[4] K. Talmadge, S. Stahl, and W. Gilbert, *Proc. Natl. Acad. Sci. U.S.A.* **77,** 3369 (1980).
[5] R. Roggenkamp, B. Kustermann-Kuhn, and C. P. Hollenberg, *Proc. Natl. Acad. Sci. U.S.A.* **78,** 4466 (1981).
[6] J. J. C. Lin, H. Kanazawa, J. Ozols, and H. C. Wu, *Proc. Natl. Acad. Sci. U.S.A.* **75,** 4891 (1978).

Use of Gene Fusions for Selection of Signal Sequence Mutations

A variety of *in vivo* and *in vitro* techniques now exist that permit the construction of fused genes that code for hybrid proteins. In the cases where gene fusions have been used in signal sequence analysis, the hybrid proteins are composed of amino-terminal segments of an exported protein and a large carboxy-terminal segment of β-galactosidase.[7] These proteins include the signal sequence plus substantial amounts of the mature portion of the exported protein. Since the amino-terminal end (up to about 30 amino acids) of β-galactosidase is not essential to its enzymatic activity, such hybrid proteins often exhibit substantial enzymatic activity. In most such fusion strains, the hybrid protein becomes embedded in the cytoplasmic membrane and is not localized to the site of the exported protein itself. For example, a hybrid between β-galactosidase (β-Gz) and maltose binding protein (MBP), a periplasmic protein, is found in the cytoplasmic membrane.

Two very different properties of strains carrying such hybrid genes have permitted direct genetic selections for signal sequence mutants.

Growth-Inhibitory Effects of Gene Fusions Allow Selection of Signal Sequence Mutants. When high levels of such a hybrid protein are made, the accumulation of this protein in the membrane inhibits cell growth. In the case of the MBP–β-GZ hybrid, high levels can be induced by the addition of maltose to the growth media. Mutant cells are isolated that are resistant to the growth-inhibitory effects of maltose. Certain of these mutants are due to mutations in the signal sequence of MBP that alter the location of the hybrid protein to the cytoplasm, thus relieving the growth-inhibitory effect.[8]

Altered Enzymatic Properties of Membrane-Bound β-Galactosidase in Gene Fusions Allow Selection for Signal Sequence Mutants. Three systems have been described in which β-GZ is fused to a protein normally localized to the periplasm or outer membrane: MBP,[9] bacteriophage λ receptor,[10] and alkaline phosphatase.[11] In those cases where the hybrid protein is localized to a *membrane* fraction, the protein exhibits reduced ability to hydrolyze lactose.[12] The result of this membrane localization is that the strains show a Lac⁻ phenotype. In contrast, when a signal sequence mutation is introduced into the hybrid gene so as to cause a *cytoplasmic* location of the hybrid

---

[7] T. J. Silhavy, P. J. Bassford, Jr., and J. R. Beckwith, *in* "Bacterial Outer Membranes" (M. Inouye, ed.), p. 203. Wiley, New York, 1979.

[8] P. J. Bassford, Jr. and J. Beckwith, *Nature (London)* **277,** 538 (1979).

[9] P. J. Bassford, Jr., T. J. Silhavy, and J. R. Beckwith, *J. Bacteriol.* **139,** 19 (1979).

[10] S. D. Emr, M. Schwartz, and T. J. Silhavy, *Proc. Natl. Acad. Sci. U.S.A.* **75,** 5802 (1978).

[11] S. Michaelis, L. Guarente, and J. Beckwith, *J. Bacteriol.* (in press).

[12] See, for example, D. Oliver and J. Beckwith, *Cell* **24,** 765 (1981).

protein, the strains are able to grow on lactose. Thus, for all three systems, beginning with a strain carrying the appropriate gene fusions, a selection for Lac⁺ derivatives yields signal sequence mutants. This latter selection is an important one, since it appears to demand only partial defects in the signal sequence, yielding a variety of different changes.

In Vitro Mutagenesis

Techniques exist for generating mutations *in vitro* in specific regions of a gene. If the region within a gene that corresponds to the signal sequence is well characterized by restriction mapping and DNA sequence, it becomes possible to obtain mutants in this region by such techniques. This has been done so far for the TEM β-lactamase (periplasm) and for the *E. coli* lipoprotein (outer membrane).

In the case of β-lactamase, the region of the gene (in single-stranded form) corresponding to the signal sequence was treated with bisulfite, a highly efficient mutagenic agent. The resulting alterations in the DNA were determined by DNA sequence analysis, and the mutated DNA was transformed back into the bacteria. Any changes in the pattern of secretion of β-lactamase could then be correlated with specific base changes.[13] This class of technique is limited at the present time in that only G-C→A-T transitions and frame shifts can be produced. This approach will probably miss many of the signal sequence changes that have strong effects on secretion, since they are most often due to transversions. However, it is anticipated that new developments in the area of *in vitro* localized mutagenesis will eventually permit the generation of transversion mutations.

In the case of the lipoprotein, mutations were generated by the use of synthetic oligonucleotides that corresponded to DNA sequences in the signal sequence portion of the gene, but differed by one base from such sequences.[14] These altered sequences could be copied into intact plasmid DNA and, thus, introduced into *E. coli* to observe their effects on lipoprotein export. Mutations affecting the charged residues near the amino terminus of this signal sequence have been obtained in this way.

Conclusion

The variety of approaches now being used and those in the developmental stage should allow a detailed analysis of signal sequences. In particular, a number of models have been proposed that suggest specific roles for various

[13] D. Shortle, D. Koshland, G. M. Weinstock, and D. Botstein, *Proc. Natl. Acad. Sci. U.S.A.* **77,** 5375 (1980).
[14] S. Inouye, T. Franceschini, K. Nakamura, X. Soberon, K. Hakura, and M. Inouye, *Proc. Natl. Acad. Sci. U.S.A.* **79,** 3438 (1982).

components or secondary structures within signal sequences. A correlation of mutational alterations and their phenotypes should allow distinction between these various models or may generate new models more consistent with the accumulated data. The *in vitro* and *in vivo* approaches should nicely complement one another. Mutations with no effect on phenotype, not easily detectable *in vivo*, can be generated *in vitro*. In contrast, the *in vivo* selections readily yield mutations in at least some important components of the signal sequence. It should be pointed out, however, that the special nature of *in vivo* selections with gene fusions may be detecting only a subclass of mutations that have deleterious effects on secretion. Such a concern has been shown to be warranted by the characterization of one signal sequence mutation in bacteriophage λ receptor isolated by a quite different approach. This mutation appears to affect both the secretion and *synthesis* of the protein and would not have been detected in the selections with gene fusions.[15]

Function of Other Components of Structural Genes for Secreted Proteins

While a role for the signal sequence in secretion is established, it is not clear what role, if any, other portions of exported proteins play in secretion. One approach to this problem is to ask whether fragments of proteins that chain-terminate prematurely are secreted. This question has been examined with nonsense mutants in the gene for the periplasmic proteins, MBP,[16] β-lactamase,[17] and arginine-binding protein[18] and for the outer membrane protein, OmpA.[19] In all cases, the nonsense protein fragments were exported across the cytoplasmic membrane, thus giving no indication that sequences distal to the signal sequence are actively involved in promoting secretion. This conclusion is limited only by the inability to detect very short nonsense fragments. However, it is consistent with the finding that all mutations obtained from gene fusions, which interfere with secretion, map in the signal sequence. (This interpretation of the latter data is subject to the reservations described above.) Also, it is known from studies with fusions to β-galactosidase, that, while distal sequences of secreted proteins may play no *active* role in secretion, attachment of the wrong kind of amino acid sequence to a signal sequence may have a *negative* effect and interfere with secretion.[20] In

[15] M. Hall and M. Schwartz, *Ann. Microbiol. (Paris)* **133A,** 123 (1982).
[16] K. Ito and J. Beckwith, *Cell* **25,** 143 (1981).
[17] D. Koshland and D. Botstein, *Cell* **30,** 893 (1982).
[18] R. T. F. Celis, *J. Biol. Chem.* **256,** 773 (1981).
[19] E. Bremer, E. Beck, I. Hindennach, I. Sonntag, and U. Henning, *Mol. Gen. Genet.* **179,** 13 (1980).
[20] K. Ito, P. J. Bassford, Jr., and J. Beckwith, *Cell* **24,** 707 (1981).

addition, it has been suggested by Hall and Schwartz[15] that amino acid sequences early within the mature protein may play a role in the secretion process.

Although it may be that very little if any of the sequence of the mature protein is required for the passage of exported proteins through the cytoplasmic membrane, these sequences may well be important in determining the ultimate location of such proteins. It has been suggested, for instance, that amino acid sequences internal to the λ receptor protein may determine its incorporation into the outer membrane.[7] Further studies on the role of the mature protein sequence in secretion can be done by engineering internal deletions within the structural gene for the protein and examining their effects. This can be done by *in vivo* techniques or by the use of the enzyme *Bal*31 *in vitro* which has been widely used in the last few years to generate deletions.

Finally, if specific sequences within the mature protein sequence are suspected to be important in secretion, the *in vitro* mutagenesis techniques described above can be used.

Secretory Apparatus

Very little is known about the cellular components involved in the machinery for secretion. While it has been hypothesized that specific ribosomal proteins, membrane receptors, and proteinaceous pores in the membrane might participate, only preliminary evidence, mainly in eukaryotic systems has been obtained.[21] One approach to characterizing this apparatus is to obtain mutants that have pleiotropic effects on secretion. These mutants may be due to alterations in structural genes for proteins that are necessary for secretion.

Mutants Pleiotropically Defective in Secretion

Various approaches have been used to detect such mutants. Some of these are based on the assumption that mutations that cause a serious defect in secretion will be lethal to the cell. For instance, a collection of conditional lethal mutants isolated by Hirota that are temperature-sensitive for growth have been screened for accumulation of precursors of secreted proteins at the high temperature without success.[22]

Another approach involved the principles of a technique developed by Schedl and Primakoff.[23] These authors were screening for conditional-lethal (temperature-sensitive) mutants in which the synthesis of stable RNA was

[21] See, for example, P. Walter, I. Ibrahimi, and G. Blobel, *J. Cell Biol.* **91**, 545 (1981).
[22] K. Ito, personal communication.
[23] P. Schedl and P. Primakoff, *Proc. Natl. Acad. Sci. U.S.A.* **70**, 2091 (1973).

defective. The basic steps were growing up mutagenized bacterial colonies on solid media at 30°, switching the plates to 42°, and spraying the media with a transducing phage that carried the gene for a stable RNA (*tyrT*). Using indicator plates to assay for the expression of this gene (which was in the form of an amber suppressor), it was possible to detect colonies defective at the high temperature in this expression without requiring the ability to grow at this temperature. In a similar fashion, Wanner *et al.*[24] used a transducing phage carrying the structural gene for alkaline phosphatase (*phoA*) to screen for conditional lethal mutants defective in the secretion of periplasmic proteins. While the mutant obtained by this technique (*perA*) had pleiotropic effects, it is not clear whether the mutation affects secretion or regulation (or both). This general approach still seems to be worthy of further application.

The characterization of this *perA* mutation and the conditional-lethal mutants above raises a general problem in screening for secretion mutants. One of the first properties to be analyzed is the accumulation of precursors of normally secreted proteins. If a precursor is found in extracts of a particular mutant, it is a good candidate for further study. However, the absence of accumulated precursors does not mean that the mutation in question is not affecting the secretory apparatus. Since there appear to be some conditions under which secretion and synthesis of a protein may be coupled,[15,21] a defect in certain components of the secretory apparatus may result in a reduction in synthesis of secreted protein but no accumulation of precursors.

A direct selection for mutants affecting the secretory apparatus has been developed based on the Lac⁻ phenotype of a strain (MM18) carrying a fusion between the genes for MBP and β-galactosidase (see above, Altered Enzymatic Properties of β-Galactosidase). As mentioned before, mutations that cause internalization of this hybrid protein restore a Lac⁺ phenotype. Further, properties of this strain indicated that only a *slight* degree of internalization would give enough β-galactosidase activity to allow growth on lactose. This finding suggested that bacteria with only *slight* defects in the secretory apparatus, which were not lethal to the cell, might be detected by selection of Lac⁺ derivatives of MM18. This proved to be the case. In fact, by selecting Lac⁺ derivatives of MM18 at 30°, some conditional-lethal (temperature-sensitive) mutants were isolated.[12]

The mutations obtained by this approach define two new genetic loci in *E. coli,* the *secA* and *secB* genes.[12,25] These mutations have pleiotropic effects on secretion, causing the accumulation of precursors of a number of

[24] B. L. Wanner, A. Sarthy, and J. Beckwith, *J. Bacteriol.* **140,** 229 (1979).
[25] D. Oliver, C. Kumamoto, M. Quinlan, and J. Beckwith, *Ann. Microbiol. (Paris)* **133A,** 105 (1982).

normally secreted proteins. In the case of *secA*, the protein product has been detected and studies are currently under way to determine its function in secretion.[26] The characterization of such proteins and their role in the secretory apparatus will involve determining the cellular location of the protein and studies on its activity in an *in vitro* secretion system.

In addition, the *secA* mutant has been used to obtain additional mutants affecting secretion. One might anticipate that some mutations in genes for proteins that interact with the *secA* protein could compensate for the defect.[27] Based on this principle, we have isolated revertants of a *secA*$_{ts}$ strain that are no longer temperature-sensitive. Among these revertants are some which are *cold-sensitive* and in which the mutation responsible is genetically unlinked to *secA*. One of these maps very close to the *prlA* mutation described in the next section.[28]

Unlinked Mutations that Restore Secretion in Signal Sequence Mutants

Another way to detect mutants in the secretory apparatus is to seek extragenic suppressors of mutations in signal sequences. Such a study has been done with a mutation in the signal sequence of λ receptor where no receptor is exported to the outer membrane. Selection for derivatives of this strain that are able to use dextrins as a carbon source (λ receptor is required for the transport of dextrins into cells) has yielded several classes of unlinked suppressor mutations. The best-studied of these are the *prlA* mutations that map within a cluster of genes specifying ribosomal proteins.[29] The finding of these mutations, which can very efficiently restore export of proteins with defective signal sequences, suggests a role in secretion for a ribosomal or ribosome-associated protein. Further studies on suppressors of various signal sequence mutants may reveal new genetic loci of this type.

Localized Mutagenesis

In those cases where conditional-lethal mutations in genes have not been found, knowing the genetic location of the gene permits a screening for conditional-lethal mutants. This involves the approach of localized mutagenesis, which would be applicable, for instance, in seeking conditional lethals in the *prlA* gene or in the *secB* gene. Ito *et al.*[30] have used this approach in the case of *prlA* and have found a temperature-sensitive mutant

[26] D. Oliver and J. Beckwith, *J. Bacteriol.* **150,** 686 (1982).

[27] J. Jarvik and D. Botstein, *Proc. Natl. Acad. Sci. U.S.A.* **72,** 2738 (1975).

[28] E. Brickman, D. Oliver, J. Garwin, and J. Beckwith, unpublished results.

[29] S. D. Emr, S. Hanley-Way, and T. J. Silhavy, *Cell* **23,** 79 (1981).

[30] K. Ito, T. Yura, and H. Nashimoto, *Ann. Microbiol. (Paris)* **133A,** 101 (1982).

defective in secretion that maps very close to *prlA*. (Remember that the original *prlA* mutations enhance rather than inhibit secretion.)

Summary

Genetic studies on the secretion process in gram-negative bacteria have made considerable progress. Within the near future, such studies should lead to a detailed understanding of the important features of signal sequences and how they function. The cloning of the structural gene for an enzyme that cleaves signal sequences from precursors of secreted proteins will allow the genetic characterization of this locus and its function.[31] Finally, the isolation and characterization of mutants that affect components of the cell's secretory apparatus are also under way. These mutants permit the detection of genes and their products that are involved in secretion. A combination of the genetic approaches and *in vitro* studies should lead to a picture of the details of passage of proteins through a membrane.

Acknowledgment

Research described in this article from the authors' laboratories was sponsored by grants from the National Institutes of Health, National Science Foundation, The American Cancer Society, and the National Cancer Institute, DHHS, under Contract No. N01-C0-75380 with Litton Bionetics, Inc.

[31] T. Date and W. Wickner, *Proc. Natl. Acad. Sci. U.S.A.* **78,** 6106 (1981).

[2] Isolation and Characterization of Mutants of *Escherichia coli* K12 Affected in Protein Localization

By THOMAS J. SILHAVY and JONATHAN BECKWITH

We have used the technique of gene fusion to study protein localization in *Escherichia coli.* With this technique, fragments of the envelope protein under study can be tagged with a label that can be monitored. By determining the cellular location of the label, we can obtain information concerning essential intragenic components. More important, fusion strains often exhibit unusual phenotypes that can be exploited to isolate mutants that are defective in the export process.

In this chapter, we discuss the techniques that we use for isolating and characterizing gene fusions. In addition, we describe various procedures that can be used to isolate mutants that exhibit an alteration in the export process. It is our belief that these techniques are sufficiently general to be applied to any gene that codes for an exported protein.

The techniques described here are genetic. It is beyond the scope of this chapter to describe recombinant DNA and cellular fractionation methods. For descriptions of these techniques the reader is referred to other sources.[1-5] Unless otherwise stated, all media should be prepared as described by J. H. Miller.[6] This reference should also be consulted for procedures for standard techniques, such as P1 transduction or Hfr mating.

Construction of Gene Fusions

During the past 5 years it has become apparent that most export information is contained at the NH_2-terminal end of the envelope protein. Accordingly, the most useful gene fusions for studying protein localization are those that produce a hybrid protein comprising an NH_2-terminal fragment of an exported protein and a COOH-terminal fragment of a protein to serve as a label for biochemical identification of the hybrid molecule. With the advent of recombinant DNA techniques, any one of a number of different proteins could be chosen to serve as a label. We have chosen the enzyme β-galactosidase.

The nature of the enzyme β-galactosidase distinguishes lactose metabolism from most other biochemical pathways and makes it easy to analyze. The enzyme will hydrolyze a wide variety of galactosides. This has permitted the synthesis of a number of chromogenic substrates. For instance, one commonly used indicator medium contains the compound 5-bromo-4-chloro-3-indolyl-β-D-galactoside (XG). When this compound is hydrolyzed, it forms a blue dye that can be used as a very sensitive test for the presence of β-galactosidase activity. This compound can be used to discriminate between strains that produce high and low levels of β-galactosidase, in general, and to detect plaque-forming transducing phages that carry the *lacZ* gene.

Another reason why β-galactosidase is particularly amenable to gene fusion techniques is the finding that the NH_2-terminal portion of the enzyme can be removed and replaced by an NH_2-terminal fragment from

[1] K. Moldave and L. Grossman, eds., "Methods in Enzymology," Vol. 60. Academic Press, New York, 1979.
[2] M. J. Osborn, J. E. Gander, E. Parisi, and J. Carson, *J. Biol. Chem.* **247**, 3962 (1972).
[3] C. A. Schnaitman, *J. Bacteriol.* **108**, 545 (1971).
[4] H. C. Neu and L. A. Heppel, *J. Biol. Chem.* **240**, 3685 (1965).
[5] M. Inouye, "Bacterial Outer Membranes." Wiley, New York, 1979.
[6] J. H. Miller, "Experiments in Molecular Genetics." Cold Spring Harbor Lab., Cold Spring Harbor, New York, 1972.

TABLE I

BACTERIA STRAINS

Strain	Genotype
Mal315	F⁻ araD139 Δ(proAB lac)X111 Δ(ara leu)7697 rpsL malT (Mu c(ts), Mu c(ts)) dII (Ampʳ, 'lac)
MC4100	F⁻ araD139 Δ(lac)U169 rpsL relA flbB ptsF
pop3186	MC4100 Φ(lamB-lacZ)hyb42-1
MM18	MC4100 [λp Φ(malE-lacZ)hyb72-47]
XPL4	F⁻ lacZ524 phoR phoA20 rpsL
S27	F⁻ araP araCam lacZ14am metB ilv::Tn10 rpsL (φ80) (φ80pSu3⁺)
SE2060	MC4100 lamBS60

any other protein, and yet the hybrid can retain β-galactosidase activity. This was first shown by Müller-Hill and Kania.[7] Fusions between lacZ and more than a dozen other genes have been constructed. The hybrid proteins range in size from smaller than β-galactosidase[8] to approximately 180,000 daltons.[9] All these hybrids retain some level of β-galactosidase activity. Apparently almost any sequence can substitute for the NH_2-terminal sequence of β-galactosidase.

A variety of *in vivo* and *in vitro* techniques for constructing lacZ fusions have been described. What follows is a detailed description of several procedures selected for their relative simplicity and general applicability. Bacterial strains and bacteriophages used in these methods are described in Tables I and II.

TABLE II

BACTERIOPHAGE

Phage	Bacterial genes carried
P1vir	—
λvir	—
φ80vir	—
φ80pSu3⁺	supF
φ80pphoA⁺	phoA⁺
φ80placZ⁺	lacZ⁺Y⁺
λcIh80 Δ(int)⁹	—
λp1(209)	::(+Mu')trp'B A'- ΔW209 lac'ZYA'
λpSG1	λp1(209) lacY::Tn9
λ561 (b221 Oam Pam cI::Tn10)	—

[7] B. Müller-Hill and J. Kania, *Nature* (London) **249,** 561 (1974).
[8] F. Moreno, A. V. Fowler, M. Hall, T. J. Silhavy, I. Zabin, and M. Schwartz, *Nature* (London) **286,** 356 (1980).
[9] S. D. Emr and T. J. Silhavy, *J. Mol. Biol.* **141,** 63 (1980).

In Vivo lacZ Gene Fusions

Historically, the target genes for fusions to *lacZ* were limited; the targets had to reside adjacent to *lac* on the chromosome or be close to a prophage attachment site. In the latter case, a transducing phage could be translocated to the appropriate *att* site in a strain deleted for wild-type *lac* genes, and thereafter *lac* could be fused to neighboring targets (i.e., *trp-lacZ*[10]). To overcome the limitations of this approach, it was necessary to develop a vector for translocating the *lac* genes to any target site on the chromosome. Such a vector, called Mu *c*(ts) dII(Amp^r, '*lac*), was developed by Malcolm Casadaban[11] and is described in Fig. 1.

Use of Mud Phages for Obtaining Gene Fusions. The ideal vector for transposition of the *lac* genes would have little if any site specificity. This lack of site specificity is a characteristic of translocons. Since the bacteriophages Mu and Mud are themselves transposons, this important criterion is fulfilled; these phages have the ability to insert more or less randomly in the chromosome, and they exhibit little target-site specificity. However, with respect to Mu DNA, the insertion event is specific. Like other translocons, Mu can insert in only one of two opposite orientations. The physical continuity of Mu is never disrupted by insertion. As described in Fig. 1, insertion of the Mud phage into a target gene destroys that gene. If the Mud phage inserts in the correct orientation and if the reading frame is correct, a hybrid gene expressing a hybrid protein will be produced. Thus, it is possible to obtain gene fusions in a single step by selecting for the loss of target gene function, resistance to ampicillin (Amp^r; the Mud phage carries a functional *bla*$^+$ gene), and ability to catabolize lactose (Lac$^+$).

The first step in the construction is to prepare a lysate that contains the Mud transducing phage. Since the phage is defective, the lysate must be prepared from a strain that also contains a nondefective helper phage. This is done by heat induction of a Mu *c*(ts) dII(Amp^r, '*lac*), Mu *c*(ts) double

FIG. 1. Genetic structure of Mu *c*(ts) dII (*Amp^r, lac*). The ends of Mu are designated *c* and *s*, *c* representing the end on which the immunity region lies. By a number of *in vivo* and *in vitro* steps, Casadaban constructed this defective (d) phage such that the *lac* genes, but not a promoter, are inserted close to the *s* end of the phage. The Mu DNA sequences remaining at the *s* end are functional for translocation; however, they do not contain transcriptional or translational stop signals. Accordingly, insertion of this phage into a target gene in the correct orientation generates a hybrid gene. If the reading frame is correct, a hybrid protein will be expressed. Thus, by using this phage, fusions can be generated in a single step.

[10] J. H. Miller, W. S. Reznikoff, A. E. Silverstone, K. Ippen, E. R. Singer, and J. R. Beckwith, *J. Bacteriol.* **104**, 1273 (1970).
[11] M. J. Casadaban and J. Chou, in preparation.

lysogen. It is important that this lysate be fresh because Mu lysates are unstable, and the titer decreases rather rapidly with time. A strain containing both prophages (Mal315, for example) is grown overnight in Luria broth (LB). The following day, 10 ml of LB are inoculated with 0.2 ml of the overnight suspension and grown at 31° with shaking to mid-log phase (OD$_{600}$, 0.3–0.4). This will require 1–2 hr. The culture is incubated at 42° for 20 min and then at 37° with shaking until the cells lyse (1–2 hr). Immediately after lysis occurs, 5–6 drops of CHCl$_3$ are added, and the mixture is vortexed. Cell debris can be removed by centrifugation at 12,000 g for 10 min. A few drops of CHCl$_3$ should be added to the supernatant fluid, and the lysate should be stored at 4°.

Gene fusions can be constructed in any *E. coli* strain. There are only two requirements. First, the strain should be deleted for *lac*. This prevents unwanted homologous recombination between the Mud phage and the chromosome, and it allows selection for Lac$^+$. Second, in order not to preclude subsequent genetic manipulations, the strain should not be a λ lysogen. (Note: In order to facilitate genetic mapping experiments, it is often desirable that the starting strain contain a genetic marker located close to the target gene, as will be discussed further.)

A suitable recipient (MC4100 for example) is grown overnight in LB. Cells are then centrifuged and resuspended in 2.5 ml of 10 mM MgSO$_4$ containing 25 μl of 0.5 M CaCl$_2$. Tenfold dilutions of the Mud lysate are prepared, up to 10^{-4}, using the same salt solution. Aliquots (0.1 ml of the cell suspension) are placed in 5 Wasserman tubes. Then 0.1 ml of undiluted lysate is added to one tube and 0.1 ml of each dilution is added to the remaining four. The mixture is incubated at 30° for 20 min without shaking. The entire mixture is then plated onto tryptone–yeast extract (TYE) agar containing ampicillin (25 μg/ml). The plates should be incubated overnight at 30°.

Results can be scored the following day. An optimum transduction will yield ∼ 2000 colonies per plate. After determining the optimum dilution of the Mud lysate, the experiment should be repeated on a larger scale, say 20 plates. This will enable the screening of ∼ 40,000 transductants.

The next step in the procedure depends upon the selection or screen to be used for identifying mutants that have lost target gene function. If, for example, we were looking for fusions of *lacZ* to the gene coding for the receptor for bacteriophage λ (*lamB*), then we would replica-plate the transductants onto TYE agar that had been spread with 10^9 λvir particles. Alternatively, if we were looking for fusions to a gene involved in maltose transport or metabolism, we would replica-plate the transductants to maltose tetrazolium agar. Mal$^-$ colonies, which will be red, can be easily scored on this medium. The mutation frequency for Mu mutagenesis is approximately 10^{-3}.

Certain selections for loss of function of envelope proteins, or resistance to colicins, for example, do not work well after replica plating. In these case, the Amp^r transductants should be scraped from the plate and resuspended in LB. This suspension should be diluted to approximately 10^7 cells per milliliter and the culture grown to mid-log phase. Aliquots containing $10^5 - 10^6$ cells from this culture can then be plated on selective agar. (Note: Mud lysogens must be grown at $30 - 31°$ to prevent heat induction of the prophage.)

Mutants that have lost target-gene function are then purified on agar containing XG. (Usually we prepare a stock solution of XG, 40 mg/ml in dimethylformamide. Aliquots of this solution (0.1 ml) are then spread on plates immediately before use.) Lac$^+$ colonies, i.e., presumptive fusion strains, should be present at $15 - 20\%$.

Gene Fusions That Are Lethal to the Cell. We have used this Mud phage to fuse β-galactosidase to a wide variety of envelope proteins. Its chief limitation stems from the fact that often such fusions kill the cell when expression of the hybrid gene is high. In practice this is not a problem provided that regulation of the gene coding for the envelope protein is understood. In these cases, fusions can be isolated under conditions in which expression of the hybrid gene is minimized. Strains carrying certain *lamB-lacZ* fusions, for example, are killed when grown in the presence of maltose (which induces high-level expression of the hybrid gene). Such strains, however, grow normally in medium that does not contain inducer.[12] Certain genes, *ompA* for example, remain refractory to this fusion approach. This gene is expressed at high levels, constitutively.[13] Repeated attempts to fuse β-galactosidase to OmpA have been unsuccessful.

We have devised a technique that should solve the problem of fusion lethality.[14] This technique utilizes a Mud phage identical to the one described earlier except that it contains the *lacZ* amber mutation *U131*. This nonsense mutation terminates polypeptide chain elongation at a position corresponding to amino acid 41 of β-galactosidase.[15] This mutation is known to relieve the lethal effects of *lamB-lacZ*[9] and *malE-lacZ*[16] fusions. Presumably, this mutation will relieve the lethal effects of other fusions as well. With the Mud *lacZU131* phage, fusions can be isolated by using the same procedure described earlier. All Mud *lacZU131* lysogens will be Lac$^-$. Fusions can be identified by replica plating onto TYE agar containing XG

[12] T. J. Silhavy, H. A. Shuman, J. Beckwith, and M. Schwartz, *Proc. Natl. Acad. Sci. U.S.A.* **74,** 5411 (1977).

[13] M. N. Hall and T. J. Silhavy, *Annu. Rev. Genet.* **15,** 91 (1981).

[14] T. Palva and T. Silhavy, unpublished data.

[15] J. K. Welply, A. V. Fowler, J. R. Beckwith, and I. Zabin, *J. Bacteriol.* **142,** 732 (1980).

[16] P. J. Bassford, T. J. Silhavy, and J. R. Beckwith, *J. Bacteriol.* **139,** 19 (1979).

that has been spread with 10^9 ϕ80pSu3$^+$ phage particles. Even if suppression of the amber mutation leads to cell death, the Lac$^+$ phenotype can be scored from the blue color reaction. The corresponding uninfected fusion strain can be recovered from the master plate from which the replica was made. By use of strains that conditionally express a Su$^+$ phenotype, these fusions can then be studied. This procedure should allow fusion of β-galactosidase to any envelope protein.

Construction of λ Derivatives of Mud Fusions. Mud fusion technology presents some problems. First, the ability of Mud to transpose and consequently affect the Lac phenotype makes direct genetic selections impossible with these strains. (For example, 2 days after Lac$^-$ Mud lysogens are streaked onto lactose MacConkey agar, hundreds of Lac$^+$ papillae are evident.) Second, the Mud phage cannot be employed to obtain specialized transducing phages that carry the entire gene fusion. This precludes the ability to perform certain types of genetic analysis, and it also makes cloning of the desired fusion onto multicopy plasmids more difficult. A solution to these problems was suggested by Komeda and Iino.[17]

Komeda and Iino took advantage of another plaque-forming transducing phage, λp1(209), which was developed by M. Casadaban, to construct a gene fusion transducing phage. This method is outlined in Fig. 2. First, the Mud gene fusion strain is lysogenized with λp1(209) (which like the original Mud carries no promoter for the *lac* genes). Next, a segregant of this lysogen that has lost the Mud phage is isolated. The final lysogen can be induced to yield plaque-forming phages that have incorporated the *lacZ* genes with the target promoters by illegitimate excision.

In practice, this procedure is done in the following way: (Note: Before using this procedure, it is important to determine that the Mud fusion candidate contains only one prophage. This test is described in the following section.) Each Mud fusion strain is grown overnight in LB at 30°. The cultures are centrifuged, and the cells are resuspended in 2.5 ml of 10 mM MgSO$_4$. Aliquots (0.1 ml) of cells are mixed with 2.5 ml of melted H-Top agar, and the mixture is then poured onto TYE agar. After the agar hardens, drops of a λpSG1 lysate (approximately 10^8 ϕ per milliliter) are spotted onto the surface of the plate. λpSG1 is a derivative of λp1(209) and is identical to the parent phage except that it carries a transposon, Tn9, which confers resistance to chloramphenicol (Cam^r) inserted in the *lacY* gene.[18] This allows direct selection of lysogens. When the spots are dry, the plates are incubated at 30° overnight. From the turbid centers of the lysed areas, cells are picked and streaked onto lactose MacConkey agar containing 25 μg of chloramphenicol per milliliter. These cultures are incubated overnight at

[17] Y. Komeda and T. Iino, *J. Bacteriol.* **139**, 721 (1979).
[18] S. Garrett, personal communication (1982).

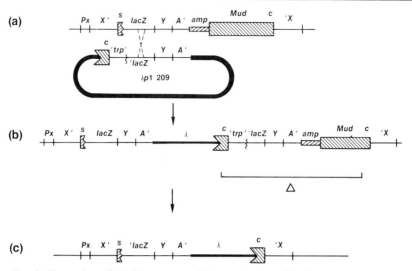

FIG. 2. Conversion of Mud lysogen to a λ lysogen by recombination. (a) A Mud lysogen (Amp^r, temperature-sensitive) infected with λp1(209) and λ lysogens are selected. λp1(209) must integrate by *lacZ* or Mu homology, since this phage has no attachment site. Integration is depicted here as occurring via *lacZ* homology. (b) The double lysogen (Amp^r, ts, λ$^+$) loses the Mud phage by recombination in Mud homology (*c* end). (c) The final λ lysogen (Amp^s, λ$^+$) can be selected by ability to grow at 42°, since the Mud prophage has a temperature-sensitive repressor. In addition to the way pictured here, integration of λ may occur by Mu homology, and excision of the Mud prophage by *lacZ* homology.

30°. Large isolated colonies, probable lysogens, are then purified by restreaking on the same medium. Depending on the position of the homologous recombination that formed the lysogen, colonies will be either Lac$^+$ or Lac$^-$. Lac$^-$ lysogens will be formed by a recombinational event that occurs promoter proximal to the Tn insertion. Lac$^+$ lysogens can occur by recombination at Mu DNA or at the *lac* DNA promoter distal to the Tn insertion. In order to ensure that the fusion strain remains Lac$^+$ when cured of Mu, it is best to pick and purify Lac$^+$ lysogens. Lac$^-$ lysogens will remain Lac$^-$ when cured of Mu (see Fig. 2). To verify that the Lac$^+$, Cam^r colonies are λ lysogens, the colonies should be cross-streaked against λvir, ϕ80vir, and λcIh80 on TYE agar at 30°. Lysogens will exhibit a λvir^s, ϕ80vir^s, λcIh80^r phenotype.

To cure Mud from the double Mud, λ lysogens, colonies should be streaked onto lactose MacConkey agar at 42°. Healthy survivors should be fusion strains containing λ, but not Mud. They should exhibit the following phenotypes: Lac$^+$, Cam^s, Amp^s, λvir^s, ϕ80vir^s, λcIh80^r, and temperature resistance.

Isolation of λ Transducing Phages Carrying Gene Fusions. In order to isolate λp (plaque-forming) fusion transducing phages, fresh colonies of strains that carry the various fusions are stabbed into TYE agar that has previously been overlaid with 2.5 ml of H-Top agar containing 0.1 ml of a stationary-phase culture of the indicator strain MC4100. By using a grid, as many as 50 fusion strains can be stabbed into a single TYE plate. The plate is placed under an ultraviolet (UV) lamp for 50 sec. This corresponds approximately to the time required to kill 50% of the cells using our UV lamp arrangement. Plates are then incubated for approximately 6 hr in the dark until visible lysis can be detected around each stab. Patches from each zone of lysis are streaked onto TYE agar containing XG, which had been prespread with 0.2 ml of a stationary-phase culture of MC4100. Plaques that appear dark blue are purified several times by restreaking on the same medium.

Isolated dark blue plaques are collected by stabbing into the agar with a Pasteur pipette. The small cylindrical piece of agar with the plaque on top is transferred to 0.2 ml of a stationary-phase culture of MC4100. Phage adsorption is allowed to occur for 10 min at room temperature. Then 2.5 ml of H-Top are added, and the suspension is poured onto TYE agar. When lysis is nearly confluent, 4 ml of LB broth are added and the top agar layer is scraped from the plate. The lysates are treated with several drops of chloroform, vortexed, and centrifuged for clarification. From a single plaque we usually obtain 5 ml of lysate with a titer of 5×10^{10} ϕ/ml.

In some cases, dark blue plaques cannot be obtained with the above method. Transducing phages that carry these fusions can be isolated by induction of an exponential liquid culture as described by Miller.[6] Dilutions of these lysates are then plated with the indicator strain on TYE agar containing XG. This allows one to look at larger numbers of plaques. Dark blue plaques are purified and lysates are prepared as described above.

Caution: Double Lysogens. One fairly frequent complication of this approach is that, during the process of forming the λ lysogen, two copies of the λ phage are integrated. This makes the subsequent detection of λ plaque-forming phages carrying the fusions difficult for the following reasons. When only a single λ is located adjacent to the fusion, there is no genetic homology available for the high-frequency excision of λ. When induced by UV radiation, λ plaque-forming phages are formed by excision caused by nonhomologous crossovers. The probability that such abnormal excision events will yield a phage carrying the fusion is relatively high. The titer of plaque-forming phages from such lysates is low (10^6 to 10^7 pfu/ml). In contrast, when two λ phages have integrated in the original insertion event, one will be adjacent to the fusion, and the second will be adjacent to the first. When induced by UV radiation, a λ plaque-forming phage is most frequently formed by recombination between homologous sequences in the two λ prophages. The titer of such lysates is considerably higher, 10^9 pfu/ml, and the λ phages carrying fusions represent a very small percentage of the total phages. They are, therefore, very difficult to detect in such lysates.

To avoid this problem, derivatives of double lysogens that are cured of one of the copies of the phage should be used. They can be obtained by using this procedure. First, double and single lysogens are selected in the following way. A lawn of a λ-sensitive indicator strain (e.g., MC4100) in H-top agar is spread onto TYE agar. Then, λ lysogens from single colonies are picked and stabbed into the TYE lawn. After overnight incubation, double lysogens will be surrounded by a halo of lysis on the indicator plate, whereas single lysogens will show no lysis. This result is attributed to the much higher frequency of spontaneous excision of active phages in the double lysogen. To detect those single lysogens that occur spontaneously as derivatives of the double lysogens, 50–100 single colonies are picked and stabbed into the lawn of the λ-sensitive indicator. At a frequency of a few percent, single lysogens will occur and such colonies will fail to give lysis on such a lawn. These single lysogens can now be used to obtain the phages carrying the *lac* fusions.

Genetic Manipulations with λ Fusion Phages. The λ transducing phages isolated by the method described above are useful for a variety of genetic manipulations. In addition, these phages are a source of DNA for subcloning the gene fusion onto a multicopy plasmid vector. With a plasmid constructed by M. Berman,[19] any fusion can be subcloned in a single step. This plasmid, pMLB524, and the scheme for subcloning fusions is described in Fig. 3. Cloning the fusion on a multicopy vector is the first step toward analyzing regulatory and export information to the level of the DNA sequence. In addition, such plasmids provide substrates for *in vitro* mutagenesis. Many fusions will be lethal on multicopy vectors because of the high levels of expression that occur. Accordingly, these cloning experiments may have to be done in mutant strains that, for one reason or another (a regulatory mutant), do not allow high-level expression from the cloned promoter. Alternatively, one may have to clone the fusion containing the *lacZU131* mutation described earlier.

Although the genetic manipulations described here may seem complicated, they are not technically difficult. All the steps from the original isolation of the fusion to the cloning can be done in less than a month.

In Vitro lacZ Gene Fusions

Gene fusions can also be generated by recombinant DNA techniques. Certain plasmid vectors have been constructed that allow the fusion of any well-characterized gene to β-galactosidase, so as to replace the amino-termi-

[19] M. Berman, in preparation.

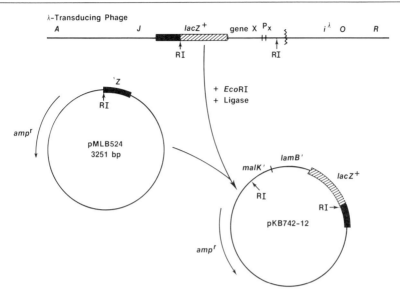

FIG. 3. Scheme for cloning gene fusions using pMLB524. In order to facilitate cloning of fusions to *lacZ* on a multicopy plasmid, the vector pMLB524 was developed.[19] A small region of pBR322 (from the *Eco*RI to the *Ava*I restriction sites) was replaced with an *Eco*RI-*Ava*I DNA fragment containing the COOH-terminal region of the *lacZ* gene (corresponding to amino acids 1004–1021). Since all gene fusions to *lacZ* contain the same *Eco*RI site at a position corresponding to amino acid 1004 of β-galactosidase, it is possible to regenerate functional *lacZ* by recloning *Eco*RI fragments from λp fusion transducing phages. Strains containing the desired plasmid will be Amp^r, LacZ^+, i.e., blue on agar spread with XG.

nal sequence of this protein.[20,21] With this approach, β-galactosidase has been fused to large amino-terminal segments of alkaline phosphatase (*phoA*), a periplasmic enzyme.[22] These fusion strains have properties analogous to fusions between β-galactosidase and bacteriophage λ receptor or maltose-binding protein. However, for genetic selections using these fusions, their existence on multicopy plasmids creates certain problems (see Application of the Approaches to Other Genes). Therefore, these gene fusions have been recombined onto λ transducing phages carrying previously isolated *phoA-lacZ* fusions.[23] This step permits the construction of strains that carry only one copy of the gene fusion as a λ prophage. If this were to be used as a general approach for other genes, it would require the prior existence of a λ transducing phage carrying the correct gene fusion.

[20] L. Guarente, G. Lauer, T. M. Roberts, and M. Ptashne, *Cell* **20**, 533 (1980).
[21] M. J. Casadaban, J. Chou, and S. N. Cohen, *J. Bacteriol.* **143**, 971 (1980).
[22] S. Michaelis, L. Guarente, and J. Beckwith, *J. Bacteriol.*, in press.
[23] A. Sarthy, S. Michaelis, and J. Beckwith, *J. Bacteriol.* **145**, 293 (1981).

Since such phages do not exist for most other genes, the utility of this approach may be limited. On the other hand, it should be possible, given the appropriate restriction enzyme sites, to clone the gene fusion from the plasmid onto a phage.

Characterization of Gene Fusion Strains

The techniques described for fusion construction were designed to minimize the occurrence of abnormal events. Nevertheless, time should not be spent working with fusion strains until their genetic structure has been verified.

Although it should be obvious, the first step in characterizing Mud fusions is to check the genotype of the strain. There is no point in characterizing a contaminant. The next step is to determine whether the Mud fusion maps at the correct site on the chromosome. This is most simply done by P1 transduction. The object is to demonstrate that the Mud fusion maps at the same place as the target gene. For these experiments, it is better to use the Mud fusion as the recipient in the transduction. It is possible to transduce a Mud fusion from one strain to another using P1. However, this transduction is complicated by zygotic induction of the Mud prophage. When this prophage enters a strain that does not contain a phage repressor, induction occurs, with the result, that, in as many as 70% of the transductants, the Mud phage has translocated to a different site on the chromosome. A much better way to do this experiment is to demonstrate that the Mud fusions show the same transductional linkage to an adjacent marker as does the target gene. For example, if we were looking for *ompC-lacZ* fusions, fusion candidates could be transduced to resistance to nalidixic acid (Nal^r) with a P1 lysate grown on an *ompC$^+$ gyrA* strain. If the Mud phage was inserted in *ompC,* then approximately 40% of Nal^r transductants should be Amp^s, Lac$^-$, OmpC$^+$.

The genes coding for many envelope proteins are not located on the chromosome next to a selectable marker. Accordingly, in many instances, it is best to select a parent strain that has a mutation located near the target gene. For example, if we were looking for *ompF-lacZ* fusions, we would chose a *pyrD* parent strain. Transduction of fusion candidates to prototrophy should yield Amp^s, Lac$^-$, temperature-resistant colonies at the expected 40% frequency. Alternatively, it is possible to employ a strain containing a transposon located next to the target gene as the donor for the transduction. Procedures for the construction of such strains are described in the Appendix.

These transductional analyses will also permit identification of multiple Mud lysogens. If no Amp^s, temperature-resistant transductants are found, it is likely that the fusion candidate contains a second Mud prophage located somewhere else on the chromosome. Such multiple lysogens can not be used in subsequent experiments. The easiest way to solve this problem is to transduce the Mud fusion into a clean genetic background. As stated earlier, these transductions are complicated by zygotic induction. Accordingly, transductants that exhibit the desired phenotypes must be characterized as carefully as any fusion candidate.

If the Mud prophage maps at the expected chromosomal location, the next step is to demonstrate that transcription and translation are initiated from the wild-type target gene signals. The simplest way to do this is to demonstrate that synthesis of the hybrid protein is regulated in a manner analogous to the target gene regulation. For example, if we were looking for *lamB-lacZ* fusions, synthesis of the hybrid protein should be induced by the presence of maltose in the growth media.

In cases where the target gene is expressed constitutively or when regulation of target gene is not understood, a different method must be employed. The technique we employ is depicted in Fig. 4. The rationale behind this method is as follows: If a mutation known to block expression of the wild-type target gene also blocks expression of the hybrid protein, it can be concluded that the hybrid protein is being expressed from target gene signals. This technique utilizes the λ fusion transducing phage isolated as

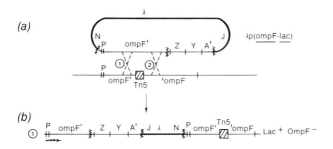

FIG. 4. Genetic verification of *ompF-lac* fusions. (a) An *ompF*::Tn5 insertion strain is lysogenized with a λ transducing phage isolated from a potential *ompF-lac* fusion strain. (b) If the transducing phage is carrying an *ompF-lac* fusion, and if the Tn5 insertion is promoter proximal to the fusion joint, two classes of lysogens are obtained depending on where the integrative recombinational event occurs. Integration promoter proximal to the Tn5 insertion (1) yields a Lac⁺, OmpF⁻ lysogen. Integration promoter distal to the Tn5 insertion (2) yields a Lac⁻, OmpF⁺ lysogen. The Lac⁻, OmpF⁺ lysogen is possible only if the transducing phage carries an *ompF-lac* fusion [M. N. Hall and T. J. Silhavy, *J. Mol. Biol.* **146**, 23 (1981)].

described earlier. A strain containing a translocon inserted into the target gene (see Appendix) is infected with the fusion transducing phage at a multiplicity of infection of 0.1. A low multiplicity of infection is best to avoid the complication of double lysogens. Phages are adsorbed by incubation at room temperature for 20 min. The infection mixtures are then diluted 10-fold with LB and incubated at 37° for 20 min with aeration. Lysogens are selected by spreading aliquots of the incubation mixture on TYE agar previously spread with 10^9 λcIh80 particles and 0.1 ml of XG (40 mg/ml in dimethylformamide). The Lac phenotype is determined by the position of the integrative recombination event (Fig. 4). Integration of the fusion phage promoter distal to the Tn insertion results in a lysogen with a Lac⁻ phenotype caused by the insertion mutation. The lysogen also should exhibit a wild-type target gene, since this gene should be regenerated by such a recombination. A Lac⁻, Target Gene⁺ lysogen can be obtained only if *lac* is fused to the target gene. This provides genetic verification of the fusion strain. This technique can be employed not only with insertion mutations,[24] but also with promoter[25] or nonsense mutations.[26]

Final verification of fusion strains requires biochemical characterization of the hybrid protein. This is best done by demonstrating that the hybrid protein can be immunoprecipitated with antisera raised against either β-galactosidase or the target-gene product. With short hybrid proteins, which contain only small amounts of the target-gene product, purification and amino acid sequencing of the NH_2-terminal portion of the molecule may be required.

A variety of cellular fractionation techniques can be used to determine the cellular location of the hybrid protein. Although we will not describe these techniques, a few general points should be made. First of all, fractionation techniques that yield consistently reproducible results with wild-type proteins often yield ambiguous results with hybrid proteins and aberrant proteins. Furthermore, different techniques may yield different results.[27,28] Thus, caution must be exercised when interpreting these data. Generally, however, if the hybrid protein is found in substantial amounts in an apparently noncytoplasmic location, the hybrid protein contains export information specified by the target gene. In most cases, strains containing such a fusion will exhibit one or more unusual phenotypes. These phenotypes can be exploited to isolate export-defective mutants.

[24] M. N. Hall and T. J. Silhavy, *J. Bacteriol.* **140,** 342 (1979).
[25] M. L. Berman and J. Beckwith, *J. Mol. Biol.* **130,** 303 (1979).
[26] M. Debarbouille and M. Schwartz, *J. Mol. Biol.* **132,** 521 (1979).
[27] K. Ito and J. R. Beckwith, *Cell* **25,** 143 (1981).
[28] I. Crowlesmith, K. Gamon, and U. Henning, *Eur. J. Biochem.* **113,** 375 (1981).

Selection Procedures Based on Gene Fusions

As described in our general overview,[29] fusion strains often exhibit unusual phenotypes that can be exploited to isolate export-defective mutants. Although we believe these phenotypes generally will be observed, it is not possible to describe a selection procedure that can be employed in all cases. In particular, the precise composition of the selection media will vary depending on the envelope protein being studied. In this section we describe examples of selection procedures that have been successfully used.

Growth Inhibitory Effects of Gene Fusions: The Isolation of lamB Signal Sequence Mutants

The expression of the *lamB-lacZ* gene fusion present in strain pop3186 is induced by maltose. A significant fraction of the hybrid protein synthesized early after induction is found in the outer membrane; however, the growth of this strain is strongly inhibited by the addition of inducer to the growth medium. This maltose-sensitive (Mal^s) phenotype is a result of the inability of the cell to export large amounts of the hybrid protein efficiently. We have shown that mutants blocked in the localization of the hybrid protein can be obtained by selecting maltose resistance (Mal^r). Unlike mutants defective in the synthesis of the hybrid protein, these mutants retain β-galactosidase activity and therefore a Lac$^+$ phenotype. We have devised the following procedure to enrich for the desired Mal^r, Lac$^+$ mutants. Independent colonies of strain pop3186 are inoculated into separate tubes of maltose minimal medium (M63), and the tubes are incubated at 37° for 48 hr or until cultures have grown to saturation. Portions (0.05 ml) of each culture are then inoculated into 5 ml of fresh maltose minimal medium, and the cultures are incubated at 37° for 24 hr. This gives rise to almost pure cultures of Mal^r cells. To select for Lac$^+$ cells present in this population, dilutions of the cultures are plated on lactose minimal agar. To ensure that all of the spontaneously occurring mutants analyzed are the result of independent events, only a single Mal^r, Lac$^+$ colony from each culture should be purified and characterized.

Nearly 50 mutants have been isolated using this procedure. All of these mutations are linked to the *lamB-lacZ* fusion. This was shown by inducing the λ fusion phage from the Mal^r, Lac$^+$ mutants. When these phages were then lysogenized into the parent strain, most of the lysogens remained Mal^r. This is possible only if the mutation conferring Mal^r is carried by the fusion phage. Subsequent deletion mapping and DNA sequence analysis have shown that all the mutations alter the LamB signal sequence. As predicted,

[29] J. Beckwith and T. J. Silhavy, this volume [1].

all the mutations prevent export of the hybrid protein. In the Mal^r mutant strains, the hybrid protein is found in the cytoplasm.

In order to determine the effect of the signal sequence mutations obtained with the Mal^r selection on the wild-type *lamB* gene product, the mutation must be recombined from the *lamB-lacZ* fusion to an otherwise wild-type *lamB* gene. Depending on the genetic system, several different techniques can be employed. The most general, and the one we use most often, involves the λ fusion phage that carries the signal sequence mutation.

Since the mutation prevents export of the hybrid protein, it probably prevents export of the wild-type LamB protein as well.[30] This would confer a λ^r phenotype, and it can be used as a selection for recombining the mutation onto the wild-type gene. A *lac* deletion strain such as MC4100 is lysogenized with phages carrying the *lamB-lacZ* fusion, and λ^r recombinants are selected. If the λ^r character resulted from a reciprocal recombination as depicted in Fig. 5, then the resulting strain should exhibit a Mal^s phenotype. Such a recombination would regenerate the original Mal^s *lamB-lacZ* fusion. We have used this procedure to select and purify merodiploids exhibiting the λ^r, Mal^s phenotype.

Evidence that the mutations were now present in gene *lamB* was obtained by selection of the reverse reciprocal recombination in these merodiploid strains (Fig. 5). Such an event, which restores the λ^s, Mal^r phenotype of the original lysogen, in fact, occurs at a high frequency in these strains, whereas the spontaneous events conferring this phenotype on strains carrying the original fusion are very rare.

In order to remove the *lamB-lacZ* hybrid gene from the merodiploid, the lysogens were cured of the respective λp transducing phages by selecting for the loss of the Mal^s phenotype in the presence of λvir (see Fig. 5). Alternatively, the fusion phage can be cured by irradiating it with UV and scoring for loss of β-galactosidase activity (see next section). Cured derivatives that retained the export-defective mutation can then be identified by scoring for the loss of target gene function.

Not all mutations that block export of a hybrid protein will completely block export of the corresponding wild-type gene product as well.[9] If these mutations are phenotypically silent when present in the wild-type gene, genetic analysis is difficult. Their presence in an otherwise wild-type gene after the fusion transducing phage is cured can be detected only by marker rescue. This is done by relysogenizing strains that are thought to carry the export-defective mutation with λ fusion transducing phages that carry the parent fusion. These lysogens will exhibit β-galactosidase activity and a phenotype identical to the parent fusion strain, e.g., Mal^s. However, recombinants that exhibit the mutant fusion phenotype, e.g., Mal^r, will appear at

[30] S. D. Emr, M. Schwartz, and T. J. Silhavy, *Proc. Natl. Acad. Sci. U.S.A.* **75**, 5802 (1978).

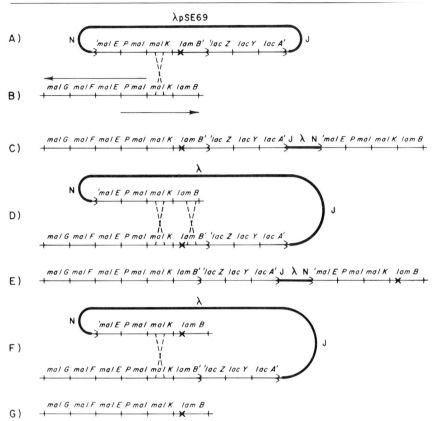

FIG. 5. Recombination of the *lamB* mutations S68 and S69 onto wild-type gene *lamB*. (A) The λp transducing phage that carries the *lamB-lacZ* fusion and the *lamB* mutation S69 (designated here with a X). (B) A portion of the chromosome of strain MC4100 depicting the two divergent operons that comprise the *malB* locus. The promoter here is designated Pmal. Because strain MC4100 is Δ(*lac*) and because the λp transducing phage is Δ(*att*), lysogenization occurs primarily via recombination between homologous *malB* DNA. Insertion of the phage as shown yields a lysogen (C) that has the phenotype λ^r, Mal^+, Lac^+. A reciprocal crossing-over as shown in (D) will recombine the S69 mutation onto the wild-type *lamB* gene and simultaneously regenerate the original *lamB-lacZ* fusion 42-1. Accordingly, the phenotype of the resulting lysogen (E) is now λ^r, Mal^s, Lac^+. Loss of the prophage by a recombinational event as depicted in (F) removes the *lamB-lacZ* fusion and thus eliminates the Mal^s phenotype of the lysogen. This can be used as a selection procedure for curing the lysogen of the prophage. This curing event yields strain SE2069, which is Mal^+, λ^r because the mutation S69 is now present in the *lamB* gene (G). It is possible to demonstrate that the steps depicted here are easily reversible; i.e., the mutation S69 can be moved from the hybrid *lamB-lacZ* gene to the wild-type gene *lamB* and vice versa by the recombination event depicted in (D). This provides evidence that the mutation present in strain SE2069 (G) is identical to the mutation present in λpSE69 (A).[30]

high frequency. In such a merodiploid, the mutant phenotype of the fusion can be rescued by the mutation present in the otherwise wild-type gene.

Mutations obtained using the Mal^r selection alter the signal sequence in all cases. Why unlinked mutations have not been obtained is not clear. This could be attributed to the fact that unlinked mutations that would relieve the Mal^s phenotype are lethal. Unfortunately, this cannot be directly tested since the selection cannot be easily adapted for the isolation of conditional lethal (e.g., temperature-sensitive) mutations. A Mal^r, temperature-sensitive strain could not survive under any conditions in the presence of inducer.

Altered Enzymatic Properties of Membrane-Bound β-Galactosidase in Fusion Strains: The Isolation of malE Signal Sequence Mutants and secA Mutants

The malE-lacZ fusion strain MM18 makes a hybrid protein that is localized to the cytoplasmic membrane. Presumably, this cellular location is the result of an abortive attempt by the cell to secrete this protein into the periplasm. Like the lamB-lacZ fusion strain pop3186, this strain also exhibits a Mal^s phenotype. (Note: Unlike lamB, the malE gene product is required for growth on maltose. In order to observe the Mal^s phenotype, a wild-type malE gene must be provided. Strain MM18 is a wild-type, Mal$^+$ strain that has been lysogenized with a λ fusion phage that carries the malE-lacZ fusion. In this merodiploid malE$^+$/malE-lacZ, the Mal^s phenotype can be observed.) These two fusion strains also exhibit another common characteristic, namely the levels of β-galactosidase present in the absence of induction are extremely low. It may be that when these hybrid proteins are embedded in a membrane at low concentrations (the uninduced state), they cannot effectively tetramerize into active enzyme. Indeed, in the absence of inducer, these strains grow very poorly on lactose. By selecting for a Lac$^+$ phenotype, mutants can be obtained in which the cellular location of the hybrid protein has been altered.[31] This selection offers the further advantage of being easily adaptable for the isolation of conditional lethal (e.g., temperature-sensitive) mutations. Since a mutation that blocks export of a variety of proteins may well be lethal, this offers a distinct advantage.

The selection procedure to be described employs lactose tetrazolium agar. Tetrazolium agar can be used for mutant screening or for mutant selection. This is unusual and requires some explanation. Tetrazolium agar is often used to screen isolated colonies for a defect in sugar catabolism. When screening isolated colonies, red indicates a sugar$^-$ phenotype. White

---

[31] D. B. Oliver and J. Beckwith, *Cell* **25**, 765 (1981).

indicates sugar utilization. Because spotting a red colony against a background of white colonies is quite easy, the medium is very useful for obtaining mutants after replica plating, for example. Tetrazolium agar can also be used to select for strains that exhibit a sugar⁺ phenotype. The reason for this is the high sugar content of the medium (1%). Cells unable to metabolize the sugar will form a white lawn on this medium. Rare sugar⁺ cells in the lawn will continue to grow after growth of the lawn has stopped because such cells can metabolize the sugar. Since the color reaction in a lawn of cells is the opposite to that seen with isolated colonies, rare sugar⁺ cells can be simply observed as red colonies growing out of a white lawn. This powerful selection enables one to screen an enormous number of cells (a confluent lawn) for rare mutants.

This selection can be employed with the *malE-lacZ* fusion strain MM18 in the following manner: Aliquots (0.1 ml) of overnight cultures are spread on lactose tetrazolium agar and incubated at 30° for 5 days. Red colonies that appeared on the white lawn are purified on the same medium again at 30°. Conditional lethal mutants can be identified by testing for growth at 42°. More than 100 export-defective mutants have been isolated using this technique. Many of the mutations appear to alter the MalE signal sequence. Such mutations are linked genetically to the *malE-lacZ* fusion. This is most simply demonstrated by inducing the λ fusion phage from the Lac⁺ mutant. If the mutation is carried by the phage, then this lysate will transduce a *lac* deletion strain to a Lac⁺ phenotype with high frequency. This can be done by nonselectively lysogenizing a *lac* deletion strain and testing lysogens for their Lac phenotype. (Note: Since the recombination event that generates the lysogen may occur between the mutation and the fusion joint, not all lysogens will exhibit a Lac⁺ phenotype; see Fig. 4.).

Mutants that exhibit a temperature-sensitive phenotype are likely to be the result of genetically unlinked mutations. This can be simply tested for by curing the fusion phage and scoring the phenotype of the nonlysogenic derivative. This is done by streaking the mutant on TYE agar spread with XG. The plate is then irradiated with UV for 30–60 sec and incubated in the dark for 24 hr at 30°. Colorless colonies, or sections, are then purified. Nonlysogenic cells should exhibit a Lac⁻, λcIh80ˢ phenotype. If these derivatives remain temperature sensitive or if they exhibit an impaired function of the corresponding wild-type gene, the mutation that caused a Lac⁺ phenotype must have been unlinked. In these cured derivatives, the effect of the mutation on protein export can then be tested. By use of this technique, the *secA* mutation was isolated. This conditional (temperature-sensitive) mutation prevents export of a variety of envelope proteins.[31] A generalized method for mapping the unlinked mutations of this type is described in the Appendix.

Application of the Approaches to Other Genes

We have already described the construction of *phoA-lacZ* fusions using *in vitro* techniques. The hybrid protein produced in such strains is localized to the cytoplasmic membrane. As with *lamB-lacZ* fusions, production of large amounts of the hybrid protein is lethal to the cell. The synthesis of the *phoA-lacZ* hybrid protein is induced by starvation for inorganic phosphate. This phosphate starvation sensitivity has been used to select for starvation-resistant mutants in anticipation that signal sequence mutants would be obtained. However, instead, we have found that among a large number of such mutants screened, all are the result of genetic rearrangements (deletions, etc.) on the plasmid. It is for this reason that we recombined the fusion onto a λ phage in order to obtain a single-copy derivative of the hybrid gene. In lysogens for the λ phage carrying the *phoA-lacZ* fusion, selection for a Lac⁺ phenotype has yielded signal sequence mutants.[32]

Because of the problems encountered in selections with the gene fusion on the multicopy plasmid, this approach may not be as practical as the *in vivo* approaches. However, the isolation of signal sequence mutants using *phoA-lacZ* fusions demonstrates further the versatility of the gene fusion approach.

In Vitro Mutagenesis

Knowledge of DNA chemistry coupled with the techniques of recombinant DNA permit one to alter certain amino acids specifically in nearly any protein. Although these techniques (*in vitro* mutagenesis) have not been extensively applied to the study of protein export, the potential is enormous.[33-36] If a particular gene or gene fusion (see Fig. 3) has been cloned and a rather complete restriction map, or better yet, the DNA sequence has been determined, a variety of *in vitro* techniques can be employed to alter a desired region specifically. These techniques permit the isolation of either deletion or point mutations.

Deletion mutations can be generated *in vitro* using the quasi-processive exonuclease *Bal*31. Linearized plasmids (prepared either by digestion with restriction enzymes that cut the plasmid at a single site or by limited digestion with restriction enzymes that cut at several sites) treated with the exonuclease for various times and then religated will yield plasmids that contain deletions of varying extents. When this technique is applied to

[32] S. Michaelis, H. Inouye, D. Oliver, and J. Beckwith, *J. Bacteriol.,* in press.
[33] D. Shortle, D. DiMaio, and D. Nathans, *Annu. Rev. Genet.* **15**, 265 (1981).
[34] M. Smith and S. Gillam, *Genet. Eng.* **3**, 1 (1981).
[35] D. Shortle, J. Pipas, S. Lazarowitz, D. DiMaio, and D. Nathans, *Genet. Eng.* **1**, 269 (1979).
[36] C. Weissmann, S. Nagata, T. Taniguchi, H. Weber, and F. Meyer, *Genet. Eng.* **1**, 133 (1979).

plasmids that carry gene fusions, the technique can be controlled so that only deletions that do not alter the reading frame can be obtained. An example of this method is shown in Fig. 6. In this particular *lamB-lacZ* fusion, digestion with the exonuclease can originate from the unique *Sma*I site near the middle of the *lamB* gene. After digestion, religation, and transformation, strains containing plasmids that carry in-frame deletions can be identified as blue colonies on TYE agar spread with XG. Strains containing plasmids with out-of-frame deletions or deletions that extend into the *lacZ* portion of the hybrid gene will not synthesize active β-galactosidase and hence will form white colonies on the medium. Deletions obtained in this manner can then be recombined onto the chromosome. This permits the analysis of the effect of the deletion on an otherwise wild-type gene. Accordingly, this technique permits one to remove specifically various portions of a particular gene. Its chief limitation concerns the availability of unique or rarely occurring restriction sites in or near the region of interest.

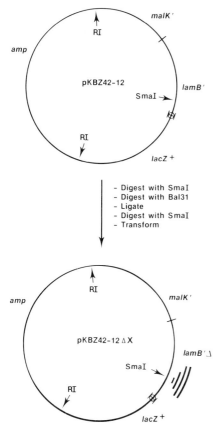

- Digest with SmaI
- Digest with Bal31
- Ligate
- Digest with SmaI
- Transform

FIG. 6. Scheme for isolating in-frame deletions. Plasmid pKB747-12 is a *lamB-lacZ* fusion that has been cloned into pMLB524 (S. A. Benson and T. J. Silhavy, *Cell,* in press) (see Fig. 3). The plasmid shown contains a unique *Sma*I site in the *lamB* portion of the hybrid gene. To create in-frame deletions, the plasmid is linearized by digestion with *Sma*I. The linearized plasmid is then treated with *Bal*31 for varying amounts of time. After nuclease treatment the plasmid is religated. To prevent reisolation of the parent plasmid, a second *Sma*I digestion is performed. Resulting DNA is then transformed into a suitable recipient. Plasmids containing in-frame deletions are scored as Amp^r, LacZ$^+$, i.e., blue in agar spread with XG.

A variety of *in vitro* techniques have been developed for the isolation of point mutations. Most methods require the generation of single-stranded gaps at the sight to be mutagenized. The region with gaps is then treated with a mutagen such as sodium bisulfite. Several different procedures can be used to generate single-stranded gaps. For example, such gaps can be produced at a restriction enzyme cleavage site by treatment of supercoiled DNA with the restriction enzyme, allowing a single-stranded nick to be made, and then inhibition of the cleavage reaction with ethidium bromide. A gap is then produced by use of the 5' to 3' exonuclease activity of DNA polymerase from *Micrococcus luteus.* After treatment with bisulfite, the gap is filled in with DNA polymerase.[37] This method is limited by the presence of restriction sites in the region of interest.

A more general method of producing single-stranded gaps utilizes the enzyme pancreatic DNase I.[38] In the presence of an optimum concentration of ethidium bromide, this enzyme induces a single nick essentially at random in a covalently closed circular DNA molecule. This nick can then be converted to a gap and mutagenized as already described. Since this method will generate mutations throughout the plasmid, it is most useful when the desired mutant exhibits an easily scorable phenotype.

A two-step method has been developed that should allow placement of a nick in any segment of DNA for which a single-stranded fragment can be isolated.[39] In this technique, the RecA protein of *E. coli* is used to produce a displacement loop (D loop) at the site in a covalently closed circular plasmid corresponding to a homologous single-stranded fragment. The single-strand-specific S1 nuclease is then used to produce a nick in the displaced DNA. The nicks are then converted to small gaps and mutagenized with bisulfite. By use of this technique, a collection of signal sequence mutations in the gene coding for the periplasmic enzyme β-lactamase has been isolated.[33,40]

The techniques described so far are limited by the fact that bisulfite is a specific mutagen causing CT→GA transitions. Thus, many potentially interesting mutations cannot be obtained. Several methods have been reported that should overcome these limitations. The first causes mutations by nucleotide misincorporation during the DNA polymerase reaction. Since a short, single-stranded gap is a good substrate for most DNA polymerases, any of the techniques described above can be used in the initial step. The gap

[37] D. Shortle and D. Nathans, *Proc. Natl. Acad. Sci. U.S.A.* **75,** 2170 (1978).
[38] D. R. Shortle, R. F. Margolskee, and D. Nathans, *Proc. Natl. Acad. Sci. U.S.A.* **76,** 6128 (1979).
[39] D. Shortle, D. Koshland, G. M. Weinstock, and D. Botstein, *Proc. Natl. Acad. Sci. U.S.A.* **77,** 5375 (1980).
[40] D. Koshland, R. T. Saver, and D. Botstein, *Cell* **30,** 903 (1982).

is then repaired in a reaction mixture in which a particular dNTP is omitted and Mn^{2+} ions are added to increase the frequency of misincorporation.[33,41]

The techniques of oligonucleotide synthesis have improved to the point where they can now be automated. Thus, in the not-too-distant future, synthetic polynucleotides will be generally available. Equipped with such reagents, it will be possible to construct specific mutants with a defined substitution at any base pair. This is done by synthesis of a short oligonucleotide (12–15 base pairs) that contains the desired substitution near the middle of the sequence. This oligonucleotide is then annealed with a plasmid (single-stranded) containing the homologous wild-type sequence. The complex is then used as a substrate for DNA polymerase. After ligation, the plasmid is transformed into cells. Replication will segregate the mutant plasmid after the first round of replication.[34] This technique has been used to generate mutations in the lipoprotein signal sequence.[42]

In studying subjects such as protein localization, it is often possible to clearly identify desired mutational changes. The techniques described in this section are likely to provide the most straightforward method for their construction. We will undoubtedly see further refinement and widespread application of this powerful methodology.

Mutant Screens

As we have already noted, screening a mutagenized cell population for export-defective mutants has not proved to be worthwhile. The probable reasons for this failure are as follows: First, even after heavy mutagenesis, the desired mutant will, at best, be present at a frequency of 10^{-3}. Unless a relatively simple screen can be devised, scoring for an export defect can require a prohibitive amount of work. Since there is no phenotypic difference between mutants that do not export and mutants that do not synthesize a particular protein, no simple screen has been devised. Techniques have, however, been described that overcome these problems at least to some degree.

Wanner et al.[43] have designed a procedure for the isolation of mutants unable to secrete periplasmic proteins. This procedure is based upon the assumption that some factor(s) exists that is necessary for the secretion of periplasmic proteins, e.g., alkaline phosphatase, but is not necessary for the synthesis of cytoplasmic proteins, e.g., β-galactosidase. Mutants are detected

[41] D. Shortle and D. Botstein, personal communication.
[42] S. Inouye, X. Soberon, T. Franceschini, K. Nakamura, K. Itakura, and M. Inouye, *Proc. Natl. Acad. Sci. U.S.A.* **79**, 3438 (1982).
[43] B. L. Wanner, A. Sarthy, and J. Beckwith, *J. Bacteriol.* **140**, 229 (1979).

on agar medium that includes indicators for both alkaline phosphatase and β-galactosidase synthesis. In addition, since mutations defective in the secretion of a number of envelope proteins may be lethal to the cell, the general approach used by Schedl and Primakoff[44] for the detection of conditional-lethal mutants affecting tRNA synthesis was adopted.

The medium used contains a tetrazolium dye (an indicator for lactose metabolism) and 5-bromo-4 chloroindolylphosphate-p-toluidine (XP; an indicator for alkaline phosphatase). This medium was initially developed by Brickman and Beckwith[45] for a different purpose. On this medium, PhoA⁻, LacZ⁻ colonies are white, PhoA⁻, LacZ⁺ colonies turn red, PhoA⁺, LacZ⁻ colonies turn blue, and PhoA⁺, LacZ⁺ colonies turn purple. *Escherichia coli* strain XPh4, which is *lacZ phoA phoR rpsL* (*strA*), is mutagenized with nitrosoguanidine and then cultured in TYE medium for 4 hr at 30°. Diluted portions are spread on indicator plates to yield about 300 colonies per plate. The plates are incubated at 30° until the colonies grow to 1–2 mm in diameter (approximately 36 hr) to allow any temperature-sensitive mutants to grow, after which the plates are incubated at 42°. One hour later each plate is sprayed with 1 ml of a prewarmed phage mixture containing 2×10^{10} ϕ80p*phoA*⁺ and 2×10^{10} ϕ80p*lacZ*⁺ phages. Incubation is continued at 42°. Sprayed colonies that turn red are picked and purified for further study. These mutants are presumed to be defective in alkaline phosphatase but not β-galactosidase synthesis. The mutation *perA* was isolated in this manner. The export of several different periplasmic and outer membrane proteins is affected by the *perA* mutation.

A second method requires preexisting knowledge of a chromosomal locus known to be involved in protein export. This technique uses local mutagenesis to increase the frequency of desired mutants. A method for doing local mutagenesis was suggested by Hong and Ames[46]; they mutagenized a suspension of the general transducing phage P22 (a *Salmonella typhimurium* generalized transducing phage) and then used it in a transduction cross, selecting for incorporation of a donor gene near the region of interest. Each donor fragment introduces a segment of the surrounding chromosome that has been heavily mutagenized. Thus, each transductant has a high probability of acquiring a mutation in any gene closely linked to the selective marker. Since only this tiny portion of the chromosome has been mutagenized, the probablity of multiple mutations is much lower than for standard mutagenesis methods. (However, multiple mutations within the region of interest may be encountered.) Use of drug-resistance elements has made this technique more widely applicable. By using tetracycline

[44] P. Schedl and P. Primakoff, *Proc. Natl. Acad. Sci. U.S.A.* **70**, 2091 (1973).
[45] E. Brickman and J. Beckwith, *J. Mol. Biol.* **96**, 307 (1975).
[46] J.-S. Hong and B. N. Ames, *Proc. Natl. Acad. Sci. U.S.A.* **68**, 3158 (1971).

resistance as a selective marker, the chromosomal region near any Tn10 insertion can be heavily mutagenized. Since this selective marker (e.g., tetracycline resistance) can be placed close to any gene (see Appendix), the technique can be applied to any region of the chromosome.

At present there are a number of ways to do this type of experiment. Either transducing phages can be mutagenized or the donor bacteria can be mutagenized before preparing a generalized transducing lysate. Generally we employ the latter technique. The details of this method are as follows. The strain to be mutagenized is grown overnight in LB. The following day, the culture is diluted 1 to 10 and it is allowed to grow to late log phase (OD_{600} ~ 1.0). Cells, 2 ml, are placed in a centrifuge tube and 50 μl of N-methyl-N'-nitro-N-nitrosoguanidine (2.5 mg/ml in 95% ethanol) are added. The mixture is incubated for 10 min at 27°. Cells are removed by centrifugation and washed once with minimal medium 63. The cell pellet is resuspended in 5 ml of LB and grown overnight.

The following day, the culture is diluted 1 to 20 and grown to mid log phase in 10 ml of LB containing 5 mM $CaCl_2$. Two drops of P1vir lysate ($\geq 5 \times 10^8 \phi$/ml) is added, and the culture is incubated with shaking until lysis occurs (2–3 hr). About 5–10 drops of chloroform are added, and the suspension is mixed throughly and then clarified by centrifugation. This lysate can be used to transduce a suitable recipient. The desired mutants should represent 0.1–0.3% of the transductants provided that the gene of interest is closely linked to the selected marker (≥ 50% cotransduction frequency).

Using a procedure similar to this, Ito et al.[47] isolated a temperature-sensitive mutant that affects protein export. The target for the local mutagenesis was the large ribosomal gene cluster, a locus known to contain the prlA mutation (see the following section). The cotransducible selected marker was aroE+. All cell manipulations were done at 30°, and transductants were screened for the inability to grow at 42°. Temperature-sensitive mutants were screened for an export defect by looking for accumulated precursor maltose-binding protein after short exposure to the nonpermissive temperature. The precursor protein was identified by antibody precipitation.

The technique of local mutagenesis has been expanded to enable the isolation of amber (UAG) mutants defective in essential genes. This expanded procedure is called the blue ghost technique.[48] In this technique a P1 lysate grown on a wild-type strain containing the amber suppressor supF is used to transduce a recipient strain containing the following relevant genetic markers: araD araC(am) lacZ(am) (ϕ80) (ϕ80pSu3+). Any marker can be used to select transductants, including drug resistance conferred by a trans-

[47] K. Ito, M. Wittekind, M. Nomura, A. Miura, K. Shiba, T. Yura, and H. Nashimoto, Cell, in press.
[48] S. Brown, E. Brickman, and J. Beckwith, J. Bacteriol. 146, 422 (1981).

locatable element. The selection is done at 30°. After transduction, colonies are replica-plated onto M63 medium containing 0.2% glycerol and M63 medium containing 0.2% glycerol, 0.5% arabinose, XG, and 1 mM isopropyl-β-D-thiogalactoside. The replica plates are incubated overnight at 30°.

The rationale behind this technique is as follows: Colonies derived from cells that contain an amber mutation in an essential gene will grow on the M63 glycerol plate since the amber mutation is suppressed by Su^{3+}. On the arabinose plate they will appear as hazy, dark blue ghosts. This is because the *araD* mutation confers sensitivity to arabinose if the strain is *araC*$^+$. In the presence of Su3$^+$, the recipient strain is sensitive to arabinose. The dark blue color is caused by the high levels of β-galactosidase that are induced by the isopropyl-β-D-thiogalactoside. Strains that do not contain an amber mutation in an essential gene will grow on M63 glycerol. On arabinose-containing medium, they will appear as large, pale-blue colonies. Since the recipient is a double $\phi80/\phi80pSu3^+$ lysogen, the phage containing the suppressor can be lost at high frequency because of homologous recombination between the two phages. The colonies that grow on arabinose-containing medium are the result of outgrowth of these frequent derivatives that have lost $\phi80pSu3^+$. They are resistant to arabinose because, in the absence of the suppressor, the strain is *araC*. The large colonies are pale blue because, in the absence of the suppressor, the strain is *lacZ*.

Given the ability to insert a drug-resistance element in any region of the chromosome (see Appendix), this technique should permit the isolation of an amber mutation in any region of the chromosome. If this technique can be used with strains carrying Su3^{ts}, export-defective mutants could be identified as those that accumulate precursors of exported proteins at the nonpermissive temperature.

Suppressors of Export-Defective Mutations

As already noted, one approach to identifying cellular components of the export machinery is to isolate mutants that are generally export defective, i.e., mutants that fail to export a number of envelope proteins. Another approach is to devise a selection of mutants in which an internalized protein is exported. The mutants we have isolated in which the precursor of the *lamB* gene product is found in the cytoplasm provide a selection for the export of an internalized protein. These mutant strains do not localize LamB because the export machinery cannot recognize the mutationally altered signal sequence. We reasoned, therefore, that it should be possible to alter the export machinery by mutation to restore recognition of the signal sequence. Such mutations would define components of the export machinery. These mutations could be obtained by selection of reversion of a signal-sequence mutation. Reversion can be selected by growth on dextrin. The mutation responsible for reversion must restore export of LamB to the outer membrane.

The inability of strains containing a mutation in the signal sequence region of the *lamB* gene to grow on maltodextrins provided a simple means to perform this genetic test. When such strains are plated on minimal maltodextrin (MS Dex) agar, pseudorevertant Dex⁺ colonies appear as a result of second-site mutations that cause an alteration of the export machinery. The major problem encountered with this simple test is to be able to distinguish "true" revertants, that is, a back mutation resulting in the regeneration of the wild-type *lamB* gene, from the desired pseudorevertants. Both appear as Dex⁺ colonies. Small deletion mutations that do not prevent synthesis and are internal to the *lamB* signal sequence region can be used to solve this problem. Because these mutations are deletions, true reversion is not possible, and the Dex⁺ colonies obtained in this way should be greatly enriched for the desired second-site pseudorevertants.

The suppressor mutations were obtained as follows: strain SE2060 which contains a deletion in the signal sequence region of the *lamB* gene is spread on TYE agar, and isolated colonies are inoculated into 5 ml LB and grown overnight at 37°, then 1 ml of cells from each tube is centrifuged and resuspended in 0.2 ml of M63 medium. The entire suspension is then spread on M63 agar containing 0.7% dextrin. Plates are then incubated at 37° for 3–4 days. One colony per plate is picked and purified to ensure that each of the spontaneous mutants is the result of an independent event. With this technique, three different classes of suppressor mutation have been isolated: *prlA*, *prlB*, and *prlC*. The *prlA* suppressor maps in the large ribosomal gene cluster near *aroE*.[49]

Conclusion

In this chapter we have described various genetic techniques that have been successfully applied to the study of protein export in *E. coli*. Most, if not all, of these techniques can be applied to any exported protein provided that suitable selections and/or screens can be devised. In view of the pronounced similarity that is observed in the mechanisms of protein export throughout biology, results obtained in *E. coli* are likely to have an immediate impact on similar studies in other organisms. Furthermore, the genetic techniques developed in *E. coli* may well provide the basis for genetic analysis of protein export in other systems, including eukaryotic cells.

---

[49] S. D. Emr, S. Hanley-Way, and T. J. Silhavy, *Cell* **23**, 79 (1981).

Appendix

Insertion of a Transposon in, or near, a Gene of Interest

A number of transposons capable or more or less random insertion are available. Generally we employ the transposon Tn10 (confers resistance to tetracycline, Tet^r). We make this choice because Tn10 insertions are stable and, more important, because selections for tetracycline sensitivity have been developed.[50,51] The ability to select for loss of the Tn is useful for a variety of genetic techniques.

Several different procedures and a variety of Tn10 vectors have been described. The procedure we use routinely is listed below. In most cases we construct a random pool of Tn insertions in a wild-type strain. From this pool, Tn10 insertions in any gene for which a positive selection exists can be recovered. Furthermore, P1 lysates grown on this pool are capable of transducing a Tn10 near any gene or in any nonessential gene.

The first step in this procedure is to translocate the Tn10 from the genome of a bacteriophage (λ561) onto the *E. coli* chromosome. Cells that have acquired the Tn10 will be recognized by their ability to form Tet^r colonies. For this experiment to work, several requirements must be met. First, the bacteriophage must not be able to lysogenize. λ561 is cI$^-$ (cI::Tn10) and Att$^-$ (b221 has removed *attP*), thus the first requirement is fulfilled. Second, the phage must not be able to replicate. This phage carries amber mutations in both the *O* and *P* genes of λ and thus cannot replicate. Furthermore, double revertants of genes O_{am} and P_{am} are uncommon and seldom present a problem in the handling of this phage. Note that in the following procedure 20 mM citrate is present throughout. Citrate is a chelating agent, and that prevents λ adsorption.

A suitable recipient, MC4100, is grown overnight in 5 ml of LB. The following day, cells are pelleted and resuspended in 2.5 ml of 10 mM MgSO$_4$. An aliquot, 0.2 ml, of this suspension is added to 20 ml of LB, and cells are grown to mid log phase (OD$_{600}$ ~ 0.5). Into 10–20 small Wasserman tubes, 1 ml of cells and 0.05 ml of λ561 (~ 10^{10} ϕ/ml) are placed. Phages are allowed to adsorb for 20 min at 32°. After adsorption, 1 ml of LB containing 40 mM citrate is added and the mixture is incubated with shaking for 1 hr at 31°. Finally, cells are pelleted and resuspended in 0.2 ml of LB containing 20 mM sodium citrate (LBC), and the entire contents are spread on TYE agar containing 25 μg of tetracycline per milliliter and 20 mM sodium citrate. Plates are incubated overnight at 37°.

The following day 1 ml of LBC is added to each plate. Colonies are

[50] B. R. Bochner, H. Huang, G. L. Schieven, and B. N. Ames, *J. Bacteriol.* **143**, 926 (1980).
[51] S. R. Maloy and W. D. Nunn, *J. Bacteriol.* **145**, 1110 (1981).

carefully loosened from the plates and pooled in a 50-ml centrifuge tube. The suspension is then centrifuged, and the supernatant is discarded. Cells are resuspended in 10–20 ml of LBC and washed once. An aliquot, 0.2–0.3 ml of the suspension, is inoculated into LB containing 25 μg of tetracycline per milliliter, and the culture is grown to mid log phase. The cells are pelleted and washed once with LB.

If the pooled population is sufficiently large ($\sim$ 20,000 independent colonies), then it should contain cells with Tn10 insertions near any gene or in any nonessential gene. At this point, aliquots of the culture can be frozen for future use. Recovering the desired insertion can be done in several ways depending upon the selections and/or screens available. For example, to find an insertion in a gene for sugar catabolism, aliquots from the pool can be plated on indicator agar; to find auxotrophs, penicillin enrichments can be performed.

In order to obtain an insertion near a particular gene, P1 lysates are prepared from the pooled population. This lysate can then be used to transduce a strain carrying a mutation in the gene of interest to a wild-type phenotype simultaneously selecting for Tet[r]. If no positive selection is available, the transduction can be performed by simultaneous selection for a linked marker and for Tet[r]. These transductants can then be scored for the desired phenotype.

General Method for Mapping Mutants with Pleiotropic Effects on Secretion

This method is particularly useful in those cases where the mutation is not conditionally lethal, although it is also generally applicable. (Conditionally lethal mutants can be mapped by using a collection of Hfrs to determine its approximate position on the chromosome.) The technique was described by Johnston and Roth[52] for the mapping of a UGA suppressor. The principles behind it are (*a*) to obtain a Tn10 transposon (carrying tetracyline resistance) inserted adjacent to the gene to be mapped; (*b*) to use the Tn10 genetic homology to direct the insert of an F' also carrying a Tn10 into the chromosome; and (*c*) determining the origin of transfer of the resulting Hfr, to locate the position of the Tn10 (and, thus, of the gene in question) on the chromosome. It has been used to map the *secB* locus.[53] Mutations in *secB*, none of which are conditionally lethal so far, restore a Lac$^+$ phenotype to the *malE-lacZ* fusion strain, MM18.

By use of techniques described in the preceding section, a collection of Tn10 insertions in strain MM18 *secB* (Lac$^+$) is prepared (Note: since our goal in the experiment is to isolate a Tn10 insertion *near secB*, the pooled

[52] H. M. Johnston and J. R. Roth, *J. Bacteriol.* **144**, 300 (1980).
[53] C. Kumanoto and J. Beckwith, unpublished data (1982).

population does not need to be large, i.e., 1000), and P1 is grown on the pooled insertions. This P1 lysate is then used to transduce MM18 simultaneously to Lac$^+$ and tetracycline resistance. Transductants with both characters are in most cases strains in which the Tn10 is adjacent to *secB*. However, this should be verified by making a P1 lysate on these latter strains and repeating the transduction of MM18.

Two F' factors, which are temperature sensitive in their replication and which carry the *lac* genes and Tn10 in one of the two possible orientations, are introduced into the Tn10 *secB* strains by conjugation. The mating is done at 30°, and Lac$^+$ sexductants are picked.[6] The Lac$^+$ colonies are purified at 30° and grown overnight in lactose minimal liquid medium. These F' strains are then grown at 42°, and Lac$^+$ derivatives are selected at the high temperature. Since the F' cannot replicate at 42°, Lac$^+$ bacteria can only arise by integration of the F' into the chromosome via Tn10, which is the only genetic homology that exists between the episome and chromosome. (The strains are deleted for the chromosomal *lac* region.) Two-hundredths milliliter of a saturated culture of the F' strain (grown at 30°) is inoculated into 5 ml of lactose minimal medium and incubated at 42° for 2 days. These cultures are diluted 1 : 10 in the same medium and grown to saturation at 42°. Dilutions of these culture are spread on lactose MacConkey agar at 42° so as to obtain single colonies. Any stable Lac$^+$ colonies at this temperature are candidates for Hfr derivatives.

The possible Hfr derivatives are then tested for transfer of various chromosomal markers. By such an analysis, it should be possible to determine precisely the origin of transfer of this Hfr, which will in turn indicate the location of the Tn10 in the chromosome. More precise mapping can then be done by using P1 transduction with known markers in the region.

Acknowledgment

Research described in this article from the authors' laboratories was sponsored by grants from the National Institutes of Health, National Science Foundation, The American Cancer Society, and the National Cancer Institute, DHHS, under Contract No. NO1-CO-75380 with Litton Bionetics, Inc.

[3] Purification and Characterization of Leader Peptidase from *Escherichia coli*

By P. B. Wolfe, C. Zwizinski, and William Wickner

Many secretory and membrane proteins of eukaryotic and prokaryotic cells are synthesized as precursors bearing an amino-terminal extension of 15–30 amino acid residues. These leader (signal) sequences are enriched in hydrophobic residues and are thought to be important for the correct localization of proteins into or across the lipid bilayer. They are also transient, being removed by an endopeptidase during or shortly after protein synthesis. Removal of the leader peptide makes these events irreversible. Furthermore, the enzyme may serve as a membrane recognition element for leader peptide-bearing nascent polypeptides and allow proteins to discriminate among membrane-bound compartments.

This laboratory has reported a 6000-fold purification and partial characterization of a leader peptidase from *Escherichia coli*.[1] We now report a greatly improved purification protocol from a strain that overproduces the enzyme and summarize the properties of the purified protein.

Leader Peptidase Assay

The activity of *E. coli* leader peptidase is determined by measuring the conversion of procoat, the precursor to the major coat protein of bacteriophage M13, to its mature form. Radiolabeled procoat is synthesized in an *in vitro* protein synthetic reaction[2] in the presence of [³⁵S]methionine using mRNA from M13-infected cells to direct the synthesis.[3] The substrate [5 μl, diluted to approximately 10,000 cpm/μl in 50 mM triethanolamine-HCl (pH 8.0), 1 mM EDTA, 1% Triton X-100], is mixed with decreasing concentrations of enzyme (5 μl, diluted in the same buffer), and the reaction is allowed to proceed for 30 min at 37°. The reaction is stopped by chilling on ice, and the samples are mixed with an equal volume of gel electrophoresis sample buffer[4] and boiled for 2 min. Ten-microliter aliquots are subjected to sodium dodecyl sulfate (SDS)-gel electrophoresis on 19.3% polyacrylamide gels as described.[1,5] The gels are fluorographed using a salicylate method of Chamberlin.[6]

One unit of leader peptidase activity is defined as the amount of enzyme required to convert radiolabeled procoat to bands of procoat and coat protein of equal intensity after fluorography (see Fig. 1).

Purification of Leader Peptidase

In earlier studies we isolated the leader peptidase from wild-type *E. coli*.[1] More recently we have developed strains containing a multicopy number

[1] C. Zwizinski and W. Wickner, *J. Biol. Chem.* **255**, 7973 (1980).
[2] L. M. Gold and M. Schweiger, this series, Vol. 20, p. 535.
[3] M. LaFarina and P. Model, *Virology* **86**, 268 (1978).
[4] U. K. Laemmli, *Nature (London)* **227**, 680 (1970).
[5] K. Ito, G. Mandel, and W. Wickner, *Proc. Natl. Acad. Sci. U.S.A.* **76**, 1199 (1979).
[6] J. P. Chamberlin, *Anal. Biochem.* **98**, 132 (1979).

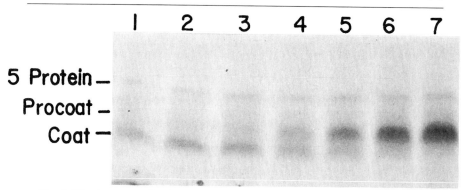

FIG. 1. Titration curve of leader peptidase activity. Serial dilutions of leader peptidase were made, and 5 μl of each dilution were incubated for 30 min in the presence of radiolabeled procoat, then subjected to gel electrophoresis in sodium dodecyl sulfate. One unit of leader peptidase activity is illustrated by lane 4 of the fluorograph.

plasmid that bears the structural gene for leader peptidase.[7] Such strains overproduce the enzyme 30- to 100-fold and are being used as starting material in the purification.

Membrane Preparation. Large-scale preparations of *E. coli* K12 strain HJM114 bearing the plasmid PS9 are grown to an A_{600} of 1.0 at 30° in L broth[8] in the fermentor. The temperature is shifted to 42° to induce transcription off the leader peptidase gene, and cell growth is continued at 42° for 1 hr. Cells are harvested by centrifugation (4°, 2800 g, 30 min), homogenized in a Waring blender in an equal volume of 50 m*M* Tris-HCl (pH 7.5) containing 10% sucrose, frozen by rapid pipetting into an excess of liquid nitrogen, and stored at −90°.

Cells are thawed in three volumes of 10 m*M* Tris-HCl (pH 8.5) containing 5 m*M* EDTA and 20% sucrose. Lysozyme and DNase I are added to final concentrations of 0.15 mg/ml and 0.015 mg/ml, respectively. The cell suspension is stirred at room temperature for 30 min, then frozen in a Dry Ice–acetone bath and thawed in a room-temperature water bath. To activate the DNase, magnesium acetate is added to a final concentration of 5 m*M*, and the suspension is stirred at room temperature for 30 min. Large membrane fragments are collected by centrifugation at 40,000 g for 30 min at 2°, suspended in 10 m*M* triethanolamine-HCl (pH 7.5) containing 10% glycerol, and centrifuged again at 40,000 g for 30 min at 2°.

Extraction in Nonionic Detergent. Washed membranes are homoge-

---

[7] T. Date and W. Wickner, *Proc. Natl. Acad. Sci. U.S.A.* **78**, 6106 (1981). See also this volume [4].

[8] J. H. Miller, "Experiments in Molecular Genetics," p. 433. Cold Spring Harbor Lab., Cold Spring Harbor, New York, 1972.

nized on ice in a Dounce homogenizer in one-fourth of the original volume in 10 mM triethanolamine-HCl (pH 7.5) containing 10% glycerol and 1% Triton X-100. The membrane extract is centrifuged at 40,000 g for 30 min at 2°.

DEAE Chromatography. The Triton X-100 extract is applied directly to a column of DEAE-cellulose (Whatman DE-52, 10 volumes of resin per volume of sample) equilibrated at 4° in 10 mM triethanolamine-HCl (pH 7.5), 5 mM magnesium chloride, 10% glycerol, and 1% Triton X-100. After application of the sample, the column is washed with two column volumes of the same buffer. A broad peak of leader peptidase emerges after the unbound material has passed through the column. Fractions containing leader peptidase activity are pooled and dialyzed overnight at 4° against two changes of 25 mM imidazole-acetate (pH 7.4) containing 1% Triton X-100 in preparation for chromatofocusing.

Chromatofocusing. The dialyzed DEAE pool is applied to a 2-ml column of Polybuffer exchanger (PBE 94, Pharmacia Fine Chemicals) equilibrated at 4° in 25 mM imidazole-acetate (pH 7.4) containing 1% Triton X-100. After application of the sample, the column is washed with two column volumes of the same buffer. The column is developed by elution with 12 column volumes of Polybuffer 74 (Pharmacia Fine Chemicals), diluted 10-fold, and made 1% in Triton X-100. Leader peptidase elutes at a pH of approximately 6.9. A summary of a typical purification is shown in Table I.

Properties of Leader Peptidase

Physical Properties. The subunit molecular weight of leader peptidase is estimated to be 37,000 by gel electrophoresis in SDS. Its isoelectric point is approximately 6.9 as judged by its elution from the chromatofocusing resin.

Stability. The isolated enzyme is stable at low ionic strength in solutions of nonionic detergents at low temperature. The pH optimum is between 8.0 and 9.0. Enzyme activity is reversibly inhibited by ionic and bile salt detergents (SDS, cholate, deoxycholate) and can be renatured, even after boiling in SDS, by dilution in Triton X-100.

TABLE I
PURIFICATION OF *Escherichia coli* LEADER PEPTIDASE

Step	Volume (ml)	Protein (mg)	Units	Specific activity	Yield (%)
Cell lysate	270	2025	3.45×10^8	1.71×10^5	100
Triton X-100 extract	86	516	2.20×10^8	4.27×10^5	63
DEAE-cellulose	208	5.4	0.66×10^8	1.23×10^7	19
Chromatofocusing	8	3.2	0.21×10^8	1.28×10^7	6

Sensitivity to Protease Inhibitors. Leader peptidase is resistant to N-tosyl-L-phenylalanylchloromethyl ketone (TPCK), N-tosyl-L-lysylchloromethyl ketone (TLCK) and phenylmethylsulfonyl fluoride (PMSF) and the metal chelators EDTA, o-phenanthroline, or 2,6-pyridinedicarboxylic acid.[9] Peptidase activity is sensitive to high concentrations of sodium chloride (> 160 mM) or magnesium ions (> 1 mM) at neutral pH. N-Ethylmaleimide, dinitrophenol, and carboxyphenanthroline are also effective inhibitors at high concentrations.[9]

Substrate Specificity. Leader peptidase has been shown to process accurately the precursor of the major capsid protein of bacteriophage M13 as well as periplasmic and outer membrane proteins[10] to their mature forms.[1] The specificity of the enzyme is apparently not solely dependent on the precise sequence of amino acids at the cleavage site. In procoat, this site is formed by the sequence Ala-Ala. Adjacent alanyl residues are also found at positions $+9$ and $+10$ in coat but are not attacked by the enzyme. When a battery of peptides were examined as potential competitive inhibitors of procoat hydrolysis by leader peptidase, none were found to be effective (unpublished observation, see Table II). In addition, a review of precursor peptide cleavage sites indicates that almost all the amino acids can form this site, although residues bearing short side chains (Ser, Gly, and Ala) are favored. Clearly, protein conformation is as important as a specific amino acid sequence in determining the site of hydrolysis by leader peptidase.

Membrane Assembly

The purification of leader peptidase to homogeneity and its characterization have led to a deeper understanding of the events occurring during membrane assembly. The correct processing of precursor proteins by the purified enzyme demonstrates that no cofactors, accessory proteins, or special environmental conditions are necessary for processing to occur. The peptidase is found in both the inner and outer membranes.[9] It is the only known protein found in both membranes, which may reflect its role in membrane assembly.

The development of a sensitive assay for leader peptidase activity has enabled us to begin genetic analyses. We have used the assay to screen the Clarke and Carbon collection for strains containing a plasmid-borne copy of the structural gene for leader peptidase.[7] With the appropriate manipulations we have been able to construct strains that overproduce the enzyme up to 100-fold.[7] This has greatly aided in the purification of large amounts of leader peptidase. In addition, we have used these strains to study the effects of leader peptidase overproduction on the assembly of bacteriophage M13

[9] C. Zwizinski, T. Date, and W. Wickner, *J. Biol. Chem.* **256** 3593 (1981).
[10] C. J. Daniels, D. Oxender, D. Perrin, and M. Hofnung, personal communication.

TABLE II

PEPTIDES TESTED FOR THEIR ABILITY TO INHIBIT PROCESSING OF M13 PROCOAT TO
COAT PROTEIN BY LEADER PEPTIDASE

Dipeptides		Tripeptides	Tetrapeptides
Ala Ala	Gly Trp	Ala Ala Ala	Ala Ala Ala Ala
Ala Asn	Gly Tyr	Ala Gly Gly	Gly Gly Gly Gly
Ala Asp	Gly Val	Ala Leu Gly	Pro Phe Gly Lys
Ala Glu			Phe Gly Phe Gly
Ala Gln	Leu Ala	Gly Ala Ala	Phe Gly Gly Phe
Ala His	Leu Phe	Gly Gly Ala	
Ala Ile	Leu Ser	Gly Gly Gly	
Ala Leu	Leu Gly	Gly Gly Ile	Others
Ala Lys		Gly Gly Leu	
Ala Met	Lys Ala	Gly Gly Phe	Ala Ala Ala Ala Ala
Ala Phe	Lys Val	Gly Gly Val	Ala Ala Ala Ala Ala Ala
Ala Pro		Gly His Gly	Gly Gly Gly Gly Gly
Ala Ser	Met Ala	Gly Phe Ala	Gly Gly Gly Gly Gly Gly
Ala Thr	Met Leu	Gly Phe Phe	
Ala Trp		Gly Pro Ala	Arg Pro Pro Gly Phe Ser Pro Phe Arg
Ala Tyr	Phe Ala	Gly Leu Tyr	
Ala Val	Phe Gly	Gly Leu Ala	
	Phe Leu		
Arg Asp	Phe Phe	Leu Gly Gly	
		Leu Gly Phe	
Gly Ala	Pro Ala	Leu Leu Leu	
Gly Asn	Pro Gly		
Gly Asp	Pro Leu	Phe Phe Phe	
Gly Glu	Pro Ile	Phe Gly Gly	
Gly Gly	Pro Met		
Gly His	Pro Phe	Ser Ser Ser	
Gly Ile	Pro Trp		
Gly Leu	Pro Tyr	Tyr Gly Gly	
Gly Met	Pro Val		
Gly Phe		Val Gly Gly	
Gly Pro	Ser Met	Val Tyr Val	
Gly Ser			
Gly Thr	Val Pro		

and its derivatives.[7,11] The isolation of conditionally lethal, temperature-sensitive mutants in leader peptidase should aid in these studies.

Purified leader peptidase has been used to reconstitute membrane assembly from purified components.[12,13] Vesicles of *E. coli* phospholipids that contain leader peptidase have been shown to bind purified M13 procoat

[11] R. Zimmermann, C. Watts, and W. Wickner, in press.
[12] P. Silver, C. Watts, and W. Wickner, *Cell* **25,** 341 (1981).
[13] C. Watts, P. Silver, and W. Wickner, *Cell* **25,** 347 (1981).

made *in vitro* and convert it posttranslationally into mature coat protein. A significant portion of the coat protein has the correct orientation in the bilayer. These results suggest that leader peptidase and a phospholipid bilayer are the only components necessary for correct binding, insertion, and processing of this precursor protein.

Future Prospects

Overproduction of leader peptidase has allowed the enzyme to be purified in very large quantities. Such amounts will be useful for determination of the physical properties of the enzyme, the amino acid composition and sequence, and the quaternary structure and in the production of antibodies. Antibodies to leader peptidase will be powerful tools for dissecting the events of membrane assembly and localizing the enzyme in the membrane.

Several questions remain concerning the role of leader peptidase in membrane assembly. What is the basis of cleavage specificity? Are there other peptidases with different specificities? Is the orientation of the enzyme in the membrane asymmetric? Does the enzyme have a role in recognition of nascent precursor polypeptides? Does it catalyze protein transfer as well as peptide bond hydrolysis? Further genetic and biochemical studies should help to answer these questions.

[4] Molecular Genetics of *Escherichia coli* Leader Peptidase

By Takayasu Date, Pamela Silver, and William Wickner

One enzymatic activity associated with the integration of proteins into membranes proteolytically removes the leader (signal) sequence from precursor molecules. This activity, termed leader peptidase, has been isolated from *E. coli* membranes and extensively characterized.[1,2] It is the only enzyme exclusively associated with membrane biogenesis and secretion to be isolated to homogeneity and studied biochemically in the purified form. It is now possible to examine membrane assembly using a genetic approach that focuses on the leader peptidase gene. Described here are methods for the identification of the leader peptidase gene and genetic manipulations to determine the role of leader peptidase in cell growth.

[1] C. Zwizinski and W. Wickner, *J. Biol. Chem.* **255**, 7973 (1980); C. Zwizinski, T. Date, and W. Wickner, *ibid.* **256**, 3593 (1981).
[2] P. Wolfe, P. Silver, and W. Wickner, *J. Biol. Chem.* **257**, 7898 (1982).

Isolation of the Leader Peptidase Gene

The clone carrying the leader peptidase gene was isolated from the Clarke and Carbon collection, which contains over 2000 *E. coli* strains. In each strain, a random segment of the *E. coli* genome is present in the multicopy plasmid ColE1.[3] Generally the identification of a specific cloned gene is carried out by mating a suitable female *E. coli* auxotrophic strain with each member of the collection and selecting directly for complementation of the host chromosomal mutation by the hybrid plasmid DNA, which is transferred during mating. This method is not applicable for the screening of the library for the clone carrying the leader peptidase gene because no mutants in the enzyme are currently available. However, one can exploit the observation that the increased copies of the plasmid can lead to overproduction of the products of the cloned genes.[3] Leader peptidase can be identified in detergent extracts of whole cells by its insensitivity to serine protease inhibitors[2] and by its ability posttranslationally to cleave M13 procoat to coat protein.[4]

Experiments

Solution

Solution A: 20% (w/v) sucrose, 1 mg/ml lysozyme, 10 mM EDTA, 10 mM Tris-HCl (pH 8.1), 1% Triton X-100, 5 μg/ml DNase, 1 μg/ml RNase

Solution B: 50 mM triethanolamine, 1 mM phenylmethylsulfonyl fluoride, 1 mM *N*-carbobenzyloxy-L-phenylethylchloromethyl ketone, 1 mM L-tosylamido-2-phenylchloromethyl ketone

1. Preparation of Crude Extracts

Each strain of the collection is cultivated in 1.3 ml of L broth[5] to $A_{600} = 0.5$. The culture is immediately chilled, and the cells are collected by centrifugation in the Brinkmann microfuge for 3 min at 4°. The cells are resuspended by vortexing in 50 μl of solution A and placed at room temperature for 30 min to allow cell lysis. The leader peptidase in the crude extracts remains fully active for up to 6 months when stored at $-20°$. One hundred extracts can easily be prepared in one day and stored for later assay.

[3] L. Clarke and J. Carbon, *Cell* **9**, 91 (1976).

[4] G. Mandel and W. Wickner, *Proc. Natl. Acad. Sci. U.S.A.* **76**, 236 (1979).

[5] J. H. Miller, "Experiments in Molecular Genetics." Cold Spring Harbor Lab., Cold Spring Harbor, New York, 1972.

2. In Vitro Assay for Leader Peptidase Overproduction

Each crude extract is diluted to 0.4 mg/ml protein with solution B, and 5 μl of each dilution are mixed with 2 μl of ^{35}S-labeled procoat (30,000 cpm) and incubated for 30 min at 37°. The reaction is terminated by addition of 7 μl of sample buffer.[1] Each sample is then subjected to SDS gel electrophoresis and fluorography[6]; 1.2 mg of crude extract from nontransformed cells typically produces equal amounts of procoat and coat (1 unit of enzyme activity), whereas only 0.3 μg of crude extract from cells transformed with pLC7-47 is required to generate the same amount of conversion of procoat to coat. Cells transformed with pTD101, an 8.8 kilobase (kb) plasmid made by subcloning the leader peptidase gene into pBR322, yield 10 times more overproduction of leader peptidase (see Fig. 1). Although this assay is specific for leader peptidase, it does not lend itself to assaying many samples, as typically required in a genetic study. However with adequate facilities, it is possible to run at least 10 gels per day (15 samples per gel).

3. In Vivo Assay of Leader Peptidase Overproduction

The conversion of procoat to coat in cells infected with mutant M13 viruses is slower than in cells infected with wt M13.[6] This altered rate of conversion is accelerated by the overproduction of leader peptidase in cells carrying the leader peptidase gene on a multicopy plasmid.

HJM114 (F' lac pro/Δlac pro)[7] is transformed with a plasmid carrying the leader peptidase gene. The transformed cells are cultivated in M9 plus glucose[5] at 37°. At $A_{600} = 0.4$ the cells are infected at a multiplicity of 100 with either M13am7 or M13am8 H1R6, a virus with three amino acid changes in its coat protein. Cells infected with either of these viruses show a delayed rate of conversion of procoat to coat protein when compared to cells infected with wild-type virus. After 1 hr of infection, 1 ml of cells is mixed in a prewarmed test tube with [^{3}H]proline (2 μCi) for 15 sec. This "pulse" is followed immediately with a "chase" of nonradioactive amino acid (200 μg). At increasing times during the chase, 0.2-ml aliquots of the reaction are precipitated with an equal volume of cold 10% trichloroacetic acid (TCA) and analyzed by SDS–polyacrylamide gel electrophoresis as described in this volume.[8] Typical results are shown in Fig. 2. Overproduction of leader peptidase in cells infected with the mutant phages relieves the delay in conversion of procoat to coat. This approach could be useful for further screening of the Clarke and Carbon library for other factors that may play a

[6] K. Ito, G. Mandel, and W. Wickner, Proc. Natl. Acad. Sci. U.S.A. 76, 1199 (1979).
[7] W. Wickner and T. Killick, Proc. Natl. Acad. Sci. U.S.A. 74, 505 (1977).
[8] W. Wickner, T. Date, R. Zimmermann, and K. Ito, this volume [5].

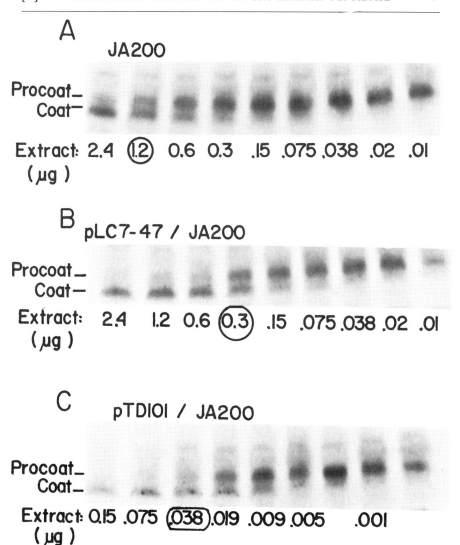

FIG. 1. Leader peptidase activity in crude extracts of (A) *Escherichia coli* JA200 and (B) this strain with plasmid pLC7-47 [T. Date and W. Wickner, *Proc. Natl. Acad. Sci. U.S.A.* **78,** 6106 (1981)] or (C) pTD101. Cultures (1 ml) of each strain were grown in 37° in L broth to $A_{600} = 0.5$ and centrifuged; the cells were resuspended in 150 μl of 10 mM Tris-Cl (pH 8.1), 10 mM EDTA, 1% (v/v) Triton X-100. Cells were disrupted by brief sonication at 0°, and the indicated amounts of cell extract were assayed for leader peptidase by posttranslational conversion of procoat to coat in the presence of mixed proteinase inhibitors, as described in the text. The levels of extract that yield procoat and coat bands of equal intensity[1] are circled. Reproduced from T. Date and W. Wickner, *Proc. Natl. Acad. Sci. U.S.A.* **78,** 6106 (1981).

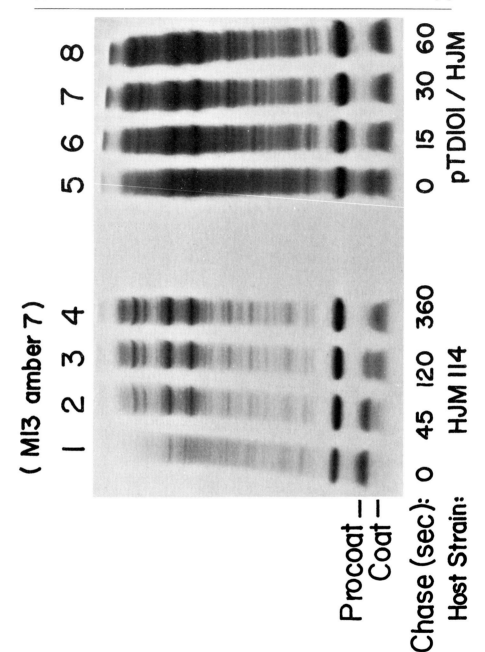

role in membrane assembly or, in general, for enzymes for which there are no corresponding mutants.

Structure and Expression of the Leader Peptidase Gene

The plasmid in the strain isolated from the Clarke and Carbon collection, pLC7-47, is 21.6 kb in length. It is composed of 15 kb of *E. coli* DNA linked to 6.6 kb of ColE1-derived DNA. This plasmid is difficult to work with. Owing to its large size it does not yield high transformation efficiency; and owing to the low copy number in transformed cells, the level of overproduction of leader peptidase is only four times higher than the level of activity found in wild-type cells. To achieve more overproduction of the enzyme, the *E. coli*-derived DNA was subcloned into pBR322 as outlined in Fig. 3. The new plasmid, pTD101, is stable, has a high transformation efficiency, and yields a 30-fold overproduction in transformed cells. The plasmid, pTD101, contains 4.8 kb of *E. coli*-derived DNA. Biochemical characterization of the purified leader peptidase protein[2] indicates that it is composed of a single 35,000 dalton polypeptide, suggesting that the gene is about 1 kb in length. Therefore about 3.5 kb of the cloned DNA in the plasmid is not leader peptidase structural gene. This section briefly describes the procedures for the determination of the essential regions of pTD101 for the expression of the leader peptidase gene.

One microgram of the plasmid DNA is digested completely with the appropriate restriction enzyme. After inactivation of the restriction enzyme, the entire reaction mixture is incubated with T4 ligase,[9] and an aliquot is used to transform JA200 (recA). Transformed colonies are selected by resistance to ampicillin. Each transformant is then cultivated in 2 ml of L broth[5] with 100 μg of ampicillin per milliliter at 37°. At $A_{600} = 0.5$, a 0.5-ml aliquot is removed and assayed for leader peptidase overproduction *in vitro* as described.[8] The remainder of the culture is used to prepare a minilysate and analyzed by agarose gel electrophoresis.[9] The results of this kind of analysis are presented in Fig. 4. There are two essential regions for the expression of the leader peptidase gene on the plasmid. One is a 1.2 kb

[9] R. W. Davis, D. Botstein, and J. R. Roth, In: "Advanced Bacterial Genetics." Cold Spring Harbor Lab., Cold Spring Harbor, New York, 1980.

FIG. 2. Effect of leader peptidase overproduction on the conversion of procoat to coat. Wild-type *E. coli* HJM114 (lanes 1–4) or overproducer strain pTD101/HJM114 (lanes 5–8) were infected by M13 *amber* 7 virus, pulse-labeled for 15 sec with [³H]proline, and chased with excess nonradioactive amino acid as previously described.[8] After the indicated intervals of chase, aliquots were harvested and analyzed by sodium dodecyl sulfate–polyacrylamide gel electrophoresis and fluorography. Reproduced from T. Date and W. Wickner, *Proc. Natl. Acad. Sci. U.S.A.* **78**, 6106 (1981).

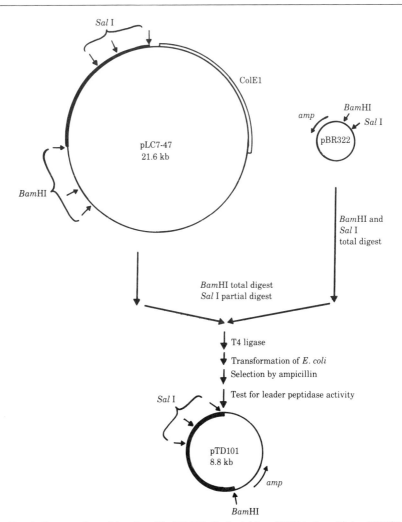

FIG. 3. Construction of the plasmid pTD101. *Escherichia coli* DNA cloned into pTD101 is indicated by the thick, dark line. Three micrograms of pLC7-47 were digested completely with *Bam*HI and partially with SalI. The digestion products were extracted with phenol, precipitated with ethanol, and resuspended in a small volume of ligation buffer.[9] pBR322 (0.8 μg) was digested to completion by *Bam*HI and *Sal*I, and the fragments were isolated in the same manner. The products of each digestion were ligated in a 60-μl reaction for 16 hr at 16°[9] and transformed [T. Date and W. Wickner, *Proc. Natl. Acad. Sci. U.S.A.* **78**, 6106 (1981)] and transformed into *E. coli* JA200. Ampicillin-resistant clones were tested for the overproduction of leader peptidase as described in the text. Reproduced from T. Date and W. Wickner, *Proc. Natl. Acad. Sci. U.S.A.* **78**, 6106 (1981).

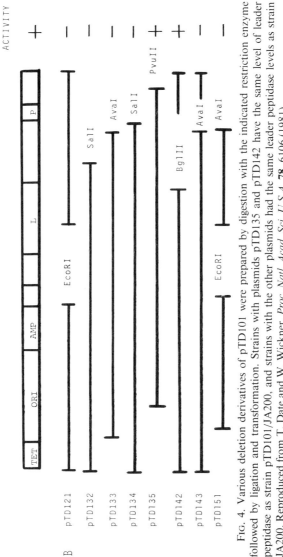

FIG. 4. Various deletion derivatives of pTD101 were prepared by digestion with the indicated restriction enzyme followed by ligation and transformation. Strains with plasmids pTD135 and pTD142 have the same level of leader peptidase as strain pTD101/JA200, and strains with the other plasmids had the same leader peptidase levels as strain JA200. Reproduced from T. Date and W. Wickner, *Proc. Natl. Acad. Sci. U.S.A.* **78**, 6106 (1981).

region, termed L, which presumably encodes the leader peptidase structural gene. The other region, termed P, is 350 base pairs in length and may act as a promoter for the overproduction of the leader peptidase. To further confirm that the P region is in fact required as a promoter, and to achieve higher levels of expression of the leader peptidase gene, a λ promoter, P_R, is inserted upstream from the leader peptidase gene. The plasmid, pBH1[10] contains an EcoRI-generated fragment of phage λC1857 cro27 bio256, which codes for a thermosensitive repressor. To place the cloned leader peptidase gene under control of either the λ promoter P_L or P_R, the cloning strategy outlined in Fig. 5 is followed. Cells (JA200) containing the new plasmid, pPS9, with the P_R substituted for the P region of the leader peptidase gene are propagated at 30°, where they are resistant to ampicillin, immune to λ phage infection, and produce wild-type levels of leader peptidase. Inactivation of the thermosensitive repressor at 42° for 1 hr results in a 100-fold increase in the expression of the leader peptidase gene.[2]

Requirement for Leader Peptidase for Cell Growth

It is important to understand the role of leader peptidase in the normal functionings of the cell. Biochemical analysis has indicated that purified leader peptidase can process several membrane protein precursors to their native size in vitro.[11] In addition, liposomes reconstituted from E. coli phospholipids and leader peptidase not only can correctly process M13 procoat to coat protein, but also can mediate the correct integration of coat protein into the lipid bilayer.[12] These findings suggest a critical role for leader peptidase in the cell. To examine its role further, the following genetic methods were employed.

Polymerase I, the product of the polA gene, is required for the replication of this plasmid DNA in transformed cells.[13] In polA⁻ cells where the plasmid can no longer replicate, hybrid plasmids containing a piece of the host chromosomal DNA can recombine at sites of homology between the plasmid and the host chromosome with a high frequency. In the case of pTD101, single recombination events between the plasmid and host DNA produce ampicillin-resistant colonies. This phenomenon is useful for examining whether the enzyme encoded by the plasmid is essential for cell growth. Deletions of the promoter (P), the structural gene (L), or both, on the plasmid and transformation into polA⁻ cells yield integrated strains whose viability reveals the necessity of the genes encoded by these regions of the chromosomes for cell growth.

[10] J. Hedgepeth, M. Ballivet, and H. Eisen, Mol. Gen. Genet. **163,** 197 (1978).
[11] C. J. Daniels, D. Oxender, D. Perrin, and M. Hofnung, personal communication.
[12] C. Watts, P. Silver, and W. Wickner, Cell **25,** 347 (1981).
[13] D. T. Kingsbury and D. R. Helinski, Biochem. Biophys. Res. Commun. **41,** 1538 (1970).

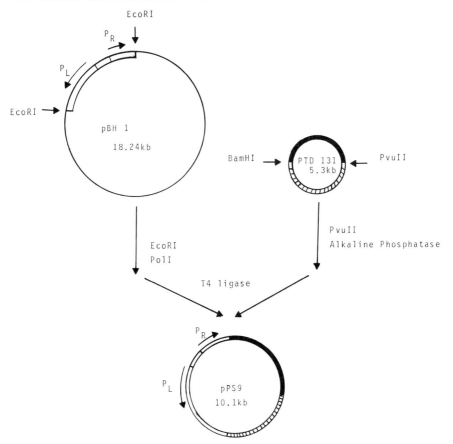

FIG. 5. Construction of plasmid pPS9. λ-derived DNA is indicated by the double light line, *Escherichia coli*-derived DNA by the thick dark line, and pBR322 derived DNA by the cross-hatched areas. Five micrograms of pBH1 were digested completely with *Eco*RI and treated with the Klenow fragment of polymerase I. After phenol extraction and ethanol precipitation, the digestion products were resuspended in a small volume of ligation buffer [T. Date and W. Wickner, *Proc. Natl. Acad. Sci. U.S.A.* **78,** 6106 (1981)]. pTD131 (2 μg) was digested with *Pvu*II, treated with bacterial alkaline phosphatase, and isolated in the same manner. The products of each digestion were ligated in a 20-μl reaction for 15 hr at 14°[9] and transformed *E. coli* JA200.

Methods

A *polA^ts* strain is used for these experiments. At 30°, the *E. coli* strain MM383[14] is resistant to methylmethanesulfonate and plasmids are capable of normal autonomous replication. Upon shifting to 42°, the cells become sensitive to methylmethanesulfonate and the plasmid can no longer repli-

[14] M. Monk and J. Kinross, *J. Bacteriol.* **109,** 971 (1972).

EFFECT OF INTEGRATION OF PLASMID ON CELL GROWTH[a]

Plasmid	Genotype	Amp^r cells (at 30°)	Amp^r cells grown at 42°	Frequency
pTD101	p^+L^+	3.4×10^5	164	4.8×10^{-4}
pTD121	p^+L^-	5.1×10^5	76	1.5×10^{-4}
pTD132	p^-L^+	5.8×10^5	64	1.1×10^{-4}
pTD151	p^-L^-	0.8×10^7	0	10^{-7}
pBR322	p^-	1.2×10^7	0	10^{-7}

[a] Strain MM383 ($polA^{ts}$) was transformed with the indicated plasmids (see Fig. 4), and the number of ampicillin-resistant colonies at 30° and 42° was scored.

cate. Instead, recombination between regions of homology in the plasmid and the host chromosome can occur.

The cells, MM383, are cultivated in 25 ml of L broth[5] at 30°. At $A_{600} = 0.4$, the cells are quickly chilled and collected by centrifugation at 8000 rpm for 10 min at 2°. The supernatant is removed with a sterile pipette. The cell pellet is suspended by vortexing in 3 ml of cold 0.1 M CaCl$_2$ and then incubated for 30 min at 0°. The cells are collected by centrifugation, suspended in 1.2 ml of cold 0.1 M CaCl$_2$, and incubated an additional 2 hr at 0°. These Ca^{2+}-treated cells are used for subsequent transformations. A 0.2-ml portion of cells is incubated with 6 μg of plasmid DNA in a 25-ml glass test tube for 30 min at 0°. The cells are then heat-shocked at 30° for 90 secs, 1 ml of L broth is added, and the cells are incubated for an additional 40 min at 30°. Transformed cells are selected by spreading on TYE plates containing 100 μg of ampicillin per milliliter. Colonies appear after 40 hr at 30°. To promote integration, transformed colonies are transferred into 4 ml of L broth, cultivated at 30°, and then spread on TYE plates plus 100 μg of ampicillin per milliliter and incubated at 42.5° for 14 hr. The results are summarized in the table. Strains constructed from plasmids lacking only the promoter region or the downstream portion of the structural gene are viable after plasmid integration. Strains transformed with plasmids that have both the promoter and the structural gene deleted are no longer viable after integration, indicating that the leader peptidase is necessary for cell growth.

The methods described above provide a means of (a) identifying the leader peptidase gene that is associated directly with membrane biogenesis; (b) determining the structure of the gene; and (c) defining its physiological role. Study of leader peptidase and its gene may elucidate the basis for leader peptidase specificity of cleavage. To further understand the role of leader peptidase and the assembly of proteins into or across membranes, conditionally lethal mutants will be of great use. Once obtained, these mutants will be particularly useful for examining directly the role of leader peptidase in protein localization and for generating second-site revertants in other genes that may be necessary for membrane assembly.

[5] Pulse-Labeling Studies of Membrane Assembly and Protein Secretion in Intact Cells: M13 Coat Protein

By WILLIAM WICKNER, TAKAYASU DATE, RICHARD ZIMMERMANN, and KOREAKI ITO

Studies of protein transfer into or across membranes have frequently employed crude, disrupted-cell systems. With the increasing number and sophistication of these experiments, it becomes imperative to relate them to studies of these processes in intact, infected cells. In turn, such studies may suggest new aspects of secretion or membrane assembly that can then be studied *in vitro.* Studies of the assembly of coliphage M13 coat protein into the plasma membrane of *E. coli* have been enriched by just this sort of interchange between the *in vivo* and *in vitro* approaches.[1] They have revealed that M13 coat protein, which spans the plasma membrane with its NH_2 terminus exposed to the periplasm, is made as a precursor (procoat) that integrates across the membrane posttranslationally in an electrical potential-dependent manner.

There are two major experimental approaches to these *in vivo* studies: (*a*) determining *where* protein is synthesized after separating the membrane-bound and unattached polysomes (as reviewed in Vol. 96); and (*b*) following the fate of a newly made, pulse-labeled protein or its precursor. This second approach is the topic of this article. The basic procedure in such experiments is to (*a*) grow cells in a medium with a low concentration of at least one amino acid; (*b*) briefly pulse-label with that amino acid; (*c*) "chase" with a large chemical excess of the amino acid; (*d*) remove portions of the culture at various intervals to an environment that will "quench" the processes of interest (protein synthesis, membrane traversal, glycosylation, proteolysis, etc.); (*e*) analyze topography where necessary by subcellular fractionation and proteolysis of closed, vesicular organelles; and (*f*) assay the pulse-labeled protein (after immunoprecipitation if necessary) by sodium dodecyl sulfate (SDS)–polyacrylamide gel electrophoresis and fluorography or autoradiography.

[1] W. Wickner, *Science* **210**, 861 (1980).

METHODS IN ENZYMOLOGY, VOL. 97

Before proceeding to the technical aspects of this work, it is important to note that there are several biological parameters that are crucial to the ease and accuracy of such a study. Each of these is favorable for the assembly of M13 coat protein, and this protein will therefore serve to illustrate these points.

1. It is useful if the protein is abundant, allowing its detection with a minimum number of purification steps. M13 coat protein accounts for 5% of the protein synthesis of *E. coli*.[2]

2. If the protein in question is not abundant, then it is crucial to have antibody; as is well known, the use of immunoprecipitation in such studies requires careful controls to allow quantitation and to avoid artifacts.

3. The cells should grow in a medium with only very small levels of the pulse-labeling amino acid. Uptake of labeled amino acid should be rapid and its incorporation proceed without dilution through large internal pools. Similarly, the addition of a large excess of nonradioactive amino acid should lead to efficient and rapid competitive dilution of the isotope incorporation. *Escherichia coli* has each of these characteristics of favorable pulse-labeling and the chase.

4. Each of the "compartments" of the cell, the membranes or the soluble phases of distinguishable protein composition that they bound, should be readily separable to allow analysis of the spatial pathway followed by pulse-labeled proteins. *Escherichia coli* has four compartments: cytoplasm, inner membrane, periplasm, and outer membrane. Each can be separated under specific, reproducible fractionation conditions.[3,4]

5. The data from pulse-labeling studies will be easily interpreted only if the assembly steps that are monitored take longer than the pulse-labeling period. This condition is *not* fulfilled for the procoat of wild-type M13 virus but is seen with procoat in cells infected by M13 *amber* 7 virus or by viruses (such as M13am8H1R1, M13am8H1R2, M13am8H1R5, or M13am8H1R6) with mutant procoat protein.[5-7]

Experimental Procedures

Cell Growth. An overnight culture of *E. coli* strain HJM114 (F' lac pro/∇*lac pro*)[7a] in minimal medium (M9 or M63 with 0.5% glucose and with

[2] H. Smilowitz, J. Carson, and P. W. Robbins, *J. Supramol. Struct.* **1**, 8 (1972).
[3] H. C. Neu and L. A. Heppel, *J. Biol. Chem.* **240**, 3685 (1965).
[4] M. J. Osborn, J. E. Gander, E. Parisi, and J. Carson, *J. Biol. Chem.* **247**, 3962 (1972).
[5] K. Ito, G. Mandel, and W. Wickner, *Proc. Natl. Acad. Sci. U.S.A.* **76**, 1199 (1979).
[6] M. Russel and P. Model, *Proc. Natl. Acad. Sci. U.S.A.* **78**, 1717 (1981).
[7] W. Wickner, unpublished observation.
[7a] W. Wickner and T. Killick, *Proc. Natl. Acad. Sci. U.S.A.* **74**, 505 (1977).

thiamin)[8] was diluted with fresh medium to $A_{600} < 0.005$ and grown at 37° in a shaking water bath. Cultures of 20 ml or less were grown in a 50-ml flask. When the A_{600} reached 0.4, the culture was infected with M13 virus or one of its mutant derivatives at a multiplicity of 100. Portions (0.5–1.0 ml) of the culture were transferred with a warmed sterile pipette to warmed 8-ml test tubes with metal closures in a rack in the shaking water bath. Growth of these infected cells was continued for 1 hr after infection, a period necessary for full viral gene expression in minimal medium.

Pulse Labeling. Radioactive amino acid ([³H]phenylalanine or [³H]proline, 1 mCi/ml in 0.01 M HCl, 100 Ci/mmol; use 40 μl per milliliter culture) was added by Pipetteman, and shaking of the tube in the bath was continued for 10–15 sec. The choice of amino acid is governed by its abundance in the protein of interest (M13 procoat has no cysteine, arginine, or histidine)[9] and relevant contaminating proteins (the abundant prolipoprotein and lipoprotein, which comigrate with procoat on SDS gels,[10] lack proline and phenylalanine[11]) as well as by cost and specific activity.

Chase. Nonradioactive amino acid (0.1 ml) was added by Pipetteman to a final concentration of 0.25 mg/ml. Drugs, such as energy poisons, were dissolved in the chase solution when desired.[12] Aliquots (0.1 ml) were removed at various times of chase (5, 30, 90, and 300 sec, for example). By using the same Pipetteman to deliver the chase and remove the first aliquot, one can reproducibly obtain a true 5-sec chase sample.

Quench. Accurate analysis is dependent on an immediate and effective quenching of the assembly and processing reactions under study. If subcellular topography is not being measured, this is most effectively accomplished by adding the 0.1-ml aliquots of labeled culture directly to 0.4 ml of ice-cold 10% trichloroacetic acid in conical 1.4-ml plastic centrifuge tubes. After 1 hr at 0°, precipitates were collected by centrifugation (2 min, 4°, Brinkmann microfuge). After careful aspiration of the supernatant, the pellets were suspended by vortexing in 0.5 ml of ice-cold acetone and collected by centrifugation (2 min, 4°). This acetone wash was repeated, and the pellets were then suspended in 50 μl of sample buffer (12.5 mM Tris-HCl, pH 6.8, 5% SDS, 10% glycerol, 20 mM dithiothreitol) by vortexing for 30 sec. Samples were placed in a boiling water bath for 3 min, then vortexed

---

[8] J. H. Miller, "Experiments in Molecular Genetics." Cold Spring Harbor Lab., Cold Spring Harbor, New York, 1972.

[9] F. von Asbeck, K. Beyreuther, H. Kohler, G. von Wettstein, and G. Braunitzer, *Hoppe-Seyler's Z. Physiol. Chem.* **350**, 1047 (1969).

[10] K. Ito, T. Date, and W. Wickner, *J. Biol. Chem.* **255**, 2123 (1980).

[11] K. Takeishi, M. Yasumura, R. Pirtle, and M. Inouye, *J. Biol. Chem.* **251**, 6259 (1976).

[12] T. Date, C. Zwizinski, S. Ludmerer, and W. Wickner, *Proc. Natl. Acad. Sci. U.S.A.* **77**, 827 (1980).

again for 30 sec. They were then ready for analysis by SDS–polyacrylamide gel electrophoresis.

If subcellular topography was to be measured, then a rapid quench was achieved by adding aliquots of the culture to twice their weight of ice in a prechilled centrifuge tube in an ice bucket.

Fractionation and Analysis. The techniques of separating the four compartments of *E. coli* have been extensively described and will therefore not be presented here in detail. Briefly, cells were converted to spheroplasts by the method of Osborn *et al.*,[4] and the protein released in this procedure is periplasmic. Separation of inner and outer membranes was as described by those workers. Where analysis demands only a separation of soluble (cytoplasmic) and membrane (inner plus outer) fractions, spheroplasts are broken by either a brief sonication[10] at low power at 0° (monitored by light microscopy) or by dilution with 4 volumes of ice-cold sucrose-free buffer. Three factors are critical to such analysis: speed of operation, strict prechilling of all components (buffers, pipettes, tubes, rotors, etc.; performing the operations in a cold room), and the concentration of divalent cations. In the presence of Mg^{2+}, many "soluble" proteins of *E. coli* are recovered with the sediment of osmotically lysed cells,[10] perhaps owing to their adherence to DNA, polysomes, or membranes in an ion-dependent manner. Cleaner separations are achieved in the presence of EDTA. It is also noteworthy with respect to M13 procoat that divalent cation has a marked effect on its solubility properties.[13]

In addition to measuring the distribution of pulse-labeled precursors among compartments, it is also important to assay the topography of their distribution across membranes. This is achieved by digestion with proteinase at 0°. Aliquots of labeled cells were quenched with ice as described above, collected by centrifugation (0°, 3 min, 10,000 g), and resuspended in their original volume of ice-cold 20% sucrose, 10 mM EDTA, 30 mM Tris-HCl, pH 8.1. This buffer destroys the outer membrane permeability barrier of M13-infected cells without the fragility and partial lysis that accompanies the formation of spheroplasts.[14] Portions were then mixed with various proteinases, incubated for 1 hr at 0°, and quenched with proteinase inhibitor. Trypsin was employed because it only has one site of cleavage in the center of procoat and is readily inhibitable by soybean trypsin inhibitor. Proteinase K was employed in other experiments for its lack of cleavage specificity and for its susceptibility to inhibition by *N*-tosyl-L-phenylala-

[13] P. Silver, C. Watts, and W. Wickner, *Cell* 25, 341 (1981).
[14] T. Date, J. Goodman, and W. Wickner, *Proc. Natl. Acad. Sci. U.S.A.* 77, 4669 (1980).

nylchloromethyl ketone. Digested samples were boiled in SDS and analyzed by gel electrophoresis. Several controls are required for meaningful results in such a study.[15] Coat protein provides a positive control for digestion at the outer surface of the plasma membrane; if infected cells are labeled with [³H]proline, then trypsin, chymotrypsin, or proteinase K will release the sole proline of coat protein (at position 6), rendering the coat band invisible. At the same time, the failure of proteinase to degrade the many pulse-labeled cytoplasmic proteins assures that the plasma membrane is intact. It is important in such mapping studies to examine a substantial range of proteinase concentrations to be able to infer that a cleavage site is on one side of the bilayer or the other. Finally, digestions should also be performed on samples intentionally lysed in order to conclude that the failure of a proteinase to cleave a protein of interest (such as procoat) is due to protection by the membrane.

Immunoprecipitation. This procedure is useful for qualitative identification of labeled bands but, when employed as a routine step in a quantitative isolation, requires great care. The abundance of M13 procoat and coat and their unusually low molecular weight make them prominent bands in fluorographs of SDS gels of whole, pulse-labeled cells. Thus, immunoprecipitation is not needed in routine analysis and will not be further reviewed here.

SDS Gels. Samples from up to 100 μl of culture were analyzed by SDS gels and autoradiography or fluorography.

Limitations of This Approach

Even if the criteria outlined above for a successful *in vivo* study are met, the study will be informative only if the kinetics of assembly are such as to reveal interesting intermediates. In employing drugs or mutations to perturb the assembly pathway, it is crucial to perform control analysis of their metabolic effects during the pulse–chase period. *In vivo* studies are inherently limited by the inability to enumerate, and control the presence of, all the potential effectors of the reaction of interest. Kinetics will also not always reveal all the mechanistically relevant intermediates. However, the lessons learned will bear the stamp of physiological relevance and can guide more biochemical approaches.

[15] C. Watts, P. Silver, and W. Wickner, *Cell* **25,** 347 (1981).

[6] Synthesis of Proteins by Membrane-Associated Polysomes and Free Polysomes

By Phang C. Tai, Michael P. Caulfield, and Bernard D. Davis

In animal cells a special role for the ribosomes bound to the endoplasmic reticulum was recognized early by Palade[1] and co-workers, who observed a parallel between the abundance of such ribosomes and the secretion of proteins. Techniques were developed for separating polysomes bound to the reticulum from unbound, free polysomes,[2] and various secretory proteins were shown to be synthesized on the bound polysomes, and various cytoplasmic proteins on the ribosomes free in the cytoplasm. The discovery of a hydrophobic N-terminal sequence on precursors of secreted proteins,[3,4] which is subsequently cleaved, lends further importance to the association with membrane, and it led to the signal hypothesis[3,5] for the cotranslational secretion of nascent peptide chains.

A similar mechanism of protein secretion has been proposed for bacteria. However, the functional significance of the association of ribosomes with membrane in bacteria has long been difficult to establish, since in thin sections of bacterial cells it has not been possible to distinguish membrane-bound and free polysomes by electron microscopy because of the dense population of ribosomes. Alternatively, gentle lysis of the cells can yield membrane-associated polysomes suitable for *in vitro* translation, and early studies revealed a greater abundance of secretory alkaline phosphatase chains on the membrane-bound than on the free polysome fraction from *Escherichia coli.*[6] Here we describe procedures for isolating active membrane-associated polysomes and free polysomes essentially free of each other; these procedures have been applied successfully to both gram-positive and gram-negative bacteria to study the site of synthesis of certain proteins.

Materials and Reagents

Buffer I: 10 mM Tris-HCl (pH 7.6), 50 mM KCl, 2 mM Mg(OAc)$_2$, 1 mM dithiothreitol

Buffer II: Buffer I, except with 10 mM Mg(OAc)$_2$

[1] G. E. Palade, *Science* **189**, 347 (1975).
[2] F. S. Rolleston, *Subcell. Biochem.* **3**, 91 (1974).
[3] C. Milstein, G. G. Brownlee, T. M. Harrison, and M. B. Mathews, *Nature (London), New Biol.* **239**, 117 (1972).
[4] I. Schechter, *Proc. Natl. Acad. Sci. U.S.A.* **70**, 2256 (1973).
[5] G. Blobel and B. Dobberstein, *J. Cell Biol.* **67**, 835 (1975).
[6] R. Cancedda and M. J. Schlesinger, *J. Bacteriol.* **117**, 290 (1974).

METHODS IN ENZYMOLOGY, VOL. 97

Buffer III: Buffer II, with 10% (v/v) glycerol

Buffer IV: Buffer II with 1 mM phenylmethylsulfonyl fluoride (PMSF) and 0.1 mM o-phenanthroline added just prior to use

Buffer V: Buffer IV with 10% glycerol

Buffer VI: Buffer IV with 1 M KCl

Buffer VII: 20 mM Tris-HCl (pH 8), 0.15 M NACl, 5 mM EDTA, 0.5% Nonidet P-40 (NP-40) or Triton X-100, 0.1% SDS, and 0.5 mg/ml ovalbumin

Energy stock solution: The stock contains, per milliliter, 0.2 mM GTP-Tris, 10 mM ATP-Tris, 50 mM potassium phosphoenolpyruvate (neutralized with Tris base), and 15 μg of pyruvate kinase

Bacterial strains: MRE600 (an $E.\ coli$ RNase I$^-$ strain) or D10 (an $E.\ coli$ K12 RNase I$^-$ strain) was used for the preparation of supernatant enzymes (S100) necessary for translation of polysomes.[7] $Escherichia$ $coli$ K12 CW3747, Met$^-$ and constitutive in the synthesis of alkaline phosphatase, and $Bacillus\ subtilis,$ a high producer for α-amylase, were from American Type Culture Collection (No. 27257 and No. 6051a, respectively); $B.\ lichenformis$ 749/C, constitutive for penicillinase, was obtained from J. Lampen; and $Vibrio\ cholerae$ 569B, a hyperproducer of cholera toxin, and $Corynebacterium\ diphtheriae$ PW8, a hyperproducer of diphtheria toxin, were from J. Murphy.

Toluene scintillation fluid: 4 g of Omnifluor dissolved in 1 liter of toluene

Sucrose stock solution made as 2.5 M in water and autoclaved

Reagents: Chloramphenicol, Tris, ATP-Tris, GTP-Tris from Sigma; pyruvate kinase, phosphoenolpyruvate (K$^+$ salt), dithiothreitol, and amino acids from Calbiochem; [^{35}S]methionine (translational grade; stored in small portions under nitrogen at $-76°$), EnHance; and Omnifluor from New England Nuclear; DNase I (RNase-free, DPFF grade) from Worthington Biochemicals; Millipore HAWP filters from Millipore Corporation or glass fiber filter GA/C from Whatman. RNase-free sucrose from Schwarz-Mann. All other chemicals (reagent grade) are commercial sources.

Isolation of Free and Membrane-Bound Polysomes

Since most ribosomes are engaged in protein synthesis in exponentially growing cells, the gently prepared lysate of such cells is a rich source of polysomes. Membrane-associated polysomes could be separated from free polysomes, based either on the lower density or on the larger size of the former. We have found that the fractionation based on differential density gave better separation (see below).

---

[7] P. C. Tai and B. D. Davis, this series, Vol. 59, p. 362.

The general principles and precautions for isolating functional polysomes were discussed earlier.[7] The most important considerations are preventing polysome runoff and preventing degradation by ribonuclease. All glassware, plastic tubes, and buffers are autoclaved, where possible, and chilled before using. We have also used heparin as a ribonuclease inhibitor.

Procedures

Step 1. Growth of Cells. When a specific protein or enzyme is under study, the optimal condition for its production by the cells should be used.

For *E. coli* alkaline phosphatase, K12CW3747 (Met⁻ and constitutive in the synthesis of alkaline phosphatase) was grown (usually 4×300 ml in 2-liter flasks) at 37° with vigorous aeration in supplemented medium A, which contains, per liter, 7 g of K_2HPO_4, 3 g of KH_2PO_4, 1 g of $(NH_4)_2SO_4$, 0.5 g of sodium citrate, 0.1 g of $MgSO_4 \cdot 7H_2O$, 5 mg of $CaCl_2$, 0.25 mg of $FeSO_4$, 0.4% glucose, and 0.2% Difco casamino acids; the last two are prepared and autoclaved in 50% and 10% solutions, respectively. Exponential phase cultures (6×10^8 cells/ml) are rapidly mixed with chloramphenicol (100 μg/ml) and chilled by pouring onto excess ice. The cells are then pelleted by centrifugation, washed once with chilled buffer I containing 100 μg of chloramphenicol per milliliter, and pelleted. This is a critical step for preserving polysomes and should take no more than 20 min. The cell pellets are resuspended in buffers, and if necessary, frozen and stored at $-76°$.

For bacterial cells that produce cell-bound or extracellular proteases (e.g., bacilli), the cells were washed once with buffer VI, containing a high salt concentration and protease inhibitors.

Step 2. Cell Lysis. In our earlier procedures for cell lysis, digestion by lysozyme or grinding with solid CO_2 were reported; the lysozyme method provides a higher yield.[7]

For gram-negative bacteria (e.g., *E. coli*) the cells from 1–2 liters of culture are resuspended in two tubes with 5 ml of Buffer I (polysomes are stable at this low Mg^{2+} and the lysis is better) containing 100 μg of chloramphenicol per milliliter, and lysozyme is added to 400 μg/ml. After two or three cycles of quick freezing and slow thawing, 100 μg of DNase and 1 mg of heparin (final 200 μg/ml) are added to each tube, and Mg^{2+} is adjusted to 10 mM. The material can then be stored at $-76°$ until further fractionation.

With gram-positive bacteria (e.g., bacilli), the cells are similarly resuspended in buffer IV containing, per milliliter, 100 μg of chloramphenicol and 400 μg of lysozyme, and warmed with shaking in a 37° water bath. As soon as the cells lyse, as shown by increased viscosity, the lysate is chilled quickly and DNase and heparin are added as above.

Cell lysis could also be achieved with a French press. Polysomes isolated from *E. coli* after lysis at 10,000 psi were as active as those prepared by the above procedure. *Corynebacterium diphtheriae* PW 8 could be lysed satisfactorily at 20,000 psi[8]; other bacteria could be lysed at 7,000–10,000 psi.

Step 3. Sucrose Gradient Fractionation. The lysates are clarified at 3000 *g* for 5 min to remove most of the unbroken cells. The supernatant is further centrifuged at 15,000 *g* for 10 min to obtain a supernatant fraction, which is used mainly to isolate free polysomes, and a pellet, which is resuspended in 6 ml of buffer II or IV with 100 μg of heparin per milliliter and is used mainly for isolating membrane–polysome complexes. This prefractionation is useful, since it gives cleaner free polysome preparations upon subsequent fractionation, which otherwise would be contaminated with unbroken cells, since both sediment through the sucrose cushion used below.

Fractions of 2.5 to 3 ml are layered on top of a block sucrose solution of 10 ml of 1.35 *M* and 12 ml of 1.8 *M* sucrose in buffer II in a Beckman polycarbonate tube in a 60 Ti fixed rotor and centrifuged at 52,000 rpm for 16 hr at 4°. (We have found that this procedure is enough to pellet 70 S ribosomes as well as short polysomes.) The membrane-associated polysomes are found in the interface and are carefully removed with a Pasteur pipette; the diffuse band on top of the gradient is membrane free of ribosomes, and the pellet from this gradient is mostly unbroken cells; these two fractions are normally discarded. However, if the pellet is quite large and recovery of membrane polysomes is a problem, the pellet, which consists in part of cells lysed during centrifugation, is resuspended in buffer II with chloramphenicol and heparin, and the suspension is put through a second gradient (see below) to recover more membrane polysome complexes.

Free polysomes from prefractionation supernatant sediment in similar gradients as a soft pellet. It is thus important to remove the sucrose solution quickly but to leave about 0.5 ml of solution for resuspension (see next paragraph). Occasionally there is a considerable amount of membrane–polysome complexes in the interface from this gradient. These could be recovered and combined with other membrane–polysome fractions.

We find it necessary to put these fractions through another similar block gradient, especially the membrane-associated polysome fraction, which otherwise would be contaminated with free polysomes and "S100" enzymes. The membrane-associated polysome fraction is diluted about 2-fold to reduce sucrose concentration to less than 1 *M* and layered again onto a 1.35 *M*–1.8 *M* sucrose block gradient as above. The free polysomes could be combined and similarly treated. The gradients are again centrifuged at 55,000 rpm for 16 hr in a 60 Ti rotor, and the fractions are reisolated. The

[8] W. P. Smith, P. C. Tai, J. R. Murphy, and B. D. Davis, *J. Bacteriol.* **141**, 184 (1980).

membrane-associated polysome fraction is diluted 5-fold with buffer II or IV containing 200 μg of heparin per milliter and centrifuged at 40,000 rpm in the 60 Ti rotor for 1 h. If it is desirable to disrupt the membrane (see the next section, Translational Activity of Polysomes), Triton X-100 is added to 1% before centrifugation. The pellets are resuspended in buffer III or V and 200 μg of heparin per milliliter. The concentration of the polysomes is determined optically (1 A_{260} = 60 μg) with samples in 2% SDS solution (to solubilize membranes that otherwise cause light scattering that interferes with the determination), and small aliquots (normally 5 A_{260}) of polysomes are stored at $-76°$.

The proportion of free and membrane-associated polysomes ranges from 1:1 (e.g., *E. coli*) to 3:1 (e.g., *Bacillus*). The yield from a 1-liter culture normally is about 100–300 A_{260} each of membrane-associated (or derived) and free polysomes. If a smaller volume of culture is fractionated, it is preferable to use the Spinco SW 50.1 rotor with swinging buckets at 45,000 rpm for 16 hr.

With gram-negative bacteria the membrane-associated polysome preparations also contain outer membranes. We have occasionally encountered difficulty recovering membrane-associated polysomes in the interface of 1.35 M–1.8 M sucrose gradients, and we have used 1.35 M–2.0 M block sucrose gradients instead.

We have also used a sieving column (Sepharose 2B or 4B), as previously described for endogenous polysomes,[7] to separate membrane-associated polysomes from free polysomes; the procedure is quicker, and the polysomes are generally more active. However, cross-contamination is greater, and we have generally used the density sucrose gradient procedures.

Translational Activity of Polysomes

Chain elongation on polysomes is carried out in a reaction mixture (0.1 ml) containing 50 mM Tris–HCl (pH 7.6), 60 mM NH$_4$Cl, 10 mM KCl, 9 mM Mg(OAc)$_2$ (for *E. coli;* optimum to be determined for other bacterial strains), 2 mM dithiothreitol, 200 μg of heparin per milliliter, 10 μl of energy stock solution (to yield 1 mM ATP-Tris, 20 μM GTP, 5 mM potassium phosphoenolpyruvate, 1.5 μg of pyruvate kinase), 20 μCi of [^{35}S]methionine supplemented with 0.5–2 μM unlabeled methionine, 19 other amino acids at 50 μM each, 2 μl of *E. coli* S100 extract (optimal amount), and 60 μg of polysomes. After incubation at 35–37° for 30 min (the incorporation is normally completed in 15 min), 10-μl samples are analyzed for incorporation activity by mixing with 2 ml of 5% trichloroacetic acid and 1 drop of 0.5% bovine serum albumin; the mixture is heated in boiling water for 15 min, chilled, filtered through Millipore (HA) or Whatman glass fiber filters, and washed with 5% trichloroacetic acid containing 0.2% casamino acids. The filters are then dried and counted with 4 ml of toluene scintillation fluid in a liquid scintillation counter. The rest of the reaction mixture is processed for immunoprecipitation or is frozen at $-76°$ until further use.

We have normally included 1 mM PMSF and 0.1 mM o-phenanthroline (with 1% ethanol final concentration) as protease inhibitors in the incorporation system; these concentrations do not inhibit incorporation. These additions may be omitted in the *E. coli* system. The optimal amount of S100 and Mg^{2+} needs to be established for each preparation. In general, polysomes from gram-positive bacteria require higher Mg^{2+} for optimal activity. We have also used heparin at 200 μg/ml as RNase inhibitor; it has sometimes increased incorporation twofold. Furthermore, at this concentration heparin virtually completely inhibits the incorporation that depends on initiation with added mRNA. Thus, although we have not rigorously tested initiation factor activity[7] in our preparations, it is likely that there is no reinitiation in the incorporating system. This conclusion has been verified also by finding lack of inhibition by kasugamycin, which specifically blocks initiation.

Membrane-associated polysomes are generally about 50% less active than free polysomes. We have found that the dissolution of membrane, by 1% Triton X-100 after a second sucrose gradient fractionation (see the preceding section), gives rise to "derived" polysomes that are almost as active as free polysomes. However, this treatment usually activates signal peptidase,[9,10] resulting in the cleavage of signal peptide from the product.

We have found that the most efficient way to remove most of the membrane lipid from derived polysomes is by first treating the vesicles with 0.1% DOC on ice from 15 min followed by centrifugation at 12,000 g for 5 min. The supernatant is removed, made 1% with Triton X-100, and then passed through a sucrose gradient as for free ribosomes above. This procedure removes ≥ 99% of the membrane lipid as compared to 90–95% by 1% Triton alone.

Identification of Specific Products Formed by Membrane-Associated and by Free Polysomes

The translational products of membrane-associated or derived polysomes are distinctly different from those of free polysomes prepared as described above. This is evident from analyses, in SDS–acrylamide gel

[9] L. L. Randall, L. G. Josefsson, and S. J. S. Hardy, *Eur. J. Biochem.* **92**, 411 (1978).
[10] G. Mandel and W. Wickner, *Proc. Natl. Acad. Sci. U.S.A.* **76**, 236 (1979).

electrophoresis, either of total incorporation products or of immunoprecipitated specific proteins.

The immunoprecipitation is carried out as follows: One microliter of 10% SDS is added to 100 μl of reaction mixture in a 1-ml Eppendorff tube; after 15 min on ice, the mixture is centrifuged in an Eppendorff or Fisher centrifuge for 4 min to remove debris. The supernatant is brought to 1 ml with immunoprecipitation buffer VII, and 10 μl of preimmune serum are added to remove nonspecifically bound proteins; the mixture is left on ice for 30 min, 100 μl of 10% *Staphylococcus aureus* (which have been washed twice and resuspended with buffer VII; 2 mg of IgG per milliliter binding capacity) is added and the mixture is incubated at 0° for 10 min and centrifuged for 2 min. The supernatant is then treated with 10 μl of serum against specific protein, incubated for 30 min on ice, and mixed with 100 μl of *Staphylococcus* A suspension and incubated at 0° for 10 min. After 30 sec of centrifugation, the pelleted cells are washed twice by resuspension in buffer VII and centrifugation, and then they are resuspended in 50 μl of sample buffer VIII and boiled for 2 min. (Total incorporation products are similarly treated for analysis by SDS–polyacrylamide gel electrophoresis.) The suspension is centrifuged for 2 min, and the supernatant is analyzed with Laemmli's gel electrophoresis system.[11]

Total incorporation products are similarly mixed with sample buffer VIII and boiled and the supernatants are analyzed. The electrophoresis is performed until the tracking dye has reached the bottom (usually 100 V for 5 hr or 20 V for 16 hr). After electrophoresis the gel is fixed with acid and processed for fluorography[12] with EnHance or salicylate[13] and exposed with Kodak X-5 film at −75°.

Preferential Synthesis by Membrane-Associated and Free Polysomes

The translational products of the membrane-associated and free polysomes thus prepared are distinctly different. By immunological assays, translational factors EFTu and EFG are found to be almost exclusively synthesized ($>$90%) by the free polysome fraction. Several secreted proteins (alkaline phosphatase in *E. coli*,[14] diphtheria toxin,[8] penicillinase of *B.*

[11] U. K. Laemmli, *Nature (London)* **227**, 680 (1970).
[12] W. M. Bonner and R. A. Laskey, *Eur. J. Biochem.* **46**, 83 (1974).
[13] J. P. Chamberlain, *Anal. Biochem.* **98**, 132 (1979).
[14] W. P. Smith, P. C. Tai, R. C. Thompson, and B. D. Davis, *Proc. Natl. Acad. Sci. U.S.A.* **74**, 2830 (1977).

licheniformis[15]) are predominantly synthesized by membrane-associated polysomes, but separation is less complete (80–90%). The higher proportion of contamination probably represents initiating ribosomes carrying short nascent peptides that are not yet firmly associated with membrane, and perhaps some polysomes that are stripped of membrane during preparation. This interpretation is supported by the finding that, on chain completion of these proteins, the free polysomes yield only precursor (making 10–20% of the total), whereas the membrane-associated polysomes yield a mixture of precursor and mature protein.

Comments

Although the procedures described are tedious, they yield free and membrane-associated polysomes with minimal cross-contamination: most proteins tests are synthesized virtually entirely on one or the other. Proteins synthesized by membrane-associated polysomes have been shown to be secreted cotranslationally, but some secreted proteins are predominantly synthesized by free polysomes, and they are probably transported posttranslationally. Thus subunit A of cholera toxin is synthesized as a larger precursor by free polysomes,[16] and in the cell it can be detected in the cytoplasm.

The cotranslational secretion of a specific protein by membrane-associated polysomes was directly demonstrated by extracellular labeling of growing nascent peptides with nonpenetrating reagents, acetylmethionyl methylphosphate sulfone[17] and diazoiodosulfanilic acid,[18] as we have previously reported for *E. coli* alkaline phosphatase,[14] *B. subtilis* α-amylase,[19] diphtheria toxin,[8] and *B. licheniformis* penicillinase.[15] Unfortunately, we have not been able to repeat the original demonstrations according to the procedures as described,[14,19] and we have been able to obtain Dr. Walter Smith's original detailed protocols. Experiments with several modifications carried out in this laboratory have confirmed the extracellular labeling of growing nascent peptides, but to a much smaller extent than was originally reported.[14,19] These procedures will be published elsewhere.

[15] W. P. Smith, P. C. Tai, and B. D. Davis, *Proc. Natl. Acad. Sci. U.S.A.* **78**, 3501 (1981).
[16] J. C. Nichols, P. C. Tai, and J. R. Murphy, *J. Bacteriol.* **144**, 518 (1980).
[17] M. S. Bretscher, *J. Mol. Biol.* **58**, 775 (1971).
[18] K. L. Carraway, *Biochim. Biophys. Acta* **415**, 319 (1975).
[19] W. P. Smith, P. C. Tai, and B. D. Davis, *Biochemistry* **18**, 198 (1979).

[7] Preparation of Free and Membrane-Bound Polysomes from *Escherichia coli*

By LINDA L. RANDALL and SIMON J. S. HARDY

At least three techniques can be used to obtain pure preparations of membrane-bound and free polyribosomes from *Escherichia coli.* Each has its advantages and disadvantages. We shall limit detailed discussion to the technique developed and routinely used in our laboratories. This technique relies on the finding that, in bacterial lysates obtained by sonication, membrane-bound ribosomes are attached to large fragments of the bacterial envelope, which sediment faster than even the largest free polysomes. The sonicate is layered onto a sucrose gradient and subjected to a short centrifugation. The membrane-bound polysomes collect on a sucrose cushion at the bottom of the gradient while the free polysomes remain in the body of the gradient. This technique has the advantages of separating the two species rapidly and of yielding clean membrane-bound ribosomes. It has the disadvantage that the fraction containing the free polysomes also contains small fragments of membrane.

A second technique, developed in the laboratory of Bernard Davis,[1] makes use of the difference in densities between the two fractions of ribosomes. Sucrose gradients are designed so that, during centrifugation, membrane-bound polysomes reach their equilibrium density whereas the more dense, free polysomes form a pellet. This technique has the great advantage of yielding clean free polysomes. The disadvantages are that separation is relatively slow, overnight centrifugation being required, and that the membrane-bound ribosomes obtained are contaminated with free monosomes and small polysomes. These small species sediment so slowly through the dense sucrose solutions required for banding the membranes that several days of centrifugation would be required to pellet them.

We have occasionally used a third technique, which is a modification of the second, designed to yield uncontaminated preparations of both kinds of ribosomes. After removal of unbroken cells, a lysate is brought to a density greater than the density of the membrane-bound ribosomes by either addition of a saturated solution of sucrose or addition of crystalline sucrose. The sample is then layered onto a cushion near the bottom of the sucrose gradient so that during centrifugation membrane-bound ribosomes move upward in the gradient while free polysomes sediment to form a pellet. Since

---

[1] W. P. Smith, P.-C. Tai, R. C. Thompson, and B. D. Davis, *Proc. Natl. Acad. Sci. U.S.A.* **74,** 2830 (1977). See also this volume [6] for a detailed discussion.

METHODS IN ENZYMOLOGY, VOL. 97

the membranes are floated away from the monosomes and small polysomes, the contamination that presents a problem in the second technique no longer occurs. However, preparation of the sample is technically difficult. Either a large volume of saturated sucrose solution must be added to the lysate, thus making it very dilute, or crystalline sucrose must be dissolved in the lysate. At the low temperatures needed for maintenance of polysome integrity and activity, the latter is a laborious process.

Preparation of Polysomes

In order to obtain good yields of active polysomes from bacteria without the use of antibiotics such as chloramphenicol, which act to trap the ribosomes on the mRNA, the cultures are harvested by rapid chilling so that elongation of nascent chains is instantaneously stopped. All subsequent operations are carried out as rapidly as possible, and precautions are taken to avoid raising the temperature of the suspension.

The major causes of failure in this preparation are associated either with sonication or with unwanted ribonuclease activity. In our original work, sonication was used to disrupt the cells in order to establish the existence of membrane-bound polysomes,[2] because it had been reported that lysozyme caused nonspecific binding of ribosomes to membranes.[3] Later we attempted to decrease the amount of sonication required by pretreating the cells with lysozyme to render them more fragile. Although 50% more ribosomes were found in the membrane fraction when cells were disrupted in this way, the protein synthetic activity of that fraction was only 67% of the activity found in the membrane fraction prepared in our usual manner by a longer period of sonication in the absence of lysozyme. An additional reason for avoiding the use of lysozyme is that until the mechanism of export is known in detail, we prefer to minimize the disruption of the structural relationship between peptidoglycan and the membranes of the cell envelope. Therefore, we have chosen to break the cells by sonication under carefully optimized conditions. Undersonication results in a low level of cell breakage, whereas oversonication causes fragmentation of polysomes. Pilot experiments to find the optimal duration of the sonic pulse must be conducted with the equipment available. Sonication should be performed by successive pulses of only a few seconds duration with periods between for cooling on ice in order to avoid heating of the sample.

The second problem is ribonuclease; every effort must be made to exclude it from the preparation. Since fingers are a source of ribonuclease (data not shown), gloves should be worn and all glassware should be

[2] L. L. Randall and S. J. S. Hardy, *Eur. J. Biochem.* **75**, 43 (1977).
[3] D. Patterson, M. Weinstein, R. Nixon, and D. Gillespie, *J. Bacteriol.* **101**, 584 (1970).

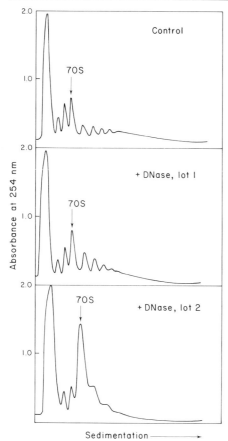

FIG. 1. Assay of ribonuclease activity. Bacterial cells and sucrose gradients can be stored frozen, providing a rapid technique for assay of RNase. Gradients are formed by freezing 3.8 ml of 17.5% (w/w) sucrose, 10 mM Tris-acetate (pH 7.6), 10 mM magnesium acetate, and 60 mM NH$_4$Cl in centrifuge tubes ($\frac{7}{16} \times 2\frac{3}{8}$ in.). Cells are grown to 5 × 10^{8}cells/ml, harvested by centrifugation, suspended in $\frac{1}{50}$th to $\frac{1}{100}$th of the original culture volume in TMN buffer, and stored, frozen, in 0.6-ml aliquots. When needed, the cells are thawed and lysed by addition of lysozyme (final concentration, 200 μg/ml) and Triton X-100 (final concentration, 1%). The DNase to be tested is added to 0.1 ml of the suspension containing the lysed cells, and the suspension is then applied to a gradient. After centrifugation at 40,000 rpm for 1 hr in the Beckman SW 60 rotor (230,000 g), the gradients are collected using an ISCO gradient fractionator with an absorbance monitor. The absorbance at 254 nm is shown for a control (no addition) and for addition of DNase (10 μg/ml) from two different commercially available lots.

heat-treated (160° for at least 2 hr). In the case of equipment that cannot be heat-treated, such as the ISCO gradient fractionator and flow cell absorbance monitor used in collection of gradients, ribonuclease contamination can be eliminated by treatment with diethyl pyrocarbonate.[4] All solutions are made with autoclaved water. Chemicals obtained from biological sources are the most likely to contain ribonuclease. Thus, even though we have obtained satisfactory polysome profiles using reagent-grade sucrose, we routinely use Schwarz–Mann ultrapure density gradient-grade sucrose. We feel that the difference in price is offset by the elimination of the risk of ribonuclease contamination. However, purchase of chemicals labeled "ribonuclease free" is not always a guarantee that no contaminating nucleases

[4] L. Ehrenberg, I. Fedorcsak, and F. Solymosy, *Prog. Nucleic Acid Res. Mol. Biol.* **16**, 189–262 (1976).

are present. The most serious problem we have had with our preparation has been due to ribonuclease activity in the purest preparation of deoxyribonuclease (DNase), which is commercially available from Sigma Chemical Company. Some lots of DNase contain sufficient ribonuclease activity to degrade almost all the free polysomes (Fig. 1). To overcome this difficulty we screen different lots of DNase in small-scale experiments as shown in Fig. 1 and purchase large quantities of the lots that are free of ribonuclease.

Procedure for Preparation of Membrane-Bound and Free Polysomes

Tiresome and unnecessary delays in the preparation of polysomes can result from unpredictable variations in the time a culture requires to enter exponential growth phase. This problem can be easily avoided by starting the cultures from an inoculum that was itself stopped in exponential growth phase on the previous day by transfer from shaking at 37° to storage in the refrigerator. For a standard preparation, 900 ml of prewarmed, minimal medium (M9 salts)[5] containing all the growth requirements of the selected strain and the desired carbon source are inoculated with 100 ml of culture containing 5×10^8 cells/ml. The diluted culture is distributed equally into four 1-liter flasks and shaken at 37° until the cell density again reaches 5×10^8 cells/ml. At this point, the cultures are harvested by rapidly pouring the contents of each flask into a centrifuge bottle containing 100 g of finely divided, frozen TMN (10 mM Tris-acetate, pH 7.6, 5 mM magnesium acetate, 60 mM NH$_4$Cl), which is made by intermittently crushing the buffer as it freezes (we use a plastic ruler for this purpose). The bottles are capped and vigorously shaken so that the temperature of the culture falls to 0° as rapidly as possible. All subsequent operations are carried out at 4° or below. The cells are pelleted in the Sorvall GS3 rotor by centrifugation for 1 min at 7500 rpm; the supernatant is discarded; the inner walls of the bottles are dried with tissues, and the pellets are suspended as rapidly as possible in a total of 12 ml of ice-cold TMN. In order to break the bacteria by ultrasonic vibration, the cell suspension is divided equally among four plastic scintillation vials, and each vial is held in a salt–ice bath (approximately −8°) during the period of sonication. Deoxyribonuclease (ribonuclease free) is added to the sonicated suspension to a final concentration of 10 μg/ml, and the suspension is then divided into six equal parts, each of which is layered onto a sucrose gradient. The gradients, made in nitrocellulose tubes ($1 \times 3\frac{1}{2}$ in.), are linear gradients of 5% to 28% (w/w) sucrose in TMN (total volume of gradient, 28 ml) poured on top of a cushion of 8 ml of 55% (w/w) sucrose,

[5] J. H. Miller, "Experiments in Molecular Genetics," p. 431. Cold Spring Harbor Lab., Cold Spring Harbor, New York, 1972.

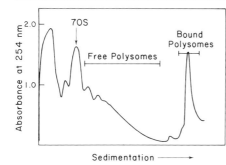

FIG. 2. Fractionation of free and membrane-bound polysomes by sucrose gradient centrifugation.

also in TMN. The gradients are centrifuged (Beckman SW 28 rotor) for 90 min at 26,000 rpm. During this centrifugation, unbroken cells pellet, the membrane-bound ribosomes collect at the cushion/gradient interface, and the free ribosomes contaminated with small fragments of membrane sediment in the body of the gradient. The gradients are collected with an ISCO gradient fractionator that injects a solution of 60% (w/w) sucrose into the bottom of the tube, floating the gradient upward through the flow cell of a recording absorbance monitor. An example of a gradient profile is shown in Fig. 2. The fractions containing the free and the membrane-bound polysomes are collected separately as shown (Fig. 2) and concentrated, after dilution with TMN, by overnight centrifugation at 30,000 rpm in a Beckman 60 Ti rotor. The pellets containing either membrane-bound or free ribosomes are suspended in TMN and used for *in vitro* protein synthesis. For our routine assays, the free ribosomes from a 1-liter culture are suspended in a total of 1.6 ml of TMN, and the membrane-bound ribosomes are suspended in 0.4 ml of TMN. The yield of polysomes will vary, but the absorbance of the suspension of free polysomes is approximately 50 A_{260}/ml. The polysome suspensions are fractionated into aliquots of 0.1 ml, rapidly frozen in liquid nitrogen, and stored at $-80°$.

Protein Synthesis *In Vitro*

Polysomes are used to direct protein synthesis *in vitro* in a system that allows elongation of nascent chains, but not initiation. All components of the system except the supernatant fraction are commercially available. The supernatant used is an S100 fraction prepared from a frozen cell paste of MRE600 (Grain Processing Corporation) by the method described previously[2] except that the buffer used is TMN. The S100 supernatant is stored in 0.1-ml aliquots at $-80°$. All other solutions are kept frozen ($-20°$) except

PROTEIN SYNTHESIS *in Vitro*

Stock solution	Added per 0.1 ml of assay mixture (ml)	Final concentration
1. H$_2$O	Variable, to bring total volume to 0.1 ml	—
2. 4 M NH$_4$Cl	Variable[a]	60 mM NH$_4$Cl
3. 1 M Tris-acetate, pH 8.2[b]	0.005	50 mM Tris-acetate, pH 8.2
4. 0.1 M Mg acetate	Variable[a]	8 mM Mg acetate
5. Amino acid mixture: 1 mM in each amino acid except methionine	0.004	40 μM in each amino acid except methionine
6. 0.5 M dithiothreitol	0.0001	0.5 mM dithiothreitol
7. Energy mixture: 10 mM ATP, 1.2 mM GTP, 50 mM phosphoenolpyruvate	0.010	1 mM ATP, 0.12 mM GTP, 5 mM phosphoenolpyruvate
8. 10 mg/ml tRNA	0.002	200 μg/ml
9. Pyruvate kinase (Sigma, P-1506), approximately 10 mg of protein/ml	0.0001	Approximately 10 μg/ml
10. Supernatant, S100 in TMN buffer	Optimal amount determined for each preparation	—
11. [^{35}S]Methionine, 5 mCi/ml; 1000 Ci/mmol	0.002	0.1 μM methionine at 1000 Ci/mmol
12. Polysome fraction in TMN buffer	0.020	—

[a] The amount of NH$_4$Cl and magnesium acetate required will vary depending on the volume of supernatant and polysome suspension [both in TMN (10 mM Tris-acetate, pH 7.6, 5 mM magnesium acetate, 60 mM NH$_4$Cl) buffer] added. Any other additions containing Mg^{2+} and/or NH$_4^+$ should be taken into consideration in calculating the amount of NH$_4$Cl and magnesium acetate required to bring the final concentrations of Mg^{2+} and NH$_4^+$ to 8 mM and 60 mM, respectively.

[b] The pH can be varied between pH 7.8 and pH 8.2. We have chosen the higher pH in order to assay proteolytic processing in the protein synthesis system.

the pyruvate kinase (Sigma Chemical Company), which is stored at 4° as a crystalline suspension in 2.2 M (NH$_4$)$_2$SO$_4$.

In making up an incorporation mixture the solutions should be mixed in the order given in the table so that the proper pH and ionic conditions are reached prior to the addition of the biologically active components. The completed mixture is incubated at 37°. Incorporation of [^{35}S]methionine proceeds linearly for 3–5 min. There is no initiation of chains in this system, since the supernatant fraction is deficient in the initiation factors. When detergent is added to membrane-bound polysomes to stimulate proteolytic processing (see below), incubation is continued for an additional 10 min.

To determine the amount of radioactivity incorporated into polypeptides, 0.01 ml of the assay mixture is precipitated with 1 ml of 10% trichloroacetic acid containing 1% casamino acids. The precipitated sample is boiled for 15 min to destroy tRNA charged with [^{35}S]methionine, and the precipitate is collected on Whatman GF/C glass-fiber filters. The filters are washed with 10 ml of 5% trichloroacetic acid and dried; the radioactivity is determined by liquid scintillation counting.

The polypeptide products can be immunoprecipitated from the assay mixtures with appropriate antisera and then analyzed by sodium dodecyl sulfate gel electrophoresis and autoradiography.

Proteolytic Processing

The products of membrane-bound polysomes include precursor forms of exported proteins. Although the membranes contain the processing enzyme leader peptidase, processing is observed only if detergent is added to the system. The ratio of membrane to detergent in the system is critical and should be determined for the individual case. Processing of two periplasmic binding proteins[6] was optimal with 0.1% Triton X-100 present in the incorporation mixture. When higher levels of detergent were present, decreased processing was observed, probably because the precursors or enzyme, or both, were released from the membrane.

The Triton X-100 can be added during protein synthesis or immediately after; however, prolonged delay (approximately 10 min) after termination of synthesis before addition of detergent results in a low level of processing.

The free polysome fraction does contain some membranes; however, there is no evidence for a functional association of polysomes with membrane, since addition of detergent does not result in proteolytic processing of precursors. Processing of precursors synthesized by free polysomes is observed when purified leader peptidase (obtained from W. Wickner) is added to the incorporation mixture in the presence of Triton X-100.

Acknowledgments

This work was supported by the Swedish Natural Sciences Research Council and the National Institutes of Health. We are indebted to C. G. Kurland for providing us with space and financial as well as moral support in the initial phase of our studies.

[6] L. L. Randall, L.-G. Josefsson, and S. J. S. Hardy, *Eur. J. Biochem.* **92**, 411 (1978).

[8] Analysis of Cotranslational Proteolytic Processing of Nascent Chains Using Two-Dimensional Gel Electrophoresis

By Lars-Göran Josefsson and Linda L. Randall

Passage of protein through membranes in both prokaryotes and eukaryotes involves an amino-terminal sequence of amino acids, the signal sequence, which plays a critical role in the export phenomenon. This sequence is proteolytically removed to generate the mature form of the protein (see Inouye and Halegoua[1] for a review). We were interested in determining when, in relation to the time course of synthesis, proteolytic processing occurred, since removal of the amino-terminal signal sequence while chains are nascent would indicate that export is initiated before the synthesis of the protein is complete. We have been able to quantify the amount of processing that occurs cotranslationally in *Escherichia coli* by determining the amount of processed and unprocessed amino termini present in a population of incomplete nascent chains of a given protein. Our approach utilizes a modification of the technique of limited proteolysis developed by Cleveland *et al.*[2] The technique could also be adapted for analysis of those covalent modifications of nascent polypeptides that result in a change in migration of the protein during sodium dodecyl sulfate gel electrophoresis.

The population of polypeptides to be analyzed is isolated by specific immunoprecipitation from a culture of exponentially growing *E. coli* that has been pulse-labeled with [³⁵S]methionine. These radioactively labeled polypeptides are separated according to molecular weight on a standard sodium dodecyl sulfate–polyacrylamide slab gel,[3] referred to as the first-dimensional gel. The lane of this first-dimensional gel that contains the polypeptides of interest is cut out along the migration dimension; the gel strip is then applied horizontally to the top of a second slab gel. The polypeptides are subjected to limited proteolysis within the slab gel as described below. The peptides generated by the proteolysis are separated according to molecular weight by electrophoresis in the second dimension. The second-dimensional gel displays simultaneously peptides derived from the completed forms of the protein (precursor and mature) and peptides derived from the population of incomplete, nascent polypeptides that mi-

[1] M. Inouye and S. Halegoua, *CRC, Crit. Rev. Biochem.* **7**, 339 (1980).
[2] D. W. Cleveland, S. G. Fischer, M. W. Kirschner, and U. K. Laemmli, *J. Biol. Chem.* **252**, 1102 (1977).
[3] L. L. Randall and S. J. S. Hardy, *Eur. J. Biochem.* **75**, 43 (1977).

METHODS IN ENZYMOLOGY, VOL. 97

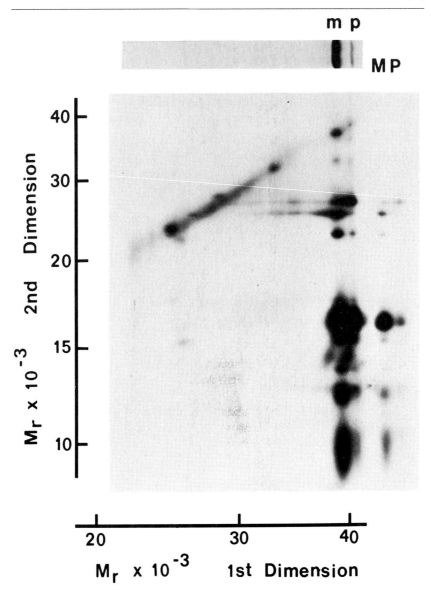

FIG. 1. Analysis of cotranslational processing of maltose-binding protein. An exponentially growing culture of *Escherichia coli* (0.5 ml containing 10^8 cells/ml) was labeled for 15 sec with 15 μCi of [^{35}S]methionine. Incorporation of the radioactive amino acid was stopped by addition of trichloroacetic acid, and the polypeptides related to maltose-binding protein were isolated by immunoprecipitation as described in the text. The immunoprecipitated polypeptides were separated by sodium dodecyl sulfate – 10% polyacrylamide gel electrophoresis (first dimension, top of figure) and visualized by autoradiography. The section of the first-dimension gel lane including the region containing precursor maltose-binding protein (p), mature maltose-bind-

grated in the first-dimensional gel at positions of lower molecular weight than the completed species (for illustration refer to Fig. 1, an analysis of the processing of maltose-binding protein, which is a protein exported into the periplasmic space of *E. coli*). The peptides derived from nascent chains of a given molecular weight appear in a vertical array beneath the position that corresponds to that molecular weight in the first-dimensional gel. Any given peptide will be present in all nascent chains that have been elongated sufficiently to contain the amino acids that comprise the peptide. Thus each peptide appears as a horizontal streak beginning at the position where nascent chains in the first-dimensional gel are first long enough to contain the entire peptide and extending up to and including the positions of the completed protein.

The presence of the amino-terminal peptide of the mature protein among peptides derived from nascent chains is evidence that some of the chains are proteolytically processed before their synthesis is complete. The amino-terminal peptide is identified by comparison of the peptide patterns of the completed mature protein and of its precursor. The amino-terminal peptide of the precursor species contains the signal sequence and thus migrates at a position of molecular weight higher than that of the corresponding peptide derived from the mature protein. The amount of cotranslational processing can be quantified if the distribution of methionine in the protein of interest is known. It must be remembered that the proteolytically matured amino-terminal peptide may contain fewer methionines than the corresponding peptide which carries the signal sequence.

A similar two-dimensional gel system which simultaneously displayed peptide patterns from fibroin and a population of proteins with lower molecular weights was used by Lizardi *et al.*[4] to demonstrate that the low molecular weight species in their system were nascent chains of fibroin.

Protease

Staphylococcus aureus V8 protease, which specifically cleaves peptide bonds on the carboxyl side of glutamic acid residues, is used in this system

---

[4] P. M. Lizardi, V. Mahdavi, D. Shields, and G. Candelas, *Proc. Natl. Acad. Sci. U.S.A.* **76**, 6211 (1979).

---

ing protein (m) as well as nascent chains was placed lengthwise on top of a sodium dodecyl sulfate–15% polyacrylamide gel (second dimension) and subjected to limited proteolysis. Reference peptide patterns from mature maltose-binding protein (lane M) and precursor maltose-binding protein (lane P) were generated by limited proteolysis of those isolated proteins excised from a gel similar to the first-dimensional gel shown. The radioactivity in the second-dimensional gel was visualized by fluorography after treatment with sodium salicylate.

because it retains activity in sodium dodecyl sulfate. If proteases with other specificities, such as trypsin or chymotrypsin, are required in a particular study, high concentrations of the protease must be used to overcome the problem of inactivation. Cleveland et al.[2] examined several proteases in their original presentation of the technique, and that publication can be referred to for details. Digestion of the protein under study should be performed initially over a wide range of concentrations of the chosen protease to determine which concentration produces the desired degree of proteolysis. This can be done rapidly on a slab gel as described by Cleveland et al.[2] We routinely use 0.05 μg of S. aureus V8 protease per 6-mm sample slot to digest individual bands excised from gels. For the digestion of samples in the 6-cm gel strips, we use 0.5 μg of protease (applied in 0.2 ml of buffer given below). Staphylococcus aureus V8 protease is available from Miles Laboratories, Inc. It is advisable to store the protease frozen, in small aliquots, since repeated freezing and thawing changes the proteolytic activity. Our enzyme stock is stored at 0.5 mg/ml in water and diluted immediately prior to use in a buffer that consists of 20 mM Tris-HCl (pH 7.8), 0.1% sodium dodecyl sulfate, 1 mM EDTA, and 20% glycerol. Ethylenediaminetetraacetic acid is included in this buffer because it inhibits a different protease, which is secreted by S. aureus cells. This enzyme is not likely to contaminate the commercial preparations; nonetheless, EDTA is included as a precaution.

Limitations of the Technique

This technique cannot be applied to proteins of low molecular weight, since the peptides generated must be large enough to be resolved in the second gel. In addition, there must be available an antiserum raised to the protein of interest, and the serum must efficiently precipitate nascent chains. We have used antisera to seven different exported proteins, which range in molecular weight from 27,000 to 45,000, and in all cases nascent chains were immunoprecipitated.[5]

The labeling conditions must be defined for each system, since it is critical that a large fraction of the incorporated label be in nascent chains. Thus the labeling must be rapidly terminated during the period of linear incorporation of the amino acid.

The immunoprecipitation must be specific. The presence of contaminants that migrate in the region of the nascent chains will complicate interpretation of the peptide patterns.

[5] L.-G. Josefsson and L. L. Randall, Cell 25, 151 (1981).

Procedure

Preparation of the Sample. We routinely label proteins with [^{35}S]methionine during a period of 10 sec to 30 sec under conditions chosen so that incorporation is linear for a period corresponding to twice the labeling time. Incorporation of radioactive amino acid is stopped by addition of trichloroacetic acid (final concentration, 5%). The sample is centrifuged and the pellet washed with 10 mM Tris (pH 7.8). The pellet is solubilized for immunoprecipitation in a small volume of buffer containing 0.5% sodium dodecyl sulfate, 10 mM Tris-HCl (pH 7.8), and 0.1 mM phenylmethylsulfonyl fluoride by incubating at 55° for 10 min. Then the sodium dodecyl sulfate is diluted at least 10-fold with 1% Triton X-100, 10 mM Tris (pH 7.8), 5 mM magnesium acetate, 60 mM NH$_4$Cl, and 0.1 mM phenylmethylsulfonyl fluoride. This step is critical in that it prevents the sodium dodecyl sulfate from denaturing immunoglobulins during immunoprecipitation. After the sample has been centrifuged to remove nonspecific aggregates, an appropriate amount of specific antiserum is added to the sample. Our experience in determination of the amount of antiserum needed is that for most antiserum (rabbit) raised to soluble proteins, approximately 10 μl of antiserum precipitates protein in the range of 0.5 – 1.0 μg.

The immunoprecipitate is isolated by centrifugation and washed extensively by successively suspending and pelleting it in 0.2 ml of buffers as follows: first, 0.15 M NaCl, 10 mM Tris-HCl (pH 7.8), 0.5% Triton X-100, 5 mM EDTA, 0.1 mM phenylmethylsulfonyl fluoride; then, the same buffer except for containing 0.5 M NaCl; then again the buffer containing 0.15 M NaCl. The salt washes are followed by two washes with 10 mM Tris-HCl, pH 7.8. Finally the pellet is suspended in sample buffer for gel electrophoresis.[6]

First Gel. The immunoprecipitate is applied to a 0.5 mm-thick slab gel that serves as the first dimension. The gel is a standard sodium dodecyl sulfate – polyacrylamide gel[6] with slight modifications as described elsewhere.[3] The choice of percentage of acrylamide to be used will depend on the molecular weight of the protein under study. The gel is fixed and stained by standard techniques[7] before it is dried and subjected to autoradiography. The lane of interest is excised from the dried gel with a razor blade. If care is exercised, the dried gel can be pulled away from most of the paper at this stage. We routinely cut out a 6-cm-long strip, beginning 2 mm above the precursor position and extending toward the bottom of the gel. The strips are rehydrated in tubes containing sufficient volume of buffer [20 mM

[6] U. K. Laemmli, *Nature* (*London*) **227**, 680 (1970).
[7] G. Fairbanks, T. L. Steck, and D. F. H. Wallach, *Biochemistry* **10**, 2606 (1971).

Tris-HCl (pH 7.8), 0.1% sodium dodecyl sulfate, 1 mM EDTA, 20% glycerol] to cover the gel completely. The tubes should be vigorously shaken several times during a 30-min incubation at room temperature to assure thorough equilibration. During this incubation bits of paper that adhered to the gel strip will come off. Any paper that remains will not interfere with the subsequent electrophoresis, although it can make insertion of the strip into the sample slot difficult.

Second Gel. The slab gel for the second electrophoresis step should be 1 mm thick to facilitate insertion of the 0.5-mm-thick sample strip. In addition, a high-percent acrylamide (15%) or a gradient of acrylamide concentration is necessary to resolve the small peptides. A sample well is formed that will accommodate the gel strip horizontally across the top of the second gel. In addition, two sample wells should be formed to the sides — one to carry molecular weight standards, and the other for a gel piece containing the protease and Bromphenol blue (to serve as a tracking dye). proteolysis in the gel and serve as references for the position of the peptides. In the original procedure of Cleveland *et al.*,[2] the gel pieces containing the proteins to be digested were overlaid with a solution containing the protease, and both the protease and sample were electrophoresed into the gel at the same time. We have found that it is easier to achieve a uniform distribution of the enzyme along the 6-cm sample slot if we first apply 0.2 ml of solution containing the protease and Bromphenol blue (to serve as a tracking dye). Electrophoresis is started and continued only until the protease solution just enters the stacking gel. One should be certain that both refractive boundaries that are present during electrophoresis are within the stacking gel when electrophoresis is stopped. The sample well is then rinsed with running buffer, the gel strip is inserted, and electrophoresis is restarted. The protein in the sample gel strip will enter the gel and overtake the enzyme. One can convince oneself that the proteins are migrating together by applying molecular weight standards twice to the same sample well — once with the enzyme and once with the gel strip.

Proteolytic Digestion

The proteins are allowed to migrate into the stacking gel until the tracking dye reaches a position approximately 1 cm above the interface of the stacking gel and separation gel. Electrophoresis is stopped for 40 min, during which time the protease digests the protein that has been eluted from the gel strip. Electrophoresis is restarted and the peptides that were generated during the digestion are separated according to molecular weight. After electrophoresis the gel is stained, destained, dried, and subjected to autoradiography or fluorography by standard methods. In many cases the level of radioactivity in the second gel will be so low that enhancement is required for detection within a reasonable time. Chamberlin[8] has described use of sodium salicylate as a fluor for enhancement of radioactivity. If sodium salicylate is used in the first gel, one must be sure to remove all salt from the gel strip before proceeding with the second electrophoresis step.

Interpretation of the Data

Interpretation of the data in terms of cotranslational processing depends on identification of peptides derived from the amino termini that are characteristic of the mature or precursor species. The peptide can be identified initially simply by comparing proteolytic digests of precursor and mature proteins. The amino-terminal peptide from the precursor species contains the signal sequence and migrates at a position of molecular weight higher than the corresponding peptide from the mature species. The identification of a shifted peptide as derived from the amino terminus can be confirmed by pulse-chase experiments similar in design to that used by Dintzis[9] in establishing the direction of protein chain elongation. In a pulse-chase experiment, the first polypeptide to be completed will be labeled at the carboxyl terminus. Polypeptide chains which were initiated during the pulse, thus carrying label in the amino terminus, will appear last. It must be kept in mind that proteolysis is not complete, therefore large fragments including the amino terminus will appear labeled before smaller amino-terminal fragments. Nevertheless, amino-terminal fragments must appear radioactively labeled after carboxy-terminal fragments. (See Hardy and Randall[10] for an example of this technique.)

Thus the critical peptide can be identified before attempting an analysis of nascent chains. However, the two-dimensional gel pattern itself provides an independent source of data that allows assignment of the peptides to positions in the protein. The schematic representation of a two-dimensional gel pattern (Fig. 2) illustrates the information that can be obtained. A peptide containing the amino terminus of the precursor protein is present in all chains that are larger than the fragment and first appears in those chains that are just long enough to contain the entire sequence. These nascent chains, which correspond to the proteolytic fragment, are not further digested by the protease and thus migrate with the same molecular weight in both the first and second gel. Thus the horizontal streak representing an amino-terminal peptide fuses with the main diagonal. Peptides that include the carboxy terminus are present, intact, only when the protein is com-

[8] J. P. Chamberlin, *Anal. Biochem.* **98**, 132 (1979).
[9] H. M. Dintzis, *Proc. Natl. Acad. Sci. U.S.A.* **47**, 247 (1961).
[10] S. J. S. Hardy and L. L. Randall, *J. Bacteriol.* **135**, 291 (1978).

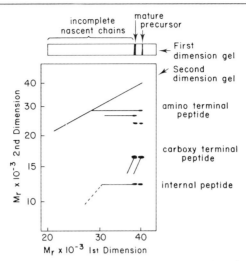

FIG. 2. Schematic representation of two-dimensional analysis of peptides shown on fluorogram in Fig. 1.

pleted. Thus, there is no horizontal streak, but a diagonal streak ending at the position of the completed polypeptide. Peptides derived from internal positions in the protein migrate in the second gel at a molecular weight lower than in the first dimension and appear as a horizontal streak that does not reach the main diagonal of the gel but ends in a secondary diagonal.

Presence of the amino-terminal peptide that is characteristic of the mature species in the proteolytic digest of nascent chains indicates that proteolytic processing has occurred while chains are nascent. The fact that the horizontal streak that represents the matured amino-terminal peptide does not fuse with the main diagonal is significant (Fig. 1). In the case of maltose-binding protein, the mature amino-terminal peptide first appears in digests of nascent chains of 30,500 molecular weight. That size corresponds to an unprocessed molecular weight of 33,000 (80% of the molecular weight of the completed precursor, 41,000). Thus it seems that a critical length must be reached to initiate processing.

In the case of proteins that are processed entirely posttranslationally, no mature amino-terminal peptide would be detected in the digests of nascent chains. On the other hand, if a protein were processed entirely cotranslationally even though no precursor species would be detected in a pulse-labeling experiment, the amino-terminal peptide carrying the signal sequence would be present in digests of nascent chains.

In studies employing this technique, one must demonstrate that the incomplete polypeptides exhibit properties of nascent chains and are not the products of degradation or premature termination. The polypeptides should become radioactively labeled during a pulse with [35S]methionine and become undetectable during a subsequent chase with excess nonradioactive methionine. The disappearance of the peptides can be shown to be the result of elongation of polypeptides to complete length and not of degradation by performing the nonradioactive chase in the presence of chloramphenicol. Under these conditions the peptides will not disappear.

Acknowledgment

The work was supported by a grant from the National Institutes of Health. We are grateful to Gerald Hazelbauer for critically reading the manuscript.

[9] Proteins Forming Large Channels from Bacterial and Mitochondrial Outer Membranes: Porins and Phage Lambda Receptor Protein

By HIROSHI NIKAIDO

Biological membranes are usually impermeable to most hydrophilic solutes, except those taken up by specific, active transport or facilitated diffusion systems. However, some membranes have an unusual location that requires a rather nonspecific permeability to a wide range of solutes: examples are the bacterial outer membrane, which is located outside the cytoplasmic membrane and even the peptidoglycan layer of the cell wall,[1,2] and the mitochondrial outer membrane. Reconstitution studies in our laboratory[3-5] showed that this nonspecific permeability is due to the presence of proteins of a special class, called porins, that produce large channels across the membrane.

Diffusion through the porin channel is too slow for solutes of certain type. In this case, the bacteria may produce a specific diffusion channel for one or more of these solutes. The phage lambda receptor protein (LamB protein) was initially shown to catalyze the diffusion of maltose and its higher homologs in intact cells,[6] and later the properties of this channel were studied by the reconstitution technique.[7]

[1] H. Nikaido and T. Nakae, *Adv. Microb. Physiol.* **20**, 163 (1979).
[2] H. Nikaido, *in* "Bacterial Outer Membranes" (M. Inouye, ed.), p. 361. Wiley, New York, 1979.
[3] T. Nakae, *J. Biol. Chem.* **251**, 2176 (1976).
[4] T. Nakae, *Biochem. Biophys. Res. Commun.* **71**, 877 (1976).
[5] L. S. Zalman, H. Nikaido, and Y. Kagawa, *J. Biol. Chem.* **255**, 1771 (1980).
[6] S. Szmelcman, M. Schwartz, T. J. Silhavy, and W. Boos, *Eur. J. Biochem.* **65**, 13 (1976).
[7] M. Luckey and H. Nikaido, *Proc. Natl. Acad. Sci. U.S.A.* **77**, 167 (1980).

The purification and *in vitro* reconstitution of these proteins into lipid bilayer membranes are described here. For work on the biogenesis of these proteins in intact cells, the readers are referred to reviews.[8,9]

General Properties of Porins and LamB Protein

Porins from Escherichia coli and Salmonella typhimurium. In these organisms, usually more than one species of porin are present. Thus the LT2 strain of *S. typhimurium* produces three porins, "36K" (coded by *ompC* gene), "35K" (coded by the *ompF* gene), and "34K" (coded by the *ompD* gene).[1,9] *Escherichia coli* K12 normally produces two porin species: Ia (or O-9, or *b*), coded by the *ompF* gene; and Ib (or O-8, or *c*), coded by the *ompC* gene.[1,8,9] In addition, phosphate starvation derepresses the production of another porin, coded for by the *phoE* gene.[10,11] *Escherichia coli* B produces only the OmpF porin.[12]

The reported molecular weights of the monomer units of these porins are centered around 36,000.[1,8,9] The OmpF porin from *E. coli* B has been sequenced,[13] and the molecular weight calculated from the sequence was 37,205. All these porins were shown to share the following properties.

1. They are all acidic proteins. More recently reported p*I* values are 5.0 and 4.8 for OmpF and OmpC porins of *E. coli* K12,[14] and around 4.8 for *S. typhimurium* porins.[15]
2. They contain a very large proportion of β-sheet structure,[12] and there is no evidence for the presence of significant amounts of α-helical region.
3. They form trimers that are stable even in sodium dodecyl sulfate (SDS).[16,17]
4. The trimers are associated with the underlying peptidoglycan sheets through noncovalent interactions.[12]
5. In the absence of other proteins, the porin trimers have a tendency to form two-dimensional, crystalline arrays of hexagonal symmetry both in the presence[12] and in the absence[16] of peptidoglycan sheets.

[8] J. M. DiRienzo, K. Nakamura, and M. Inouye, *Annu. Rev. Biochem.* **47**, 481 (1978).
[9] M. J. Osborn and H. C. P. Wu, *Annu. Rev. Microbiol.* **34**, 369 (1980).
[10] M. Argast and W. Boos, *J. Bacteriol.* **143**, 142 (1980).
[11] J. Tommassen and B. Lugtenberg, *J. Bacteriol.* **143**, 151 (1980).
[12] J. P. Rosenbusch, *J. Biol. Chem.* **249**, 8019 (1974).
[13] R. Chen, C. Kramer, W. Schmidmayr, and U. Henning, *Proc. Natl. Acad. Sci. U.S.A.* **76**, 5014 (1979).
[14] C. J. Schmitges and U. Henning, *Eur. J. Biochem.* **63**, 47 (1976).
[15] H. Tokunaga, M. Tokunaga, and T. Nakae, *Eur. J. Biochem.* **95**, 433 (1979).
[16] A. C. Steven, B. ten Heggeler, R. Muller, J. Kistler, and J. P. Rosenbusch, *J. Cell Biol.* **72**, 292 (1977).
[17] M. Tokunaga, H. Tokunaga, Y. Okajima, and T. Nakae, *Eur. J. Biochem.* **95**, 441 (1979).

LamB Protein from E. coli K12. Based on the nucleotide sequence of *lamB* gene,[18] the molecular weight of the mature protein should be 47,392. This protein shares all the properties of the porins listed above.[19-23] There are some differences, however. For example, the interaction between porin trimers and the peptidoglycan can be broken by 0.5–1 M NaCl,[24] whereas the association between the LamB protein and the peptidoglycan can be disrupted by lower concentrations of NaCl[22] or with EDTA.[7]

Porins from Other Sources. Porins from *Pseudomonas aeruginosa*[25,25a] and *Phaseolus aureus* mitochondrial outer membrane[5] have been purified. The former protein can be purified as functional monomers in SDS.[25a] The latter protein occurs also as oligomers, but the quaternary structure is more easily destroyed by sodium dodecyl sulfate (SDS) than with *E. coli* or *S. typhimurium* porins. The mitochondrial porin is a weakly basic protein.[5]

More recently, porins have been purified from a number of gram-negative bacteria.[26,27] In all cases so far examined, the apparent molecular weights of the monomer are in the range of 30,000–45,000, and some of them can be purified as oligomers in SDS. They are usually acidic proteins, and in many cases show association with the peptidoglycan layer.

Purification of Porins from *E. coli* or *S. typhimurium*

Unless *E. coli* B or its derivative is used as the starting material, the outer membrane contains multiple species of porin, and separation of one porin from another is difficult. The best strategy here is to start from mutant strains that produce only one species of porin: such strains are available both in *E. coli* K12[28,29] and in *S. typhimurium* LT2.[30] Even in the wild-type strains, one can sometimes repress the production of one or more porins specifically through manipulation of culture conditions.[24,29,31] Most purifica-

---

[18] J. M. Clement and M. Hofnung, *Cell* **27**, 507 (1981).
[19] P. Bavoil and H. Nikaido, unpublished observation.
[20] J. P. Rosenbusch, personal communication.
[21] E. T. Palva and P. Westermann, *FEBS Lett.* **99**, 77 (1979).
[22] J. Gabay and K. Yasunaka, *Eur. J. Biochem.* **104**, 13 (1980).
[23] H. Yamada, T. Nogami, and S. Mizushima, *J. Bacteriol.* **147**, 660 (1981).
[24] K. Nakamura and S. Mizushima, *J. Biochem.* (*Tokyo*) **80**, 1411 (1976).
[25] R. E. W. Hancock, G. M. Decad, and H. Nikaido, *Biochim. Biophys. Acta* **554**, 323 (1979).
[25a] F. Yoshimura, L. S. Zalman, and H. Nikaido, *J. Biol. Chem.* **258**, 2308 (1983).
[26] J. T. Douglas, M. D. Lee, and H. Nikaido, *FEMS Microbiol. Lett.* **12**, 305 (1981).
[27] L. S. Zalman and H. Nikaido, in preparation.
[28] T.-J. Chai and J. Foulds, *J. Bacteriol.* **135**, 164 (1978).
[29] M. N. Hall and T. J. Silhavy, *J. Mol. Biol.* **146**, 23 (1981).
[30] M. Nurminen, K. Lounatmaa, M. Sarvas, P. H. Makela, and T. Nakae, *J. Bacteriol.* **127**, 941 (1976).
[31] C. A. Schnaitman, D. Smith, and M. F. de Salsas, *J. Virol.* **15**, 1121 (1975).

tion methods for porins are based on the fact that they produce SDS-resistant trimers that remain associated with peptidoglycan. The following is a minor modification of that described by Tokunaga et al.[17]

Materials

L broth: Add 10 g of Bacto-yeast extract (Difco), 10 g of Bacto-tryptone (Difco), and 5 g of NaCl to a liter of distilled water; after autoclaving, add a sterile solution of D-glucose up to a final concentration of 0.5%. Glucose is added to repress the production of LamB protein.

NaCl buffer: 50 mM Tris-HCl (pH 7.7), 0.4 M NaCl, 1% SDS, 5 mM EDTA, 0.05% mercaptoethanol, 3 mM NaN$_3$

Procedure

Preparation of the Cell Envelope Fraction. Twenty-five milliliters of an overnight culture of the bacterial strain in L broth is diluted into 1 liter of prewarmed, sterile, L broth in a 6-liter Erlenmeyer flask; the incubation is continued with shaking at 37°. When the culture has attained a Klett reading (red filter) of 130, the cells are harvested by centrifugation. They are washed once with cold 50 mM Tris-HCl buffer, pH 7.2, then resuspended in 5 ml of 50 mM Tris-HCl buffer, pH 7.7; the suspension is passed through a chilled French pressure cell three times at 9000 psi. The extract is centrifuged for 10 min at 1000 g to remove unbroken cells, and the supernatant is then centrifuged at 100,000 g for 1 hr. The pellet corresponds to "cell envelope," which contains the outer and inner (cytoplasmic) membranes as well as the peptidoglycan layer.

Preparation of Peptidoglycan-Associated Porin. The cell envelope fraction is resuspended in 5 ml of 2% SDS–10 mM Tris-HCl, pH 7.7; the suspension is kept at 32° for 30 min and then centrifuged at 100,000 g for 30 min at 20°. The pellet should consist of peptidoglycan sheets with covalently linked lipoprotein and noncovalently associated porin.[12] This can be checked by solubilizing a small portion of the sample by boiling it in the Laemmli sample buffer[32] and analyzing it by slab polyacrylamide gel electrophoresis.[33] If significant amounts of nonporin proteins are detected, the extraction with 2% SDS–10 mM Tris-HCl, pH 7.7, is repeated.

Elution and Purification of Porin Trimers. The peptidoglycan-associated porin preparation is now resuspended in 2 ml of NaCl buffer (see above) and extracted at 37° for 2 hr. The suspension is then centrifuged at 100,000 g for 30 min at 25°, and the supernatant is applied to a column (1.6 × 90 cm) of

[32] U. K. Laemmli, *Nature (London)* **227**, 680 (1970).
[33] B. Lugtenberg, J. Meijers, R. Peters, P. van der Hoek, and L. van Alphen, *FEBS Lett.* **58**, 254 (1975).

Sephacryl S-200 superfine (Pharmacia), which has been preequilibrated with the NaCl buffer. The column is eluted at 25° with the NaCl buffer at the flow rate of about 4 ml/hr. The porin trimers are eluted as a symmetrical peak after about 75 ml, or with a K_d of 0.06. The fractions containing the porin are pooled, then dialyzed against 3 mM NaN$_3$ first at room temperature and then at 4° for 7–10 days. The yield usually is 3–4 mg of porin, which is at least 98% pure on the basis of the scanning of the Coomassie Blue-stained slab gels. The final preparation stays in solution if the protein concentration is low (usually less than 0.5 mg/ml). Its pore-forming activity remains constant for several months at 4°.

Comments. This procedure produces reasonably monodisperse trimers,[15] whereas earlier procedures tended to produce higher aggregates of trimers.[3] The final preparation contained very little SDS, less than 0.1% of the weight of the protein.[3] Although the bulk of lipopolysaccharides (LPS) and phospholipids is removed during the first SDS extraction, it is difficult to obtain a preparation completely free of LPS contamination. A more recent study[34] showed that the leading edge of the porin peak is LPS-free.

Purification of LamB Protein from *E. coli*

In the earlier procedures,[7,22] contamination by porin was a major problem. A more selective and rapid method has been developed[35]; this is based on the specific interaction between the LamB protein and immobilized maltose-binding protein (MBP).

Materials

MBP. This is purified from cells of any Mal$^+$ strain of *E. coli* K12, such as HfrG6, according to the method of Ferenci and Klotz,[36] and is dialyzed for at least 5 days against 10 mM Tris-HCl, pH 7.2, and then overnight against 0.5 M NaCl–0.1 N NaHCO$_3$, pH 8.1.

MBP–Sepharose column.[35] MBP (10 mg) is coupled to 2 g (dry weight) of CNBr-activated Sepharose 6 MB (Pharmacia) as follows. The beads are swollen in 1 mM HCl, washed with 0.5 M NaCl–0.1 N NaHCO$_3$, pH 8.1, and then allowed to react with MBP in 30 ml of the same buffer for 20 hr at 4°. Remaining binding sites are blocked by incubation with 1 M ethanolamine, pH 9, for 1 hr at room temperature. The beads are then washed four times, each time by the alternate application of 0.5 M NaCl–0.1 M acetate buffer, pH 4, and 0.5 M NaCl–0.1 N borate

[34] H. Schindler and J. P. Rosenbusch, *Proc. Natl. Acad. Sci. U.S.A.* **78**, 2302 (1981).
[35] P. Bavoil and H. Nikaido, *J. Biol. Chem.* **256**, 11385 (1981).
[36] T. Ferenci and U. Klotz, *FEBS Lett.* **88**, 71 (1978).

buffer, pH 8. The beads are poured into a Pharmacia K9/15 column, and the column is stored at 4° in 0.1% Triton X-100–0.01% NaN_3–10 mM Tris-HCl, pH 7.4.

Procedure

Escherichia coli cells are grown in a medium that would cause the maximal induction of the LamB protein, such as mineral medium containing maltose as the sole carbon source, or L broth with 0.8% (w/v) maltose. The cells are harvested at the late exponential phase, and cell envelope fraction (see above), or preferably, the outer membranes,[37] are prepared. The outer membranes (5–10 mg of protein) are extracted with 2% Triton X-100–10 mM EDTA–20 mM Tris-HCl, pH 7.4 (total volume, 2 ml) for 30 min at 4°. (To this and all other solutions used in subsequent steps, 0.2 mM PMSF is added.) Mild sonication may be used to make certain that the material is finely dispersed. The mixture is centrifuged at 20,000 g for 1 hr, and the supernatant fluid is applied to the MBP-Sepharose column described earlier. The column is washed, at 4°, with at least five column volumes of 0.1% Triton X-100–10 mM Tris-HCl, pH 7.4, and is then eluted with 0.2 M NaCl–0.1% Triton X-100–10 mM Tris-HCl, pH 7.4. The LamB protein (yield 300–700 μg) is eluted very close to the solvent front. It is usually at least 90% pure, and contains less than 5% porin. In this preparation, LamB protein exists as trimers, as shown by SDS–polyacrylamide gel electrophoresis. For reconstitution, NaCl is removed by dialyzing the sample overnight against 0.1% Triton X-100–10 mM Tris-HCl, pH 7.4–0.01% NaN_3. The pore-forming activity is stable for months if the sample is stored at 4°.

Purification of Mitochondrial Porin[5]

"Bean sprouts" (hypocotyls of *Phaseolus aureus*) are a convenient source of mitochondrial outer membrane of rather simple protein composition. To a 0.45-ml suspension of *P. aureus* mitochondrial outer membrane (containing 2 mg of protein)[38] in 10 mM phosphate buffer, pH 7, 0.05 ml of 20% Triton X-100 is added; after 15 min at room temperature the mixture is centrifuged at 130,000 g for 1 hr at 4°. The pellet is then resuspended in 0.5 ml of 2% Triton X-100–1 M NaCl–10 mM phosphate buffer, pH 7, and the suspension is centrifuged in a similar manner after 15 min. The

---

[37] If a French pressure cell is available, the procedure of J. Smit, Y. Kamio, and H. Nikaido [*J. Bacteriol.* **124**, 942 (1975)] is recommended, as it is highly reproducible and is usable in large-scale preparations. Otherwise, the procedure of M. J. Osborn, J. E. Gander, E. Parisi, and J. Carson [*J. Biol. Chem.* **247**, 3962 (1972)] may be used.

[38] C. A. Manella and W. D. Bonner, Jr., *Biochim. Biophys. Acta* **413**, 213 (1975).

supernatant contains predominantly the 30,000-dalton porin, although it is still contaminated by a small amount of 54,000-dalton protein. If further purification is desired, one can use a rate centrifugation in a sucrose gradient in 1 M NaCl–0.05% Triton X-100–10 mM phosphate buffer, pH 7.

Purification of Porins from Other Bacteria

A method involving ion-exchange chromatography has been published for the purification of *Pseudomonas aeruginosa* porin.[25] However, the yield was rather poor and contaminating proteins could not be removed completely. Our experience indicates that it is best to utilize the two widespread properties of porins: (*a*) they are extracted only when EDTA or high concentrations of salts are present in addition to detergents; and (*b*) in their oligomeric forms they have a large size matched by few other outer membrane proteins. Thus differential extraction with cholate (or SDS if this does not dissociate the oligomer) plus NaCl or EDTA, and fractionation by gel filtration, appears to be a universally applicable procedure that produces homogeneous preparations of porins from various bacteria, including *P. aeruginosa,* at high yields.[25a,26,27]

Liposome Reconstitution for Evaluation of Exclusion Limits

In this method, multilayered liposomes are reconstituted from phospholipids and pore-forming protein (and usually also LPS) in the presence of two radiolabeled solutes.[3,4] One solute is chosen so that it will not diffuse through the protein pores; dextran is most often used. The other solute, labeled with a different radioisotope, is of much lower molecular weight. The reconstitution mixture is then passed through a column of Sepharose 4B (Pharmacia). In this column, even large dextrans are significantly "included" and retarded, and liposomes that come off at the void volume become completely separated from constituents of the extravesicular fluid. More important, if the low molecular weight solutes can diffuse through the pores, they will diffuse out of the vesicles during the gel filtration process, so that the ratio of the two radioisotopes in the vesicles recovered from the column would become very different from the ratio in the original reconstitution mixture. The extents of efflux of various solutes will allow the determination of "exclusion limits" of the pore.

Materials

Pore-forming proteins. Their purification has been described already. Although it is best to remove as much detergent as possible before reconstitution, the system can tolerate fairly large amounts of detergents. Thus, up to 1 mol of Triton X-100 could be added to 7 mol of phospholipids without affecting the results significantly.[5,25]

Phospholipids. In most experiments we have used crude lipids from *E. coli* or *S. typhimurium*.[3,4] These are extracted according to Folch *et al.*[39] from late exponential phase cells of these bacteria, grown with aeration at 37° in L broth (see above; glucose is omitted). The lipid extract is washed twice with Folch's "theoretical upper phase" mixture, dried with anhydrous Na_2SO_4, and kept at $-70°$. In some experiments, especially those with mitochondrial porin,[5] acetone-extracted[40] "asolectin" (soybean lipids) has been used. However, so far there is little evidence for requirement of any specific lipids in the reconstitution assay described here.

Lipopolysaccharides (LPS). The LPS is prepared from late exponential phase culture of *S. typhimurium* HN202 grown in L broth at 37°, by the method of Galanos *et al.*[41] The LPS is quantitated by a colorimetric assay for heptose.[42] In our earlier papers,[3,4] we assumed that LPS molecule is a trimer, containing 6 molecules of heptose. However, more recent studies have shown that LPS exists as monomers, and we now express the molar quantities of LPS based on the assumption that 1 mol of LPS contains 2.5 mol of heptose. Some batches of LPS inhibit the efficient production of liposomes. Probably this is due to the nature of bound cations present. With these preparations, electrodialysis[43] usually converts the LPS into a form active in the reconstitution assay. With some LPS preparations it was necessary to reduce the Mg^{2+} concentration in the reconstitution mixture (see below) to 1 mM or even to 0 nM (see also Luckey and Nikaido[7]).

Radiolabeled solutes. Nonelectrolytes are easier to use because they are not affected by the pH gradient or membrane potential and are not adsorbed electrostatically to membrane surface. Various monosaccharides and disaccharides are available commercially as [14]C- and/or ^{3}H-labeled compounds. Preparation of other labeled saccharides has been described.[44,45]

Procedure[3,4]

About 1 μmol of *E. coli* or *S. typhimurium* phospholipids is added to a test tube. We use thick-walled, 20 × 120 mm tubes. The solvent is removed

[39] J. Folch, M. Lees, and G. H. S. Stanley, *J. Biol. Chem.* **226**, 497 (1957).
[40] Y. Kagawa and E. Racker, *J. Biol. Chem.* **246**, 5477 (1971).
[41] C. Galanos, O. Luderitz, and O. Westphal, *Eur. J. Biochem.* **9**, 245 (1969).
[42] M. J. Osborn, *Proc. Natl. Acad. Sci. U.S.A.* **50**, 499 (1963).
[43] C. Galanos and O. Luderitz, *Eur. J. Biochem.* **54**, 603 (1975).
[44] G. Decad and H. Nikaido, *J. Bacteriol.* **128**, 325 (1976).
[45] T. Nakae and H. Nikaido, *J. Biol. Chem.* **250**, 7359 (1975).

under a stream of N_2, the lipid is dissolved in 0.5 ml of benzene, and the solvent is evaporated again. Finally, the lipid is dissolved in about 0.2 ml of ethyl ether, and the solvent is removed under a nitrogen stream while the tube is shaken rapidly; this produces a thin, even film of phospholipids. The tubes are dried further in a desiccator, which is evacuated with an oil pump for at least 30 min.

An aqueous suspension of porin (if it is *E. coli* porin, 10–20 μg should suffice) and LPS (0.2 μmol) (in 0.3 ml total volume) is then added, and the lipid is dispersed first by manual shaking and then by sonication in a bath-type sonicator (we use Bransonic 12, Branson Cleaning Equipment Co., Shelton, Connecticut). If the system is "in tune" by having the correct water level and by having the tube positioned at the right place, it will take less than 30 sec for the suspension to turn translucent, indicating the conversion of most liposomes into the vesicles of very small sizes. The tube is connected to a gas-drying jar containing anhydrous $CaSO_4$, and the jar is in turn connected to an oil pump. The tube is immersed in a water bath at 45°, and the vacuum is applied while the tube is shaken manually. It takes usually less than 2 min to complete the drying process. The tubes are then dried in an evacuated desiccator for at least 1 hr. These tubes may be kept overnight at room temperature in the dark.

The lipid–protein mixture is then resuspended, first by manual shaking, in 0.2 ml of a solution containing 10 mM HEPES-NaOH buffer, pH 7.4, 0.1 M NaCl, 10 mM $MgCl_2$, 0.15 mM chloramphenicol, [^{3}H]dextran (2×10^5 cpm), and [^{14}C]sucrose (or any other compound to be tested) (2×10^5 cpm). The suspension is then sonicated as above, but for only 2–3 sec, just to break up large aggregates and to produce an even suspension. It is then heated for 30 min in a water bath at 45°; finally, after turning off the water bath, it is allowed to cool slowly for 2 hr.

The suspension is applied to a column (9 × 95 mm) of Sepharose 4B equilibrated with 10 mM HEPES-NaOH buffer, pH 7.4–0.1 M NaCl–0.1 mM $MgCl_2$–3.3 mM NaN_3, and the column is eluted with the same buffer. Liposomes that come off the column at the void volume are collected, and the ^{3}H and ^{14}C content of this preparation is determined by the use of a liquid scintillation spectrometer. If the low molecular weight solute, labeled with ^{14}C, can diffuse out through the protein pores, the ^{14}C:^{3}H ratio in the liposomes should be significantly lower than the ratio in the control liposomes made without the protein. (In some experiments with porin, the ratio decreased to 1–2% of the control.) Furthermore, the ^{14}C:^{3}H ratio in the control liposome made without the addition of protein should be about the same as in the reconstitution mixture itself: a significantly lower value in the former suggests the formation of "leaky" (presumably unstable) bilayer. Finally, the presence of protein should not decrease significantly the total amount of [³H]dextran trapped inside the liposome. Usually this amounts to 1–2% of what is added to the reconstitution mixture.

Comments. In some experiments we have used filtration through membrane filters[4] instead of gel filtration. This alternative method is more convenient, but it is not recommended for general use, as its success depends on many factors and is not easily predictable.

Addition of LPS is required if the assay is carried out as described, but this is not necessarily related to the possible effect of LPS on the functions of porins. In this system, Mg^{2+} is present. Without LPS, which binds divalent cations,[46] Mg^{2+} will precipitate the acidic phospholipids and inhibit the formation of liposomes. The addition of LPS also improves the assay by increasing the electrostatic repulsion between phospholipid bilayers and thereby favoring the production of unilamellar vesicles. However, the assay can be successfully performed without the addition of LPS, at least for mitochondrial porins,[5] if the Mg^{2+} concentration is kept at 1 mM or lower and if more phospholipids are used to counteract the decrease in intravesicular space.

With some solutes, practically complete efflux or complete retention is observed. However, with other solutes an intermediate degree of retention is often seen. When the solute is not monodisperse in terms of its size, perhaps the sieving effect is contributing to this phenomenon. But the phenomenon is also seen with a homogeneous solute, and in this case must mean either that the efflux process is too slow to be completed during the gel filtration process or that the pore size is not totally uniform (see Schindler and Rosenbusch[34]).

Liposome Reconstitution for Evaluation of Penetration Rates

The assay described above has a poor resolution in the time domain, as the diffusion occurs during the slow gel filtration step, requiring 15–20 min. In contrast, the "swelling assay"[7] gives precise information on the *rates* of diffusion of solutes through the porin or LamB protein channels.[47]

---

[46] M. Schindler and M. J. Osborn, *Biochemistry* **18**, 4425 (1979).

[47] An alternative method for measuring the penetration rates through the porin channel has been described [M. Tokunaga, H. Tokunaga, and T. Nakae, *FEBS Lett.* **106**, 85 (1979)]. Very small unilamellar vesicles containing hydrolytic enzymes are made, and the penetration rate through the porin channel is determined by measuring the rate of hydrolysis of substrates added to the extravesicular medium. In this method, however, calculation shows that often only a fraction of the vesicles are expected to contain porin. Thus the method sometimes might be measuring the statistical distribution of porin-containing vesicles rather than measuring the actual rates of diffusion through the porin channels.

The method is based on that used to measure the permeability of phospholipid bilayers of liposomes.[48] Thus multilayered liposomes containing the pore-forming protein are made so that large solute molecules incapable of penetration through the pore are trapped within the liposomes. They are then diluted into isotonic solutions of test solutes. If the test solute does not diffuse through the pore, the system will remain at an osmotic equilibrium. If, however, the pore is permeable to the solute, an influx of the solute molecules will occur because of the chemical potential gradient of this solute, and this will be followed by the influx of water. This will lower the average refractive index of the liposomes[49] and will reduce the light scattering, and therefore the optical density of the liposome suspension. Although the entire time course of the swelling of the multilayered liposomes is complex, we can assume that the *initial rate* of swelling must reflect the solute penetration rate into the outermost layer.

Materials

Phospholipids. Phospholipids from *E. coli* or *S. typhimurium* (see above) are excellent because they produce bilayers with extremely low permeability to nonelectrolytes and therefore give very low "background" swelling rates. On the other hand, they are sometimes difficult to resuspend, and preparation of large amounts of these phospholipids is time-consuming.

Commercial egg phosphatidylcholine (type IX-E, Sigma) can be obtained in large amounts and resuspended very easily. However, this preparation usually gives very high background permeability to nonelectrolytes, and must be extracted with acetone.[40] The crude phosphatidylcholine (2.5 g) is crushed into fine powders with mortar and pestle, and then 50 ml of acetone containing 2 mM mercaptoethanol is added. The extraction is continued for 3 hr at room temperature with occasional crushing and kneading of phospholipid with a pestle. The acetone extract is discarded, and the phospholipid is partitioned in the Folch system[39] with the addition of 1 mM EDTA instead of water, washed once with the "theoretical upper phase mixture,"[39] and stored, after drying with anhydrous Na_2SO_4, at $-70°$.

[48] J. De Gier, J. G. Mandersloot, and L. L. M. van Deenen, *Biochim. Biophys. Acta* **150**, 666 (1968).

[49] This is obvious when the intravesicular fluid contains such solutes as dextran and has a higher refractive index than the extravesicular solution. However, this statement is valid even when the intra- and extravesicular fluids have similar refractive indices, as the refractive index of phospholipid bilayers is far higher than that of water and the pulling apart of successive layers of multilayered liposomes is expected to lower the averaged refractive index of these liposomes. [See C. S. Chong and K. Colbow, *Biochim. Biophys. Acta* **436**, 260 (1976).]

Procedure

Preparation of Liposomes. About 2 μmol of phospholipids (either *E. coli* phospholipids or egg phosphatidylcholine) are added to a large test tube, the solution is dried, and the film is resuspended in about 0.3 ml of aqueous suspension of pore-forming protein. The procedure follows what has been described for the preparation of phospholipid–LPS–protein liposomes except that LPS is usually not added and the amounts of porin added are much smaller (see below). The suspension is dried, again as described above, and is finally resuspended in 0.6 ml of 17% (w/v) Dextran T-20 (Pharmacia) containing 5 mM Tris-HCl, pH 7.5. When *E. coli* or *S. typhimurium* phospholipids are used, best results are obtained by rotating the tube slowly to wet the film, and then leaving the tube in a 30° water bath without agitation for at least 1 hr, and then shaking by hand to complete the resuspension process.[50]

Alternatively, the liposomes may be resuspended in stachyose or monosodium nicotinamide adenine diphosphonucleotide (Na-NAD$^+$)[51] if *E. coli* or *S. typhimurium* porins are used.[51a] Since in this case the refractive index of the intravesicular fluid is not much higher than that of the solutions into which the vesicles will be diluted, the extent of light scattering is low, and large amounts of lipids must be used in reconstitution. We usually use about 10 μmol of acetone-extracted, egg phosphatidylcholine per tube, and prepare liposomes in the manner described above except that the lipid-protein film is resuspended in 0.3 ml of 20 mM stachyose–2 mM Tris-HCl, pH 7.4, or of 10 mM Na-NAD$^+$–2 mM Tris–NAD$^+$, pH 7.4.

Swelling Assay. To a "semimicro" spectrophotometric cell (total volume 1.2–1.5 ml), 0.6 ml of an isotonic solution (see below) of a test solute is added. The solution always contains the same concentration of the buffer that is present in the liposome suspension. To this cuvette, usually 0.02 ml of the liposome suspension is added, the cuvette is covered with a piece of Parafilm, the content is mixed rapidly, and the recording of the optical density is started in a spectrophotometer connected with a recorder. Since the sensitivity becomes higher at shorter wavelengths, we normally use 400 nm. Because the *initial* rates are important, the recording should be begun as soon as possible after the addition of the liposomes; with practice, the interval can be shortened to about 10 sec.[52]

[50] M. Luckey and H. Nikaido, *Biochem. Biophys. Res. Commun.* **93**, 166 (1980).
[51] Although the standard biochemical abbreviation for this compound shows a positive charge on the nicotinamide moiety, the compound as a whole has one net negative charge.
[51a] H. Nikaido and E. Y. Rosenberg. *J. Bacteriol.* **153**, 241 (1983).
[52] With photometers using modulated light (for example, Brinkmann PC600), one can add liposomes to the test solution while the latter is being stirred in a lighted room and observe the turbidity changes in the early (2–10 sec) portion of the curve. The results, however, were qualitatively similar to what was obtained by the determination of the slope in the 10–20 sec range, except with very rapidly swelling liposomes (H. Nikaido, unpublished results).

With stachyose liposomes, the isotonic concentration lies quite close to 20 mM, but can be slightly higher. With dextran liposomes, the isotonic concentration is typically in the range of 40 mM. This is far higher than is expected from the molar concentration of dextran, and the isotonic concentration furthermore shows rather large tube-to-tube and day-to-day variation. The most probable explanation of this phenomenon is that the dextran has a few charged groups, and these charges attract a number of small ions through the Donnan effect. Because of this, it is necessary to determine the isotonic concentration of the particular liposome suspension to be used. One thus dilutes portions of liposomes into solutions, at various concentrations, of a compound that penetrates through the pore only with negligible rates. (For example, stachyose solutions are convenient for liposomes containing $E. coli$ porin.) Isotonicity is indicated when the optical density trace is flat from the beginning. A rapid increase (or decrease) in optical density during the first 10–20 sec indicates that the solution is hypertonic (or hypotonic).

Once the isotonic concentration is established, the liposomes are diluted into other solutions of the same osmolality and the initial slopes of the curves are determined. If the test solutions have similar refractive indexes, then the slopes themselves could be used as rough indicators of the permeability. However, if the refractive indexes are significantly different, or if accurate values are desired, the following two steps must be followed.

1. Make the refractive indexes of all solutions equal by adding appropriate amounts of Ficoll 4000 to solutions of lower refractive index.
2. Then the slope of the initial portion of the curve, $d(OD)/dt$, must be converted to the rate of change of $1/(OD)$. Since $d[1/(OD)]/dt = [1/(OD)^2][d(OD)/dt]$, the initial slope of the OD versus time tracing is divided by the square of the initial turbidity of the suspension.[53]

Comments. Although the swelling method is simple and rapid, some care is needed in its execution. The solution used must be precisely isotonic. This is sometimes difficult to achieve, because many biochemicals contain unspecified amounts of water and activity coefficients are different from one compound to another. For solutions at fairly high tonicity (30–50 milliosmolar), measurements with vapor pressure or freezing point osmometers may give the most reliable results. For more dilute solutions a more accurate method is the use of "control" liposomes, which do not contain pore-forming proteins, as an osmometer. Thus they are diluted into varying concentrations of the test solute as well as a "nonpenetrating" solute, such as stachyose. If the test solution at the nominal concentration of 21 mM and 20 mM stachyose both gave flat tracings, then the correction factor is $21/20 = 1.05$. Therefore, if the porin-containing liposomes gave a flat

[53] A. D. Bangham, M. W. Hill, and M. G. A. Miller, *Methods Membr. Biol.* **1,** 1 (1974).

tracing upon dilution into 23 mM stachyose, the test solute should be used at the nominal concentration of $1.05 \times 23 = 24.15$ mM.

If the test solute is an organic anion, and if an alkali metal salt of this compound is used, the swelling assay with bacterial porins will be essentially measuring the diffusion rate of the organic anion, as small cations are expected to penetrate through these pores much more rapidly.[1,54] In this case, however, the liposomes should contain the same alkali metal salt of a nonpenetrating anion, because with nonelectrolyte-containing liposomes (a) the alkali metal ion outside with rapidly flow in following its own concentration gradient, accompanied by the influx of permeable anions in the mixture (such as Cl^- in the Tris-HCl buffer) against their concentration gradient; and (b) a large membrane potential will be generated. For example, in order to measure the diffusion rate of gluconate anion, the liposomes are made in Na-NAD$^+$ and they are diluted into an isotonic solution of sodium gluconate.

One assumption in the swelling assay is that the liposomes are far more permeable to water than to the test solute, and therefore the rate of swelling is not limited by the permeability to water. It is important to operate within the range where this assumption is valid. Thus one should measure the swelling rate limited by water permeability, by diluting liposomes into water. In comparison with this slope, the slopes obtained upon dilution into test solutions should be far less steep. For some solutes this requires that very small amounts of porin be used; for measurement of arabinose penetration rates using *E. coli* OmpF porin, less than 0.5 μg per 10 μmol phospholipid will suffice.

The turbidity of liposome suspension is sensitive to temperature, and the cell compartment of the spectrophotometer should be temperature-regulated.

With some lipid preparations, liposomes tend to aggregate together, especially upon dilution into electrolyte solutions. Washing the lipid with EDTA (see above) reduces this tendency. It is also beneficial to filter the liposomes, before dilution, through Millipore filters (pore size 3 μm) to remove large clumps that nucleate the aggregation process.

Reconstitution into Planar Films

The channel-forming proteins have been successfully reconstituted into "black lipid films" and into "planar lipid bilayers" that are devoid of hydrocarbon solvents. Thus with the former system, *E. coli* porin[54,55] and *P.*

[54] R. Benz, K. Janko, and P. Lauger, *Biochim. Biophys. Acta* **551**, 238 (1979).
[55] R. Benz, K. Janko, W. Boos, and P. Lauger, *Biochim. Biophys. Acta* **511**, 305 (1978).
[56] R. Benz and R. E. W. Hancock, *Biochim. Biophys. Acta* **646**, 298 (1978).

aeruginosa porin[56] were reconstituted into black lipid films of oxidized cholesterol or monoolein, by adding the SDS solutions of porin into an aqueous compartment. In the planar lipid bilayer system, both *E. coli* porin[34,57] and mitochondrial porins have been reconstituted.[58] In these cases the porin was either added to the aqueous phase[58] or mixed with phospholipids, and then the mixtures were transformed first into monolayers and finally into bilayers.[34,57,58]

The planar film technique has several important advantages. Owing to the sensitivity of the electrical measurements, one can observe the penetration through individual porin channels. Also, the electrolyte penetration and the effect of membrane potential are more easily studied than in the liposome system.

Since this technique requires specialized equipment frequently unavailable in biochemical laboratories, the reader is referred to the original articles[34,54–58] for the details of the procedure.

Properties of the Pore

Porin Channels. The rates of diffusion of solutes through the porin channel are affected by at least three gross physiochemical properties of the solute.

1. The larger the size of the solute, the slower the rate of diffusion. This size dependence is seen well within the "exclusion limit" of the pore. Gross heterogeneity in pore size cannot be the explanation, because planar lipid bilayer experiments[34,57–58] showed that the individual channels have exactly the same size.[59] Thus the collision at the rims of the pore and the drag exerted by the pore wall are the most likely causes of this effect; comparison with theoretical models suggests the pore diameter of about 11–12 Å for *E. coli* porins[51a,60] and in the range of 20–25 Å for *P. aeruginosa* porin.[1,25a] The interpretation of the single-channel conductivity data is controversial,[1] but we believe that these data with *E. coli* porin are consistent with the notion that a porin trimer contains three channels, a conclusion derived from the detailed analysis of the conductance, presumably caused by the insertion of a single trimer into the bilayer,[34,57] and from electron microscopy.[16]

2. The diffusion through the pore is retarded by the hydrophobicity of the solute.[1,51a,61]

[57] H. Schindler and J. P. Rosenbusch, *Proc. Natl. Acad. Sci. U.S.A.* **75,** 3571 (1978).

[58] M. Colombini, *J. Membr. Biol.* **53,** 79 (1979).

[59] However, an *E. coli* porin produced channels of apparently different size, depending on the mode of lateral aggregation.[34]

[60] H. Nikaido and E. Y. Rosenberg, *J. Gen. Physiol.* **77,** 121 (1981).

[61] H. Nikaido, E. Y. Rosenberg, and J. Foulds, *J. Bacteriol.* **153,** 232 (1983).

3. With *E. coli* porins normally produced (OmpC and OmpF), there is a preference for cations, and negative charges tend to slow down the penetration rate.[1,51a,54] In contrast, the *E. coli* PhoE porin channel does not retard the penetration of solutes with multiple negative charges,[51a,61,62] and it is hypothesized to act as a preferential pore for phosphate and phosphorylated compounds.[10,11] The mitochondrial porin also shows a preference for anions.[58]

Although there have been several reports claiming that these channels are substrate-specific, most of them are marred by technical or conceptual flaws.[63] The PhoE porin channel, however, apparently is "clogged" by low cencentrations of polyphosphate and many contain a specific binding site.[62]

4. In planar bilayer models, the porin channels become reversibly closed whenever a large electrical potential was applied across the bilayer.[34,57] The physiological significance of this phenomenon is not clear at present.

LamB Channel. When reconstituted, this channel shows some nonspecific pore activity, and allows the rapid diffusion of metal cations, $Tris^+$, Cl^-, hexoses, etc.[7,55] However, as the size of the solute becomes larger, the configurational specificity of the channel becomes more apparent, and sucrose diffuses with only 2.5% of the rate of maltose, for example.[7] Also the penetration rates of oligosaccharides of maltose series do not decrease with size as drastically as with porin, and even maltoheptaose penetrates with significant rates.[7] The channel appears to possess specific binding site(s) for maltodextrins, as shown by inhibition studies[50] and by the direct binding of fluorescein-modified amylopectin to *E. coli* cell surface.[64]

[62] N. Overbeeke and B. Lugtenberg, *Eur. J. Biochem.* **126,** 113 (1982).
[63] H. Nikaido, M. Luckey, and E. Y. Rosenberg, *J. Supramol. Struct.* **13,** 305 (1980).
[64] T. Ferenci, M. Schwentorat, S. Ullrich, and J. Vilmart, *J. Bacteriol.* **142,** 521 (1980).

[10] Phage λ Receptor (LamB Protein) in *Escherichia coli*

By Maxime Schwartz

The phage λ receptor is an outer membrane protein of *Escherichia coli.*[1] In addition to being a phage receptor, it is a component of the transport system for maltose[2,3] and maltodextrins.[4,5] Its structural gene, called *lamB,* is

[1] L. Randall-Hazelbauer and M. Schwartz, *J. Bacteriol.* **116,** 1436 (1973).
[2] S. Szmelcman and M. Hofnung, *J. Bacteriol.* **124,** 112 (1975).
[3] G. L. Hazelbauer, *J. Bacteriol.* **124,** 119 (1975).
[4] S. Szmelcman, M. Schwartz, T. Silhavy, and W. Boos, *Eur. J. Biochem.* **65,** 13 (1976).
[5] C. Wandersman, M. Schwartz, and T. Ferenci, *J. Bacteriol.* **140,** 1 (1979).

located in one of the three maltose operons.[6-8] In accordance with the usual nomenclature for outer membrane proteins, it is also called the LamB protein.

Assay

The LamB protein is most easily assayed by its ability to neutralize host-range mutants of λ *in vitro*.[1] The conditions are such that the concentration of receptor molecules is several orders of magnitude greater than that of phage particles. Therefore the neutralization follows pseudo-first-order kinetics, and the rate of neutralization is directly proportional to the concentration of receptor.

λ *Phage*. It is best to use λV, a virulent strain of λ because it gives clear and large plaques.[9] Host-range mutants of λ can be selected by plating this phage on strain CR63,[10] which carries a particular missense mutation in gene *lamB*. The phage used in the author's laboratory is λVho. A stock[11] of this phage, grown on CR63, is diluted in 4 mM MgSO$_4$ in such a way that the final dilution contains between 6×10^3 and 10^4 plaque-forming units per milliliter.

Indicator Strain. Strain CR63, or any other strain sensitive to λVho, is grown at 37° in a complete medium such as LB[12] (yeast extract, 5 g/liter; Bacto-tryptone, 10 g/liter; NaCl, 5 g/liter; pH 7.4) The cells are harvested in late exponential or early stationary phase, centrifuged, and resuspended in 10 mM MgSO$_4$, at a density of about 5×10^8 cells per milliliter (optical density 1.0 at 600 nm).

Solid Medium for Plating. Petri dishes are filled with 30–40 ml of tryptone agar (Bacto-tryptone, 10 g/liter; NaCl, 2.5 g/liter; agar, 10 g/liter) that has been autoclaved for 20 min at 115°. Soft agar is the same as above, but with only 6 g of agar per liter.

The solution containing the λ receptor is serially diluted in 10 mM Tris-HCl, pH 7.5. When the solution contains a detergent (1–3% sodium dodecyl sulfate, sodium cholate, or Triton X-100) and/or millimolar amounts of ethylenediaminetetraacetic acid (EDTA), dilutions of 1:10 or greater must be used because these compounds would otherwise interfere with the assay. (Extracts such as those described below will generally have to

[6] M. Schwartz, *Ann. Inst. Pasteur, Paris* **113**, 685 (1967).
[7] J. P. Thirion and M. Hofnung, *Genetics* **71**, 207 (1972).
[8] O. Raibaud, M. Roa, C. Braun-Breton, and M. Schwartz, *Mol. Gen. Genet.* **174**, 241 (1979).
[9] F. Jacob and E. L. Wollman, *Ann. Inst. Pasteur, Paris* **87**, 653 (1954).
[10] R. K. Appleyard, J. F. McGregor, and K. M. Baird, *J. Virol.* **2**, 565 (1956).
[11] J. H. Miller, "Experiments in Molecular Genetics," p. 37. Cold Spring Harbor Lab., Cold Spring Harbor, New York, 1972.
[12] J. H. Miller, "Experiments in Molecular Genetics," p. 433. Cold Spring Harbor Lab., Cold Spring Harbor, New York, 1972.

be diluted 10^3- or 10^4-fold.) Equal volumes of the λ receptor dilution and of the phage dilution are mixed, then incubated at 37°. One hundred-microliter samples are withdrawn after various incubation periods, usually up to 15 or 30 min, and transferred into tubes containing 0.4 ml of the suspension of CR63. [In these tubes the concentration of Mg^{2+} ions (see below) and the respective concentrations of receptor and bacterial cells are such that the rate of adsorption to the CR63 cells is much greater than the rate of inactivation by the receptor.] After a further incubation of about 5 min, 3 ml of soft agar are added to these tubes, and the contents are plated on tryptone agar plates. The number of plaques, which corresponds to the number of surviving phage particles at the time of sampling, is counted after at least 6 hr of incubation at 37°. The pseudo-first-order rate of the reaction (K) can be determined by plotting the logarithm of the number of surviving particles as a function of the time of sampling. It is the inverse of the time at which the number of particles is 37% (e^{-1}) of the input.

Since the inactivation reaction is stoichiometric, it is theoretically possible to assay quantitatively the number of receptor molecules by a titration procedure.[1] However, this assay is tedious to do routinely and requires the use of a phage stock in which the proportions of plaque-forming particles and inactive particles are known.[13] This titration has been performed on several extracts and thus allows the following empirical formula[13] to be established:

$$[R] = K/(4 \times 10^8)$$

where [R] is the molarity of receptor, and K is the pseudo-first-order rate of phage inactivation, expressed in sec^{-1}.

Extraction. Any λ-sensitive strain of *E. coli* K12 can in principle be used as a source of λ receptor. However, for purification it is convenient to use an *ompR1* mutant[14] (previously called *ompB* mutant[15]) because such mutants synthesize greatly reduced levels of the OmpC and OmpF proteins, two major outer membrane proteins that tend to copurify with the λ receptor. The strain is generally grown in a minimal salt medium, supplemented with maltose as a carbon source and as an inducer of *lamB* gene expression.

The λ receptor is an integral membrane protein and is bound to the peptidoglycan layer by noncovalent bonds (see a later section). Its complete solubilization requires the use of a combination of the adequate detergent or organic solvent, and of agents (EDTA or high concentrations of salt) that release it from the peptidoglycan. Apparently complete solubilization is obtained[16] when exponentially growing cells are harvested at a density of

[13] M. Schwartz, *J. Mol. Biol.* **99**, 185 (1975).
[14] M. N. Hall and T. J. Silhavy, *J. Mol. Biol.* **151**, 1 (1981).
[15] V. Sarma and P. Reeves, *J. Bacteriol.* **132**, 23 (1977).
[16] C. Chapon, *J. Bacteriol.* **150**, 722 (1982).

5×10^8 per milliter, washed once with 10 mM Tris-HCl, pH 7.5, and resuspended at the same cell density in 10 mM Tris-HCl (pH 7.5) containing 3% Triton X-100 and 5 mM EDTA, and incubated for 30 min at 37°. The amount of λ receptor found in such extracts corresponds to 3 to 8 $\times$ 10^4 active molecules per cell.[17] When sodium cholate is used instead of Triton X-100, lower values are obtained,[1,13,18] presumably because the protein is incompletely released from peptidoglycan[19] and only partially solubilized.

Purification

Several purification procedures have been published,[1,20-28] but only one will be described here. It is probably the simplest, but it yields a protein that is still contaminated with lipopolysaccharide.[19,26] A special mention should be made of the procedure of Neuhaus.[27] It takes more time and involves the use of an expensive detergent, but it yields a protein that is free of lipopolysaccharide and can be crystallized.[29]

The *ompR1* mutant pop1130 is grown in 10 liters of medium M63 [KH$_2$PO$_4$, 13.6 g/liter; (NH$_4$)$_2$SO$_4$, 2 g/liter; FeSO$_4$ · 7H$_2$O, 0.5 mg/liter; MgSO$_4$ · 7H$_2$O, 1 ml of a 1 M solution added after autoclaving] supplemented, per liter, with 0.1 g of L-histidine, 1 mg of thiamine, and 4 g of maltose. The cells are harvested in exponential phase when the culture reaches an optical density of 1.0 at 600 nm. The cells are then washed once with 500 ml of 10 mM Tris-HCl, pH 7.4, and resuspended in 500 ml of extraction buffer (10 mM Tris-HCl, pH 7.4, containing 2% sodium dodecyl sulfate, 10% glycerol, 2 mM MgCl$_2$). After a 30-min incubation at 60°, the suspension is centrifuged at 100,000 g for 1 hr at 20°. Most of the cellular proteins are then found in the supernatant. The pellet, which contains the peptidoglycan and 30–40% of the λ receptor, is washed twice with 500 ml of 2 mM MgCl$_2$. Quantitative dissociation of the λ receptor from the pepti-

[17] V. Braun and H. J. Krieger-Brauer, *Biochim. Biophys. Acta* **469**, 89 (1977).
[18] M. Débarbouillé, H. A. Shuman, T. J. Silhavy, and M. Schwartz, *J. Mol. Biol.* **124**, 359 (1978).
[19] J. Gabay and K. Yasunaka, *Eur. J. Biochem.* **104**, 13 (1980).
[20] R. Endermann, I. Hindennach, and U. Henning, *FEBS Lett.* **88**, 71 (1978).
[21] H. Sandermann, P. Bavoil, and H. Nikaido, *FEBS Lett.* **95**, 107 (1978).
[22] E. T. Palva and P. Westermann, *FEBS Lett.* **99**, 77 (1979).
[23] B. A. Boehler-Kohler, W. Boos, W. Dieterle, and P. Benz, *J. Bacteriol.* **138**, 33 (1979).
[24] T. Nakae, *Biochem. Biophys. Res. Commun.* **88**, 774 (1979).
[25] M. Luckey and H. Nikaido, *Proc. Natl. Acad. Sci. U.S.A.* **77**, 167 (1980).
[26] J. Gabay and M. Schwartz, *J. Biol. Chem.* **257**, 6627 (1982).
[27] J. M. Neuhaus, *Ann. Microbiol. (Paris)* **133A**, 27 (1982).
[28] P. Bavoil and H. Nikaido, *J. Biol. Chem.* **256**, 11385 (1982).
[29] R. M. Garavito, J. A. Jenkins, J. M. Neuhaus, A. P. Pugsley, and J. P. Rosenbusch, *Ann. Microbiol. (Paris)* **133A**, 37 (1982).

doglycan is then achieved by resuspending the washed pellet in 80 ml of 10 mM Tris-HCl buffer, pH 7.4, containing 2% Triton X-100 and 5 mM EDTA. After a 30-min incubation at 37°, the suspension is centrifuged at 100,000 g for 1 hr, and the λ receptor is dialyzed three times for 4 hr against 1 liter of 10 mM Tris-HCl (pH 7.4), 1% Triton X-100, 0.1 M NaCl. It is then deposited onto a 30-ml column (2.5 × 20 cm) of quaternary aminoethyl (QAE)-Sephadex A-50 equilibrated with the same buffer. The column is washed with 100 ml of the buffer, and a 200-ml salt gradient, from 0.1 to 0.8 M NaCl, is applied. The λ receptor is the major protein eluted from the column, at about 0.2 M salt. The final yield is approximately 5 mg of λ receptor, which gives a single band upon gel electrophoresis in the presence of sodium dodecyl sulfate. However the protein is still contaminated with lipopolysaccharide, and phospholipids (1–5 mol of phosphate per mole of protein). The final yield is close to 30%, the only significant loss occurring at the first step.

A possible modification of the above procedure involves the following steps (S. Schenkman, personal communication):

1. Elution from the peptidoglycan using 2% octyl polyoxyethylene,[27] which does not absorb around 280 nm, instead of Triton X-100.

2. A 1 hr treatment of the resulting solution with 1 μg/ml PMS-treated deoxyribonuclease I (Worthington) in the presence of 50 mM MgCl$_2$. This reduces the viscosity of the sample, but also induces a precipitation of the λ receptor. The solution is then centrifuged for 1 hr at 100,000 g, and the λ receptor is solubilized in 10 ml of 10 mM Tris-HCl, pH 7.4, containing 2% octyl polyoxyethylene and 5 mM EDTA.

3. Further purification can be achieved either by using the above-described chromatography on QAE Sephadex, or more conveniently by performing an affinity chromatography on Sepharose 4B covalently coupled to a monoclonal antibody[26] directed against the LamB protein. In the latter case purification of 1 mg of LamB protein is performed on a 2 ml column containing 7 mg of immunoglobulin. Elution is then obtained with a 0.1 M glycine-HCl buffer, pH 2.8, containing 1% octyl polyoxyethylene.

Physicochemical Properties

The molecular weight of the denatured protein, estimated by polyacrylamide gel electrophoresis in the presence of sodium dodecyl sulfate (SDS)[13,17,20-23,30] or by ultracentrifugation in the presence of 6 M guanidine hydrochloride[27,31] was generally found to be between 45,000 and 50,000. The exact molecular weight of the polypeptide, calculated from the amino

[30] P. W. Kühl FEBS Lett. **96**, 385 (1978).
[31] J. N. Ishii, Y. Okajima, and T. Nakae, FEBS Lett. **134**, 217 (1981).

acid composition (421 residues), which was itself deduced from the gene sequence,[32] is 47,392.

The nondenatured protein, as obtained in mild detergents, or even after solubilization in SDS at a temperature lower than 60°, has a mobility in SDS–polyacrylamide gel electrophoresis that corresponds to a molecular weight of 80,000 to 100,000.[27,30,31] However this value is erroneous because of the low amount of detergent bound to the protein.[27,31] More recently the molecular weight of the protein, purified according to two different procedures, was determined by ultracentrifugation. Ishii *et al.*[31] performed sedimentation equilibrium studies in a solution containing the homogeneous nonionic surfactant octaethylene glycol dodecyl ether and D_2O. They reported a value of 135,600. Neuhaus[27] performed the ultracentrifugation experiments in octyl tetraoxyethylene, and reported a value of 138,000 ± 8000. Both of these values suggest that the mature protein is a trimer, as was previously indicated by cross-linking data.[22] The results of complementation experiments, involving various *lamB* mutants, are also best interpreted by assuming that the LamB protein is a trimer.[33]

The UV-absorption spectra of monomer and trimer show maximum absorption peaks at 280–282 nm, and their absorption coefficients in a 0.1% solution at 280 nm (A_{280} 0.1%) were found to be 2.48 and 2.47, respectively.[31] A value obtained from amino acid analysis[27] was A_{282} 0.1% = 2.45.

The circular dichroism (CD) spectrum of the trimer in solutions of SDS or octaethylene glycol dodecyl ether is characteristic of a β-sheet structure.[31] The molar ellipticity at 217 nm was calculated to be −4.3 and −5.1 × 10^3 deg cm^2 dmol^{-1} in these two solutions, respectively. The CD spectra of the monomers in SDS and in 6 M guanidine-HCl showed α-helix and random coil-like structures, respectively. Similar conclusions were reached by Neuhaus.[27] Secondary structure prediction derived from the amino acid sequence originally suggested that the protein contained almost equal proportions of α-helical (0.25), β-extended (0.23), random coil (0.32), and β-turn (0.20) regions. These predictions have now been modified, after taking into account the results of the CD studies.[34]

The λ-Receptor Is a Transmembrane Protein

The mere fact that the λ receptor is a phage receptor (details in next sections) indicates that part of the molecule must be accessible at the cell surface. In addition, antibodies directed against the protein react with the

[32] J. M. Clément and M. Hofnung, *Cell* **27**, 507 (1981).
[33] C. Marchal and M. Hofnung, *EMBO J.* **2**, 81 (1983).
[34] J. M. Clément, E. Lepouce, C. Marchal, and M. Hofnung, *EMBO J.* **2**, 77 (1983).

cell surface, since they can be used to label the cells by immunofluores-cence[26,35] and to direct the cytolytic action of complement.[36]

Evidence that the λ receptor also protrudes on the inner aspect of the outer membrane is also quite strong. First, as mentioned earlier, the λ receptor remains strongly associated with the underlying peptidoglycan layer when whole cells or envelopes are dissolved in SDS at 60°.[20,22,24,26] Once eluted from the peptidoglycan it can reassociate with it, in the presence of lipopolysaccharide, to form a hexagonal lattice.[37] Second, the λ receptor is involved in a functional interaction with the periplasmic maltose binding protein (see later section). Finally, and more directly, immuno electron microscopic studies have shown that the inner face of the outer membrane can be labeled with polyclonal and monoclonal antibodies directed against the λ receptor (S. Schenkman, E. Couture, and M. Schwartz, submitted).

The path of the polypeptide within the membrane is mostly unknown. Of the 421 residues in the polypeptide, 89 are charged and these are scattered throughout the sequence.[32] There are no long hydrophobic regions of the molecule that would be obvious candidates for spanning the membrane. Tentative conclusions regarding portions of the polypeptide that are accessi-ble at the cell surface have been obtained by localizing, through DNA sequencing, mutations that seem to affect the primary interaction of the protein with bacteriophages.[34,38] Hence it seems likely that two regions of the polypeptide, one including residues 151–155, and the other residue 382, are exposed at the cell surface. Another approach to the same problem has involved the use of monoclonal antibodies directed against the λ receptor.[26] The portion of the polypeptide located beyond residue 351 was concluded to be, at least in part, exposed at the outer surface of the membrane because it carries antigenic determinants recognized by two monoclonal antibodies that react with the cell surface.[39] This conclusion, which fits with the above-mentioned finding that residue 382 plays an essential role in phage adsorption, is also consistent with the most recent secondary structure predictions (mentioned in Ref. 34).

A detailed knowledge of λ receptor three-dimensional structure will require an analysis, by high-resolution electron microscopy, X-ray diffrac-tion, or other physical techniques, of the crystalline or quasi-crystalline forms of the protein that were recently described.[29,37]

[35] M. Schwartz and L. Le Minor, J. Virol. 15, 679 (1975).
[36] J. Gabay, FEMS Microbiol. Lett. 2, 61 (1977).
[37] H. Yamada, T. Nogami, and S. Mizushima, Ann. Microbiol. (Paris) 133A, 43 (1982).
[38] M. Roa and J. M. Clément, FEBS Lett. 121, 127 (1980).
[39] J. Gabay, S. Benson, and M. Schwartz, J. Biol. Chem. 258, 2410 (1983).

Interaction with Phage λ

The λ receptor fails to neutralize wild-type phage λ *in vitro* unless a saturating amount of chloroform[1] or 10–20% ethanol[35] is added to the reaction mixture. In the absence of these solvents the protein does interact with the phage, but in a reversible manner.[13,40] Two favorable circumstances allowed the determination of the kinetic and equilibrium parameters defining this interaction.[13] One was the extreme stability of the phage receptor complex at low concentrations of cations. The other was the availability of a variant λ receptor (see later section) that did inactivate wild-type phage λ in the absence of solvent, and could therefore be used to neutralize free phage in mixtures of normal receptor and wild-type λ The complexed phage could then be assayed after dissociating the complex in the presence of higher concentrations of cations. Monovalent and divalent cations affect both the rate of association and of dissociation of the complex. The highest affinity was observed in the presence of 2 mM (Mg^{2+}) ions, the equilibrium constant being about 5×10^{-12} M. Under the same conditions the kinetic constants for association and dissociation were $k_a = 3.7 \times 10^8$ M^{-1} sec^{-1} and $k_d = 3 \times 10^{-4}$ sec^{-1}, respectively. (It should be mentioned that the above constants were determined using a cholate–EDTA extract of bacteria. Since the λ receptor present in such extracts is now known to be imperfectly solubilized, slightly different constants might be obtained if one used fully solubilized λ receptor, as obtained with Triton X-100 and EDTA.)

Three instances have been described where the interaction of λ with the λ receptor leads to phage neutralization; (*a*) when chloroform or ethanol is added to the reaction mixture; (*b*) when one uses one of the above-mentioned variants of receptor, as can be extracted from some wild strains of *E. coli* or *Shigella;* and (*c*) when one uses a host-range mutant of λ, as can be selected by plating the wild-type phage on some *lamB* missense mutants. The first step in the neutralization reaction seems to be, in all these cases, a reversible interaction having the same characteristics as those described for the interaction between wild-type λ and *E. coli* K12 receptor, in the absence of solvents.[13] In particular the effect of cations is the same on the neutralization reaction as it is on the reversible interaction, with a sharp maximum at 2 mM (Mg^{2+}). Under adequate experimental conditions the neutralization reaction is followed by an ejection of phage DNA.[40-42] Since the neutralization of phage by the receptor is stoichiometric, the receptor must be inactivated also in the reaction that leads to phage neutralization. Receptor

[40] M. Roa and D. Scandella, *Virology* **72,** 182 (1976).
[41] D. J. MacKay and V. Bode, *Virology* **72,** 167 (1976).
[42] M. Roa, *FEMS Microbiol. Lett.* **11,** 257 (1981).

inactivation may merely correspond to a very strong binding to the inactive phage particles and the empty ghosts, but could also correspond to some chemical modification.

Early studies clearly established that the LamB protein is a necessary component of the λ receptor. They did not establish, however, whether a λ binding site corresponded to a single polypeptide chain and whether lipopolysaccharide, which strongly interacts with the LamB protein, is also a necessary component of the receptor. The latter question was recently answered[27] by the demonstration that lipopolysaccharide-free LamB protein still inactivates λVh. As for the first question the results indicate that each LamB polypeptide may bear one λ binding site. Indeed the number of LamB polypeptide chains per cell, as can be estimated from the yield and recovery of the purification described earlier (45,000 molecules per cell) correlates well with the number of "active" λ-receptor molecules that can be assayed after complete solubilization of the bacteria (30,000 to 80,000 molecules per cell). In addition, assays performed on the most active preparations of receptor, purified according to Randall-Hazelbauer and Schwartz,[1] yielded one binding site for 65,000 daltons (unpublished results). Therefore each LamB protein trimer would seem to bear three λ binding sites. No λ-inactivating monomer has as yet described, but this is not surprising in view of the harsh treatments required to obtain such monomers.

Interaction with Other Phages

Genetic studies have shown that the LamB protein is required for the adsorption of several phages other than λ. Phage K10 is one of them[43,44] and has the same morphology as λ. It interacts reversibly with the λ receptor *in vitro*. Competition experiments have shown that, in 2 mM MgSO$_4$, the affinity of the λ receptor for K10 is approximately 200 times less than for λ. Presumably because of the lower affinity, the rate of adsorption of phage K10 to sensitive cells depends much more upon the density of receptor at the cell surface[45] than does that of λ.[46] Other phages that can use the λ receptor for their adsorption[47-49] have the same morphology as T4 and have the peculiarity that they can "choose" between using the LamB protein or another outer membrane protein (the OmpF or OmpC proteins, or both) for

[43] R. E. Hancock and P. Reeves, *J. Bacteriol.* **127**, 98 (1976).
[44] M. Roa, *J. Bacteriol.* **140**, 680 (1979).
[45] M. Schwartz, M. Roa, and M. Débarbouillé, *Proc. Natl. Acad. Sci. U.S.A.* **78**, 2937 (1981).
[46] M. Schwartz, *J. Mol. Biol.* **103**, 521 (1976).
[47] C. Wandersman and M. Schwartz, *Proc. Natl. Acad. Sci. U.S.A.* **75**, 5636 (1978).
[48] F. Moreno and C. Wandersman, *J. Bacteriol.* **144**, 1182 (1980).
[49] M. G. Beher and A. P. Pugsley, *J. Virol.* **38**, 372 (1981).

their adsorption. In these cases, however, there is no evidence that the phage interacts directly with the LamB protein.[26]

Role in the Transport of Maltose and Maltodextrins

Wild-type (*lamB*[+]) strains of *E. coli* K12 grow equally well on maltose and on maltodextrins containing up to 6 or 7 glucose residues.[5] They take up maltose at a high efficiency even when this sugar is present at a very low concentration in the medium, the K_m of the transport system being 1 μM. Mutants devoid of λ receptor (*lamB* nonsense mutants) still grow normally on 1–10 mM maltose, but they fail to grow on maltodextrins containing more than three glucose residues.[4,5] The K_m of their transport system for maltose is increased to 100–500 μM, the V_{max} remaining unchanged.[2-4] These results led to the speculation that the λ receptor might function as a pore, or a channel, which would facilitate the diffusion across the outer membrane of maltose and maltodextrins, and perhaps of other substrates of similar or smaller size.[4] Studies performed *in vivo*[50] as well as *in vitro*[23,24,27] demonstrated that, indeed, like the porins coded by genes *ompC* and *ompF*, the λ receptor allows the diffusion of low molecular weight compounds (mono- and disaccharides, amino acids, nucleosides, ions) across natural or artificial membranes. However, unlike the porins, the λ receptor displays a strong preference for $\alpha(1 \rightarrow 4)$-linked oligoglucosides.[25,51] Using a "liposome swelling assay," Luckey and Nikaido demonstrated that maltose diffuses 40 times faster than sucrose into λ receptor-containing liposomes, while the maximum difference in diffusion rates among disaccharides was only about three-fold in OmpF-containing liposomes. A higher oligosaccharide, such as maltoheptaose, diffused at a significant rate (2.5% of the maltose flux) through the λ-receptor pore, but not detectably through the other porins.

Although the notion of a "stereospecific pore" is not absolutely clear, it would seem to imply that the pore is, or contains, a binding site specific for maltose and maltodextrins. Ferenci *et al.*[52] demonstrated that a fluorescent analog of amylopectin binds to *lamB*[+] cells, but not to *lamB* nonsense mutants. Maltose and maltodextrins inhibited this binding, and this provided a way to measure, by competition, their affinity constants. The affinity for maltose is rather poor ($K_d = 14$ mM), and this explains why it was not detected by other techniques.[4] The affinities for longer maltodextrins are much higher ($K_d = 75$ μM for maltohexaose). These results strongly suggest that the LamB protein possesses a binding site for maltose and maltodextrin. However, the exact relation between this binding site and the "stereospecific

[50] K. von Meyenburg and H. Nikaido, *Biochem. Biophys. Res. Commun.* **78**, 1100 (1977).
[51] T. Nakae and J. Ishii, *J. Bacteriol.* **142**, 735 (1980).
[52] T. Ferenci, M. Schwentorat, S. Ullrich, and J. Vilmart, *J. Bacteriol.* **142**, 521 (1980).

pore" remains to be determined. No data are available regarding the number of pores and of binding sites per LamB protein trimer. By analogy with results obtained with a porin,[53] it is tempting to hypothesize that each monomer constitutes one channel.

A possibility considered in several laboratories[2,20,28,54] is that part of the substrate specificity of the LamB protein might result from the existence of a specific interaction with the periplasmic maltose-binding protein (*malE* product). This hypothesis is now supported by the following lines of evidence: (*a*) the phenotype of some *malE* mutants is most easily interpreted as the result of a lack of interaction between the LamB and MalE proteins[5]; this interpretation obviously implies that this interaction exists in the wild-type strain; (*b*) the nonspecific pore properties of the LamB protein are more pronounced in the absence than in the presence of MalE protein, suggesting that the MalE protein might "plug" the LamB pore[54]; (*c*) immunoelectron microscopy suggests the existence of a LamB protein-dependent association of the MalE protein with the outer membrane[55]; (*d*) LamB protein is specifically retained on Sepharose columns containing covalently bound MalE protein.[28]

In summary, the present model would be that the LamB protein constitutes a stereospecific pore, plugged on the periplasmic side by the MalE protein. Once into the pore, maltose or maltodextrin molecules would be in a position to combine with the maltose-binding protein. This would induce a conformational change in the maltose-binding protein,[4] which would be released with its bound ligand into the periplasmic space.

Variants of the LamB Protein in *E. coli* and Other Enterobacteriaceae

Many *LamB* mutants have been isolated on the basis of their resistance to λ,[6,7,8,56] K10,[44] or TP1,[47] or on their inability to take up maltodextrins[5,57] or to bind to immobilized starch.[58] Some of these mutants synthesize a modified LamB protein that is altered in its interaction with phages and/or its ability to form stereospecific channels in reconstituted membranes.[13,59,60] A search for λ-receptor activity has been made in extracts of several species of Enterobacteriaceae.[35] All strains of *E. coli* and most Mal⁺ strains of *Shigella*

[53] H. Schindler and J. Rosenbusch. *Proc. Natl. Acad. Sci. (U.S.A.)* **78**, 2302 (1981).
[54] M. W. Heuzenroeder and P. Reeves. *J. Bacteriol.* **141**, 431 (1980).
[55] W. Boos and L. Staehelin, *Arch. Mikrobiol.* **129**, 240 (1981).
[56] M. Hofnung, A. Jezierska, and C. Braun-Breton, *Molec. Gen. Genet.* **145**, 207 (1976).
[57] C. Wandersman and M. Schwartz, *J. Bacteriol.* **151**, 15 (1982).
[58] T. Ferenci and K. S. Lee, *J. Mol. Biol.* **160**, 431 (1982).
[59] C. Braun-Breton and M. Hofnung, *J. Bacteriol.* **148**, 845 (1981).
[60] M. Luckey and H. Nikaido, *J. Bacteriol.* **153**, 1056 (1983).

had λ receptor, but no activity was found in extracts of *Salmonella, Klebsiella*, and *Enterobacter* strains. The receptor found in most strains of *E. coli* and *Shigella* differed from that found in *E. coli* K12 in that it inactivated wild-type λ in the absence of added solvents. (When the *lamB* gene of a *Shigella* strain was transduced into *E. coli* K12, the receptor of the resulting transductant had this same property.[40])

PROPERTIES OF THE λ RECEPTOR FROM *Escherichia coli* K12

Property[a]	Value
Cellular amount[26] (maltose-grown cells)	30,000 to 80,000 LamB polypeptide chains per cell
Purification[26]	Yield: 0.35 mg/g of bacteria (wet weight); recovery 30%
Native quaternary structure[22,27,31]	Trimer
Molecular weight of the monomer[32] (calculated from sequence)	47,392 (421 amino acids)
Secondary structure	
Predicted from sequence[32]	25% α-helical, 23% β-extended, 32% random coil, 20% β-turn
Predicted from CD spectra[27,31]	β-Sheet
Maximum absorption coefficient	
Calculated from amino acids[27]	$A_{282}^{0.1\%} = 2.45$
Measured[31]	$A_{280}^{0.1\%} = 2.47$
Cellular location[1,19,26,28]	Outer membrane; most probably spans the entire membrane
Interaction with	
Phage λ+, reversibly[13b]	In 2 mM MgSO$_4$, $K_d = 5 \times 10^{-12}\ M$, $k_a = 3.7 \times 10^8\ M^{-1}\ sec^{-1}$; $k_d = 3 \times 10^{-4}\ sec^{-1}$
Phage λh, irreversibly[1,13]	The phage is inactivated with a pseudo-first-order rate constant of $4 \times 10^8\ sec^{-1}$ in a 1.0 M solution of receptor
Phage K10, reversibly[44b]	In 2 mM MgSO$_4$; $K_d \simeq 5 \times 10^{-10}\ M$
α(1 → 4)-linked oligoglucosides,[25,52] i.e, (glucose)$_n$[c]	$n = 2, K_d = 1.4 \times 10^{-2}\ M$ $n = 4, K_d = 3.0 \times 10^{-4}\ M$ $n = 6, K_d = 7.5 \times 10^{-5}\ M$ $n = 10, K_d = 5.7 \times 10^{-5}\ M$ Maltose diffuses 40 times faster than sucrose through the λ receptor "pore."
Peptidoglycan[19,20,22,24,26,37]	Binding requires 5 mM Mg^{2+}
Periplasmic maltose-binding protein[5,28,54,55]	Binding characteristics unknown

[a] Superscript numbers refer to text footnotes.
[b] The molarity of λ receptor is expressed in phage binding sites. Available data suggest that there is one binding site per LamB protomer.
[c] The affinity for oligoglucosides was measured *in vivo*.

Another difference is that in most wild strains of *E. coli* and *Shigella* the λ receptor is not accessible at the cell surface, probably because it is buried under a luxuriant polysaccharide structure. These bacteria fail to adsorb λ or to react with anti λ-receptor antibodies.

Some bacterial species that fail to display any λ-receptor activity, *in vivo* or as measured in cell-free extracts, still possess an outer membrane protein that seems to have the same function in maltose and maltodextrin transport. This is the case in particular for *Salmonella typhimurium*[22] and *Klebsiella pneumoniae*.[61]

Summary and Conclusion

The main properties of the λ receptor are summarized in the table. Because these can be studied by a combination of genetic, biophysical, and biochemical techniques, the λ receptor now appears to represent one of the best systems for study of structure–function relationships in a membrane protein. In addition, as explained in this volume,[62,63] it also constitutes a good system for study of the export of proteins to extracytoplasmic locations.

[61] G. Wöhner and G. Wöber, *Arch. Microbiol.* **116**, 311 (1978).
[62] J. Beckwith and T. Silhavy, this volume [1].
[63] T. Silhavy and J. Beckwith, this volume [2].

[11] Synthesis and Assembly of the Outer Membrane Proteins OmpA and OmpF of *Escherichia coli*

By IAN CROWLESMITH and KONRAD GAMON

The kinetics of synthesis and assembly[1] of outer membrane proteins in gram-negative bacteria has been investigated in several laboratories over the past few years,[2-12] both because these proteins are often present at high concentrations (numbers in excess of 10^5 molecules per cell are typical for the major outer membrane proteins, including OmpA and OmpF), and

[1] Assembly is defined as the integration of newly synthesized protein into the outer membrane and implies the adoption of a molecular conformation that is operationally indistinguishable from that of stably integrated older protein molecules.
[2] N. Lee and M. Inouye, *FEBS Lett.* **39**, 167 (1974).
[3] K. Ito, T. Sato, and T. Yura, *Cell* **11**, 551 (1977).
[4] L. De Leij, J. Kingma, and B. Witholt, *Biochim. Biophys. Acta* **512**, 365 (1978).
[5] L. De Leij, J. Kingma, and B. Witholt, *Biochim. Biophys. Acta* **553**, 224 (1979).

METHODS IN ENZYMOLOGY, VOL. 97

especially after early studies suggested that these proteins might be synthesized unusually slowly,[3,6,7] yet without the appearance of transient intermediates either in the form of precursor molecules[2-4,6,7] or as a pool of newly synthesized molecules (whether processed or not) in the inner membrane.[2-6] However, mature lipoprotein has been detected in the inner membrane,[7] and recent work[8,10-12] has shown that both OmpA and OmpF are synthesized at the same rate as other cell proteins,[8,12] and as precursor molecules[10,11] that associate first with the inner membrane[11] and are subsequently processed rapidly to mature protein.[10,12] The kinetics of processing suggest that there may be as many as three processing pathways,[9,10,12] the relative contribution of which may be temperature dependent.[9,12]

The techniques used for some of this work[11,12] are described here. Key points are: (a) the use of very short pulses (5–10 sec) of very high specific activity [^{35}S]methionine; (b) low growth temperatures (25°) to slow the kinetics to a manageable rate; (c) the avoidance, where possible, of cell fractionation procedures, such as sonication and centrifugation, that might permit either posttranslational processing (e.g., due to local heating at the tip of the sonication probe) or the selective loss of intermediates of interest (as discarded supernatants); and (d) the use of specific antibodies to purify labeled proteins. Quantitative analysis requires, in addition, the use of a double-labeling methodology and a knowledge of the primary sequence (number and location of methionine residues) of the protein(s) under investigation.[12] The techniques described differ from those of Josefsson and Randall[9] with regard to both methodology and the kind of information derivable (see this volume [8]).

Growth of Bacteria and Pulse-Labeling Conditions

Reagents

M9 Minimal salts medium. Mix the following sterile ingredients: 100 ml of 10× M9 salts (60 g of anhydrous Na_2HPO_4, 30 g of anhydrous KH_2PO_4, 5 g of NaCl, and 10 g of NH_4Cl per liter), 10 ml of 0.1 M $MgSO_4$, 10 ml of 10 mM $CaCl_2$, 10 ml of 40% glucose (as carbon source), 10 ml of each required supplement, made up in advance as 100× concentrated solutions, and sterile deionized water

[6] J. J. C. Lin and H. C. Wu, *J. Biol. Chem.* **255**, 802 (1980).
[7] J. J. C. Lin, C. Z. Giam, and H. C. Wu, *J. Biol. Chem.* **255**, 807 (1980).
[8] A. Boyd and I. B. Holland, *J. Bacteriol.* **143**, 1538 (1980).
[9] L. G. Josefsson and L. L. Randall, *Cell* **25**, 151 (1981).
[10] K. Ito, P. J. Bassford, and J. Beckwith, *Cell* **24**, 707 (1981).
[11] I. Crowlesmith, K. Gamon, and U. Henning, *Eur. J. Biochem.* **113**, 375 (1981).
[12] I. Crowlesmith and K. Gamon, *Eur. J. Biochem.* **124**, 577 (1982).

to a final volume of 1 liter. For strain P400.6 the supplements used are L-leucine, L-threonine, L-arginine, L-proline (each at a final concentration of 50 μg/ml) and thiamin (5 μg/ml).
L-Methionine, freshly diluted to 0.2 M in M9 salts
[^{35}S]Methionine, 1000 Ci/mmol
[^{3}H]Lysine, 100 Ci/mmol

Procedure. Bacteria are grown at 25° with vigorous aeration to a density of 3 to 5 × 10^8 cells/ml (A_{600} = 0.5) in M9 minimal salts medium. A ^{3}H-labeled amino acid that is rapidly taken up into trichloroacetic acid-pre-cipitable material (for strain P400.6 [^{3}H]lysine is suitable[12]) is added at a final concentration of 2 – 5 μCi/ml, and the culture is harvested 10 – 30 min later, when the cell density has reached 5 × 10^8 bacteria per milliliter (A_{600} = 0.5 to 0.7). The cells are chilled rapidly to 4°, centrifuged, and resuspended to a density of 5 to 6 × 10^9 bacteria per milliliter in fresh M9 minimal salts medium at 4°. Pulse-labeling is carried out by reincubating the cells at 25° for 5 min and adding high-specific-activity [^{35}S]methionine. In a typical experiment 0.05 ml of [^{35}S]methionine (5 mCi/ml) is added with rapid mixing to 4.5 ml of concentrated cells. The pulse is terminated (and the chase begun) after 5 – 10 sec by the addition of 0.5 ml of 0.2 M L-methio-nine (in M9 salts, prewarmed to 25°). Samples of 0.2 ml are withdrawn at 5-sec intervals and rapidly injected into glass test tubes held at – 78° in a solid CO_2 – acetone bath. Samples are kept small in order to ensure that they freeze instantaneously on contact with the cold glass.

The kinetics of incorporation of [^{35}S]methionine into nascent whole-cell protein are measured in order to permit quantitative analysis of the kinetics of labeling of OmpA and OmpF. The frozen pulse-chase samples are allowed to thaw at 0° (*not* at room temperature), and 5-μl aliquots are boiled for 15 min in 5% trichloroacetic acid – 1% casamino acids (as carrier) to ensure the hydrolysis of any methionyl-tRNA^{Met} and the precipitation of labeled nascent protein. This is collected on glass fiber filters (e.g., Schleicher & Schüll No. 6), washed, dried, and counted in a scintillation counter. A typical result is shown in Fig. 1. The dashed line (0 – 5 sec) represents the expected rate of incorporation during the pulse phase, determined from control experiments with an uninterrupted pulse. At this experiment at 25° shows, the chase may only slowly become fully effective. However at 37°, or in control experiments at 25° using lower concentrations of [^{35}S]methionine (5 μCi/ml), the chase is complete within 5 – 10 sec.[12] The reasons for these differences are unclear, but they do not affect the validity of the results provided they are taken fully into consideration for quantitative analysis (see below).

Alternative Procedures and Optimization of Labeling Conditions. The procedure described is suitable for a double-label pulse-chase experiment in

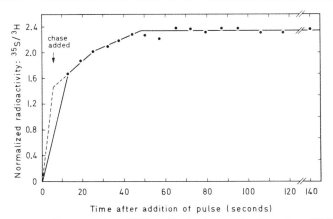

FIG. 1. Kinetics of incorporation of [³⁵S]methionine into total cell protein at 25°. See text for details. From Crowlesmith and Gamon.[12]

which high incorporation of radioactivity must be balanced against the need for extremely short pulse times and small samples. If high incorporation of radioactivity is not essential, cells need not be concentrated prior to labeling with [³⁵S]methionine. On the other hand, if large numbers of cells are needed (e.g., 10^{11} cells may be required for experiments in which they are lysed in the French press[11]), then a pulse or pulse-chase experiment may be effectively terminated by pouring 10–50 ml of concentrated, labeled cells into a prechilled flask (0°) containing an equal volume of crushed ice. This variation may not permit samples to be taken every 5 sec, but suffices for a pulse-chase experiment of somewhat lower resolution. Another method of terminating the chase is to mix samples with an equal volume of ice-cold 10% trichloroacetic acid.[10] This method, however, has not been used in the experiments on which the present chapter is based.[11,12]

An alternative chase procedure might be to add a protein synthesis inhibitor, such as chloramphenicol (0.2 mg/ml), tetracycline or oxytetracycline (1 mg/ml). The chase in each case is as effective as when unlabeled methionine is used (incorporation of [³⁵S]methionine, 5 μCi/ml, into nascent protein ceases within 5 sec of addition of the chase or inhibitor), but the use of inhibitors carries with it the potential disadvantage that other physiological reactions such as processing may be affected: chloramphenicol, for example, has been reported to stimulate processing,[13] and in our own laboratory we have observed variations in the ratio of mature to precursor protein in experiments with tetracyclines.

For experiments in which the specific radioactivity of labeled protein

[13] P. J. Bassford, *Ann. Microbiol.* (*Paris*) **133A**, 91 (1982).

must be maximized, it is important to choose concentrations of cells and [^{35}S]methionine such that methionine incorporation into nascent protein is not limited by its uptake into cells. For strain P400.6, 5 μCi/ml [^{35}S]methionine (1000 Ci/mmol) is incorporated into nascent protein with maximal kinetics only at cell concentrations greater than 5×10^8 bacteria/ml. Hence 5×10^9 cell/ml are necessary to maximize the rate of incorporation of 50 μCi/ml [^{35}S]methionine (the pulse-labeling conditions described in this chapter).

Immunoprecipitation and Sodium Dodecyl Sulfate (SDS) – Polyacrylamide Gel Electrophoresis

Reagents

Antisera: OmpA and OmpF antisera are raised in rabbits by intramuscular injection of the purified proteins, preemulsified in phosphate-buffered saline (pH 7.2) plus Freund's adjuvant, at multiple sites.[14] The purified antigens are prepared using published procedures[15] involving the denaturation of outer membrane proteins in boiling 2% SDS and separation of the denatured proteins by gel chromatography in the presence of 1.5% SDS. It may not be possible to avoid a minute contamination of OmpF protein by OmpA (and hence the presence of OmpA antibodies in OmpF antiserum),[12] but this need not matter provided the two proteins are well separated after SDS-polyacrylamide gel electrophoresis.

Tris–SDS buffer: 12% SDS (w/v) in 0.3 M Tris-HCl, pH 7.5

Tris buffer: 10 mM Tris-HCl, pH 7.5

Wash buffer: 10 mM Tris-HCl, pH 7.5 containing NaCl (0.9%), EDTA (0.1 mM), and SDS [0.02%, 0.2%, or 2% (w/v)]

NaCl, 2 M

Electrophoresis reagents: 12% slab gels are prepared using a modification[15] of the Laemmli system[16] in which N,N,N',N'-tetramethylenediamine (TEMED) is used at 0.15% instead of 0.25%. The gels are prepared up to a week in advance and stored in a cold, moist atmosphere (e.g., in a closed box containing wet filter paper in the cold room). All gels are run in the cold room at 4 – 8°. Electrophoresis sample buffer consists of 62.5 mM Tris-HCl, pH 6.8, containing glycerol (10%, w/v), SDS (2%, w/v), and mercaptoethanol (5%, v/v).

Staphylococcus aureus cells (Cowan I strain): 1 – 2 ml of a solution of heat-killed, formalin-fixed cells,[17] maintained as a stock 10% (w/v)

[14] U. Henning, H. Schwarz, and R. Chen, *Anal. Biochem.* **97**, 153 (1979).
[15] M. Schweizer, I. Hindennach, W. Garten, and U. Henning, *Eur. J. Biochem.* **82**, 211 (1978).
[16] U. K. Laemmli, *Nature (London)* **222**, 293 (1970).
[17] S. Kessler, *J. Immunol.* **115**, 1617 (1975).

suspension in phosphate-buffered saline (pH 7.2) at $-50°$, are thawed, diluted with 10 ml of wash buffer containing 2% SDS (see above), and boiled for 5 min. The boiled cells are washed twice in wash buffer containing 0.2% SDS (10,000 rpm, 5 min) and resuspended to the original volume (1–2 ml) in 10 mM Tris-HCl, pH 7.5.

Procedure. The frozen pulse-chased cell samples are thawed at 0° and 0.1-ml aliquots are added to 0.02 ml of ice-cold Tris-SDS buffer in 1.5-ml plastic Eppendorf tubes, yielding suspensions of 4 to 5 × 10^8 cells in solutions containing 2% SDS. The SDS will probably have crystallized out, but this does not matter. The tubes are closed and plunged directly from the ice bath into a vigorously boiling water bath. The rapid temperature change and small volumes ensure that the samples are heated almost at once beyond the physiological temperatures at which posttranslational processing might occur. After boiling for 5 min, the samples (0.12 ml) are cooled and mixed with a solution containing, per sample, 0.24 ml of antiserum and 0.12 ml of 2 M NaCl (final NaCl concentrations, 0.5 M). These mixtures are incubated for 60 min at 37°, under which conditions immune complexes are formed, and are then diluted with 0.5 ml of water and mixed with 0.15 ml of washed *S. aureus* cells. After further incubation for 30 min at 37°, the cell-bound immune complexes are precipitated by centrifugation in an Eppendorf microcentrifuge (2–4 min, room temperature) and washed twice with 1 ml of wash buffer (10 mM Tris-HCl, pH 7.5, containing 0.9% NaCl, 0.1 mM EDTA, and 0.02% SDS) prior to resuspension in 0.05 ml of electrophoresis sample buffer [62.5 mM Tris-HCl, pH 6.8, 10% (w/v) glycerol, 2% SDS, and 5% (v/v) mercaptoethanol]. Resuspension of the precipitate at each stage of this washing procedure is tedious and is best accomplished using an Eppendorf tube vibrator (allow 30 min for complete resuspension). Alternatively, the precipitate can be resuspended by vigorous agitation with a robust wire held between thumb and forefinger (using a rapid strumming motion of the finger).

Samples are boiled for 5 min to denature and solubilize the immune complexes, spun briefly to precipitate the *S. aureus* cells, and loaded as 10–20-μl samples into every other well of an SDS–polyacrylamide gel, which minimizes the chance of cross-contamination between the radioactive protein bands of neighboring samples. After electrophoresis, which can be performed in a cold room (4–8°), the gels are fixed in methanol–water–acetic acid (45:45:10 by volume) for 30 min and stored in 7% acetic acid to await fluorography.

Factors Affecting Immunoprecipitation. The procedure described depends ostensibly upon the ability of protein A-bearing *S. aureus* cells (Cowan I strain) to adsorb and precipitate immune complexes. In fact a good immunoprecipitate can also be obtained, particularly for OmpA, even in the absence of *S. aureus* cells. However, for OmpF the conditions described are necessary if the yield of immunoprecipitated protein is to be maximized. Note especially that the concentration of SDS has been chosen to ensure a ratio of SDS to bacterial protein of 4:1 (w/w), yet at the same time to minimize the ratio of SDS to antiserum in the immunoprecipitation reaction.

Quantification of Radioactive Proteins. Fixed SDS–polyacrylamide gels are prepared for fluorography by soaking either in En³Hance (New England Nuclear) or in 1 M sodium salicylate,[18] and are dried in a slab gel drier (e.g., Hoefer Scientific Instruments). The corners of the paper onto which the gel has been dried are marked with radioactive dye (Coomassie Blue mixed with any surplus radioisotope is convenient), and the gel is exposed to preflashed film[19] for up to a week. The developed film is relaid over the gel in its original position (by alignment with the radioactive corner markers), fixed into place along all four sides with adhesive tape, and the radioactive protein bands are cut out with a scalpel. This can be done extremely accurately if the bands are outlined in pencil on the film (the outlines should be covered with transparent adhesive tape to prevent smudging), and the gel is first cut into strips, each strip carrying the separated proteins from one sample. All free edges are secured with adhesive tape after each cut. The dried pieces of gel are prised away from the backing paper using the tip of the scalpel and swollen overnight at 37° in sealed glass scintillation vials containing 30% H_2O_2–0.2 M NH_4OH. The gel pieces disintegrate entirely in this solution provided they are not too large, and a homogeneous scintillation mixture can be prepared by the addition of 10 ml of Unisolve scintillant (Zinsser) or other suitable scintillation cocktail. The samples are counted for double label (^3H, ^{35}S) with appropriate quench corrections. Background radiation is measured with nonradioactive pieces of the same gel carried through the same procedure; background levels higher than expected may arise if, for example, radioactive molecules diffuse into the gel fixing solution and are subsequently reabsorbed at random into the gel.

Theoretical Analysis of Pulse-Chase Data

Kinetics of OmpA and OmpF Synthesis. The elongation rate is determined by matching experimental data for the incorporation of [^{35}S]methionine into a given protein against a series of theoretical curves. This technique is illustrated in Fig. 2, in which a hypothetical protein with 450 amino acids, including methionine residues at the indicated positions, is synthesized at 10 amino acids per second. The [^{35}S]methionine pulse of 5 sec is followed by a chase of unlabeled methionine, whose effectiveness is

[18] J. P. Chamberlain, *Anal. Biochem.* **98**, 132 (1979).
[19] R. A. Laskey and A. D. Mills, *Eur. J. Biochem.* **56**, 335 (1975).

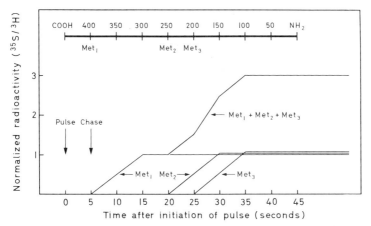

FIG. 2. Kinetics of synthesis of a hypothetical protein. The protein is 450 amino acids long, contains 3 methionine residues, and in this example is synthesized at 10 amino acids per second. See text for details.

monitored as described in Fig. 1. In this example we assume for simplicity that the chase has no effect for 5 sec, at which time it instantaneously becomes 100% effective. Hence the *effective* pulse time is 10 sec. After this time no further radioactivity is incorporated into nascent protein molecules; however, radioactivity is detected in these molecules only when their synthesis is completed. Therefore, three curves can be drawn (labeled Met_1, Met_2, and Met_3 in Fig. 2) that represent the kinetics of appearance of completed protein molecules containing [^{35}S]methionine incorporated at positions 400, 250, and 200, respectively, in the primary sequence. The sum of these curves ($Met_1 + Met_2 + Met_3$) represents the theoretical rate of appearance of [^{35}S]methionine in newly synthesized full-length protein molecules. Similar curves can be drawn for other elongation rates, and comparison of the experimental data with these curves yields the rate of synthesis of the protein in question to an accuracy of $\pm$ 10%.

Two key factors required for the above analysis are (*a*) that the chase with unlabeled methionine becomes 100% effective at some time during the experiment; and (*b*) that the protein under investigation contains methionine residues of known location within the amino acid sequence. Alternative methods for measuring elongation rates sometimes assume a uniform methionine distribution, and this may lead to erroneous results.

Note that no distinction is made between precursor and mature protein: both must be counted. However, since nascent precursor molecules lacking 15–25 C-terminal amino acids may be expected to comigrate on SDS–polyacrylamide gels with mature protein, such molecules will contribute significantly to the apparent radioactivity incorporated into mature protein at early times, before much of the latter has accumulated. Therefore early data points should be expected to lie somewhat above the optimal theoretical curve.[12] On the other hand, labeled methionine residues in the precursor signal sequence may be neglected, since such residues would not appear in full-length precursor molecules until well after the time at which most such molecules are already processed.[12]

Kinetics of Pro-OmpA and Pro-OmpF Processing. The kinetics of processing are determined from the same data already used to calculate the rates of OmpA and OmpF elongation. The method entails matching the data to a series of theoretical curves, in an analogous manner to that described above, and depends upon a stepwise solution of the differential equation $ds/dt = k - \lambda s$, where k is the rate of incorporation of [35S]methionine, s, into full-length precursor molecules, and λ is the processing rate constant.[12] It is assumed that λ is constant (i.e., all precursor molecules are equally likely to be processed) and that nascent precursor is not a substrate for the processing reaction. Other assumptions can equally well be made.[12] Figure 3 shows typical results obtained for OmpF protein and provides evidence for two

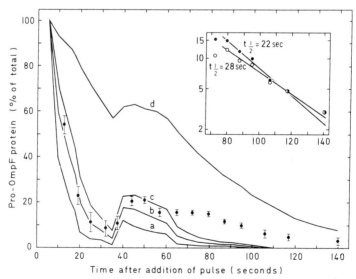

FIG. 3. Kinetics of processing of the pro-OmpF protein at 25°. The data are derived from the experiment illustrated in Fig. 1 and from kinetics of synthesis data analogous to those shown in Fig. 2. Theoretical curves are derived assuming pro-OmpF half-lives ($t_{1/2}$) of 1.5 sec (curve a), 3 sec (curve b), 4.5 sec (curve c), and 25 sec (curve d). The inset is a semilogarithmic plot of the right-hand half of the main figure and shows the relative stability of pro-OmpF protein at late chase times. Data points are replotted (●) without correction, or (○) after subtraction of pro-OmpF radioactivity attributable to a theoretical curve with $t_{1/2} = 4$ sec. From Crowlesmith and Gamon.[12]

processing reactions: a fast reaction ($t_{1/2} = 3-5$ sec) represented by curves *b/c*, and a slow reaction (inset) that gives rise to a divergence between the data points and the theoretical curves at late chase times.

Immunoreplicate Electrophoresis of OmpA and OmpF

Reagents

Tris–SDS–glycerol buffer: 0.3 M Tris-HCl, pH 7.5, containing 12% SDS (w/v) and 30% glycerol (w/v)

Gel buffer: 10 mM Tris-HCl, pH 7.5, containing NaCl (150 mM) and SDS (0.1% or 0.01%)

Agarose, 1% in gel buffer. This is prepared as a molten solution at 46°: mix 0.1 g of agarose with gel buffer containing 0.1% SDS to a final volume of 10 ml, boil for 20 min with continuous stirring to dissolve the agarose, and cool in a water bath held at 46°.

Agarose, 0.2%: Dissolve 20 mg of agarose in 10 ml of deionized water by boiling for 20 min as above; cool to 46° in a water bath.

Antiserum: Prepare as described above (see preceding section).

Sodium salicylate, 1 M

Procedure. The basic pulse–chase experiment with [³⁵S]methionine is as described above, except that cells are not prelabeled with [³H]lysine. The pulse-chased samples (10^9 cells in 0.2 ml) are prepared as usual by freezing at $-78°$ and are subsequently thawed at 0°; 0.1 ml of each sample is mixed with 0.02 ml of ice-cold Tris–SDS–glycerol buffer (0.3 M Tris-HCl, pH 7.5, containing 12% SDS and 30% glycerol), and the mixtures are boiled for 5 min to denature and solubilize total cell proteins. The boiled samples are cooled, spun in an Eppendorf microcentrifuge to remove any precipitate, and loaded onto 12% SDS–polyacrylamide slab gels (as 10-µl samples, each containing ca. 4×10^7 cell equivalents, per well).

While the gels are running, preparations are made for immunoreplication; the method described is a modification of that of Showe *et al.*[20] Glass plates of the size routinely used for slab gel electrophoresis are thoroughly cleaned, dried, and overlaid with molten 0.2% agarose until the surface is just covered. When the agarose has solidified, the plates are heated at 37° or (preferably) 60° to evaporate the water and leave a crust of dried agarose covering the entire surface of each plate. This layer subsequently ensures a firm seal between the glass plate and the agarose gel that will be poured over its surface. Thick adhesive tape (preferably of the kind used to label laboratory equipment; *not* transparent Scotch tape or Sellotape) is wound around the sides of each plate so as to form an enclosed space with the tape projecting about 0.5 cm above the surface of the plate. The plate is reheated

---

[20] M. K. Showe, E. Isore, and L. Onorato, *J. Mol. Biol.* **107**, 55 (1976).

briefly in a 60° oven (to bring its temperature to 40–50°) and placed on a level surface; onto it is poured 6 ml of an antiserum–agarose mixture prepared by mixing 1 ml of antiserum (warmed briefly to 37°) and 5 ml of molten 1% agarose in gel buffer containing 0.1% SDS, held at 46°. The plate must be poured quickly but without creating air bubbles and should be tilted from side to side while pouring to ensure that the molten mixture forms an even layer over the whole surface of the plate. The adhesive tape is removed when the agarose has set.

The unfixed polyacrylamide gel carrying electrophoresed proteins is now placed on top of the agarose gel (care must again be taken to exclude air bubbles) and overlaid with a heavy glass plate, the purpose of which is to ensure that the two gels are pressed firmly together. The entire sandwich of glass plates and gels is put into an airtight box, kept humid with moist filter paper, and left overnight at room temperature to allow proteins to effuse from the polyacrylamide gel into the agarose gel. The polyacrylamide gel is then carefully removed and discarded, while the agarose gel (still attached to its glass base plate) is washed 3–4 times with a stirred solution of 500 ml of gel buffer containing 0.01% SDS (changed at 5–15-hr intervals over a total of at least 24 hr). It is important to avoid shaking the gel or stirring the solution too vigorously; otherwise the gel may become dislodged from the glass plate. A suitable arrangement is to support the ends of the glass plate, creating a space *beneath* it for a magnetic stirring rod.

The washed gel (still attached to the glass plate) is prepared for fluorography by immersing it in 1 M sodium salicylate[18] for 30 min and drying it at 37° overnight or until the agarose is completely dry. The dried gel is overlaid with a preflashed film[19] to expose radioactive protein bands that have been fixed by the specific OmpA or OmpF antiserum. Typical results for OmpA protein are shown in Fig. 4. In this experiment one portion of the pulse-chased samples was heated at 50° (instead of 100°) in the same buffer containing SDS at a final concentration of 2%, in order to analyze the heat-modifiability of newly synthesized OmpA protein. This is a property of mature OmpA protein that results in an apparent M_r of 28,000 (instead of the usual 33,000) on SDS–polyacrylamide gels when the protein is denatured at 50° instead of 100°. In Fig. 4 it is evident that, whereas late chase samples of OmpA protein heated at 50° run as expected with an apparent M_r of 28,000, early chase samples (lanes 3 and 4) do not exhibit this property of heat-modifiability. The experiment suggests that newly synthesized OmpA protein undergoes a conformational change, resulting in the adoption of a native (heat-modifiable) confirmation, 1–2 min after synthesis.[12]

It should be noted that no pro-OmpA protein is detected using this technique, even though control experiments confirm the presence of this protein in early chase samples. This is apparently an artifact of the immu-

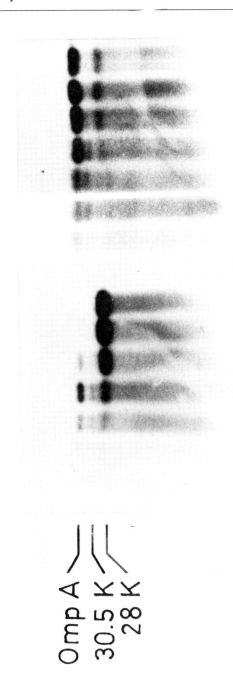

FIG. 4. Pulse-chase analysis of OmpA protein by immunoreplicate electrophoresis. Cells were pulse-labeled with [³⁵S]methionine for 10 sec at 25° and chased with nonradioactive methoinine for up to 30 min. Whole-cell extracts were electrophoresed on a 12% SDS–polyacrylamide gel and replicated onto a 1% agarose gel containing OmpA-specific antiserum. Lanes 1–7: cell extracts solubilized at 50°; lanes 8–14: cell extracts solubilized at 100°. Chase times for samples in lanes 1–7 and 8–14 were: 7, 15, 30, 50, and 110 sec, 15 min, 30 min.

noreplication method, since pro-OmpF protein is also not detected in comparable experiments using OmpF-specific antiserum. The protein band at 30,500 (30.5 K, Fig. 4) may be a biosynthetic intermediate of OmpA protein.[12]

[12] Isolation of Mutants of the Major Outer Membrane Lipoprotein of *Escherichia coli* for the Study of Its Assembly

By JACK COLEMAN, SUMIKO INOUYE, and MASAYORI INOUYE

Gram-negative bacteria are surrounded by two distinct membranes, the inner or cytoplasmic membrane and the outer membrane. Between these two membranes there is the peptidoglycan layer (for review, see Inouye[1]). The outer membrane contains a few major proteins; among them the lipoprotein is the numerically most abundant protein in *Escherichia coli* (for review, see Inouye[2]). The lipoprotein has a very peculiar amino-terminal structure; a glycerylcysteine [*S*-(propane-2′,3′-diol)-3-thioaminopropionic acid] with two fatty acids attached by ester linkages and one fatty acid attached by an amide link. Furthermore, one-third of the mature lipoprotein is bound to the peptidoglycan through an ε-amino group of the carboxy-terminal lysine linkage to the carboxyl group of the *meso*-diaminopimelic acid of the peptidoglycan replacing the D-alanine residues found normally at this position (see Fig. 1). Structural features of the lipoprotein provide an excellent system to study posttranslational modification systems, and lipoprotein biogenesis and assembly into the outer membrane.

Use of Globomycin for Isolation of Lipoprotein Mutants

Globomycin, a unique cyclic peptide antibiotic, was isolated from *Streptomyces*,[3] and its structure is shown in Fig. 2. It has been shown to block the signal peptidase of the prolipoprotein, accumulating prolipoprotein III (see Fig. 1) in the cytoplasmic membrane,[4] which results in lysis of

[1] M. Inouye, *in* "Bacterial Outer Membranes: Biogenesis and Functions" (M. Inouye, ed.), p. 1. Wiley, New York, 1979.
[2] M. Inouye, *Biomembranes* **10**, 141 (1979).
[3] M. Inukai, M. Nakamima, M. Osawa, T. Haneieshi, and M. Arai, *J. Antibiot.,* **31,** 421 (1978).
[4] M. Hussain, S. Ichihara, and S. Mizushima, *J. Biol. Chem.* **255**, 3707 (1980).

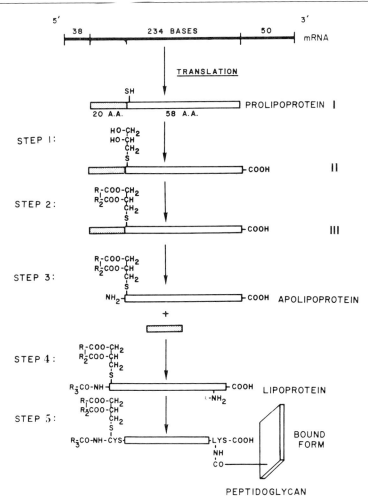

FIG. 1. A proposed mechanism for the biogenesis of the lipoprotein. All the lipoprotein precursors shown have been detected with the exception of prolipoprotein II. A.A., amino acids.

cells. Thus, mutants that lack the lipoprotein[5] or some mutants that produce defective lipoproteins[6] become globomycin resistant.

Isolation of Mutants. Isolation of globomycin-resistant mutants can be carried out as follows: An overnight culture of an *E. coli* K12 strain in L

[5] L. J. Zwiebel, M. Inukai, K. Nakamura, and M. Inouye, *J. Bacteriol.* **145**, 654 (1981).

[6] J. S. Lai, W. M. Philbrick, S. Hayashi, M. Arai, Y. Hirota, and H. Wu, *J. Bacteriol.* **145**, 657 (1981).

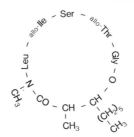

FIG. 2. Structure of globomycin.

broth[7] is diluted 100- and 10,000-fold with L broth. One hundred micro-liters of each dilution and the original culture are each plated on L-broth agar plates containing 25 μg of globomycin per milliliter. Globomycin is not commercially available, but it can be obtained from Dr. Mamoru Arai (Sankyo Co. Ltd., 2-58, 1-Chrome, Hiromachi, Shinagawa-ku, Tokyo 140, Japan).

The globomycin-containing plates can be prepared by pouring 15 ml of L-broth agar into a 15 cm in diameter petri dish containing 18.8 μl of 20 mg/ml globomycin solution in 95% ethyl alcohol. The plates should be incubated overnight at 37° before use. Single colonies isolated from the plates are further purified by streaking on L-broth plates (no globomycin to conserve the drug) with a sterile loop. Purified single colonies from each of the globomycin-resistant colonies on the original plates are grown overnight in L broth, and their globomycin resistance is retested by spotting 10 μl of each culture on globomycin-containing plates, after a 100-fold dilution. For each plate, 50–60 spots can be placed. The concentration of globomycin necessary may vary from 5 μg/ml for *E. coli* B strains to as high as 100 μg/ml for high globomycin-resistant mutants of *E. coli* K12 strains.

Site-Specific Mutagenesis

The majority of the globomycin-resistant mutants isolated above are *lpp⁻* as a result of deletion mutations in the *lpp* gene. Therefore, it is more desirable to use guided site-specific mutagenesis, by which one can generate various different structural gene mutants on the basis of the known DNA sequence of the gene.

The most precise and versatile way to introduce specific mutations in DNA is through the use of short synthetic oligonucleotides as site-specific mutagens.[8] By use of this technique an oligonucleotide is synthesized that

---

[7] J. H. Miller, "Experiments in Molecular Genetics." Cold Spring Harbor Lab., Cold Spring Harbor, New York, 1972.

[8] R. B. Wallace, P. F. Johnson, S. Tanaka, M. Schold, K. Itakura, and J. Abelson, *Science* **209**, 1396 (1980).

contains the desired mutation (point mutation, deletion, or insertion). The oligonucleotide is used as a primer to direct the synthesis of DNA on a single-stranded circular DNA template. The oligonucleotide is thereby incorporated into the resulting circular DNA molecule forming a hetero-duplex.

Transformation into *E. coli* followed by DNA replication resolves this heteroduplex into mutant and parental genes. It is important to use an inducible system for this mutagenesis, since some mutations thus obtained may be lethal to the cells. For example, the wild-type sequence of the amino-terminal region of the signal peptide of the prolipoprotein, Met-Lys-Ala-Thr-Lys-(with a $+2$ charge) has been changed to Met-Lys-Asp-Thr-Lys-(charge of $+1$), Met-Ala-Thr-Lys-(charge of $+1$), Met-Asp-Thr-Lys-(charge of 0), and Met-Glu-Asp-Thr-Lys-(charge of -1), using this method.[9]

In this fashion, one can introduce at the site of interest various desired mutations that cause amino acid replacement, deletion, or insertion of amino acid residues. Thus, mutations can be introduced not only in the signal sequence region, but also in the entire lipoprotein structural region. Analysis of these mutant lipoproteins will provide us important information as to the function of the signal peptide for protein secretion across the membrane and the assembly mechanism of the lipoprotein in the outer membrane.

Preliminary Characterization of Mutants

A preliminary characterization of the mutants can be done immunolog-ically by use of the Ouchterlony method[10] with anti-lipoprotein serum.[11] The Ouchterlony requires 10 μl of cells made from a 1-ml overnight culture, resuspended in 100 μl of 1% SDS and boiled for 5 min.

This method is useful only to determine whether the lipoprotein is made in near normal amounts or if the mutation is due to reduced production or total lack of the lipoprotein. Further characterizations, such as isolation of membranes and polyacrylamide gel electrophoresis, are necessary for the study of the mutants that produce near normal amounts of the lipoprotein. These characterization methods will be discussed later.

Analysis of Mutants

Mutations in the *lpp* gene can be determined by determining DNA sequence of the mutant *lpp* genes.[9] In order to characterize the mutant

[9] S. Inouye, X. Soberon, T. Franceschini, K. Nakamura, K. Itakura, and M. Inouye, *Proc. Natl. Acad. Sci. U.S.A* **79**, 3438 (1982).

[10] O. Ouchterlony, *Ark. Kemi* **I**, 43 (1949).

[11] S. Inouye, K. Takeishi, N. Lee, M. DeMartini, A. Hiroshima, and M. Inouye, *J. Bacteriol.* **27**, 555 (1976).

lipoproteins in terms of their secretion across the membrane and assembly in the outer membrane, the cells carrying a mutant *lpp* gene are pulse-labeled with [^{35}S]methionine for 10–30 sec. A fraction of the pulse-labeled cells are also chased in the presence of a large excess of nonradioactive methionine for 2–5 min.[9] The cells are then sonicated, and the membrane and the soluble fractions are separated by differential centrifugation.[12] These fractions are treated with anti-lipoprotein serum, and the immunoprecipitates are analyzed by two different gel electrophoresis systems to separate most of the intermediates shown in Fig. 1, as described below.

When it is found that the mutant lipoproteins or their precursor molecules exist in the membrane fraction, it is then important to determine whether they are localized in the cytoplasmic membrane or in the outer membrane. This can be achieved either by physical separation of the membranes using a sucrose density gradient[13] or by solubilizing only the cytoplasmic membrane using Sarkosyl, a detergent.[14] It is desirable to use both methods for the separation of both membranes, because some intermediates distribute in the different membrane fractions depending upon the separation methods used.[14a]

Separating Proteins on Gels

Two gel systems have to be used to separate most of the different forms of the lipoprotein. The first gel system is a Laemmli[15] 17.5% acrylamide gel as modified by Anderson *et al.*[16] except that the Tris-HCl [tris(hydroxy methyl)aminomethane hydrochloride] for the separating gel (in a 1 M solution) contains 0.12 M NaCl. This extra salt produces sharper bands for the lower molecular weight proteins. In the absence of a reducing agent, most of prolipoprotein I (see Fig. 1) runs as a dimer, well above the mature lipoprotein (see Fig. 3). The monomer of prolipoprotein I runs in the same position as the normal mature lipoprotein. Prolipoprotein III runs slightly above the mature lipoprotein, and apolipoprotein (see Fig. 1) runs slightly below the mature lipoprotein. The second type of gel system is a phosphate-buffered gel.[9] This type of gel system offers a clearer separation of prolipoprotein I monomer from the mature lipoprotein. Prolipoprotein III runs at the same position as the prolipoprotein I monomer.

[12] M. Inouye and J. P. Guthree, *Proc. Natl. Acad. Sci. U.S.A.* **64,** 957 (1969).
[13] M. J. Osborn, J. E. Gander, E. Parisi, and J. Carson, *J. Biol. Chem.* **247,** 3962 (1972).
[14] C. Filip, G. Fletcher, J. Wulff, and C. F. Earhart, *J. Bacteriol.* **115,** 717 (1973).
[14a] M. Inukai and M. Inouye, *Eur. J. Biochem.* **130,** 27 (1983).
[15] U. K. Laemmli, *Nature (London)* **227,** 680 (1970).
[16] C. W. Anderson, P. R. Baum, and R. F. Gesteland, *J. Virol.* **12,** 241 (1973).

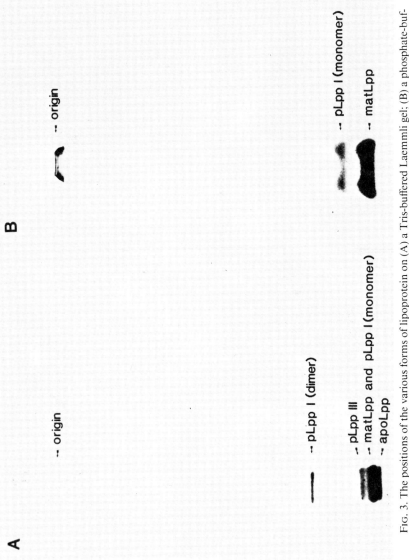

FIG. 3. The positions of the various forms of lipoprotein on (A) a Tris-buffered Laemmli gel; (B) a phosphate-buffered gel. pLpp = prolipoprotein, apoLpp = apolipoprotein, matLpp = mature lipoprotein (see Fig. 1 for a description of the various forms of the lipoprotein).

[13] Analysis of M13 Procoat Assembly into Membranes *in Vitro*

By Colin Watts, Joel M. Goodman, Pamela Silver, and
William Wickner

The diverse transmembrane topologies of different membrane proteins and the finding that most are insoluble in aqueous media stimulated the current interest in how these proteins get into membranes. Investigations are now under way in many laboratories into the mode of transfer of specific polar regions across the lipid bilayer and whether this requires a net input of energy. Information on the relative timing of synthesis, insertion, and proteolytic processing of precursor forms is now available for both prokaryotic and eukaryotic membrane and secreted proteins. The nature and number of components in the target membrane necessary for correct assembly can, in principle, be elucidated by reconstructing the assembly process *in vitro,* using purified components. Here, we describe specific experimental procedures that have extended our understanding of the assembly of one particular membrane protein.

The major coat protein of the coliphage M13 becomes a transmembrane component of the inner *Escherichia coli* membrane prior to its incorporation into new virus particles. It has proved to be very useful in studies of membrane protein assembly *in vitro.*

Assembly of Coat Protein into *E. coli* Inner Membrane Vesicles

The preparation of fractions for cell-free protein synthesis and the preparation of inverted inner membrane vesicles is described elsewhere.[1,2] The major products following the addition of either M13 replicative form (RF-I) DNA or mRNA isolated from M13-infected *E. coli* to a cell-free transcription – translation reaction are the products of gene 5 (DNA single-strand binding protein) and gene 8 (the major coat protein). The latter is made as a higher molecular weight precursor, termed procoat, with 23 extra amino acids at the N terminus. When inverted inner membrane vesicles from uninfected *E. coli* are present during this synthetic reaction, there is a partial conversion of procoat to authentic coat protein, which is integrated in the vesicle bilayer, as originally shown by Chang *et al.*[2] This occurred only

[1] L. M. Gold and M. Schweiger, this series, Vol. 20, p. 537.
[2] N. C. Chang, P. Model, and G. Blobel, *Proc. Natl. Acad. Sci. U.S.A.* **76,** 1251 (1979).

very inefficiently when the vesicles were added after 1 hr of protein synthesis. We have identified an additional process that occurs *in vitro:* namely, the conversion of procoat from a form that is competent to assemble to a form no longer able to undergo this reaction.[3] The half-life of this active form (approximately 2 min) was of the same order as that observed *in vivo* for pulse-labeled procoat synthesized in infected *E. coli.* We believe this to be an important limitation to identifying a cotranslational rather than posttranslational mechanism based on the relative efficiency of membrane assembly after long synthesis times *in vitro.*

When membranes are added *during* synthesis, the level of total protein synthesis is depressed. This also complicates the quantitation of assembly and processing efficiency, since less protein is made in the presence of membranes than would have been made in their absence. It is therefore difficult to show that the extent of processing of a precursor form to its mature product is dependent on the concentration of membranes present during translation; the amount of mature form produced might actually decrease with increasing membrane levels as a result of diminishing synthesis of the precursor. The *relative* amounts of the two forms made when membranes are added following, rather than during, the synthesis of the same amount of protein is the best index of efficiency. Membrane addition after a shorter time of synthesis will be required in this latter case, since the extent of synthesis is greater in the absence of membranes.

Criteria for Transmembrane Insertion of Proteins

The analysis of the association of newly synthesized proteins with both inner membrane vesicles and reconstituted vesicles (discussed below) involves measuring the extent of processing of precursor forms by the vesicles and localizing the proteins with respect to the plane of the lipid bilayer. Rather than give specific experimental details, which may be found in references cited in footnotes 2–4, we wish in this section to discuss these assays in a general way.

1. The extent of *in vitro* conversion of precursors to the mature form is assessed by SDS–polyacrylamide gel electrophoresis of the labeled proteins and fluorography of the dried gel. Loss of radioactive amino acids as a result of removal of the leader (signal) peptide should be taken into account when comparing the relative amounts of precursor and product. For example, two of the three methionine residues in M13 procoat are in the leader peptide. Thus the detection by fluorography of equal-intensity bands of ^{35}S-labeled

---

[3] J. Goodman, C. Watts, and W. Wickner, *Cell* **24,** 437 (1981).
[4] C. Watts, P. Silver, and W. Wickner, *Cell* **25,** 347 (1981).

procoat and ^{35}S-labeled coat indicates that 75% of the precursor was processed.

2. To determine whether the interaction of the labeled proteins with the added membranes is integral or peripheral, the vesicles are washed by centrifugation and resuspension in buffers of high and low ionic strength at mildly alkaline pH. These conditions were originally used to define the two types of protein association with red blood cell membranes.[5] In many instances integral membrane proteins were shown by other means to span the phospholipid bilayer. These treatments at nonphysiological pH and ionic strength should be regarded merely as giving a preliminary idea of whether or not a protein is inserted into the bilayer.

3. The association of the *in vitro* synthesized proteins with the added membranes is assessed further by proteolysis with, for example, proteinase K or Pronase. This is performed in the presence or the absence of detergent or following some physical stress such as freeze–thawing, which should destroy the integrity of the lipid bilayer. Proteins bound to the vesicle periphery are expected to be degraded under all conditions, whereas proteins inserted into the bilayer or sequestered within the vesicles are expected to be at least partially resistant except under conditions where the bilayer has been destroyed.

In our studies of M13 procoat assembly *in vitro,* we found it useful to employ a variety of proteinase concentrations. Following digestion we analyzed *all* the remaining labeled proteins by gel electrophoresis, i.e., without immunoprecipitation. In this way we were able to determine the proteinase concentrations that degraded peripherally bound substrates, such as the product of gene 5, compared with procoat and coat protein. In Fig. 1 we see that the gene 5 protein, which remains membrane associated, is completely degraded by proteinase K levels as low as 8 μg/ml after digestion for 20 hr. The procoat protein is also almost completely degraded at these levels, whereas the coat protein, labeled with [³H]proline at residue 6 from the N terminus, is sensitive only in the presence of detergent. At 10-fold higher proteinase K or Pronase concentrations, some of the coat protein becomes degraded. This probably is the result of vesicle lysis. The resistance of coat protein to digestion is attributed to its insertion into the vesicle bilayer. As a positive control for the digestion pattern shown in Fig. 1, we exposed inverted membrane vesicles isolated from M13-infected *E. coli* to a similar range of proteinase concentrations. These vesicles are derived from membranes that have been shown by analysis of the intact cell to have coat protein inserted through the bilayer.[6,7] The same insensitivity of the coat protein to a range of proteinase concentrations was observed.

[5] T. L. Steck and J. Yu, *J. Supramol. Struct.* **1**, 220 (1973).
[6] W. Wickner, *Proc. Natl. Acad. Sci. U.S.A.* **73**, 1159 (1976).
[7] R. E. Webster and J. S. Cashman, *in* "The Single Stranded DNA Phages" (D. T. Denhardt,

FIG. 1. Inverted *Escherichia coli* membrane vesicles (0.58 OD_{280} unit) were included in an *in vitro* synthesis reaction directed by M13 RF-I DNA (final volume 185 μl). [^{3}H]Proline was the labeled amino acid. After 1 hr of synthesis at 37°, the vesicles were pelleted through a cushion of 50 mM Tris-Cl (pH 8.0), 10 mM EDTA, 1 mM dithiothreitol, 0.5 M NaCl, and 0.5 M sucrose[2] in the Beckman Airfuge at 160,000 g for 5 min at 24°. The pellets were resuspended in 75 μl of 50 mM Tris-HCl (pH 8.0), 10 mM $CaCl_2$. Portions (3 μl) were added to 3 μl of proteinase K (Sigma type XI) dissolved in the same buffer to give the final concentrations shown. Where indicated, 1% Triton X-100 was present. After 20 hr, 6 μl of gel-electrophoresis sample buffer containing 10 mM phenylmethylsulfonyl fluoride was added; after heating to 100° for 3 min, the samples were analyzed by gel electrophoresis and fluorography.[12]

The coat protein associated with both vesicle types displayed a slightly greater mobility following exposure to higher proteinase concentrations. We used an antibody specific for the N terminus[6] and the fact that proline, the labeled amino acid, is present only as the sixth residue in the molecule to show that the N terminus was intact. The shift in mobility is in fact due to removal of residues from the C-terminal hydrophilic domain as originally shown by Chang *et al.*[2] This partial sensitivity to proteinase is expected for a membrane protein that spans the bilayer once, displaying portions of its sequence on both sides. Localizing the protein in relation to the bilayer provides the most rigorous criteria for deciding whether a membrane protein assembly event has been constructed *in vitro.*

We have used these assays to demonstrate that procoat made *in vitro*

D. Dressler, and D. S. Ray, eds.), p. 557. Cold Spring Harbor Lab., Cold Spring Harbor, New York, 1978.

can assemble posttranslationally into inverted *E. coli* membrane vesicles.[3] We have used the same assays and criteria to evaluate our reconstitution of membrane protein assembly from purified components.

Reconstitution of Vesicles Containing Highly Purified Leader Peptidase

In order to understand the interaction of membrane protein precursors with their target membrane and the events that lead to correct, asymmetric membrane assembly, we have attempted to define the required components. We prepared detergent extracts of purified *E. coli* inner membrane vesicles. Removal of the detergent by dialysis or by gel filtration results in the formation of vesicular structures. These may then be tested as target membranes for membrane protein precursors made *in vitro*. A soluble extract for our purposes was defined as the supernatant from a centrifugation at 100,000 *g* for 1 hr. Before removal of the detergent, proteins present in the extract may be fractionated (in the presence of detergent) by chromatographic methods. The separate fractions, either singly or in combinations, can then be mixed with lipid and freed of detergent to yield vesicular structures. We were able to detect an activity in a Triton X-100 extract of *E. coli* membrane that we felt certain would be relevant to the assembly of M13 procoat into membranes. This was the proteolytic activity that processed M13 procoat to authentic coat protein. Purification of this enzyme (leader peptidase) on the basis of its ability to make this cleavage in Triton X-100 containing solution is described in this volume [3].

We compared crude detergent extracts of *E. coli* membranes, partially purified preparations of leader peptidase, and highly purified leader peptidase. It was found that when crude extracts were made in the presence of the nonionic detergent octyl glucoside they could be reconstituted by dialysis into sedimentable vesicles that would still process procoat to authentic coat protein. Extracts prepared with deoxycholate and cholate were also tested but, following detergent removal, the vesicles had only weak processing activity. Our encouraging results with octyl glucoside and the difficulty of efficiently and rapidly removing Triton X-100 suggested that this is the detergent of choice. The initial step in the purification of leader peptidase employs chromatography in Triton X-100 on the anion-exchange resin DE-52 (Whatman).[8] We were able to exchange the detergent by rebinding the enzyme to a DE-52 column at low ionic strength, washing with buffer containing octyl glucoside, and eluting the enzyme by raising the ionic strength. This procedure and the subsequent reconstitution of the enzyme into liposomes are described below.

[8] C. Zwizinski and W. Wickner, *J. Biol. Chem.* **255,** 7973 (1980).

Exchange of Leader Peptidase into Octyl Glucoside

A small (1.0 × 3 cm) column of DEAE-cellulose (DE-52) was equilibrated with 0.01 M triethanolamine-Cl (pH 7.5), 1% (v/v) Triton X-100, 10% glycerol. Complete equilibration is essential, as leader peptidase binds only weakly to this resin. Leader peptidase fractions were dialyzed extensively against the same buffer and then applied to the column. The column was then washed with 8 ml of 0.01 M triethanolamine-HCl (pH 7.5), 10% glycerol, 1% octyl glucoside (Calbiochem) and eluted with the same buffer plus 0.1 M KCl. Fractions (0.5 ml) were collected throughout the application, washing, and elution of the protein. Triton X-100 removal was monitored by its absorbance at 278 nm. Fractions containing leader peptidase were identified by their ability to catalyze the conversion of M13 procoat, made *in vitro* in the presence of Triton X-100, to coat protein.[8] These fractions were pooled and stored at −80°.

Phospholipids

Phospholipids were extracted from *E. coli* by the method of Bligh and Dyer.[9] Removal of neutral lipid and separation of phosphatidylethanolamine from the acidic phospholipids, phosphatidylglycerol, and cardiolipin, was achieved by chromatography on silicic acid.

The lipid was applied in chloroform, and the column was washed with three column volumes of the same solvent. Phospholipids were eluted with chloroform–methanol (7:3, v/v), and fractions were assayed by thin-layer chromatography on silica gel 60 plates (Merck) in chloroform–methanol–acetic acid, 65:25:10 (v/v/v). Fractions containing phosphatidylethanolamine alone and a fraction containing phosphatidylglycerol and cardiolipin were obtained in this way. For reconstitution these were generally recombined so that there was 1 mol of phosphatidylethanolamine phosphate per mole of phosphatidylglycerol–cardiolipin phosphate.

Reconstitution of Leader Peptidase into Liposomes

The phospholipids were dried thoroughly as a film and resuspended at 23° in 10 mM triethanolamine-HCl, pH 7.5, by vortexing in the presence of a few glass beads (1 mm in diameter). The suspension was flushed with N_2 and placed in a bath sonicator for 2 min at 23°. A mixed micellar solution of lipid (5–10 mg/ml), octyl glycoside (17 mg/ml), and leader peptidase (40–400 μg/ml depending on the purity of the fraction) was prepared in 0.01 M triethanolamine-HCl, pH 7.5. Leader peptidase was diluted 2- to

[9] E. Bligh and W. J. Dyer, *Can. J. Biochem. Physiol.* **37,** 911 (1959).

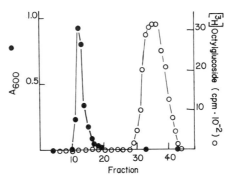

Fig. 2. Formation of liposomes containing leader peptidase by detergent removal on a gel-filtration column. A column (1.5 × 17 cm) of Sephadex G-25 was equilibrated at 23° in 10 mM triethanolamine-HCl, pH 7.5. The column was prerun with lipid (4 mg) dissolved in 10 mM triethanolamine-HCl, pH 7.5, 10% glycerol, and 1% octyl glucoside, applied to the column, and eluted with column buffer. The sample to be reconstituted contained 4 mg of *Escherichia coli* phospholipids, approximately 40 μg of leader peptidase, and 10 mg of octyl glucoside in a final volume of 1.2 ml of 10 mM triethanolamine-HCl, pH 7.5; 1 μCi of [14C]octyl glucoside was included in the mixture. The sample was applied to the column and eluted with 10 mM triethanolamine, pH 7.5. Fractions (0.5 ml) were collected, and the turbidity ($A_{600\,nm}$) and [14C]octyl glucoside content were measured.

3-fold on making this mixture, so that the final KCl and glycerol levels are approximately 40 mM and 4%, respectively.

The removal of octyl glucoside and the formation of active vesicles has been achieved by both dialysis and gel filtration.[10] Each method removed over 99.8% of the detergent. Using 14C-labeled octyl glucoside we found that 1 molecule of detergent remained after dialysis per 60 phospholipid molecules. When the detergent was removed by filtration through a Sephadex G-25 column, only one molecule of detergent remained per 300 phospholipid molecules. As shown in Fig. 2, the detergent is almost completely resolved from the reconstituted lipid. $CaCl_2$ was added to the fractions which eluted at the void volume. These were turbid owing to small phospholipid vesicles. The addition of $CaCl_2$ produced a dense, white precipitate. Addition of 15 mM EDTA caused the formation of large unilamellar vesicles as described by Papahadjopoulos and colleagues.[11] These were collected by centrifugation (50,000 g, 15 min, 4°), and the pellet was resuspended in 2 ml of 10 mM triethanolamine-HCl, pH 7.5, 1 mM EDTA. The suspension was centrifuged as above and the vesicles were resuspended in 0.5 ml of 10 mM triethanolamine-HCl, pH 7.5, and frozen in liquid N_2.

[10] J. Brunner, P. Skrobal, and H. Hauser, *Biochim. Biophys. Acta* **455**, 322 (1976).
[11] D. Papahadjopoulos, W. J. Vail, K. Jacobson, and G. Poste, *Biochim. Biophys. Acta* **394**, 483 (1975).

Leader peptidase reconstituted into phospholipid vesicles by the above method is able to support the assembly and processing of M13 procoat *in vitro* as determined by the assays discussed earlier. Like the inverted inner membrane vesicles, the reconstituted system is able to process and assemble M13 procoat posttranslationally in the absence of ribosomes.[4] The amount of coat protein resistant to digestion when proteinase K is added to the vesicles after exposure to M13 procoat is somewhat less in the reconstituted vesicles than in the inner membrane vesicles, where almost all of the coat protein produced is resistant (Fig. 1). We believe this to be due to some instability peculiar to the reconstituted vesicles, though we cannot rule out the possibility that some of the reconstituted peptidase mediates processing without insertion or that partial or nonasymmetric insertion occurs.

The behavior of the reconstituted vesicles was not affected when the detergent-to-lipid ratio in the starting mixture was varied from 1 : 1 to 5 : 1 (w/w). Increasing the phospholipid to protein ratio did produce an effect in that less coat protein was made by the reconstituted vesicles as the amount of phospholipid was increased. Under these conditions, the proportion resistant to digestion remained constant. At high phospholipid-to-leader peptidase ratios, vesicles may be formed that lack active peptidase molecules but still support procoat binding. The reconstitution of leader peptidase did not appear to be particularly sensitive to ionic strength. When 0.2 M NaCl or NaSCN was included in the column elution buffer or in the dialysis buffer, liposomes were formed that showed processing of the procoat made *in vitro* and protection of the coat protein.

Isolation of Radiochemically Pure M13 Procoat

Our studies on the assembly of M13 coat protein *in vitro* have also focused on the properties of the precursor, procoat, which may be isolated in a soluble form from M13-infected *E. coli* and which assembles and is processed posttranslationally *in vivo*[12,13] and *in vitro*.[3] These findings alone did not rule out a possible role in the assembly reaction for ribosomes, soluble proteins, or other cofactors. The apparent instability of procoat *in vitro*[3] and its short half-life *in vivo* makes a study of the solubility properties of isolated procoat particularly interesting. We have described the isolation of this precursor in radiochemically pure form following its solubilization and denaturation in detergent and organic solvents.[14] The procoat, which is obtained from an *in vitro* protein synthetic reaction, is resolved from larger proteins and smaller molecules, such as detergents, by chromatography on

[12] K. Ito, T. Date, and W. Wickner, *J. Biol. Chem.* **255**, 2123 (1980).
[13] T. Date, J. M. Goodman, and W. Wickner, *Proc. Natl. Acad. Sci. U.S.A.* **77**, 4669 (1980).
[14] P. Silver, C. Watts, and W. Wickner, *Cell* **25**, 341 (1981).

Sephadex LH-60 in formic acid–ethanol, 3:7 (v/v). When the organic solvents were removed from the fractions containing procoat it was found that the protein could be resolubilized in aqueous buffers at low ionic strength and mildly alkaline pH. Remarkably, the resolubilized protein was processed to coat protein by either inverted inner membrane vesicles or by liposomes containing purified leader peptidase. The coat protein was inserted into the bilayer by the criteria discussed above. These assembly reactions and the solubility of the isolated procoat were strongly affected by Mg^{2+}, which was required for binding to membranes, but which, in the absence of membranes, rendered the procoat insoluble.[14]

This radiochemically pure form of procoat has also been used as a substrate in a rapid and quantitative assay for leader peptidase and for leader peptide hydrolase, an enzyme (or set of enzymes) that degrades the leader peptide after its cleavage from procoat. The assay exploits the generation by leader peptidase and leader peptide hydrolase of acid-soluble material from acid-insoluble procoat. This approach might allow the isolation of leader peptidases from other organisms that recognize procoat, and the isolation of other radiochemically pure precursors might allow other peptidases with different specificities to be isolated.

[14] Insertion of Proteins into Bacterial Membranes

By PETER MODEL and MARJORIE RUSSEL

This chapter deals with methods for the study of prokaryotic membrane and certain periplasmic proteins, both *in vivo* and *in vitro*. Our main interest has been in the synthesis and membrane insertion of the coat proteins of bacteriophage f1, which enter the plasma membrane of the infected cell as an early step in the biosynthesis of the phage. We have also worked with β-lactamase, alkaline phosphatase, lipoprotein, and a variety of other *E. coli* membrane proteins.[1-7] There are two main problems to be faced when biosynthesis and membrane insertion are to be studied. The first difficulty is mainly an *in vivo* problem — one keenly feels the lack of reagent(s) that will

[1] C. N. Chang, G. Blobel, and P. Model, *Proc. Natl. Acad. Sci. U.S.A.* **75**, 361 (1978).
[2] C. N. Chang, P. Model, and G. Blobel, *Proc. Natl. Acad. Sci. U.S.A.* **76**, 1251 (1979).
[3] J. D. Boeke, M. Russel, and P. Model, *J. Mol. Biol.* **144**, 103 (1980).
[4] M. Russel and P. Model, *Proc. Natl. Acad. Sci. U.S.A.* **78**, 1717 (1981).
[5] M. Russel and P. Model, *Cell* **28**, 177 (1982).
[6] C. N. Chang, H. Inouye, P. Model, and J. Beckwith, *J. Bacteriol.* **141**, 726 (1980).
[7] J. D. Boeke, P. Model, and N. D. Zinder, *Mol. Gen. Genet.* **186**, 185 (1982).

METHODS IN ENZYMOLOGY, VOL. 97

specifically inhibit membrane insertion and processing by a well-defined mechanism. The second problem is general in the characterization of membrane proteins, i.e., the difficulty in determining the nature of the association of a particular protein with the membrane, and will be discussed where appropriate.

In Vivo

Cells are grown in an appropriate medium: either M9[8] (Na_2HPO_4, 6 g; KH_2PO_4, 3 g; NaCl, 0.5 g; NH_4Cl, 1 g per liter) or Vogel–Bonner minimal salts[9] (per liter of 50× concentration: $MgSO_4 \cdot 7H_2O$, 10 g; citric acid, 100 g; K_2HPO_4, 500 g; $NaNH_4HPO_4 \cdot 4H_2O$, 175 g). Ingredients are dissolved in the order listed, and the solution is autoclaved. It is diluted with sterile tap water as needed. Both media are made up with sterile tap water, since this will supply needed minor components; this will not always be possible, but depends on the quality of the local water. We include a carbon source (usually glucose) at 0.2%, all amino acids at 1–2 mM except that with which labeling is to be carried out, except when carbonyl cyanide m-chlorophenyl-hydrozone (CCCP) is added (see below). Cells are grown from an overnight culture made up in the appropriate minimal medium by diluting at least 1:20 and shaking in a water bath at 37° (some precursor proteins are more stable at 30°). They are used when they reach a density of 2 to 4 × 10^8/ml, as measured by apparent optical density (OD) at 650 nm with the use of a calibration curve derived from viable plate counts vs apparent OD. We grow cells in small flasks and, just before labeling, transfer 200 μl into 12 × 100 mm tubes, continue shaking, and then carry out the labeling protocol.

Labeling with [^{35}S]methionine is useful because of its availability at high specific activity (10^6Ci/mol); particularly with membrane fractions, the ability rapidly to incorporate much label into a small volume of cell culture is useful for subsequent cell fractionation, since gels of membrane fractions are particularly sensitive to overloading. In the case of many proteins, precursors can be detected if the labeling time is short enough — say less than 10 sec, and often preferably less than 5 sec. Samples (a few microliters) are taken into cold 5% trichloroacetic acid. Incorporation of the label can be monitored by subsequent filtration, and the proteins can be analyzed by solubilization in sample buffer (final concentration: 10% glycerol, 5% 2-mercaptoethanol, 3% sodium dodecyl sulfate, 0.0625 M Tris-HCl, pH

[8] J. H. Miller, "Experiments in Molecular Genetics." Cold Spring Harbor Lab., Cold Spring Harbor, New York, 1972.
[9] H. J. Vogel and D. M. Bonner, *J. Biol. Chem.* **218**, 97 (1965).

6.8), boiling for 2 – 5 min, and analysis on an appropriate SDS-containing gel system.

It is often difficult to resolve precursors from mature proteins for low molecular weight proteins, such as lipoprotein or the bacteriophage f1 major coat protein. A particularly effective gel system for doing so contains 20% acrylamide, 0.077% bisacrylamide, and 0.09 M NaCl as well as 0.40 M Tris-HCl (pH 8.8), 0.1% SDS, and 6 M urea. A standard stacking gel, but containing 6 M urea, is used in this system. Gels are easily prepared if a stock solution containing 40% (w/v) acrylamide and 0.154% (w/v) bisacrylamide and a 3 M Tris-HCl stock are first prepared. Small adjustments in the bis-acrylamide concentrations may be needed to achieve optimal separation. These gels must not be allowed to stand at temperatures lower than 20 – 22°. If they are at this temperature for any length of time, microcrystals of urea, almost invisible, will form and the resolving power of the gel will be lost.

Proteins that are present in control lanes, which do not contain the protein of interest, are run on the same gel. It is often necessary to carry out immune precipitation, if the protein of interest is not abundant. Precipitation is carried out after solubilization in SDS by diluting the protein into an excess of Triton-containing buffer. This is believed to lead to the formation of mixed micelles of SDS-Triton, which at the relative concentrations indicated do not seriously interfere with specific precipitation by the antiserum. We use the following procedure. Samples that can be derived from trichloroacetic acid precipitates to reduce the sample volume are brought to 4% SDS and boiled for 1 min. Those derived from NaOH pellets (see below) are sonicated for 10 sec to ensure dispersal of the pellet. The samples (6 – 10 μl) are diluted to <0.2% SDS with 1% Triton buffer [150 mM NaCl, 50 mM Tris-HCl (pH 7.5), 5 mM EDTA, and 1% Triton X-100] and then brought to 1.5% Triton X-100. Trasylol (250 units/ml) and excess antibody are added, and the samples are incubated overnight at room temperature. Packed protein A – Sepharose CL-4B (Pharmacia) is added (25 – 50 μl); after 30 – 60 min of gentle shaking at room temperature, the samples are centrifuged and washed three times with 1 ml of 1% Triton buffer, using a drawn-out Pasteur pipette to avoid disturbing the Sepharose beads. Finally, the beads are resuspended in 50 μl of 4% SDS and heated in a boiling water bath for 2 min. Sample buffer (5×; 12.5 μl but lacking SDS) is added, and the samples are heated in a boiling water bath for 1 min shortly before electrophoresis. We have found it important to ensure that we are in antibody excess, by performing appropriate control experiments, and to ensure that the molar ratio of SDS to Triton in the precipitation buffer does not exceed what is recommended above. The products of the immune precipitation are subjected to electrophoresis on acrylamide gels and detected by autoradiography.

Examination of the products of labeling allows one to determine the processing of precursors to products, but does not give one an opportunity to study the cellular location of the precursors to membrane or periplasmic proteins, since processing (and presumably membrane insertion) continue when cells are harvested (even in the cold) whether or not metabolism continues. We have used two methods to get around this problem. First, one can transfer the cells into 0.1 M NaOH[5]; 0.2 ml of culture is added to 1 ml of 0.1 M NaOH, freshly diluted from a stock solution of 40% (w/v; approximately 10 M). The samples are vortexed vigorously. If subsequent proteolysis is to be carried out (see below) the samples are neutralized at this point. If not, the material is centrifuged for 15 min in an Eppendorf microfuge at 4°. The supernatant is removed with a drawn-out Pasteur pipette so as not to remove the gelatinous, translucent pellet, which occasionally detaches from the wall of the tube. For direct analysis on SDS–polyacrylamide gels, one-tenth volume of 100% (w/v) trichloroacetic acid is added to the supernatant and the pellet is resuspended in 5% trichloroacetic acid. These samples are left on ice for 20 min. After centrifugation, both fractions are washed once with a small volume of acetone to remove trichloroacetic acid, dried, and resuspended in sample buffer. This method stops processing and leads to isolation of the integral membrane proteins in the pellet fraction. The supernatant contains most of the cytoplasmic proteins and the nonintegral membrane-associated proteins. This fractionation is quick and reproducible, and it should stop enzymatic processes. The disadvantages of this procedure are that nonintegral membrane proteins are released and that one must do a specific control to ensure that the protein of interest is not precipitated by the base. Very few proteins are. β-Galactosidase and cellular (presumably DNA-bound) RNA polymerase precipitates in the base. Purified RNA polymerase, when tested by itself, does not precipitate. In any event there was little evidence for precipitation of cytoplasmic proteins of size less than 100,000 daltons. Proteolysis of the NaOH-treated material can be carried out by careful neutralization followed by the addition of the protease.

In a typical protocol 200-μl portions of cells are treated with an equal volume of 0.2 N NaOH and then brought to pH 8.4 with Tris-HCl. As controls, to show that the proteins of interest are susceptible to protease, portions are brought to 2% Triton X-100 or 0.2% SDS. These, together with a sample not treated with detergent are brought to 100 μg/ml with autodigested Pronase. Samples are incubated at 37° for 30 min, then centrifuged; the pellets and supernatant are independently precipitated with trichloroacetic acid. The centrifugation separates the protease-resistant portion of the integral member proteins from the digestion products of the remainder of the cell proteins. At this Pronase concentration, all of the cytoplasmic proteins in the supernatant fraction are digested. We have found it impossible to centrifuge first and then treat the pellets with protease; in the absence of detergent, the pellets are almost impossible to disaggregate after the centrifugation step.

A method by which precursors can be stabilized is by the addition of inhibitors of processing, such as CCCP[10] or procaine.[11] We have found CCCP to be effective, but cessation of processing occurs only in minimal medium which does not contain amino acids, and only if the CCCP is added shortly (at least 5 sec) before label is added. When this is done, however, protein synthesis is strongly affected, presumably as the ATP pools of the cell decline. Insertion of proteins into the cell membrane is partially, but not completely affected. The mechanism by which CCCP affects processing is not understood (the uncoupler does not inhibit purified signal peptidase[10]), and it is not clear whether the distribution of membrane proteins in CCCP-treated cells is normal or aberrant. To carry out a CCCP experiment the drug is conveniently dissolved in DMSO and is added to a culture, grown in M9 minimal medium without amino acids to a final concentration of 60 μM.

When the dynamic aspects of insertion into or transport through the cell membrane are not of particular interest the cells can be fractionated by conventional means into periplasm, membrane, and cytoplasm. Cells (500 μl) are labeled as appropriate, chilled on ice, and centrifuged. The pellet is resuspended in 500 μl of freshly mixed 0.1 M Tris-HCl, pH 8.1, 18% sucrose, 10 mM EDTA containing 4×10^9 carrier cells; 100 μg/ml (final concentration) of lysozyme is added, and the cells are incubated on ice for 10 min and centrifuged in an Eppendorf microfuge for 2 min. The supernatants contain periplasmic proteins (e.g., > 95% of the β-lactamase activity). The rather sticky pellet is resuspended in 100 μl of Tris-sucrose with the use of a rounded glass rod and made 10 mM in $MgCl_2$; DNase I is added to a final concentration of 50 μg/ml. The cells are lysed by the addition of 400 μl of cold water and by brief sonication (Heat Systems; microtip, setting 1 – 2, 5 sec). (The sonication step is very sensitive to the power of the sonicator, but also to the cell concentration. We find it almost impossible to get reproducible results at very low cell concentrations, hence the carrier cells.) After addition of EDTA (10 mM final concentration) 100 μl of the lysate are layered onto 12 ml of 15% sucrose – 3 mM EDTA over a cushion of 0.5 ml of 70% sucrose and centrifuged in a Beckman SW 40.1 rotor for 2 hr at 40,000 rpm at 0°. When the gradient is collected, the membrane proteins are found on the bottom, on the sucrose shelf, and the cytoplasmic proteins in the uppermost fraction or two. If the sonication has been too intense, the

[10] T. Date, C. Zwizinski, S. Ludmerer, and W. Wickner, *Proc. Natl. Acad. Sci. U.S.A.* **77**, 827 (1980).
[11] C. Lazdunski, D. Baty, and J.-M. Pages, *Eur. J. Biochem.* **96**, 49 (1979).

membrane proteins may trail back from the bottom. Once good sonication conditions have been worked out, we find the centrifugation in sucrose to be virtually unecessary; the same separation can be achieved by short (2 – 5 min) centrifugation in the Eppendorf microfuge. The membrane fraction can further be divided into inner and outer membrane fractions, either by centrifugation on sucrose-density gradients[12,13] or by selective extraction with nonionic detergents.[14] Osmotic shock can be used to release proteins from the periplasmic compartment, but this method is not very efficient and varies in efficiency for various proteins. Thus it is mainly useful in indicating in a qualitative way whether a particular protein is likely to be found in the periplasmic compartment.

One technique that we have found to be extremely useful is the plasmolyis of the outer membrane by sucrose, which permeabilizes the outer membrane of fl-infected cells and some other cells sufficiently so that proteolytic enzymes (trypsin, chymotrypsin, Pronase) are admitted. These will then digest the susceptible portions of proteins exposed on the outer face of the inner membrane and help to establish their location. Proper controls are absolutely essential, since a number of periplasmic proteins and some membrane proteins are highly resistant to protease, so that susceptibility must be established before membrane protection, and thus insertion, can be inferred. Cells (200 μl) are rapidly chilled, centrifuged, and resuspended in an equal volume of Tris-HCl (0.03 M, pH 8.1)–sucrose (20%)–EDTA (10 mM). Pronase is added to 100 μg/ml (final); after 2 hr on ice, the samples are precipitated with trichloroacetic acid.

In Vitro

Protein-synthesizing extracts prepared by any of several standard methods[15-17] make the precursors to most membrane and/or periplasmic proteins upon addition of the appropriate template (either DNA or RNA). Examples include prelipoprotein, prealkaline phosphatase, and the precursors to two of the fl coat proteins, pre-gene VIII protein and pre-gene III protein. Other examples abound. If a membrane preparation (see below) is added to such an extract, processing of the preprotein to the mature form is observed, and in a number of instances this cleavage has been shown to be

[12] K. Ito, T. Sato, and T. Yura, *Cell* **11**, 551 (1977).
[13] M. J. Osborn and R. Munson, this series, Vol. 31, p. 642.
[14] D. A. White, W. J. Lennarz, and C. Schnaitman, *J. Bacteriol.* **109**, 686 (1972).
[15] L. M. Gold and M. Schweiger, this series, Vol. 20, p. 537.
[16] G. Zubay, *Annu. Rev. Genet.* **7**, 267 (1973).
[17] R. E. Webster, D. L. Engelhardt, N. D. Zinder, and W. Konigsberg, *J. Mol. Biol.* **29**, 27 (1967).

an endoproteolytic cut at the position expected for the conversion of the preprotein to the mature form. The efficiency of the conversion varies from preparation to preparation and can range from a few percent to more than 50% conversion. Beyond a certain point, addition of more membrane vesicles does not raise the fraction of newly synthesized product that is cleaved, but only inhibits protein synthesis. Zwizinski and Wickner[18] (also see this volume [3]) have purified an activity from uninfected *E. coli* that can carry out the cleavage reaction. We find that the vesicles must be added at the time of synthesis in order for the cleavage reaction to occur; adding them after synthesis is concluded leads to little or no conversion. Goodman *et al.*[19] have examined the kinetics of decay of "insertibility" in some detail and find that after about 1 min the capacity of the system to process precoat protein of phage M13 is lost.

With the use of inverted vesicles (see below) it is possible to demonstrate insertion of f1 phage precoat protein into the membrane. When these vesicles are reisolated on sucrose gradients following translation, and subjected to proteolysis, the products obtained show that the N terminus of the mature protein, together with the membrane-spanning domain, survives and that the C-terminal portion has been attacked by the protease. The protein susceptibility to protease mimics that found for protein made *in vivo* and found in inverted vesicles prepared from infected cells.

A disadvantage of this procedure is the variability of the system from preparation to preparation. Processing activity (as distinct from membrane insertion and sequestration) can reproducibly be observed if vesicles are added to the *in vitro* system together with a nonionic detergent. The most effective detergent is Nikkol (Nikko Chemical, Tokyo) although Triton too can be used. The capacity of membranes to catalyze processing as assayed in the presence of Nikkol is much more uniform than in its absence, and always greater. The method of preparing these vesicles has been shown[20,21] to lead to the formation of "inside-out" vesicles; we believe that the signal peptidase is on what was originally the outside (luminal side) of the plasma membrane, and is on what becomes the inside of the vesicles. Thus processing will occur only after the newly synthesized protein has been inserted in the vesicle membrane or by those vesicles that are not sealed.

Nonionic detergents disrupt the membrane so as to expose the protein to the action of the signal peptidase. In the presence of nonionic detergents most protein synthesis extracts will show processing activity to some degree even if no added membranes are added. This activity is probably due to

[18] C. Zwizinski and W. Wickner, *J. Biol. Chem.* **255**, 7973 (1980).
[19] J. M. Goodman, C. Watts, and W. Wickner, *Cell* **24**, 437 (1981).
[20] M. Futai, *J. Mol. Biol.* **15**, 15 (1974).
[21] T. Tsuchiya and B. P. Rosen, *J. Biol. Chem.* **250**, 7687 (1975).

contaminating membrane fragments and can be reduced by carefully washing the ribosomes and centrifuging the supernatant.

Preparation of Membrane Vesicles[1,2]

It is preferable to use strains that lack RNase I (MRE600, BL15, Q13,
A19, AB301 are such), since the membranes tend to be enriched in this
enzyme when it is present. Cells are grown in tryptone broth (10 g of
tryptone, 1 g of yeast extract, 1 g of glucose, 6 g of NaCl) with good aeration
at 37°. They are harvested at early mid-log phase (about 2 to 4 × 10^8/ml,
chilled, and washed once in 50 mM triethanolamine-HCl, pH 7.5, 5 mM
magnesium acetate, 1 mM dithiothreitol (buffer A). The cells are resuspended in 2–4 volumes of buffer A, passed once through a French
pressure cell at a pressure below 2000 psi, and centrifuged at 10,000 g for 10
min. The supernatant is layered in 8-ml portions over 2 ml of 20% sucrose
and centrifuged in a type 40 rotor at 100,000 g for 1 hr. The pellet is
resuspended in a small volume of buffer A and loaded onto a linear gradient
of 20 to 50% sucrose in buffer A. After centrifugation in a swinging-bucket
rotor at 180,000 g at 4° for 20 hr, a zone containing turbid material sedimenting between the middle and the bottom of the tube is withdrawn (with a
needle, puncturing the side of the tube). This material is sedimented
(100,000 g, 1 hr, type 40 rotor) and is then suspended in buffer B (50 mM
triethanolamine-HCl, pH 7.5 – 10 mM EDTA – 1 mM dithiothreitol) at a
concentration of 10 A_{280}/ml; 8-ml portions are layered over 2 ml of 20%
sucrose in buffer C (50 mM triethanolamine-HCl, pH 7.5 – 5 mM
Mg(OAc)$_2$—1 mM dithiothreitol). Centrifugation is carried out for 2 hr at
100,000 g in a type 40 rotor; the pellet is resuspended in buffer C at a
concentration of 75 A_{280}/ml and stored at − 80° or in liquid nitrogen. A_{280} is
measured by solubilization in 3% SDS at room temperature.

The ability of preparations such as this to sequester proteins and process
(without added detergent) is somewhat variable. We have substituted Tris
for triethanolamine and have omitted the pelleting between the 20-hr
gradient and the EDTA step—these preparations were inferior to the one
described above, but it is not clear whether the cause was the substitutions.

Processing can be assayed in a suitable *in vitro* system to which the
messenger for f1 precoat (or another suitable precursor) is added. Titration
is carried out by adding a range of membrane vesicle concentrations, so that
between 1 : 50 and 1 : 10 of the incorporation mix is made up of the vesicle
preparation. Subsequent electrophoresis of the products on polyacrylamide
gels will give an indication of the optimal range.

If processing only (as distinct from membrane insertion) is to be studied,
the system usually is stimulated by the addition of a nonionic detergent.
Nikkol (octaethylene glycol dodecyl ether; from Nikko Chemicals Co.,

Tokyo, Japan) has proved to be most suitable. Maximal activation of the processing activity is usually observed at about $1-2$ mM Nikkol; a considerable excess (12 mM) inhibits neither cleavage nor protein synthesis.

Acknowledgment

Work in this laboratory is supported by grants from the National Science Foundation and the National Institutes of Health.

[15] Influence of Membrane Potential on the Insertion and Transport of Proteins in Bacterial Membranes

By ROBERT C. LANDICK, CHARLES J. DANIELS, and DALE L. OXENDER

Models for protein secretion that emphasize either the importance of the primary structure of the signal sequence[1] or the three-dimensional conformation of the secretory protein precursor[2] have been proposed. Studies on *Escherichia coli* have led to the suggestion that electrochemical potentials across the inner membrane may be an equally important factor in protein secretion.[3-5] The role of membrane potential in protein secretion may be to align the electrical dipoles of the signal sequence α-helix in the transmembrane electrical field so that the processing site becomes accessible to the leader peptidase on the outer surface of the inner membrane. As shown in Fig. 1, this alignment requires that portions of the mature protein sequence also span the inner membrane. After processing, the secreted protein vectorially refolds from the high-energy transmembrane conformation to a stable, mature periplasmic protein. This suggestion is based on the observation that membrane perturbing agents, such as 2-phenylethanol (PEA), and ionophores, such as carbonyl cyanide m-chlorophenylhydrazone (CCCP) and valinomycin, which alter electrochemical gradients across the inner membrane, are effective inhibitors of protein processing and secretion. Because both secretion and processing are inhibited by these compounds, it is possible to detect the precursor forms of many secreted proteins by

[1] G. Blobel, *Proc. Natl. Acad. Sci. U.S.A.* **77**, 1496 (1980).

[2] W. Wickner, *Annu. Rev. Biochem.* **48**, 23 (1979).

[3] H. G. Enequist, T. R. Hirst, S. Harayama, S. J. S. Hardy, and L. Randall, *Eur. J. Biochem.* **116**, 227 (1981).

[4] C. J. Daniels, D. G. Bole, S. C. Quay, and D. L. Oxender, *Proc. Natl. Acad. Sci. U.S.A.* **78**, 5396 (1981).

[5] T. Date, J. M. Goodman, and W. Wickner *Proc. Natl. Acad. Sci. U.S.A.* **77**, 4669 (1980).

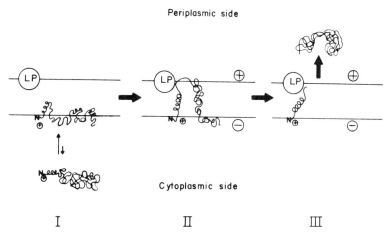

Periplasmic side

Cytoplasmic side

I   II   III

FIG. 1. A model for the role of membrane potential in secretion of periplasmic proteins. Panel I: In the absence of a membrane potential, the precursor of the secreted protein partitions between the cytoplasm and the inner membrane with the equilibrium usually in favor of the membrane-bound form. Panel II: When a membrane potential is present, the dipoles in the signal sequence α-helix align in the electrical field, causing the signal sequence to span the membrane. The positive charge at the N terminus of the signal sequence helps anchor it on the cytoplasmic side of the inner membrane. Negative charge near the processing site may assist in orienting the protein in the membrane. Panel III: After the leader peptidase (LP) has removed the signal sequence, the mature protein uses refolding energy, and possibly the membrane potential, to translocate vectorially into the periplasmic space.

sodium dodecyl sulfate–polyacrylamide gel electrophoresis (SDS-PAGE). By varying the concentration of secretion inhibitors, it is possible to determine the sensitivity of a given secretory protein to the inhibitor. The concentration required to produce 50% maximal inhibition is different for different proteins[4] and therefore reflects a requirement for different levels of membrane potential for a given protein to be effectively processed and secreted.

In this chapter we describe assays for CCCP inhibition of secretion in EDTA-treated whole cells and for valinomycin inhibition of secretion in spheroplasts. The action of both inhibitors can be reversed by resuspending labeled cells in fresh media for a 60-min chase period. These assays are useful for studying the role of membrane potential in protein secretion and for studying precursor–product relationships of secreted proteins. Finally, we have found that the CCCP assay can be a useful tool for detecting the expression of secretory protein gene mutants, which often produce altered proteins that are rapidly degraded in untreated cells.

Reagents

SDS: Sodium dodecyl sulfate (Bio-Rad)

PAGE: Polyacrylamide (Sigma) gel electrophoresis
EDTA: Disodium ethylenediaminetetraacetic acid (Mallinckrodt)
MOPS: 3-(N-Morpholino)propanesulfonic acid (Sigma)
Tris: Tris(hydroxymethyl)aminomethane (Sigma, reagent grade)
CCCP: Carbonyl cyanide m-chlorophenylhydrazone (Sigma)
PEA: 2-Phenylethanol (Sigma)
DNase: Bovine pancreatic deoxyribonuclease I (Sigma)
Valinomycin (Sigma)
L-[^{35}S]Methionine (New England Nuclear) specific activity > 800 Ci/mmol
Goat (rabbit IgG) antisera (Miles)
Lysozyme (Miles)

Cell Growth of Protein Processing Assays

A variety of growth media may be used for cultures to be assayed. The choice of an optimum medium is dependent on the particular protein to be analyzed. Most simply, if the protein of interest is constitutively expressed or if expression can be induced by adding an inducer to the culture (e.g., maltose for *mal* gene expression or isopropylthiogalactoside for proteins expressed from the *lac* promoter), cells can be grown in Luria broth.[6] For the expression of a secretory protein that requires the cells to be grown in the absence of a given nutrient (e.g., leucine for *liv* genes or tryptophan for proteins expressed from the *trp* promoter), a synthetic rich medium such as MOPS-rich medium is preferable.[7]

These protocols have been used successfully with a variety of *E. coli* K12 strains that have been derived from CGSC strain 4273, also designated JC1552. *Escherichia coli* strain RGC123 carrying a lesion in the *lon* (*capR9*) gene[8] has proved to be particularly useful because the absence of the ATP-dependent membrane protease reduces proteolysis of protein precursors lodged in the inner membrane that occurs after CCCP or valinomycin treatment.

Unambiguous identification of a given protein and its unprocessed precursor is greatly facilitated by overproduction of the protein from a plasmid vector. Plasmids that express the secretory protein β-lactamase provide an internal control for processing experiments, since the precursor and processed protein bands can be identified in a whole-cell labeling electrophoretic pattern without immunoprecipitation.

[6] S. E. Luria and J. W. Burrows, *J. Bacteriol.* **74**, 461 (1957).
[7] F. C. Neidhardt, P. L. Bloch, and D. F. Smith, *J. Bacteriol.* **119**, 736 (1974).
[8] C. H. Chung and A. L. Goldberg, *Proc. Natl. Acad. Sci. U.S.A.* **78**, 4931 (1981).

CCCP Assay

CCCP collapses bacterial inner membrane electrochemical potentials by acting as a proton ionophore.[9] In spheroplasts and membrane vesicles, CCCP completely inhibits proton gradient formation at low concentrations; CCCP is less effective when used with whole cells because the outer membrane acts as a permeability barrier. The permeability problem of whole cells can be partially overcome by treating the cells with buffered 10 mM EDTA at 37° which disrupts the permeability of the outer membrane (see Fig. 2). Tris-HCl (100 mM, pH 7.2) and 100 mM potassium phosphate (pH 7.2) are equally effective as buffers for the EDTA treatment; CCCP stock solution are prepared in ethanol at 100 times the desired final concentration for each level to be used in the assay. These solutions may be stored at −20° for later use.

Procedure

1. Harvest cells at an OD$_{600}$ of 0.8 by centrifugation at 3000–5000 g for 5–10 min.
2. Resuspend the cells in 100 mM Tris-HCl (pH 7.2), 10 mM EDTA and incubate at 37° for 5 min.

[9] H. R. Kaback, J. P. Reeves, S. A. Short, and F. J. Lombardi, *Arch. Biochem. Biophys.* **160**, 215 (1974).

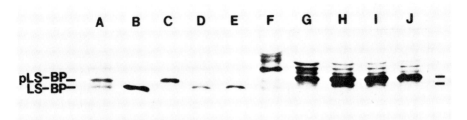

FIG. 2. Inhibition of processing by ionophores and reversal of inhibition by dilution into fresh medium. Lane A shows the precursor of the leucine-specific binding protein (pLS-BP) and the mature form of the leucine-specific binding protein (LS-BP) synthesized in the presence of 10 μM CCCP. Lane B shows the LS-BP synthesized in the absence of CCCP, and lane C shows the pLS-BP synthesized in the presence of 32 μM CCCP. Lane D shows the reversal of processing inhibition by incubation of cells in medium containing 100 μg of chloramphenicol per milliliter and lacking CCCP after synthesis in the presence of 32 μM CCCP. Lanes E and F show shock fluid and cell pellet, respectively, derived from cells treated as those shown in lane D. Lane G shows the pLS-BP synthesized in spheroplasts in the presence of 5 μM valinomycin. Lanes H, I, and J show the reversal of valinomycin inhibition by resuspension of spheroplasts treated as in lane G in fresh media containing 5 μM valinomycin and 0, 1, or 5 mM KCl. All lanes show only a portion of SDS–11% polyacrylamide gels containing pLS-BP and LS-BP immunoprecipitated with rabbit anti-LS-BP antisera and goat anti-rabbit-IgG antisera.

3. Harvest by centrifugation as above. Resuspend the cells in methionine-free, MOPS-rich medium to the original cell density.
4. Add 100-μl aliquots of the cell suspension to 1.5-ml plastic Eppendorf tubes. Add 1 μl of CCCP stock solutions to the cell suspensions. The precise CCCP concentrations chosen will depend on the particular experimental protocol. In general, we have found that 100 μM CCCP is adequate to inhibit the processing completely.
5. Incubate at 23° for 10 min. Add 10–20 μCi of L-[^{35}S]methionine to each cell suspension.
6. Incubate at 37° for 30 min.
7. Centrifuge in a table-top microfuge and resuspend the cells in 100 μl of 0.36% SDS in 125 mM Tris-HCl (pH 6.8) for immunoprecipitation or resuspend directly into SDS-PAGE sample buffer for electrophoresis. It is also possible to prepare osmotic shock fluid from the labeled cells using the procedure of Neu and Heppel[10] prior to addition of SDS.
8. Place the tubes in boiling water bath for 2–5 min.
9. For immunoprecipitation, cool the tubes and add 900 μl of 1% Triton X-100 in 125 mM Tris-HCl (pH 6.8) and appropriate antisera. The antibody preparations should be filtered to produce the optimum precipitation conditions. Incubate this preparation at 37° for 1 hr, after which goat antiserum directed against rabbit IgG is added and the incubation is continued for 4 hr. The immune complexes are recovered by centrifugation for 15 sec in a microfuge, washed vigorously with 1% Triton X-100 in 125 mM Tris-HCl (pH 6.8), and then dissolved in SDS-PAGE sample buffer for electrophoresis.

Reversal of Inhibition by Dilution. The CCCP inhibition of processing can be reversed by resuspension of CCCP-treated cells in fresh CCCP-free medium. To demonstrate reversal, continued incorporation of L-[^{35}S]methionine into protein must be blocked by addition of unlabeled methionine and/or chloramphenicol (100 μg/ml) to the resuspension medium. Reproducible reversal of CCCP inhibition has been accomplished by incubation at 37° for 60 min prior to immunoprecipitation and electrophoresis.

Valinomycin Assay

Valinomycin is a potassium ionophore that can alter the inner membrane transmembrane potential by facilitating free diffusion of potassium ion through the membrane. The outer membrane of *E. coli* effectively excludes valinomycin from access to the inner membrane. Even partial

---

[10] H. C. Neu and L. A. Heppel, *J. Biol. Chem.* **240**, 3685 (1965).

disruption of the outer membrane with EDTA, however, does not allow effective concentrations of valinomycin to reach the inner membrane. Spheroplasts must, therefore, be employed for these experiments. Use of spheroplasts also offers the added advantage of allowing separation of the secreted proteins from proteins in the cytoplasm and membrane by simple centrifugation. The suitability of using valinomycin as an inhibitor of processing obviously depends on whether a given bacterial strain can be effectively converted to spheroplasts.

Procedure

1. Harvest cells at an OD_{600} of 0.8 by centrifugation at 3000–5000 g for 5–10 min.
2. Resuspend the cells in one-third volume of ice cold 0.75 M sucrose in 10 mM Tris-HCl (pH 8.0).
3. Hold on ice for 5 min. Add lysozyme to a final concentration of 150 μg/ml.
4. Hold on ice for 5 min.
5. While vortexing the cells at low speed, slowly add two-thirds volume of 1.5 mM EDTA (pH 8.0) dropwise. Watch the suspension to ensure that the EDTA solution is mixing with the heavier sucrose layer.
6. Add DNase I to a final concentration of 5 μg/ml. Hold on ice for 10 min.
7. Examine the suspension under a phase contrast microscope to verify that >90% of the rod-shaped cells have been converted to oval spheroplasts.
8. Harvest the spheroplasts by centrifugation at 10,000 g for 15 min.
9. Resuspend the spheroplasts to 20% of the original culture volume in MOPS-rich medium containing 12% (w/v) sucrose and lacking L-methionine.
10. Aliquot 100 μl of spheroplast suspension to an Eppendorf tube for each assay condition. Add 1 μl of 100× valinomycin in ethanol.
11. Hold at 23° for 10 min and then shift to 37° and add 10–20 μCi of L-[^{35}S]methionine (specific activity > 800 Ci/mmol). Incubate at 37° for 30 min.
12. Chill to 4° and centrifuge in a microfuge for 10–15 min to pellet spheroplasts. Remove the supernatant fluid.
13. To assay the labeling medium for the presence of secreted proteins, add 10 μl of 4% SDS in 1.4 M Tris-HCl (pH 6.8) to the 100 μl of supernatant fluid obtained from step 12. Proceed to step 8 of the CCCP assay.
14. Add 100 μl of 0.36% SDS in 125 mM Tris-HCl (pH 6.8) to the spheroplast pellet. Proceed to step 8 of the CCCP assay.

Reversal of Inhibition by Dilution. To demonstrate reversal of valinomycin inhibition of processing, spheroplasts labeled in the presence of valinomycin (40 μM has worked well for the leucine binding proteins) are pelleted and resuspended in 125 mM Tris-HCl (pH 6.8) containing 12% w/v sucrose, 5 μM valinomycin final concentration, and various concentrations

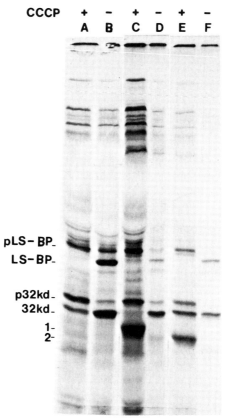

FIG. 3. Detection of deletion mutants of the LS-BP expressed *in vivo.* Lane A shows the immunoprecipitate of proteins synthesized from a strain containing a plasmid (pOX7) coding for the leucine-specific binding protein (LS-BP) in the presence of 32 μM CCCP [D. L. Oxender, J. J. Anderson, C. J. Daniels, R. Landick, R. P. Gunsalus, G. Zurawski, and C. Yanofsky, *Proc. Natl. Acad. Sci. U.S.A.* **77**, 2005 (1980)]. Lane B is the same as lane A without CCCP. Lanes C and E show immunoprecipitates of two different pLS-BP deletions mutants labeled with [^{35}S]methionine in the presence of 32 μM CCCP (bands 1 and 2). Lanes D and F show immunoprecipitates of the same LS-BP deletion mutants labeled in the absence of CCCP. Immunoprecipitation was performed as described in the legend for Fig. 2. 32 kd refers to an outer membrane protein (probably *ompA* protein) that coprecipitates with the immunoprecipitate because it is insoluble in 1% Triton X-100.

of KC1 from 0 to 40 mM. These suspensions are incubated at 37° for 60 min and then immunoprecipitated as described above.

Detection of Unstable Secreted Mutant Proteins

Advances in molecular genetics have produced a large variety of techniques, collectively referred to as directed mutagenesis, that can be used to alter specifically the amino acid sequence of proteins in a controlled fashion.[11] These techniques are powerful tools for addressing the role played by different portions of a protein in transmembrane secretion. Specific amino acids or sets of amino acids can be altered or deleted to ask whether or not these residues are required for secretion. One complication in this approach is that changes in a protein's sequence may alter the protein's conformation such that it is recognized and degraded by cellular proteases.[12,13] In some cases, altered proteins may be degraded so rapidly that it is difficult to detect their expression. We have found that such proteins can be detected *in vivo* by treating cells with CCCP to inhibit processing and secretion (see Fig. 3). In CCCP-treated cells, the altered proteins are often trapped in the inner membrane in precursor form and are degraded much more slowly than if they are allowed to be processed and secreted into the periplasm. Destruction of the membrane potential of *E. coli* cells by ionophore treatment can be used both to study protein secretion and to detect mutant proteins, which would otherwise be rapidly degraded.

Acknowledgments

This work was supported by Grant GM 11024 to D. L. O. and by a Horace H. Rackham Predoctoral Fellowship and dissertation grant to R. C. L.

[11] D. Shortle, D. DiMaio, and D. Nathans, *Annu. Rev. Genet.* **15**, 265 (1981).
[12] K. H. Sveedhara Swamy and A. L. Goldberg, *J. Bacteriol.* **149**, 1027 (1982).
[13] K. Talmadge and W. Gilbert, *Proc. Natl. Acad. Sci. U.S.A.* **79**, 1830 (1981).

[16] Penicillinase Secretion *in Vivo* and *in Vitro*

By JENNIFER B. K. NIELSEN

The study of penicillinase secretion has contributed extensively to knowledge of secretory processes in bacteria. Pencillinases comprise class A of β-lactamases, classified by Ambler[1] on the basis of size, substrate speci-

[1] R. P. Ambler, *Philos. Trans. R. Soc. London, Ser. B* **289**, 321 (1980).

ficity, and sequence homology. They are found in gram-negative organisms, where the best-known example is the constitutive R6K β-lactamase (formerly RTEM) encoded by the plasmid pBR322 in *Escherichia coli*. Those in gram-positive organisms are the β-lactamases of *Bacillus licheniformis, Bacillus cereus* (penicillinase type I, to distinguish it from the thiometallocephalosporinase, type II), both chromosomally encoded and inducible, and the inducible plasmid encoded penicillinase of *Staphylococcus aureus*. As with other secreted enzymes the penicillinases are synthesized with signal peptides, sequenced now for all but the *B. cereus* case.

In gram-negative bacteria the R6K penicillinase is found in the periplasm; its secretion, and the particular defects in secretion when carboxylterminal deletions are present, have been studied by Koshland and Botstein.[2] The gram-positive penicillinases occur in two forms, hydrophilic exoenzyme and hydrophobic membrane-bound enzyme, encoded by the same gene. We have identified all three as lipoproteins of the thioether type,[3] a structure first described by Hantke and Braun[4] for the major outer membrane protein of *E. coli*. This chapter concerns itself with the synthesis and isotopic labeling *in vivo* of the lipoprotein membrane-bound forms of gram-positive penicillinases, and the synthesis both in permeablized cells and in cell-free extracts of full-length translation products. The labeling procedures, though developed for penicillinases in gram-positive organisms, are applicable to the study of thioether lipoproteins in a wide range of bacteria.

Growth, Induction, and Isotopic Labeling of Membrane Forms of Gram-Positive Penicillinases

Bacillus licheniformis 749 and 6346 and their magnoconstitutive mutants 749/C and 6346/C (see the table) are grown at 34° in the following media:

L broth (per liter: 10 g of Bacto-tryptone, 5 g of yeast extract, 5 g of NaCl, pH 6.9) for growth without labeling or for labeling with [^{3}H]palmitate (specific activity 10–30 Ci/mmol; from New England Nuclear) at 10–60 μCi/ml. L broth gives higher yields of both exo and membrane penicillinases than the following media.

CH/S (1% Bacto acid-hydrolyzed casamino acids, 20 mM KH$_2$PO$_4$, 10 mM MgCl$_2$, 0.1% Pollock's salts,[5] pH 6.5) for [^{35}S]cysteine labeling (specific activity 1000 Ci/mmol; from New England Nuclear) at

[2] D. Koshland and D. Botstein, *Cell* **20**, 749 (1980).
[3] J. B. K. Nielsen and J. O. Lampen, *J. Biol. Chem.* **257**, 4490 (1982).
[4] K. Hantke and V. Braun, *Eur. J. Biochem.* **34**, 284 (1973).
[5] M. R. Pollock, *Biochem. J.* **94**, 666 (1965).

SOURCE FOR ORGANISM STOCK

Organism	Strain[a]	Original source
Bacillus licheniformis	749, BGSC 5A20	M. W. Pollock
	749/C, ATTC 25972, BGSC 5A20	M. W. Pollock
	6346	J. F. Collins (Edinburgh)
	6346/C	J. F. Collins (Edinburgh)
Bacillus cereus	569, ATCC 27348, BGSC 6A3	N. Citri (Jerusalem)
	569/H	N. Citri (Jerusalem)
	5/B, ATCC el3061	R. A. Day
Staphylococcus aureus	All strains	R. P. Novick (New York Public Health)

[a] Bacillus Genetic Stock Center, Department of Microbiology, The Ohio State University, 484 W. 12th Avenue, Columbus, Ohio 43210. ATCC, American Type Culture Collection.

50–200 μCi/ml. CH/S may be used for palmitate labeling, but in our hands only the diglyceride (*O*-acyl) fatty acids, not the *N*-acyl residue, were labeled in CH/S, whereas all three were labeled in brain heart infusion or the veal medium used by Lai *et al.*[6] CH/S may be used also for 2-[^{3}H]glycerol labeling (specific activity 10 Ci/mmol; from New England Nuclear) at 20–100 μCi/ml.

Minimal medium (per liter: 2 g of $(NH_4)_2SO_4$, 6 g of KH_2PO_4, 14 g of K_2HPO_4, 1 g of sodium citrate, 0.1 g of thiamin, 1 ml of Pollock's salts, 1% sodium glutamate, 0.5% glucose (autoclaved separately), and amino acids (lacking methionine) at 0.1 mg each per milliliter for [^{35}S]methionine labeling.

In all media the yield of membrane penicillinase is maximal before the onset of stationary phase, when membrane is largely converted to exoenzyme. Prior to this there is very little turnover of the membrane form,[7] and the amount of labeled lipoprotein increases with the labeling time, (usually 2–3 generation times). Cultures of 2.5 ml provide a convenient amount of penicillinase for analysis on 1 or 2 gel lanes.

Bacillus cereus 569, 5B6, and magnoconstitutive 569/H are grown at 34°. CH/S medium can be used for all purposes except labeling with methionine, when minimal medium as above, is used. L broth gives lower enzyme yields.

Staphylococcus aureus pI 524 inducible for PCl penicillinase, pII 147 inducible for PCII, and the constitutive mutants of both[8] are grown at 30° in CY medium.

[6] J.-S. Lai, M. Sarvas, W. J. Brammar, K. Neugebauer, and H. C. Wu, *Proc. Natl. Acad. Sci. U.S.A.* **78**, 3506 (1981).

[7] L. J. Crane, G. E. Bettinger, and J. O. Lampen, *Biochem. Biophys. Res. Commun.* **50**, 220 (1973).

[8] R. P. Novick, *J. Gen. Microbiol.* **33**, 121 (1963).

CY medium: $0.12\ M$ sodium β-glycerophosphate, 1 mM MgSO$_4$, 0.02 ml/liter trace metal solution (0.5% CuSO$_4 \cdot$ 5H$_2$O, 0.5% ZnSO$_4 \cdot$ 7H$_2$O, 0.5% FeSO$_4 \cdot$ 7H$_2$O, 0.2% MnCl$_2 \cdot$ 4H$_2$O), 1% yeast extract, 1% casamino acids, 0.8% glucose, $5 \times 10^{-5}\ M$ Cd(NO$_3$)$_2$ (to maintain presence of the plasmid, which also bears a cadmium resistance gene). This medium is used for growth and for palmitate or glycerol labeling. For labeling with methionine, the level of yeast extract is reduced to 0.2%.

For all three organisms penicillinase production is induced by adding the pseudo gratuitous inducer 2-(2'-carboxyphenyl)benzoyl-6-aminopenicillanic acid (Sigma) at 5 μM to a culture in early exponential phase.[9] Assay is performed as described by Sargent,[10] except that the tubes are siliconized and 0.1% gelatin[11] is present in the assay buffer to minimize adsorption of *B. cereus* penicillinase to glass. Membrane penicillinase is not cryptic in grampositive cells and may be measured by direct assay of a washed cell suspension.

Labeling is terminated by adding an excess of the appropriate unlabeled compound. Cells are harvested, washed, and converted to protoplasts as described for the bacilli[12] and for *S. aureus*.[3]

The protoplasts are disrupted, after the addition of protease inhibitors [1 mM diisopropylfluorophosphate (DFP) and 1 mM phenyl-methanesulfonyl fluoride (PMSF) in 0.05 M Tris-HCl buffer, pH 7.5]. Membranes are pelleted at 15 g for 45 min in a microfuge, extracted twice with CHCl$_3$–CH$_3$OH, 2:1,[13] dissolved in 2$\times$ electrophoresis buffer for direct analysis on 10% SDS–polyacrylamide gels[14] or in antibody wash buffer (1% Triton X-100, 50 mM Tris-HCl, pH 7.5, 5 mM EDTA, 0.15% NaCl) and clarified by centrifugation at 15 g for 30 min. Antibody precipitation, with or without *Staphylococcus* cells[15] as a second adsorbent, is carried out by standard methods and the precipitates are analyzed on SDS–polyacrylamide gels by autoradiography or by fluorography[16] in case of [^{3}H]glycerol or [^{3}H]palmitate.

The formation of glycerylcysteine sulfone from oxidation of gel-purified [^{35}S]cysteine-labeled penicillinases is assayed by a modification[3] of a pre-

[9] G. E. Bettinger and J. O. Lampen, *J. Bacteriol.* **104**, 283 (1970).
[10] M. G. Sargent, *J. Bacteriol.* **95**, 1493 (1968).
[11] M. R. Pollock, *J. Gen. Microbiol.* **15**, 154 (1956).
[12] J. B. K. Nielsen, M. P. Caulfield, and J. O. Lampen, *Proc. Natl. Acad. Sci. U.S.A.* **78**, 3511 (1981).
[13] E. G. Bligh and W. J. Dyer, *Can. J. Biochem. Physiol.* **37**, 911 (1959).
[14] U. K. Laemmli, *Nature (London)* **227**, 680 (1970).
[15] S. W. Kessler, *J. Immunol.* **117**, 1482 (1976).
[16] J. P. Chamberlain, *Anal. Biochem.* **98**, 132 (1979).

viously published procedure.[17] This is a useful way of distinguishing for any palmitate-labeled protein the diglyceride thioether type of lipoprotein from the *O*-acyl type described by Schlesinger[18] and is readily performed on [^{35}S]cysteine-labeled proteins cut from gels stained with Coomassie Blue and dried.

In Vitro Methods for Analysis of Processing

mRNA-Directed E. coli S-30 System.[19] Full-length translation product can be isolated on a gel and converted to processed hydrophilic forms upon the addition of membrane vesicles from either *E. coli* or *B. licheniformis.*[20] No modification reactions have been demonstrated *in vitro,* and no evidence of cotranslational processing with protection from proteolysis, as shown for many proteins, has been seen. The hydrophilic products of processing can, however, be sequenced, unlike the initial translation product, which carries the blocking *N*-formylmethionyl group.

DNA-Directed Synthesis. Cloned penicillinase DNA[21] can be used as template in supercoiled DNA or restriction fragment-directed systems prepared by a modification[22] of Zubay's method. Synthesis, again of blocked protein, is considerably more efficient and processing was demonstrated as in the RNA system.

Synthesis Using Phenethyl Alcohol-Treated Cells. The unique value of this system, a modification[12] of that described by Halegoua and Inouye,[23] is that full length precursor is made highly efficiently with the *N*-formyl group largely removed. Sequencing is then possible. The time of exposure to phenethyl alcohol will be that for penicillinase in *B. licheniformis,* but a small-scale trial should be run testing a concentration range from 0.2% to 1.2% phenethyl alcohol to find conditions yielding maximal amounts of precursor.

Bacillus licheniformis 749/C cells are grown in minimal medium (see above in section on growth) to a cell density of about 0.85 mg dry weight per milliliter (A_{540} of 1.2), harvested, resuspended in 0.25 volume of cold 10 m*M* Tris-HCl (pH 7.5), 50 m*M* NH$_4$Cl, 10 m*M* MgCl$_2$. Phenethyl alcohol is added to 0.7%, and the cells are mixed well and kept for 10 min in ice. An

[17] J. J. C. Lin and H. C. Wu, *J. Bacteriol.* **125**, 892 (1976).
[18] M. J. Schlesinger, *Annu. Rev. Biochem.* **50**, 193 (1981).
[19] M. W. Nirenberg and J. H. Matthaei, *Proc. Natl. Acad. Sci. U.S.A.* **47**, 1588 (1961).
[20] C. N. Chang. J. B. K. Nielsen, K. Izui, G. Blobel, and J. O. Lampen, *J. Biol. Chem.* **257**, 4340 (1982).
[21] W. J. Brammar, S. Muir, and A. McMorris, *Mol. Gen. Genet.* **178**, 217 (1980).
[22] H.-L. Yang, L. Ivashkiv, H.-Z. Chen, G. Zubay, and M. Cashel, *Proc. Natl. Acad. Sci. U.S.A.* **77**, 7029 (1980).
[23] S. Halegoua and M. Inouye, *J. Mol. Biol.* **130**, 39 (1979).

equal volume of $2\times$ reaction mix is added to give a final concentration of 2 mM ATP, 0.2 mM GTP, 10 mM MgCl$_2$, 50 mM NH$_4$Cl, 50 mM Tris-HCl (pH 8.0), 5 mM dithiothreitol, 0.5 mM amino acids excluding the labeled amino acid, and labeled amino acid at 5–500 μCi/ml. Incorporation is generally linear at 20–22° for only 5–10 min, although the cells can be incubated for 20 min without decrease in incorporation. At the end of this period excess cold amino acid is added, and membrane proteins are isolated for immunoprecipitation and gel analysis as above.

[17] Lactose Permease of *Escherichia coli*

By J. K. WRIGHT, R. M. TEATHER, and P. OVERATH

Lactose permease,[1] also called lactose carrier or "M-protein," is an integral protein of the cytoplasmic membrane of *E. coli,* which functions as a galactoside-proton symporter.[2] Within the framework of a classical carrier model, the carrier is considered to bind one molecule of galactoside and one proton on the external face of the membrane. The resulting ternary complex reorients to the internal face and dissociates. The cycle is completed by reorientation of the empty carrier from the inside to the outside. In the absence of an electrochemical proton gradient ($\Delta\tilde{\mu}_{H^+} = 0$) the carrier merely catalyzes the equilibration of substrate across the membrane (facilitated diffusion). In the presence of an electrochemical proton gradient ($\Delta\tilde{\mu}_{H^+} \ll 0$) or one of its components (electrical potential difference, $\Delta\psi$, outside positive, or a pH difference, ΔpH, outside acid) as driving force, the coupling of galactoside and proton fluxes via the carrier results in active transport of galactosides into the cell or vesicle. Results on the energetics and kinetics of this transport system have been described.[3-12]

[1] H. V. Rickenberg, G. N. Cohen, G. Buttin, and J. Monod, *Ann. Inst. Pasteur, Paris* **91**, 829 (1956).
[2] I. C. West and P. Mitchell, *Biochem. J.* **132**, 587 (1973).
[3] H. Hirata, K. Altendorf, and F. M. Harold, *J. Biol. Chem.* **249**, 2939 (1974).
[4] S. Ramos and H. R. Kaback, *Biochemistry* **16**, 854 (1977).
[5] J. L. Flagg and T. H. Wilson, *J. Membr. Biol.* **31**, 233 (1977).
[6] D. Zilberstein, S. Schuldiner, and E. Padan, *Biochemistry* **18**, 669 (1979).
[7] S. Ahmed and I. R. Booth, *Biochem. J.* **200**, 583 (1981).
[8] G. J. Kaczorowski and H. R. Kaback, *Biochemistry* **18**, 3691 (1979).
[9] G. J. Kaczorowski, D. E. Robertson, and H. R. Kaback, *Biochemistry* **18**, 3697 (1979).
[10] D. E. Robertson, G. J. Kaczorowski, M. Garcia, and H. R. Kaback, *Biochemistry* **19**, 5692 (1980).
[11] J. K. Wright, I. Riede, and P. Overath, *Biochemistry* **20**, 6404 (1981).
[12] J. K. Wright, *Biochim. Biophys. Acta* (submitted for publication).

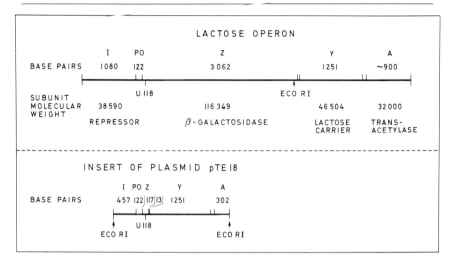

Fig. 1. Lactose operon (upper part) and *lac*Y-containing DNA-fragment (lower part) of the plasmids pTE18 or pGM21. The DNA fragment is flanked by cleavage sites of the restriction endonuclease *Eco*RI and carries an intact promoter-operator region (PO) and the lactose carrier gene (Y) but only parts of the *I, Z,* and *A* genes. The fragment is integrated into the single *Eco*RI site of either pBR322 (pTE18) or pACYC184 (pGM21) as a vector; in both cases the orientation of the fragment is the same as that of the *lac* operon on the *Escherichia coli* chromosome [see R. M. Teather, J. Bramhall, I. Riede, J. K. Wright, M. Fürst, G. Aichele, V. Wilhelm, and P. Overath, *Eur. J. Biochem.* **108**, 223 (1980) and M. Mieschendahl, D. Büchel, H. Bocklage, and B. Müller-Hill, *Proc. Natl. Acad. Sci. U.S.A.* **78**, 7652 (1981) for further details].

Lactose permease was first identified as a protein [apparent $M_r = 31{,}000$ by sodium dodecyl sulfate (SDS)–polyacrylamide gel electrophoresis] by a specific labeling technique using the thiol reagent N-ethylmaleimide[13] and by double-labeling experiments with radioactive amino acids.[14] The amino acid sequence of the protein is now known from the DNA sequence (417 residues, M_r 46,500).[15,16] N-Ethylmaleimide labels a cysteinyl residue at position 148 in the protein sequence.[17] The cloning of the structural gene for the protein (*lac*Y gene of the lactose operon, cf. Fig. 1 and Teather *et al.*[18,19]) on multicopy plasmid vectors enabled the amplication of the protein in the

[13] C. F. Fox and E. P. Kennedy, *Proc. Natl. Acad. Sci. U.S.A.* **54**, 891 (1965).

[14] T. H. D. Jones and E. P. Kennedy, *J. Biol. Chem.* **244**, 5981 (1969).

[15] D. E. Büchel, B. Gronenborn, and B. Müller-Hill, *Nature (London)* **283**, 541 (1980).

[16] R. Ehring, K. Beyreuther, J. K. Wright, and P. Overath, *Nature (London)* **283**, 537 (1980).

[17] K. Beyreuther, B. Bieseler, R. Ehring, and B. Müller-Hill, *in* "Methods in Protein Sequence Analysis." Humana Press, Clifton, New Jersey, p. 139 (1982).

[18] R. M. Teather, B. Müller-Hill, U. Abrutsch, G. Aichele, and P. Overath, *Mol. Gen. Genet.* **159**, 239 (1978).

[19] R. M. Teather, J. Bramhall, I. Riede, J. K. Wright, M. Fürst, G. Aichele, U. Wilhelm, and P. Overath, *Eur. J. Biochem.* **108**, 223 (1981).

cytoplasmic membrane by 10- to 20-fold. Substrate binding experiments with such membranes suggest that the carrier has one type of binding site and that the stoichiometry of carrier substrate interaction is close to one.[11,19,20] Cytoplasmic membranes with elevated carrier level provide suitable starting material for studies on reconstitution and purification of permease.[21-23]

Substrates

Lactose permease recognizes and translocates a large number of α- and β-galactosides.[24] Lactose, the physiological substrate, is most commonly used for transport experiments (half-saturation constant for active transport, $K_T = 0.08$ mM in cytoplasmic membrane vesicles, prepared according to Kaback,[25] and $K_T = 0.27$ mM in cells), but has a very low affinity for binding (equilibrium binding constant, $K_D = 14$ mM). β-D-Galactosyl-1-thio-β-D-galactopyranoside (GalSGal) has a high affinity both for binding ($K_D = 0.053$ mM) and for active transport ($K_T = 0.044$ mM). Both substrates have a negligible tendency to bind unspecifically to membranes or phospholipids. Therefore, GalSGal is a good alternative substrate to p-nitrophenyl α-D-galactopyranoside (NpαGal) when binding is measured, for instance, to reconstituted samples with a very high lipid-to-protein ratio. NpαGal has a high affinity for the carrier ($K_D = 0.022$ mM) and is most useful for estimating carrier levels by equilibrium binding down to 0.1–0.2 nmol per milligram of membrane protein.

The rate of transport of o-nitrophenyl β-D-galactopyranoside (NpβGal) in cells containing an excess of β-galactosidase activity provides a convenient and reliable optical assay for the determination of carrier activity ($K_T = 0.91$ mM).[26-28] A series of fluorescent substrates can be used to measure binding or active transport in membrane vesicles.[20,29] Among these 2'-(N-dansyl)aminoethyl 1-O-β-D-galactopyranoside (DnsEtOGal) and

[20] P. Overath, R. M. Teather, R. D. Simoni, G. Aichele, and U. Wilhelm, *Biochemistry* **18,** 1 (1979).

[21] J. K. Wright, H. Schwarz, E. Straub, P. Overath, B. Bieseler, and K. Beyreuther, *Eur. J. Biochem.* **124,** 545 (1982).

[22] M. J. Newman, D. L. Foster, T. H. Wilson, and H. R. Kaback, *J. Biol. Chem.* **256,** 11804 (1981).

[23] J. K. Wright and P. Overath, *Eur. J. Biochem.* (submitted for publication).

[24] See Wright *et al.*[11] for a recent summary of binding and translocation data. This paper also gives references to earlier work.

[25] H. R. Kaback, this series, Vol. 22, p. 99.

[26] L. A. Herzenberg, *Biochim. Biophys. Acta* **31,** 525 (1959).

[27] J. R. Carter, C. F. Fox, and E. P. Kennedy, *Proc. Natl. Acad. Sci. U.S.A.* **60,** 725 (1968).

[28] P. Overath, F. F. Hill, and I. Lamnek-Hirsch, *Nature (London), New Biol.* **234,** 264 (1971).

[29] S. Schuldiner, G. K. Kerwar, H. R. Kaback, and R. Weil, *J. Biol. Chem.* **250,** 1361 (1975).

6'-(*N*-dansyl)aminohexyl 1-thio-β-D-galactopyranoside (DnsHxSGal) are both of high affinity (K_D = 0.017 m*M* and K_D = 0.007 m*M*, respectively), and their synthesis is straightforward.[30]

All substrates can be radioactively labeled in good yield and high specific radioactivity by enzymatic oxidation to the aldehyde at carbon atom six of the galactosyl residue by galactose oxidase (EC 1.1.3.9) and subsequent reduction with sodium boro[³H]hydride.[31] It should be noted that whenever galactosides with an *O*-β-glycosidic linkage are employed for binding or transport experiments the use of strains defective in the synthesis of β-galactosidase is required. Membranes isolated from β-galactosidase-containing strains are always contaminated by this enzyme.

Escherichia coli Strains

A series of bacterial strains are listed in Table I that are commonly employed to study various aspects of lactose permease. Strain ML30 and its derivatives are suitable both for transport studies in cells and, in particular, in derived cytoplasmic membrane vesicles.[25] Lactose permease overproducing strains of the ML series, which carry the *lacY* gene on the hybrid plasmid pGM21 (Fig. 1), have become available.[32] The K12 strains T192 and T206 carry plasmid pGM21 in the *lacZ*⁺ background of strain T28RT or the *lacZ*⁻ background of strain T184, respectively. Strain T206 is the preferred strain for the isolation of cytoplasmic membranes with an elevated level of permease.

Growth of Bacteria and Isolation of Cytoplasmic Membranes

Several methods are available for the preparation of membranes from *E. coli*. For active transport experiments, cytoplasmic membrane vesicles prepared according to the procedure of Kaback are most useful as their orientation is the same as in cells.[25] For certain transport experiments membrane vesicles of inverted orientation have been employed. These can be prepared either from cells by passage through a French pressure cell[33] or by brief sonication of right-side-out vesicles.[34] For routine binding experiments a crude envelope preparation containing inner and outer membranes is obtained by sonication of cells.[11] The preparation of a cytoplasmic

[30] DnsHxSGal is readily obtained by dansylation of 6'-aminohexyl 1-thio-β-D-galactopyranoside[29] (commercially available from Calbiochem GmbH, Giessen, F.R.G.).

[31] E. P. Kennedy, M. K. Rumley, and J. B. Armstrong, *J. Biol. Chem.* **249**, 33 (1974).

[32] V. Weigel (unpublished experiments).

[33] J. R. Lancaster, Jr. and P. Hinkle, *J. Biol. Chem.* **252**, 7657 (1977).

[34] R. M. Teather, O. Hamelin, H. Schwarz, and P. Overath, *Biochim. Biophys. Acta* **467**, 386 (1977).

TABLE I
Escherichia coli Strains

Strain	lac genotype[a]			Lac phenotype	Other markers[b]			Notes
	Chromosome	F'[c]	Plasmid[d]		Chromosome	F'	Plasmid	
ML30	$I^+O^+Z^+Y^+$	—	—	$I^+O^+Z^+Y^+$	—	—	—	e
ML308-225	$I^-O^+Z^-Y^+$	—	—	$I^-O^+Z^-Y^+$	—	—	—	f
ML3	$I^+O^+Z^+Y^-$	—	—	$I^+O^+Z^+Y^-$	—	—	—	e
T240	$I^+O^+Z^-(Y^+)$	—	—	$I^+O^+Z^-Y^-$	—	—	—	g: derivative of strain ML30
T4UW	$I^+O^+Z^-Y^-$	—	$\Delta(I)O^+\Delta(Z)Y^+$	$I^+O^+Z^+Y^+$	—	—	tet^R	g: derivative of strain ML3
T7UW	$I^+O^+Z^-(Y^+)$	—	$\Delta(I)O^+\Delta(Z)Y^+$	$I^+O^+Z^-Y^+$[i]	—	—	tet^R	g: derivative of strain T240
T28RT	$I^+O^+Z^+Y^-$	$I^qO^+Z^+Y^-$	—	$I^qO^+Z^+Y^-$	$rpsL\ pro^-\ met^-thr^-$ $recA\ hsdM\ hsdR$	pro^+	—	h
T184	$I^+O^+Z^-Y^-$	$I^qO^+Z^{U118}(Y^+)$	—	$I^qO^+Z^-Y^-$	(T28RT)	pro^+	—	h
T192	$I^+O^+Z^+Y^-$	$I^qO^+Z^+Y^-$	$\Delta(I)O^+\Delta(Z)Y^+$	$I^qO^+Z^+Y^+$	(T28RT)	pro^+	tet^R	h: derivative of strain T28RT
T206	$I^+O^+Z^-Y^-$	$I^qO^+Z^{U118}(Y^+)$	$\Delta(I)O^+\Delta(Z)Y^+$	$I^qO^+Z^-Y^+$	(T28RT)	pro^+	tet^R	h: derivative of strain T184

[a] Designation of the genes of the *lac* operon: I, repressor [I^+, inducible, I^- constitutive, I^q, mutation causing overproduction of repressor; cf. B. Müller-Hill, L. Crapo, and W. Gilbert. *Proc. Natl. Acad. Sci. U.S.A.* **59**, 1259 (1968)]; O, operator; Z, β-galactosidase; Y, permease [(Y^+), defective synthesis of permease caused by polar mutation in Z gene]. Δ before a bracketed symbol indicates that the gene is (partially) deleted.

[b] For genetic designations, see B. J. Bachmann, K. B. Low, and A. L. Taylor, *Bacteriol. Rev.* **40**, 116 (1976). Strains ML30 through T7UW are prototrophic; strains T28RT through T206 require methionine, threonine, and thiamin for growth.

[c] F' refers to an F factor that carries the *pro* (AB) genes in addition to the *lac* operon. Z^{U118} refers to a polar mutation in the Z gene.

[d] These strains carry the DNA-fragment shown in Fig. 1 integrated into the single cleavage site for the restriction enzyme *Eco*RI of the plasmid vector pACYC 184 [A. C. Y. Chang and S. N. Cohen, *J. Bacteriol.* **134**, 1141 (1978)]. Strains carrying this plasmid are resistant to tetracycline (tet^R).

[e] H. V. Rickenberg, G. N. Cohen, G. Buttin, and J. Monod, *Ann. Inst. Pasteur, Paris* **91**, 829 (1956).

[f] H. H. Winkler and T. H. Wilson, *J. Biol. Chem.* **241**, 2200 (1966).

[g] V. Weigel (unpublished experiments).

[h] R. M. Teather, J. Bramhall, I. Riede, J. K. Wright, M. Fürst, G. Aichele, U. Wilhelm, and P. Overath, *Eur. J. Biochem.* **108**, 223 (1980).

[i] I^+ refers to partial derepression of the *lac* operon due to the high copy number of the plasmid.

membrane fraction by sucrose density-gradient centrifugation of a soni-cated lysate of EDTA–lysozyme spheroplasts[35] yields starting material suitable for reconstitution and purification of the carrier. This type of membrane preparation is described for the plasmid-harboring strain T206.[19]

Strain T206 (Table I) is grown in the dark in Cohen–Rickenberg mineral salts medium[36] containing glycerol (0.5%), methionine (50 μg/ml), threonine (50 μg/ml), thiamin (10 μg/ml), streptomycin (100 μg/ml), and tetracycline (3.7 μg/ml[37]). An exponential culture (A_{420} 0.4 – 0.8) grown with shaking at 37° is diluted into 20 liters of medium in an aerated bottle to give an initial absorbance of 0.001 – 0.002. After growth overnight to an absor-bance of 0.2, the culture is induced by addition of 0.6 g of isopropyl 1-thio-β-D-galactopyranoside. Growth continues at an exponential rate (generation time 2 hr) to an absorbance of 1.5 – 2.0. The culture is then cooled and quickly centrifuged (5 min, 13,000 g). Cells (20 g wet weight) are suspended at 0 – 5° in 150 ml of 0.75 M sucrose, 10 mM Tris-HCl, pH 7.8, and stirred in an ice bath. After the immediate addition of 150 ml of cold 20 mM sodium-EDTA containing 15 mg of lysozyme (EC 3.2.1.17), stir-ring is continued for up to 45 min; 70 – 90% of the cells are converted to spheroplasts; some lysis of the cells is unavoidable. The suspension is sonicated at 0 – 5° for 4 min in 150-ml batches with stirring, using a Branson sonifier equipped with the standard horn at a power setting of 90 – 100 W. Residual intact cells and spheroplasts are removed by centrifugation (20 min at 1500 g), and the membranes are pelleted by centrifugation (90 min at 150,000 g). The membranes are suspended to a final volume of 18 ml in 25% sucrose, 0.5 mM sodium-EDTA, pH 7.5, and applied to six gradients containing, 55% (1.5 ml), 50%, 45%, 40%, 35%, 30% sucrose, 0.5 mM sodium-EDTA, pH 7.5 (6.3 ml each). The gradients are centrifuged at 150,000 g for 4.5 hr in a Beckman V50 Ti rotor at 0 – 5°. The brownish fraction of cytoplasmic membrane (combined $L_1 + L_2$ fraction,[35] about 2.5 – 4.5 cm from the top of the tubes) is aspirated with a Pasteur pipette. The suspension is diluted threefold with 50 mM potassium phosphate, pH 6.6, and centrifuged (2 hr, 150,000 g). The membranes are washed once with 100 ml of phosphate buffer. They are finally resuspended in 50 mM potassium phosphate, pH 6.6, and stored in liquid nitrogen. The procedure yields 50 – 70 mg of protein. The carrier content as assessed by flow dialysis amounts to 2 – 4 nmol of substrate binding sites per milligram of membrane protein.

---

[35] M. J. Osborn, J. E. Gander, E. Parisi, and J. Carson, J. Biol. Chem. **247**, 3962 (1972).
[36] Y. Anraku, J. Biol. Chem. **242**, 793 (1967).
[37] We use 10 μg of Reverin per milliliter (Hoechst, Frankfurt/Main), which contains 0.37 mg of pyrrolidinomethyltetracycline per milligram dry weight.

Estimation of Permease Content by Substrate Binding

The carrier content of membrane preparations can be determined by equilibrium binding experiments using radioactively labeled high-affinity substrates such as NpαGal or GalSGal.[11,20,38] Alternatively, a fluorometric method using dansylgalactosides is available that is convenient for the rapid estimation of the relative carrier content.[20]

Equilibrium Binding Determined by Flow Dialysis. A flow dialysis cell consists of two compartments separated by a dialysis membrane.[39] The upper compartment is the reaction vessel. Free ligand in the upper compartment diffuses slowly across the membrane into the lower compartment, which is flushed with buffer at a constant rate. The effluent of the lower compartment is collected by a fraction collector. Thus, changes in the concentration of the free ligand in the upper compartment are monitored by changes in the steady-state concentration of ligand in the effluent of the lower compartment.[40]

A suspension of membranes (0.5 ml, 5–10 mg protein per milliliter for cytoplasmic membranes from strain T206 or twice that concentration of a total cell envelope preparation) in 50 mM potassium phosphate, pH 6.6, 1 mM MgSO$_4$, and 10 mM NaN$_3$ (optional) is placed at 22° in the stirred reaction vessel of a flow dialysis cell. [6′-^{3}H]NpαGal (5 µl, 1 mM, 2.7 MBq/ml) is added, and sufficient fractions are taken so that the concentration of [6′-^{3}H]NpαGal in the effluent is in the steady state ($t > 2$ min; cf. Fig. 2). The specifically bound ligand is released from the carrier by addition of a freshly prepared solution of *p*-(chloromercuri)benzene sulfonate (ClHgBzSO$_3$, 5 or 10 µl, 20 mM). Another 5–10 fractions are collected. Aliquots of the fractions (1 ml) are mixed with 5 ml of scintillation cocktail (e.g., Quickszint-212, Zinsser, Frankfurt am Main) and counted. Figure 2 shows a typical experiment. The data are evaluated as indicated in the figure legend, employing an equilibrium dissociation constant, $K_D = 22$ µM. Using 10–15 mg of protein per milliliter, but otherwise identical conditions, the ligand [6,6′-^{3}H$_2$]GalSGal (5 µl, 2.5 mM, 3.1 MBq/ml, $K_D = 53$ µM) can be used instead of [6′-^{3}H]NpαGal.

The assay can be used not only to estimate the binding activity of various

[38] E. P. Kennedy, M. K. Rumley, and J. B. Armstrong, *J. Biol. Chem.* **249**, 33 (1974).
[39] S. P. Colowick and F. C. Womack, *J. Biol. Chem.* **244**, 774 (1969).
[40] A flow dialysis cell is commercially available from Bel Arts Products. In order to increase the response time of the cell, the volume of the lower compartment should be as small as possible. We routinely use a cell similar to that described by K. Feldmann [*Anal. Biochem.* **88**, 225 (1978)] (upper compartment, diameter 2 cm, height 1 cm; the lower compartment is a spiral groove of approximately 20 µl volume; Union Carbide dialysis tubing, flow rate 1.1 ml/min, fraction volume 1.1 ml). The lower compartment should always be rinsed with or stored in H$_2$O to prevent depositing of the buffer.

membrane preparations, but also to determine binding constants for both a labeled ligand by stepwise addition of the unlabeled ligand or for a second unlabeled ligand by competition experiments.[11] Solutions of NpαGal should be protected from light by wrapping in aluminum foil.

Fluorometric Assay.[20] A fluorometer[41] with a cuvette holder equipped with a magnetic stirrer is used. The fluorescence signal is recorded as a function of time on an XY-recorder. A cytoplasmic membrane preparation of strain T206 (1 mg of protein) in 50 mM potassium phosphate, pH 6.6, and 10 mM MgSO$_4$ containing 20 μM DnsEtOGal (total volume 2 ml, $T = 25°$) and a small stirring bar are placed in a cuvette. The fluorescence is monitored as a function of time at an emission wavelength of 500 nm (excitation at 340 nm). The constant signal observed is the sum of three components: ligand in the aqueous phase, ligand unspecifically bound to the vesicles, and ligand specifically bound to the lactose permease. The specifically bound ligand can be released by addition of 10 μl of ClHgBzSO$_3$ (20 mM) or the competing ligand GalSGal (20 μl, 0.2 M). In both cases a rapid ($t_{1/2} \leq 3$ sec) decrease in fluorescence intensity is observed. The change in fluorescence intensity is a semiquantitative, relative measure of the carrier content of the sample. Suitable control experiments are membranes from uninduced strain T206 or carrier-containing membranes pretreated with sulfhydryl reagents. The permease concentration in membranes from haploid *E. coli* strains (ML308-225; cf. Table I) is at the limit of sensitivity of this assay. This assay can also be used for determining the binding constants of fluorescent ligands or for competing nonfluorescent ligands.[11,20]

Radioactive Labeling of Permease

Three methods are available for radioactive labeling of lactose permease. The first method uses radioactive amino acids.[14,18,20] An inducible strain is grown in the presence of a ^{14}C-labeled amino acid mixture in the absence of inducer or in the presence of a ^{3}H-labeled mixture in the presence of inducer. Differential labeling of permease can be evaluated after electrophoretic separation of the cytoplasmic membrane proteins of a mixture of the two cell populations. Although this method gives an objective estimate of the relative carrier content, it is tedious to perform. The second method uses 4-nitro[2-^{3}H]phenyl α-D-galactopyranoside as a photoaffinity reagent.[42] In membrane vesicles of strain ML308-225, 25–50% of the incorporated label is specifically associated with the carrier. For membranes from

[41] We use a Hitachi Perkin–Elmer Spectrofluorometer, Model MPF3.
[42] G. J. Kaczorowski, G. LeBlanc, and H. R. Kaback, *Proc. Natl. Acad. Sci. U.S.A.* **77**, 6319 (1980).

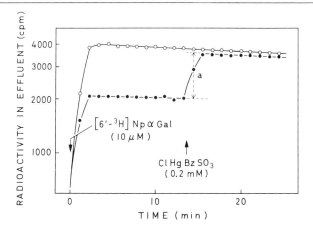

FIG. 2. Determination of substrate binding by flow dialysis. The binding of galactosides to the carrier is described by the law of mass action, which can be used to calculate the number of binding sites from a single flow dialysis. In general, the amount of NpαGal bound to the carrier (nanomoles of NpαGal per milligram of protein) under the experimental conditions is given by

$$\bar{v} = (C_g/C_p)(a/b) \qquad (1)$$

where C_g is the total concentration of ligand (μM), C_p is the protein concentration in the sample (mg/ml), a is the segment shown in the figure, and b is the radioactivity in the effluent after addition of ClHgBzSO$_3$. The free ligand concentration before addition of ClHgBzSO$_3$ (C_f) is given by

$$C_f = C_g[1 - (a/b)] \qquad (2)$$

The binding at saturation (nmol permease or NpαGal/mg protein, n, is then given by

$$n = \bar{v}[(K_D + C_f)/C_f] \qquad (3)$$

where K_D is the equilibrium dissociation constant (22 μM). The flow dialysis experiment above can be evaluated as follows.

At time $t = 0$, 5 μl of a 1 mM solution of [6'-^{3}H]NpαGal was added to a stirred suspension of cytoplasmic membrane vesicles from strain T206, which was not induced (O——O) or was induced (●——●) for the synthesis of the lactose carrier (both at 5 mg of protein per milliliter). After approximately 2 min, a constant level of radioactivity is detected in the effluent form the lower chamber of the flow dialysis cell. The slight decrease in radioactivity with time is due to the loss of [6'-^{3}H]NpαGal from the upper into the lower chamber. The lower level of radioactivity in the sample containing the lactose carrier corresponds to the lower concentration of free (unbound) galactoside due to binding of the substrate to the carrier. Addition of ClHgBzSO$_3$ (0.2 mM) to this vesicle suspension causes release of the bound galactoside reflected in the increase of the radioactivity in the effluent. Addition of ClHgBzSO$_3$ to the suspension of vesicles lacking the lactose carrier does not evoke any change in the radioactivity in the effluent.

After the addition of ClHgBzSO$_3$, both samples are characterized by the same level of radioactivity in the effluent, because the specific binding of NpαGal in the sample of carrier-containing vesicles has been completely inhibited. For this example, the total concentration of NpαGal must be corrected for loss due to dialysis. Extrapolating the curve for the carrier-containing vesicles after the addition of ClHgBzSO$_3$ to $t = 0$, the theoretical radioactivity in the effluent would be 3820 cpm, which corresponds to 10 μM [6'-^{3}H]NpαGal; when ClHgBzSO$_3$

overproducing strains, the method is highly specific.[22,43] The third method is based on the fact that permease substrates such as GalSGal protect an essential sulfhydryl moiety against attack by *N*-ethylmaleimide (MalNEt). Membranes are first treated with unlabeled MalNEt in the presence of substrate and subsequently with MalN[^{3}H]Et in the absence of substrate. The following protocol is a modified version[44] of the original procedure.[13,45]

Cytoplasmic membranes from strain T206 (3–4 nmol of substrate binding activity per milligram of protein) are washed[46] once in 50 m*M* potassium phosphate, pH 7.0, and resuspended in this buffer at a concentration of up to 8 mg of protein per milliliter. GalSGal (25 µl of an 0.2 *M* aqueous solution per milliliter of suspension) is added to the membranes. The mixture is briefly sonicated and incubated at 28°. After addition of 10 µl of 0.1 *M* MalNEt/ml, the suspension is again sonicated and left at 28° for 15 min. A freshly prepared solution of iodoacetamide (IAcNH$_2$, 50 µl/ml, 0.1 *M*) is added; after brief sonication, incubation is continued for 5 min. A freshly prepared solution of cysteine (75 µl/ml, 0.1 *M*, pH 7) is added, and the sample is sonicated at room temperature for 5 min. The membranes are then centrifuged, washed three times with 50 m*M* potassium phosphate, pH 6.3, and resuspended to a protein concentration of about 10 mg/ml.

The membrane suspension (1 ml) pretreated with MalNEt and IAcNH$_2$ is now added to 0.93 MBq MalN[1-^{3}H]Et (17.5 GBq/mmol, dried from pentane or benzene in a test tube), sonicated for 5 min, and incubated for 40 min at 28°. After addition of cysteine (50 µl/ml, 0.1 *M*, pH 7) and brief sonication, the sample is washed three times with 50 m*M* potassium phosphate, pH 6.3, or 10 m*M* Tris, pH 7.5, resuspended in 2 ml of this buffer, and stored in 50-µl aliquots at −20°.

The pretreated membranes retain 71 ± 5% of the original substrate

---

[43] A drawback of this method is that 4-nitro[2-^{3}H]phenyl α-D-galactoside is at present not commercially available.

[44] J. K. Wright, unpublished experiments.

[45] C. F. Fox, J. R. Carter, and E. P. Kennedy, *Proc. Natl. Acad. Sci. U.S.A.* **57**, 698 (1967).

[46] Throughout this procedure all centrifugations are for 90 min at 150,000 *g*; membrane pellets are dispersed by brief sonication in a bath sonicator, i.e., Sonorex RK 510 (Bandelin, Berlin FRG). "Washing" refers to centrifugation and resuspension.

---

was added, this theoretical value had decreased to 3610, so that the total concentration of galactoside at the time of ClHgBzSO$_3$ addition was 9.45 µ*M*. Addition of ClHgBzSO$_3$ to the sample containing the carrier causes an increase in radioactivity in the effluent of $a = 3610 - 2010 = 1600$ cpm, corresponding to the concentration of galactoside bound to the carrier. The total concentration of galactoside corresponds to 3610 cpm, so that the fraction of the total galactoside bound to the carrier before the addition of ClHgBzSO$_3$, (a/b) is $1600/3610 = 0.443$. Using values of $C_g = 9.45$ µ*M*, $C_p = 5.0$ mg/ml, and Eqs. 1–3 above, the number of binding sites for NpαGal (i.e., carrier molecules) is 4.3 nmol per milligram of protein.

binding activity. After separation of the membrane proteins on a 10.8% polyacrylamide gel,[47,47a] 80–90% of the radioactivity is associated with a single broad band of apparent $M_r = 31,000$, which is characteristic for lactose permease.

Purification and Reconstitution

For protein chemistry a purification of denatured, inactive permease is available[16,44] that relies on the tendency of this protein to aggregate. Two procedures for the purification of active permease have been described.[21-23, 48,48a] The method of Newman *et al.*[22] and Foster *et al.*[48a] uses a total envelope fraction from strain T206 prepared by passage of the cells through a French press. The crude membranes are extracted sequentially by urea and cholate and then solubilized in octyl β-D-glucopyranoside in the presence of additional *E. coli* phospholipids. Passage of this extract through a DEAE-Sepharose column yields a product that shows a single major band (M_r 33,000) on polyacrylamide gels and has an amino acid composition close to that expected from the DNA sequence of permease. Reconstitution of the purified permease by a dilution regimen yields vesicles that show countertransport activity, active transport in response to a potassium diffusion gradient in the presence of valinomycin, and lactose-induced proton movements. At present, a rigorous characterization of the purified permease in terms of specific binding activity and catalytic constants is lacking. In the following section, the procedures for the purification of inactive permease as well as the purification and reconstitution of active permease used in our laboratory are described (see also Addendum).

Purification of Inactive Permease[44]

Cytoplasmic membrane vesicles from strain T206 containing 100 Bq per milligram of total protein MalN[³H]Et-labeled permease as a tracer are diluted to a protein concentration of 1 mg/ml in 10 mM Tris-HCl, 1 mM EDTA, pH 8.8, briefly sonicated, and pelleted by centrifugation (90 min, 150,000 g, 4°). The pellet is suspended at a concentration of 12–13 mg/ml in 50 mM sodium phosphate–5 mM EDTA, pH 7.4. A 10% (w/w) solution of Triton X-100 is added to a final concentration of 0.2% at room temperature, and the sample is sonicated[46] for 10 min at $T \leq 20°$. After centrifugation (90 min, 150,000 g, 4°), the pellet is resuspended in 0.5 M NaCl,

[47] U. K. Laemmli and M. Favre, *J. Mol. Biol.* **80,** 575 (1973).
[47a] Samples should not be heated in "sample buffer" before application to polyacrylamide gels as this treatment causes irreversible aggregation of permease (cf. Ref. 18).
[48] M. J. Newman and T. H. Wilson, *J. Biol. Chem.* **255,** 10583 (1980).
[48a] D. L. Foster, M. L. Gracia, M. J. Newman, L. Patel, and H. R. Kaback, *Biochemistry* **21,** 5634 (1982).

50 mM sodium phosphate, 10 mM EDTA, 30% (v/v) glycerol, pH 7.5 (1 ml/5 mg of protein before the Triton treatment). The sample is sonicated at $T \leq 22°$ and then diluted with an equal volume of 0.2 M octyl β-D-glucoside dissolved in the above buffer. After an additional 5-min sonication the sample is kept at room temperature for 45 min. The now turbid sample is centrifuged (12,000 g, 15 min, 15°), and the brownish supernatant is carefully removed. The pellet is dissolved in 150–250 μl of 90 mM lithium dodecyl sulfate, briefly sonicated, and centrifuged (10,000 g, 10 min, 15°). The supernatant is applied to a 2.5 × 90 cm Ultrogel AcA-34 (LKB, France) column equilibrated with 10 mM sodium dodecyl sulfate, 0.4 M urea. Permease is eluted around fraction 70 (3 ml/fraction, flow rate 7 min/fraction).

Purification is followed by counting aliquots and by polyacrylamide gel electrophoresis (Fig. 3). The overall yield is 56% in terms of MalN[³H]Et-labeled permease (87% of the radioactivity is found in the pellet after the octyl glucoside extraction; this product may be sufficient for many purposes). The purity of the product is ≥ 90% as judged by gel electrophoresis and amino acid composition.[16] Comparison of the protein yield and the substrate binding activity of the starting vesicle preparation suggests a stoichiometry of carrier–substrate interaction of 1 : 1 for a subunit molecular weight of 46,500.[19]

Partial Purification and Reconstitution of Active Permease[21,23]

Solubilization. Cytoplasmic membrane vesicles from strain T206, containing 60 Bq MalN[³H]Et-labeled carrier per milligram of protein are washed[46] in 15 mM Tris, 0.1 mM dithiothreitol, 1.0 mM sodium EDTA, pH 7.5, and resuspended in this buffer by bath sonication. The suspension (4 ml, 5 mg of protein per milliliter) is stirred on a Vortex mixer, and 4 ml of freshly prepared dodecyl O-β-D-maltoside (DodOMalt, Calbiochem, Giessen, FRG; 15 mg/ml in 15 mM Tris-HCl, 0.1 mM DTT, 0.05 mM sodium EDTA, pH 7.5) are rapidly added (3–4 sec) at room temperature. The sample is briefly sonicated and centrifuged for 30 min at 150,000 g ($T = 10°$).[49] The clear, brown supernatant will be referred to as the DodOMalt extract.

ECTEOLA Column Chromatography. ECTEOLA (prepared by the treatment of alkaline cellulose with epichlorhydrin and triethanolamine; ECTEOLA 23, Servacel; Serva, Heidelberg, FRG) in H$_2$O is titrated to pH 7.5 with NaOH and allowed to swell overnight at 2–4°. In order to prevent channeling and to obtain reproducible flow rates, the resin is dispersed by

---

[49] Since DodOMalt dissolves the membrane completely, centrifugation for 15 min at 12,000 g is sufficient.

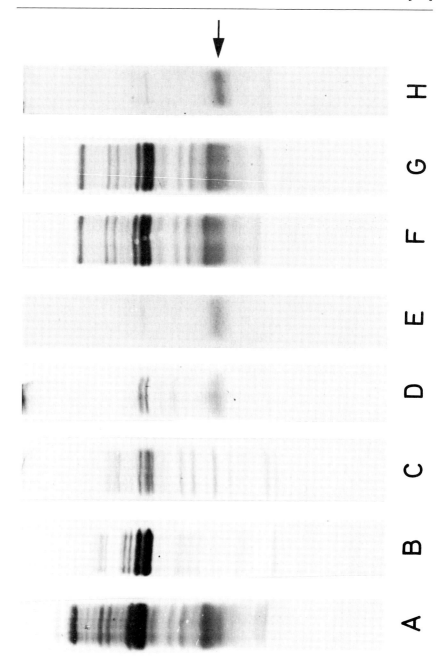

magnetically stirring the suspension of ECTEOLA in 15 mM Tris, 1 mM EDTA for 1 hr in a beaker. The pH is adjusted to 7.5 with NaOH as required. Fine particles are removed by suspending the ECTEOLA twice in 15 mM Tris, 1 mM EDTA, pH 7.5, and twice in 15 mM Tris, 1 mM EDTA, 20 mg of glycerol per milliliter, pH 7.5. The pH in each case is adjusted to 7.5, if necessary, the resin is allowed to settle, and the supernatant is decanted. The resin should be washed in the glycerol-containing buffer only on the day of use to avoid bacterial growth in the slurry. A 12 × 2 cm (38 ml) column is poured and equilibrated with 10 ml of freshly prepared 20 mg/ml glycerol, 15 mM Tris, 1 mM EDTA, 0.1 mM DTT, pH 7.5; then 7 ml of freshly prepared 1 mg/ml DodOMalt, 20 mg/ml glycerol, 15 mM Tris, 1 mM EDTA, 0.1 mM DTT, pH 7.5, at about 20° are allowed to run into the bed. Thereafter, the DodOMalt extract (8 ml; 20 mg of protein) is applied to the ECTEOLA column at room temperature and washed into the bed with 1–2 ml of the previous buffer. The column is developed with this same buffer, and fractions of 3–4 ml per 4 min are collected. The lactose carrier appears in 4–5 fractions around fraction 8. The appearance of the lactose carrier can be assessed most rapidly and conveniently by monitoring the absorbance of the effluent at 280 nm. This rises to 0.5–1.0 in the peak fraction. The fractions with $A_{280} > 0.1$ are collected and made 10 mM in sodium EDTA and 1 mM in DTT (see below).[50]

The ECTEOLA column can be regenerated by removing the bound proteins with 0.5 M NaCl, 0.1% Triton X-100 and subsequently washing with 10 mM Tris, pH 7.5, until the absorbance of the effluent at 280 nm sinks to zero. However, we find it more practical simply to prepare new columns from fresh ECTEOLA.

Reconstitution. Sodium EDTA (final concentration 10 mM) and DTT (final concentration 1 mM) are added to the DodOMalt extract and thereafter 20 mg of solid dodecyl polyoxyethylene ($n = 9.5$) ether (DodO(EtO)$_{10}$H, Lubrol PX, Sigma, Munich, FRG; i.e., 1 mg/mg protein). The sample is then briefly sonicated. Alternatively, the combined permease-containing fractions from the ECTEOLA column are similarly supplemented with

---

[50] The method for the purification of the permease can be scaled down. As little as 150 μl of the DodOMalt extract can be loaded on a 1–2-ml ECTEOLA column (e.g., in a Pasteur pipette), which may make this method useful in screening large numbers of samples.

FIG. 3. Polyacrylamide gel electrophoresis[47a] of fractions throughout the purification of inactive (lanes A–E) and active (lanes F–H) permease. A, cytoplasmic membrane from strain T206; B, supernatant after treatment with Triton X-100; C, supernatant after octyl β-D-glucoside treatment; D. pellet after octyl β-D-glucoside treatment; E, purified permease after Ultrogel AcA-34 column chromatography; F, cytoplasmic membrane from strain T206; G, DodOMalt extract; H, partially purified permease after ECTEOLA-column chromatography.

EDTA (10 mM), DTT (1 mM), and last $DoDO(EtO)_{10}H$ (1 mg/mg protein in the initial DodOMalt extract).

Eight milliliters of either preparation are added to 100 mg of *E. coli* phospholipids.[51] The sample is sonicated[46] for 5 – 10 min at T 20°, stirred with a glass rod until clear, and applied at room temperature to a styrene – divinylbenzene copolymer column (BioBeads SM2, 20 – 50 mesh; Bio-Rad Laboratories, Munich, FRG, 2.5 × 20 cm, 100 ml) equilibrated with 10 mM Tris, 10 mM EDTA, 25 mM NaCl, 1.3 M glycerol, pH 7.4, and covered by a plastic net to prohibit floating of the beads. Fractions of 6 ml are collected (flow rate 1 ml/min). Turbid fractions are combined and diluted 1 : 1 with water; the vesicles are pelleted by centrifugation (2 hr, 150,000 g, 4°).[52] The vesicles are suspended by sonication in 20 ml of 50 mM potassium – hydrogen phosphate buffer, pH 6.6, and centrifuged for 2 hr at 150,000 g, 4°. Finally, the vesicles are resuspended in about 1 ml of this buffer, and substrate binding in the presence of 10 μM [6'-³H]NpαGal is measured. The purification is followed by polyacrylamide gel electrophoresis, (Fig. 3), by substrate binding, and the yield of both Mal N[³H]Et-labeled carrier and protein (Table II). Permease corresponds to a broad band at $M_r = 31,000$ (cf. Fig. 3H). There are traces of impurities in the region of $M_r = 70,000$. Table II shows that the carrier is purified in a yield of 39 ± 5% in terms of Mal N[³H]Et-labeled protein and 32 ± 3% in terms of binding sites for NpαGal, indicating that at least 83% of the carrier molecules are active in binding substrate.

Transport Activity of Reconstituted Vesicles. A convenient, qualitative optical test for carrier-mediated transport is the lactose-driven countertransport of a fluorescent galactoside, such as DnsHxSGal. The reconstitution sample is taken up in 10 – 20 ml of 35 mM potassium – hydrogen phosphate, 40 mM lactose, pH 6.6, and sonicated (the sample from the binding experiment may be used for this experiment after washing in 40 ml of buffer containing 0.25 mM 2-mercaptoethanol to release $ClHgBzSO_3$).

---

[51] Lipids are extracted from cells of strain ML308-225 with chloroform – methanol according to G. F. Ames [*J. Bacteriol.* **95**, 833 (1968)], and a total phospholipid fraction is prepared by silica gel chromatography [cf. P. Overath and H. Träuble, *Biochemistry* **12**, 2625 (1973)] and stored as benzene solution at −20°. For the reconstitution, the desired amount of lipid is transferred to a glass vessel and dried to a thin film under a stream of N_2. The lipid is taken up in diethyl ether, dried to a film with a stream of N_2 (about 2 mg of lipid per square centimeter,) and kept for 1 hr at 37°. To facilitate dispersion of the lipid in the reconstitution sample and to minimize sonication time, the lipid was loosened from the wall of the vessel with a glass rod.

[52] To regenerate the column, the BioBeads are first washed with 50 ml of H_2O. Bound detergent is eluted with 150 ml of 50% (v/v) methanol. The column is further washed with 100 ml of H_2O and stored in this form. Immediately before use, the column is washed slowly with 100 ml of equilibration buffer.

TABLE II
PARTIAL PURIFICATION AND RECONSTITUTION OF THE LACTOSE CARRIER

| | Lactose carrier | | |
Sample	MalN[³H] Et-carrier (%)	NpαGal binding sites (%)	Protein (%)
Cytoplasmic-membrane vesicles[a]	100	100	100[b]
DodOMalt extract	89 ± 7	—	91 ± 5
ECTEOLA fractions	56 ± 3	—	10 ± 2
Reconstitution sample	39 ± 5	32 ± 3	8 ± 1

[a] The original sample contained 61 nmol of lactose carrier determined by NpαGal binding, 1.2 KBq MalN[³H]Et-labeled lactose carrier and 20 mg of protein.

[b] Protein was determined by the method of G. L. Peterson, *Anal. Biochem.* **83**, 346 (1977). For the reconstitution sample, a standard curve in the presence of *Escherichia coli* phospholipids was determined.

This sample is frozen at − 18° and then allowed to thaw slowly at room temperature. This freeze–thaw cycle causes the small vesicles obtained from the BioBead column to fuse. Countertransport is more pronounced in larger vesicles.[21] The large vesicles are collected by centrifugation for 20 min at 37,000 g, 10°. The pellet is suspended in 1 ml of 40 mM lactose, 35 mM potassium-hydrogen phosphate, pH 6.6, by stirring with a glass rod and by gently drawing the suspension up in a pipette several times. To 2 ml of 50 mM potassium-hydrogen phosphate, pH 6.6, containing 6 μM DnsHxSGal 5 – 10 μl of vesicle suspension is added. The time-course of the transient increase in the fluorescence of DnsHxSGal (excited at 350 nm, monitored at 500 nm) corresponds to the transient accumulation of DnsHxSGal due to the lactose gradient. Given that all galactosides share a common binding site, the ability of the reconstituted lactose carrier to cause a transient accumulation of DnsHxSGal is merely equated with the ability of the system to catalyze facilitated diffusion. Countertransport is suppressed when the vesicles are diluted into 40 mM lactose, 35 mM potassium-hydrogen phosphate, pH 6.6 containing 6 μM DnsHxSGal.

The partial purification of the lactose carrier described here is attractive for several reasons. The procedure is simple and rapid, and the yield is relatively high. The partially purified permease is in a form amenable to further study and is active after reconstitution with respect to galactoside binding and countertransport. Although the quantitation of the reconstituted carrier in terms of the rates of facilitated and active transport still poses methodological difficulties, substrate binding is a convenient and reliable assay in the reconstituted system.

Addendum

The purification of the lactose carrier and measurement of active transport of the reconstituted carrier have recently been improved by the following additions and modifications.[23]

Purification. Cytoplasmic membrane vesicles are first washed twice in 0.25 M sodium 5-sulfosalicylate, 10 mM Tris, 10 mM Na-EDTA, 1 mM DTT, pH 7.4. This treatment removes the impurities that appear in the M_r 70,000 region on the gels. The vesicles are washed in 15 mM Tris, 1 mM Na-EDTA, pH 7.5, twice to lower the ionic strength before proceeding with the extraction and the ECTEOLA column. The traces of residual impurities are removed by an immunoabsorbent. Cytoplasmic membrane vesicles lacking the permease from strain T184 (Table I) or noninduced strain T206 are extracted as described above and applied to the ECTEOLA column. The fractions containing the impurities were mixed with complete Freund's adjuvant and used to immunize rabbits. The IgG fraction is isolated from the serum and coupled to cyanogen-bromide activated agarose. The permease-containing fractions from the ECTEOLA column are gently shaken with the absorbant for 1–3 hr in the presence of 1 mM DTT.

Reconstitution. The previous procedure for the preparation of an *E. coli* phospholipid dispersion is modified by the addition of Lubrol PX and valinomycin. For a purification starting from 20 mg of cytoplasmic membrane vesicles a typical preparation is 1 ml *E. coli* phospholipids, (100 mg/ml, in benzene) 40 μl 1 mM valinomycin in benzene, and 20 mg Lubrol PX. The benzene is removed under N_2, the residue is dissolved in diethyl ether and dried under N_2 to a thin film by rotating the vessel (a 40 ml test tube is convenient). The sample is held for 1 hr at 37° to remove traces of solvents. The ECTEOLA fractions are made 1 mM in DTT, 10 mM in Na EDTA, and added to the lipid–detergent–valinomycin mixture. Reconstitution follows as described.

Active Transport. The proteoliposomes from the Bio-Bead column are diluted with one volume of 0.4 M potassium phosphate, pH 7.0, and sonicated for 15 min at 20°. Vesicles are collected as described. The pellet is frozen at −20° and subsequently allowed to thaw at room temperature in the air. The vesicles are suspended in 0.2 M potassium phosphate, pH 7.0, sonicated briefly (10–30 sec) to disperse flocculent material, kept for 2 hr on ice to equilibrate the inner compartment, and frozen for a second time at −20°. The sample is thawed at room temperature, dispersed by drawing up into a pipette several times, and the large vesicles (cf. Ref. 21) are harvested by centrifugation for 10 min at 13,000 g. Active transport can be achieved by diluting this vesicle suspension, made 1 mM in $MgSO_4$, 100-fold in 0.2 M sodium phosphate, 1 mM $MgSO_4$, pH 6.0, containing substrate. The stop buffer must contain 1–2 mg/ml protamin sulfate to cause the vesicles to flocculate and 2 mM HgCl$_2$ to inhibit the carrier. NpαGal binding is measured on a part of the sample so that the maximal velocity of transport may be expressed as a turnover number: nanomole of substrate transported per second per nanomole of lactose carrier. By following this modified procedure, the maximal velocity of lactose active transport has been increased 50-fold from a value of 0.085 sec^{-1} in the original reconstitution product from the previous protocol to 4.3 sec^{-1} in the proteoliposomes obtained above. While this figure is similar to the turnover number in vesicles (2.9 sec^{-1}; Ref. 11), the turnover number for lactose in EDTA-treated ML308-225 cells is considerably higher, 50 sec^{-1}. At present the reason for the discrepancy between vesicles and cells has not been elucidated.

The 50-fold increase in the maximal velocity of the reconstituted lactose carrier due to changes in the reconstitution procedure might be interpreted as follows: First, the mixing of the valinomycin with the phospholipids in benzene produces a homogeneous mixture. Every vesicle should now contain valinomycin. Commonly, valinomycin is added to a suspension of proteoliposomes which may not lead to a homogeneous distribution into all vesicles. Obviously, the transport activity of lactose carrier in liposomes containing little or no valinomycin would be low. Second, by adding Lubrol PX to the lipid, the carrier is presented with micellar lipid. This might facilitate a better mixing of lactose carrier, phospholipid, and detergent leading to a homogeneous product with respect to the composition of the individual vesicles after reconstitution. Third, as shown previously, the efficiency of carrier-catalyzed countertransport in reconstituted proteoliposomes is enhanced by a freeze-thaw cycle.[21] This effect was tentatively ascribed to the increase in the size of the vesicles. In large vesicles, driving gradients are more stable and the product does not accumulate as rapidly. In the modified protocol, two freeze-thaw cycles are included. Additionally, the suspension is enriched in the largest vesicles by centrifuging at only 13,000 g and discarding the supernatant. Whereas this last step results in loss of carrier in the smaller vesicles (the carrier in the supernatant exhibits normal NpαGal binding), the larger vesicles are more efficient in transport. Fourth, since liposomes are highly impermeable to many ionic compounds, e.g., potassium phosphate, equilibration of the internal space can be best achieved by sonication of the vesicles at some stage and incubation at 0°.

[18] Cloning of the Structural Genes of the Escherichia coli Adenosinetriphosphatase Complex

By David A. Jans and Frank Gibson

The structural genes (*unc* genes[1]) encoding the eight proteins of the adenosinetriphosphatase complex of *Escherichia coli* are arranged in an operon at about 83.5 min on the *E. coli* chromosome (Fig. 1). Five of these genes, *uncA, D, G, H,* and *C,* code for the α, β, γ, δ, and ϵ subunits of the soluble portion of the complex (F_1-ATPase), respectively, and the remaining three genes, *uncB, E,* and *F,* encode the insoluble portion (F_0).

A number of approaches to the problem of cloning the structural genes of the *unc* operon have been used. Downie *et al.*[2] used a combination of methods to clone the two restriction endonuclease *Hin*dIII fragments specifying these genes. An *E. coli Hin*dIII gene pool inserted into the F-mobilizable plasmid pGM706 was used to complement a mutation in the *uncB* gene selecting for growth on succinate medium. The *Hin*dIII fragment causing this complementation was recloned into plasmid pACYC184 to give plasmid pAN51 carrying the *uncB, E, F, H,* and *A* genes. The entire operon was not derived from the plasmid pools, and an RP4 hybrid plasmid carrying the structural genes of the operon was constructed, using Mu phage-mediated transposition of the chromosomal *unc* genes onto RP4. Plasmid pAN45 (Fig. 2) was ultimately derived from the RP4 hybrid plasmid and plasmid pAN51.

Other groups used specialized λ transducing phages as enriched sources of the *unc* genes from which to clone into multicopy plasmid vectors, using restriction endonuclease *Hin*dIII as well as other enzymes. Futai's group,[3] as well as that of Walker,[4] used λ *asn* 5, whereas von Meyenburg and co-workers[5] used λ *asn*-105 as the source of *unc* DNA. A detailed restriction

---

[1] The following designations have been suggested for the genes coding for the ATPase complex of *E. coli: unc:* J. D. Butlin, G. B. Cox, and F. Gibson, *Biochem. J.* **124,** 75 (1971). *bcf:* B. P. Rosen, R. N. Brey, and S. M. Hasan, *J. Bacteriol.* **134,** 1030 (1978). *atp:* K. von Meyenburg and F. G. Hansen, *ICN–UCLA Symp. Mol. Cell. Biol.* **19,** 137 (1980). *pap:* H. Kanazawa, K. Mabuchi, T. Kayano, F. Tamura, and M. Futai, *Biochem. Biophys. Res. Commun.* **100,** 219 (1981).

[2] J. A. Downie, L. Langman, G. B. Cox, C. Yanofsky, and F. Gibson, *J. Bacteriol.* **143,** 8 (1980).

[3] H. Kanazawa, F. Tamura, K. Mabuchi, T. Miki, and M. Futai, *Proc. Natl. Acad. Sci. U.S.A.* **77,** 7005 (1980).

[4] N. J. Gay and J. E. Walker, *Nucleic Acids Res.* **9,** 2187 (1981).

[5] J. Nielsen, F. G. Hansen, J. Hoppe, P. Friedl, and K. von Meyenburg, *Mol. Gen. Genet.* **184,** 33 (1981).

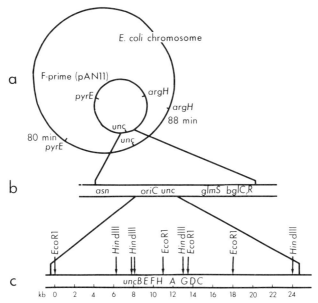

FIG. 1. Schematic representation (not to scale) of (a) the partial diploid strain AN862, (b) the genes close to the *unc* genes, and (c) the *Hind*III and *Eco*RI restriction sites around the structural genes of the *unc* operon. It has been shown[21,22] that there is an open reading frame between the *uncB* gene and the putative *unc* operator, but this does not code for one of the structural genes for the known subunits of the ATPase complex.

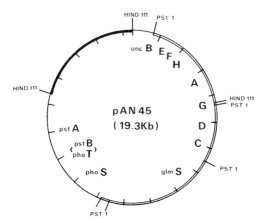

FIG. 2. Map of plasmid pAN45 showing *Hind*III and *Pst*I restriction sites. Each division represents 1 kilobase. The heavy line represents the vector portion of the plasmid.

map of the *unc* region, mapping at ~ 83.5 min on the *E. coli* chromosome, is presented in Fig. 1.

A simple approach to cloning the *unc* structural genes uses an F' plasmid isolation as the enriched DNA source for insertion of the operon into a multicopy plasmid vector using standard restriction and ligation procedures. A partial diploid (strain AN862) that duplicates the entire *unc* operon of *E. coli* is used as a source of the DNA to be cloned. The F' plasmid in strain AN862 can be isolated and used as an enriched source of the *unc* genes. The Hansen and Olsen[6] Triton lysis–alkali pulse plasmid preparation method can be used, but because the F' involved is large, and shears easily, the yield of DNA is low (although the *unc* operon has been successfully cloned using this method[7]). However, a rapid isolation method that does not involve cesium chloride gradients can be used if the problem of contamination of plasmid DNA by chromosomal fragments is negated by the use of the appropriate partial diploids. The procedure described is modified from that of Portnoy and White as described by Crosa and Falkow.[8]

After restriction of the isolated F' plasmid and ligation with the cloning vector, the ligation mix can be used to transform an appropriate recipient strain, and selection is made for the antibiotic resistance determinant of the plasmid vector and for the ability to grow on succinate minimal medium (the Unc+ phenotype). All, or parts of, the *unc* operon can be selected for as cloned inserts by using the appropriate *recA* derivatives of *unc* mutant strains. Transformants can be tested for maintenance of a plasmid of the correct size by using miniscreen analysis. The plasmid DNA is then isolated for restriction analysis, to confirm the identity of the cloned DNA and the orientation of the insert in the cloning vector. Transformation into other *unc* mutant strains, selecting plasmid antibiotic resistance, and subsequent screening of the transformed strains for growth on succinate minimal medium, can verify that the desired structural genes of the *unc* operon have been successfully cloned.

F' Plasmid Isolation

Growth Media. Minimal medium comprises 60 mM K$_2$HPO$_4$, 40 mM NaH$_2$PO$_4$ · 2H$_2$O, and 15 mM (NH$_4$)$_2$SO$_4$, to which 1 ml of a trace-element solution (14 mM ZnSO$_4$, 1 mM MnSO$_4$, 4.7 mM H$_3$BO$_3$, 0.7 mM CaSO$_4$,

[6] J. B. Hansen and R. H. Olsen, *J. Bacteriol.* **135**, 227 (1978).
[7] D. A. Jans, unpublished.
[8] J. H. Crosa and S. Falkow, *in* "Manual of Methods for General Bacteriology" (P. Gerhardt, R. G. E. Murray, R. N. Costilow, E. W. Nester, W. A. Wood, N. R. Krieg, and G. B. Phillips, eds.), p. 269. Am. Soc. Microbiol., Washington, D.C., 1981.

2.5 mM CaCl$_2$ and 1.8 mM FeCl$_3$) is added per liter. After sterilization, 1 ml of a sterile 1 M MgSO$_4$ solution is added per liter, together with other growth supplements, as required, to the following final concentrations: 20 mM D-glucose, 0.2 μM thiamin hydrochloride, 40 μM 2,3-dihydroxybenzoate, 0.8 mM L-arginine hydrochloride, 0.2 mM uracil, 0.15 mM adenine hydrochloride, 0.3 mM L-valine, 0.8 mM L-threonine, 0.8 mM L-leucine, 0.3 mM L-isoleucine. All media are supplemented with 0.05% acid-hydrolyzed casein (Difco, Detroit, Michigan).

Solid media are prepared by addition of 2% (w/v) agar to the medium described above. In agar plates containing succinate as the sole carbon source, glucose was replaced by sodium succinate (30 mM).

Materials

Organism: *E. coli* strain AN862 (see the table[1,9–11]), but any strain, carrying the plasmid F111[12] would be suitable.

TE: 0.05 M Tris-HCl, 0.01 M EDTA, pH 8.0

Lysis buffer: 4% w/v sodium dodecyl sulfate in TE, pH 12.4[13]

Tris-HCl, 2.0 M, pH 4.7

NaCl, 5 M

Isopropanol

Distilled phenol, equilibrated with 0.01 M Tris, pH 8.0

Chloroform

Diethyl ether

Sodium acetate, 3 M

Ethanol, absolute

Ethanol, 70% (v/v) plus 0.1 mM EDTA

Method

1. Grow 100 ml of culture of strain AN862 overnight at 37° on a shaker bath. The absence of isoleucine-valine, arginine, and uracil in the growth medium (20 mM D-glucose, 0.05% casamino acids, 0.15 mM adenine hydrochloride) selects for maintenance of the F-plasmid pAN11, covering the *pyrE$^+$-unc$^+$-ilvC$^+$-argH$^+$* region of the *E. coli* chromosome (about 81–89 min; see Fig. 1).
2. Spin down the cells at 6000 rpm for 10 min at 4°.
3. Resuspend the pellet in 100 ml of TE.
4. Spin at 6000 rpm, 10 min, 4°.

[9] B. J. Bachmann, K. B. Low, and A. L. Taylor, *Bacteriol. Rev.* **40**, 116 (1976).

[10] R. P. Novick, R. C. Clowes, S. N. Cohen, R. Curtiss, N. Datta, and S. Falkow, *Bacteriol. Rev.* **40**, 168 (1976).

[11] G. B. Cox, J. A. Downie, F. Gibson, and J. Radik, *Biochem. J.* **170**, 593 (1978).

[12] K. B. Low, *Bacteriol. Rev.* **36**, 587 (1972).

[13] The closer the pH of the lysis buffer to 12.5 the better, but a pH of greater than 12.5 irreversibly denatures the plasmid DNA.

5. Resuspend the pellet in 2 ml of TE.
6. Add 30 ml of lysis buffer (freshly prepared) to the cell suspension, mixing gently but well. Leave for 25 min at 37°.
7. Add 2 ml of 2 M Tris-HCl(pH 4.7), inverting until a change in viscosity is perceptible. This step returns the pH to ~ 8.0.
8. Add 12 ml of 5 M NaCl to bring the salt concentration to 1 M. Mix by gentle inversion and leave on ice for at least 1 hr. Although about 4 hr are required for complete removal of chromosomal DNA, maximal plasmid purity is not essential, and 1 hr on ice is sufficient.
9. Spin at 12,500 rpm for 10 min.
10. Decant the supernatant into a new tube and add 30 ml of isopropanol. Keep at $-20°$ for 30 min.
11. Spin at 12,500 rpm for 5 min and then, discarding the supernatant, dry the pellet under vacuum.
12. Resuspend the pellet in 0.5 ml of TE and transfer to a microfuge tube.
13. Add 0.5 ml of distilled phenol. Mix well.
14. Add 0.5 ml of chloroform. Mix well.
15. Spin to separate phases.
16. Remove upper aqueous phase into a new microfuge tube with a Pasteur pipette.
17. Continue phenol extractions of aqueous layer, filling the tube with phenol each time, until the aqueous phase is clear.
18. Fill tube with diethyl ether and mix.
19. Spin for 30 sec in microcentrifuge.
20. Remove the upper organic phase and discard it.
21. Repeat the ether extractions (steps 18–20) twice more.
22. Remove half of the aqueous layer to a new microfuge tube.
23. Add 30 μl of 3 M sodium acetate and 1 ml of ice-cold absolute ethanol to both tubes and mix. Keep at $-70°$ (Dry Ice and 95% ethanol) for 5 min.
24. Spin for 5 min in the microcentrifuge.
25. Discard the supernatant and add 1 ml of ice cold 70% ethanol + 0.1 mM EDTA, taking care not to disturb the pellet. Keep for 5 min at $-70°$. This is intended as a wash step to remove metal ions, but DNA can dissolve in 70% EtOH + 0.1 mM EDTA if the pellet is resuspended.
26. Spin for 5 min in the microcentrifuge.
27. Discard the supernatant and add 1 ml of ice-cold absolute ethanol, again being careful to avoid resuspension of the pellet. Keep at $-70°$ for 5 min.
28. Spin for 5 min in the microcentrifuge.
29. Discard the supernatant and dry the pellet under vacuum.

Restriction and Ligation

Materials

Restriction endonuclease *Hin*dIII (New England BioLabs, Beverly, Massachusetts)

T4 DNA-ligase: New England BioLabs, Beverly, Massachusetts

Plasmid pACYC184[14] DNA solution: about 0.05 µg/ml

*Hin*dIII buffer, 10 × strength: 0.6 M NaCl, 70 mM MgCl$_2$, 0.1 M Tris-HCl, pH 7.4, and 1 mg of bovine serum albumin (sterile) per milliliter

MgCl$_2$ 130 mM (sterile)

ATP 20 mM (sterile)

DL-Dithiothreitol (DTT), 200 mM (Sigma Chemical Co., St. Louis, Missouri) (sterile)

Tris, 1 M (sterile)

Water (sterile)

Method. All steps should be performed aseptically.

1. Resuspend pellets individually in 18 µl of sterile water and pool in one microfuge tube.
2. Add 15 µl of plasmid pACYC184 solution (approximately 1 µg of DNA), 6 µl of 10× *Hin*dIII buffer, and 3 µl of restriction endonuclease *Hin*dIII (about 10 units)
3. Incubate overnight at 37°.
4. Hold at 65° for 15 min to denature the *Hin*dIII restriction endonuclease.
5. Cool slowly to room temperature.
6. Add 7 µl of sterile water, 5 µl of 130 mM MgCl$_2$, 5 µl of 20 mM ATP, 5 µl of 200 mM DTT, 5 µl of 1 M Tris, and 3 µl of T4 DNA-ligase.
7. Ligate by incubation overnight at 12°.

Transformation and Selection for *unc* Inserts

Materials

Organism: *E. coli* strain AN1363 (*uncG, recA*) (see the table)

Glu-Luria broth: 1.5 ml of 2 M D-glucose in Luria broth[15] (sterile)

MgSO$_4$, 10 mM (sterile)

CaCl$_2$, 50 mM (sterile)

Succinate minimal agar plates, 30 mM, with 100 µg of chloramphenicol per milliliter

Minimal medium (sterile)

---

[14] A. C. Y. Chang and S. N. Cohen, *J. Bacteriol.* **134,** 1141 (1978).

[15] S. E. Luria and J. W. Burrous, *J. Bacteriol.* **74,** 461 (1957).

Bacterial strain[a]	Relevant genetic loci[b]	Reference footnote
AN862	F' (pAN11-*arg*+, *unc*+, *ilv*+, *pyrE*+), *argH, ilvC, pyrE, purE, recA*	11
AN1363	*uncG428, thr⁻, leu⁻, rk⁻, mk*+, *recA srl*∷Tn10	1

[a] These strains may be obtained from Dr. B. J. Bachmann, *E. coli* Genetic Stock Center, Department of Human Genetics, Yale University School of Medicine, 310 Cedar Street, New Haven, Connecticut 06510.

[b] Gene designations are as listed by Bachmann *et al.*,[9] and plasmid nomenclature is that of Novick *et al.*[10]

Method. This procedure is a slight modification of the standard DNA transformation method of Lederberg and Cohen.[16]

1. Inoculate 3–5 colonies from an agar plate growth of strain AN1363 (*uncG, recA*) into 20 ml of Glu-Luria broth, or alternatively add 1 ml of an overnight Glu-Luria culture into 20 ml of Glu-Luria broth. Strain AN1363 (*uncG, recA*) is used in order to select a plasmid containing the full complement of *unc* structural genes. As the restriction endonuclease *Hin*dIII cuts the *uncG* gene, the two *Hin*dIII fragments comprising the *unc* structural genes (Fig. 2) are thus selected for as transformants on succinate minimal medium.[17]
2. Grow to Klett 100 (about 1.3×10^8 cells/ml) at 37° in a shaker.
3. Chill the cells for 5 min at 4°.
4. Spin down the cells at 4000 rpm for 10 min at 4°.
5. Resuspend the cells in 20 ml of ice-cold 10 mM MgSO$_4$ and keep on ice for 30 min.
6. Spin at 4000 rpm, 10 min, 4°.
7. Resuspend the cells in 10 ml of ice-cold 50 mM CaCl$_2$ and keep on ice for 15 min.
8. Spin at 4000 rpm, 10 min, 4°.

[16] E. M. Lederberg and S. N. Cohen, *J. Bacteriol.* **119**, 1072 (1974).

[17] Alternatively, strain AN887 [F. Gibson, J. A. Downie, G. B. Cox, and J. Radik, *J. Bacteriol.* **134**, 728 (1978)], a polarity mutant in which all of the structural genes of the *unc* operon are affected, could be used to achieve the same end result. Strain AN1363 is preferred because it transforms at an approximately 10-fold higher frequency. A further alternative would be to transform appropriate *recA* derivatives of *unc* mutant strains affected in genes either side of the *uncG* gene with portions of the ligation mix. Either *Hin*dIII fragment of the operon can be individually selected, and subsequently *Hin*dIII restricted and ligated to yield a plasmid analogous to pAN45.

9. Resuspend the cells in 1 ml of ice-cold 50 mM CaCl$_2$, using a cold pipette.
10. Add 0.25 ml of cell suspension to sterile 5-ml tubes. (Include a control with no DNA.)
11. Add ligation mix (90 μl).[18]
12. Place on ice for 15 min, mixing gently at intervals.
13. Transfer to a beaker at 42° for 2.5 min.
14. Remove to a rack at room temperature and let stand for 10 min.
15. Add 5 ml of Glu-Luria broth to each tube and transfer the contents to sterile flask. Incubate at 37° for 90 min, shaking slowly.
16. Spin down the cells at 4000 rpm for 10 min at room temperature.
17. Resuspend the pellet in 0.5 ml of minimal medium and then plate out 0.1 ml/plate on succinate minimal agar plus chloramphenicol. Appropriate controls of untransformed AN1363 cells on glucose minimal agar plus chloramphenicol, and succinate minimal agar, as well as on succinate minimal agar plus chloramphenicol plates, should be included.
18. Incubate the plates 48 hr at 37°.

Single-Colony Lysate Miniscreen of Transformants

Materials

Plates: succinate minimal agar plus 100 μg of chloramphenicol per milliliter

Agarose, type 2 (Sigma Chemical Co., St. Louis, Missouri)

Agarose gel electrophoresis buffer: 40 mM Tris acetae (pH 7.8), 20 mM sodium acetate, and 10 mM sodium EDTA

Ethidium bromide solution, 10 mg/ml

Plasmid size marker DNA solutions; plasmid pACYC184 and another marker about 20 kilobases in size

SDS-xylene cyanole mix: 50% (v/v) glycerol, 5% (w/v) sodium dodecyl sulfate, 0.2% (w/v) xylene cyanole, 0.1 M EDTA

Method

1. Purify the transformants on succinate minimal medium plus chloramphenicol.
2. Sector the transformants on the same medium; incubate for 48 hr at 37°.
3. Suspend a full loop of bacterial culture of each transformant to be assayed in 75 μl of agarose gel electrophoresis buffer.

---

[18] It is advisable to retain at least 10 μl of the ligation mix. If no transformants are obtained, this can be used to check the DNA and ligation efficiency using agarose gel electrophoresis.

4. Add 40 μl of SDS–xylene cyanole mix (premelted at 37°).
5. Vortex for 10 sec, and then incubate at 65° for 20 min.
6. Vortex for 30 sec.
7. Load about 70 μl on a 1% agarose gel with appropriate plasmid size markers.
8. Run samples into the gel for 10 min at 100 mA and then run for 3–4 hr at 120 mA, or overnight at 30 mA.
9. Stain the gel in 500 ml of distilled water plus 100 μl of ethidium bromide at room temperature for 20 min.
10. Observe the gel under ultraviolet light (and photograph if required using Polaroid type 55 P/N film and red filter). Plasmids bearing the appropriate HindIII inserts should be about 20 kilobases in size.

Isolation of Multicopy Plasmid

Materials
Spectinomycin hydrochloride: purchased as Trobicin (Upjohn Co., Kalamazoo, Michigan)
25% Sucrose, 50 mM Tris, 20 μM dodecylamide: Add 0.1 ml of 20 mM dodecylamide to 100 ml, as required.
Lysozyme solution: 20 mg/ml in 0.25 M Tris, pH 8.0
EDTA, 0.5 M, pH 8.0
Triton buffer: 0.3% (w/v) Triton X-100, 50 mM Tris, 62.5 mM EDTA, pH 8.0
Cesium chloride (British Drug Houses Chemicals Ltd., Poole, England)
Protease from *Streptomyces griseus,* type XIV (Sigma Chemical Co., St. Louis, Missouri), 20 mg/ml in TE, autodigested at 80° for 10 min
Paraffin
Isopropanol: NaCl saturated; stored over TE
TE/10: TE diluted one in ten.
Method[19]

1. Inoculate culture of transformant into 100 ml of glucose minimal broth containing 0.5% casamino acids and 100 μg of chloramphenicol per milliliter to retain the multicopy plasmid.
2. Grow to Klett 120 (about 3 × 10⁸ cells/ml) on a shaker at 37° and amplify by addition of 0.03 g of solid spectinomycin.
3. Shake overnight at 37°.
4. Spin down the cells at 6000 rpm for 10 min at 4°.

---

[19] The quantity of plasmid DNA isolated by this method is sufficient to carry out restriction analysis and DNA transformation experiments. If larger amounts are required, as, for example, in DNA sequencing, the method of E. Selker, K. Brown, and C. Yanofsky [*J. Bacteriol.* **129**, 388 (1977)] is recommended.

5. Resuspend the cells in 1 ml of sucrose – Tris – dodecylamide solution. Keep for 5 min on ice.
6. Add 0.1 ml of lysozyme solution, mixing gently but well. Keep for 5 min on ice.
7. Add 0.1 ml of 0.5 M EDTA, mixing gently but well. Keep for 5 min on ice.
8. Add 1.6 ml Triton buffer, mixing gently but well. Keep for 15 min on ice with occasional swirling.
9. Spin at 15,000 rpm for 20 min at 4°.
10. Decant supernatant carefully from the viscous pellet of cell debris.
11. Add 20 μl of protease and digest at 37° for 1 hr.
12. Make volume up to 4.9 ml with TE and add to preweighed 5 g of cesium chloride in Beckman polyallomer centrifuge tube ($\frac{5}{8}$ inch diameter $\times$ 3 inches). Add 300 μl of 10 mg/ml ethidium bromide.
13. Dissolve the cesium chloride, fill the tube with paraffin, and then centrifuge for 43 hr at 39,000 rpm at 15°.
14. Visualize DNA bands under ultraviolet light and remove lower plasmid band by piercing tube and sucking out the band with an 18-gauge needle (1.2 $\times$ 38 mm) and 1-ml syringe.
15. Inject syringe contents into two volumes of NaCl-saturated isopropanol.
16. Mix the phases and allow to stand until the phases separate.
17. Discard the upper organic phase and repeat isopropanol extractions until all red coloration (ethidium bromide) has been removed.
18. Transfer the aqueous phase to a small-bore dialysis sac and dialyze overnight against TE/10 at 4°.
19. Confirm plasmid identity by restriction analysis. Approximately 15 μl (0.5 μg of DNA) of plasmid preparation should be sufficient for an agarose gel track. The restriction patterns obtained using restriction endonucleases *Eco*RI, *Hin*dIII, and *Pst*I are presented in Fig. 3.

Further confirmation of successful cloning of the structural genes of the adenosinetriphosphatase complex of *E. coli* can be provided by transformation, selecting chloramphenicol resistance, into appropriate *recA* derivatives of other *unc* mutant strains, and then screening for ability to grow on succinate minimal medium.[20]

By using the appropriate *recA unc* mutant strains, therefore, part or the

[20] Complementation of growth on succinate in strain AN727 (*uncB402*)[11] indicates the presence of the proximal *Hin*dIII fragment; complementation of strain AN819 (*uncD409*)[11] demonstrates the presence of the promoter distal fragment; and complementation of strain AN888 (*unc-416*::Mu) [G. B. Cox, J. A. Downie, L. Langman, A. E. Senior, G. Ash, D. R. H. Fayle, and F. Gibson, *J. Bacteriol.* **148**, 130 (1981)] shows successful cloning of all structural genes of the ATPase complex.

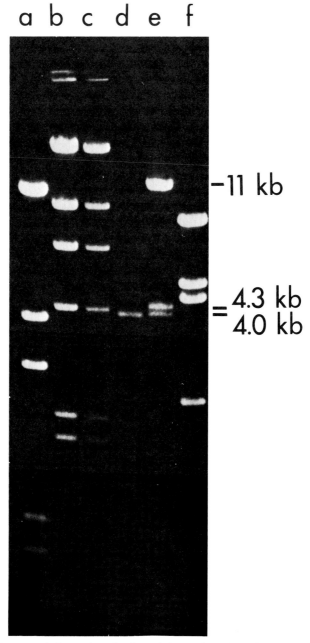

FIG. 3. Agarose gel electrophoresis of restriction endonuclease fragments of DNA. (a) pAN45 + *Pst*I; (b) and (c) lambda DNA + *Hin*dIII; (d) pACYC184 + *Hin*dIII; (e)

entire complement of the F_1F_0-ATPase structural genes can be selected as a cloned insert into the multicopy plasmid vector. The same procedure has been used to clone mutant *unc* alleles onto multicopy plasmids following transfer of mutant alleles to F′ plasmids.[7]

By using the above procedure of F′ plasmid isolation, but using other restriction endonucleases such as *Eco*RI or *Pst*I, it is reasonable to assume that the entire *unc* operon, including the open reading frame preceding the *uncB* gene,[21,22] could be cloned. *Eco*RI cuts the *unc* operon into three fragments (Fig. 1), but attempts to use this enzyme have thus far resulted in cloned inserts with unexpected restriction patterns.[7] This is possibly due to the fact that the large 11 kb *Eco*RI fragment, including the promoter and promoter-proximal region of the *unc* operon, also includes *oriC*—the origin of replication of the *E. coli* chromosome. *Pst*I may be suitable, especially if *Pst*I-cut sites occur between the *unc* promoter and the *oriC* locus. The outlined procedure of F′ plasmid isolation, as an enriched source of DNA from which to clone genes of interest, would seem to have widespread application.

Acknowledgments

We are indebted to Lyndall Langman for assistance in developing the experimental procedure and to both Lyndall Langman and Graeme Cox for helpful discussions.

[21] N. J. Gay and J. E. Walker, *Nucleic Acids Res.* **9**, 3919 (1981).
[22] H. Kanazawa, K. Mabuchi, T. Kayano, T. Noumi, T. Sekiya, and M. Futai, *Biochem. Biophys. Res. Commun.* **103**, 613 (1981).

pAN45 + *Hin*dIII; (f) pAN45 + *Eco*RI. Samples of DNA (about 0.5 μg) in either *Hin*dIII buffer (60 m*M* NaCl; 7 m*M* MgCl$_2$; 10 m*M* Tris-HCl, pH 7.4; 100 μg of bovine serum albumin per milliliter), *Pst*I buffer (50 m*M* NaCl; 6 m*M* Tris-HCl, pH 7.4; 6 m*M* MgCl$_2$; 6 m*M* 2-mercaptoethanol; 100 μg of bovine serum albumin per milliliter), or *Eco*RI buffer (100 m*M* Tris-HCl, pH 7.5; 50 m*M* NaCl; 5 m*M* MgCl$_2$; 100 μg of bovine serum albumin per milliliter) were digested at 37° with the respective restriction endonuclease *Hin*dIII, *Pst*I, or *Eco*RI for 2 hr. Estimated sizes (in kilobases) of the *Hin*dIII DNA fragments of plasmid pAN45 are indicated.

[19] Biogenesis of an Oligomeric Membrane Protein Complex: The Proton Translocating ATPase of *Escherichia coli*

By WILLIAM S. A. BRUSILOW, ROBERT P. GUNSALUS, and ROBERT D. SIMONI

The relative locations of genes on a chromosome can be determined in many ways using genetic techniques that depend on the availability of genetic exchange mechanisms. Such approaches can determine the location of various genes but alone cannot establish gene–polypeptide relationships, which must be obtained with supplemental biochemical or physiological information. Most frequently this can be accomplished with an enzyme assay or other physiological assay. Thus the general approach is to localize a gene by genetic techniques and determine the biochemical consequence of a lesion in that gene by biochemical or physiological techniques. While this approach has been the foundation of biochemical genetics, it has several limitations. Most significantly it depends on a functional measure of each gene product.

The proton translocating adenosinetriphosphatase (ATP synthetase, EC 3.6.1.3) of *E. coli* provides a good example of the limitations of the traditional approaches to determining the gene–polypeptide relationships. This complex, located in the cytoplasmic membrane of *E. coli,* utilizes an electrochemical gradient of protons to synthesize ATP and can also hydrolyze ATP with resultant generation of proton gradient. The complex consists of eight types of polypeptides. Five polypeptides, α, β, γ, δ, and ϵ, comprise the F_1 portion of the complex, which is peripheral to the membrane and has the catalytic activity of ATP synthesis and hydrolysis. Three polypeptides, a, b, and c, comprise the F_0 portion of the complex, which is integral to the membrane, and function as the proton conducting channel.[1,2] The structure of the complex is further complicated by an unusual subunit stoichiometry, which is probably $\alpha(3)$, $\beta(3)$, $\gamma(1)$, $\delta(1)$, $\epsilon(1)$, $a(1-2)$, $b(2)$, and $c(6-10)$.[3,4]

Genetic analysis of mutants in the ATPase complex determined that the genes that specify the eight polypeptides are all located in one region of the *E. coli* chromosome.[5] It was difficult, however, to determine which polypep-

[1] D. L. Foster and R. H. Fillingame, *J. Biol. Chem.* **254,** 8230 (1979).
[2] P. Friedl, C. Friedl, and H. U. Schairer, *Eur. J. Biochem.* **100,** 175 (1979).
[3] D. L. Foster and R. H. Fillingame, *J. Biol. Chem.* **257,** 2009 (1982).
[4] J. Nielson, F. G. Hansen, J. Hoppe, P. Friedl, and K. von Meyenburg, *Mol. Gen. Genet.* **184,** 33 (1981).
[5] J. A. Downie, F. Gibson, and G. B. Cox, *Annu. Rev. Biochem.* **48,** 103 (1979).

tides were altered by each mutation and thus establish the gene – polypeptide relationships. Because there was no functional assay available for the individual polypeptides of the complex, the traditional approaches alone were not adequate.

Described here in a general procedure for both mapping genes and determining the gene – polypeptide relationships when the function of the polypeptides are not measurable or are unknown.

Principle. The basic method is to clone the genes into plasmids, then use restriction mapping, subcloning of restriction fragments, and *in vitro* protein synthesis to determine which regions of DNA code for certain polypeptides. Because the relative locations of the restriction sites are known, this analysis can determine the relative locations of genes coding for the polypeptides synthesized *in vitro*. In order absolutely to establish gene – polypeptide relationships, and thus gene order, there must be a method of identifying the polypeptides of interest. For the ATPase the identities of the *in vitro* products were established by coelectrophoresis with authentic ATPase and by antibody precipitation.[6] Even without such identification, however, the relative locations of polypeptide-coding regions can be established by this method.

Methods

DNA isolation and restriction mapping will not be discussed in detail, since they are covered in other sections of this volume. It is of practical advantage to start by isolating DNA significantly enriched for the gene or genes of interest. For ATPase, this was accomplished by isolating λ transducing phage for this region of the *E. coli* chromosome[7,8] or by using F′ plasmid known to carry ATPase genes.[9] At the outset of this work, it was not clear how many or which ATPase genes were located at the same position of the chromosome. The use of λ transducing phage offered a considerable advantage in the studies on ATPase because it could be shown that the transducing DNA also carried flanking genes, thus ensuring that all ATPase genes in this region of the chromosome were contained within this piece of DNA.

Once the λ transducing phage DNA was isolated, a restriction map was constructed. As a matter of strategy it was efficient to begin with restriction

[6] W. S. A. Brusilow, R. P. Gunsalus, E. C. Hardeman, K. P. Decker, and R. D. Simoni, *J. Biol. Chem.* **256**, 3141 (1981).

[7] T. Miki, S. Hiraga, T. Nagata, and T. Yura, *Proc. Natl. Acad. Sci. U.S.A.* **75**, 5099 (1978).

[8] K. von Meyenberg, F. G. Hansen, L. D. Nielsen, and E. Riise, *Mol. Gen. Genet.* **160**, 287 (1978).

[9] G. B. Cox, J. A. Downie, F. Gibson, and J. Radik, *Biochem J.* **170**, 593 (1978).

endonucleases that infrequently cut the DNA and progress to those for which the sites appear more frequently. With the restriction site information, suitable plasmid vectors were selected that permitted cloning of the desired DNA fragments. A suitable *E. coli* strain was transformed with these recombinant plasmids. In order that the isolation not be biased by specific selections, transformants were selected for the antibiotic resistance carried by the vector and then screened for the specific size DNA inserts. Because only antibiotic resistance was selected for, the vector DNA was phosphatased before ligating in the fragments, thus ensuring that most antibiotic-resistant bacteria carried plasmids containing inserts. Alternatively, of course, a strain carrying a specific genetic defect could be transformed and a scheme designed to select only those plasmids carrying an insert that genetically complements a specific mutation. Transformants were isolated and grown for isolation of plasmid DNA, from which the size of the DNA insert and the gene products were determined as follows.

Rapid Isolation of Plasmid DNA

   Solutions[10]
   Solution I: 50 mM glucose, 10 mM EDTA, 25 mM Tris-HCl (pH 8).
      Store at 0°, and before use add egg white lysozyme to 2 mg/ml.
   Solution II: 0.2 N NaOH with 1% sodium dodecyl sulfate (SDS)
   Solution III: 3 M sodium acetate, pH 4.8
   Solution IV: RNase A is dissolved in 50 mM Tris-HCl buffer (pH 8.0)
      to 10 mg/ml and heated to 80° for 20 min to destroy any DNase
      activity.

   Isolation Procedure. Each transformant to be screened is grown overnight in tubes containing 5 ml of L broth including 20 μg/ml of the appropriate antibiotic. We normally screen 16–24 transformants at a time. The cells from each overnight culture are isolated by centrifugation in plastic tubes. Cells are resuspended in 200 μl of solution I and allowed to stand on ice for 30 min, allowing the lysozyme to digest the cell wall. Then 400 μl of solution II are added, and the suspension is incubated on ice for an additional 5 min, during which the cells lyse, releasing their DNA. Solution III, 300 μl, is added with gentle mixing, and the mixture is allowed to remain on ice for an additional 60 min. The mixture is then centrifuged for 20 min at 27,000 g in the SS34 rotor of a Sorvall centrifuge. The supernatant solution, usually about 0.8 ml, is carefully transferred to 1.5-ml Eppendorf microcentrifuge tubes. Isopropanol (0.5 ml) is added, and precipitation of the DNA occurs by allowing the mixture to remain for 10 min in the Dry Ice–ethanol bath. The plasmid DNA is isolated by centrifugation for 5 min

[10] H. C. Birnbaum and J. Doly, *Nucleic Acids Res.* 7, 1513 (1979).

in the microfuge in the cold. The pellet is then redissolved in 100 μl of 0.3 M sodium acetate. Two volumes of ethanol are added, and the solution again is cooled in Dry Ice to reprecipitate the plasmid DNA. The DNA is isolated by centrifugation, and the pellets are dried by lyophilization for 15–30 min. The dried DNA is resuspended in 100 μl of TE buffer (10 mM Tris, 1 mM EDTA, pH 7.9). This preparation is now suitable for both gel analyses of the size of the insert and *in vitro* protein synthesis.

The size of the DNA insert is determined by gel analysis of the plasmid, which has been digested with the same restriction enzyme used in the cloning. The plasmid preparation (2–5 μl) is incubated with restriction enzyme under appropriate conditions for that enzyme and with 1 μl of RNase A solution (IV). The RNase is important, since large amounts of RNA can inhibit the restriction endonuclease activity and also smear on the gel and obscure the restriction fragments. Once the size of the insert is determined, it is then related to the restriction map, and the position of the cloned fragment in the larger DNA is deduced.

In Vitro Protein Synthesis

The coupled transcription–translation system originally described by Zubay[11,12] is a convenient way to determine which polypeptides are encoded by each cloned piece of DNA. There are some potential limitations to testing cloned DNA in this way, since the structural genes may be separated from the transcription start region (promoter). In our experience this problem has not arisen, presumably because there are promoters in the vector DNA from which transcription can proceed.

Solutions for in Vitro Protein Synthesis

Incubation mixture for protein synthesis: Mix A, 2.5 μl; mix B, 2.5 μl; mix C, 2.5 μl; PEG 6000, 25%, 2.5 μl; DNA template 1–10 μl; Tris-acetate buffer, 10 mM pH 8.0 to 25 μl; S-30 extract, 5 μl; [³⁵S]methionine (1000 Ci/mmol), 0.5 μl/reaction

Composition of mixes

Mix A: Immediately before use, add 0.165 ml of mix A-1 to 0.05 ml of mix A-2. Shake well to suspend.

Mix A-1: 1.0 M Tris-acetate, pH 8.2, 8.80 ml; 2.75 M potassium acetate, 4.00 ml; 2.0 M ammonium acetate, 2.60 ml. Store at 4°. Excellent long-term stability.

Mix A-2: 20 mM amino acids (all amino acids, except Met, Asn, Gln, Cys), 2.20 ml; 20 mM amino acids (Asn, Gln, Cys), 2.20 ml; DTT,

[11] G. Zubay, D. A. Chambers, and L. D. Cheong, *In* "The Lactose Operon" (J. R. Beckwith and D. Zipser, eds.), p. 375. Cold Spring Habor Lab., Cold Spring Harbor, New York, 1971.
[12] H. Zalkin, C. Yanofsky, and C. L. Squires, *J. Biol. Chem.* **249**, 465 (1974).

1.0 M, 0.28 ml. Store at $-20°$ in 0.05 ml portions. Excellent long-term stability. Use once or twice, then discard unused portion.

Mix B: Na_3PEP (0.83 M), 0.51 ml; calcium leucovorin (2.7 mg/ml), 0.20 ml; $E.\ coli$ tRNA (10 mg/ml), 0.20 ml; TPN (9 mg/ml), 0.06 ml; FAD (5 mg/ml), 0.108 ml; ATP (0.44 M), 0.10 ml; CTP (0.11 M), 0.10 ml; GTP (0.11 M), 0.10 ml; UTP (0.11 M), 0.10 ml; H_2O, to 2.00 ml. Store frozen in 0.4-ml aliquots. Activity may decline a few percent per month upon repeated freeze–thawing, but appears to be very stable for long periods when not thawed.

Mix C: Store 0.14 M magnesium acetate and 0.148 M calcium acetate frozen and mix 1:1 before use. Mix C has been optimized for production of ATPase proteins. The Mg^{2+} ion concentration should be optimized for production of any other proteins.

[35S]methionine: New England Nuclear or Amersham, translation grade, > 1000 Ci/mmol.

Preparation of the S-30 Extract. The $E.\ coli$ strain used is called A 19.[13] It is RNase I^- 19 Hfr met$^-$ lambda$^+$ del ton-trp. It therefore requires methionine and tryptophan. In our experience, however, any $E.\ coli$ strain can be used. The growth medium is as follows: K_2HPO_4 (289 g/liter), 90 ml; KH_2PO_4 (56 g/liter), 90 ml; 10% yeast extract, 90 ml; 0.01 M $FeCl_3$, 18 ml; B_1 (15 mg/ml), 0.9 ml; 50% glucose (autoclave separately), 18 ml; L-tryptophan, 50 mg; L-methionine, 50 mg. Add H_2O to 1 liter.

Grow cells with vigorous shaking at 30° to a Klett reading (red filter) of 200–300. Harvest cells and wash twice with [0.01 M Tris-acetate buffer (pH 8.2), 0.014 M MgOAc, 0.06 M KOAc, 1 mM DTT]. Suspend cells in 4 ml of the same buffer for each gram of cells. Break cells in a French press at 6000–8000 psi. Immediately after breaking the cells, add 1.0 M DTT to a 1 mM final concentration. Centrifuge the extract twice at 30,000 g for 30 min, saving the supernatant fraction each time. This extract is now preincubated at 37° to remove endogenous mRNA.

Preincubation mix 1 (prepare fresh from frozen stocks): 1 M Tris-acetate (pH 7.8), 10.0 ml; 0.14 M MgOAc, 2.0 ml; 1.0 M DTT, 0.03 ml; 20 mM amino acids I, 0.05 ml; 20 mM methionine, 0.05 ml.

Preincubation mix 2 (prepare fresh): 0.2 M ATP (pH 7), 0.4 ml; 0.075 M NaPEP, 12.0 ml; pyruvate kinase, 1 mg; 20 mM amino acids II, 0.05 ml.

Combine 0.49 ml of mix 1 + 0.50 ml of mix 2 and add 2.46 ml per 10 ml of extract. Incubate at 37° for 80 min in the dark, then dialyze against 50 volumes of the buffer used to make the extract. Change the solution every 2 hr. Total dialysis time is about 6–8 hr. Store in 1-ml portions in liquid N_2. The protein concentration is about 30–34 mg/ml (biuret). The activity of

13 R. Gesteland, *J. Mol. Biol.* **16**, 67 (1966).

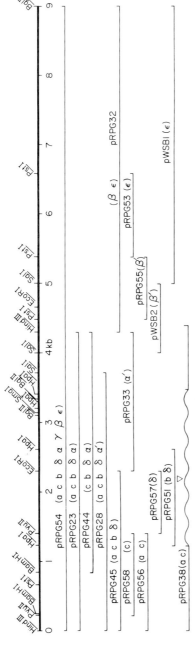

Fig. 1. Restriction map of plasmids containing various regions of the *unc* operon. The heavy line shows the location of the restriction endonuclease recognition sites with the distance given in kilobases. Regions of the *unc* operon present on the various plasmids are indicated. The ends of the lines indicate the restriction sites within the operon used for cloning. Individual H+-ATPase polypeptides produced *in vitro* from each plasmid are indicated, as are partial translation products (e.g., α′). Details of the individual clonings and resultant plasmids are given by Gunsalus *et al.*[14]

an S-30 declines with freezing and thawing but remains useful for 4–5 freeze–thaw cycles.

Conducting the reaction. Mix the three mixes and the PEG with the hot methionine, and pipette 10.5 μl of this solution into each reaction tube. Add buffer and DNA totaling 10 μl, then add 5 μl of the S-30 extract, so that the

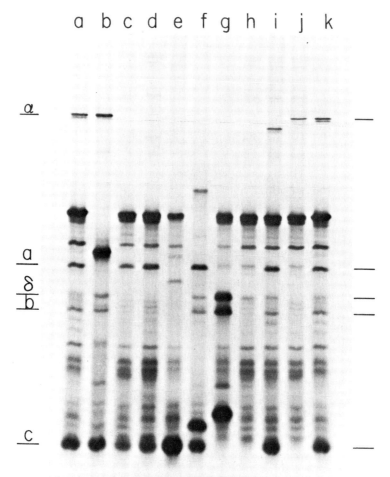

FIG. 2. *In vitro* protein synthesis products of plasmids constructed to determine the order of the first five genes of the *unc* operon. *In vitro* DNA-directed protein synthesis was carried out as described above. Proteins were labeled with [35S]methionine, separated on a polyacrylamide gel, and located by autoradiography. The locations of the *unc* polypeptides α, a, δ, b, and c are indicated alongside the autoradiogram and were determined by running pure F_1F_0 alongside. All other proteins are either vector-encoded or partial ATPase polypeptides. Lanes: a, pRPG23; b, pRPG44; c, pRPG38; d, pRPG56; e, pRPG58; f, pRPG45; g, pRPG51; h, pRPG57; i, pRPG28; j, pRPG33; and k, pRPG23.

final volume is 25–26 μl. The DNA concentration should be optimized for each DNA. Usually 1 μg per reaction is a good starting concentration. Incubate the reaction mixture at 37° for 20–45 min. Different S-30 preparations have different optimal synthesis times. The reactions are stopped by diluting each tube with 25 μl of Tris-acetate buffer (10 mM, pH 8.0) and precipitating the proteins with 1 ml of acetone, followed by 10 min on ice. The precipitated protein is centrifuged in a microfuge and then dissolved in sample buffer for gel electrophoresis.

Comments

The information presented in Fig. 1 summarizes the restriction map and cloning of parts of the *unc* operon of *E. coli*,[14] while the data in Fig. 2 demonstrate the polypeptides synthesized from each of the plasmids described in Fig. 1. This technique of ordering genes by cloning and *in vitro* protein synthesis has been used not only to determine the relative locations of genes in the *unc* operon, but also to determine the relative locations of protein-coding regions near the origin of DNA replication of *V. harveyi*.[15]

[14] R. P. Gunsalus, W. S. A. Brusilow, and R. D. Simoni, *Proc. Natl. Acad. Sci. U.S.A.* **79**, 320.
[15] J. W. Zyskind, J. M. Cleary, W. S. A. Brusilow, N. E. Harding, and D. W. Smith, *Proc. Natl. Acad. Sci. U.S.A.* **80**, 1164.

[20] Analysis of *Escherichia coli* ATP Synthase Subunits by DNA and Protein Sequencing

By JOHN E. WALKER and NICHOLAS J. GAY

The genes encoding the eight subunits of the *Escherichia coli* ATP synthase complex form a single transcriptional unit, the *unc* operon[1–3] found at about minute 83 on the 100-min linkage map.[4] Gene–product relationships have been established, and an order for the genes has been deduced from both genetic[1,2,5] and *in vitro* analyses.[5–7] A more precise

[1] J. A. Downie, F. Gibson, and G. B. Cox, *Annu. Rev. Biochem.* **48**, 103 (1979).
[2] F. Gibson, *Proc. R. Soc. London Ser. B* **215**, 1 (1982).
[3] R. H. Fillingame, *Curr. Top. Bioenerg.* **11**, 35 (1981).
[4] B. J. Bachman and B. K. Low, *Microbiol. Rev.* **44**, 1 (1980).
[5] G. B. Cox, J. A. Downie, L. Langman, A. E. Senior, G. Ask, D. H. R. Fayle, and F. Gibson, *J. Bacteriol.* **148**, 30 (1981).
[6] R. P. Gunsalus, W. S. A. Brusilow, and R. D. Simoni, *Proc. Natl. Acad. Sci. U.S.A.* **79**, 320 (1982).
[7] K. von Meyenburg and F. G. Hansen, *ICN–UCLA Symp. Mol. Cell. Biol.* **19**, 139 (1980).

METHODS IN ENZYMOLOGY, VOL. 97

description of this region of the *E. coli* chromosome is provided by the DNA sequence and the independent identification of the genes by protein sequence analysis[8-13] and of the promoter by *in vitro* experimentation.[14] The DNA sequence analysis has been determined by primed synthesis in the presence of dideoxy chain terminators of DNA fragments cloned into bacteriophage M13, as first described by Sanger *et al.*[15,16] The *unc* promoter (the region at which RNA polymerase binds adjacent to the point of initiation of transcription) has been identified by making runoff transcripts from appropriate DNA fragments. Some of the genes have been identified from protein sequences determined with the aid of a novel solid-phase microsequencer on ATP synthase subunits eluted from polyacrylamide gels. The methodologies employed in these three aspects of the work are described below.

I. DNA Sequence

Principles of DNA Sequence Analysis

DNA is a polymer of deoxyribose units linked via phosphodiester bonds at their 5′ and 3′ positions; to the 1′ position of the deoxyribose monomers is attached one of the four nucleotide bases thymine (T), cytosine (C), guanine (G), or adenine (A). The sequence of DNA is the order of nucleotide bases along the sugar–phosphate backbone. Natural DNA molecules are very long chains (the *E. coli* chromosome is approximately 4×10^6 bases in length). Therefore, prior to sequence analysis the chain is broken up, and the fragments are both purified and amplified by cloning into a plasmid or viral vector. After amplification, the inserted DNA often is released by digestion of the vector with restriction enzymes. The cloning and amplification of the *unc* operon is described below.

The enzymic method of DNA sequencing is usually performed with

[8] N. J. Gay and J. E. Walker, *Nucleic Acids Res.* **9**, 2187 (1981).

[9] N. J. Gay and J. E. Walker, *Nucleic Acids Rev.* **9**, 3919 (1981).

[10] M. Saraste, A. Eberle, N. J. Gay, M. J. Runswick and J. E. Walker, *Nucleic Acids Res.* **9**, 5287 (1981).

[11] J. E. Walker, A. Eberle, N. J. Gay, P. Hanisch, M. Saraste, and M. J. Runswick, *in* "Methods in Peptide and Protein Sequence Analysis" (M. Elzinga, ed.), p. 337. Humana, Clifton, New Jersey, 1982.

[12] J. E. Walker, A. D. Auffret, A. Carne, A. Gurnett, P. Hanisch, D. Hill, and M. Saraste, *Eur. J. Biochem.* **123**, 253 (1982).

[13] J. E. Walker, M. Saraste, and N. J. Gay, *Biochim. Biophys. Acta* (in press).

[14] N. J. Gay and J. E. Walker, *J. Mol. Biol.*, submitted for publication.

[15] F. Sanger, S. Nicklen, and A. R. Coulson, *Proc. Natl. Acad. Sci. U.S.A.* **74**, 5463 (1977).

[16] F. Sanger, A. R. Coulson, B. G. Barrell, A. J. H. Smith, and B. A. Roe, *J. Mol. Biol.* **143**, 161 (1980).

single-stranded DNA templates produced by cloning into bacteriophage M13. In the M13 shotgun strategy (employed for sequencing fragments R1 and R2, see Fig. 1), DNA to be sequenced is broken into smaller fragments by digestion either with restriction enzymes or random nucleases[17] or alternatively physically sheared by sonication.[18] The DNA fragments are cloned into the unique restriction sites of the replicative form of M13mp5, M13mp7[19] or M13mp8. The replicative form (RF) is then used to transform a male host. The transformed cells are mixed with fresh cells and spread on an agar plate. They excrete phage and serve as foci for phage infection, leaving small areas of retarded growth (plaques) in the lawn of cells. In order to distinguish plaques of recombinant phage from the original phage, a complementation assay is used. The vectors M13mp2-9 contain a portion of the *lacZ* gene of *E. coli* including the operator, the uv5 promoter, and the N-terminal 145 amino acid residues of β-galactosidase (the α-peptide): when induced by isopropyl-β-D-thiogalactoside (IPTG) this part of β-galactosidase is not a functional enzyme. Host *E. coli* (JM101) has no chromosomal β-galactosidase gene but has a defective β-galactosidase, from which amino acid residues 11–41 are missing on an episome. When M13mp2-9 infect the host, the two defective β-galactosidase polypeptides complement and produce a functional β-galactosidase. Plaques derived from infected cells producing a functional β-galactosidase can be identified by incorporation of a lactose analog 5-bromo-4-chloro-3-indolyl-β-D-galactopyranoside (BCIG) into the agar plate; this colorless compound is hydrolyzed by β-galactosidase to release a blue dye, bromochloroindole. This confers a blue color on plaques of infected cells. By contrast, plaques containing recombinant phage are usually colorless; the inserted DNA interrupts the α-peptide DNA and so prevents complementation of the defective episomal β-galactosidase. The recombinant phage is harvested by sticking a toothpick into the white plaque; the toothpick is then used to inoculate fresh medium, and a small culture is grown. Phage is secreted into the medium and is obtained from the supernatant after first centrifuging the cells. The protein coats of the phage are removed with phenol, and the single-stranded recombinant DNA is used as a template for sequencing.

In the enzymic method of DNA sequencing[15] the oligonucleotide primer is hybridized to a single-stranded DNA template adjacent to the recombinant insert to be sequenced. Several universal primers complementary to the region of M13mp7 at the 3' end of the inserted DNA are available either as synthetic primers[20] (P-L Biochemicals, Collaborative Research, or Be-

[17] S. Anderson, *Nucleic Acids Res.* **9**, 3015 (1981).

[18] P. Deininger, manuscript in preparation.

[19] J. Messing, R. Crea, and P. Seeburg, *Nucleic Acids Res.* **9**, 309 (1981).

[20] M. L. Duckworth, M. J. Gait, P. Goelet, G. F. Hong, M. Singh, and R. C. Titmas, *Nucleic Acids Res.* **9**, 1691 (1981).

thesda Research Laboratories) or as a short restriction fragment.[21] The primer is extended through the recombinant insert by DNA polymerase in the presence of both [^{32}P]dATP deoxynucleotide triphosphates (dNTPs) and dideoxynucleoside triphosphates (ddNTPS) to make a radioactively labeled complementary strand; the ddNTPs compete for incorporation with the analogous dNTPs, but when incorporated into the transcript their lack of a 3′-hydroxyl group on the ribose moiety prevents further polymerization and terminates the chain. Four separate reactions are performed, each containing a different ddNTP. This generates four sets of fragments corresponding to terminations at T, C, G, or A positions in the original DNA sequence. These are separated in denaturing polyacrylamide gels that can resolve molecules differing in length by one nucleotide. This gives rise to four ladders revealed by autoradiography, from which the sequence can be read.

Cloning the unc Operon

Sequence analysis of the *unc* operon has been helped by the availability of a lambda transducing phage containing about 30 kilobases (kb) of the *E. coli* chromosome including the loci *asnA, oriC, unc,* and *glmS* (see Fig. 1a).[22] A number of restriction sites had been identified in the *unc* region,[7,23] thereby providing a simple means of obtaining DNA fragments that would cover at least part of the *unc* operon (Fig. 1a and b). They include *Eco*RI fragments R1 and R2 (Fig. 1c). These were cloned into a plasmid pACYC184 to amplify them, excised and sequenced using the M13 shotgun dideoxy sequence strategy. Other regions were cloned directly into bacteriophage M13 (see Fig. 1c).

Materials

Polyethylene glycol, M_r 6000 (PEG 6000) purchased from Koch Light Laboratories, Colnbrook, Bucks, England

BACTERIAL STRAINS

Strain	Genotype
JM101[24]	Δ(*lac, pro*), *sup*E, *thi*-, F′ *tra* D36, *pro* AB⁻, *lac*I^q, M15
HB101[25]	*rec*A13, *hsd*520 (r_B⁻, m_B⁻), *ara*14, *pro*A2, *lac*Y1, *gal*K2, *rps*L20, *xy*15, *mtl*1, *sup*E44
KY7485[26]	*asn*₃₁⁻, *thr*⁻, *rif*ʳ (lambda *asn*⁺5, lambda *cI857S7*)

[21] S. Anderson, M. J. Gait, L. Mayol, and I. G. Young, *Nucleic Acids Res.* **8**, 1731 (1980).
[22] T. Miki, S. Hiraga, T. Nagata, and T. Yura, *Proc. Natl. Acad. Sci. U.S.A.* **75**, 5099 (1978).
[23] H. Kanazawa, T. Miki, F. Tamura, T. Yura, and M. Futai, *Proc. Natl. Acad. Sci. U.S.A.* **76**, 1126 (1979).
[24] J. Messing, *Recomb. DNA Tech. Bull.* **2**, 43 (1979).

PLASMIDS

Plasmid	Genotype
pACYC184[27]	*Cam^r Tet^r*, ColEl replicon
pBR322[28]	*Amp^r Tet^r*, ColEl replicon

Media

2 × TY: 16 g of tryptone, 10 g of yeast extract, 5 g NaCl, 1 liter distilled water, adjusted to pH 7.4

TYE: 15 g of agar, 8 g of NaCl, 10 g of Bacto-tryptone, 5 g of yeast extract, 1 liter of distilled water

H-top agar: 8 g of Bacto-agar, 10 g of Bacto-tryptone, 8 g of NaCl, 1 liter of distilled water

Lambda dil: 10 ml of 10 mM Tris, pH 7.4, 5 ml of 5 mM MgSO$_4$, 200 ml of 0.2 M NaCl, 1 g of 0.1% gelatin, 1 liter of distilled water

Antibiotics were incorporated into media as required, usually at the following concentrations: chloramphenicol, 25 μg ml^{-1}, tetracycline, 15 μg ml^{-1}, ampicillin, 25 μg ml^{-1}.

Preparation of DNA from Transducing Phage lambda asn5

An overnight culture of the lysogen KY7485 was diluted 1 : 100 in 1 liter of 2 × TY + 10 mM MgSO$_4$ and grown to OD$_{590}$ 0.6. Then 1 liter of 2 × TY preheated to 52° was added, and the cells were incubated at 42° for 4 hr. Chloroform (5 ml) was added, and the culture was shaken for 5 min. Cells were pelleted (7500 g, 10 min), and the supernatant was decanted off. Phage particles were precipitated for 12 hr at 4° by addition of NaCl to 2% and PEG 6000 to 8% (final concentrations). Then phage was centrifuged (5500 g, 30 min) and resuspended in 28 ml of phage buffer consisting of 50 mM MgCl$_2$, 50 mM NaCl, 20 mM Tris-HCl, pH 7.4. Excess PEG 6000 was removed by centrifugation (8000 g, 30 min). A cesium chloride step gradient was prepared in a 30-ml Beckman polyallomer tube from 2-ml layers of three cesium chloride solutions in phage buffer with refractive indexes at 1.4, 1.38, and 1.36. The phage suspension was layered on the uppermost and centrifuged in a Beckman SW 27 rotor (113,000 g, 2 hr, 8°). The phage forms a band at the interface of the two layers with refractive indexes of 1.4 and 1.38.

[25] H. W. Boyer and D. Roulland-Dussoix, *J. Mol. Biol.* **41**, 459 (1969).

[26] H. Kanazawa, F. Tamura, K. Mabuchi, T. Miki, and M. Futai, *Proc. Natl. Acad. Sci. U.S.A.* **77**, 7005 (1980).

[27] A. C. Y. Chang and S. N. Cohen, *J. Bacteriol.* **134**, 1141 (1978).

[28] F. Bolivar, R. L. Rodriguez, P. J. Greene, M. C. Betlach, H. L. Heyneker, H. W. Boyer, J. H. Crosa, and S. Falkow, *Gene* **2**, 95 (1977).

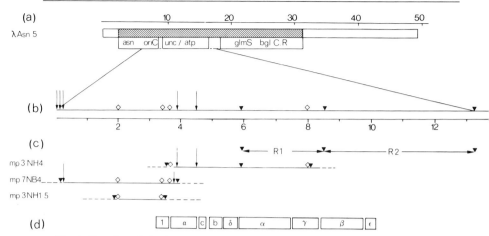

FIG. 1. Genetic and physical maps of the *Escherichia coli* chromosome in the vicinity of the *unc* operon. (a) The extent of the *E. coli* chromosome (hatched) in λ *Asn*5 showing genetic markers. The scale is in kilobases. (b) Restriction map of part of this region redrawn from Kanazawa *et al.*[23] Arrows denote restriction sites for *Bam*HI, ▼, *Eco*RI; and ◇, *Hin*dIII. (c) Alignment of *Eco*RI fragments R₁ and R₂ and primary fragments used for sequence analysis in clones M13mp5.NH4, mp7.NB4, and mp5.NH1.5. (d) Alignment of the nine genes of the *unc* operon with restriction map. Transcription is from left to right.

It was collected with a Pasteur pipette and recentrifuged in a second linear cesium chloride gradient (60 Ti rotor, 123,000 g, 24 hr, 8°). The lowest band was collected and dialyzed against 1 liter of phage buffer (with three changes). DNA was obtained by extraction of the contents of the dialysis bag, twice with equal volumes of neutralized water-saturated phenol and once with diethyl ether. DNA was ethanol precipitated, washed, dried *in vacuo* and redissolved in TE buffer.

Plasmid and M13 RF Preparation

The alkaline sodium dodecyl sulfate (SDS) method,[29] devised for plasmids, was used. Single colonies were inoculated into 50 ml 2× TY supplemented with the appropriate antibiotic. A culture was grown overnight and diluted 100-fold with 2× TY broth (50 ml) and grown with shaking to $OD_{590} = 0.4$ at 37°. Then chloramphenicol (150 μg/ml) was added, and the cells were incubated for a further 18 hr.[30] For M13 RF preparations, single plaques were inoculated into 2× TY broth (50 ml) to which had been added a drop of an overnight culture of JM101. This culture was grown with shaking at 37° to $OD_{590} = 0.8$. Cultures were centrifuged in 50-ml polycar-

[29] H. C. Birnboim and J. Doly, *Nucleic Acids Res.* **7**, 1513 (1979).
[30] D. B. Clewell and D. R. Helinski, *Biochemistry* **9**, 4428 (1970).

bonate tubes at 1200 g for 10 min. The supernatants were discarded, and the cells were resuspended in 1 ml of buffer containing 25 mM Tris- HCl, pH 8.0, 10 mM EDTA, 50 mM glucose, 2 mg ml^{-1} lysozyme. After incubation at 0° for 30 min, 2 ml of solution containing 0.2 M NaOH and 1% SDS were mixed thoroughly with it. After a further 5 min, 3 M sodium acetate, pH 5.0 (1.5 ml) was added and the mixture was incubated at 0° for 90 min. Cell debris and precipitated chromosomal DNA was removed by centrifugation (28,000 g, 10 min). The supernatant was removed, and the plasmid was precipitated by addition of 95% ethanol (10 ml). The solution was kept at −25° for 15 min, and the precipitate was collected by centrifugation (12,000 g, 10 min).

The pellet was redissolved in 2 ml of buffer containing 0.1 M sodium acetate, 1 mM EDTA, 0.1% SDS, 40 mM Tris-HCl, pH 8.0 and extracted once with phenol–chloroform (1 : 1 w/v). The aqueous phase was removed, and the phenol phase was reextracted with a further 1 ml of sodium acetate buffer. DNA in the aqueous phase was precipitated by addition of 95% ethanol (7.5 ml) and then centrifuged (12,000 g, 10 min). The pellet was redissolved in H$_2$O (400 μl), and 60 μl of 1 M sodium acetate, pH 8.0, and 1 ml of 95% ethanol were added. The solution was kept at −20° for 5 min and then centrifuged for 2 min in an Eppendorf centrifuge. This step was repeated, and the final pellet was redissolved in 200 μl of H$_2$O. Pancreatic ribonuclease (20 μl, 1 mg ml^{-1}) heated at 100° for 5 min to remove DNase activity was added, and the solution was incubated at 37° for 30 min. Then 4 M sodium acetate (pH 6.0, 7.5 μl) and 95% ethanol (300 μl) were added. After incubation at 25° for 10 min, purified plasmid or RF was pelleted, washed with 95% ethanol (1 ml), dried *in vacuo,* and redissolved in TE buffer (200 μl; 10 mM Tris-HCl, pH 7.8, 0.1 mM EDTA). Often DNA was purified further by three additional ethanol precipitations, two extractions with aqueous phenol (neutralized with 1 M Tris-HCl, pH 8.0) followed by extraction with diethyl ether and further ethanol precipitation and washing steps. For 400-ml cultures, the above procedure was scaled up.

Analytical and Preparative Agarose Gel Electrophoresis

Agarose gel electrophoresis (AGE) was performed either in a 10 cm × 10 cm "submarine" mini-gel apparatus (Uniscience, Cambridge, England) using TBE buffer (90 mM Tris, 90 mM boric acid, 2.5 mM EDTA) or in a 20 cm × 20 cm flat-bed apparatus in TAE buffer (40 mM Tris-acetate, pH 8.0, 20 mM sodium acetate, 2.0 mM EDTA). Generally, analytical gels were made from 1% high-gelling-temperature agarose (Sea-Kem HGT, FMC Corporation, Marine Colloids Division, Rockland, Maine). Preparative gels were made with 1% low-gelling-temperature agarose (LGT) (Marine Colloids). Minigels were run for 30 min at 80 mA (constant current); large gels were run for 12–18 hr at 80–100 V. After staining with ethidium bromide (ca. 7.5 mg ml⁻¹) DNA was detected with long-wave ultraviolet light.

Recovery of DNA Restriction Fragments from Preparative AGE

As described previously,[16] bands were excised from the LGT agarose gels and the gel slice was melted at 80°. The molten agarose was extracted four times with neutralized, water-saturated phenol (0.5 volume), and the final aqueous phase was extracted with diethyl ether (5 volumes). Then DNA was precipitated by addition of 0.1 volume of 3 M sodium acetate, pH 5.0, and 2.5 volumes of 95% ethanol. The solution was frozen in a Dry Ice–ethanol slurry and centrifuged for 10 min in an Eppendorf 5412 centrifuge.

Digestion of DNA with Restriction Endonucleases

All restriction endonucleases were from New England BioLabs (Beverly, Massachusetts) with the exceptions of *Taq*1, a gift from M. J. Runswick, and *Eco*RI, kindly donated by A. R. Coulson. Digestions were usually carried out at 37° in a buffer containing 10 mM Tris-HCl, pH 7.8, 10 mM MgCl$_2$, and 50 mM NaCl, with the exception of *Eco*RI (digestion in 100 mM Tris-HCl, pH 7.4, 50 mM NaCl, 10 mM MgCl). Digestions with *Taq*1 were performed at 65°.

Ligation of DNA Fragments to Vectors

Ligations were performed in a solution containing 50 mM Tris-HCl, pH 7.8, 10 mM dithiothreitol, 10 mM MgCl$_2$, 1.0 mM ribo-ATP. T4 DNA ligase was a gift from D. L. Bentley. Five units of DNA ligase per microgram of DNA was added for cohesive-ended ligation, and 250 units/μg for blunt ends. Typically a fivefold molar excess of fragment with respect to the vector was added. Incubations were performed at 15° for 1–3 hr in the case of cohesive-ended reactions and for 18 hr for blunt-ended ligation.

Treatment of Vectors with Calf Alkaline Phosphatase

This treatment removes the 5′-phosphate group from DNA, thereby preventing the vector from recircularizing.[31] This reduces the background by eliminating transfections from religated vector. Calf alkaline phosphatase (CAP) was purified as described elsewhere.[32] Restriction digests of vector were heated (80°, 10 min) and then adjusted to 100 mM Tris-HCl,

---

[31] R. K. Morton, *Biochem. J.* **60**, 573 (1955).
[32] A. Efstratiadis, J. N. Vournakis, H. Donis-Keller, G. Chaconas, D. K. Dougall, and F. C. Kafatos, *Nucleic Acids Res.* **4**, 4165 (1977).

pH 9.5, with an appropriate volume of 1 M Tris-HCl, pH 9.5. CAP (10 units) was added, and the solution was incubated at 37° for 30 min. Then the solution was extracted twice with neutralized water-saturated phenol (0.5 volume) and once with diethyl ether (5 volumes). Finally, DNA was precipitated with ethanol as described above.

Transformation and Transfection of E. coli Cells

As described elsewhere[33] fresh overnight cultures E. coli were diluted 50-fold into 2× TY broth (50 ml) and grown to $OD_{590} = 0.5$. Cultures were transferred to 50-ml Corning centrifuge tubes and pelleted (750 g, 5 min, 4°). Cells were resuspended in ice-cold $CaCl_2$ (50 mM, 25 ml). The suspension was placed on ice for 30 min, and then the cells were centrifuged (750 g, 5 min, 4°). The pellet was resuspended in ice-cold 50 mM $CaCl_2$ (2 ml). After a further 30 min at 4°, cells are competent for transformation or transfection and remain so for about 76 hr. Ligation mixtures were added to 300-μl portions of cells and incubated at 0° for 30 min. Cells were then heated at 44° for 2 min. For plasmid transformation 2× TY broth (2 ml) was added, and the culture was incubated for 30 min at 37° and then plated on TYE agar supplemented with the appropriate antibiotic. For M13 transfection, 200–300 μl of transfected cells were added to H-top (3 ml) agar supplemented with IPTG (25 μl, 25 mg/ml), BCIG (25 μl, 25 mg/ml), and 200 μl of an early log phase culture of E. coli JM101. This mixture was overlaid on H-agar plates and incubated at 37° for 18 hr.

Preparation of Single-Stranded M13 Template DNA

Colorless (recombinant) plaques were transferred with toothpicks into 1.5 ml of an early log phase culture of E. coli JM101. This was grown at 37° with shaking for 4.5–5.5 hr. These cultures were transferred into 1.5-ml Eppendorf tubes, and cells were centrifuged (5 min, Eppendorf 5412). Supernatants were mixed with 150 μl of a solution containing 20% PEG 6000 and 2.5 M sodium chloride· in a second 1.5-ml Eppendorf tube. Precipitated phage particles were pelleted (5 min, Eppendorf 5412), and the supernatant was decanted. The tubes were recentrifuged briefly, and traces of PEG 6000 were removed with a drawn-out capillary. The pellet was redissolved in TE buffer (100 μl) and extracted once with neutralized, water-saturated phenol (0.5 volume). To the aqueous phase were added 3 M sodium acetate (10 μl) and 95% ethanol (250 μl). The solution was cooled in a Dry Ice–ethanol slurry, and the precipitated DNA was centrifuged (10 min, Eppendorf 5412). The pellets were washed with 95% ethanol (1 ml), dried in vacuo, and redissolved in TE buffer (40 μl).

[33] S. N. Cohen, A. C. Y. Chang, and L. Hsu, Proc. Natl. Acad. Sci. U.S.A. **69**, 2110 (1972).

DNA Sequencing

Materials for DNA Sequencing

Primer, template and DNA polymerase (Klenow) fragment, 0.5 U/μl (Boehringer). M13 recombinant templates were prepared from 1.5 ml of culture and dissolved in 40 μl of TE buffer. pSP14 primer was derived from an *Eco*RI and *Bam*HI digest of pSP14 plasmid (100 μg).[21] Alternatively a synthetic primer was employed.[20]

Buffers and mixes

Stock solutions

Tris-HCl, 1 M, pH 7.5

MgCl$_2$, 1 M

NaCl, 1 M

EDTA, 100 mM, pH 7.5

Dithiothreitol (DTT), 100 mM

ddNTP, 10 mM

dNTP, 20 mM

DTT, ddNTP, and dNTP solutions are stored frozen, the nucleotide triphosphates having been dissolved in 5 mM Tris-HCl, 0.1 mM EDTA, pH 8.0.

10× annealing buffer: Tris-HCl, 1 M, pH 8.0 (100 μl) MgCl$_2$, 1 M (100 μl) H$_2$O (800 μl). This is stored frozen.

ddNTP working stocks: The appropriate level of ddNTP is determined by trial and error. The following concentrations are a good starting point: ddTTP, 0.8 mM; ddCTP, 0.1 mM; ddGTP, 0.2 mM; ddATP, 0.1 mM. These are stored frozen in 5 mM Tris-HCl, 0.1 mM EDTA, pH 8.0.

dNTP working stocks: 0.5 mM working stocks are made freshly by dilution into 0.1 mM EDTA.

dNTP$^\circ$ mixes

dNTP$^\circ$ MIXES

Mix	dTTP$^\circ$	dCTP$^\circ$	dGTP$^\circ$	dATP$^\circ$
dTTP, 0.5 mM	1 μl	20 μl	20 μl	20 μl
dCTP, 0.5 mM	20 μl	1 μl	20 μl	20 μl
dGTP, 0.5 mM	20 μl	20 μl	1 μl	20 μl
Tris-HCl, 50 mM				
EDTA, 1 mM pH 8.0	5 μl	5 μl	5 μl	5 μl

These mixes keep indefinitely when frozen.

[^{32}P]dATP (400 Ci/mmol, 1 mCi ml^{-1}) (Amersham International Ltd., Amersham, Bucks, England)

Tracking dyes: 100 ml of formamide are stirred with 5 g of Amberlite MB1 for 30 min, then the resin is filtered off. Xylene cyanol FF (0.3 g), Bromphenol blue (0.3 g), and Na$_2$ EDTA (20 mM) are dissolved in formamide.

Repelcote: for siliconization of glass (Hopkin & Williams)

Stock solution of 40% acrylamide: Acrylamide (190 g) and bisacrylamide (10 g) are made up to 500 ml in H$_2$O. The solution is stirred gently with Amberlite MB1 (5 g) for 30 min. Then the resin is filtered off, and the solution is stored at 4°.

10× TBE buffer: Tris, 108 g; H$_3$BO$_3$, 55 g; Na$_2$EDTA·2H$_2$O, 9.5 g; H$_2$O, to 1 liter

10% ammonium persulfate: ammonium persulfate, 10 g; H$_2$O, 100 ml

TEMED: N,N,N',N'-tetramethylethylenediamine

Urea: Ultra-pure (Bethesda Research Laboratories)

Other Convenient Materials

Melting point capillaries, 10 cm × 1 mm i.d. (Gallenkamp-Griffin, MFB 210-538L)

Micropets, 1–5 μl, calibrated (Clay-Adams). Micropets should not contain alkaline deposits and should be washed if necessary.

Siliconized glass tubes (1 cm × 5 cm bacteriological test tubes). After siliconization with Repelcote, tubes are washed with distilled water and then baked at 150° for 4 hr.

Slab gel apparatus, 40 cm (Raven Scientific Ltd., Haverhill, Suffolk, England)

Mirrow-glass plates (20 cm × 40 cm × 4 mm); back plate with ears (Raven Scientific Ltd.)

Side spacers and comb with 18 teeth (each tooth is 3 mm wide and separated from adjacent tooth by a 1.5-mm gap). Spacers and combs are cut from 0.3 mm Plastikard (Raven Scientific Ltd).

Yellow tape: Clipper Canadian Technical Tape Ltd. or Tuck Tape

Dow Saran Wrap plastic film (other makes are usually not satisfactory).

Steel plate (43 cm × 34 cm × ⅛th inch) for weighting down X-ray film.

Hybridization

Single-stranded M13 template and primer are denatured together by boiling and are annealed. The procedure given below is for two clones but can be extended to more. It can be performed in either a glass capillary, as described, or a capped Eppendorf tube.

A capillary is drawn out by making the glass pliable in a small Bunsen flame, removed from the flame, and pulled gently to give a neck of about 3 cm. The neck is scored lightly with a diamond pencil and broken to give flush ends. It is calibrated by blowing into the drawn out tip of the capillary with a micropipet. The hybridization mixture contains 5 μl of M13 clone, 1 μl of 10× sequencing buffer, 1 μl of primer, and 3 μl of water. These are mixed by blowing the sample up and down onto a piece of Parafilm. The blunt end of the capillary is sealed in a Bunsen flame; as the glass cools the sample should draw back from the capillary tip. Then the tip is sealed. The capillary or Eppendorf tube is left for 2 min in a boiling water bath and then cooled on the bench to room temperature (15–30 min). When the synthetic primer is employed in Eppendorf tubes, the denaturation is performed at about 70° in a water bath or heated oven for 1 hr.

Gel Preparation[34]

Urea (21 g) is dissolved by stirring in 7.5 ml of 40% acrylamide stock solution, 5 ml of 10× TBE, and water to 50 ml. This gives a 6% gel. One of a pair of glass plates is siliconized, and both are swabbed with ethanol. Plastikard spacers (0.3 mm thick) are placed between the plates at the edges; then they are taped together with yellow tape, and the plates are clamped with two Bulldog clips. TEMED (40 μl) and 10% ammonium persulfate (300 μl) are mixed with the acrylamide solution. The glass plates are held at about 30° from the horizontal and the acrylamide solution injected from a 50-ml disposable syinge.

Formamide Sequencing Gels. Sometimes these are useful for removal of compressions (see Reading a Sequence, below). The gel mix contains 25% deionized formamide, 5.8% acrylamide, 0.2% N,N'-methylenebisacrylamide, 8 M urea, 1× TBE. After electrophoresis, 20% ethanol is added to the fixing solution to help removal of formamide from the gel.

Polymerization

In the following procedure capillaries can be replaced by Eppendorf tubes, and solutions can be mixed in an Eppendorf centrifuge.

Four aliquots (2 μl) of [α-^{32}P]deoxyadenosine triphosphate (specific activity 400 Ci/mmol) are dried *in vacuo* in siliconized glass tubes. Eight capillaries are drawn out, and each is rested in a siliconized glass tube. It is convenient to arrange the tubes in a 4 × 2 matrix. Each hybridized clone is distributed in 2-μl aliquots along each row of the four drawn-out capillaries; i.e., row 1 = clone 1, row 2 = clone 2. Four capillaries are drawn out and rested in the four tubes containing dried-down [α-^{32}P]deoxyadenosine triphosphate; to the first capillary is added 2 μl of dTTP^0 mix and 2 μl of ddTTP; to the second, 2 μl of dCTP^0 mix and 2 μl of ddCTP; to the third, 2 μl of dGTP^0 mix and 2 μl of ddGTP; to the fourth, 2 μl of dATP^0 mix and 2 μl of ddATP. It is convenient to arrange these capillaries as a third row of the matrix of clones; i.e. row 1 = clone 1, row 2 = clone 2, row 3 = dNTP^0

[34] F. Sanger and A. R. Coulson, *FEBS Lett.* **87**, 107 (1978).

mix + ddNTP + dried-down [α-^{32}P]dATP. The mixture of dTTP^0 and ddTTP is blown out into the glass tube to take up the radioactive label. The solution is sucked into the capillary. (Transfer of radioactivity can be checked with a Geiger counter.) Klenow polymerase (0.4 μl) is added to the capillary, and the liquid is blown out again into the glass tube. To ensure thorough mixing of the enzyme with the triphosphates, the drop of liquid is rolled around the bottom of the tube. The solution is taken back into the capillary, and 2-μl aliquots of it are added to the first capillary of each row of clones (the T column). A stopwatch is started ($t = 0$), and the primer/template and triphosphates/enzyme mixed thoroughly by blowing them up and down the capillary.

The procedure is now repeated except that additions are made in turn to the second, third, and fourth capillaries of each row of clones with the dCTP^0 mix + ddCTP, then with dGTP^0 mix + ddGTP, and finally with dATP^0 mix + ddATP. This should take about 5 min.

After $t = 15$ min, 1 μl of 0.5 mM dNTP chase is added to each of the clones in the T column. The content of each capillary is blown out into its glass tube, and the capillary is discarded. Then the chase is repeated with the C column, G column, and A column: in each column a 15-min reaction time should elapse before the chase is added.

After $t = 30$ min, the reaction in the T column is terminated by addition of formamide dyes (4 μl). This procedure is repeated with the C, G, and A columns: before addition of the formamide dyes to each column, a full 15-min chase reaction time should have elapsed. The open tubes are placed in a boiling water bath for 3 min to reduce the volume of the samples.

Gel Electrophoresis

The comb is removed from the gel and the slots are flushed out with running buffer to remove unpolymerized acrylamide. The tape at the bottom of the gel is cut, and the gel is clamped into the gel apparatus. Buffer (1 liter of 1× TBE) is distributed equally between the top and bottom reservoirs. The slots are washed out with buffer. Meanwhile the samples are removed from the boiling water bath and allowed to cool on the bench, *not in ice.* Portions of the samples (2 μl) are sucked into a drawn-out capillary and loaded at the bottom of the slots. Immediately prior to loading, each slot is again flushed out with running buffer. It is inadvisable to try to load more than about 1.5 μl of each sample from the capillary.

The gel is run for 1.5 hr at 30 mA (about 1.1 kV) when the Bromphenol blue dye should have reached the bottom. The plates are removed from the apparatus and prized apart. The gel should adhere to the nonsiliconized plate: if it adheres to both, the plates can be separated under 10% acetic acid and the upper plate carefully lifted off. A piece of plastic mesh is placed over the gel, which is immersed in a bath of 10% acetic acid for 10 min; the mesh prevents the gel from floating from away. Then the gel is taken from the bath, and excess acetic acid is removed by rolling a 25-ml pipette over the surface. Several layers of tissue paper are placed on top of the gel, and the roller is used again to squeeze out surplus liquid. The tissue paper is carefully removed, and the gel is covered with Saran Wrap. Alternatively the gels are transferred to Whatman 3 MM paper and dried with a Bio-Rad 224 vacuum gel drier. Gels are autoradiographed for 12–24 hr with Kodak X-Omat film.

Reading a Sequence

A number of characteristics help in the interpretation of sequencing gels. For example, bands in each sequencing track are usually not of equal intensity, and for doublets the following rules apply: the upper C is always *more* intense than the lower C; the upper G is always *more* intense than the lower G if the Gs follow a T; the upper A is often *less* intense than the lower A. It is advisable when reading a gel to read the spaces as well as the band and to ensure that the proposed sequence fits exactly into the space available.

In addition a number of artifacts occur, the most serious being "pileups" or "compressions."

1. A pileup seems to arise during synthesis when Klenow polymerase reaches a persistent secondary structure in the template, thus causing the enzyme to stop. The problem may be overcome by sequencing recombinant M13 template in which the opposite strand of the DNA has been cloned. Alternatively, the secondary structures in the template appear to be destabilized by minimizing the ionic strength of the sequencing reaction mixtures (M. D. Biggin and T. J. Gibson, unpublished work).

2. The compression may arise from the persistence of local secondary structure in the copied DNA that is not denatured on the urea–polyacrylamide gel. These structures seem to result in compression of the spacing between bands and are usually followed by a slight expansion of spacing above; in extreme cases these compressions appear as bands in more than one track. One remedy is to sequence the opposite strand of the DNA, when the compression should be displaced by a few residues. Alternatively, denaturation is achieved sometimes by the use of a formamide sequencing gel (see above).

Compilation and Analysis of Sequences

DNA sequence data are best stored and manipulated with the aid of a computer. For this purpose a range of computer programs have been written[35-38]; some are described briefly below.

[35] R. Staden, *Nucleic Acids Res.* **4,** 4037 (1977).

BATIN: Sequences deduced from autoradiograms of gels (gel readings) are entered into the computer and given file names.

DBCOMP: Compares sequences entered in BATIN with preexisting data and identifies any homologous matches. The different gel readings are also compared among themselves.

DBUTIL: This program keeps a record of preexisting gel readings and the relationships between them, the data base. It allows new gel readings to be entered into the data base and joined onto the preexisting data. If any gel readings overlap two preexisting blocks of data ("contigs" = contiguous sequences) the program allows them to be joined together. Both preexisting contigs and incoming data may be corrected by editing. The sequence of any contig may be displayed, showing all the individual gel readings comprising that piece of sequence. Incoming DNA sequences can be complemented to conform with the preexisting orientation.

DBAUTO[38]: This is a method for automated comparison of new gel readings with each other and existing readings. The data base and consensus are automatically updated accordingly.

CONSEN: Sets up a file containing a sequence that is a consensus of all gel readings in any contigs, allowing for any ambiguities. This file may be used in subsequent manipulations.

TRANMT: Translates DNA sequence in a consensus file into protein sequence using a prescribed or the universal genetic code.

SQRVCM: Computes the *reverse* and complement of a consensus file.

TRANDK: Sets up a file of protein sequence data that is derived from a consensus file.

MWCALC: Uses a protein sequence file set up in TRANDK to calculate the amino acid composition, molecular mass, and polarity index of a protein.

CODSUM: Computes the codon usage in any given gene.

HAIRGU: Searches for and reports any secondary structures in a consensus sequence. The user defines the maximum loop size and minimum base-paired stem. This includes G:U base pairs.

SEQLST: Lists a sequence in any named file.

SEQEDT: Allows editing of a consensus file.

DIAGON[39]: This is a user-interactive program for comparison of two nucleic acid or protein sequences. It displays signficant matches on a diagonal matrix. The parameters within which a match is considered

---

[36] R. Staden, *Nucleic Acids Res.* **5**, 1013 (1978).

[37] R. Staden, *Nucleic Acids Res.* **6**, 2601 (1979).

[38] R. Staden, *Nucleic Acids Res.* **10**, 4731 (1982).

[39] R. Staden, *Nucleic Acids Res.* **10**, 2951 (1982).

significant are defined by the operator. The protein comparison version is based on a proportional matching option.[40] This involves calculating a score at each position in the comparison matrix by summing points found when looking forward and backward along a diagonal line of given length. This length is called the span. The algorithm uses a score matrix for every possible pair of amino acid substitutions. This matrix, MDM78, was calculated from accepted point mutations of 71 families of related proteins and found to be the most powerful score matrix for the detection of distant relationships.[41]

Results

The DNA sequence of the *unc* operon determined as described above has been summarized elsewhere.[13] With protein sequences (determined as described below) and other information (summarized by Walker *et al.*[13]) the order of genes shown to scale in Fig. 1d has been established. The first gene in the operon (gene 1) appears not to code for a structural component of the enzyme complex. It may influence assembly.[13]

II. *In Vitro* Transcription Methods

Initially, the promoter for the *unc* operon was identified tentatively by comparison of the DNA sequence with the canonical sequence for prokaryotic promoters.[42] Experimental verification for this proposal was obtained by *in vitro* transcription of DNA fragments, as described below.[14] These results have been recently confirmed independently.[42a]

Methods

Escherichia coli RNA polymerase was a gift from Dr. A. A. Travers. Reaction mixtures[43] contained the following: 0.04 M Tris-HCl, pH 7.9, 0.01 M MgCl$_2$, 0.075 M KCl, 0.006 M 2-mercaptoethanol, 0.0001 M EDTA, 0.002 M ATP, 0.001 M GTP, 0.00025 M CTP, 0.000004 M [α-^{32}P]UTP (50 Ci/mmol, Amersham International), and 5 nM DNA fragment. Reaction mixtures were preincubated at 32° for 5 min, and RNA synthesis was started by the addition of RNA polymerase holoenzyme (100 nM). The reactions were performed at 32° for 30 min and terminated by the

[40] A. D. McLachlan, *J. Mol. Biol.* **61**, 409 (1971).
[41] M. O. Dayhoff, *in* "Atlas of Protein Sequence and Structure," Vol. 5, Supple. 3, p. 356. National Biomedical Research Foundation, Washington, D.C., 1978.
[42] M. Rosenberg and D. Court, *Annu. Rev. Genet.* **13**, 319 (1979).
[42a] H. Kanazawa, K. Mabuchi, and M. Futai, *Biochem. Biophys. Res. Commun.* **107**, 568 (1982).
[43] A. A. Travers, *J. Mol. Biol.* **141**, 91 (1980).

addition of an equal volume of formamide dye mix (see Section I). The products separated by PAGE in the presence of urea as denaturant and autoradiographed as described above. For end-labeling of RNA transcripts with [α-^{32}P]GTP or ATP (14 Ci/mmol) the reaction conditions were as above except that the nucleotide concentrations were 0.05 mM for ATP or GTP and 0.125 mM for GTP and UTP.

Results

An active promoter in the region preceding gene 1 was demonstrated by *in vitro* transcription of a restriction fragment encompassing it [Fig. 2 E(i)]. The length of the transcript was approximately 135 bases (Fig. 3) in agreement with its predicted length [Fig. 2 E(i)]. The size of transcript made from a longer restriction fragment covering the region also agreed with the predicted length [Fig. 2 E(ii)].[14] No *in vitro* transcript was made from a DNA fragment covering the region of the gene 1 and protein a junction [Fig. 2 E(iii)], although a weak promoter has been reported.[44] The construction of a deletion mutant in gene 1[45] still able to make ATPase eliminated the possibility of a promoter in this region, as proposed by Kanazawa *et al.*[46]

III. N-Terminal Protein Sequence Analysis of F₁ Subunits

In order to determine the starts of the genes for F$_1$ subunits, we devised a simple procedure for N-terminal protein microsequence analysis of material recovered from polyacrylamide gels. After electrophoresis, the proteins are stained briefly with a dye and eluted by electrophoresis. Prior to sequence analysis, the protein is selectively attached to a porous-glass derivative. Contaminating salts or dyes do not attach and are removed in a simple washing step. The immobilized protein is then sequenced in a solid-phase microsequencer. This technique was applied successfully to the *E. coli* F$_1$ α, β, γ, and ϵ subunits.[12]

Materials and Methods

The *p*-phenylene diisothiocyanato derivative of 3-aminopropyl glass (DITC-glass) was prepared from controlled pore glass (7.5 nm pore size).[47] F$_1$-ATPase was prepared from *E. coli* membranes and then purified further

[44] J. Nielsen, B. B. Jorgensen, F. G. Hansen, P. E. Petersen, and K. von Meyenburg, *EBEC Rep.* **2**, 611 (1982).

[45] N. J. Gay, *J. Mol. Biol.*, submitted for publication.

[46] H. Kanazawa, K. Mabuchi, T. Kayano, T. Noumi, T. Sekiya, and M. Futai, *Biochem. Biophys. Res. Commun.* **103**, 613 (1981).

[47] E. Wachter, W. Machleidt, H. Hofner, and J. Otto, *FEBS Lett.* **35**, 97 (1973).

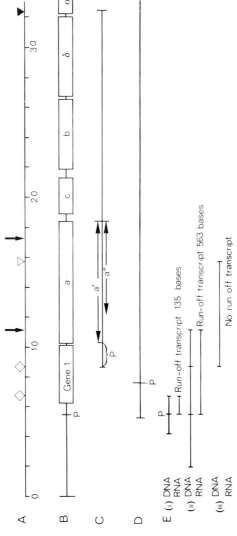

FIG. 2. Experiments to detect the promoter for the *unc* operon. (A) Physical map. ◇ *Hind*III, ↓ *Bam*HI, ▽ *Hae*III, ▶ *Eco*RI restriction sites. (B) Genes drawn to scale. P represents the site of initiation of transcription proposed by Gay and Walker[9] and verified as described. (C) P represents the location of a weak promoter proposed by Nielsen *et al.*[44] (D) P is the position of a promoter proposed by Kanazawa *et al.*[46] and disproved by Gay.[45] (E) (i)–(iii) Predicted lengths of *in vitro* transcripts from various DNA fragments.

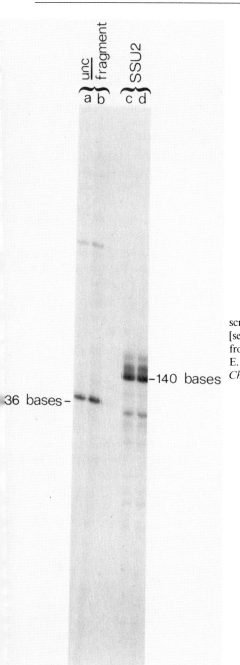

FIG. 3. Autoradiograph of gel showing *in vitro* transcript of DNA fragment containing proposed promoter [see Fig. 2B and 2E(i)] compared with transcript made from the SSU2 promoter [M. J. Ryan, B. Ramamoorthy, E. L. Brown, H. J. Fritz, and H. G. Khorana, *J. Biol. Chem.* **254**, 5802 (1979)].

TABLE I
PROGRAM FOR THE SOLID-PHASE EDMAN DEGRADATION EMPLOYING A
MODIFIED MICROSEQUENCER[a]

Time (min)	Function	Reagent
0–17	Coupling	Phenyl isothiocyanate (10%) in acetonitrile and buffer
17–19	Washes	Benzene and methanol alternately
19–25	Cleavage	Trifluoroacetic acid
25–26	Washes	Methanol

[a] All reactions were carried out at 58°. The buffer used in coupling was methanol containing 0.5% triethylamine. In the cleavage reaction the first 100 μl were collected for conversion.

by gel filtration through Ultrogel AcA-34.[48] Reagents employed in protein sequencing were prepared and purified as follows:

Buffer: 2% triethylamine (v/v) in 72% aqueous methanol. Triethylamine (Pierce, Sequenal grade) was refluxed over ninhydrin for 1 hr and then double-distilled immediately prior to being made up into buffer.

Trifluoroacetic acid: Koch Light trifluoroacetic acid, 1 kg, was refluxed over $K_2Cr_2O_7$ for 1 hr and then distilled. The distillate was stored over neutral alumina (dried) either overnight or prior to the second distillation. Then the anhydrous trifluoroacetic acid was decanted from alumina, refluxed with dithiothreitol (DTT) (approximately 50 mg/100 ml) for ca. 1 hr and then distilled. Double-distilled trifluoroacetic acid was used in the sequencer without further addition of DTT.

Coupling reagents

Phenyl isothiocyanate was used as a 10% solution in acetonitrile (Fison's HPLC grade). The reagent was obtained from Rathburn Chemicals Ltd., Walkerburn, Scotland and was used either without further purification or after redistillation *in vacuo* at 0.5–1.0 mM Hg.

Acetonitrile: Use without further purification.

Solvents

Methanol and benzene: dry, analytical grade solvents from Merck were used without further purification.

Dimethylformamide is used as solvent in the activation of aminopropyl glass with diphenyl isothiocyanate. It is deionized by passage through a column of dried, mixed (acid:base 1:1, w/w) alumina (Woelm).

Diphenylisothiocyanate (Sigma) and 3-aminopropyltriethoxysilane (Pierce). Both were used without further purification.

[48] M. Saraste, M. J. Runswick, J. Deatherage, M. A. Naughton, and J. E. Walker, *Biochemistry,* submitted for publication (1982).

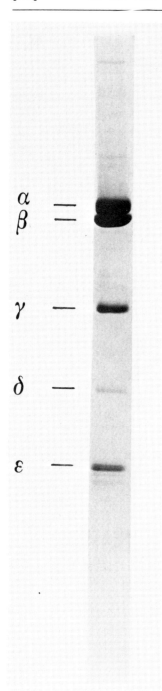

FIG. 4. The *Escherichia coli* F$_1$-ATPase complex. The apparent molecular weights of subunits are as follows: α, 57×10^3; β, 52×10^3; γ, 32×10^3; δ, 20–22×10^3; and ϵ, 12×10^3.[3] For sequencing experiments, about 5 mg of F$_1$ protein were applied to the gel in 10 lanes. This corresponds to about 10 nmol each of α and β and 3 nmol of the γ and ϵ proteins. The recoveries, however, were not determined.

TABLE II. PROTEIN SEQUENCES IN THE AMINO-TERMINAL REGIONS

```
α Subunit
  Protein                                    M    Q    L    N    S    T    E
  DNA       G  G [G  G  A] C  T [G  G  A  G] C [ATG] CAA  CTG  AAT  TCC  ACC  GAA
β Subunit
  Protein                                    A    T    G    K    I    V
  DNA       T  A [G  A  G  G] A  T  T  T  A  A  G [ATG] GCT  ACT  GGA  AAG  ATT  GTC
γ Subunit
  Protein                                    A    G    A    K    E    I
  DNA       T [G  A  G  G  A  G] A  A  G  C  T  C [ATG] GCC  GGC  GCA  AAA  GAG  ATA
ε Subunit
  Protein                                    A    M    T    Y    X    L
  DNA       A  A  T  C [G  G  A  G  G  G] G  T  G  A  T [ATG] GCA  ATG  ACT  TAC  CAC  CTG
```

[a] Protein sequences were derived from material eluted from bands from 20 slots of a gel run as in Fig. 4. ATG codons in boxes are proposed initiation codons and are preceded by boxed sequences complementary to the end of ribosomal RNA [J. Shine and L. Dalgarno, *Proc. Natl. Acad. Sci. U.S.A.* **71**, 1342 (1974)]. The one-letter code is used for amino acids in the

Gel Electrophoresis and Recovery of Proteins

Electrophoresis was performed in 15 to 30% acrylamide gradient slab gels (15 × 15 cm × 1.5 mm thick) or in 25% acrylamide gels surmounted by a stacking gel with 20 loading wells. Gels and buffers were prepared according to Laemmli.[49] The sample was dissolved in five parts of 10% SDS, with five parts of H_2O and one part of sample buffer. Sample buffer contained 50% sucrose, 90 mM Tris, 90 mM boric acid, 2.5 mM EDTA, 0.3% xylene cyanole FF, and 0.3% Bromphenol blue. Electrophoresis was performed until the Bromphenol blue reached the bottom of the gel. Proteins were detected by staining for 10 min in a solution of 0.2% Coomassie Blue in 50% methanol, 7% acetic acid, and destained in a solution of the same composition without dye. Bands were excised and cut into conveniently sized pieces with a razor blade. It is important not to macerate them (e.g., by pushing the gel through a syringe needle, since this releases contaminating materials that interfere with subsequent chemical manipulations. The gel slices were then put into tubes 15 cm long × 7 mm internal diameter, with one end tapered down to 1.5 mm, made from disposable plastic pipettes. A piece of dialysis tubing 1 cm across × 8 cm long was pushed onto the tapered end to give a tight seal and clipped off at its open end with a Spectrum Medical Industries plastic closure. Smaller polypeptides were collected in appropriate low molecular weight cutoff tubing (Spectrapor). The plastic tube with attached dialysis tubing was introduced

[49] U. K. Laemmli, *Nature (London)*, **227**, 680 (1970).

OF SUBUNITS OF *Escherichia coli* F₁-ATPase[a]

I	X	E	L									
ATC	AGC	GAA	CTG									

Q	V	I	G	A	V	V	D	V	E	F	P	Q	D
CAG	GTA	ATC	GGC	GCC	GTA	GTT	GAC	GTC	GAA	TTC	CCT	CAG	GAT

X	S	K	I	A	S	V	Q	N
CGT	AGT	AAG	ATC	GCA	AGC	GTC	CAG	AAC

D	V	V	S	A	E
GAC	GTC	GTC	AGC	GCA	GAG

protein sequences; in the DNA sequences, the d representing deoxy and the hyphens representing phosphodiester linkages have been omitted.

into a disc gel apparatus and filled with elution buffer. This was either 50 mM sodium bicarbonate, 0.1% SDS, 50 mM Tris-acetate, pH 8.0, in 0.1% SDS or electrophoresis tank buffer.[49] The bottom compartment, containing the dialysis sac, also contained the anode. After electrophoresis for 18 hr at 100–200 V, the dialysis sac containing the eluted protein was removed and dialyzed in 10% ethanol for 24–36 hr and then sometimes freeze-dried. At this stage the protein solution contained SDS and Coomassie Blue, which do not interfere with coupling of the protein to porous-glass derivatives.

Covalent Attachment of Protein to DITC-Glass

Freeze-dried eluted proteins were usually dissolved in 0.1 M sodium hydrogen carbonate containing 10% 1-propanol (Pierce). Redissolved proteins were then added to DITC-glass (25 mg), and the volume was adjusted by addition of more buffer until the glass beads were just covered by the liquid.

Sequence Analysis

This was performed with a computer-controlled solid-phase microsequencer of construction and design to be described in detail elsewhere. Briefly, the main features of the instrument are that dead spaces in the reagent delivery system and the column size have been minimized. All functions including the pumps are computer-controlled, and the volumes of reagents and solvents used per cycle and reaction times have been drastically reduced. The instrument is now operated with a 26-min degradation cycle,[50] described in Table I. These combined features serve to minimize exposure of proteins to reagents (and minor contaminants therein). This has the effect of reducing side reactions of the protein and helps to eliminate contaminants in the phenylthiohydantoins released, which would otherwise interfere with their identification. Phenylthiohydantoins were identified by reverse-phase chromatography on Zorbax-ODS as described earlier.[51] In some experiments lysine and phenylalanine were not separated. Back hydrolysis and amino acid analysis were employed to confirm assignments.[52]

Results

N-Terminal Sequences. The F_1-ATPase preparation contained low amounts of δ subunit, which was not recovered from the gel (see Fig. 4). The N-terminal sequences derived for the other F_1 subunits allowed their corresponding gene starts to be identified, as illustrated in Table II.

Acknowledgments

We thank G. Winter, M. D. Biggin, T. J. Gibson, and A. R. Coulson for helpful comments. Nicholas J. Gay was supported by an MRC Studentship.

[50] J. E. Walker, R. Blows and I. Fearnley, manuscript in preparation.
[51] C. J. Brock and J. E. Walker, *Biochemistry* **19**, 2873 (1980).
[52] E. Mendez and C. Y. Lai, *Anal. Biochem.* **68**, 47 (1975).

[21] Biogenesis of Purple Membrane in Halobacteria

By Dorothea-Ch. Neugebauer, Horst-Peter Zingsheim, and Dieter Oesterhelt

The biogenesis of purple membranes in halobacteria proceeds in several steps: the biosynthesis of the membrane constituents—lipids and protein—similar to other biological membranes; the production of crystalline arrays of bacteriorhodopsin forming the purple membrane patches within the cell membrane[1,2]; and the growth of the purple membrane patches under continued bacteriorhodopsin synthesis. Only this last process is described here.

Purple membranes are isolated as round sheets of an approximate

[1] A. E. Blaurock and W. Stoeckenius, *Nature (London), New Biol.* **233**, 152 (1971).
[2] P. N. T. Unwin and R. Henderson, *J. Mol. Biol.* **94**, 425 (1975).

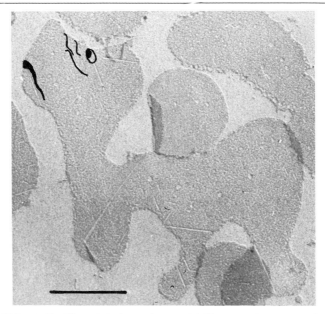

FIG. 1. "Monster" sniffing at electron microscopist. These membranes were isolated after 50 hr of growth. Note the cracks that mark the lattice orientation.[1,5] There are also two young, round membranes and another large irregularly shaped membrane in the right top corner. Ta/W shadowed. The bar marks 0.5 μm.

diameter of 0.5 μm or less from cells in which bacteriorhodopsin synthesis has been induced by limited aeration.[3] Much larger purple membranes of irregular shape, which cover most of the cell surface (up to 80%), are found after prolonged growth (Fig. 1). These large membranes, like the small ones, are monocrystalline. Two alternative explanations for the mode of their formation are possible: (a) addition of bacteriorhodopsin molecules to a single nucleation site or (b) fusion and recrystallization of several membrane patches during their growth. Freeze fracturing of whole cells revealed several sites of nucleation, since often more than one patch of purple membrane per cell is found[1] (Fig. 2).

Fusion and recrystallization of purple membranes can be demonstrated if (a) intermediate stages in the formation of large patches are detected and recognized[4]; and (b) a method is used that is sensitive enough to detect not only the overall crystallinity, but that of very small regions within a single purple membrane patch.[5,6]

[3] D. Oesterhelt and W. Stoeckenius, this series, Vol. 31, p. 667.

[4] D.-Ch. Neugebauer, H. P. Zingsheim, and D. Oesterhelt, *J. Mol. Biol.* **123**, 247 (1978).

[5] D.-Ch. Neugebauer and H. P. Zingsheim, *J. Mol. Biol.* **123**, 235 (1978).

[6] D.-Ch. Neugebauer, this series, Vol. 88, p. 235.

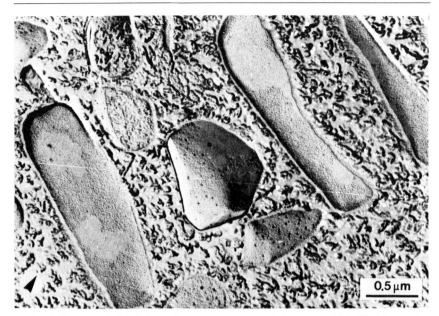

FIG. 2. Freeze-etched halobacteria, spray frozen [L. Bachmann and W. W. Schmitt-Fumian, *in* "Freeze Etching Techniques and Applications" (E. L. Benedetti and P. Farard, eds.), p. 73. 1973], shadowed with Ta/W). Note that the bacteria contain more than one patch of purple membrane.

Electron microscopic screening of purple membranes from bacteria grown under carefully controlled conditions satisfies requirement (*a*),[4] and metal decoration provides a sensitive method for observing the crystallinity of small areas (*b*).[5,6] Confidence in the results is strengthened by the ability to distinguish purple membrane from other membrane areas by electrostatic ferritin labeling[7] and by successful recrystallization *in vitro*.[7,8]

Methods

Growth of Bacteria and Isolation of the Purple Membrane

Halobacterium halobium strain R_1M_1 is grown in a MicroFerm fermentor (New Brunswick Scientific, New Brunswick, New Jersey) with illumination by the bank of daylight fluorescent tubes of the MicroFerm. The medium contains, in 1000 ml: NaCl, 250.0 g; $MgSO_4 \cdot 7H_2O$, 20 g; trisodium citrate $\cdot 2H_2O$, 3.0 g; KCl, 2.0 g; Oxoid Bacteriological Peptone L 37

[7] D.-Ch. Neugebauer, D. Oesterhelt, and H. P. Zingsheim, *J. Mol. Biol.* **125**, 123 (1978).
[8] K. A. Fischer, K. Yanuyimoto, and W. Stoeckenius, *J. Cell Biol.* **77**, 611 (1978).

(Oxoid Deutschland GmbH, 4230 Wesel), 10 g. The medium is prefiltered through soft filter paper and autoclaved in the fermenting vessel for 20 min at 120°. The culture is aerated with 120 liters of air per hour and stirred at 40° with 300 rpm. The inoculum (400 ml) consists of bacteria from an agar plate cultivated in a Fernbach flask for 40–68 hr at 40° on a linear shaker (105 cycles/min, amplitude 2.7 cm).

Careful preparation of the inoculum is important if purple membrane biogenesis is to be studied. Bacteria in the stationary phase are not suitable for an inoculum because such bacteria usually contain large amounts of purple membrane, which would contribute a rather high portion of the total purple membrane in the early stages. Thus, early developmental forms would be obscured.

Beginning 32 hr after the start of the culture, samples of 400 ml are drawn every 4–5 hr, and purple membranes are isolated. The cells are spun down at 10,000 g for 15 min, and the pellet is dissolved in 3 ml of distilled water containing 1 mg of DNase (Rinderpankreas, Roth OHG Karlsruhe, Federal Republic of Germany). The clear lysate is directly layered on a linear 30% to 50% sucrose density gradient with a 2-ml 60% sucrose cushion at the bottom in tubes of a Spinco L2-65 SW 27 rotor. After centrifugation for 17 hr at 100,000 g and 15°, the material of the purple band is collected with a Pasteur pipette and dialyzed against 0.02% (w/v) NaN_3. Dialysis is preferred to centrifugation for the exchange of medium in order to keep the forces exerted on the purple membrane as low as possible (see below). Membranes are then stored at 4° at a concentration of about 6 mg of protein per milliliter. Protein concentration is determined by absorption measurement at the bacteriorhodopsin's maximum at 568 nm. The molar extinction is 63,000 M^{-1} cm^{-1}, and the molecular weight is 26,000.

Ferritin Labeling

Membranes at a protein concentration of 280 μg/ml are incubated in a total volume of 250 μl with horse spleen ferritin (560 μg/ml in 5 mM sodium acetate buffer, pH 4.4–5.2) for 30 min at room temperature. The suspension is then centrifuged at 10,000 g for 10 min, washed twice by centrifugation in 0.5 ml of buffer, and resuspended in 50–100 μl of the same buffer. The membrane–ferritin complexes form aggregates that can, however, be sufficiently resuspended by passing the suspension through a fine Pasteur pipette until no aggregates are visible to the naked eye.

Electron Microscopy

A drop of the suspension containing purple membrane or membrane–ferritin complex is allowed to stand for 30–60 min. During this time, incompletely resuspended aggregates sediment. Then a support, i.e., a carbon-coated grid (thickness of the carbon-coated film is 3–4 nm) is brought into contact with the suspension for 1–2 sec, washed with a few milliliters of buffer followed by distilled water or by floating the specimens on three drops of buffer and then three drops of distilled water for 5 min each time. The air-dried specimens are shadowed with or without previous decoration with silver. For specimens to be decorated, it is useful to chose the conditions for adsorption such that the extracellular face is preferentially absorbed,[8,9] since only the intracellular face reveals the crystalline pattern of the protein lattice.[10] Thorough washing of the specimens is important, since silver decoration is extremely sensitive to contamination of small amounts of organic materials, such as sugar or Tris.[5,6]

Silver Decoration and Shadowing

For silver decoration the metal is vacuum-deposited at right angles to the specimen surface. With small amounts of metal no continuous film is formed, but the metal collects at nucleation sites. This effect is particularly pronounced with a low-melting metal, such as gold and silver.[11]

On the cytoplasmic face of the purple membrane, the silver grains mark the lattice of bacteriorhodopsin.[10] For optimal results the amount of metal is critical. With the purple membrane an average thickness of the metal film of 0.6 nm (measured by a Balzers quartz crystal thin film monitor, QSG 201) is recommended.[5,6]

Shadowing at an angle of 45° with platinum–carbon or with tantalum–tungsten by electron beam evaporation (average thickness 0.6–0.7 nm) enhances the contrast and reveals the surface morphology of the membranes.

Silver-decorated specimens are not stable. Therefore a thin film of carbon (1–2 nm) should be deposited immediately after decoration and shadowing. Nevertheless, since the silver keeps moving at a slow rate, the specimens should be photographed within 24 hr.[6]

Optical Diffraction

For contrast enhancement the original micrographs are contact copied on FO 81 p plane film (Agfa) followed by a second copying on FO 71. This film material is chosen because it lacks intrinsic phase structures that would produce spurious diffraction. The optical diffraction patterns are recorded

[9] K. A. Fischer, this series, Vol. 88, p. 230.
[10] H. P. Zingsheim, R. Henderson, and D.-Ch. Neugebauer, *J. Mol. Biol.* **123,** 275 (1978).
[11] G. A. Basset, J. W. Menter, and D. W. Pashley, *in* "Structure and Properties of Thin Films" (C. A. Neugebauer, J. B. Newkirk, and D. A. Vermileya, eds.), p. 11. Wiley, New York, 1959.

on Ilford FP4 (35 mm) film, and the angles between the reflections are measured with a low-power light microscope equipped with a graded rotatable stage. Alternatively, a graded rotatable stage can be used on the optical diffractometer.

In Vitro Recrystallization

A membrane suspension (100 μl containing 1–2 mg of protein per milliliter) isolated after 40 hr of growth is heated in a water bath to 50° at a rate of 20° per minute, kept at 50° for 10 min, and cooled to room temperature. The suspension is used without further dilution. No denaturation of bacteriorhodopsin takes place under these conditions.

Results

The bacteria grow logarithmically from the beginning for about 60 hr if an inoculum as described is used (Fig. 3a). Purple membranes appear after about 32 hr, well before the end of the logarithmic growth phase. Membranes isolated at such an early stage are almost ideally round, about 0.5 μm in diameter, and rather homogeneous in size (Figs. 3b, 4a).

As the total amount of purple membrane increases, the proportion of round membranes decreases rapidly with a concomitant increase in irregularly shaped larger membranes, which we call "old monsters" (Figs. 3b, 1). Their outlines suggest that they may have formed by fusion of smaller membranes. Investigation of their crystallinity reveals in most of them a monocrystalline lattice. However, two further species are found in addition

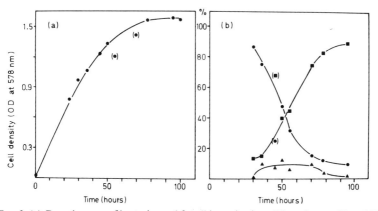

FIG. 3. (a) Growth curve of bacteria used for all investigations. The values at 55 and 78 hr are probably erroneous owing to some sedimentation of the cells before they are measured. (b) Decrease of round membranes (●), increase of "old monsters" (■), and intermediate appearance of "young monsters" (▲) during growth.

to the small round membranes and the "monsters." As expected for intermediate stages, these species reach their maximal concentration in the cell at the crossover point of the decrease of round membranes and the increase of "old monsters" (Fig. 3b). Both species have the characteristic outline of monsters, but contain several crystalline regions oriented randomly to one another.

One of these species is called "young monster" (Fig. 4b, c top,c bottom). It is most readily visualized by electrostatic labeling with native ferritin[7] at low pH.[10] "Young monsters" contain regions that bind ferritin in higher amounts than does either face of the purple membrane. Strong ferritin binding occurs not only around the edges of all purple membranes, but also on dividing lines running across membranes (Fig. 4) or on entrapped areas

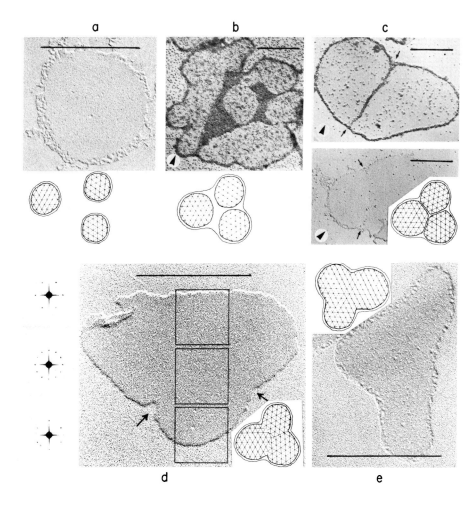

a   b   c

d   e

of fused patches (Fig. 4b). This suggests that the strong binding of the ferritin to the rim of the purple membrane is not just an edge effect. Further, shadowed membranes without bound ferritin often show fuzzy rims (Fig. 4a,e). Comparable areas are sometimes seen as lines or patches on the "young monsters" (Fig. 4c bottom). The fuzzy rim is fragile and can easily be removed by centrifuging the purple membranes in distilled water at high g forces. From this we conclude that areas that bind more ferritin are not purple membrane, but belong to the surrounding cell membranes, which can be entrapped during fusion of purple membranes.

The second intermediate stage between small, round sheets and "old monsters" are "middle-aged monsters." They differ from "young monsters" in the region separating the different lattices. This region is very sharp, not more than 2 unit cells wide, and cannot be labeled with ferritin (Fig. 2). Apparently membrane constituents other than those of purple membrane are removed from between the purple membrane patches during the transformation of "young monsters" into "middle-aged monsters." Finally, lattices of different orientation recrystallize to form the "old monsters" (Fig. 4e).

The recrystallization can also be effected *in vitro* by gently heating a membrane suspension containing a high proportion of "middle-aged monsters." This proportion of "middle-aged monsters" then decreases by 50%. At the same time their shape changes from very irregular to mostly oval[4] (Fig. 5). This occurs also during recrystallization *in vivo*.

Thus it appears that the biogenesis of purple membrane in the bacterium starts by nucleation of the crystallization of bacteriorhodopsin, giving rise to a number of patches to which bacteriorhodopsin is added, so that round

---

FIG. 4. All specimens are shadowed with Ta/W; those of panels d and e are also decorated with silver. The bar in each micrograph marks 0.5 μm. Underneath each developmental stage is a schematic drawing of the purple membrane surrounded by cell membrane. The protein lattice on the purple membrane is indicated. (a) Purple membrane isolated after 38 hr of culture. Note the fuzzy rim. (b) "Young monster" incubated with ferritin at pH 4.4. Not all originally separated patches are entirely outlined by ferritin, but comparatively large thickly decorated areas are left on the face, indicating a different composition of the underlying membrane from the rest of the membrane. (c, top) "Young monster" with only a thin line decorated with ferritin. (c, bottom) "Young monster" incubated with ferritin at pH 6.0. At pH 6.0 no ferritin binds electrostatically to the purple membrane or the rim. Note the fuzzy rim and the dividing line between the arrows. (d) "Middle-aged monster." The line between the fused membranes can be visualized neither by shadowing nor by labeling with ferritin. Decoration with silver shows that the two patches have different lattice orientations. The angle between them is 30°. The arrows mark the beginning and the end of the boundary between different lattice orientations. The diffraction patterns are from the indicated areas. (e) "Old monster" from a culture grown for about 100 hr. The uniform lattice is revealed by silver decoration (not visible in the micrograph). See also Fig. 1.

(a)   (b)

FIG. 5. Two typical examples showing the shape of membranes before (a), and after (b) heating to 50°.

membranes are formed. As these grow, they come into contact with each other, their relative lattice orientations being random. The intervening membrane is gradually eliminated. Originally separate membrane areas can still be discerned by silver decoration, which shows that their lattice orientations are still random. Finally, the bacteriorhodopsin recrystallizes—a process that does not require metabolic energy—to form a large single crystal[4] (Fig. 4).

Although bacteriorhodopsin is functionally active in the monomeric state,[12] its arrangement in purple membrane patches was shown to be much more efficient in transducing light energy compared to the same cellular concentration of bacteriorhodopsin in the nonpurple membrane state.[13] However, the question arises why the purple membrane should need to recrystallize after fusion. So far, this seems to have no physiological effect. Recrystallization *in vitro* occurs within 10 min at 50°, and the bacteria live at 40°. An obvious explanation seems to be that recrystallization is merely a consequence of the properties of bacteriorhodopsin and its environment.

[12] N. A. Dencher and M. P. Heyn, *FEBS Lett.* **108**, 307 (1979).
[13] R. Hartmann, H.-D. Sickinger, and D. Oesterhelt, *FEBS Lett.* **82**, 1 (1977).

[22] Isolation of the Bacterioopsin Gene by Colony Hybridization

By HEIKE VOGELSANG, WOLFGANG OERTEL, and DIETER OESTERHELT

Bacteriorhodopsin in the cell membrane of halobacteria is one of the best-studied membrane proteins. It occurs in two-dimensional crystalline arrays called the purple membrane (see reviews[1-3]) and its primary structure

[1] W. Stoeckenius and R. A. Bogomolni, *Annu. Rev. Biochem.* **51**, 587 (1982).
[2] M. Eisenbach and S. R. Caplan, *Curr. Top. Membr. Transp.* **12**, 165 (1979).

has been elucidated by protein and DNA sequencing.[4-6] Models of bacteriorhodopsin's spatial structure have been proposed,[7] and three-dimensional structure analysis will become possible with the help of bacteriorhodopsin crystals.[8] Studies on the function of this light-driven proton pump include chemical modification of its amino acid side chains, the use of various retinal analog compounds, and the complete set of modern biophysical methods. Therefore it can be assumed that in principle the structure–function relationship in bacteriorhodopsin can be elucided at atomic resolution. However, chemical modification is limited by specificity of reagents and availability of functional groups on the amino acid side chains, and usually it is difficult to interpret its effects on function. Thus it is desirable to modify the bacteriorhodopsin molecule by means of genetic manipulation. This would allow removal, insertion, or exchange of individual amino acids or stretches of the protein's sequence for functional studies. The necessary steps toward this goal are isolation, modification, and expression of the bacterioopsin gene. Our procedure for the isolation of the gene is described in this chapter. It involves the establishment of a gene library of the halobacterial genome in *Escherichia coli* using the plasmid pBR322, transformation of *E. coli* with insert carrying plasmid DNA, and identification of the bacterioopsin gene carrying clones by hybridization with a synthetic oligonucleotide.

Materials

Chemicals

Standard laboratory chemicals were from Merck and of analytical grade. Chemicals were obtained from Boehringer: nucleotide triphosphates (rATP and dNTP), polyribocytidylic acid, T4 DNA ligase, restriction endonucleases *Bam*HI, *Kpn*I, *Bgl*I, tetracycline, chloramphenicol, dithioerythritol, and the molecular weight marker DNAs; and from P-L Biochemicals (St. Goar, West Germany): restriction endonuclease *Mbo*I, terminal deoxynucleotidyltransferase, and the 14-base-long oligonucleotide (15′

[3] M. Ottolenghi, *Adv. Photochem.* **12**, 97 (1980).
[4] H. G. Khorana, G. E. Gerber, W. C. Herlihy, C. P. Gray, R. J. Anderegg, K. Nihei, and K. Biemann, *Proc. Natl. Acad. Sci. U.S.A.* **76**, 5046 (1979).
[5] Y. A. Ovchinnikov, N. G. Abdulaev, M. Yu. Feigina, A. V. Kiselev, and N. A. Lobanov, *FEBS Lett.* **100**, 219 (1979).
[6] R. Dunn, J. McCoy, M. Simsek, A. Majumdar, S. H. Chang, U. L. RajBhandary, and H. G. Khorana, *Proc. Natl. Acad. Sci. U.S.A.* **78**, 6744 (1981).
[7] D. M. Engelman, R. Henderson, A. D. McLachlan, and B. A. Wallace, *Proc. Natl. Acad. Sci. U.S.A.* **77**, 2023 (1980).
[8] H. Michel and D. Oesterhelt, *Proc. Natl. Acad. Sci. U.S.A.* **77**, 1283 (1980).

CTGGGCCTGCGATA 3') (custom synthesized). Endonuclease *Bst*EII was from New England BioLabs (Schwalbach, West Germany); bacterial alkali phosphatase (BAPF) from Worthington (Freehold, New Jersey); ampicillin from Sigma (St. Louis, Missouri); ethidium bromide and pancreatic RNase, analytical grade, from Serva (Heidelberg, West Germany); agarose from SeaKem (Rockland, Maine); cesium chloride from Baker (Gross-Gerau, West Germany); [γ-^{32}P]rATP and [α^{32}P]dCTP (3000 Ci/mmol) from Amersham (Braunschweig, West Germany); peptone and nutrient broth from Difco (Detroit, Michigan); peptone L-37 for halobacterial growth from Oxoid (Wesel, West Germany), and nitrocellulose filters from Schleicher & Schüll (Dassel, West Germany).

Strains

Escherichia coli 5K ($r_K^- m_K^+$ *leu⁻ thr⁻ thi⁻*)[9] was obtained from H. Schrempf (Würzburg). *Halobacterium halobium* NRL R_1M_1 (*rub⁻ vac⁻*) is a derivative of the wild-type strain NRC 34020[10] and was prepared from R1 in our laboratory. The *E. coli* vector plasmid pBR322[11] was obtained from W. Goebel (Würzburg) and can also be purchased from Boehringer (Mannheim) or other commercial sources.

Standard and Stock Solutions

Unless mentioned otherwise, all buffers and reaction mixtures were prepared from the following stock solutions.
4 *M* NaCl
30% (w/v) and 0.5 *M* sodium dodecyl sulfate (SDS)
Sodium phosphate, 0.5 *M*, pH 7.0; Tris-HCl 1 *M*, pHs 7.5 and 8.2
Potassium phosphate 1 *M*, pH 7.0
Sodium cacodylate, 1 *M*, pH 7.0 (freshly prepared)
$CaCl_2$, 1 *M*; $MgCl_2$, 1 *M*; sodium EDTA, 1 *M*, pH 8.2
Potassium-EDTA, 1 *M*, pH 7.5
$CoCl_2$, 0.1 *M*
Dithioerythritol (DTE), 0.1 *M*;
$Na_4P_2O_7$, 0.1 *M*;
Sodium-EGTA, 0.1 *M*, pH 7.5
TE buffer: 50 m*M* Tris-HCl, pH 7.5, 1 m*M* sodium EDTA
The following stock solutions were stored frozen at −20°: 50 m*M*, each:

---

[9] I. Hubacek and S. W. Glover, *J. Mol. Biol.* **50,** 111 (1970).
[10] W. Stoeckenius, R. H. Lozier, and R. A. Bogomolni, *Biochim. Biophys. Acta* **505,** 215 (1979).
[11] F. Bolivar, R. L. Rodriguez, P. J. Greene, M. C. Betlach, H. L. Heyneker, and H. W. Boyer, *Gene* **2,** 95 (1977).

rATP, dATP, dGTP, dCTP, dTTP; calf thymus DNA, 10 mg/ml (sonicated); poly(rC), 10 mg/ml; bovine serum albumin (BSA), 10 mg/ml; and phenol (freshly distilled and saturated at 60° with 1 M Tris-HCl, pH 7.5, and 50 mM sodium EDTA).

Media

> *Escherichia coli* complete medium "enriched nutrient broth" (ENB) containing, per liter: 5 g of Bacto-peptone, 1 g of glucose, 1.5 g of KH_2PO_4, 3.5 g of Na_2HPO_4, 5 g of NaCl, 5 mg of thiamin, and, if indicated, the antibiotics ampicillin (100 μg/ml), tetracycline (20 μg/ml), or chloramphenicol (100 μg/ml).
>
> *Halobacterium* complex medium, containing per liter[12]: 250 g of NaCl, 20 g of $MgSO_4 \cdot 7H_2O$, 3 g of sodium citrate, 2 g of KCl, 10 g of Oxoid peptone (code L-37). Adjusted to pH 7.2 before sterilizing.
>
> Difco agar, 15 g/liter, used for plating.

Methods

Digestion with Restriction Endonucleases

Restriction endonucleases *Mbo*I, *Bam*HI, *Bst*EII, *Kpn*I, and *Bgl*I were used according to the recommendations of the manufacturer without modification.

Agarose Gel Electrophoresis and "Southern" Blotting

DNA samples were analyzed by electrophoresis on 1% agarose gels using TBE buffer (90 mM Tris-OH, 2.7 mM sodium-EDTA, 60 mM boric acid; adjusted to pH 8.3 by addition of glacial acetic acid) or Tris–phosphate buffer (40 mM Tris-OH, 20 mM NaH_2PO_4, 18 mM NaCl, 2 mM sodium-EDTA).[13,14] After electrophoresis the gel was stained with ethidium bromide (2.5 μg/ml) and photographed under UV light (254 nm) using an orange filter (Heliopan, Gräfelfing, West Germany; 590 nm). For analysis of the gel pattern by hybridization, the DNA was blotted to a sheet of nitrocellulose filter according to the procedure of Southern[15] and hybridized to a radioactive DNA- or oligonucleotide probe.[16] Then an X-ray film (Linhardt X-ray 90) was exposed to the nitrocellulose filter using an intensifier screen.

[12] D. Oesterhelt and W. Stoeckenius, this series, Vol. 31, p. 667.

[13] J. A. Meyers, D. Sanchez, L. P. Elwell, S. Falkow, *J. Bacteriol.* **127,** 1529 (1976).

[14] G. G. Hayward, *Virology* **49,** 342 (1972).

[15] E. M. Southern, *J. Mol. Biol.* **98,** 503 (1975).

[16] R. W. Davis, B. Botstein, and J. R. Roth, eds., "Advances in Bacteriol Genetics." Cold Spring Harbor Lab., Cold Spring Harbor, New York, 1980.

Isolation of DNA

Total DNA from Halobacterium halobium

Halobacterium halobium was grown aerobically in 1.5 liters of *Halobacterium* complex medium at 40° to early stationary growth phase ($OD_{578} = 1.0$), harvested by centrifugation (12,000 g for 15 min), and resuspended carefully in 5 ml of basal salt (i.e., complex medium without peptone, pH 7.0). This cell suspension was added dropwise into 150 ml of lysis buffer (0.1 M Tris, pH 8.2, 1 mM sodium EDTA, 3%, w/v, sodium dodecyl sarcosinate, 200 μg of Pronase per milliliter) and incubated for 2 hr at 60°. The resulting lysate was extracted three times with the same volume of buffer-saturated phenol in an Erlenmeyer flask, which was moderately rotated on a laboratory shaker. To the aqueous phase, 15 ml of 10 M LiCl and 375 ml of ethanol were added, and the precipitating DNA fibers were recovered by spooling on a glass rod. After washing with 70% ethanol containing 0.3 M LiCl (1×) followed by absolute ethanol (2×), the DNA was dried under vacuum and redissolved in 30 ml of TE buffer, avoiding shearing. This sample was incubated for 90 min with 75 μg of pancreatic RNase at 60°, extracted three times with phenol as described above, and precipitated with LiCl–ethanol again.

The dry DNA was dissolved in 40 ml of TE buffer and, after addition of exactly 1.3 g of cesium chloride per milliliter, centrifuged to equilibrium in a fixed-angle rotor (115,000 g for 48 hr at 20°). The position of the DNA in the density gradient was localized by measuring the optical density of gradient fractions at 260 nm. Fractions containing DNA were diluted 5× with TE buffer; the DNA was precipitated as described above and redissolved in 20 ml of TE buffer.

The purity was monitored by measuring the UV spectrum. The size was determined to be larger than 30 kilobase pairs (kbp) by agarose gel electrophoresis.

Plasmid DNA from E. coli

Plasmids from *E. coli* were isolated and purified essentially as described by Oertel *et al.*[17] or by Kahn *et al.*,[18] omitting the amplification step.

Pretreatment of the Vector Plasmid for Ligation

Isolated pBR322 (15 μg)[17] was cleaved to completion by endonuclease *Bam*HI in 100 μl of buffer. The DNA was precipitated as described above

[17] W. Oertel, R. Kollek, E. Beck, and W. Goebel, *Mol. Gen. Genet.* **171**, 277 (1979).
[18] M. Kahn, R. Kolter, C. Thomas, D. Figershi, R. Meyer, E. Remcent, and B. R. Helsinki, this series, Vol. 68, p. 269.

and redissolved in 100 μl of 50 mM Tris-HCl buffer, pH 8, containing 5 mM magnesium chloride. The completion of cleavage was verified by analyzing an aliquot by gel electrophoresis. To remove the 5'-phosphate groups, the sample was incubated with 100 μg of bacterial alkaline phosphatase at 60° for 1 hr. After inactivation of the enzyme by addition of 50 mM EGTA and further incubation for 10 min at 60°, the protein was carefully extracted with the same volume of buffer-saturated phenol at 60° (5 times) and the DNA was finally recovered from the aqueous phase by precipitation with lithium chloride–ethanol as described above.

Digestion of Halobacterial DNA

Appropriate conditions for partial digestion, yielding fragments of the desired size range, were determined by trial experiments varying the time of incubation and the proportion of DNA and enzyme. The samples were analyzed by agarose gel electrophoresis in the presence of size markers. For preparative digestion, 320 μg of halobacterial DNA were treated with 90 units of endonuclease MboI[19] for 60 min at 37°, and the degree of cleavage was confirmed. The DNA fragments were separated by band sedimentation (210,000 g for 4.5 hr at 20° in a swinging-bucket rotor)[20] in a sucrose gradient [5 to 20% (w/v); in 50 mM Tris-HCl, pH 7.5, 1 M NaCl, and 1 mM EDTA]. The distribution of the DNA within the gradient was determined by measuring optical density (260 nm) in the fractions collected from the bottom of the tubes. After size determination, four classes of fragments were pooled; the DNA was concentrated by precipitation and redissolved in a small volume of TE buffer.

Ligation

Ten micrograms of pretreated pBR322 DNA and 50 μg of partially MboI-digested halobacterial DNA fragments (5–10 kbp) were coprecipitated by lithium chloride–ethanol as described above, dissolved in 200 μl of ligase buffer (20 mM Tris-HCl, pH 7.6, 10 mM MgCl$_2$, 10 mM DTE, 0.6 mM ATP) and incubated for 14 hr at 7° after addition of 12 units of T4 DNA ligase.

Transformation of E. coli

Aliquots of 100 μl of E. coli cells (5 K) rendered competent for transformation by the method described by Mandel and Higa[21] with the modifica-

[19] K. A. Nasmyth and S. I. Recel, *Proc. Natl. Acad. Sci. U.S.A.* **77**, 2119 (1980).
[20] T. Maniatis, R. C. Hardison, E. Lacy, J. Lauer, C. O'Connel, and D. Quon, *Cell* **15**, 687 (1978).
[21] M. Mandel and A. Higa, *J. Mol. Biol.* **53**, 159 (1970).

tions described by Pfeifer[22] were mixed at 0° with the same volume of transformation buffer (10 mM Tris-HCl, pH 8.0, 10 mM CaCl$_2$, 10 mM MgCl$_2$, 10 mM NaCl, 10% (v/v) formamide) containing 0.5 – 2 μg of ligase-treated DNA at the bottom of a sterile centrifuge glass tube. The samples were incubated for 40 min at 0°, followed by a heat shock at 42° for 120 sec, and placed immediately back into an ice bath. After another 10 min at 0° the sample was diluted with 2 ml of transformation medium (10 g of Bacto-tryptone, 1 g of yeast extract, 1 g of glucose, 8 g of sodium chloride, 0.3 g of calcium chloride, 5 mg of thiamin per liter) and incubated in a shaker for 45 min at 37°. Aliquots (0.1 ml) of this suspension were plated on ENB-ampicillin agar plates. Transformants resistant to the antibiotic were selected from the plates and checked for tetracycline sensitivity by replica plating. A total of 5000 tetracycline-sensitive clones were placed in a regular pattern on agar plates for convenient replication.

Maintenance of the Genomic Library

The individual clones were transferred to Microtiter plates containing 150 μl of ENB-ampicillin medium in each well and grown to stationary phase at 37°. For long-term storage 50 μl of 60% (v/v) sterile glycerol were added to each well and the plates were frozen at −70°.

Terminal Labeling of the Oligonucleotide Probe

Deoxyribonucleotide (5′ CTGGGCCTGCGATA 3′) (40 pmol) in 100 μl of a buffer containing 100 mM sodium-cacodylate, pH 7; 7.5 mM KPO$_4$, pH 7.5; 8 mM MgCl$_2$; 0.2 mM DTE; 0.5 mM CoCl$_2$; 150 μg of BSA per milliliter, and 5 μM [α^{32}P]dCTP (500 Ci/mmol) was incubated for 3 hr at 37° with 30 units of terminal deoxyribonucleotidyltransferase. The reaction was stopped by addition of 5 μl 1 M EDTA solution, and the product separated from unreacted [α^{32}P]dCTP by gel filtration on a Sephadex G-25 column (1 cm in diameter and 25 cm long) in 10 mM Tris-HCl, pH 7.5, containing 0.1 mM EDTA. The fractions of the void volume containing the labeled oligonucleotide were identified by measuring Cerenkov radiation in a scintillation counter, pooled, and stored frozen.

The analysis of the product was performed by acrylamide gel electrophoresis (24% polyacrylamide – urea gel[23]). The results showed an average elongation of the nucleotide by about 5 – 8 dC residues.

[22] F. Pfeifer, Doctoral thesis, University of Würzburg, 1981.
[23] A. M. Maxam and W. Gilbert, *Proc. Natl. Acad. Sci. U.S.A.* **74**, 560 (1977).

Identification of Bacterial Clones Containing DNA Complementary to the Oligonucleotide Probe

Colony Hybridization

Preparation of the Filters for Colony Hybridization.[24] This was essentially carried out as described by Davis *et al.*[16] Five thousand colonies containing plasmids with a halobacterial DNA insert were grown on agar plates in a regular pattern (88 each) and were then replica-plated onto nitrocellulose filters (NC filters, Schleicher & Schüll, BA 85/21, diameter 82 mm) placed onto agar plates containing ampicillin. After incubation overnight at 37°, the nitrocellulose filters were transferred to agar plates containing chloramphenicol for amplification of the plasmids[25] and further incubated for 24 hr at 37°. The filters were placed upside up for 10 min on three layers of Whatman 3 MM paper soaked before in 0.5 M sodium hydroxide and 1.5 M sodium chloride in order to lyse the cells and were then dried by transfer to dry filter paper. After being placed on a second set of Whatman 3 MM paper wetted with 0.5 M Tris-HCl, pH 7.5, and 1.5 M sodium chloride for neutralization, the filters were dried again by blotting with filter paper and shaken for 1 hr at room temperature in a glass petri dish filled with chloroform. This procedure decreases the background radioactivity on the filter by removal of lipids (R. Schnabel, personal communication). The filters were kept for 2 hr at 80° and could then be stored for several months in the refrigerator.

Hybridization Procedure. The filters were prehybridized for 2–4 hr at 40° with 100 μg of sonicated and freshly denatured calf thymus DNA per milliliter in 5× SSPE buffer containing 0.3% SDS (1× SSPE: 0.18 M sodium chloride, 10 mM sodium phosphate, pH 7, 1 mM sodium EDTA; a 20-fold concentrated stock solution is prepared). For hybridization with the radioactive oligonucleotide, three filters, separated from each other by filter paper soaked in the preincubation buffer, were placed in a plastic bag that was then heat-sealed. Preincubation buffer (10 ml) was injected into the plastic bag by a syringe, and finally 10^6 cpm of the elongated oligonucleotide in a small volume were added. Care was taken that no air was trapped in the plastic bag. After incubation for 48 hr at 45° the filters were removed from the plastic bag and washed for 10–15 min with 2× SSPE buffer containing 0.2% SDS (3 times) followed by one wash with 2× SSPE alone. The filters were dried at 80° for 10 min, and a Linhardt X-ray 90 film was exposed for 1–2 days. Colonies showing a positive hybridization signal were checked

[24] M. Grunstein and D. S. Hogness, *Proc. Natl. Acad. Sci. U.S.A.* **72,** 3961 (1975).
[25] D. B. Clevell, *J. Bacteriol.* **110,** 667 (1972).

twice by oligonucleotide hybridization. Out of the 5000 clones, 12 positive clones were found.

Analysis of Plasmid DNA Derived from Positive Clones

Restriction Analysis. The plasmids were isolated from 200-ml cultures of the clones by the standard procedure (see above), digested with various restriction endonucleases, and analyzed by agarose gel electrophoresis in order to compare their cleavage patterns. After double digestion with *Bam*HI and *Bst*EII an approximately 1.6 kbp DNA fragment was found, which hybridized strongly to the oligonucleotide probe in Southern blot-experiments[15] and was shown to be common in 7 of the 12 clones. This fragment is most likely of the same size as a *Bam*HI – *Bst*EII fragment found previously[6] that contained the bacterioopsin gene.

Sequence Analysis. For preparative isolation of the 1.6 kbp fragment, 75 μg of plasmid DNA were cut by endonuclease *Bam*HI (150 units) for 2 hr and subsequently with *Bst*EII (150 units) for 12 hr. The DNA fragments were separated on agarose gels[13,14] and stained with ethidium bromide. The piece of the gel containing the 1.6-kbp fragment was cut out and dissolved in saturated potassium iodide solution containing 0.1 M Tris-HCl, pH 8.2, 10 mM sodium bisulfite, and 1 mM EDTA and purified from the agarose by adsorption and desorption to hydroxyapatite as described by Oertel *et al.*[17]

For determination of the nucleotide sequence, 4 μg of the purified 1.6-kbp DNA fragment containing the bacterioopsin gene were cleaved with the endonuclease *Kpn*I using the conditions proposed by the vendor. The two resulting fragments (about 600 and 1000 base pairs long) were separated by agarose gel electrophoresis. After elution of the fragments according to Young *et al.*[26] the smaller DNA piece was labeled at the 5' ends with ^{32}P, using polynucleotide kinase, and cleaved with *Bgl*I (see Scheme 2). The nucleotide sequences of the two resulting 5'-labeled subfragments were partially determined using the technique of Maxam and Gilbert.[23]

Biosafety Conditions

The experiments were carried out under EK 1/L 2 conditions according to the National Institutes of Health (USA) guidelines and the rules of the Zentrale Kommission für biologische Sicherheit (West Germany).

Results and Comments

Scheme 1 summarizes the individual steps for construction of the gene library of *Halobacterium halobium* of *E. coli.* The plasmid pBR322 is

---

[26] R. C.-A. Young, J. Lis, and R. Wu, this series, Vol. 68, p. 176.

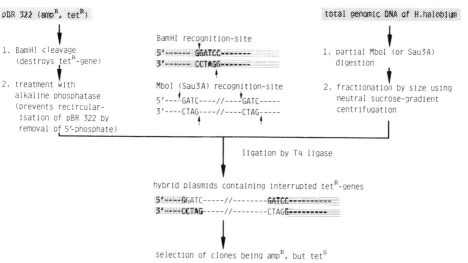

SCHEME 1. Construction of the gene library of *Halobacterium halobium* in pBR322.

certified by the National Institutes of Health for DNA cloning experiments and is most commonly used because *E. coli* cells produce it in high copy numbers after amplification in the presence of chloramphenicol. The plasmid's nucleotide sequence is known, and it contains two antibiotic resistance genes including several unique restriction endonuclease cleavage sites into which DNA can be inserted, inactivating one of the antibiotic resistance markers. Transformed cells carrying recombinant plasmid DNA are selected as ampicillin-resistant, tetracycline-sensitive clones according to Scheme 1. This selection procedure is unnecessary if vectors are used that allow positive selection of recombinant plasmid DNA carrying cells[27] (M. Betlach, personal communication).

The tetracycline resistance gene of pBR322 was cleaved with *Bam*HI because *Mbo*I (or *Sau*3AI) fragments can be inserted into this site, and the sticky ends created by both enzymes are identical. The bacterial alkaline phosphatase treatment removes the terminal 5'-phosphate and therefore prevents the recircularization of the linearized plasmid DNA during the DNA ligase reaction. Only ligation of the linearized plasmid DNA with a halobacterial DNA fragment that carries a 5'-phosphate group leads to a closed-ring structure with two nicks. The percentage of clones containing a halobacterial DNA fragment increased from about 5% to maximally 90% as a response to bacterial alkaline phosphatase treatment. Dephosphorylation was performed at 60° because at this temperature exonucleases are inactive and the efficiency of phosphatase treatment is increased. At higher tempera-

---

[27] H. Hennecke, I. Günther, and F. Binder, *Gene* **19**, 231 (1982).

ture (60°) the double-stranded DNA termini are in equilibrium with their denatured (i.e., single-stranded) form and can be attacked more easily by the enzyme that prefers single strands. Dephosphorylation was stopped by EGTA, which complexes zinc, the cofactor of alkaline phosphatases. The enzyme was removed by extraction with phenol. This step has to be carried out very carefully in order to avoid reactivation of the alkaline phosphatase in the ligase reaction mixture; otherwise, dephosphorylation of the insert-DNA and the DNA ligase-cofactor rATP prevents ligation.

Halobacterial total genomic DNA was isolated according to standard procedures[28] adapted to the particular properties of halobacteria as described in the Methods section. Special care was taken to avoid mechanical and UV damage. Complete cleavage of the DNA with an enzyme like *Bam*HI, even if creating fragments of appropriate size, has the disadvantage that the chosen enzyme may eventually cut within the gene to be isolated. Therefore, quasi-statistical partial digestion of the DNA using a frequently cleaving restriction endonuclease is preferable. *Mbo*I or *Sau*3AI are particularly well suited for this purpose because their recognition site is in part the same as for *Bam*HI, but less specific, requiring only the sequence GATC instead of GGATCC. Cleavage of DNA by each of the mentioned endonucleases creates the 5'-protruding end 5'-P-GATC. *Mbo*I is commercially available at a relatively low price and acts on halobacterial DNA, indicating that adenine residues are not methylated. Since the endonuclease *Sau*3A has the same specificity and works also on methylated restriction sites, e.g., in *E. coli* DNA, partial digestion by *Sau*3AI is more generally applicable. Fragments of 5–10 kbp corresponding to the average length of about 2–4 genes of the bacterioopsin gene size are preferable.

The conditions for partial digestion of DNA have to be optimized for each new batch of DNA and restriction endonuclease, since the enzyme (especially *Mbo*I) is very sensitive to minor contaminants in the DNA preparations. Furthermore, the enzyme activities given by the supplier are not very accurate and, in addition, may vary with the time of storage. Although the DNA fragments obtained after preparative partial digestion can be used directly, it is advisable to subfractionate the mixture into several fractions containing narrower size classes using preparative sucrose gradient centrifugation. Otherwise one would select preferentially the smallest pieces of the DNA mixture during the DNA ligase reaction and transformation.

Restriction analysis after ligation of the linearized plasmid with the insert DNA (see Fig. 2) demonstrates the presence of a *Bam*HI site in the recombinant DNA. This means that either the old *Bam*HI site was regenerated, an event that occurs with a probability of 25%, or the halobacterial DNA introduced a new *Bam*HI site.

[28] J. Marmur, *J. Mol. Biol.* **3**, 208 (1961).

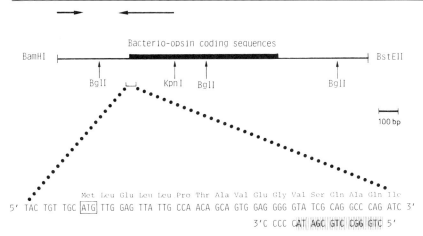

SCHEME 2. Physical map of the halobacterial DNA fragment carrying the bacterioopsin gene. Only relevant restriction endonuclease cleavage sites used for the isolation of the gene and its identification by sequence analysis are shown. The part of the sequence hatched in the graph was used as the hybridization probe.

In the steps of transformation and selection, we followed standard procedures but increased the calcium chloride treatment at 0° from 3 hr to 15 hr, thereby increasing the efficiency of transformation considerably.[29] After selection of ampicillin-resistant and tetracycline-sensitive colonies, 5000 clones were kept on Microtiter plates as the gene library.

Efficient and specific colony hybridization requires an oligonucleotide long enough to allow stable annealing to its target sequence at a temperature at which partially complementary subsequences of this probe do not form stable hybrids with other areas of the genome. As shown in Scheme 2, we selected a tetradecanucleotide near the beginning of the bacterioopsin gene. This stretch of DNA is followed by the sequence 5′ CCCCCTCC 3′. The tetradecanucleotide was custom synthesized by P-L laboratories and could be elongated enzymatically by at least 5 additional highly radioactive C residues to yield a completely complementary probe 19 bases long, or by 8 residues to yield a 22-base-long oligonucleotide containing 1 mismatch. The [α^{32}P]dCTP used for the terminal transferase reactions was purchased in 50% ethanol. This allowed concentration of the label without enrichment of substances, e.g. Tricine buffer, which might inhibit the enzyme action. It is necessary to keep the reaction volume at a minimum because for reasons of economy the radioactive nucleotide has to be used in a suboptimal concentration below its K_m value. Unlabeled dCTP was added to an amount that

[29] M. Dagert and S. D. Ehrlich, *Gene* **6**, 23 (1979).

optimizes saturation of the enzyme with dCTP and allows the average chain extension to be at least 5 residues. At the same time these conditions created an oligonucleotide substrate of a specific radioactivity high enough to allow easy detection of hybridizing colonies.

Hybridization was carried out with the DNA from colonies grown on nitrocellulose filters. The cells were lysed and the DNA adsorbed to the filter after denaturation. After heating to 80° in order to fix the DNA irreversibly, the filters were preincubated with denatured calf thymus DNA and poly(dC), which prevents unspecific hybridization of the oligonucleotide probe to G-rich regions of the halobacterial DNA. The hybridization temperature should be as high as possible for increase in specificity. We chose 45° without formamide (which reduces the hybridization temperature) because this is well above the melting temperature of any duplex that may form by less specific interaction of the oligonucleotide with other regions of the halobacterial DNA than the bacterioopsin gene, but within the range where one would expect a G-C-rich 14-mer (or longer) stably to anneal. These conditions cannot be generalized for all oligonucleotides of a particular size. A lower temperature of hybridization is advisable for AT-rich, purine- or pyrimidine-containing oligonucleotides or if the probe may contain mismatches. A higher temperature should be chosen if one expects secondary structures of the oligonucleotide or target sequences, e.g., hairpin structures. Figure 1 shows the result from a typical hybridization experiment with two positive clones on two of the filters. They are distinguishable beyond doubt from the background radioactivity of nonhybridizing clones. Out of the 5000 E. coli clones tested, we found 12 positive ones from which recombinant plasmid DNA was isolated. The size of recombinant plasmids with halobacterial insertions ranges from 6.3 to 14 kbp. Since it is known from other laboratories[6] that digestion with BamHI and BstEII cuts a 1.6 kbp fragment from the halobacterial genome carrying the bacterioopsin gene, cleavage with this enzyme combination was used to demonstrate the presence of the opsin gene within the various recombinant plasmids. As a result, 7 out of the 12 clones carry the characteristic 1.6-kbp fragment. The plasmid DNA from one clone (18/34) was then analyzed further by DNA sequence analysis.

The isolated plasmid DNA separates on agarose gels into four A forms. Two strong bands (trace a, thick arrows) and two much weaker ones (trace a, thin arrows) can be seen in Fig. 2. After BamHI treatment only one band is seen (trace b) as expected as the result of the linearization of the circular DNA upon restriction endonuclease treatment. As an established fact, the linear form of a plasmid is always found in a position above the supercoiled, and below the nicked relaxed, ring form. This means that the two weak bands in trace a amounting to less than 5% represent the monomeric forms of the DNA, whereas the two strong bands must represent a tandem dimer.

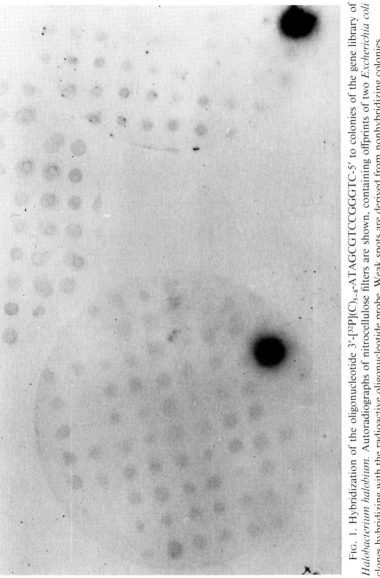

FIG. 1. Hybridization of the oligonucleotide 3'-[^{32}P](C)$_{5-8}$-ATAGCGTCCGGGTC-5' to colonies of the gene library of *Halobacterium halobium*. Autoradiographs of nitrocellulose filters are shown, containing offprints of two *Excherichia coli* clones hybridizing with the radioactive oligonucleotide probe. Weak spots are derived from nonhybridizing colonies.

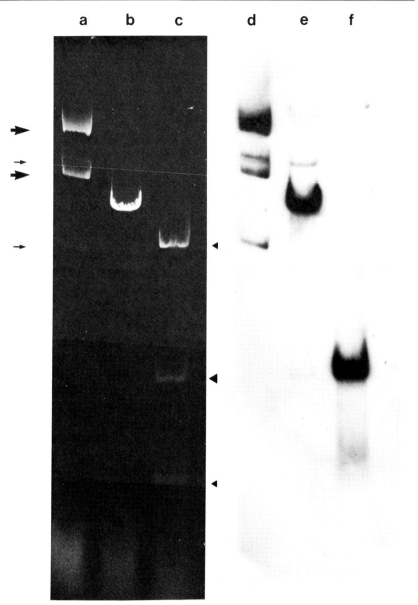

FIG. 2. Identification of the bacterioopsin gene in a 1.6-kbp fragment of halobacterial DNA by agarose gel electrophoresis and hybridization. Traces a–c: Fluorogram of ethidium bromide-stained plasmid DNA, clone 18/34 (trace a), cleaved with BamHI (trace b), and BamHI/BstEII (trace c). Traces d–f: Hybridization of the radioactive bacterioopsin sequence-specific oligonucleotide probe to the "Southern" blot of samples a–c.

The Southern blot of the DNA fragments against the oligonucleotide probe in traces d–f visualizes the monomeric bands more clearly (trace d) because the smaller monomeric DNA molecules are transferred more efficiently to the nitrocellulose than the dimeric molecules. In trace b a small band due to incomplete *Bam*HI digestion is seen beside the main band representing the linearized monomer. Cleavage of the recombinant DNA by *Bam*HI and *Bst*EII yields a large fragment containing the complete pBR322 vector and two halobacterial DNA fragments (1.6 kbp and 900 bp) only one (1.6 kbp) of them containing the bacterioopsin gene as shown by the hybridization with the oligonucleotide probe (trace f).

A *Bst*EII/*Bam*HI fragment of probably the same size has been reported to contain the bacterioopsin gene.[6] For definite proof that the 1.6-kbp fragment in trace f of Fig. 2 is identical to the one described by Dunn *et al.*,[6] this fragment was cleaved with *Kpn*I as indicated in Scheme 2, and the smaller of the resulting fragments was cut after isolation with *Bgl*I. Sequence analysis (→, ←, Scheme 2) confirmed the nucleotide sequence of the bacterioopsin gene reported by Dunn *et al.*[6]

Acknowledgment

The authors are indebted to H. Priess for sequencing the DNA fragments.

Section II

Mitochondria

[23] Assessing Import of Proteins into Mitochondria: An Overview

By Susan M. Gasser and Rick Hay

Mitochondria depend upon the nucleocytoplasmic system of gene expression for most of their polypeptides.[1] This poses two problems for the eukaryotic cell. First, the cell must possess some mechanism for sorting polypeptides to ensure that the appropriate molecules ultimately reach a mitochondrion rather than another intra- or extracellular compartment. Second, once a polypeptide arrives at a mitochondrion, it must be recognized, taken up, and routed to its proper intramitochondrial location.

Experiments on the interaction between proteins and mitochondria in the yeast *Saccharomyces cerevisiae* have shown that recognition, import, and maturation of cytoplasmically synthesized mitochondrial proteins differ in important respects from the corresponding steps in the production of extramitochondrial proteins.[2-4] Mitochondria utilize not only one mechanism, but at least three different ones, for acquiring and correctly inserting cytoplasmically made proteins.

This chapter outlines a general approach for examining the import of a given mitochondrial protein and emphasizes some of the rationale and the methods currently available for doing so.

Background

Earlier work on mitochondrial protein import has produced the following findings.

1. Many, but not all, imported mitochondrial proteins of yeast and other species are initially translated on 80 S ribosomal polysomes as precursors, which are up to 10 kilodaltons (kd) larger than the corresponding mature proteins.[2,5,6]

2. None of the mitochondrial proteins tested so far appears to be

---

[1] G. Schatz and T. L. Mason, *Annu. Rev. Biochem.* **43**, 51 (1974).

[2] M.-L. Maccecchini, Y. Rubin, G. Blobel, and G. Schatz, *Proc. Natl. Acad. Sci. U.S.A.* **76**, 343 (1979).

[3] N. Nelson and G. Schatz, *Proc. Natl. Acad. Sci. U.S.A.* **76**, 4365 (1979).

[4] S. M. Gasser, G. Ohashi, G. Daum, P. C. Böhni, J. Gibson, G. A. Reid, T. Yonetani, and G. Schatz, *Proc. Natl. Acad. Sci. U.S.A.* **79**, 267 (1982).

[5] A. Lewin, I. Gregor, T. L. Mason, N. Nelson, and G. Schatz, *Proc. Natl. Acad. Sci. U.S.A.* **77**, 3998 (1980).

[6] W. Neupert and G. Schatz, *Trends Biochem. Sci.* **6**, 1 (1981).

METHODS IN ENZYMOLOGY, VOL. 97

imported by the obligately cotranslational "vectorial translation" pathway that has been demonstrated for the transport of proteins across the endoplasmic reticulum. Recognition and import of precursors by mitochondria can occur posttranslationally both *in vivo* and *in vitro*.[2-4,7]

3. Translocation of the precursor forms of polypeptides destined for the inner membrane or the matrix requires an electrochemical potential across the mitochondrial inner membrane.[2,7] The precursors are cleaved to their mature forms by a protease located in the soluble matrix space. This protease is inhibited by divalent cation-chelating agents.[8]

4. Precursor forms of two major cytochromes found within or protruding into the intermembrane space are processed to their mature forms in two discrete steps.[4] The first step, apparently identical to that required for import across the inner membrane, converts the precursor to an intermediate form whose electrophoretic mobility is between that of the precursor and that of the mature form. The intermediate form remains bound to the outer face of the inner membrane. The second, energy-independent step involves cleavage of the intermediate to the mature form by a second protease most likely bound to the outer face of the inner membrane. This cleavage is inhibited by membrane disruption, but not by chelators.[9]

5. The major proteins of the yeast mitochondrial outer membrane appear to be synthesized with their mature molecular weights. The initial translation products are then posttranslationally inserted into the outer membrane independent of proteolysis or of an electrochemical potential.[9a]

6. Some yeast mitochondrial proteins may require additional, nonproteolytic modifications for proper import and processing to occur. For example, covalent attachment of heme to the apocytochrome c_1-intermediate form is necessary before the final maturation step can occur.[10]

Experimental Approaches

In examining the import of a particular mitochondrial protein, nine questions must be answered. These will be discussed below.

1. Is the Protein Initially Synthesized as a Larger Precursor?

Most, but not all of the cytoplasmically synthesized mitochondrial proteins studied so far are initially translated as larger precursors. Since no

[7] S. M. Gasser, G. Daum, and G. Schatz, *J. Biol. Chem.* **257**, 13034 (1982).
[8] P. C. Böhni, S. Gasser, C. Leaver, and G. Schatz, *in* "The Organization and Expression of the Mitochondrial Genome" (A. M. Kroon and C. Saccone, eds.), p. 423. Elsevier/North-Holland, Amsterdam, 1980.
[9] G. Daum, S. M. Gasser, and G. Schatz, *J. Biol. Chem.* **257**, 13075 (1982).
[9a] S. M. Gasser and G. Schatz, *J. Biol. Chem.* **258**, 3427 (1983).
[10] A. Ohashi, J. Gibson, I. Gregor, and G. Schatz, *J. Biol. Chem.* **257**, 13042 (1982).

general feature seems to distinguish those proteins that have larger precursors from those that do not, each case has to be examined individually. This can be done in two ways.

In Vitro Protein Synthesis. RNA isolated from yeast grown under suitable conditions is translated *in vitro* with [^{35}S]methionine in a hemin- and mRNA-stimulated rabbit reticulocyte lysate system as described elsewhere.[10a] Translation is stopped by the addition of unlabeled methionine, protease inhibitors, a reducing agent, and hot sodium dodecyl sulfate (SDS), followed by heating of the mixture to 95°. These precautions minimize the likelihood of proteolysis of the translation products and nonenzymatic radiolabeling of proteins in the reaction mixture.[11] Insoluble material should be removed by centrifugation, since it may interfere with subsequent immunoprecipitation. The samples are subjected to immunoprecipitation, SDS-gel electrophoresis, and fluorography as described elsewhere.[10a]

The specificity of the antiserum toward the mature antigen must be rigorously established beforehand as outlined in this volume [29].

The radioactive products should be compared directly to a radiolabeled mature protein standard that has been immunoprecipitated from yeast cells grown overnight in the presence of $^{35}SO_4^{2-}$. In comparing the electrophoretic mobilities of the *in vitro* product with the authentic mature protein, it is essential that the two samples be coelectrophoresed on the same gel slab, both individually in adjacent lanes and together in the same lane. This precaution is necessary because the electrophoretic mobility of a polypeptide in an SDS–polyacrylamide gel slab often appears to increase as the amount of protein in the polypeptide band under investigation is increased. For this reason, the radioactive band of an immunoprecipitated *in vitro* translation product (containing a vanishingly small amount of protein) may appear to migrate more slowly than the corresponding mature protein standard (which is usually labeled to a lower specific radioactivity and, hence, represents a larger amount of protein).

Finally, the structural relatedness of the precursor and mature forms should be confirmed by peptide mapping[5,12] or by radiochemical amino acid sequencing.

In Vivo Accumulation of Precursors. Precursors destined to be imported into the matrix, the inner membrane, and the intermembrane space accumulate in the cytoplasm if yeast spheroplasts or intact cells are pulse-labeled in the presence of the uncoupler carbonyl cyanide *m*-chlorophenylhydrazone (CCCP). Although the stability of different mitochondrial protein precursors in the cytoplasm varies considerably, all the precursors appear to

[10a] S. M. Gasser, this volume [32].
[11] M. Suissa, *Anal. Biochem.* **115,** 67 (1981).
[12] D. W. Cleveland, S. G. Fischer, M. W. Kirschner, and U. K. Laemmli, *J. Biol. Chem.* **252,** 1102 (1977).

be sufficiently stable to be detected in a pulse-labeling experiment. The main advantages of accumulating precursors in cells are the convenience of studying their posttranslational import *in vivo*,[13] and the possibility of using $^{35}SO_4^{2-}$ (instead of the more expensive [^{35}S]methionine) as the radiolabel.

2. Can the Polypeptide Be Taken Up by Mitochondria Posttranslationally?

Many cytoplasmically synthesized mitochondrial proteins have been shown to be imported and processed to their mature forms posttranslationally, in contrast to the strictly cotranslational segregation of secretory proteins into the lumen of the endoplasmic reticulum.[2,6] Posttranslational import of yeast mitochondrial proteins can be assayed both *in vitro* and in whole cells.

In Vitro Import Assay. An efficient and reproducible assay for the import of *in vitro* synthesized polypeptides by isolated yeast mitochondria is described by Gasser.[10a] After translation of yeast mRNA *in vitro,* protein synthesis in the reticulocyte lysate is stopped by one or more of the following four methods: removal of polysomes by centrifugation, chilling, gel filtration to remove amino acids and cofactors for protein synthesis, or addition of cycloheximide. The lysate is then incubated with isolated mitochondria, and import of radiolabeled *in vitro* products is checked by the "processing assay" or the "protection assay." Any import detected under these conditions must have occurred posttranslationally. The efficiency of import should be quantified as described by Gasser.[10a]

"Pulse-Chase" Experiments with Spheroplasts or Whole Cells. Evidence for posttranslational import of mitochondrial precursors in yeast cells can be obtained by pulse-labeling yeast spheroplasts or whole yeast cells in the presence of 20 μM CCCP, stopping translation with cycloheximide (100 $\mu g/ml$), and subsequently inactivating the uncoupler with 7 mM 2-mercaptoethanol.[13] After a suitable chase period in the presence of unlabeled methionine, the radiolabeled polypeptide under investigation is analyzed by immunoprecipitation, SDS–gel electrophoresis, and fluorography. If the precursor is imported and processed posttranslationally, it will be converted to the mature polypeptide during the chase. Quantitation of the conversion must take into account the loss of radioactivity due to the cleavage of the prepiece (see the discussion of cytochrome *c* oxidase subunit VI by Reid in this volume [31]). In addition, fractionation of spheroplasts pulse-labeled in the presence of CCCP is necessary to show that the precursor indeed accumulates outside the mitochondria. This is not an easy task, since careful controls have to be included to assess leakage of proteins from

---

[13] G. A. Reid and G. Schatz, *J. Biol. Chem.* **257**, 13062 (1982).

the mitochondria during cell fractionation.[14] Also, proteolysis of the precursors by the matrix-located processing protease or by vacuolar proteases should be minimized by adding 1 mM phenylmethylsulfonyl fluoride (PMSF) and 1 mM o-phenanthroline to all solutions and by keeping all fractions close to 0°.

If no precursor to a mitochondrial polypeptide can be detected by *in vivo* labeling, and if the *in vitro* synthesized precursor cannot be taken up by isolated mitochondria posttranslationally, it is possible that the protein is only inserted into mitochondria cotranslationally. (Such a case has not yet been observed, but it must always be considered.) Isolated mitochondria or microsomes can be added to the reticulocyte lysate reaction mixture[2,15] during the translation of yeast mRNA to check for cotranslational insertion. The mitochondria should be reisolated and subjected to proteolysis and quantitative immunoprecipitation as described for the *in vitro* import assay.

3. Is Import of the Protein Mediated by Binding to a Mitochondrial Outer Membrane Receptor?

Although a receptor for mitochondrial precursors has not yet been isolated, there is evidence to suggest that mitochondria must have at least one type of receptor located on the outer membrane.[16,16a]

Specific binding of a mitochondrial precursor to mitochondrial outer membrane vesicles from yeast has been demonstrated.[16a] Outer membranes are isolated,[17] incubated with gel-filtered reticulocyte lysate, reisolated by centrifugation, dissociated in hot SDS, and subjected to immunoprecipitation. Whereas mitochondrial outer membranes bind precursors with high affinity and specificity, vesicles derived from nonmitochondrial membranes (such as the endoplasmic reticulum) show only a low level of nonspecific binding.

The presence of more than one receptor specific for different groups of mitochondrial polypeptides may be tested by selectively blocking import or binding of some proteins through modification of a receptor. For example, if mitochondria or outer membrane vesicles are treated with 10 μg of trypsin per milliliter for 10 min at 0° both preparations lose the ability to bind precursors to F_1-ATPase β subunit and cytochrome b_2, and the mitochondria can no longer import these precursors.[7] However, *in vitro* insertion of the major 29-kilodalton outer membrane protein is not blocked by this trypsin treatment.[9a] Another approach for selective modification of outer

[14] G. A. Reid, T. Yonetani, and G. Schatz, *J. Biol. Chem.* **257**, 13068 (1982).
[15] G. Blobel and B. Dobberstein, *J. Cell Biol.* **67**, 852 (1975).
[16] R. Zimmermann, B. Hennig, and W. Neupert, *Eur. J. Biochem.* **116**, 455 (1981).
[16a] H. Riezman, R. Hay, C. Witte, N. Nelson, and G. Schatz, *EMBO J.*, in press.
[17] G. Daum, P. C. Böhni, and G. Schatz, *J. Biol. Chem.* **257**, 13028 (1982).

membrane receptors is to saturate the binding sites with a specific, unlabeled precursor, such as apocytochrome c. When *Neurospora crassa* mitochondria are preincubated with an excess of apocytochrome c prior to the *in vitro* import assay, the uptake of radiolabeled apocytochrome c is blocked, but that of the ADP--ATP carrier protein is unaffected.[16]

Precautions should be taken to show that (*a*) binding is specific for precursors, but not for mature forms of mitochondrial polypeptides; (*b*) binding is saturable and of high affinity; (*c*) nonmitochondrial proteins do not bind or compete for binding; and (*d*) binding is neither energy-dependent nor accompanied by proteolytic cleavage of the precursor.

Definitive evidence for receptor(s) must await isolation of the components and their functional reconstitution into liposomes.

4. Does Import of the Polypeptide Require an Energized Inner Membrane?

The import of polypeptides into the mitochondrial matrix, the inner membrane, or the intermembrane space requires an electrochemical gradient across the mitochondrial inner membrane.[7] If an imported mitochondrial polypeptide is made as a larger precursor, then the energy dependence of import can be conveniently checked by demonstrating accumulation of this precursor in yeast cells pulse-labeled in the presence, but not in the absence, of 20 μM CCCP (see above). This assay is unsuitable for precursors that have the same molecular weight as the mature protein. It should also be kept in mind that CCCP may have side effects, particularly in whole yeast cells. For this reason, the energy requirement of import should also be tested with the *in vitro* import assay. Gel filtration of the radiolabeled reticulocyte lysate used for the assay removes nucleoside triphosphates, substrates for oxidative phosphorylation, and other potential energy sources so that import becomes strictly dependent on an added oxidizable substrate or on added ATP (ATP can generate a membrane potential via the ATPase system). Preincubation of the mitochondria with valinomycin (2 μM) and K^+ (50 mM), or with oligomycin (70 μM) prior to incubation with the labeled lysate will inhibit the uptake of those precursors that require a potential across the mitochondrial inner membrane. By testing different inhibitors in various combinations, complications from side effects of the inhibitors can be minimized.

If protein import can occur in the absence of an energy source and if preincubation with valinomycin and K^+ does not affect import, then a requirement for other cofactors should be explored. Stimulation of uptake by the addition of metal ions, or inhibition of uptake by chelators (EDTA, EGTA), can also be assessed with the *in vitro* import assay. Controls must be done to ascertain that the "inhibitor" is not merely reacting with the precursor and altering its conformation or charge, and that the reagent does not inhibit the processing enzyme rather than the translocation step itself.

5. Does Maturation of the Protein Involve the Chelator-Sensitive Matrix Protease?

All imported yeast mitochondrial proteins that are synthesized as larger precursors appear to require cleavage after or during their translocation across the inner membrane by the soluble, chelator-sensitive matrix protease described by Böhni and Daum [30]. Translocation across a membrane is not a prerequisite for proteolysis *in vitro*, since *in vitro*-synthesized precursors can be efficiently processed posttranslationally by the partially purified matrix protease. This posttranslational cleavage does not require the presence of detergent.[8]

Some chelators, such as *o*-phenanthroline, penetrate mitochondrial membranes and may be used to inhibit the matrix protease both *in vivo* and in isolated mitochondria. However, the concentrations of chelator needed to affect the protease *in vivo* are quite high (5–10 mM) and may well perturb other aspects of import and processing.

Susceptibility of a precursor to the matrix protease can best be tested *in vitro* by examining the effects of a partially purified preparation of protease upon *in vitro* translation products or *in vivo* accumulated precursors under various conditions, and comparison of the products with a mature standard by SDS–polyacrylamide gel electrophoresis and fluorography.

Demonstration that a specific precursor is cleaved by the matrix protease should include controls to establish (*a*) that precursors to nonmitochondrial proteins are not cleaved by the protease; (*b*) that the mature form of the mitochondrial protein studied is not affected by the protease; (*c*) that cleavage is inhibited by chelators and that this inhibition is reversed by Co^{2+} and Zn^{2+}; (*d*) that cleavage is not affected by serine-protease inhibitors; (*e*) that failure to observe precursor cleavage does not result from chemical modification of the precursor by a protease inhibitor, e.g., covalent binding of *N*-ethylmaleimide or *N*-tosyl-L-phenylalanylchloromethyl ketone (TPCK); and (*f*) whether additional cleavages may be required, as in the case of two-step processing for components of the intermembrane space, fully to convert precursors to their mature forms observed *in vivo*.

6. Are Additional Proteolytic Events Involved in Processing?

Earlier work in our laboratory has demonstrated that the matrix protease generates "intermediate forms" from precursors to two cytochromes that occupy or protrude into the intermembrane space.[4] The intermediate is bound to the outer face of the inner membrane.[9] A second protease (probably located on the outer face of the mitochondrial inner membrane) converts the intermediate to the mature form, which can then assume its proper location in the intermembrane space. The second protease in not inhibited by either serine-protease inhibitors or agents chelating divalent metal ions. It can, however, be selectively inhibited *in vitro* by pretreatment of mitoplasts with detergents or trypsin.[9] Under these conditions, isolated mitochondria incubated with *in vitro* synthesized cytochrome b_2 precursor transiently accumulate the cytochrome b_2 intermediate. With untreated mitochondria, the intermediate is detectable only after very brief periods of incubation.

As least one additional protease may participate in processing a particular precursor if the product of *in vitro* treatment with the purified matrix protease remains larger than the mature protein. Whether maturation of this product involves a second protease may be tested by two procedures.

1. The precursor is subjected to *in vitro* import at $4°$ and the accumulation of an intermediate form is tested by immunoprecipitation and SDS-polyacrylamide gel electrophoresis. Any intermediate found should be chased to the mature form upon stopping further import with CCCP and shifting the temperature back to $28°$.

2. Yeast cells are allowed to accumulate precursor in the presence of CCCP and [^{35}S]methionine; incorporation of label is stopped with unlabeled methionine; CCCP is inactivated with 2-mercaptoethanol, and the time course of conversion of the precursor to the mature form is followed by denaturing aliquots of the cells with trichloroacetic acid and subjecting the extracts to immunoprecipitation. If an intermediate form is found, it should exhibit kinetics consistent with a maturation sequence precursor → intermediate → mature form.[14]

7. Is the Protein Properly Located within the Mitochondria?

Once a radiolabeled protein has been taken up by isolated mitochondria, it is essential to check whether it has been transported to its correct intramitochondrial location. If this location is one of the two membranes, it should be determined whether the imported radiolabeled protein is oriented in the membrane in the same way as its unlabeled counterpart.

Proper intramitochondrial localization can be ascertained by fractionating the mitochondria after *in vitro* import into matrix, intermembrane space, and membrane fractions (see this volume [30]). The distribution of the imported polypeptide is then determined by quantitative immunoprecipitation,[7] SDS-polyacrylamide gel electrophoresis, fluorography, and either scanning of the fluorograms or counting of excised gel slices.[7]

Difficulties in this procedure arise in trying to isolate inner and outer membranes after *in vitro* import. Although a clean fraction of inner membrane can be obtained, the recovery of outer membrane is drastically reduced, probably as a result of mitochondrial exposure to the reticulocyte lysate.

To check whether an *in vitro* synthesized outer membrane protein has been correctly inserted into the outer membrane, one can exploit the fact that some outer membrane proteins are only partially cleaved if intact mitochondria are subjected to mild proteolysis (e.g., 10 μg of trypsin per milliliter for 10 min at 0°, 2 mg of yeast mitochondrial protein per milliliter in 0.6 M mannitol, 20 mM HEPES-KOH, pH 7.4). Trypsin activity is then stopped by PMSF, trypsin inhibitor, 2-mercaptoethanol, and hot SDS; the outer-membrane polypeptide is isolated by immunoprecipitation, and its apparent molecular weight is determined by SDS–polyacrylamide gel electrophoresis. If an imported, radiolabeled protein of the outer membrane shows a shift in mobility identical to that shown by the unlabeled protein, then it is probably correctly inserted into the membrane. If the protein is insensitive to a gentle protease treatment of whole mitochondria, but sensitive after the mitochondria have been converted to mitoplasts (see this volume [30]), then the membrane protein is located either on the inner side of the outer membrane, or on the outer side of the inner membrane. Similar assays can be used to determine whether the protein is soluble or membrane-bound, as has been done for the localization of the cytochrome b_2 intermediate form.[9,14]

8. Are Additional, Nonproteolytic Modifications of the Protein Required for Import and Processing?

With yeast cells, the covalent, nonproteolytic modifications of mitochondrial protein import can be detected by checking (*a*) import in strains lacking the presumed modification pathway; and (*b*) import in wild-type cells treated with a specific inhibitor of that pathway.

Both approaches have been used to show that covalent attachment of heme to apocytochrome c_1 is required for complete maturation of the protein.[10] A mutant unable to make heme and wild-type cells grown in the presence of levulinic acid [an inhibitor of δ-aminolevulinic acid dehydratase (porphobilinogen synthase)] were pulse-labeled with [^{35}S]methionine in the presence of CCCP in order to accumulate labeled precursor to apocytochrome c_1. Upon inactivation of the uncoupler with 2-mercaptoethanol, the precursors were converted to the intermediate form. Subsequent conversion to the mature form of cytochrome c_1 occurred only after heme synthesis had been restored by addition of a suitable heme precursor. This shows that heme attachment occurs after the first intramitochondrial proteolytic cleavage, and that it must occur before the second cleavage can take place.

Similar approaches might be applied to determine whether other ligands, such as carbohydrates or fatty acids, are attached to particular proteins during their import and maturation.

9. Does a Polypeptide Imported in Vitro Become Functional inside the Mitochondria?

This question is the most difficult to answer. In fact, it has not yet been answered for any polypeptide imported into mitochondria, although a positive result must be obtained before one can conclude that correct import has been achieved *in vitro*. Promising techniques include immunoprecipitation of oligomeric enzymes by antisera directed against subunits other than those whose import is being studied, chemical cross-linking to adjacent subunits, and examination of changes in proteolytic sensitivity of hemoproteins as a result of oxidation or reduction.[18]

[18] G. Basile, C. Dibello, and H. Taniuchi, *J. Biol. Chem.* **255**, 7181 (1980).

[24] Molecular Cloning of Middle-Abundant mRNAs from *Neurospora crassa*

By Adelheid Viebrock, Angela Perz, and Walter Sebald

In exponentially growing hyphae of *Neurospora crassa* three abundancy classes of mRNA have been found,[1] similar to those in other cells. Many nuclear-coded mitochondrial proteins, e.g., the adenine nucleotide translocator, as well as subunits of the ATP synthase, the cytochrome oxidase, and the bc_1 complex, represent a sizable portion of total cellular protein in *Neurospora*.[2-4] The mRNAs of these proteins most likely belong to the middle-abundancy class, and accordingly should occur at a frequency range of around 1 per 1000. After cloning total polyadenylated RNA as cDNA plasmids in *E. coli* there is a high probability of finding one of these middle-abundant species if a few thousand clones are analyzed. Screening such a number of clones based entirely on the translation of hybridization-selected mRNA appears to be feasible. Cell-free translation of mRNA selected by hybridization to immobilized DNA represents an established

[1] L.-J. Wong and G. A. Marzluf, *Biochim. Biophys. Acta* **607**, 122 (1980).
[2] W. Sebald and G. Wild, this series, Vol. 55, p. 344.
[3] H. Weiss and W. Sebald, this series, Vol. 53, p. 66.
[4] H. Weiss, B. Juchs, and B. Ziganke, this series, Vol. 53, p. 99.

technique for the identification of coding sequences in DNA fragments.[5,6] Hybridization to cDNA plasmids immobilized on nitrocellulose filters has been used to isolate cDNAs complementary to, e.g., the mRNA of interferon[7] and of histocompatibility antigen.[8] Large numbers of cDNA plasmids can be screened in the described experiments, since pooled cDNA plasmids are used for the selection of mRNA.

Principle

The procedure for the identification of a certain cDNA involves four steps: (*a*) the preparation of an ordered cDNA clone bank from total polyadenylated RNA; (*b*) the construction of a cDNA-plasmid bank consisting of DNA covalently coupled to diazobenzoxymethyl paper; (*c*) the isolation of mRNA by hybridization; and (*d*) cell-free translation of selected mRNA, immunoprecipitation of translation products with antibodies against a certain protein, and analysis of translation products and immunoprecipitates by SDS–gel electrophoresis.

The hybridization selection of mRNA is specific enough, and cell-free translation of mRNA is sensitive enough so that many cDNA plasmids can be combined, and bound together to one paper. Thus, hundreds of clones can be analyzed in one experiment. If a plasmid pool has been identified containing a cDNA complementary to the desired mRNA, individual clones can be analyzed. Once a bank of paper-bound cDNA plasmids has been prepared, it can be used several times for the screening for different mRNAs. In the following a protocol is described that allowed the isolation of cDNA encoding the major part of the mRNA of the proteolipid subunit of the mitochondrial ATP synthase from *Neurospora crassa*.[9]

Procedures

Construction of a cDNA Clone Bank

Isolation of Polyadenylated RNA. Hyphae of *Neurospora crassa* SL 74A (FGSC stock No. 987) are grown under aeration in Vogel's minimal me-

[5] J. G. Williams, *in* "Genetic Engineering" (R. Williamson, ed.), p. 1. Academic Press, New York, 1981.

[6] J. R. Parnes, B. Velan, A. Felsenfeld, L. Ramanathan, U. Ferrini, E. Appella, and J. G. Seidman, *Proc. Natl. Acad. Sci. U.S.A.* **78**, 2253 (1981).

[7] S. Nagata, H. Taira, A. Hall, L. Johnsrud, M. Stenli, J. Ecsödi, W. Boll, K. Cantell, and C. Weissmann, *Nature (London)* **284**, 316 (1980).

[8] S. Kvist, F. Bregegere, L. Rask, B. Carni, H. Garoff, F. Daniel, K. Wiman, D. Larhammar, J. P. Abastado, G. Gachelin, P. A. Peterson, B. Dobberstein, and P. Kourilsky, *Proc. Natl. Acad. Sci. U.S.A.* **78**, 2772 (1981).

[9] A. Viebrock, A. Perz, and W. Sebald, *EMBO J.* **1**, 565 (1982).

dium plus 2% sucrose at 30° to a wet weight of 10–15 g/liter.[10] They are harvested by filtration, washed with water, immediately frozen in liquid nitrogen, and pulverized under liquid nitrogen in a Waring blender. From 100 g wet weight of cells, about 750 mg of total RNA are extracted using a phenol medium.[11] After chromatography on poly(U)-cellulose,[12] about 3 mg of enriched polyadenylated RNA are obtained, which are stored in 70% ethanol at −20°.

Preparation of cDNA Recombinant Plasmids. Single-stranded cDNA is synthesized from total polyadenylated RNA according to Friedman and Rosbash[13] omitting actinomycin D and using 1 unit of reverse transcriptase (avian myeloblastosis virus) per microgram of RNA. Yields of cDNA are between 3 and 5%. The second strand is synthesized at 25° for 4 hr in the presence of 1 mM of each of the four dNTPs.[14] Thereafter, the DNA is treated with S1 nuclease with a recovery of about 40% of the input cDNA. The double-stranded cDNA is tailed with dC.[15] Plasmid pBR322 is cleaved with *Pst*I, and the linearized form is purified by CsCl gradient centrifugation. After tailing with dG[16] the plasmid DNA is annealed with dC-tailed double-stranded cDNA at a weight ratio of 5:1. Transformation of *E. coli* 5K is performed according to Dagert and Ehrlich.[17] Employing 12 ng of the annealed DNAs per 0.1 ml of competent *E. coli* cells, 800–1600 tetracycline-resistant cells are obtained after plating directly on nitrocellulose filters. This corresponds to 400–800 colonies per nanogram of input cDNA. The tailed plasmid annealed in the absence of cDNA yields 10–20 tetracycline-resistant cells, i.e., <2%. Cells of single colonies are inoculated in 50 μl of LB medium (1% tryptone, 0.5% yeast extract, 1% NaCl) plus tetracycline in the 96 wells of a Microtiter plate. After overnight growth, 25 μl of glycerol are added, and the sealed plates are stored at −20°. Analysis of a random sample of 96 clones shows that the cloned cDNA inserts range in size from about 100 bp to 1400 bp.

cDNA-Plasmid Bank

Preparation of Plasmid DNA. Single colonies from the clone bank are inoculated individually into 50 ml of LB medium plus tetracycline (5 μg/ml) and grown overnight at 37° on a rotatory shaker. Each time twelve cultures

[10] W. Sebald, W. Neupert, and H. Weiss, this series, Vol. 55, p. 144.
[11] R. Michel, E. Wachter, and W. Sebald, *FEBS Lett.* **101,** 373 (1979).
[12] R. Sheldon, C. Jurale, and J. Kates, *Proc. Natl. Acad. Sci. U.S.A.* **69,** 417 (1972).
[13] E. G. Friedman and M. Rosbash, *Nucleic Acids Res.* **4,** 3455 (1977).
[14] D. Woods, J. Crampton, B. Clarke, and R. Williamson, *Nucleic Acids Res.* **8,** 5157 (1980).
[15] H. Land, M. Grez, H. Hauser, W. Lindenmaier, and G. Schütz, *Nucleic Acids Res.* **9,** 2251 (1981).
[16] J. H. J. Hoeijmakers, P. Borst, J. Van den Burg, C. Weissmann, and G. A. M. Cross, *Gene* **8,** 391 (1980).
[17] M. Dagert and S. D. Ehrlich, *Gene* **6,** 23 (1979).

are combined, supercoiled plasmid DNA is extracted following a modified protocol[18] of Birnboim and Doly,[19] and is purified by banding once in an 11-ml CsCl gradient. Yields from twelve 50-ml cultures are 50–200 μg of DNA that is slightly contaminated by RNA.

Plasmid DNA (50–100 μg) is partially depurinated by an incubation for 10 min at 25° in 200 μl of 50 mM HCl. The DNA is broken and dissociated by adding 0.2 ml of 1 N NaOH for 30 min at 37°. The DNA is precipitated with 1 ml of ethanol plus 0.08 ml of 3 M sodium acetate, pH 4.8. After a 70% ethanol wash, the sediment is desiccated and dissolved in 10 μl of water. The solution is heated for 2 min at 100°. Then 40 μl of dimethyl sulfoxide plus 0.5 μl of 3 M sodium acetate, pH 4.8, are added. The solution is added to two 1 cm^2 circles of activated diazobenzyloxymethyl paper[20] that has been equilibrated with 60 mM sodium acetate, pH 4.8, in 80% dimethyl sulfoxide. After incubation at room temperature overnight, the loaded papers are washed once with water and four times for 30 min with 0.5 N NaOH at 37°. The papers are rinsed with water until the pH is neutral and stored at 4° in hybridization buffer [50% formamide, 5× SSC = 750 mM NaCl plus 75 mM sodium citrate, 0.1% SDS, 100 μg of poly(A), and 100 μg of yeast tRNA/ml].

Selections of mRNA by Hybridization

Two circles of DNA paper are soaked with 80 μl of 1.25-fold concentrated hybridization buffer, and 50 μg of polyadenylated RNA dissolved in 20 μl of H$_2$O are added. Hybridization proceeds for 15 hr at 37°. The papers are then washed twice with hybridization buffer (5 ml), twice with 50% formamide, 0.2× SSC, 30 mM sodium phosphate, pH 7.3, 0.1% SDS at 37°. The bound RNA is eluted at 65° with four times 0.1 ml of 98% formamide, containing 5 μg of yeast tRNA per milliliter, 1 mM EDTA, 0.1% SDS, and 10 mM Tris-HCl, pH 7.5. Particles—mainly disintegrated paper—are removed by centrifugation for 5 min at 18,000 g. RNA is precipitated with 1 ml of ethanol after addition of 40 μl of 2 M sodium acetate, pH 6. RNA is reprecipitated from 400 μl of water and dissolved in 10 μl of water: 5-μl aliquots are analyzed in a 15-μl wheat germ assay.

Cell-Free Translation of mRNA and Immunoprecipitation

A cell-free protein-synthesizing system is prepared from commercial wheat embryos[21] using an acetate medium.[22] Assay conditions are as de-

[18] F. G. Grosveld, H. M. Dahl, E. de Boer, and R. A. Flavell, *Gene* **13**, 227 (1981).

[19] A. C. Birnboim and J. Doly, *Nucleic Acids Res.* **7**, 1513 (1979).

[20] M. L. Goldberg, R. P. Lifton, G. R. Stark, and J. G. Williams, this series, Vol. 68, p. 206.

[21] B. E. Roberts and B. M. Paterson, *Proc. Natl. Acad. Sci. U.S.A.* **70**, 2330 (1973).

[22] J. W. Davies, A. M. J. Aalbers, E. J. Stuik, and A. Van Kammen, *FEBS Lett.* **77**, 265 (1977).

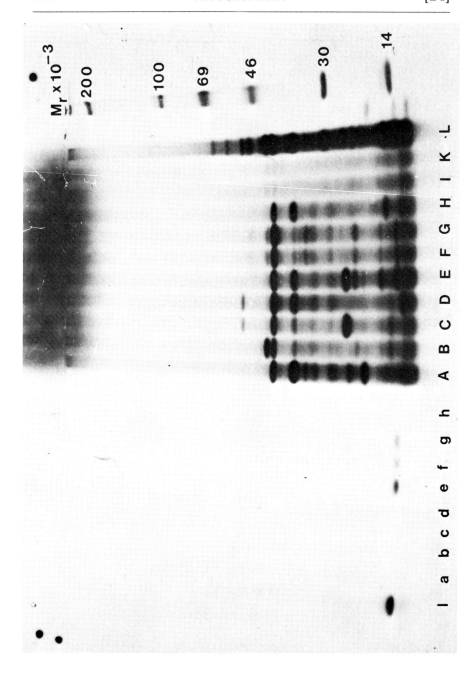

scribed[11] employing 1 mCi of [^{35}S]methionine per milliliter. Using 15-μl assays, the system corresponds linearly to added polyadenylated RNA up to 100 μg/ml. Incorporation into hot trichloroacetic acid-insoluble material is 1 to 3 × 10^6 cpm per microgram of added RNA corresponding to a 10- to 30-fold stimulation over endogenous incorporation.

Protein synthesis is stopped by adding 15 μl of 2% SDS and by heating the mixture for 2 min at 100°. Aliquots (about 10%) are removed for analysis of total translation products. The rest is diluted with 0.5 ml of 0.5% Triton X-100, 0.15 M NaCl, 10 mM sodium phosphate, pH 7.5. Specific immuno-globulins are added for 1–4 hr at 0°. Immunoglobulins are adsorbed to protein A-Sepharose CL-4B for another 1–15 hr. After washing twice with 1 ml of 0.5% Triton X-100, 0.15 M NaCl, 10 mM sodium phosphate, pH 7.5, twice with 0.15 M NaCl, 10 mM sodium phosphate, pH 7.5, and once with 10 mM Tris-HCl, pH 7.5, the adsorbed protein is released by boiling for 1 min in 62.5 mM Tris-HCl, pH 6.8, 20% glycerol, 2% SDS, 2% 2-mercap-toethanol. Immunoadsorbed proteins and total translation products are analyzed electrophoretically on 12 to 20% concave gradients of SDS–poly-acrylamide gels,[23] and radioactivity is visualized by fluorography.[24]

Results

Figure 1 shows an SDS-gel of the total translation products synthesized in response to 8 mRNA populations selected by hybridization to DNA papers each containing 12 pooled cDNA plasmids. The corresponding immunoprecipitates obtained with antibodies to the proteolipid subunit of the mitochondrial ATP synthase from *Neurospora crassa* have been separated on the same gel.

The mRNA from three pools directed the synthesis of a protein that reacted with proteolipid antibodies and exhibited a molecular weight similar to that of the preproteolipid. The twelve clones comprising the positive pool Pl-E have been grown up individually, and single cDNA plasmids have been analyzed as described above. Proteolipid mRNA is specifically selected by

[23] U. K. Laemmli, *Nature* (*London*) **227**, 680 (1970).
[24] W. M. Bonner and R. A. Laskey, *Eur. J. Biochem.* **46**, 83 (1974).

Fig. 1. Translation of mRNAs selected by hybridization to 12 pooled cDNA plasmids. A fluorography of a sodium dodecyl sulfate gel is shown after electrophoretic separation of total cell-free translation products (lanes A–L) and of immunoprecipitates (lanes 1, a–h) obtained with antibodies against the ATP synthase proteolipid subunit. The [^{35}S]methionine-labeled proteins were synthesized in response to total polyadenylated RNA (lanes L, 1), to water (lane K), and to mRNAs selected by 8 pools of cDNA plasmids Pl-A to Pl-H (lanes A–H, and a–h). Each pool contained 12 cDNA plasmids. Lane I shows the translation products synthesized in response to mRNA selected by a paper containing only plasmid pBR322.

the cDNA in clone Pl-E2.[9] (Similarly, one positive clone could be identified among the positive pool Pl-G. This clone, as well as the positive pool Pl-F, have not been further analyzed.) The cDNA insert from clone Pl-E2 represents a copy of about 150 bp of the 3′ end of the proteolipid mRNA including a short poly(dA) tail. Using this insert as a probe, further cDNA clones encoding proteolipid mRNA were identified by colony filter hybridization at a frequency of about 0.2%. One isolated cDNA represented the major part of the proteolipid mRNA. The nucleotide sequence of this cDNA showed 243 bases corresponding to the known amino acid sequence of the mature proteolipid and, in addition, 178 bases coding for an amino-terminal presequence. Noncoding sequences of 48 bases at the 5′ end and of 358 bases at the 3′ end plus a poly(A) tail were determined.[9] The long presequence of 66 amino acids is very polar, in contrast to the lipophilic mature proteolipid, and includes 12 basic and no acidic side chains. It is suggested that the presequence is specifically designed to solubilize the proteolipid for posttranslational import into the mitochondria.

Prospects

The identification of a cloned mRNA by techniques based on the translation of hybridization-selected mRNA depends on few prerequisites, provided that an immunological or another specific test for a protein is available. Thus, this method can be applied in systems where enrichment of mRNA or of cDNA clones is impossible, and where, as in yeast, genetic complementation of defined mutants by cloned genomic DNA cannot be applied. The screening of large numbers of cDNA clones appears to be possible. Several thousand cDNA plasmids in groups of 96 pooled clones were covalently bound to paper, and in several of these pools a cDNA complementary to the mRNA of the adenine nucleotide translocator from *Neurospora* was identified (H. Arends and W. Sebald, unpublished results). The adenine nucleotide translocator[25] and the ATP synthase proteolipid are on a molar basis the most abundant mitochondrial proteins in *Neurospora*. The cloned mRNA of these two proteins is therefore most easily isolated. It is a distinct possibility, however, that the cloned mRNAs of other mitochondrial or cellular proteins occurring at a molar ratio up to 10-fold lower can be identified by the described approach. It is hoped that once a bank of paper-bound cDNA plasmids has been prepared, it can be utilized for several rounds of hybridization for various proteins.

[25] H. Hackenberg, P. Riccio, and M. Klingenberg, *Eur. J. Biochem.* **88**, 373 (1978).

[25] Biogenesis of Cytochrome c in Neurospora crassa

By Bernd Hennig and Walter Neupert

Cytochrome c is an exceptionally well-studied protein with respect to its primary and tertiary structure, function in the mitochondrial respiratory chain, topological arrangement,[1] and molecular genetics and evolution.[2] This protein is therefore particularly suitable for investigating the molecular mechanism of mitochondrial assembly. Several steps of the biogenesis of cytochrome c have already been elucidated.[3] Its structural gene(s) are contained in nuclear DNA and transcribed into poly(A)-mRNA. The mRNA is translated on free cytoplasmic ribosomes. The primary translation product is an extramitochondrial precursor which, unlike most other mitochondrial precursor proteins, does not carry an N-terminal or C-terminal transient sequence. Apparently it is identical to apocytochrome c, which can be prepared from holocytochrome c by chemical means. Apocytochrome c binds to specific sites on the surface of mitochondria. During the transfer into mitochondria, the heme group becomes covalently linked to the apoprotein via two thioether bonds. The mature protein, holocytochrome c, binds to the inner membrane and can reversibly dissociate into the intermembrane space. Antibodies can be obtained against apo- and holocytochrome c that do not cross-react, presumably because apocytochrome c differs strongly from holocytochrome c in its conformation. These antibodies allow one to trace precursor and mature protein during the assembly process.

A further important aspect is that practically unlimited amounts of apocytochrome c can be chemically prepared, in contrast to the minute amounts of precursors, which can be obtained by synthesis in cell-free systems. This is a great advantage in investigating (a) the specificity, affinity, and identity of the putative receptor sites on the mitochondrial surface; (b) the mechanism of translocation of apocytochrome c across the membrane; and (c) the conversion to holocytochrome c by the putative enzyme, cytochrome c heme lyase.

[1] R. Timkovich, *in* "The Porphyrins" (D. Dophin, ed.), p. 241. Academic Press, New York, 1979.

[2] E. Margoliash, *Proc. Int. Conf. Theor. Phys. Biol. 3rd, 1971* p. 175 (1973).

[3] B. Hennig and W. Neupert, *Horiz. Biochem. Biophys.* 7 (in press).

METHODS IN ENZYMOLOGY, VOL. 97

Isolation and Purification of Cytochrome c from Neurospora crassa

Neurospora cells are grown in liquid culture according to established procedures.[4] Vogel's minimal medium supplemented with 2% sucrose is used. A maximal yield of 0.4 mg of cytochrome c per gram of protein of crude cell extract is obtained when wild-type (No. 262, Fungal stock center, Arcata, California) is harvested at mid log phase (Fig. 1). This stage (optical density (OD) at 410 nm = 1.2) is reached 14–16 hr after inoculating the culture with 1×10^9 conidia per liter (OD at 410 nm = 0.1) and growth under vigorous aeration at 25°. Longer growth leads to a decline of the cytochrome c content in cells to about half the value obtained at mid log phase. Cells are harvested by suction filtration on cellulose filter paper in a

[4] R. H. Davis and F. J. Serres, this series, Vol. 17A, p. 79.

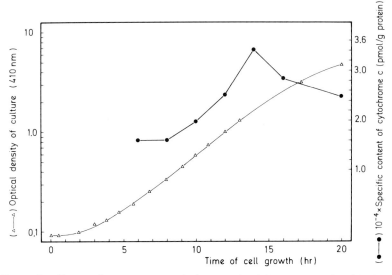

FIG. 1. Specific cytochrome c content during growth of *Neurospora* cells. Eight bottles containing 1 liter of Vogel's minimal medium supplemented with 2% sucrose were inoculated with 10^9 conidia and aerated at 25°. Samples were taken from one culture at the indicated times, and growth was determined by measuring the optical density of the culture at 410 nm. The other cultures were harvested at the indicated times, and the cytochrome c content of cells was determined in the following way. Cells were suspended in 150 mM KP$_i$ and disrupted with a homogenizer and by sonication as described in the text. The homogenate was centrifuged in a SS-34 rotor for 20 min at 20,000 rpm. In the supernatant, total protein of the crude cell extract was determined. After high-spin centrifugation in a Ti-50 rotor for 60 min at 50,000 rpm, cytochrome c was determined in the resulting supernatant by difference spectroscopy of the dithionite-reduced vs ferricyanide-oxidized samples ($E_{550\,nm-533\,nm} = 21.6 \times 10^3 cm^2/mmol$).

Büchner funnel and washed with distilled water. They can then be kept frozen at $-20°$ for up to several months.

Two different procedures are recommended for isolation of cytochrome *c*. Procedure A is preferred for small amounts of cells (less than 100 g wet weight); procedure B should be employed with large amounts of cells (1–2 kg).

Procedure A. This procedure allows nearly quantitative recovery of cytochrome *c* and has been successfully used to purify amounts of less than 1 mg. It is a modified version of the microscale preparation described by Hennig.[5]

Cells are thawed in 2 ml of 0.15 M KP$_i$, pH 7.2, per gram of cell wet weight. Then 5 μl of phenylmethylsulfonyl fluoride (PMSF) (from a 200 mM stock solution in absolute ethanol) are added per milliliter of homogenate, and cells are homogenized with an Ultra-Turrax (Janke and Kunkel, Staufen, Federal Republic of Germany) for 1–5 min at 0°. After addition of 1 μl of 2-mercaptoethanol per milliliter, the homogenate is sonicated at maximum power in an ice bath. Temperature is kept below 10° by frequent interruptions of sonication. Sonication is continued until cell breakage is complete as judged by phase contrast microscope. Usually, complete breakage requires about 10 min of total sonication time for a 50-ml portion. The crude extract is prepared by centrifugation in a Sorvall SS-34 rotor at 18,000 rpm for 20 min. The supernatant is mixed with 0.05 g of wet BioRex 70 (200–400 mesh, Bio-Rad, Richmond, California) per milliliter preequilibrated with 15 mM KP$_i$, pH 7.0. The suspension is diluted with constant stirring by addition of 9 ml of cold distilled water per milliliter. Adsorption of cytochrome *c* to the resin is allowed to proceed under constant stirring at 4° for 60 min. The resin is separated from the homogenate by filtration on cellulose filter paper, and resuspended in 0.7 ml of 10 mM KP$_i$, pH 7.0, per gram of wet resin. The filtrate is incubated for a second time with newly added resin and treated as described before.

The combined slurry of BioRex is transferred into a column of appropriate size (for isolation of cytochrome *c* from 50 g of cells, a column size of 0.9 × 30 cm is required). The packed column is thoroughly washed with 10 mM KP$_i$, pH 7.0, at a flow rate of 100 ml/hr per square centimeter (at least 10 column volumes are run through). Cytochrome *c* is eluted from the column with 0.5 M NaCl, 10 mM KP$_i$, pH 7.0, and the fractions absorbing at 410 nm are collected.

The pooled fractions (about 5 ml) are applied onto a column of Sephadex G-75 (1.6 × 86 cm) and eluted with 0.2 M NH$_4$HCO$_3$. The eluate is monitored for absorption at both 280 nm and 410 nm. The cytochrome

[5] B. Hennig, *Eur. J. Biochem.* **55**, 167 (1975).

c-containing fractions separate well from a large 280 nm peak. They are collected and adjusted to pH 6.5 by addition of 5 M HCl.

Then the solution is pumped into a short precolumn (1.6 × 1 cm) of CM-32-cellulose (Whatman, Maidstone, England) equilibrated with 5 mM KP$_i$, pH 6.5. Stable binding of cytochrome c to the CM-cellulose requires 10-fold dilution of the applied sample. To avoid loss of cytochrome c due to adsorption on the glassware, this dilution should occur concomitantly with application of the sample to the column by means of a three-way valve. The precolumn is then washed with at least 100 ml of 5 mM KP$_i$, pH 6.5. Cytochrome c is oxidized by an additional wash with 5 ml of 1 mM K$_3$Fe(CN)$_6$, 20 mM KP$_i$ pH 6.5, followed by washing with 10 ml of 20 mM KP$_i$, pH 6.5. The precolumn is connected to the fractionation column (0.9 × 46 cm) of CM-32-cellulose preequilibrated with 20 mM KP$_i$, pH 6.5. The interconnected columns are eluted with a linear gradient of 0.02 M to 0.3 M KP$_i$, pH 6.5, at 28 ml/hr per square centimeter. The slope of the gradient is about 0.5 mM KP$_i$ per hour per cubic centimeter of column volume.

Cytochrome c-containing fractions are detected by their absorption at 410 nm. They are pooled, lyophilized, dissolved with distilled water to a concentration of 10 mg of cytochrome c per milliliter, and dialyzed in Visking cellulose tubing 18/32 (Union Carbide Corporation, Chicago, Illinois) against the desired final buffer (we use either 100 mM NaP$_i$, pH 7.0, or 0.25 M NH$_4$HCO$_3$). Cytochrome c thus obtained is pure as judged by its absorption spectrum (A_{410}/A_{280} of the oxidized cytochrome c = 4.7), by electrophoresis on SDS–polyacrylamide gels, and by amino acid analysis.

Procedure B. This procedure allows isolation of 20–100 mg of cytochrome c by a modified version of a previously described method,[6] but it is not designed for quantitative recovery of cytochrome c. Cells (1 kg, wet weight) are suspended in 3 liters of distilled water and homogenized with a Waring blender (Dynamics Corp., New Hartford, Connecticut) at 4°. The resulting slurry is passed three times through a grind mill[7] to break the cells. The homogenate is adjusted to pH 9.8 with 25% NH$_4$OH and stirred for 1 at 4°. It is then centrifuged in a Sorvall GS-3 rotor for 10 min at 8000 rpm. The resulting pellet is discarded. The supernatant is adjusted to pH 7.4 by addition of glacial acetic acid. The mixture is centrifuged as above for 50 min, and the pellet is discarded. The supernatant is diluted with water to 35 liters, and 0.875 g of NaN$_3$ and 1.75 g of K$_3$Fe(CN)$_6$ are added and dissolved by stirring. Then 52.5 g of wet BioRex 70 (Bio-Rad, Richmond, California) is added while stirring with a motor-driven rotor blade, and stirring is continued for 12 hr at 4°. The resin is preequilibrated with 20 mM NH$_4$P$_i$, pH 7.0.

[6] J. Heller and E. L. Smith, *J. Biol. Chem.* **241**, 3158 (1966).
[7] H. Weiss, G. von Jagow, M. Klingenberg, and T. Bücher, *Eur. J. Biochem.* **14**, 75 (1970).

The resin with adsorbed cytochrome c is collected by filtration on cellulose filter paper in a Büchner funnel. The filtrate is supplemented with an additional 1.75 g of $K_3Fe(CN)_6$ and 45 g of preequilibrated BioRex 70, and stirring is continued for 4 hr. Again, the resin is collected by filtration, and the two batches of resin are combined. The resin is washed twice with distilled water, once with 50 mM NH$_4$P$_i$, pH 7.4, and is collected by filtration after each wash. Then a column (2.6 × 40 cm) is filled to a height of 5 cm with BioRex 70 preequilibrated with 20 mM NH$_4$P$_i$, pH 7.4, and the sample of BioRex containing the bound cytochrome c is layered on top; 100 ml of 50 mM NH$_4$P$_i$, pH 7.4, are passed through the column, followed by 60 ml of 0.1 M NH$_4$P$_i$, pH 7.4. Upon this treatment, the red band of cytochrome c starts to migrate down the column. It is eluted with 0.5 M NH$_4$P$_i$, pH 7.4, and is recovered in about 10–15 ml of the eluate.

A column of CM 32-cellulose (1.6 × 5 cm) is equilibrated with 20 mM KP$_i$, pH 6.5. The sample of cytochrome c is diluted with distilled water by 1:25 and is pumped over the column. Cytochrome c is retained. Then 5 ml of 1 mM K$_3$Fe(CN)$_6$ are passed through the column, followed by 20 mM KP$_i$, pH 6.5, for 12 hr at a rate of 20 ml/hr. A linear gradient of 20 mM to 300 mM KP$_i$, pH 6.5, is applied, and the eluate is directly passed over a second column of CM-32-cellulose (2.6 × 20 cm) equilibrated with 20 mM KP$_i$, pH 6.5. The gradient has a total volume of 500 ml and is run for 16 hr. Fractions of 5 ml are collected.

Cytochrome c is detected by its red color and is recovered in about five fractions. These are combined, diluted with distilled water by 1:25 and pumped onto a BioRex 70 column (1.5 × 1.5 cm) equilibrated with 20 mM KP$_i$, pH 7.0. Cytochrome c binds to the column. It is washed first with 5 ml of 1 mM K$_3$Fe(CN)$_6$, 10 mM KP$_i$, pH 7.0, and then with 100 ml of 10 mM KP$_i$, pH 7.0 (flow rate 10 ml/hr). A gradient of 50 mM KP$_i$ to 300 mM NaCl, 50 mM KP$_i$, pH 7.0 (500 ml total volume) is applied. The eluate is lead directly to a second column of BioRex 70 (2.6 × 20 cm) equilibrated with 50 mM KP$_i$, pH 7.0. The cytochrome c-containing fractions are combined and lyophilized. The dry material is dissolved with 5 ml distilled water and dialyzed in Visking cellulose tubing 18/32 against 0.1 M NaP$_i$, pH 7.0. The total yield is 30–40 mg of cytochrome c. It is pure according to its absorption spectrum and gel electrophoresis.

Comments. Cytochrome c strongly adsorbs to glass walls, dialysis tubing, gel matrix, etc., at low ionic strength. For quantitative recovery, cytochrome c should be kept in buffers of sufficiently high ionic strength (at least $I = 0.1$); 0.25 M NH$_4$HCO$_3$ is recommended as a volatile solvent. The methylated and unmethylated forms of cytochrome c[8] can be separated by chromatography on BioRex 70 or Amberlite CG-50 using the gradient described for the resin in procedure B. Visking cellulose tubing, 18/32, was found to be

---

[8] W. A. Scott and H. K. Mitchell, *Biochemistry* **8**, 4282 (1969).

particularly suitable for dialysis of cytochrome c, since it is completely impermeable for this small protein. The tubing is presoaked in a solution containing 50% (v/v) ethanol, 10 mM EDTA for several hours and washed with distilled water.

Preparation of Apocytochrome c

The recommended procedure is a modified version of that described for bacterial cytochromes c.[9] After dialysis against 0.25 M NH$_4$HCO$_3$, holocytochrome c is lyophilized. A solution of 8 M urea and 0.1 M NaCl is prepared and adjusted to pH 2.0 by addition of 5 M HCl; 10 mg of cytochrome c and 50 mg of HgCl$_2$ are dissolved each in 1 ml of this solution. They are mixed and incubated at room temperature for 16 hr in the dark. The mixture is applied on a column of Sephadex G-25 (0.9 × 84 cm) and eluted with 0.25 M NH$_4$OAc, pH 6.0. Fractions containing the protein are collected, combined, and lyophilized. The dry sample is dissolved in 1 ml of 8 M urea, 0.25 M NH$_4$HCO$_3$; 1 mg of solid dithiothreitol is added, and incubation is performed for 30 min at 37° in order to remove protein-bound Hg. The sample is applied on a column of Sephadex G-25 (0.9 × 84) and eluted with 0.25 M NH$_4$HCO$_3$. The fractions containing the protein are collected and lyophilized. To renature the protein, it is dissolved in 1 ml of 8 M urea, 0.25 M NH$_4$HCO$_3$, 1 mM dithiothreitol and dialyzed in Visking cellulose tubing, 18/32, at 4° against 0.25 M NH$_4$HCO$_3$, 0.01% 2-mercaptoethanol for 16 hr. The exact quantity of apocytochrome c can be determined by amino acid analysis. The measured value of arginine is a reliable basis for rapid calculation. The protein contains 3 mol of arginine per mole of protein.[10] The yield is higher than 90%. The protein has no absorption at 410 nm, reflecting complete absence of heme.

For the preparation of small amounts of apocytochrome c, the following method is more convenient: 100 μg of cytochrome c are treated with HgCl$_2$ as described above, but in a total volume of 20 μl of 8 M urea, 0.1 M NaCl, pH 2.0. After incubation for 16 hr, the mixture is transferred into a 50-μl microdialysis chamber (Schütt, Göttingen, Federal Republic of Germany) and dialyzed for 16 hr against 0.25 M NH$_4$OAc, pH 6.0, at 4°. Then dialysis is continued against a freshly prepared solution containing 8 M urea, 0.25 M NH$_4$HCO$_3$, 1 mM dithiothreitol for 30 min at 37°. Finally, dialysis is carried out against 0.25 M NH$_4$HCO$_3$, 0.01% 2-mercaptoethanol for 16 hr at 4°. The yield is about 95%.

Comment. Apocytochrome c is easily denatured. This is in contrast to the stability of holocytochrome c. Solutions of apocytochrome c should be

[9] R. P. Ambler and M. Wynn, *Biochem. J.* **131**, 485 (1973).
[10] J. Heller and E. L. Smith, *J. Biol. Chem.* **241**, 3165 (1966).

stored in aliquots at $-20°$. They should be thawed immediately before use. Repeated thawing and freezing must be avoided, since this denatures the protein. Denatured protein can be renatured by dialysis first against 8 M urea, 0.25 M NH$_4$HCO$_3$, 1 mM dithiothreitol and then against urea-free NH$_4$HCO$_3$ containing 0.01% 2-mercaptoethanol.

Preparation of Antibodies against Holocytochrome *c* and Apocytochrome *c* and Immunoprecipitation

Holocytochrome *c* or apocytochrome *c* is dialyzed against 0.1 M NaP$_i$, pH 7.0, and adjusted to a concentration of 0.6 mg of protein per milliliter; 2 μl of 25% glutaraldehyde are added per milliliter, and the mixture is kept for 1 hr at room temperature.[11] Then 18 mg of solid lysine hydrochloride are added per milliliter, and incubation is continued for 1 hr. The mixture is dialyzed in Visking 18/32 tubing against 0.1 M NaP$_i$, pH 7.0, for 16 hr at 4°. Aliquots containing 0.2 mg of protein in 250 μl of 0.1 M NaP$_i$, pH 7.0, are stored at $-20°$. Rabbits are injected first with a mixture of 250 μl of cytochrome *c* solution and 250 μl of Freund's complete adjuvant (Behring-werke, Marburg, Federal Republic of Germany) into the neck region at four positions. Injections are repeated 4 times at weekly intervals with 250 μl of cytochrome *c* solution and 250 μl of Freund's incomplete adjuvants. After 6 weeks, 40–60 ml blood are drawn from the ear vein. The rabbits can be boostered after 3–4 weeks by two injections with the same amount of protein mixed with incomplete Freund's adjuvant. Blood is collected after a further week. The rabbit can be boostered repeatedly. Serum is collected from the blood samples after clotting for 1 hr at room temperature. It is centrifuged at 10,000 rpm in a Sorvall SS-34 rotor for 30 min. Aliquots of 1 ml are kept frozen at $-20°$. Higher titers can be obtained when the protein is coupled to keyhole limpet hemocyanin (Calbiochem, Lahn-Giessen, Federal Republic of Germany). One milligram of cytochrome *c* and 10 mg of hemocyanin in 1 ml of 0.1 M NaP$_i$, pH 7.0, are polymerized with glutaraldehyde as described above; 0.1 mg of cytochrome *c* is applied per injection. Immune response and cross-reactivity are checked by double immunodiffusion, according to Ouchterlony.[12] The titers of the sera are determined by direct immunoprecipitation from ^{35}S-labeled mitochondria for holocytochrome *c*, and by immunoabsorption employing Sepharose-bound protein A (Pharmacia, Uppsala, Sweden) from [^{3}H]leucine-labeled cell-free translation mixtures (see below) for apocytochrome *c*.

Specific antibodies directed against holocytochrome *c* or apocytochrome *c* are isolated by affinity chromatography on Sepharose containing the

[11] M. Reichlin, A. Nisonoff, and E. Margoliash, *J. Biol. Chem.* **245**, 947 (1970).
[12] O. Ouchterlony, *Acta Pathol. Microbiol. Scand.* **32**, 231 (1953).

covalently bound antigens. The affinity matrix is prepared by coupling 1 mg of pure protein to 0.2 g (dry weight) of preswollen CNBr-activated Sepharose 4B (Pharmacia, Uppsala, Sweden). The coupling reaction is carried out following the instructions of the manufacturer. The affinity matrix is transferred into a column (0.6 × 5 cm) and washed with 50 mM Tris-HCl, pH 7.5 (adjusted at 20°), 0.3 M KCl. Thirty milliliters of serum are thawed, subjected to a clarifying spin (20 min at 10,000 rpm, Sorvall SS-34 rotor), and pumped over the column with a flow rate of 10 ml/hr at 4°. Then the column is washed with 20 ml of the same buffer used for prewashing. Elution is performed with 0.2 M acetic acid, 0.5 M KCl (pH 2.8), and the eluate (3 ml) is adjusted immediately to pH 7 with 6 M KOH. Aliquots of 100 μl are kept frozen at −20°.

Immunoprecipitation of holocytochrome c from mitochondria is carried out by dissolving mitochondria (1 mg of mitochondrial protein) in 1 ml of "lysis buffer" (1% Triton X-100, 0.3 M KCl, 10 mM Tris-HCl, pH 7.5, adjusted at 20°) and adding about 0.25 ml of serum or 20 μl of specific antibodies (according to the measured titer). The mixture is incubated in Eppendorf tubes for 15 hr at 4°. The immunoprecipitate is collected by centrifugation in a Sorvall SS-34 rotor using plastic adaptors for 10 min at 10,000 rpm. The pellet is washed twice with 1.5 ml of lysis buffer and three times with 0.01 M Tris-HCl, pH 7.5 (adjusted at 20°). Then the pellet is dissolved in 40 μl of 2% SDS, 10 mM Tris-HCl, pH 7.5, and 0.35 M 2-mercaptoethanol and heated to 95° for 5 min.

For immunoprecipitation of apocytochrome c from mitochondria, pellets are lysed with "lysis buffer." For immunoprecipitation from postribosomal supernatant fraction, 110 μl of 3 M KCl, 10 mM Tris-HCl, pH 7.5, and 60 μl of 20% Triton X-100 are added to 1-ml samples. Then about 75 μl of antiserum or 10 μl of specific antibodies are added (according to the measured titers), and the samples are incubated for 20 min at 4°. Ten milligrams of protein A-Sepharose (Pharmacia, Uppsala, Sweden) or protein A-Agarose (Sigma Chem. Corp., St. Louis, Missouri) are swollen for 15 min in 100 μl of 10 mM Tris-HCl, pH 7.5, and then added to the samples. They are shaken for 20 min at 4°. The beads are collected by centrifugation in an Eppendorf centrifuge and washed as described above for the direct immunoprecipitation of holocytochrome c. Finally, the beads are suspended in 50 μl of "dissociation buffer" (2% SDS, 10 mM Tris-HCl, ph 7.5, 0.35 M 2-mercaptoethanol). The suspension is heated for 5 min at 95°; after centrifugation for 3 min in an Eppendorf centrifuge, the supernatant is collected.

The dissolved immunoprecipitates are subjected to SDS-gel electrophoresis either on horizontal slab gels[13] or on vertical slab gels.[14]

[13] H. Korb and W. Neupert, Eur. J. Biochem. **91**, 609 (1978).
[14] U. K. Laemmli, Nature (London) **227**, 680 (1970).

Comments. The antibodies against holocytochrome *c* and apocytochrome *c* prepared by these procedures did not cross-react either on Ouchterlony plates or upon direct or indirect immunoprecipitation (Fig. 2). Heating of apocytochrome *c* before immunoprecipitation, however, should be avoided, since under this condition cross-reaction with holocytochrome *c* antibodies may occur. Small amounts of unlabeled cytochrome *c* can be analyzed by blotting the gel to nitrocellulose membranes and subsequent treatment with the specific antibodies and radioactive protein A.[15]

Synthesis of Apocytochrome *c* in Cell-Free Systems

Synthesis of apocytochrome *c* can be performed *in vitro* with similar efficiency either in a homologous cell-free translation system prepared from *Neurospora* or in the heterologous rabbit reticulocyte lysate system. The homologous system uses endogenous mRNA; the heterologous system requires isolation of *Neurospora* mRNA.

Homologous System. Neurospora cells (WT 74 A) are grown to mid log phase. Cells labeled *in vivo* with [^{35}S]sulfate are grown in 1 liter of Vogel's medium with unlabeled sulfate reduced to 0.08 mM in the presence of 250 μCi of [^{35}S]sulfate. Growth is stopped by pouring a 1-liter culture into 1 liter of ice-cold wash buffer (100 g crushed ice suspended in 10 mM MgCl$_2$, 30 mM sucrose, pH 7.4, adjusted with triethylamine). Cells are collected by suction on cellulose filter paper, resuspended in 500 ml of ice-cold wash buffer, and filtered again. Four grams of the resulting cell pad are mixed in a sterile mortar with 6 g of washed quartz sand (Riedel de Haen, Hannover), which has been dried and sterilized at 200°. The mixture is ground for 5 min at 4° with four successive additions of 2 ml of medium A. Medium A contains 0.3 M sucrose, 30 mM KCl, 60 mM triethanolamine-HCl, 30 mM triethanolamine, 2 mM EDTA, 5 mM MgCl$_2$, 20 mM KH$_2$PO$_4$, 0.25 mM each of the L-amino acids except the labeled one, 0.007% 2-mercaptoethanol, adjusted to pH 7.4, with 5 M KOH. It can be stored at $-20°$.

The homogenate is centrifuged twice in a Sorvall SS-34 rotor for 5 min at 4000 rpm. The pellets are discarded, and the supernatant is centrifuged in an SS-34 rotor for 12 min at 12,000 rpm to sediment the mitochondria. The supernatant is centrifuged once more in the SS-34 rotor for 12 min at 12,000 rpm to obtain the postmitochondrial supernatant. This is used for cell-free translation, and the following additions are made: 4 mM ATP (20 μl of 200 mM stock solution per milliliter), 1 mM GTP (20 μl of 50 mM stock solution per milliliter), 8 mM creatine phosphate (3.6 mg/ml), 50 μg of creatine kinase per milliliter (10 μl of 0.5% stock solution in 50% glycerol per milliliter). Usually labeling is carried out with [^{3}H]leucine. Then

[15] W. N. Burnette, *Anal. Biochem.* **112**, 195 (1981).

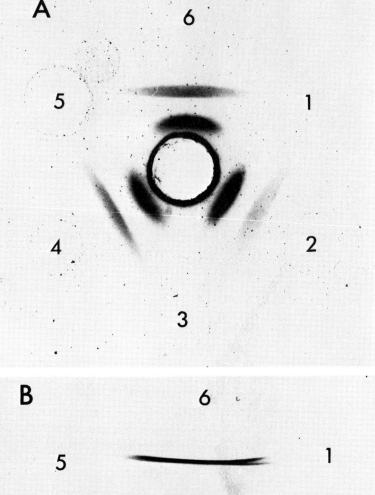

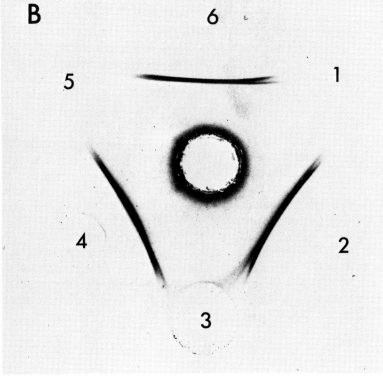

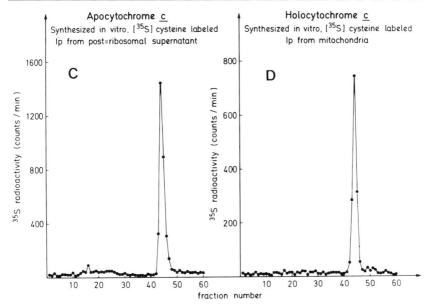

FIG. 2. Specificity of immunoprecipitation by antibodies directed against apocytochrome *c* and holocytochrome *c*. (A) and (B) Immunodiffusion according to Ouchterlony. Center well in (A): antiserum against apocytochrome *c* (20 μl); outer wells: 2, 4, 6, apocytochrome *c* (0.5 μg), and 1, 3, 4, holocytochrome *c* (0.5, 1,2 μg, respectively). Center well in (B): antiserum against holocytochrome *c* (20 μl); outer well: 2, 4, 6, holocytochrome *c* (0.5 μg), and 1, 3, 5, apocytochrome *c* (0.5, 1, 5 μg, respectively). (C) and (D) SDS-PAGE analysis of radiolabeled immunoprecipitates. (C) 35S-labeled apocytochrome *c* was synthesized *in vitro* and immunoprecipitated with antibodies directed against apocytochrome *c*. (D) 35S-labeled holocytochrome *c* labeled *in vitro* immunoprecipitated with antibodies directed against holocytochrome *c*.

0.5 mM [^{3}H]leucine (specific radioactivity 50 Ci/mmol, New England Nuclear Corp. (NEN), Boston, Massachusetts in 0.5 ml of 2 mM KCl (adjusted to pH 7.4) is added per milliliter of incubation mixture. When radioactive cysteine is to be incorporated, 0.5 mCi of [^{35}S]cysteine (1000 Ci/mmol, NEN, Boston, Massachusetts) in 0.5 ml of 2 mM KCl is added per milliliter of incubation mixture. Incorporation of the radioactive amino acids is performed for 10 min at 25°. Cycloheximide (0.1 mg/ml) is added to the incubation mixture after 10 min, and the sample is cooled to 0°. Centrifugation in Eppendorf tubes is carried out in a Beckman 50 Ti rotor using plastic adaptors for 60 min at 50,000 rpm. The postribosomal supernatant contains the labeled apocytochrome *c*. The average incorporation into total protein is 20 to 30 × 10^6 dpm per milliliter of incubation mixture. Incorporation into apocytochrome *c* is roughly 60 × 10^3 dpm per milliliter of incubation mixture.

Heterologous System. Preparation of reticulocyte lysates and cell-free protein synthesis directed by *Neurospora* RNA is performed as described.[16] After protein synthesis for 60 min, a postribosomal supernatant is prepared by centrifugation for 60 min at 50,000 rpm in a Beckman 50 Ti rotor. This contains about 0.1 pmol of labeled apocytochrome c per milliliter.

Preparation of Labeled Apocytochrome c by Reductive Methylation

Apocytochrome c, 600 μg, is dialyzed in Visking 18/32 cellulose tubing against 0.1 M KP$_i$, pH 7.0, and adjusted to a volume of 120 μl. Then at a temperature of 25°, 50 μl of 0.1 M KP$_i$, pH 7.0, containing 125 μCi of [^{14}C]formaldehyde (specific activity 40–60 Ci/mmol; NEN, Boston, Massachusetts) are added, followed by 60 μl of 0.1 M KP$_i$, pH 7.0, containing 6 mg of NaBH$_3$CN per milliliter (Serva, Heidelberg, FRG).[17] The mixture is kept for 1 hr at 25°. It is dialyzed first against 0.1 M KP$_i$, pH 7.0, at 4°, then against 0.25 M NH$_4$HCO$_3$. Specific activity of ^{14}C-methylated apocytochrome c is about 2.0×10^6 dpm/nmol as determined by amino acid analysis.

Comment. Labeling of holocytochrome c with this procedure followed by the removal of the heme group yields apocytochrome c that does not bind to mitochondria and is not converted to holocytochrome c.

Transfer of Apocytochrome c into Mitochondria and Conversion to Holocytochrome c

To study import of apocytochrome c into mitochondria *in vitro,* mitochondria are incubated either with the postribosomal supernatant of the cell-free system containing labeled apocytochrome c or with appropriate buffers containing apocytochrome c labeled by reductive methylation. Mitochondria are isolated by grinding 1 g (wet weight) of *Neurospora* cells harvested at mid log phase with 1.5 g of quartz sand and 2 ml of "sucrose medium" containing 0.3 M sucrose, 2 mM EDTA, 30 mM Tris-HCl, pH 7.2 (adjusted at 20°). Sand and cell membranes are removed by centrifugation in a Sorvall SS-34 rotor for 5 min at 5000 rpm. Mitochondria are sedimented by centrifugation in the SS-34 rotor for 12 min at 12,000 rpm. Mitochondria are then washed twice with "sucrose medium," or further purified by density gradient centrifugation.

For further purification, a mitochondrial suspension is layered on top of a linear sucrose gradient (0.96 M to 1.8 M sucrose in 10 mM Tris-HCl, 2 mM EDTA, pH 7.2, and centrifuged in a Beckman SW 60 rotor for 60 min at 60,000 rpm.[18] The mitochondrial fraction bands in the lower half

[16] H. R. B. Pelham and R. J. Jackson, *Eur. J. Biochem.* **67**, 247 (1976).
[17] D. Dottavio-Martin and J. M. Ravel, *Anal. Biochem.* **87**, 562 (1978).
[18] D. J. L. Luck, this series, Vol. 12A, p. 465.

of the gradient. It is recovered by use of a Pasteur pipette, diluted with 0.2 M sucrose, 10 mM Tris-HCl, 2 mM EDTA, pH 7.2, and centrifuged in an SS-34 rotor for 12 min at 12,000 rpm. The yield of mitochondria is about 1 mg of mitochondrial protein per milliliter of original cell homogenate. The isolated mitochondria are resuspended in the postribosomal supernatant of the cell-free translation system at a concentration of about 1 mg of protein per milliliter. Incubation is performed at 25°. Transfer of apocytochrome *c* into mitochondria is practically complete after 15–20 min (Fig. 3). Mitochondria are sedimented by centrifugation in an SS-34 rotor for 12 min at 12,000 rpm and washed once with 1 ml of "sucrose medium." Apo- and holocytochrome *c* are immunoprecipated from the lysed mitochondria and from the first postmitochondrial supernatant. About 80–90% of the radioactive apocytochrome *c* originally present is recovered as holocytochrome *c* in the mitochondrial fraction. Conversion of apocytochrome *c* to holocytochrome *c* in this system can be inhibited by addition of deuteroheme; 10 nmol of deuteroheme per milliliter inhibits conversion by about 70%, and 100 nmol/ml by more than 95%. (For details see the next section.)

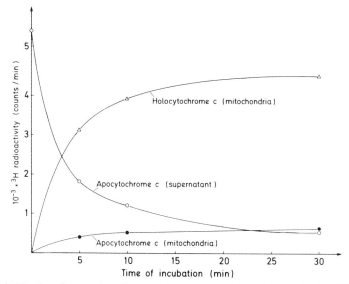

FIG. 3. Kinetics of conversion of apocytochrome *c* to holocytochrome *c* *in vitro*. ^{3}H-labeled apocytochrome *c* was synthesized in a cell-free homologous system in the presence of [^{3}H]leucine. The postribosomal supernatant was prepared and mixed with mitochondria. The mixture was incubated at 25° for the indicated periods of time. Postmitochondrial supernatant and mitochondria were separated by centrifugation. Apocytochrome *c* and holocytochrome *c* were precipitated from the supernatant and from mitochondria employing specific antibodies. The immunoprecipitates were separated by electrophoresis on SDS gels. The radioactivities in the cytochrome *c* peaks were determined.

Comments. The quality of mitochondria is important for binding and import of apocytochrome *c*. Swollen or sonicated mitochondria are unable to convert apocytochrome *c* to holocytochrome *c*. Rapid purification of small amounts of mitochondria can be performed in Eppendorf tubes with density gradients using Percoll (Pharmacia, Uppsala) according the described procedure.[19] The amount and stability of mitochondria can be controlled when mitochondria are used that are labeled *in vivo* with [^{35}S]sulfate.

Binding of Apocytochrome *c* to Mitochondrial Receptor Sites

Mitochondria (about 1 mg protein) are suspended in 100 μl of Sucrose-MOPS buffer (250 mM sucrose, 10 mM MOPS, 1 mM EDTA, 0.1% bovine serum albumin, pH 7.2) containing 10–100 nmol of deuterohemin (Porphyrin Products, Logan, Utah).[20] Deuterohemin is added from a stock solution prepared as follows. One milligram of deuterohemin is dissolved in 10 μl in 1 M KOH, then 1 ml of Sucrose–MOPS buffer is added, and the pH is adjusted to 7.2 by addition of 1 M HCl. The mixture is incubated for 5 min at 25°, then 1 ml of postribosomal supernatant containing [^{3}H]leucine-labeled apocytochrome *c* is added; incubation is performed for further 15 min. Mitochondria and supernatant are separated by centrifugation in a Sorvall SS-34 rotor for 12 min at 12,000 rpm.

For binding of isolated apocytochrome *c* labeled by reductive methylation, mitochondria (about 1 mg of protein) are suspended in 50 μl of Sucrose-MOPS buffer containing 10–100 nmol of deuteroheme. The mixture is incubated for 5 min at 25°. Then 1 ml of additional Sucrose-MOPS buffer is added, and aliquots of 50 μl are transferred into glass minivials (300 μl). Appropriate amounts of ^{14}C-labeled apocytochrome *c* are added, and binding is allowed to occur for 15 min at 25°. Then mitochondria are spun down and washed twice with 200 μl of Sucrose-MOPS buffer. Radioactivity is determined after dissolving mitochondria in 100 μl of 0.1% SDS.

Comments. Results obtained with binding of ^{14}C-labeled apocytochrome *c* labeled by reductive methylation allow a more accurate determination of the number and affinity of binding sites. Conditions and requirements of binding can be studied better in this simple system.

[19] J. R. Mickelson, M. L. Graeser, and B. B. Marsh, *Anal. Biochem.* **109,** 255 (1980).
[20] B. Hennig and W. Neupert, *Eur. J. Biochem.* **121,** 203 (1981).

[26] Biosynthesis and Assembly of Nuclear-Coded Mitochondrial Membrane Proteins in *Neurospora crassa*

By RICHARD ZIMMERMANN and WALTER NEUPERT

The vast majority of mitochondrial proteins are coded for by nuclear genes and synthesized on cytoplasmic ribosomes. These proteins must be transported into the different mitochondrial compartments: the outer membrane, the intermembrane space, the inner membrane, and the matrix space. Three major problems are currently being investigated in order better to understand the molecular basis of this transport process: What determines the specificity of the process, i.e., how is a mitochondrial protein selectively directed into a mitochondrion? How are the newly synthesized proteins translocated across the lipid bilayer of one or two mitochondrial membranes? How are the proteins assembled, so that they assume their specific orientation in the membranes and are put together to multisubunit structures?

Experimental approaches to these problems can be made by studying the various steps *in vivo,* or *in vitro* in reconstituted systems. Both of these approaches have been demonstrated to yield useful information.[1] The transport into mitochondria has been found to be a posttranslational process involving the occurrence of extramitochondrial precursor proteins. Synthesis of precursor proteins and transfer into the mitochondria are processes that are not, at least not obligatorily, coupled.[2-7] These soluble precursors in most cases have an additional peptide sequence that is not present in the mature membrane-bound protein. Specific receptors on the mitochondrial surface appear to function in binding mitochondrial precursors to the mitochondria.[8-11] We describe here methods to analyze, *in vivo,*

[1] W. Neupert and G. Schatz, *Trends Biochem. Sci.* **6,** 1 (1981).
[2] G. Hallermayer, R. Zimmermann, and W. Neupert, *Eur. J. Biochem.* **81,** 523 (1977).
[3] M. A. Harmey, G. Hallermayer, H. Korb, and W. Neupert, *Eur. J. Biochem.* **81,** 533 (1977).
[4] H. Korb and W. Neupert, *Eur. J. Biochem.* **91,** 609 (1978).
[5] M.-L. Maccechini, Y. Rudin, G. Blobel, and G. Schatz, *Proc. Natl. Acad. Sci. U.S.A.* **76,** 343 (1979).
[6] Y. Raymond and G. S. Shore, *J. Biol. Chem.* **254,** 9335 (1979).
[7] M. Mori, S. Miura, M. Tatibana, and P. P. Cohen, *Proc. Natl. Acad. Sci. U.S.A.* **77,** 7044 (1980).
[8] R. Zimmermann and W. Neupert, *Eur. J. Biochem.* **109,** 217 (1980).
[9] R. Zimmermann, B. Hennig, and W. Neupert, *Eur. J. Biochem.* **116,** 455 (1981).
[10] B. Hennig and W. Neupert, *Eur. J. Biochem.* **121,** 203 (1981).
[11] M. Schleyer, B. Schmidt, and W. Neupert, *Eur. J. Biochem.* **125,** 109 (1982).

METHODS IN ENZYMOLOGY, VOL. 97

the kinetics of transfer of newly synthesized proteins into mitochondria and to identify precursors of mitochondrial proteins after labeling of intact cells. Furthermore, methods with *in vitro* systems are described that can be employed to investigate the following problems: (*a*) structure and properties of extramitochondrial precursor proteins; (*b*) identification of ribosomes involved in the synthesis of mitochondrial precursor proteins; (*c*) transfer of precursor proteins in reconstituted systems and dissection of the process into different steps; (*d*) energy requirements of the transfer process.

For these studies it is desirable to investigate individual, well-defined mitochondrial proteins; preferentially those with known function, structure, and submitochondrial topology. Unfortunately, only a few mitochondrial proteins are currently characterized in such detail.

The methods described here deal mainly with the study of integral proteins of the inner mitochondrial membrane of *Neurospora crassa,* the ADP–ATP carrier,[12] and subunit 9 of the oligomycin-sensitive ATPase (also called DCCD-binding protein).[13] Similar methods have been used to study the various subunits of the cytochrome bc_1 complex.[14,15] Elsewhere[16] we describe methods for the study of the biogenesis of cytochrome *c*.

Antibodies against Mitochondrial Membrane Proteins

Immunological procedures are necessary to identify the extremely small amounts of proteins that are being transported *in vivo* and *in vitro*. Antibodies against the purified mature protein of interest can be raised in rabbits by any of a number of established procedures. For this purpose soluble, Triton-solubilized, or SDS-solubilized protein may be used. The antibodies obtained usually precipitate not only the appropriate mature proteins, but also recognize the precursor forms. There have been exceptions, however, to this generalization. For example, antibodies prepared against holocytochrome *c* did not recognize the precursor apocytochrome *c*, and antibodies obtained against chemically prepared apocytochrome *c* recognized apocytochrome *c* synthesized *in vivo* or *in vitro,* but not holocytochrome *c*.[4,16] Antibodies against ATPase subunit 9 only weekly interacted with the precursor form.

Kinetic Analysis of Transport of Proteins into Mitochondria in Whole Cells

The rate of appearance of newly synthesized proteins in mitochondria can be measured by pulse labeling whole *Neurospora* cells with [^{3}H]leucine.

[12] H. Hackenberg, P. Riccio, and M. Klingenberg, *Eur. J. Biochem.* **88,** 373 (1978).
[13] W. Sebald, T. Graf, and H. B. Lukins, *Eur. J. Biochem.* **93,** 587 (1979).

Such labeling studies are carried out at 6–9°, since at normal growth temperature assembly is too rapid to be resolved into its various steps. Furthermore, it is useful to employ cells that are grown in the presence of [^{35}S]sulfate to have an internal measure for the recovery of mature proteins in subsequent analyses. Incorporation of added leucine at 8° is complete within about 6 min, but shorter, more defined periods of radiolabeling can be achieved by adding, after the desired labeling period, cycloheximide plus millimolar concentrations of leucine or a chase of leucine alone. In the first case, labeled nascent chains remain attached to polysomes, whereas chain completion and release occurs when labeling is stopped with leucine alone. Incorporation of [^{3}H]leucine into polypeptide chains is inhibited in both cases in less than 5 sec.

Hyphae of *Neurospora crassa* are grown in cultures of 1–2 liters in Vogel's minimal medium[17] containing 2% sucrose. Cultures are vigorously aerated for 13 hr at 25°, at which time the cells are in mid-log phase. To allow labeling with [^{35}S]sulfate, the concentration of sulfate in the medium is reduced to 80 μM (added as MgSO$_4$). Further reduction to 8 μM is possible without affecting growth of cells for 13–14 hr. Together with the inoculum of 2 × 10^6 conidia per milliliter of culture, sodium [^{35}S]sulfate (NEN, Boston, Massachusetts; specific radioactivity 10–1000 Ci/mol) is added at a concentration of 0.25–1 mCi/liter. After 13 hr, the culture is made 1 mM in MgSO$_4$ and incubated further for 1 hr. Then the temperature is shifted to 8° and maintained there for 1–2 hr prior to leucine labeling.

For pulse labeling, [^{3}H]leucine (NEN, Boston, Massachusetts; specific radioactivity 40–60 mCi/mol) is added at a concentration of 2 mCi/liter. After incubation for 1–4 min, cycloheximide (0.1 mg/ml) and leucine (final concentration 10 mM) are added. After various times of further incubation at 8°, aliquots of 100–200 ml are removed and poured into 2 volumes of ice water containing cycloheximide (0.1 mg/ml) and leucine (10 mM). Cells are immediately harvested by filtration, using a funnel specially constructed to keep the temperature at 0°. This apparatus is shown in Fig. 1; as can be seen, it allows the filtration to be performed at room temperature. This helps minimize the time between chilling and harvesting and between harvest and subsequent steps, since all the components of the experiment may be placed in close proximity to each other.

The mycelial pad (0.5 g wet weight) is scraped from the filter paper with a cooled spatula and transferred to a small beaker containing 5 ml of homogenization buffer [0.44 M sucrose, 5 mM NH$_4$Cl, 30 mM Tris-HCl,

[14] H. Weiss and H. J. Kolb, *Eur. J. Biochem.* **99**, 139 (1979).
[15] M. Teintze, M. Slaughter, H. Weiss, and W. Neupert, *J. Biol. Chem.* **257**, 10364 (1982).
[16] B. Hennig and W. Neupert, this volume [25].
[17] H. J. Vogel, *Am. Nat.* **98**, 435 (1964).

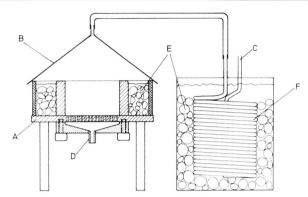

FIG. 1. Filtration apparatus to harvest *Neurospora* cells at 0°. A, aluminum filter plate; B, plastic funnel; C, air inlet; D, outlet to water aspirator; E, ice water; F, copper spiral.

pH 7.4, 1 mM 2-mercaptoethanol, 1 mM p-hydroxymercuribenzoate, 1 mM phenylmethanesulfonyl fluoride (PMSF)]. Hyphae are resuspended for 6 sec in an Ultra-Turrax homogenizer (Jahnke u. Kunkel, Stauffen, Germany) whose metal sheath was precooled to 0°. Then cells are broken in a small-scale grind mill, which is so designed that the passage of the cells occurs within about 5–10 sec.[18] The mill is kept in a freezer adjusted to −5°. The time between harvesting the cells and collection of the homogenate should not exceed 2–3 min, and care must be taken that the temperature is always close to 0°.

The crude homogenate is subfractionated by differential centrifugation according to standard procedures.[2,18] [3]H and ^{35}S radioactivities are determined in proteins precipitated with trichloroacetic acid. Immunoprecipitation from the various fractions lysed in Triton X-100 can be used to follow the kinetics of assembly of individual proteins.[2,9]

Comments. Kinetic analyses of this type give information on the pool sizes of extramitochondrial precursors. Figure 2A shows the labeling kinetics of various cellular fractions. Labeling of mitochondria occurs with a lag not only relative to total cellular protein, but also relative to cytosolic proteins. Various mitochondrial (mt) proteins (Fig. 2B) immunoprecipitated from isolated mitochondria show widely differing labeling kinetics with different lag periods, indicating different pool sizes of extramitochondrial precursors. Redistribution of proteins during cell fractionation, however, is a possible source of artifacts in this type of analysis and should be carefully checked.

Identification of Precursor Proteins by Labeling of Whole Cells

Many precursor proteins cannot be analyzed by the fractionation procedure described above. This can be due to a very short half-life of the

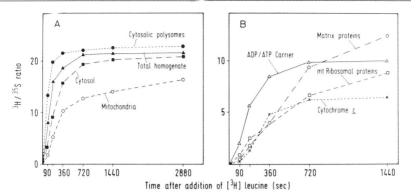

FIG. 2. Kinetics of pulse labeling of mitochondrial proteins in *Neurospora* cells. Cells were grown in the presence of [^{35}S]sulfate and pulsed with [^{3}H]leucine at 8°. Specific labeling of (A) various cellular fractions (trichloroacetic acid-precipitable protein) and (B) mitochondrial proteins immunoprecipitated from mitochondria. From Hallermayer *et al.*[2]

precursor or to the precursor being particularly sensitive to nonspecific proteolytic degration. Therefore it is necessary to stop any proteolytic reactions as rapidly and completely as possible. This can be achieved by immediately solubilizing the harvested cells in SDS followed by immunoprecipitation to detect precursor proteins.

Cells are pulse labeled, cooled to 0° by mixing with iced water (see above), and filtered on a cooled funnel. The hyphae are removed from the filter and immersed in a boiling solution containing 1 mM EDTA, 1 mM EGTA, 3% SDS with the pH adjusted to 7.5. For 0.1 g of cells (wet weight), 1 ml of this solution is used. The cells are homogenized with an Ultra-Turrax homogenizer for 10 sec. This mixture is kept in a boiling water bath for 5 min, then 9 ml of ice-cold Triton buffer (1% Triton, 0.3 M NaCl, 5 mM EDTA, 10 mM Tris-HCl, pH 7.5) are added, and PMSF and *o*-phenanthroline are added to a final concentration of 0.1 mM each. Centrifugation is carried out for 15 min at 39,000 g, and the supernatant is collected. This procedure was adapted for *Neurospora* by modifying a method described for yeast spheroplasts.[19] An alternative procedure is to transfer cells immediately after harvesting into a mortar containing liquid N$_2$ and grinding the cells under liquid N$_2$ for about 5 min. Then the liquid N$_2$ is allowed to evaporate, and a solution containing 3% SDS, 10 mM Tris-HCl, pH 7.5, 2 mM EDTA, 1 mM 2-mercaptoethanol, and 1 mM PMSF is added to the cells (5 ml/g wet weight). The frozen powder is mixed carefully with the SDS solution and allowed to thaw. The mixture is shaken at 4° for 15 min and

[18] W. Sebald, W. Neupert, and H. Weiss, this series, Vol. 55, p. 144.
[19] N. Nelson and G. Schatz, *Proc. Natl. Acad. Sci. U.S.A.* **76**, 4365 (1979).

centrifuged for 15 min at 39,000 g. The supernatant is then diluted with Triton buffer as described above for cells extracted with boiling SDS buffer.

Immunoprecipitation from the extracts is carried out employing *Staphylococcus aureus*.[20]

Comments. This procedure is useful to analyze the properties of precursor proteins that differ from the mature forms in apparent molecular weight, i.e., in electrophoretic mobility on SDS gels. Information can be obtained on the pool size of such precursors, on their mechanism of processing, and on the energy dependence of processing. As an example, Fig. 3 shows the synthesis and processing of cytochrome c_1. This protein is synthesized as a precursor protein with an extension of about 7000 M_r (see also the table). It is processed in two steps, as is cytochrome c_1 in yeast.[21] The intermediate form accumulates in *Neurospora* when the cells are kept at 8° and is processed when growth temperature is raised to 25°. Both processing steps are halted when cells are poisoned with the protonophore carbonyl cyanide *m*-chlorophenylhydrazone. The drawback of the method described above is that there is no way to determine the subcellular localization of precursor and mature forms. The extraction procedure involving grinding of cells under liquid N_2 is useful in those cases where boiling alters the protein in question in such a way that it cannot no longer be immunoprecipitated. One such example is apocytochrome *c*.

Analysis of Transfer of Mitochondrial Precursor Proteins *in Vitro*

Transport of proteins into mitochondria is a multistep process, and an understanding of the individual steps requires *in vitro* systems in which the complex overall reaction can be separately studied. The steps that can be resolved so far are (*a*) synthesis of precursor proteins on cytoplasmic ribosomes; (*b*) transfer through cytosol to the mitochondria; (*c*) binding to specific sites on the mitochondria; and (*d*) translocation across inner or outer membrane, or both, accompanied in most cases by proteolytic cleavage or by some other covalent modification.

Synthesis of Precursor Proteins in Cell-Free Systems

Mitochondrial precursor proteins are synthesized in reticulocyte lysates according to standard procedures.[22-24] Either total RNA from *Neurospora*

[20] S. W. Kessler, *J. Immunol.* **115,** 1617 (1975).
[21] G. Schatz and A. Ohashi, personal communication.
[22] E. H. Allen and R. W. Schweet, (1962) *J. Biol. Chem.* **237,** 760 (1962).
[23] T. Hunt and R. J. Jackson, *in* "Modern Trends in Human Leukaemia" (R. Neth, R. C. Gallo, S. Spiegelmann, and F. Stohlman, eds.), p. 300. J. F. Lehmans Verlag, Munich, 1974.
[24] H. R. B. Pelham and R. J. Jackson, *Eur. J. Biochem.* **67,** 247 (1976).

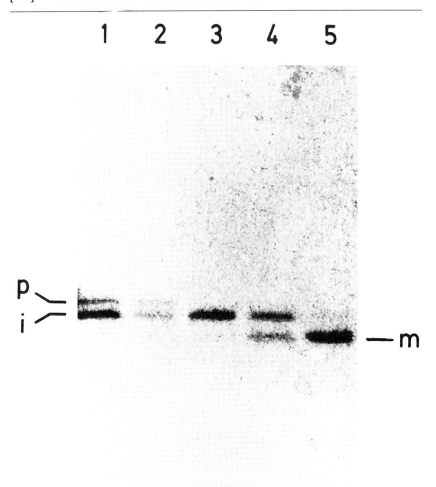

FIG. 3. Labeling of cytochrome c_1 precursor in whole *Neurospora* cells. *Neurospora* cells were pulse-labeled at 8° with [³H]leucine. They were cooled to 0°, harvested, and extracted with hot SDS-containing medium. After dilution with Triton buffer, immunoprecipitation was carried out using antibodies against cytochrome c_1 and Sepharose-bound protein A. Immunoprecipitates were analyzed by sodium dodecyl sulfate – polyacrylamide gel electrophoresis and fluorography. Lanes: 1, Pulse for 3 min, then cycloheximide (CHI) and a leucine chase were added, and the sample was immediately cooled to 0° and harvested; 2, Pulse for 3 min, then CHI and chase were added together with 12.5 μM carbonyl cyanide *m*-chlorophenylhydrazone (CCCP), and the sample was kept at 8° for 7 min, followed by cooling to 0° and harvesting; 3, as for lane 2, but without addition of CCCP; 4, pulse for 10 min at 8°, then addition of CHI, chase, and CCCP and further incubation for 5 min at 25°, then cooling to 0° and harvesting; 5, as for lane 4, but without addition of CCCP. p, Precursor; i, intermediate form; m, mature cytochrome c_1.

cells or poly(A)-containing RNA can be used to program reticulocyte lysates to synthesize mitochondrial proteins. RNA is isolated by phenol extraction by the following procedure. *Neurospora* hyphae (25 g wet weight, grown for 14 hr) are harvested by filtration and rapidly frozen with liquid nitrogen. The frozen hyphae are ground in a mortar cooled with liquid nitrogen to obtain a fine powder until at least 50% of the hyphae are broken, as determined by microscopic examination.

For extraction of nucleic acids, the powder is transferred to a bottle containing 100 ml of extraction medium A [phenol medium[25] equilibrated with 10 mM Tris-HCl, pH 8.2; detergent medium according to Leaver and Ingle[26]; chloroform medium: 49 ml of chloroform, 1 ml of isoamylalcohol; phenol medium, detergent medium, and chloroform medium are mixed in _ ¬atio of 2:1:1 (v/v/v)]. Solid sodium dodecyl sulfate is added to a final concentration of 1% (w/v), and the mixture is shaken vigorously for 15 min at room temperature. The organic phase is then separated from the aqueous phase by centrifugation at 4° in polypropylene tubes for 15 min at 30,000 g, and the aqueous phase is saved. The organic phase including the white precipitate at the interface is extracted once more with 100 ml of medium A for 15 min. Again the phases are separated by centrifugation, the organic phase and interface being discarded.

The aqueous phases are combined and 3 volumes of extraction medium B [phenol medium and chloroform medium mixed in a ratio of 2:1 (v/v)] are added. After shaking for 15 min at room temperature, centrifugation is performed as described above. The aqueous phase is collected, and 3 volumes of chloroform medium are added. After shaking for 10 min and centrifugation, the organic phase is discarded. To precipitate nucleic acids from the aqueous phase, two volumes of absolute ethanol are added and the mixture is held overnight at −20°. Precipitated nucleic acids are collected by centrifugation at −20° for 20 min at 12,000 g. The pellet is washed twice with 66% ethanol (v/v) at −20° and dried over $CaCl_2$ *in vacuo*. The yield is about 1000 OD_{260} units per 10 g of hyphae. Separation of poly(A)-containing RNA is carried out according to Aviv and Leder.[27] The yield is about 5 OD_{260} units per 10 g of hyphae.

Comments. The table lists the precursors of *Neurospora* mitochondrial proteins so far identified.[1,4,8,9,11,15,28-34] Four proteins have been found to

[25] J. H. Parish and K. S. Kirby, *Biochim. Biophys. Acta* **129**, 554 (1966).
[26] C. J. Leaver and J. Ingle, *Biochem. J.* **123**, 235 (1971).
[27] H. Aviv and P. Leder, *Proc. Natl. Acad. Sci. U.S.A.* **69**, 1408 (1972).
[28] H. Freitag, W. Neupert, and R. Benz, *Eur. J. Biochem.* **123**, 629 (1982).
[29] H. Freitag, M. Janes, and W. Neupert, *Eur. J. Biochem.* **126**, 197 (1982).
[30] R. Zimmermann, U. Paluch, and W. Neupert, *FEBS Lett.* **108**, 141 (1979).
[31] R. Zimmermann, U. Paluch, M. Sprinzl, and W. Neupert, *Eur. J. Biochem.* **99**, 247 (1979).
[32] B. Schmidt and W. Neupert, unpublished.

MITOCHONDRIAL PRECURSOR PROTEINS IN *Neurospora crassa*

Sample	Location	Apparent molecular weight × 10⁻³		Reference
		Pre-cursor[a]	Mature protein	
Mitochondrial porin	Outer membrane	31	31	28,29
Cytochrome c	Inter-membrane space	11.9*	11.9[b]	4,30
ADT–ATP carrier	Inner membrane	32	32	8,31
Cytochrome bc_1 complex				
Subunit I (core I)		51.5	50	15
II (core II)		47.5	45	15
IV (cytochrome c_1)		38*	31	1,15
V (Fe/S protein)		28	25	15
VI		14	14	15
VII		12	11.5	15
VIII		11.6	11.2	15
ATPase complex				
α-Subunit		62	58	32
Subunit 9		15.4*	9.5	9,11,33
Citrate synthase	Matrix	47.5*	45	34

[a] Precursor proteins indicated by an asterisk have also been detected after labeling of whole cells.

[b] Molecular weight of holocytochrome c without the heme group.

have no detectable additional sequence. The additional sequences of the other proteins varied between 0.4×10^3 and 7×10^3 apparent molecular weight.

Transfer of Precursors Synthesized in Reticulocyte Lysates into Isolated *Neurospora* Mitochondria

Postribosomal supernatants of reticulocyte lysates after synthesis of precursor proteins are prepared by centrifuging the cooled lysates at 0° for 1 hr at 160,000 g. This is conveniently done in Eppendorf tubes that are placed in water-filled adaptors (made from polystyrol) in a 50 Ti Beckman rotor. The supernatants are made 0.3 M in sucrose and 0.05 mM in the amino acid used for radioactive labeling.

[33] R. Michel, E. Wachter, and W. Sebald, *FEBS Lett.* **101**, 373 (1979).
[34] M. A. Harmey and W. Neupert, *FEBS Lett.* **108**, 385 (1979).

Mitochondria are isolated from *Neurospora* hyphae converted to spheroplasts. To prepare spheroplasts, 10 g of hyphae (wet weight) are resuspended in 50 ml of conversion medium (1 M sorbitol, 4.5 mM sucrose in Vogel's minimal medium[17]). Then 1 ml of a solution containing 5.2 units of β-glucuronidase and 4.5 units of arylsulfatase from *Helix pomatia* (Boehringer, Mannheim) are added. The mixture is shaken gently for 30 min at 25° then cooled to 0° and centrifuged for 10 min at 5000 g in a Sorvall refrigerated centrifuge. The pellet is washed once with conversion medium by gentle resuspension and by centrifugation. The final pellet is transferred to a Dounce homogenizer with 10 ml of the following medium: 0.3 M sucrose, 20 mM KH$_2$PO$_4$, 10 mM MgCl$_2$, 2 mM EDTA, 30 mM KCl, 30 mM triethanolamine, 60 mM triethanolamine-HCl; the pH is adjusted with 5 M KOH to 7.4. Homogenization is carried out by 10 strokes with a loosely fitted pestle. The homogenate is centrifuged twice for 5 min at 4000 g, then mitochondria are pelleted by centrifugation for 12 min at 17,300 g and held at 0° as a pellet until use.

Mitochondria are gently resuspended in the postribosomal supernatant of the reticulocyte lysate at a concentration of 0.5–1 mg of mitochondrial protein per milliliter. After incubation at 25°, aliquots are removed, cooled to 0°, and analyzed for transfer of proteins (Fig. 4).

For analysis of transfer efficiency, mitochondria and supernatant are separated again by centrifugation for 12 min at 17,300 g. The mitochondrial pellet is washed once with sucrose buffer (0.3 M sucrose, 2 mM EDTA, 10 mM Tris-HCl, pH 7.8), resuspended in the same buffer, and treated with proteinase K for 60 min at 4°. The concentration of proteinase K required to digest precursor, but not the mature assembled protein, should be determined for each different protein. In the case of the ADP–ATP carrier, for example, a concentration of 10 μg/ml is sufficient to degrade ADP–ATP carrier that is not transferred, whereas ADP–ATP carrier transferred *in vitro* or the mature preexisting form are largely resistant to concentrations up to 260 μg/ml. Discrimination of extramitochondrial and transferred form can also be made by incubating the whole transfer mixture with proteinase K.

Comments. Isolated mitochondria show the ability to bind precursor proteins in a saturable fashion. Binding can be studied as a separate step when the mitochondrial membrane potential is dissipated by protonophores such as CCCP, FCCP, or dinitrophenol, by ionophores such as valinomycin/K$^+$ or by a combination of oligomycin and the respiration inhibitors antimycin A or KCN (Fig. 5). The electrical membrane potential has been identified as the primary energy form to drive import of most precursor proteins in *Neurospora*.[11]

Correct insertion of precursors in cell-free systems is difficult to demonstrate owing to the very small amounts of protein transferred. Nevertheless,

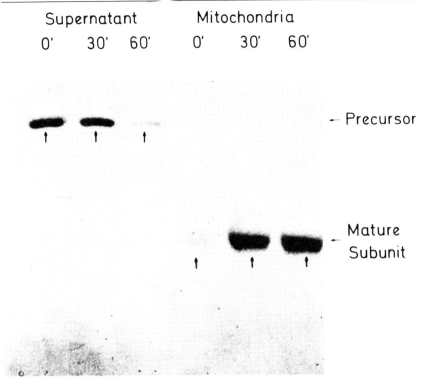

FIG. 4. Transfer and processing *in vitro* of ATPase subunit 9. A reticulocyte lysate was programmed with *Neurospora* RNA in the presence of [^{35}S]methionine. After incubation for 60 min, the postribosomal supernatant was prepared and incubated with mitochondria isolated from *Neurospora* spheroplasts. After various times of incubation, mitochondria and supernatant were reisolated. Subunit 9 was immunoprecipitated from mitochondria by direct immunoprecipitation and from the supernatant employing Sepharose-bound protein A. Immunoprecipitates were analyzed by sodium dodecyl sulfate–polyacrylamide gel electrophoresis and autoradiography.

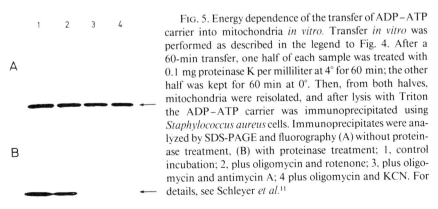

FIG. 5. Energy dependence of the transfer of ADP–ATP carrier into mitochondria *in vitro*. Transfer *in vitro* was performed as described in the legend to Fig. 4. After a 60-min transfer, one half of each sample was treated with 0.1 mg proteinase K per milliliter at 4° for 60 min; the other half was kept for 60 min at 0°. Then, from both halves, mitochondria were reisolated, and after lysis with Triton the ADP–ATP carrier was immunoprecipitated using *Staphylococcus aureus* cells. Immunoprecipitates were analyzed by SDS-PAGE and fluorography (A) without proteinase treatment, (B) with proteinase treatment; 1, control incubation; 2, plus oligomycin and rotenone; 3, plus oligomycin and antimycin A; 4 plus oligomycin and KCN. For details, see Schleyer *et al.*[11]

for a number of proteins it is possible to demonstrate properties specific for the assembled proteins. The ADP–ATP carrier is known to bind carboxyatractyloside only in the membrane-integrated dimeric form. Binding of this inhibitor protects the carrier form proteolytic fragmentation.[35] This was also demonstrated with the carrier transferred *in vitro,* indicating its assembly into a functional position.[36] For subunit 9 of ATPase, it has been found that, unlike the precursor form, the transferred and processed protein acquired the property of being soluble in chloroform–methanol and becomes precipitable with an antibody against F_1-ATPase.

The *in vitro* system can be used to determine how universal the transport mechanism is, i.e., whether species well separated in evolution are capable of taking up and processing each others precursor proteins. *Neurospora* precursors have been found to be imported and processed by yeast and rat liver mitochondria.[11,15] This indicates a degree of universality of the components of the import machinery.

[35] M. Klingenberg, H. Aquila, and P. Riccio, this series, Vol. 56, p. 407.
[36] M. Schleyer and W. Neupert, unpublished.

[27] Isolation and Properties of the Porin of the Outer Mitochondrial Membrane from *Neurospora crassa*

By Helmut Freitag, Roland Benz, and Walter Neupert

The outer mitochondrial membrane is freely permeable to low molecular weight components such as metal ions, substrates, and nucleotides.[1,2] Therefore pores in the outer membrane have been postulated. The size of these pores is not clearly defined. Molecules with apparent molecular weights up to 2000–8000 were found to permeate through these channels. Detergent extracts of mitochondrial outer membranes or purified pore proteins were incorporated into liposomes or bilayers. Formation of nonspecific pores with a diameter of about 20 Å was observed.[3-8] The pore-

[1] R. L. O'Brien and G. Brierley, *J. Biol. Chem.* **240**, 4527 (1965).
[2] E. Pfaff, M. Klingenberg, E. Ritt, and W. Vogell, *Eur. J. Biochem.* **5**, 222 (1968).
[3] M. Colombini, *Nature (London)* **279**, 643 (1979).
[4] M. Colombini, *J. Membr. Biol.* **53**, 79 (1980).
[5] L. S. Zalman, H. Nikaido, and Y. Kagawa, *J. Biol. Chem.* **255**, 1771 (1980).
[6] H. Freitag, W. Neupert, and R. Benz, *Eur. J. Biochem.* **123**, 629 (1982).
[7] N. Roos, R. Benz, and D. Brdiczka, *Biochim. Biophys. Acta* **686**, 204 (1982)
[8] H. Freitag, G. Genchi, R. Benz, F. Palmieri, and W. Neupert, *FEBS Lett.* **145**, 72 (1982).

METHODS IN ENZYMOLOGY, VOL. 97

forming protein, called mitochondrial porin, is a major component of outer mitochondrial membranes. It has an apparent molecular weight of about 31,000. In many respects the mitochondrial porin resembles the porins of outer membranes from gram-negative bacteria.

Here we describe two different isolation procedures for this protein and discuss some of its properties and its reconstitution into artificial bilayers.

Isolation of Mitochondrial Porin

Growth of *Neurospora crassa* hyphae (wild-type 74A) for 17 hr in 8-liter cultures (inoculum 1×10^6 conidia/ml) in the presence of 2% sucrose and disruption of cells are carried out according to standard procedures.[9] Mitochondria are isolated in isolation medium (0.44 M sucrose, 2 mM EDTA, 30 mM Tris-HCl, pH 7.6, adjusted at 4°) with the addition of 0.5 mM phenylmethylsulfonyl fluoride (PMSF; final concentration), freshly dissolved in ethanol. The following centrifugation steps are carried out at 4°, employing a Sorvall RC 5B centrifuge and a GS-3 rotor. The cell debris are removed by centrifugation for 20 min at 2000 g. The resulting supernatant is centrifuged for 45 min at 17,000 g. The upper layers of the resulting sediment (containing the mitochondria) are washed twice by resuspending them in a glass–Teflon potter and centrifuging for 45 min at 17,000 g. The final mitochondrial pellet is the starting material for the porin isolation. Under typical conditions, from an 8-liter culture of *Neurospora crassa* 180 g of hyphae (wet weight) are obtained and the yield of mitochondria is 230 mg protein.

Porin Isolation from Outer Membranes[6]

Principle

Mitochondrial outer membranes are isolated by swelling–shrinking of the mitochondria and subsequent separation by a sucrose step centrifugation. Differential detergent extraction of the outer membrane followed by DEAE-cellulose chromatography gives pure porin, which forms active pores after reconstitution into artificial bilayers.

Reagents

Swelling medium: 2.5 mM potassium H_2PO_4, adjusted to pH 7.3 with KOH

PMSF stock solution: 100 mM PMSF in ethanol, freshly prepared

Shrinking medium: 2.2 M sucrose, 8 mM $MgCl_2$, 8 mM ATP, adjusted to pH 7.3 with KOH

[9] W. Neupert and G. D. Ludwig, *Eur. J. Biochem.* **19**, 523 (1971).

Isolation medium: 0.44 M sucrose, 2 mM EDTA, 30 mM Tris-HCl, pH 7.5

Sucrose step medium: 0.95 M sucrose, 10 mM Tris-HCl, pH 7.5

Tris buffer: 10 mM Tris-HCl, pH 7.5

Phosphate buffer: 100 mM potassium H_2PO_4, adjusted to pH 7.3 with KOH

Octyl glucoside solution: 60 mM octyl-β-D-glucoside, 5 mM Tris-HCl, pH 7.5 (17.5 mg of octyl glucoside per milliliter)

Genapol solution: 1% Genapol X-100 (Farbwerke Hoechst, Frankfurt, FRG), 5 mM Tris-HCl, pH 7.5. Genapol X-100 contains polyethoxy chains like Triton X-100, but has isotridecyl groups instead of aromatic rings and therefore shows no absorption at 280 nm

DEAE-cellulose column: A column (15 ml, 1 cm diameter) is filled with DEAE-cellulose (DE-52, Whatman, Maidstone, England) and equilibrated with the Genapol solution. The flow rate was 10 ml/hr. In all solutions, the pH was adjusted at 4°.

Procedure

Isolation of Outer Membranes. Swelling solution, 100 ml containing 0.5 mM PMSF is used for 100 mg of mitochondrial protein. The mitochondrial pellet is suspended in the swelling medium with a glass–Teflon potter. The suspension is stirred for 15 min at 4°, and then 70 ml of the shrinking medium are added while stirring. During shrinking the suspension becomes more turbid. It is further stirred for 10 min. Then 60-ml portions of the suspension are transferred into a glass–teflon potter, and the pestle, driven by a motor at 500–700 rpm, is moved up and down five times. The membranes are then pelleted by centrifugation for 45 min at 98,000 g in a Beckman ultracentrifuge using a 45 Ti rotor. The pellets are suspended in isolation medium with a glass–Teflon potter, then 35 ml of the sucrose step medium are filled into a Beckman SW 25.2 centrifuge tube and overlayered with 20 ml of the mitochondrial suspension, containing the membranes from about 100 mg of mitochondria. Centrifugation is performed for 60 min at 70,000 g. The small band of red outer membranes on top of the 0.95 M sucrose solution is collected with a bended Pasteur pipette. The suspension is diluted with an equal volume of Tris buffer and centrifuged for 30 min at 166,000 g in a Beckman 50 Ti rotor. The outer membrane pellets are suspended in phosphate buffer with the addition of 8% dimethyl sulfoxide (v/v) at a protein concentration of 3.5 mg/ml and stored in −20°. The yield of outer membranes is about 0.5% of protein with respect to whole mitochondria, which is about 25% of the theoretical amount. Samples should be stored not longer than 1 week, because prolonged storage alters the behavior of the membranes during the subsequent extraction procedures.

Isolation of Porin. The thawed suspension of outer membranes is first extracted with octyl glucoside, which removes most of the outer membrane proteins, except the porin.[6] For this extraction it is important to adjust the octyl glucoside concentration to 30 mM, and the octyl glucoside-to-protein ratio to 10 (w/w), to obtain reproducible results. One milliliter of outer membrane suspension is first diluted with phosphate buffer to 2 ml, then 2 ml of 60 mM octyl glucoside solution are added. The mixture is kept at 4° for 45 min. Then it is transferred into Eppendorf tubes, which are placed in water-filled plastic adaptors and centrifuged for 30 min at 150,000 g in a Beckman 50 Ti rotor. The membrane pellets are suspended in the Genapol solution by adding 1 ml per milligram of the original outer membrane protein. They are shaken for 30 min at 4°. After centrifugation in an Eppendorf centrifuge for 5 min, the supernatant is subjected to chromatography on a DEAE-cellulose column, equilibrated with the Ganapol solution. Elution is followed by monitoring the absorbance at 280 nm. The protein that elutes with the void volume is pure porin and can be stored for several months at −20°.

Porin Isolation from Whole Mitochondria[8]

Principle

Whole mitochondria are lysed with Genapol X-100 and chromatographed on hydroxyapatite, Celite, and DEAE-cellulose. This procedure gives pure and functionally active porin in high yield and can be easily scaled up.

Reagents

Buffer A: 2.5% Genapol X-100 (Farbwerke Hoechst, Frankfurt, Federal Republic of Germany), 50 mM KCl, 10 mM potassium H_2PO_4, 1 mM EDTA, 10 mM Tris adjusted to pH 7.0 with HCl
Genapol solution: 1% Genapol X-100, 5 mM Tris-HCl, pH 7.5
Dry hydroxyapatite (BioGel HTP, Bio-Rad, Richmond, California)
Dry Celite (Celite 535, Roth, Karlsruhe, Federal Republic of Germany).

Procedure

For this isolation procedure, frozen mitochondria can be used. Two Pasteur pipettes plugged with cotton wool, one filled with 0.6 g of hydroxyapatite and the other filled with 0.6 g of a dry mixture of equal weights of hydroxyapatite and Celite, are prepared. To 4–6 mg of protein of a mitochondrial pellet, 0.7 ml of buffer A are added, mixed, and kept for 30 min at 4°, followed by centrifugation for 15 min at 27,000 g. The supernatant is applied to the hydroxyapatite pipette. When the solution has entered the column, further buffer is applied until the eluate has reached the original volume (0.7 ml). This eluate is transferred to the hydroxyapatite–Celite pipette. Again 0.7 ml of the eluate is collected, which already contains the largely enriched porin.[8] The eluate is dialyzed overnight against 20 volumes of the Genapol solution and subjected to chromatography on DEAE-cellulose as described before.

There are two ways to scale up the isolation procedure. For up to 200 μg of porin it is recommended to use 10 Pasteur pipettes in parallel and to combine the eluates for one DEAE-cellulose column of the size described above. In order to prepare porin in larger amounts, larger columns should be used. Mitochondria (1.2 g of protein) are dissolved in 130 ml of buffer A and chromatographed on columns containing 600 g of hydroxyapatite or hydroxyapatite–Celite mixture (diameter 3 cm) and over a DEAE-cellulose column of 40 × 3 cm. The yield is 3–4 mg of porin.

Comment

The porin contains an acid-labile bond.[8] Precipitation of porin from detergent-containing solutions with acids should be carried out in the cold. Add one-tenth volume of 3 M trichloroacetic acid and one-third volume of methanol (to prevent precipitation of the detergent). After 1 hr at 0–4° the porin can be separated by centrifugation.

Properties of the Mitochondrial Porin

General Properties

The mitochondrial porin (M_r 31,000) is the major protein of the outer membrane. It represents about 20% of total outer membrane protein and 0.4% of total mitochondrial protein. Upon isoelectric focusing, and purified porin gives two bands with pIs of 7.7 (major band) and 7.8.[8] The reason for this heterogeneity is not known, but porins from other sources also show this behavior.[10,11]

From protease digestion experiments[6] and from x-ray diffraction data,[12] it was concluded that the porin is deeply embedded into the membrane and does not extend into the aqueous phase.

It has been demonstrated that the mitochondrial hexokinase binding protein, which binds brain hexokinase[12a] or hexokinase from fast growing tumor cells,[13] is identical with the mitochondrial porin.[11,14]

[10] H. Tokunaga, M. Tokunaga, and T. Nakae, *Eur. J. Biochem.* **95**, 433 (1979).
[11] M. Linden, P. Gellerfors, and B. D. Nelson, *FEBS Lett.* **141**, 189 (1982).
[12] C. A. Mannella, *Biochim. Biophys. Acta* **645**, 33 (1981).
[12a] J. E. Wilson, J. L. Messer, and P. L. Felgner, this volume [42].

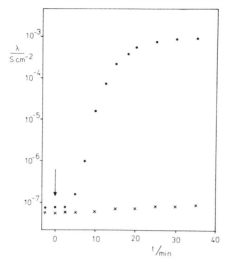

FIG. 1. Specific membrane conductance λ as a function of time after addition of 100 ng of mitochondrial porin from *Neurospora crassa* per milliliter to a black lipid bilayer membrane from asolectin–*n*-decane (arrow). The crosses represent a control experiment where only 10 μg of Genapol X-100 per milliliter were added to another membrane; 1 M KCl, pH 6; $T = 25°$.

Biosynthesis of the Porin[15]

Mitochondrial porin is synthesized on cytoplasmic free polysomes without an additional presequence. In contrast to the biogenesis of mitochondrial inner membrane or matrix proteins,[16] assembly of the porin is not dependent on energization of the mitochondria. By digitonin treatment it could be shown that the porin precursor in an *in vitro* transfer assay is incorporated into the outer membrane. Binding to the outer membrane precedes the incorporation into this membrane.[15]

Reconstitution of the Porin in Lipid Bilayer Membranes

The addition of small quantities of mitochondrial porin to the aqueous phase bathing a black lipid bilayer membrane prepared according to established procedures[17] results in a strong increase of the membrane conductance. A typical experiment is given in Fig. 1. Mitochondrial porin was added in a final concentration of 100 ng/ml to a black membrane from

[13] E. Bustamente, H. P. Morris, and P. L. Pedersen, *J. Biol. Chem.* **254,** 8699 (1981).

[14] C. Fiek, R. Benz, N. Roos, and D. Brdiczka, *Biochem. Biophys. Acta* **688,** 429 (1982).

[15] H. Freitag, M. Janes, and W. Neupert, *Eur. J. Biochem.* **126,** 197 (1982).

[16] M. Schleyer, B. Schmidt, and W. Neupert, *Eur. J. Biochem.* **125,** 109 (1982).

[17] R. Benz, K. Janko, W. Boos, and P. Läuger, *Biochim. Biophys. Acta* **511,** 305 (1978).

asolectin – n-decane. After an initial lag of about 5 min, presumably caused by the diffusion of the protein through unstirred layers, the conductance increased by about four orders of magnitude within about 30 min. Only a slight further increase occurred after that time. The conductance increase was approximately linearly dependent on the protein concentration in the aqueous phase if the same times after addition of the protein were considered (usually 20 – 30 min).

A considerable asymmetry of the action of the voltage on the single conductive unit was observed, if the protein was added to only one side of the membrane.[6,7] This indicates an asymmetric insertion of the protein into the membrane.

Figure 2 shows fluctuations of current observed with mitochondrial porin inserted into a asolectin membrane in the presence of 1 M KCl. Most of the fluctuations were directed upward, and terminating steps were only rarely observed at a membrane voltage of 5 – 10 mV membrane potential.[6] This indicated that the lifetime of the pores was long at low voltage and usually exceeded 5 min.

The single conductance increment was fairly uniform in size at a membrane voltage of 5 – 10 mV (Fig. 3).

Lifetime and single-channel conductance were strongly dependent on the membrane potential. Whereas the single channel conductance Λ was about 4.5 nS at 5 mV (1 M KCl), it decreased at 50 mV to about 0.7 nS.[6]

The conductance of the mitochondrial porin is comparable to that of the bacterial porins under otherwise identical conditions. Assuming that the pores are filled with a solution of the same specific conductivity σ as the

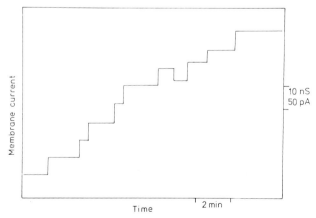

FIG. 2. Stepwise increase of the membrane current in the presence of 5 ng of mitochondrial porin per milliliter from *Neurospora crassa* added to the aqueous phase containing 1 M KCl; $T = 25°$. The membrane was formed from asolectin – n-decane; $V_m = 5$ mV.

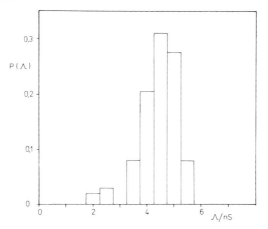

FIG. 3. Histogram of the conductance fluctuations observed with membranes from asolectin–n-decane in the presence of mitochondrial porin from *Neurospora crassa*. The aqueous phase contained 1 M KCl; pH 6; $T = 25°$. Applied voltage was 5 mV; the mean value of all observed conductance fluctuations was 4.5 nS for 253 single events.

external solution and assuming a pore length 7.5 nm (corresponding to the thickness of the outer membranes of bacteria and mitochondria), the average pore diameter $d(= 2\ r)$ and the cross section can be calculated according to the equation $\Lambda = \sigma\pi r^2/l$. The table shows the diameter and the cross section for mitochondrial porins from *Neurospora crassa* and rat liver, as well as for the bacterial porins from different gram-negative bacteria.[6,7,17–21]

[18] R. Benz and R. E. W. Hancock, *Biochim. Biophys. Acta* **646**, 298 (1981).
[19] R. E. W. Hancock, G. M. Decad, and H. Nikaido, *Biochim. Biophys. Acta* **554**, 323 (1979).
[20] H. Nikaido and T. Nakae, *Adv. Microbiol. Physiol.* **20**, 163 (1979).
[21] R. Benz, J. Ishii, and T. Nakae, *J. Membr. Biol.* **56**, 19 (1980).

COMPARISON OF THE PORES FORMED BY MITOCHONDRIAL AND
BACTERIAL PORINS IN LIPID BILAYER MEMBRANE[a]

Porin	Λ (nS)	d (nm)	Area (nm²)	Reference footnote
Mitochondrial				
Neurospora crassa	4.5	2.0	3.1	6
Rat liver	4.3	1.9	2.9	7
Bacterial				
Escherichia coli	1.9	1.3	1.3	17
Salmonella typhimurium	2.4	1.4	1.6	21
Pseudomonas aeruginosa	5.6	2.2	3.8	18

[a] The pore diameter d was calculated from the pore conductance in 1 M KCl according to $\Lambda = \sigma\pi r^2/l$ (using $\sigma = 110$ mS cm⁻¹ and length (l) = 7.5 nm).

The diameter of the pore from *Pseudomonas aeruginosa* protein F is about the same as that of the mitochondrial porin pores.[18] This is consistent with the observations on the pores in reconstituted vesicles, which in both cases were permeable to hydrophilic solutes of molecular weights up to 6000.[5,19]

[28] Synthesis and Assembly of Subunit 6 of the Mitochondrial ATPase in Yeast

By SANGKOT MARZUKI and ANTHONY W. LINNANE

The assembly of the mitochondrial H^+-translocating ATPase (ATP-synthetase, EC 3.6.1.3) has been the subject of intensive investigations in several laboratories (see recent reviews[1-3]) because of its unique property, shared only by three other enzyme complexes of the inner mitochondrial membrane, in that the enzyme complex is assembled from protein subunits that are synthesized in the mitochondria as well as those imported from the extramitochondrial cytoplasm. Like H^+-ATPase from other sources, the yeast mitochondrial ATPase can be divided functionally into two sectors: the F_1 sector, which contains the catalytic site for the synthesis and hydrolysis of ATP, and the F_0 sector, which is an integral part of the mitochondrial inner membrane and thought to act as a proton channel, linking the transmembrane proton gradient to the synthesis of ATP on the F_1 sector. Unlike the relatively hydrophilic F_1 sector, which is fairly well characterized, the subunit composition and the structure of the mitochondrial F_0 sector is still not well understood. Studies on the assembly (as well as the function) of the yeast mitochondrial F_0 sector, therefore, are currently focused on the only two polypeptides that have been shown conclusively to be subunits of the F_0 sector.[4] Both are products of the mitochondrial protein synthesis: subunit 6 [apparent molecular weight in sodium dodecyl sulfate (SDS)–polyacrylamide gel 20,000] and subunit 9 (a proteolipid, M_r 7600). In this chapter, methods that are routinely used in our laboratory to study the synthesis and assembly of subunit 6 are described. This subunit is a functional component of the mitochondrial H^+-ATPase. Although it appears that subunit 6 is not required for the formation of the enzyme proton channel, its presence is essential for the coupled ATP synthesis activity of the

[1] R. S. Criddle, R. F. Johnston, and R. J. Stack, *Curr. Top. Bioenerg.* **9**, 89 (1979).

[2] A. Tzagoloff, G. Macino, and W. Sebald, *Annu. Rev. Biochem.* **48**, 419 (1979).

[3] S. Marzuki, H. Roberts, and A. W. Linnane, *in* "Membranes and Transport" (A. N. Martonosi, ed.), Vol. 1, p. 491. Plenum, New York, 1982.

[4] J. M. Orian, M. Murphy, and S. Marzuki, *Biochim. Biophys. Acta* **652**, 234 (1981).

[5] W. M. Choo and S. Marzuki, unpublished observation (1982).

H^+-ATPase.[5] Furthermore, a mutant of yeast with a lesion in the structural gene of subunit 6 has been shown to be defective in oxidative phosphorylation or $ATP-P_i$ exchange activities, although the mutant mitochondria contain an oligomycin-sensitive mitochondrial H^+-ATPase, indicating that the F_0 and F_1 sectors of the enzyme complex are properly assembled.[6]

Comments on Growth Conditions

It is a common practice in many laboratories to use, for biochemical studies, yeast cells that have been grown in liquid batch cultures in a semisynthetic medium containing yeast extract, a salts mixture, and a carbon energy source (ethanol, glucose, or galactose), or in a rich medium containing yeast extract, peptone, and the carbon energy source (see this series, Vol. 56 [4], for example). While this growth procedure is quite satisfactory for most studies, it can create a complication when used to grow yeast cells which, as the result of a mutation or a physiological manipulation, can no longer grow by oxidative metabolism. These cells have to be grown in growth media containing a fermentable energy source such as glucose, which induces catabolite repression; frequently, it is very difficult to distinguish the effects of catabolite repression from the primary effects of the mitochondrial mutation or the physiological manipulation being studied. For example, several mutations in various regions on the mitochondrial genome, including those in the structural gene of mtATPase subunit 6, have been found to affect indirectly the activity of the cytochrome oxidase complex.[7-9] When these mutants are grown in batch cultures with glucose as an energy source, it is impossible to determine whether the reduction in the cytochrome oxidase activity is due to the mutations or to catabolite repression.

Several laboratories try to minimize the effect of catabolite repression by using galactose as an alternative fermentable energy source.[10,11] However,

[6] M. Murphy, H. Roberts, W. M. Choo, I. Macreadie, S. Marzuki, H. B. Lukins, and A. W. Linnane, *Biochim. Biophys. Acta* **592**, 431 (1980).

[7] A. W. Linnane, A. Astin, S. Marzuki, M. Murphy, and S. C. Smith, *in* "Extrachromosomal DNA" (D. J. Cummings, P. Borst, I. B. Dawid, S. M. Weissman, and C. F. Fox, eds.), p. 249. Academic Press, New York, 1979.

[8] A. W. Linnane, S. Marzuki, P. Nagley, H. Roberts, M. W. Beilharz, W. M. Choo, G. S. Cobon, M. Murphy, and J. M. Orian, *in* "The Plant Genome" (D. R. Davies and D. A. Hopwood, eds.), p. 99. John Innes Charity, Norwich, England, 1980.

[9] A. W. Linnane and S. Marzuki, *in* "New Horizons in Biological Chemistry" (M. Koike, T. Nagatsu, J. Okuda, and T. Ozawa, eds.), p. 55. J. Sci. Soc. Press, Tokyo, 1980.

[10] This series, Vol 56 [4].

[11] P. Pajot, M. L. Wambier-Kluppel, Z. Kotylak, and P. P. Slonimski, *in* "Genetics and Biogenesis of Choroplasts and Mitochondria" (T. Bucher, W. Neupert, W. Sebald, and S. Werner, eds.), p. 443. North-Holland Publ., Amsterdam, 1976.

not all strains of yeast can grow on galactose, and even in mutants that can grow on galactose, the problem is not completely solved, because galactose also induces catabolite repression although not as severe as that of glucose. Several mitochondrial mutants have been found to be more sensitive to catabolite repression.[11,12]

To avoid these problems, we routinely grow mutant cells for studies of mitochondrial functions in glucose-limited chemostat cultures. As has been discussed in this series (Vol. 56 [52]), in chemostat cultures, the steady state concentration of the growth-limiting substrate is determined solely by the dilution rate of the culture according to the equation

$$S = K_s D / (\mu_{max} - D)$$

where S is the steady-state concentration of the growth-limiting substrate (in this case glucose); μ_{max} (hr^{-1}) is the maximum specific growth rate (a constant), i.e., the growth rate of the cells in the growth medium if glucose is not limiting; D is the dilution rate (hr^{-1}) of the culture, which, in the steady state, equals the specific growth rate μ of the cells; and K_s is the concentration of the growth-limiting substrate that allows the cells to grow at half of μ_{max} (also a constant). It has been shown that at a $D = 0.1$ hr^{-1} the steady-state glucose concentration in cultures of most strains of yeast is lower than that which induces catabolite repression (this series, Vol. 56 [52]).

Labeling of Mitochondrial Translation Products

Since mitochondria can synthesize only a very limited number of proteins (eight in yeast), mtATPase subunit 6 can be easily identified among these products by a variety of separation methods, as discussed in the subsequent sections. Mitochondrial translation products are labeled with [^{35}S]sulfate in the presence of cycloheximide essentially as described by Douglas and Butow (this series, Vol. 56 [6]). However, our laboratory wild-type strain (J69-1B), from which most of our mutants have been derived, can incorporate [^{35}S]sulfate into mitochondrial translation products at only a very slow rate. To improve the incorporation of [^{35}S]sulfate, we have utilized an observation that was initially reported by Tzagoloff,[13] that mitochondrial protein synthesis can proceed to a greater extent in the presence of cycloheximide if, previous to the actual labeling step, the cells are preincubated in the presence of a mitochondrial protein synthesis

[12] G. S. Cobon, D. J. Groot Obbink, R. M. Hall, R. Maxwell, M. Murphy, J. Rytka, and A. W. Linnane, in "Genetics and Biogenesis of Chloroplasts and Mitochondria" (T. Bucher, W. Neupert, W. Sebald, and S. Werner, eds.), p. 453. North-Holland Publ., Amsterdam, 1976.
[13] A. Tzagoloff, M. S. Rubin, and M. F. Sierra, Biochim. Biophys. Acta 301, 71 (1973).

inhibitor. The preincubation presumably allows the accumulation of some cytoplasmically synthesized mitochondrial proteins that, because of the absence of mitochondrial protein synthesis, cannot be assembled into the inner membrane enzyme complexes. The size of this pool is thought to determine the extent to which mitochondrial protein synthesis can proceed when the cytoplasmic translation is subsequently inhibited.

The preincubation is carried out as follows: Cells, harvested from continuous or batch cultures, are resuspended at 2 mg of cells, dry weight, per milliliter in low-sulfate medium (this series, Vol. 56 [6]) containing chloramphenicol (2 mg/ml) and glucose (6 mg/ml). The cells are then incubated in a rotary shaker for 2 hr at 28°. After the incubation the cells are washed free of chloramphenicol by centrifugation and resuspension in the low-sulfate medium three times. The cells are then resuspended at a cell density of 12 mg of cells, dry weight, per milliliter in low-sulfate medium containing cycloheximide (0.5 mg/ml) and glucose (20 mg/ml) for labeling with [^{35}S]sulfate as previously described (this series, Vol. 56 [6]).

The 2-hr preincubation time was selected because up to 2 hr the extent of the incorporation of [^{35}S]sulfate into mitochondrial translation products is directly related to the length of the preincubation period (data not shown). Preincubation for longer than 2 hr, however, does not increase the extent of mitochondrial protein synthesis any further.

Analysis of mtATPase Subunit 6 by One-Dimensional Polyacrylamide Gel Electrophoresis of Total Mitochondrial Translation Products

Procedures for the analysis of hydrophobic proteins of the mitochondrial membrane by high-resolution polyacrylamide gel electrophoresis have been described in detail in this series (Vol. 56 [6]) and elsewhere.[14,15] By using these procedures, it is sometimes possible to separate the mtATPase subunit 6 from subunit III of the cytochrome oxidase complex, which in yeast have almost similar mobilities in SDS–polyacrylamide gels (Fig. 1a). The conditions of gel electrophoresis that we use to obtain the separation are essentially as described by Studier[16] and Laemmli[17] with the following modifications.

Separating and Stacking gels. The separating gel employed for electrophoresis (16 × 16 × 0.15 cm) contains 12.5% acrylamide, 0.34% *N,N'*-

[14] M. G. Douglas and R. A. Butow, *Proc. Natl. Acad. Sci. U.S.A.* **73**, 1083 (1976).
[15] M. Murphy, K. B. Choo, I. Macreadie, S. Marzuki, H. B. Lukins, P. Nagley, and A. W. Linnane, *Arch. Biochem. Biophys.* **203**, 260 (1980).
[16] F. W. Studies, *J. Mol. Biol.* **79**, 237 (1973).
[17] U. K. Laemmli and M. Favre, *J. Mol. Biol.* **80**, 575 (1973).

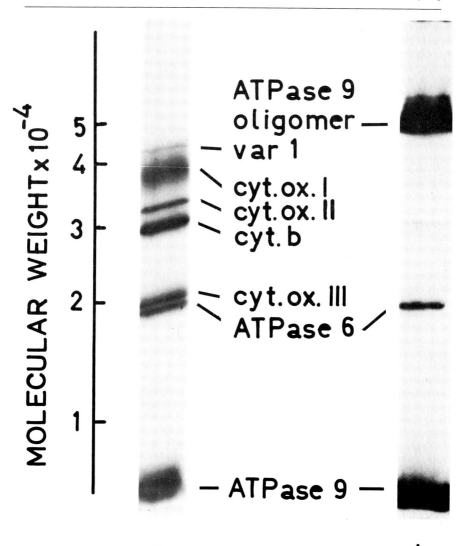

FIG. 1. Analysis of mitochondrial (mt) ATPase subunit 6 by one-dimensional polyacrylamide gel electrophoresis of (a) total mitochondrial translation products and (b) mtATPase immunoprecipitate. Mitochondrial translation products of a wild-type strain of *Saccharomyces cerevisiae* (J69-1B) were labeled with [^{35}S]sulfate, and mtATPase was isolated by immunoprecipitation with a rabbit antiserum against the F_0F_1-ATPase as described. Mitochondrial translation products were then analyzed by sodium dodecyl sulfate–polyacrylamide gel electrophoresis and visualized by direct autoradiography (a) or by fluorography (b). ATPase 6 and 9 are subunits 6 and 9 of the H^+-ATPase. Cyt. ox. I, II and III are subunits I, II and III of the cytochrome oxidase complex. Cyt. *b* is the apoprotein of cytochrome *b*. Var 1 is a variant protein associated with the small subunit of the mitochondrial ribosome.

methylenebisacrylamide, 0.1% SDS, and 0.4 M Tris-HCl (pH 8.9). Polymerization is initiated with 0.08% $N,N,N'N'$-tetramethylethylenediamine and 0.04% ammonium persulfate. A stacking gel containing 5% acrylamide, 0.6% N,N'-methylenebisacrylamide, 0.1% SDS, and 75 mM Tris-HCl (pH 6.7) is layered on top of the separating gel.

Preparation of Samples. Samples for electrophoresis are resuspended at 2–4 mg of protein per milliliter in 120 mM Tris-HCl buffer (pH 6.7), containing 2% SDS, 1% 2-mercaptoethanol, 15% sucrose, 0.025% bromphenol blue, and protease inhibitors.[18] After 2 min of incubation in a boiling water bath, 40–50 μl of each sample are then loaded onto the gel.

Running Conditions. Electrophoresis is carried out at 4° at 30 mA until the tracking dye front reaches the bottom of the gel. The running buffer contains 50 mM Tris, 0.38 M glycine, and 0.1% SDS.

Detection of Radioactively Labeled Mitochondrial Translation Products. For the analysis of mtATPase subunit 6, ^{35}S-labeled mitochondrial translation products in the polyacrylamide gels are detected by direct autoradiography. The separation between mtATPase subunit 6 and cytochrome oxidase subunit III is more difficult to observe when the mitochondrial translation products are detected by fluorography (using eith PPO-DMSO[19] or salicylate[20] as the scintillating agent). This is due to the more diffuse bands obtained in fluorography, presumably as the result of the gel pretreatment before the exposure of the X-ray film.

Two-Dimensional Electrophoresis Analysis of Mitochondrial Translation Products

A two-dimensional electrophoresis technique that resolves proteins according to their charge in the first dimension and molecular weight in the second dimension allows a more reproducible separation of the mtATPase subunit 6 and subunit III of cytochrome oxidase. Such a technique could also bring additional information with regard to molecular microheterogeneity that can arise as the result of, e.g., minor posttranslational processing events or missense mutations that escape the sensitivity limit of the more conventional one-dimensional SDS–polyacrylamide gel electrophoresis. The two-dimensional electrophoretic method as developed originally by O'Farrell,[21] however, proved to be of limited applicability to the study of the mitochondrial translation products for two main reasons: (*a*) because it does not cover the alkaline pH range essential for the separation of mitochondrial translation products; and (*b*) because of the solubility problems arising from

[18] I. J. Ryrie, *Arch. Biochem. Biophys.* **184,** 464 (1977).
[19] W. M. Bonner and R. A. Laskey, *Eur. J. Biochem.* **46,** 83 (1974).
[20] J. P. Chamberlain, *Anal. Biochem.* **98,** 132 (1979).

the high hydrophobicity of the mitochondrial translation products. The following procedure is a modification of the O'Farrell method,[21] developed in our laboratory[22] specifically for the separation of yeast mitochondrial translation products. Detailed description of the development of this procedure has been published elsewhere.[22]

Composition and Preparation of Isoelectric Focusing Gels

Isoelectric focusing gels (cylindrical, 11 cm long and 3 mm in diameter) are cast and polymerized essentially according to the method of O'Farrell.[21] The final gel composition is: 3.77% (w/v) acrylamide, 0.22% (w/v) N,N'-methylenebisacrylamide (both electrophoresis purity grade), 8 M urea (ultrapure-grade, Schwarz–Mann, Orangeburg, New York), 2% (w/v) Triton X-100, Ampholines (LKB, Bromma, Sweden), 0.6% (w/v) pH 3.5–10, 0.6% (w/v) pH 2.5–4, 0.6% (w/v) pH 9–11. The gel mixture is degassed under vacuum for 2 min, and ammonium persulfate (final concentration 0.87 mM) and $N,N,N'N'$-tetramethylenediamine (final concentration 6.7 mM) are added to start polymerization. The gels are overlaid with a small volume of 8 M urea during polymerization for 24 hr at room temperature.

Sample Preparation for Isoelectric Focusing

^{35}S-Labeled mitochondrial samples (50–100 μg of protein and 2 to 10 $\times$ 10^5 dpm) are resuspended in 25–50 μl of solubilizing buffer containing 8 M urea, 0.4 mM phenylmethylsulfonyl fluoride, 4 mM ϵ-aminocaproic acid, 4 mM p-aminobenzamide-HCl, 1% (v/v) 2-mercaptoethanol (final concentrations) and SDS in a ratio of SDS to protein of 2 (w/w). The sample mixtures are then incubated for 30–60 min at room temperature.

Isoelectric Focusing Running Conditions

The buffers used are 0.66 M ethylenediamine–8 M urea in the cathode compartment (top) and 0.1 M citric acid–8 M urea in the anode compartment (bottom). The gels are prerun for 10 min at a constant current of 1 mA/tube (LKB 2103 power supply), to remove the excess ammonium persulfate[23] and cyanate that might have been formed from the decomposition of urea.[21,24]

Samples prepared as above are adjusted to pH 3 with 100 mM citric acid immediately before application and then applied under the anode buffer

[21] P. O. O'Farrell, J. Biol. Chem. 250, 4007 (1975).
[22] G. Stephenson, S. Marzuki, and A. W. Linnane, Biochim. Biophys. Acta 609, 329 (1980).
[23] P. Righetti and J. W. Drysdale, Biochim. Biophys. Acta 263, 17 (1970).
[24] R. C. Warner, J. Biol. Chem. 142, 705 (1942).

and overlaid with 10 μl of a solution containing 6 M urea, 2% (w/v) Triton X-100, and 10 mM citric acid. After the sample application, the current is maintained at 1 mA/tube until the voltage reaches 600 V (40 min). Thereafter the voltage is kept constant at 600 V for another 240 min, during which time the current drops to a lower and constant level of about 0.18 mA/tube. At the end of the run the voltage is increased to 800 V for another 15 min in order to sharpen the electrofocused bands. To minimize the "cathode drift," longer running times and higher voltages are avoided.[25] The isoelectric focused gels are recovered from the glass tubes and frozen immediately at −80° until run on the second dimension.

Second Dimension — Polyacrylamide Gel Electrophoresis in the Presence of SDS

The composition of the slab gels (18 × 14 × 0.15 cm) and the running buffer used for the second dimension is as for the one-dimensional SDS–polyacrylamide gel electrophoresis described in the preceding section. The plates used for casting the gels are made of Perspex and are specially designed to accommodate the cylindrical isoelectricfocusing gel as well as two lateral samples that are electrophoresed at the same time. The previously frozen electrofocused cylindrical gels are shaken at room temperature for 2–10 min in 8 ml of equilibrating buffer containing 2% (w/v) SDS and 75 mM Tris-HCl, pH 6.7. Each equilibrated electrofocused gel is then sealed on top of the slab gel using a warm solution of 1% (w/v) agarose (60°) in 75 mM Tris-HCl buffer, pH 6.7, containing 0.1% (w/v) SDS.

The second-dimension gels are run at 30 mA constant current, until the dye from the lateral samples reaches the bottom of the gel (4$\frac{1}{2}$–5 hr). The electrophoresed slab gels are fixed overnight in a solution containing 11.5% (w/v) trichloroacetic acid, 3.45% (w/v) sulfosalicylic acid, and 30% (w/v) methanol and are subsequently treated for fluorography.[19,20]

All the products of mitochondrial protein synthesis are well separated in the two-dimensional gel (Fig. 2). These include: the oligomeric form of mtATPase subunit 9, the var1 protein, cytochrome oxidase subunits I, II, and III, mtATPase subunit 6, and cytochrome b apoprotein. It is interesting to note that subunit 6 of the mtATPase, which in the two-dimensional gel can be separated from subunit III of cytochrome oxidase, resolves into two spots of similar apparent molecular weight (Fig. 2). These two spots have previously been shown to be simultaneously absent in several mit^- mutants, which, as shown by immunoprecipitation of their mtATPase complex, do not synthesize the mtATPase subunit 6.[26] Furthermore, in one mutant

[25] A. Chrambach, P. Doerr, G. R. Finlayson, L. E. M. Miles, R. Sherins, and D. Rodbard, *Ann. N.Y. Acad. Sci.* **209**, 44 (1973).

[26] G. Stephenson, S. Marzuki, and A. W. Linnane, *Biochim. Biophys. Acta* **636**, 104 (1981).

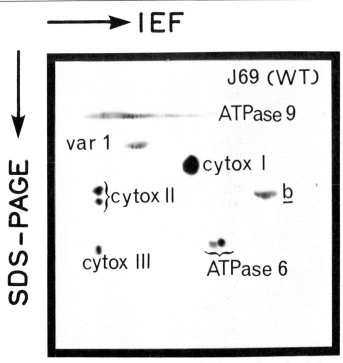

FIG. 2. Analysis of mitochondrial ATPase subunit 6 by two-dimensional electrophoresis. Mitochondrial translation products were labeled with [³⁵S]sulfate and analyzed by two-dimensional electrophoresis as described. Subunits of the H⁺-ATPase and other enzyme complexes of the mitochondrial inner membrane as previously identified[22] are indicated on the two-dimensional map: ATPase 6 and 9 are subunits 6 and 9 of the H⁺-ATPase. Cyt. ox. I, II, and III are subunits I, II and III of the cytochrome oxidase complex. *b* is the apocytochrome *b*, and var 1 is the var 1 protein associated with the small subunit of the mitochondrial ribosome. IEF, isoelectrofocusing—first dimension; SDS–PAGE, polyacrylamide gel electrophoresis in the presence of sodium dodecyl sulfate—second dimension.

strain, the disappearance of these two spots is accompanied by the appearance of two new mitochondrial translation products. Thus, although the possibility that the two subunit 6 spots are due to a technical artifact cannot be ruled out, it is possible that subunit 6 is in fact composed of two slightly different polypeptides, which are genetically related. One, for example, can arise from posttranslation modification of the other.

Immunoprecipitation Technique

Immunochemical procedures are very useful in the study of the assembly of subunit 6. For example, immunoprecipitation of the ATPase complex by using antisera raised against F_0F_1-ATPase, F_1-ATPase or by using mono-specific antisera against purified subunits of the enzyme complex can reveal valuable information on whether subunit 6 is assembled to the ATPase complex in a particular mutant or under a particular physiological manipulation. To avoid the problems associated with the use of conventional immunoprecipitation methods in the analysis of altered mitochondrial membrane proteins (see Schatz[10]), we routinely employ protein A-carrying *Staphylococcus aureus* as an antibody adsorbant.

ATPase subunits are immunoprecipitated from Triton extract of mitochondria, obtained by incubating the mitochondrial samples on ice for 15 min in 5.4 mM Tris-acetate buffer, pH 7.5, containing 62 mM sucrose, 0.5% (w/v) Triton X-100, and protease inhibitors,[18] at a concentration of 6 mg of mitochondrial protein per milliliter, followed by centrifugation at 105,000 g for 20 min. Cells of *S. aureus* Cowan strain I are grown, heat-killed, formaldehyde-fixed, and stored as described by Kessler.[27] Immediately before immunoprecipitation, formaldehyde-fixed cells (10 ml packed cell volume per 100 ml of suspension) are collected by centrifugation, resuspended to the same cell concentration in 4 mM Tris-acetate buffer, pH 7.5, containing 2 mM ATP, 1 mM EDTA, protease inhibitors,[18] and 62 mM sucrose (immunoprecipitation buffer) and incubated at room temperature for 15 min in a shaking incubator. The cells are washed once with the immunoprecipitation buffer and resuspended again in the same buffer.

To minimize nonspecific precipitation by the *S. aureus* cells, the Triton extract (50 μl diluted to 200 μl with the immunoprecipitation buffer) is mixed vigorously with 100 μl of *S. aureus* cell suspension treated as above. The cells are collected by centrifugation, then the supernatant is mixed with 100 μl of antiserum and incubated at 4° for 1–6 hr. *Staphylococcus aureus* cell suspension (100 μl) is then added to the antigen–antibody mixture, mixed, incubated for 30–60 min, and centrifuged. The cells are washed two to three times with the immunoprecipitation buffer.

For electrophoretic analysis of SDS–polyacrylamide gels (Fig. 1b) the immunoprecipitate is dissociated at 100° for 2 min in 50 μl of 120 mM Tris-HCl buffer, pH 6.7, containing a 2% SDS, 1% 2-mercaptoethanol, 15% (w/v) sucrose, 0.025% bromphenol blue, and protease inhibitors.[18] After removing the *S. aureus* cells by centrifugation, the supernatant is analyzed. For two-dimensional electrophoretic studies, the immunoprecipitate is incubated for 15 min at room temperature, in a medium containing (final concentrations): 1% (w/v) SDS, 8 M urea, and 1% (w/v) 2-mercaptoethanol. *Staphylococcus aureus* cells are removed by centrifugation, and the supernatants, adjusted to pH 3.0 with 100 mM citric acid, are applied to the positive end of the isoelectric-focusing gel as previously described.

[27] S. W. Kessler, *J. Immunol.* **115**, 1617 (1975).

Cytochrome Oxidase Subunit III-less *oxi2 mit⁻* Mutant as a Source of mtATPase Subunit 6

An increasing number of micro procedures have been developed allowing detailed structural analysis of microgram quantities of proteins that can be recovered from polyacrylamide gels after separation by electrophoresis. Some of these procedures, which include the determination of amino acid composition, peptide mapping, analysis of N-terminal amino acid, and partial amino acid sequencing, have been applied successfully to the analysis of mitochondrial translation products in yeast (see, for example, Hanson *et al.*[28] and Werner *et al.*[29]).

For most of these products, the application of the micro procedures is quite simple because yeast mitochondria synthesize only a limited number of polypeptides that can be labeled specifically in the presence of an inhibitor of cytoplasmic protein synthesis and easily separated by SDS – polyacrylamide gel electrophoresis. For mtATPase subunit 6, however, the analysis is complicated by the fact that it is very difficult to obtain this subunit free from contamination by cytochrome oxidase subunit III. As mentioned earlier, one-dimensional gel electrophoresis in the presence of SDS does not give enough separation between the two subunits. While two-dimensional gel electrophoresis allows good separation of these two mitochondrial translation products, the procedure is tedious, and the amount of labeled subunit 6 that can be recovered from each gel is too low for the gel to be of any practical use for semipreparative purposes.

We found that the best and the most convenient source of radioactively labeled mtATPase subunit 6 for microanalysis is *oxi2 mit⁻* mutants that do not synthesize cytochrome oxidase subunit III. Several of these mutants have previously been reported[30] and can be used to obtain the normal mtATPase subunit 6. For the analysis of altered subunit 6 in mutants with lesions in its structural gene, however, double mutants first have to be constructed; these carry an *oxi2 mit⁻* mutation (from a cytochrome oxidase subunit III-less strain) as well as the mutations of interest.

To avoid the difficulties in the detection of the presence of two mitochondrial mutations in a diploid strain, a mtDNA-less (*rho⁰*) strain of yeast carrying the *Kar1-1* mutation is routinely employed in our laboratory in the construction of the double mutants. Haploid cells carrying the *Kar1-1* mutation form diploids at only a low frequency[31]; the majority of cells

[28] D. K. Hanson, D. H. Miller, H. R. Mahler, N. J. Alexander, and P. S. Perlman, *J. Biol. Chem.* **254**, 2480 (1979).

[29] S. Werner, W. Machleidt, H. Bertrand, and G. Wild, *in* "The Organization and Expression of the Mitochondrial Genome" (A. M. Kroon and C. Saccone, eds.), p. 399. Elsevier/North-Holland, Amsterdam, 1980.

resulting from such a cross are heterokaryons, which subsequently segregate haploid cells containing one of the parental nuclei with the mixed cytoplasm from both parents. The following is an example of the strategy used in the construction of a double mutant that allows the analysis of the mtATPase subunit 6 from an oligomycin-resistant mutant (strain D273-10B/A48 α met [oli2-r]). For this purpose, an oxi2 mit⁻ mutant strain 2008 that does not synthesize subunit III of the cytochrome oxidase complex (α adel his[oxi2])[30] is used. The mitochondrial DNA carrying the oxi2 mutation is first transferred from strain 2008 to strain BT2-1 a leu1 Kar1-1 [rho⁰] by crossing the two haploid strains.[32] Colonies having the BT2-1 nuclearly determined auxotropic growth requirement (leu) are first selected. These were then screened for the presence of mtDNA by crossing to one or more petite deletion mutant(s) which carries the normal oxi2 allele (see Linnane and Nagley[33]). Colonies that produced respiratory competent cells in these crosses contain the BT2-1 nucleus and strain 2008 mtDNA. One of these colonies can then be used in the subsequent cross with strain D273-10B/A48, which carries the oli2 mutation. In this cross, respiratory incompetent colonies having the parental nucleus genetic marker (met) are first selected. These strains are then analyzed by marker rescue analysis using a variety of petite mutants to ensure that they contain the oxi2 mutation, and to detect the presence of the oli2 mutation (see Linnane and Nagley[33]).

---

[30] G. Stephenson, S. Marzuki, and A. W. Linnane, *Biochim. Biophys. Acta* **653**, 416 (1981).

[31] J. Conde and G. R. Fink, *Proc. Natl. Acad. Sci. U.S.A.* **73**, 3651 (1976).

[32] J. M. Orian and S. Marzuki, *J. Bacteriol.* **146**, 813 (1981).

[33] A. W. Linnane and P. Nagley, *Arch. Biochem. Biophys.* **187**, 277 (1978).

[29] Preparation and Use of Antibodies against Insoluble Membrane Proteins

By Mordechai Suissa and Graeme A. Reid

Antibodies are extremely useful for the detection and characterization of proteins. We describe here some special problems encountered in preparing and using antibodies directed against membrane proteins. Most of these problems arise from the poor solubility of membrane proteins in aqueous media. In addition, many membrane proteins are poor immunogens, and antigen–antibody reactions are then relatively weak. Fortunately, the resulting difficulties can be largely overcome by slightly modifying the procedures normally used with soluble antigens.

Preparation of Antibody

The first step in the purification of a membrane protein is almost always the isolation of the membrane with which the protein is associated. For example, purification of yeast cytochrome c oxidase would start with the isolation of mitochondria.[1,2]

Because of the insolubility of most membrane proteins in aqueous media, their isolation must be performed in the presence of suitable detergents. Purification of the membrane protein should be thorough to avoid later problems; because many membrane proteins are weak immunogens, the presence of even very small amounts of a more antigenic contaminant in the preparation used for immunization will lead to a strong, unwanted side reaction of the resulting serum. It is not at all uncommon that the major reaction of the antiserum is against a contaminant. This possibility should be checked as carefully as possible (see below) but is best avoided by purifying the immunogen to homogeneity.

In many cases one may wish to raise antibodies against a native protein, particularly if these antibodies are to be used with intact membranes or to inhibit the activity of a membrane enzyme. When the antigen is a single subunit of a complex protein then the complex must be dissociated to allow separation of the subunits; this is most readily done with SDS, but gentler methods may be found in particular instances (e.g., see Douglas *et al.*[3]).

Most of the antisera raised in our laboratory against proteins of yeast mitochondrial membranes are directed against sodium dodecyl sulfate (SDS)-denatured antigens. The final step in the purification of these proteins is usually SDS–polyacrylamide gel electrophoresis. The stained gel gives an indication of the purity of the antigen, which is simply excised from the gel and, if required, reelectrophoresed. The protein can then be electrophoretically eluted from the gel slices and prepared for immunization.[4,5] However, the use of gel electrophoresis to purify antigens deserves a warning. Even when a single protein band is observed after electrophoresis, a serum raised against this preparation may react not only with the polypeptide of interest but also with one or more others, usually of higher molecular weight. This almost invariably results from the presence in the antigen preparation of proteolytic fragments derived from larger polypeptides. Such fragments may comigrate on SDS–polyacrylamide gels with the antigen of interest. To avoid this problem, steps should be taken throughout the purification

[1] T. L. Mason, R. O. Poyton, D. C. Wharton, and G. Schatz, *J. Biol. Chem.* **248**, 1346 (1973).
[2] W. Birchmeier, C. Kohler, and G. Schatz, *Proc. Natl. Acad. Sci. U.S.A.* **73**, 4334 (1976).
[3] M. G. Douglas, Y. Koh, E. Ebner, E. Agsteribbe, and G. Schatz, *J. Biol. Chem.* **254**, 1335 (1979).
[4] N. Nelson, D. W. Deters, H. Nelson, and E. Racker, *J. Biol. Chem.* **248**, 2049 (1973).
[5] G. Daum, P. C. Böhni, and G. Schatz, *J. Biol. Chem.* **257**, 13028 (1982).

procedure to limit proteolysis by employing protease inhibitors and working as quickly as possible.

To illustrate this problem, we describe our experience with preparing different batches of rabbit sera directed against the β-subunit of yeast F_1-ATPase. The antigen was prepared by first isolating F_1-ATPase using standard procedures[6] such that the final preparation was extremely pure. The ATPase was then subjected to SDS–polyacrylamide gel electrophoresis, and the gel was stained with Coomassie Blue. The band corresponding to the β-subunit was excised and reelectrophoresed on a second SDS–polyacrylamide gel slab. The single band obtained upon reelectrophoresis was excised and electroeluted; the eluted polypeptide preparation was used to immunize a rabbit. All the antisera obtained so far react well with the F_1-ATPase β-subunit, but some react strongly also with the larger α-subunit of the F_1-ATPase. (The α-subunit is particularly sensitive to proteolysis.) This problem can be overcome by isolating mitochondria and ATPase in the presence of 1 mM phenylmethylsulfonyl fluoride (PMSF).

Still, it is always advisable to check the purity of an antigen excised from an SDS–polyacrylamide gel by another procedure, such as isoelectric focusing in the presence of urea and nonionic detergent.[7] This is particularly important if the desired polypeptide tends to be contaminated by highly antigenic contaminants (not necessarily representing proteolytic fragments), which comigrate with the antigen under study in SDS–polyacrylamide gels.

Some membrane proteins may be quickly and easily purified in a single step, by subjecting purified membranes to SDS–polyacrylamide gel electrophoresis, if the desired protein is relatively abundant and no other polypeptides migrate too closely to it on the gel. This procedure has been used to prepare the major polypeptide (M_r 29,000) of the yeast mitochondrial outer membrane; injection of this preparation into rabbits produced a strong and monospecific antiserum.[5]

One may be able to circumvent some of the problems involved in purifying membrane proteins by preparing monoclonal antibodies[8] against a crude mixture of polypeptides, e.g., a purified membrane fraction; a monospecific antibody is obtained by subsequent cloning of antibody-producing cells. This approach has been used in our laboratory, but, so far, the resulting antibodies have not proved to be as useful as expected. In our case, purified yeast mitochondrial outer membrane was used as the antigen. The mouse serum reacted strongly with several polypeptides of that membrane,

[6] M. G. Douglas, Y. Koh, M. Dockter, and G. Schatz, *J. Biol. Chem.* **252,** 8333 (1977).

[7] F. Cabral and G. Schatz, this series, Vol. 56, p. 602.

[8] R. H. Kennett, T. J. McKearn, and K. B. Bechtol, eds., "Monoclonal Antibodies." Plenum, New York, 1980.

but few of the cloned hybridomas produced antibodies that reacted sufficiently well to allow immune precipitation of individual polypeptides from solubilized mitochondrial outer membranes. Since the maintenance and screening of the hybridoma lines requires a great deal of effort, the purification of a cell line producing antibody against a specific antigen may prove to be more difficult than purifying the antigen before immunization. There are many other reasons why one may or may not wish to use hybridoma techniques,[8] and these must also be taken into consideration.

General methods for preparing antisera and monoclonal antibodies are well described elsewhere[8-10]; the only special point worth noting here is that membrane proteins are often poor immunogens. The optimal immunizing dose thus depends on the antigen used. The initial immunization should be performed with Freund's complete adjuvant or other adjuvant mixtures (see Refs. 9 and 10) to potentiate the immune response against the antigen.

Characterization of Antisera

Before using new antibodies experimentally, two important properties should be investigated; (*a*) specificity — whether the antibody react specifically with the antigen of interest; (*b*) titer — how much antigen is bound per milliliter of serum (or per milligram of purified immunoglobulin).

Specificity of Antibody. Monospecificity of an antiserum is best demonstrated by showing that the antiserum reacts with only a single component when challenged with a crude cell homogenate. This test is normally performed by double diffusion in agarose gels according to Ouchterlony[11]; in this test monospecific antisera will give a single precipitin line. This test, however, may be unreliable or even useless if the antigen is an insoluble membrane protein; under the conditions of double diffusion, such a protein will tend to aggregate not only with itself, but also with other proteins. As a result, a monospecific antiserum may yield several precipitin lines, since the antigen may migrate in the gel as several different aggregates. Conversely, an antiserum directed against several membrane proteins may yield a single precipitin line if all the antigens recognized by the serum form a tight complex. Addition of 1% Triton X-100 to the agarose gels lessens, but does not overcome, this serious problem. Double diffusion according to Ouchterlony is thus unsuitable for testing the specificity of an antiserum raised against membrane proteins.

[9] C. A. Williams and M. W. Chase, eds., "Methods in Immunology and Immunochemistry," Vol. 1. Academic Press, New York, 1967.
[10] B. A. L. Hurn and S. M. Chantler, this series, Vol. 70, p. 104.
[11] O. Ouchterlony, *Acta Pathol. Microbiol. Scand.* **32**, 231 (1953).

In our laboratory antisera are, therefore, routinely characterized by the immune replica technique,[12] which clearly reveals the presence of contaminating antibodies. It also indicates the apparent molecular weight of all polypeptides that are recognized by the antiserum. This is perhaps the most decisive advantage of this method over the Ouchterlony technique. The immune replica technique is, however, not foolproof. It will not detect a contaminating antibody against a polypeptide that comigrates in SDS–polyacrylamide gels with the polypeptide antigen to be studied. Indeed, the reaction of a serum against the desired antigen may be insignificant compared to the reaction against the contaminant. In order to show that an antiserum is directed against a particular enzyme or protein, one should be able to show an inhibition of that protein's biochemical activity by the antiserum. Also, an antibody might be missed (or its concentration underestimated) if the corresponding polypeptide antigen transfers inefficiently to the nitrocellulose test strip. Ways to overcome this are discussed by Haid and Suissa.[12]

Antibody Titer. The titer of an antibody preparation must be known in order to optimize the conditions of immunoprecipitation or immune replication. In principle, the method designed to estimate antibody titer consists of a titration of serial dilutions of antiserum against constant amounts of antigen, or vice versa. Several methods have been described in detail,[13] but two in particular should be mentioned here. Constant amounts of antigen are bound to the wells of a Microtiter dish and challenged with serial dilutions of antiserum.[14,15] The bound antibody is exposed to ^{125}I-labeled protein A, and the bound radioactivity is measured by cutting out the wells and counting them individually in a gamma counter. The principle of this method is the same as that of the immune replica technique[8] except that the antigen is bound directly to a solid surface rather than being subjected first to electrophoresis. The Microtiter assay is accordingly simpler and faster, but it often suffers from high background. Also it is sometimes difficult adequately to fix membrane preparations to the surface of the wells (Triton X-100 should be avoided since it interferes with the binding of proteins to the wells). More information can be obtained by testing serial dilutions of the serum by the immune replica assay, since the immune reaction against a specific polypeptide can be determined even when the serum reacts with more than one protein.

[12] A. Haid and M. Suissa, this series, Vol. 96 [13].
[13] C. A. Williams and M. W. Chase, eds., "Methods in Immunology and Immunochemistry," Vol. 3. Academic Press, New York, 1971.
[14] R. J. Farr, *J. Infect. Dis.* **103**, 329 (1958).
[15] J. E. Hopper, A. B. McDonald, and A. Nisonoff, *J. Exp. Med.* **131**, 42 (1970).

Special Problems in Immunoprecipitation of Membrane Proteins

Since membrane proteins are generally insoluble in aqueous media, they must be solubilized with detergents and incubated with antibody under conditions where the antigen remains soluble and does not bind to other proteins. On the other hand, the antibody–antigen interaction must still be possible. This can be tricky, and each antibody–antigen system has to be optimized individually if efficient immunoprecipitation is essential. One general approach is first to solubilize the antigen with SDS, then to add an excess of Triton buffer (1% Triton X-100, 150 mM NaCl, 1 mM EDTA, 50 mM Tris-HCl, pH 8.0), so that the Triton is in at least a fivefold weight excess over SDS.[16] Under these condition much of the SDS will be incorporated into the Triton micelles so that antibodies are able to react well with antigens.

Since the antigen(s) of interest will later be collected by precipitation, it is essential to ensure that the diluted sample is a solution, containing no particulate material that might contaminate the immunoprecipitate. This is achieved by centrifuging the sample in Triton buffer at 45,000 g for 1 hr. The antibody is then added to the sample and incubated either at room temperature for 4 to 6 hr or at 4° overnight. (The whole can be shaken). The antibody–antigen complex is then immoblized on glutaraldehyde-fixed *Staphylococcus aureus* cells,[17] by adding a 10% (w/v) suspension of the fixed *S. aureus* (5 to 15 times the volume of antiserum) and incubation is continued at room temperature for 1 hr with shaking. The cells are collected by centrifugation for 10 min at 6000 g at room temperature. The pellets are then washed about 4 or 5 times by thoroughly resuspending the cells in Triton buffer. The bound proteins are eluted from the final pellet by resuspending the cells in buffer containing SDS (and, if desired, 2-mercaptoethanol) and heating. The cells are then removed by centrifugation, the eluted material can be subjected to SDS–polyacrylamide gel electrophoresis and the radioactive polypeptides can be examined by fluorography.[18,19]

If immunoprecipitation is found to be inefficient, one should try milder conditions for dissociating the antigen or increase the amount of antiserum. If the immunoprecipitates are heavily contaminated, one can try harsher conditions of dissociation, use purified IgG fractions instead of crude serum, preincubate the sample with fixed *S. aureus* cells (0.1 ml per 10 ml of sample for 1 – 3 hr at room temperature; remove the cells by centrifugation)

[16] N.-H. Chua and F. Blomberg, *J. Biol. Chem.* **254,** 215 (1979).
[17] S. W. Kessler, *J. Immunol.* **115,** 1617 (1975).
[18] R. A. Laskey and A. D. Mills, *Eur. J. Biochem.* **56,** 335 (1975).
[19] J. P. Chamberlain, *Anal. Biochem.* **98,** 132 (1979).

before adding antibody[20] or dissolve the immunoprecipitate in SDS and subject it to a second immunoprecipitation.

Comments

In this chapter we have described the immunoprecipitation technique as one of the methods for detecting antigen. This method and others, such as the immune replica technique and binding assays, can be used when antibodies are employed in research. The antibody can serve as a tool for many types of investigation, some of which are suggested here: purification of membrane proteins; the detection of contaminating antigens in a membrane fraction, such as in the subfractionation of mitochondria; the quantitation of antigens; the location of the protein of interest in the cell by fluorescence microscopy or immunoelectron microscopy, the sidedness of membranes, the synthesis and degradation of proteins, evolutionary relations by checking immune cross-reaction with similar antigens in different species, the screening of mutants defective in the formation of a functional protein; or activity, such as the binding of proteins to receptors.

[20] A. Ohashi, J. Gibson, I. Gregor, and G. Schatz, *J. Biol. Chem.* **257**, 13042 (1982).

[30] Processing of Mitochondrial Polypeptide Precursors in Yeast

By Peter C. Böhni and Günther Daum

Most of the polypeptides imported into mitochondria from the cytosol are initially made as precursors that are several thousand daltons larger than the corresponding mature polypeptides.[1] Processing to the mature forms is effected by mitochondrial proteases.[2] Yeast mitochondria contain at least two proteases involved in processing of imported precursors. One is located in the soluble matrix,[2] and the other is probably membrane-bound.[3]

The matrix-localized protease processes precursor polypeptides destined

[1] G. Schatz, *FEBS Lett.* **103**, 201 (1979).
[2] P. C. Böhni, S. M. Gasser, C. Leaver, and G. Schatz, *in* "Organization and Expression of the Mitochondrial Genome" (A. M. Kroon and C. Saccone, eds.), p. 423. Elsevier/North-Holland, Amsterdam, 1981.
[3] G. Daum, S. M. Gasser, and G. Schatz, *J. Biol. Chem.* **257**, 13075 (1982).

METHODS IN ENZYMOLOGY, VOL. 97

to be imported into the matrix or into the mitochondrial inner membrane. The protease has been partially purified.[4]

The second mitochondrial processing protease is less well characterized. It is probably associated with the mitochondrial inner membrane, but its exact intramitochondrial location has not been determined. It catalyzes the two-step processing[3,5-7] of precursors of intermembrane space proteins and of cytochrome c_1. These precursors are first imported across both mitochondrial membranes so that they become exposed to the chelator-sensitive protease in the matrix. In a first step, this protease converts the precursors to an intermediate form that remains bound to the inner membrane, protruding from the outer face of that membrane into the intermembrane space.[3,7] The apparent molecular weights of the intermediate forms are between those of the corresponding precursors and mature forms, respectively. In a second step, these membrane-bound intermediate forms are processed to the mature polypeptides by a membrane-bound protease that is insensitive to chelators.[3] It is not yet clear whether the processing pathway demonstrated for cytochrome b_2, cytochrome c peroxidase, and cytochrome c_1 applies to other imported mitochondrial proteins.

This chapter describes the partial purification, the assay, and some properties of the chelator-sensitive protease from the yeast mitochondrial matrix and summarizes the current, rather limited information on the membrane-bound protease. Still another protease is probably involved in the processing of polypeptide precursors synthesized by the mitochondrial genetic system.[8,9]

Substrates. Mitochondrial precursor polypeptides are prepared by translating total yeast mRNA[10] in a reticulocyte lysate[11] in the presence of [^{35}S]methionine.[12] The translated lysate can be quick-frozen in liquid N_2 and stored in small aliquots at $-80°$. Just before use, each aliquot is quickly thawed in a 30° water bath.

[4] P. C. Böhni, G. Daum, and G. Schatz, *J. Biol. Chem.* (in press).
[5] S. M. Gasser, A. Ohashi, G. Daum, P. C. Böhni, J. Gibson, G. A. Reid, T. Yonetani, and G. Schatz, *Proc. Natl. Acad. Sci. U.S.A.* **79**, 267 (1982).
[6] A. Ohashi, J. Gibson, I. Gregor, and G. Schatz, *J. Biol. Chem.* **257**, 13042 (1982).
[7] G. A. Reid, T. Yonetani, and G. Schatz, *J. Biol. Chem.* **257**, 13068 (1982).
[8] K. A. Sevarino, and R. O. Poyton, *Proc. Natl. Acad. Sci. U.S.A.* **77**, 142 (1980).
[9] S. Werner, W. Machleidt, H. Bertrand, and G. Wild, *in* "Organization and Expression of the Mitochondrial Genome" (A. M. Kroon and C. Saccone, eds.), p. 399. Elsevier/North-Holland, Amsterdam, 1981.
[10] M.-L. Maccecchini, Y. Rudin, G. Blobel, and G. Schatz, *Proc. Natl. Acad. Sci. U.S.A.* **76**, 343 (1979).
[11] H. R. B. Pelham and R. J. Jackson, *Eur. J. Biochem.* **67**, 247 (1976).
[12] L. V. Crawford and R. J. Gesteland, *J. Mol. Biol.* **74**, 627 (1973).

Reagents. Phospholipid vesicles are prepared by sonicating 40 mg of acetone-washed[13] L-α-phosphatidylcholine from soybean (type II-S, Sigma), in 1 ml of 50 mM HEPES, pH 7.4, until the solution becomes clear. 1-*O*-*n*-Octyl-β-D-glucopyranoside (octyl glucoside) is available from Sigma and was kept as 10% stock solution at 4°.

Partial Purification of the Matrix-Localized Processing Protease

For exploratory experiments, a crude preparation of the enzyme can be obtained as follows: *Saccharomyces cerevisiae* cells (such as the haploid strain D-273-10B; ATCC 25657) are grown in rich medium containing 1% galactose to mid-logarithmic phase (170–190 Klett units; 8–10 g of cells wet weight per liter), harvested by centrifugation, and converted to spheroplasts with Zymolyase (see Reid [31] in this volume). Mitochondria are isolated from the spheroplasts as described by Reid except that all solutions employed for homogenizing the spheroplasts and washing the mitochondria contain 1 mM phenylmethylsulfonyl fluoride (PMSF) and 1 mM N-tosyl-L-lysylchloromethyl ketone (TLCK). The mitochondria are suspended to 5 mg of protein per milliliter in 10 mM Tris-HCl, pH 7.4. The suspension is frozen twice by successive immersion in a solid CO_2–acetone bath and a 28° water bath and centrifuged for 120 min at 125,000 g to remove membranes. The soluble extract contains virtually all of the chelator-sensitive matrix protease. It is still contaminated with other proteases, but most of these can be inhibited by protease inhibitors (see below).

The partially purified protease is prepared from 1 kg of yeast cells (grown in 100 liters of medium in a 150-liter fermentor). The cells are converted to spheroplasts, and mitochondria are prepared as described above for the small-scale preparation.

Step 1. Isolation of Soluble Matrix Contents. All steps are performed at 0–4°. The mitochondria (4.5 g of protein) are suspended in 220 ml of 0.6 M mannitol/10 mM Tris-HCl, pH 7.4–1 mM PMSF–1 mM TLCK, and the stirred suspension is diluted sixfold (i.e., to a mannitol concentration of 0.1 M) by adding ice-cold 10 mM Tris-HCl, pH 7.4–1 mM PMSF–1 mM TLCK with a peristaltic pump (at a rate of 30 ml/min). The diluted suspension is stirred very gently on a magnetic stirrer at 0° for 20 min and centrifuged in a Beckman rotor 19 for 25 min at 50,000 g. The supernatant (diluted intermembrane space) is discarded, and the pellet (mitoplasts) is resuspended in 500 ml of 10 mM Tris-HCl, pH 7.4–1 mM PMSF–1 mM TLCK with a tight-fitting Dounce homogenizer. The suspension is frozen and thawed as described above, disrupted twice at 12,000 rpm

---

[13] Y. Kagawa and E. Racker, *J. Biol. Chem.* **246**, 5477 (1971).

for 30 sec with a Polytron homogenizer (Kinematica) equipped with PTA 20N accessory, and centrifuged at 115,000 g for 120 min. The supernatant represents the matrix fraction (0.7 mg/ml). It can be frozen at $-70°$ for at least several months without significant loss of processing activity. For further purification it is convenient to use 50 ml of this fraction.

Step 2. Batchwise Adsorption to Hydoxyapatite. The soluble matrix fraction (50 ml; 35 mg of protein) is mixed with 5 g (dry weight) of hydroxy-apatite (BioGel HTP, manufactured by Bio-Rad Corporation) which had been equilibrated with 10 mM Tris-HCl, pH 7.4. The mixture is gently stirred for 20 min and then centrifuged for 3 min at 1500 g. The supernatant is discarded, and the sedimented hydroxyapatite is washed twice with 50 mM KP$_i$ + 10 mM Tris-HCl, pH 7.4. The matrix protease is eluted with two consecutive aliquots of 30 ml of 225 mM KP$_i$– 10 mM Tris-HCl, pH 7.4. The combined eluates (containing approximately 60% of the protein from the crude extract) are concentrated to approximately 8 ml with an Amicon ultrafiltration cell equipped with a PM-10 membrane.

Step 3. Gel Filtration. The concentrated hydoxyapatite eluate is passed through a 1.6 × 100 cm Sephacryl S-200 column. The column is equilibrated with 100 mM NaCl– 10 mM Tris-HCl, pH 7.4 and developed with the same solution. Fractions of 2 ml each are collected and assayed for their processing activity as described below. The activity elutes with an apparent molecular weight of 115,000. The active fractions are pooled.

Step 4. Ion-Exchange Chromatography. The pooled fractions from step 3 are diluted with 10 mM Tris-HCl, pH 7.4, to 30 mM NaCl and adsorbed to a 1 × 3 cm DEAE-cellulose column (Whatman DE-52) equilibrated with 30 mM NaCl– 10 mM Tris-HCl, pH 7.4. The column is washed with 6 ml of 50 mM NaCl– 10 mM Tris-HCl, pH 7.4, and the enzyme is then eluted with 6 ml of 150 mM NaCl– 10 mM Tris-HCl, pH 7.4. Fractions of 0.5 ml each are collected. Activity is recovered in fractions 4–7. The active fractions (combined volume: 2 ml) are pooled.

Step 5. Sucrose Gradient Centrifugation. The pooled fractions from step 4 are adjusted to 0.2% octyl glucoside and 0.1% phospholipid and divided into four aliquots (approximately 0.5 ml each). Each aliquot is layered onto a 10-ml linear sucrose gradient (7–30% sucrose, 50 mM HEPES, pH 7.4, 0.2% octyl glucoside, 0.1% phospholipid) and the gradients are centrifuged for 16 hr at 36,000 rpm in a Beckman SW 41 rotor at 2°. Ten equal fractions are collected from each gradient and assayed for processing activity. Most of the activity sediments just behind the turbid phospholipid band and is recovered in the seventh fraction from the bottom. This fraction can be stored for several days at $-70°$ without significant loss of activity. Its specific processing activity (see below) is approximately 100 times greater than that of the starting mitochondria.

Assay of the Matrix-Localized Processing Protease

The fraction to be checked for processing activity (total volume up to 400 μl) is incubated for 20 min at 21° in 10 mM Tris-HCl, pH 7.4 – 1 mM PMSF – 1 mM TLCK – 30 μM pepstatin – 50 μM CoCl$_2$ – 50 μM ZnCl$_2$. The protease inhibitors block most contaminating proteases and Co^{2+} and Zn^{2+} are added to reactivate enzyme that might have lost its putative metal cofactor during purification. The assay solution is prepared just before use. One hundred microliters of freshly thawed, prelabeled reticulocyte lysate (see above) is added, and the mixture is incubated in a 28° water bath. After 60 min, 20% sodium dodecyl sulfate (SDS) is added to a final concentration of 3%, and the mixture is heated for 4 min in a 95° water bath. Immunoprecipitation (e.g., for the F$_1$-ATPase β-subunit) is then carried out as described by Gasser (this volume [32]). The immunoprecipitate is subjected to SDS–polyacrylamide gel electrophoresis in a 10% polyacrylamide gel slab.[14] In order to speed up the assay, the gels (1.6 × 100 × 140 mm) can be run at 50 mA in the cold room (+ 4°).

Electrophoresis is then completed in only 3.3 hr without serious distortion of the protein bands. The slab is treated with sodium salicylate,[15] dried, and fluorographed. The intensities of the precursor and mature forms of the F$_1$-ATPase β-subunit (or of another suitable imported mitochondrial protein) are quantified by scanning the X-ray films or by excising the labeled bands from the dried gel and counting them for ^{35}S radioactivity.[16] One unit is defined as the amount of enzyme that produces precursor- and mature bands of equal intensity under the defined assay conditions.[2] Control experiments showed that, for a given limiting amount of enzyme, the maximal extent of processing of F$_1$-ATPase β-subunit precursor is already reached after 10 min. Other precursors are processed more slowly. Nevertheless, the assay is linear with the amount of enzyme added[2] as long as the substrate measured is the F$_1$-ATPase β-subunit precursor and as long as the percentage of processing is below 80%. Specific activity is defined as units per milligram of protein (see the table).

Properties of the Matrix-Localized Protease

The enzyme is located in the soluble matrix space (Fig. 1)[4,17] and converts a variety of imported mitochondrial precursor proteins to a poly-

---

[14] M. Douglas and R. A. Butow, *Proc. Natl. Acad. Sci. U.S.A.* **73**, 1083 (1976).

[15] J. P. Chamberlin, *Anal. Biochem.* **98**, 132 (1979).

[16] G. S. P. Groot, W. Rouslin, and G. Schatz, *J. Biol. Chem.* **247**, 1735 (1972).

[17] G. Daum, P. C. Böhni, and G. Schatz, *J. Biol. Chem.* **257**, 13028.

Fraction	Protein (mg)	Specific activity (units/mg)	Total units	Purification (-fold)	Yield (%)
Mitochondria	450	6.7	3000	1	100
Crude extract	35	71	2500	11	83
Hydroxyapatite eluate	22	83	1833	12	61
Sephacryl S-200 eluate	7.5	222	1667	33	56
DEAE-cellulose eluate	2.4	476	1142	71	38
Sucrose gradient peak	1.2	666	800	100	27

[a] See text for details.

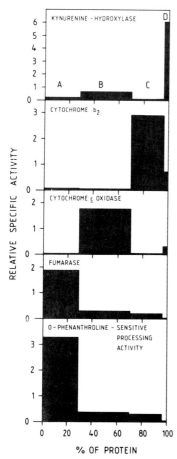

FIG. 1. The chelator-sensitive processing protease is located in the soluble matrix.[4] Yeast mitochondria were subfractionated,[17] and the distribution of typical marker enzymes was compared to that of the processing enzyme (determined with the precursor of the F_1-ATPase β-subunit as substrate). Relative specific activity denotes the specific activity of an enzyme in a given subfraction divided by the corresponding specific activity in whole mitochondria. A, matrix fraction; B, inner membrane; C, intermembrane space; D, outer membrane. From Böhni et al. (in press).[4]

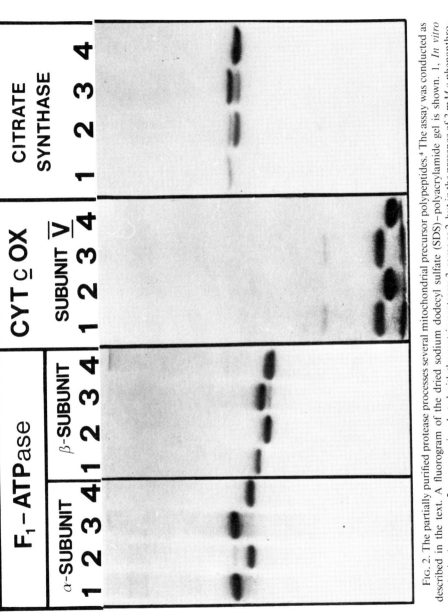

FIG. 2. The partially purified protease processes several mitochondrial precursor polypeptides.[4] The assay was conducted as described in the text. A fluorogram of the dried sodium dodecyl sulfate (SDS)–polyacrylamide gel is shown. 1, *In vitro* synthesized precursor; 2, precursor incubated with the matrix protease; 3, same as 2, but in the presence of 2 mM o-phenanthroline; 4, mature polypeptide immunoprecipitated from yeast cells grown overnight in the presence of $^{35}SO_4^{2-}$. Cyt c OX = cytochrome c oxidase. From Böhni *et al.* (in press).[4]

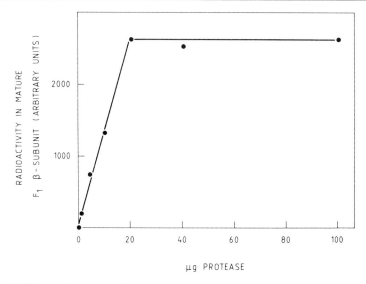

FIG. 3. The matrix protease cleaves the precursor to the F_1-ATPase β-subunit to a polypeptide similar to the mature subunit, but no further.[4] The assay was carried out as described in the text except that the prelabeled reticulocyte lysate was incubated with increasing amounts of the matrix protease (purified through step 3). The radioactive mature protein band was excised from the gel and counted for radioactivity.

peptide resembling the corresponding mature form (Fig. 2). The processed polypeptides are resistant to further proteolytic degradation by the enzyme (Fig. 3). Precursors to soluble intermembrane space proteins, such as cytochrome b_2, are processed to a polypeptide migrating between the precursor and the mature form (Fig. 4). The protease can be inhibited by chelators such as o-phenanthroline (Fig. 2), EDTA, or GTP (see also this volume [32]). No inhibition is observed with inhibitors of serine proteases, antipain, chymostatin, leupeptin, pepstatin, or chymotryptic and tryptic inhibitors isolated from dog pancreas, chicken egg white, or soybeans. None of several nonmitochondrial proteins checked was cleaved by the enzyme.

The partially purified enzyme still yields twelve major bands on SDS–polyacrylamide gel electrophoresis.[4] It is stimulated by Co^{2+} and Zn^{2+} (50 μM). After inhibition by chelators, up to 80% of the activity can be restored by either Co^{2+} or Zn^{2+}. The enzyme thus appears to be a metal-containing protease. It is inactivated by most detergents and is also sensitive to sonication. This suggests that the active enzyme may be a lipoprotein, similar to the signal peptidase from dog pancreas.[18] Chelator-sensitive processing proteases have also been detected in hypotonic extracts from mitochondria

[18] R. C. Jackson and W. R. White, *J. Biol. Chem.* **256**, 2545 (1981).

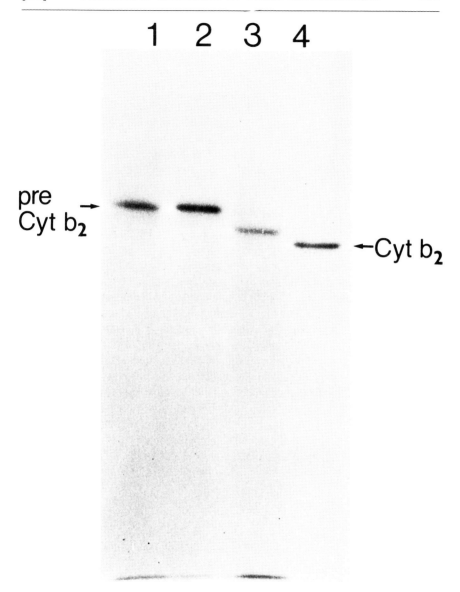

FIG. 4. The matrix protease cleaves the *in vitro* synthesized cytochrome b_2 precursor to an intermediate form.[5] The assay was carried out as described in the text. 1, *In vitro* synthesized precursor; 2, precursor incubated with matrix protease in the presence of 2 mM o-phenanthroline; 3, precursor incubated with matrix protease; 4, mature cytochrome b_2 immunoprecipitated from yeast cells grown overnight in the presence of $^{35}SO_4^{2-}$. From Gasser *et al.*[5]

of rat heart, rat liver, and maize.[2] These extracts process mitochondrial polypeptide precursors from yeast to their mature size.

Processing of Intermembrane Space Proteins

The membrane-bound mitochondrial protease that catalyzes the second maturation step of some intermembrane space proteins is difficult to assay, since no defined substrates are yet available. In principle, it should be possible first to prepare an intermediate form (e.g., that of cytochrome b_2) by processing the *in vitro* synthesized precursors with the purified matrix protease and then to incubate the resulting intermediate form (cf. Fig. 4) with mitochondria or mitochondrial subfractions. This approach has not yet succeeded, perhaps because the intermediate form generated *in vitro* does not exist in the correct conformation.

In order to assay the membrane-bound protease, yeast mitochondria are allowed to import *in vitro* synthesized yeast precursors (see this volume [32]) for 40 min at 5°.[3] [Under these conditions, the membrane-bound protease is apparently more strongly slowed down than the matrix-localized protease, resulting in the accumulation of intermediate forms (cf. Fig. 5).] Carbonyl cyanide *m*-chlorophenylhydrazone is then added to 2.5 nM to inhibit further import.[19,20] After 3 min, the mixture is warmed to 28° to allow conversion of the intermediate to the mature polypeptide by the membrane-bound protease. This conversion is unaffected by *o*-phenanthroline or any other protease inhibitor tested so far. Although it is inhibited by low concentrations of digitonin or Triton X-100, this inhibition may reflect an action on the conformation of the substrate rather than an inhibition of the enzyme protein. The protease is still present in "mitoplasts" and is, thus, not a soluble intermembrane space component. Since it releases cytochrome b_2 into the intermembrane space, it should act outside the inner membrane. Presumably, it is bound to the outer face of the inner membrane. This protease can also be studied (albeit less conveniently) by pulse-labeling experiments with intact yeast cells[7]: if yeast cells are first allowed to accumulate labeled cytochrome b_2 precursor in the presence of the uncoupler carbonyl cyanide *m*-chlorophenylhydrazone and then treated with 2-mercaptoethanol to remove the uncoupler, they convert the precursor first to the intermediate and only then to mature cytochrome b_2 (Fig. 6). Yet another possibility is offered by the fact that the second processing step of cytochrome c_1 maturation requires heme; if a heme-deficient yeast mutant[21]

[19] N. Nelson and G. Schatz, *Proc. Natl. Acad. Sci. U.S.A.* **76**, 4365 (1979).
[20] S. M. Gasser, G. Daum, and G. Schatz, *J. Biol. Chem.* **257**, 13034.
[21] E. G. Gollub, K. Liu, J. Dayan, M. Adlersberg, and D. B. Sprinson, *J. Biol. Chem.* **252**, 2846 (1977).

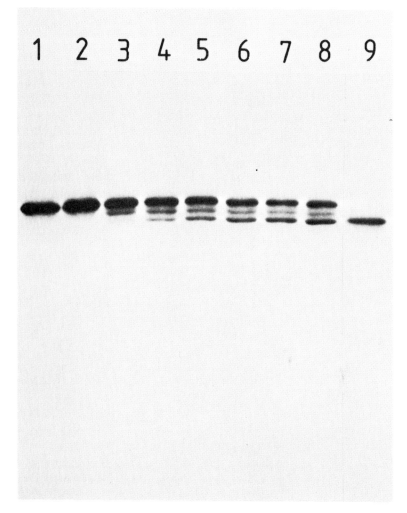

FIG. 5. Accumulation of the cytochrome b_2 intermediate during *in vitro* import at 5°. Yeast mitoplasts (0.3 mg) were incubated with 0.1 ml of a labeled and gel-filtered reticulocyte lysate at 5° for 0 (lane 2), 20 (lane 3), 40 (lane 4), 60 (lane 5), 80 (lane 6), 100 (lane 7), and 120 (lane 8) minutes. Lane 1: cytochrome b_2 precursor; lane 9: mature cytochrome b_2 standard. From Daum *et al.*[3]

is labeled with [^{35}S]methionine, the only labeled cytochrome c_1 species is the intermediate form. This form can then be chased to mature cytochrome c_1 upon addition of a suitable heme precursor, which restores heme sufficiency to the mutant.[6]

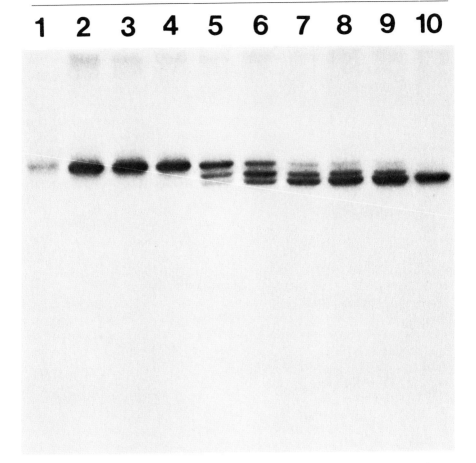

Fig. 6. Two-step processing of cytochrome b_2 *in vivo*. Yeast cells were labeled for 10 min at 28° with [^{35}S]methionine in the presence of 20 μM carbonyl cyanide *m*-chlorophenylhydrazone[2] and then chased for 0, 5, 10, 15, 20, 30, 40, and 60 min (lanes 2–9, respectively) in the presence of 2 mM unlabeled methionine and 7 mM 2-mercaptoethanol. The cells were denatured by trichloroacetic acid and extracted with SDS. Cytochrome b_2 polypeptides were immunoprecipitated from the extracts and analyzed by SDS-polyacrylamide gel electrophoresis. Lane 1: *in vitro* synthesized cytochrome b_2 precursor; lane 10: mature cytochrome b_2 standard (cf. Fig. 4). From Reid *et al.*[7]

Figure 7 summarizes how the two-step processing of cytochrome b_2 and cytochrome c_1 might work. This model is still largely hypothetical but makes the predictions that (*a*) precursor, intermediate, and mature forms should differ in the amino termini, but not their carboxy termini; (*b*) the intermediate forms should span the inner membrane. These predictions can be tested.

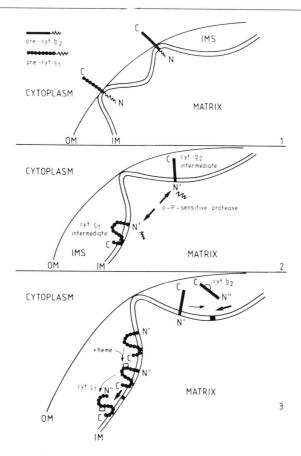

FIG. 7. Hypothetical model for two-step processing of cytochrome b_2 and cytochrome c_1. The zigzag line signifies the (presumably N-terminal) peptide extension of the precursor, and the arrows signify proteolytic cleavages. Noncovalently and covalently attached heme is represented by an open half-circle or box, respectively. With both cytochromes, the first proteolytic cleavage of the precursor in the matrix would generate a new N terminus (N'). The second proteolytic cleavage of the membrane-bound intermediate is presumed to occur close to the outer face of the inner membrane (IM) and would generate the N terminus of the mature protein (N''). This cleavage would release cytochrome b_2 in a soluble form into the intermembrane space (IMS). In contrast, cytochrome c_1 would remain attached to the inner membrane via its hydrophobic carboxy terminus. OM, Outer membrane. From Daum et al.[3]

[31] Pulse Labeling of Yeast Cells and Spheroplasts

By GRAEME A. REID

The biosynthetic pathway of yeast proteins may be followed *in vivo* by incorporating radioactive amino acids biosynthetically into newly synthesized polypeptides and following the fate of the radioactivity by suitable analytical techniques. For example, many cytoplasmically synthesized mitochondrial polypeptides have been shown to be made first as larger precursors *in vivo* by pulse-labeling yeast spheroplasts with [35S]methionine, isolating individual polypeptides by immunoprecipitation, and analyzing these polypeptides by sodium dodecyl sulfate (SDS)–polyacrylamide gel electrophoresis and fluorography.[1-3]

In principle, pulse-labeling experiments are technically simple, but several variables must be carefully controlled. These variables will now be discussed in detail.

Choice of Radioactive Precursor

It is always useful to know the amino acid composition of the polypeptide to be studied. Clearly, the specific radioactivity of the completed polypeptide will depend not only on the specific radioactivity of the amino acid pool in the cell, but also on the abundance of that amino acid in the polypeptide chain. While membrane proteins lacking leucine (a commonly used radiotracer) are extremely rare, membrane proteins may lack methionine. An example is subunit VI of yeast cytochrome *c* oxidase.[4] This protein cannot be labeled with [35S]methionine. In contrast, the subunit VI precursor can be labeled since it contains methionine in the peptide extension.[5] Unless this is known, pulse–chase data obtained for this protein may be difficult to interpret. All of the common amino acids present in proteins are commercially available in radioactive form. Our laboratory routinely uses [35S]methionine prepared biosynthetically[6] with *Escherichia coli* from

[1] M.-L. Maccecchini, Y. Rudin, G. Blobel, and G. Schatz, *Proc. Natl. Acad. Sci. U.S.A.* **76,** 343 (1979).
[2] N. Nelson and G. Schatz, *Proc. Natl. Acad. Sci. U.S.A.* **76,** 4365 (1979).
[3] G. A. Reid, T. Yonetani, and G. Schatz, *J. Biol. Chem.* **257,** 13068 (1982).
[4] I. Gregor and A. Tsugita, *J. Biol. Chem.* **257,** 13081 (1982).
[5] A. S. Lewin, I. Gregor, T. L. Mason, N. Nelson, and G. Schatz, *Proc. Natl. Acad. Sci. U.S.A.* **77,** 3998 (1980).
[6] L. V. Crawford and R. J. Gesteland, *J. Mol. Biol.* **74,** 627 (1973).

METHODS IN ENZYMOLOGY, VOL. 97

$^{35}SO_4^{2-}$ at about 100 Ci/mmol; ^{35}S is well suited to fluorographic detection of labeled polypeptides.[7]

Growth of Cells and Preparation of Spheroplasts

Reagents and Solutions

Growth medium: 0.3% (w/v) yeast extract, sulfate-free salts (containing, per liter: 1 g of KH_2PO_4, 0.98 g of NH_4Cl, 0.4 g of $CaCl_2$, 0.5 g of NaCl, 0.6 g of $MgCl_2 \cdot 6H_2O$, 0.5 g of NaCl, 5 mg of $FeCl_3$), 2% (w/v) galactose, glucose, or sodium lactate, pH 6.0

Spheroplast buffer: 1.3 M sorbitol, 40 mM KP_i, pH 7.4, 0.05% yeast extract, 1% (w/v) galactose, glucose, or sodium lactate, pH 7.4

Spheroplast labeling buffer: 1.3 M sorbitol, sulfate-free salts (as above), 1% galactose, glucose, or sodium lactate, pH 6.0

Zymolyase 5000: Extract from *Arthrobacter luteus* with yeast lytic activity, obtained from Kirin Brewery Company, Tokyo, Japan.

Procedure. If yeast cells are to be labeled with [^{35}S]methionine, they should be grown on a sulfur-deficient medium so that the cells do not accumulate large pools of methionine or its sulfur-containing precursor(s). To the growth medium described under Reagents and Solutions is added a suitable carbon source, normally galactose, glucose, or lactate. The cells are grown at 30° with vigorous shaking (to ensure adequate aeration) and harvested in mid-exponential phase. With 2% galactose as carbon source, this corresponds to a turbidity of 120–150 Klett units, using the red filter. If possible, pulse-label experiments should be performed with whole cells because the observed metabolism most closely reflects that of the growing yeast, and experimental manipulation is minimized. However, intact cells are sometimes unsuitable experimental subjects, particularly if subcellular fractions are to be analyzed after labeling; because of their thick cell wall, yeast cells can be disrupted only under conditions that severely damage cellular organelles such as mitochondria. This problem can be circumvented by first enzymatically removing the rigid cell wall, converting the yeast to spheroplasts. Most published procedures for preparing yeast spheroplasts entail long incubation of the harvested cells in an osmotically stabilized buffer under nongrowing conditions. After such treatment, the spheroplasts synthesize protein only very slowly, but they can be "regenerated" by subsequent incubation in osmotically stabilized growth medium. However, it is much simpler and faster to supplement all incubation media with the carbon and energy source used for growing the cells. In this way, the spheroplasts are formed under conditions close to those of growth. As shown below, these spheroplasts incorporate amino acid into protein at a high rate.

[7] J. P. Chamberlin, *Anal. Biochem.* **98**, 132 (1979).

The harvested cells are suspended to 0.1 g wet weight per milliliter in 0.1 M Tris-SO$_4$, pH 9.4, containing 10 mM dithiothreitol and shaken at 30° for 5 min (this step enhances the subsequent digestion of the cell wall). The cells are pelleted in a bench-top centrifuge and resuspended to 0.1 g wet weight per milliliter in spheroplast buffer containing the desired carbon source, normally the same as was present during growth. The pH of this medium is higher than during growth to optimize the activity of the cell wall-digesting enzymes. Zymolyase 5000 is added to 1 mg/ml, and the suspension is shaken at 30°. The formation of spheroplasts can be judged from their osmotic sensitivity[8] and is normally complete after 30–60 min.

Pulse-Labeling of Cells and Spheroplasts

Reagents and Solutions

Labeling buffer for cells: 40 mM KP$_i$, pH 6.0, containing a suitable carbon and energy source. It is often best to use the same carbon and energy source that had been present in the growth medium. If the cells had been grown on glucose, they might require considerable adaptation time to metabolize galactose or L-lactate efficiently.

Labeling buffer for spheroplasts: 40 mM KP$_i$, pH 6.0, sulfate-free salts (cf. preceding section), a carbon and energy source (cf. above), and 1.3 M sorbitol as osmotic support

Procedure. In order to utilize as much of the added radioactive tracer as possible, pulse-labeling should be performed under conditions where protein synthesis is rapid. One important factor here is to transfer the cells or spheroplasts from growth medium to labeling medium rather quickly; only 3–4 min are necessary for centrifugation and resuspension. The labeling medium contains no amino acids, but small amounts may be carried over from the growth medium of spheroplast buffer since the cells and spheroplasts are not washed before suspending in labeling buffer. These small amounts do not significantly compete with [35S]methionine for incorporation into protein, but may provide pools of other amino acids necessary to sustain protein synthesis.

The cells or spheroplasts are suspended in the appropriate labeling buffer at 0.1 g wet weight per milliliter, and the suspensions are shaken at 30°. Cells can be shaken very vigorously to maximize aeration, but spheroplasts are more fragile and should be shaken gently. Labeling should commence as soon as possible by adding the desired amount of radioactive amino acid.

[8] G. Schatz and L. Kováč, this series, Vol. 31a, p. 627.

Determination of Incorporation of Radioactivity into Protein

Very reproducible estimates of the protein-bound radioactivity in pulse-labeled samples can be obtained by the following method. Samples (5 μl) of the labeled suspension are diluted into 0.5–1 ml of 5% (w/v) trichloroacetic acid and filtered through a glass fiber disk (Whatman GF/A). The precipitate on the filter is washed successively with 5% trichloroacetic acid, ethanol, ethanol–diethyl ether (1:1), and, finally, diethyl ether (about 1–2 ml each). After drying the filter, it is transferred into a scintillation vial, covered with scintillation fluid, and counted for ^{35}S radioactivity.

Kinetics of Incorporation of Radioactivity into Protein

Figure 1 gives a typical time course of incorporation of [^{35}S]methionine into proteins of intact wild-type yeast cells. After a short lag, incorporation is rapid and reaches a plateau after a few minutes. Less than 100% of the ^{35}S is incorporated into protein, probably because some of the labeled methionine had been modified by oxidation. Such degradation can be minimized by storing the radioactive methionine at −80° in the presence of 1 mM dithiothreitol and by avoiding unnecessary thawing and refreezing. Figure 1 also shows that the incorporation of radioactive methionine into protein can be very rapidly stopped by addition of an excess of unlabeled methionine (2 mM).

The incorporation of [^{35}S]methionine into spheroplasts prepared by different methods is compared in Fig. 2. When our standard protocol is

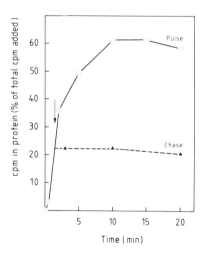

FIG. 1. Pulse-labeling kinetics of yeast cells. Wild-type *Saccharomyces cerevisiae* (D273-10B) was grown with 2% (w/v) galactose and pulse-labeled with [^{35}S]methionine (0.2 mCi/ml) as described. At various times 5-μl samples were removed to measure incorporation of radioactivity into protein (O——O). After 1.5 min (arrow) unlabeled methionine (2 mM) was added to a fraction of the suspension (▲---▲) and the incubation was continued.

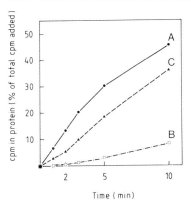

FIG. 2. Pulse-labeling kinetics of yeast sphero-plasts. Wild-type *Saccharomyces cerevisiae* (D273-10B) was grown with 2% lactate and harvested in mid-exponential phase. A portion of the cells was converted to spheroplasts and labeled immediately as described in the text (A). The remainder was treated similarly except that the spheroplast buffer was 1.3 M sorbitol, 40 mM KP$_i$, pH 7.4. The spheroplasts were centrifuged, and one portion was resuspended in lactate-containing labeling buffer and labeled immediately (B). The remainder was suspended in labeling buffer containing 0.05% (w/v) yeast extract and regenerated at 30° for 60 min before labeling in fresh labeling buffer (C). Each suspension was labeled with 0.1 mCi of [^{35}S]methionine per milliliter, and the ^{35}S radioactivity in protein was determined in 5-μl samples after various times.

used, incorporation (Fig. 2A) is almost as rapid as in intact cells. When cells are converted to spheroplasts in a simple buffer (1.3 M sorbitol, 40 mM KP$_i$, pH 7.4) incorporation is extremely slow (B) and is not fully restored even after 1 hr of regeneration in osmotically stabilized growth medium (C).

Analysis of Pulse-Labeled Polypeptides

When total protein is to be analyzed, all cellular reactions can be stopped rapidly by adding trichloroacetic acid to 15–20% (w/v). After breaking the cells or spheroplasts,[9] labeled polypeptides are extracted into buffer containing SDS. They may then be subjected to immunoprecipitation or otherwise analyzed. If spheroplasts are to be fractionated before analysis, posttranslational modification of the labeled polypeptides can be minimized or even prevented by rapidly chilling the suspension to 0°. This is most readily achieved by adding ice-cold buffer.

Comments

Pulse–chase experiments with yeast cells may be complicated by the fact that the cells may accumulate large intracellular amino acid pools that exchange only slowly, or incompletely, with externally added amino acids. As a consequence, addition of unlabeled amino acid will not immediately stop further incorporation of the radioactive amino acid. It is, therefore,

[9] A. Ohashi, J. Gibson, I. Gregor, and G. Schatz, *J. Biol. Chem.* **257**, 13042 (1982).

essential to document in each case that the chase was effective (see Fig. 1). If this is not the case, then the conditions of cell growth or the composition of the labeling medium might have to be modified until a clear-cut "chase" can be obtained with the yeast strain under study.

It should also be remembered that the intracellular amino acid pool sizes may be considerably affected by the presence of inhibitors or by the nature of the carbon source. Such effects might well influence the incorporation of the radioactive tracer and be erroneously interpreted as an altered rate of protein synthesis. A similar complication might arise if the uptake of the radioactive tracer into the cells was modified. Quantitative interpretation of pulse-labeling data obtained with whole yeast cells or yeast spheroplasts must therefore be based on measurements of the cellular amino acid pools.

[32] Import of Polypeptides into Isolated Yeast Mitochondria

By Susan M. Gasser

Isolated yeast mitochondria can take up *in vitro* synthesized mitochondrial precursor polypeptides into their internal compartments by a process that is dependent on energy, independent of protein synthesis, and usually accompanied by the proteolytic "maturation" of the precursors.[1-3] Recent work has confirmed that the import of polypeptides *in vitro* reflects the process as it occurs in living cells: up to 70% of a given *in vitro* synthesized precursor can be processed and protected from proteases added to the incubation mixture, the imported polypeptides are localized in their correct submitochondrial compartment, and the uptake requires an electrochemical potential across the mitochondrial inner membrane.[4] This efficient and reproducible assay for the uptake of precursors by isolated mitochondria has allowed a detailed study of the mechanisms by which polypeptides are segregated into their correct locations within mitochondria. For example, it was found[5] that cytochrome b_2 (a soluble component of the intermembrane space) and cytochrome c_1 (a component bound to the outer face of the inner

[1] G. Schatz, *FEBS Lett.* **103**, 201 (1979).
[2] W. Neupert and G. Schatz, *Trends Biochem. Sci.* **6**, 1 (1981).
[3] M.-L. Macchecchini, Y. Rudin, G. Blobel, and G. Schatz, *Proc. Natl. Acad. Sci. U.S.A.* **76**, 343 (1979).
[4] S. M. Gasser, G. Daum, and G. Schatz, *J. Biol. Chem.* **257**, 13034 (1982).
[5] S. M. Gasser, A. Ohashi, G. Daum, P. C. Böhni, J. Gibson, G. A. Reid, T. Yonetani, and G. Schatz, *Proc. Natl. Acad. Sci. U.S.A.* **79**, 267 (1982).

METHODS IN ENZYMOLOGY, VOL. 97

membrane) are imported and processed in two distinct steps. The *in vitro* import assay to be described here also provides a means of searching for "import receptors" on the mitochondrial surface and of testing whether such receptors are specific for different groups of precursor polypeptides.

Reagents and Solutions

Growth medium: A *Saccharomyces cerevisiae* wild-type strain, such as D273-10B (ATCC 25657) is grown on medium containing, per liter: 3 g of yeast extract (Difco), 1 g of glucose, 1 g of KH_2PO_4, 1 g of NH_4Cl, 0.5 g of $CaCl_2 \cdot H_2O$, 0.5 g of NaCl, 0.6 g of $MgSO_4 \cdot H_2O$, 0.3 ml of 1% $FeCl_3$, and 22 ml of 90% lactic acid. The final pH is adjusted to 5.5 with NaOH before autoclaving.

Solutions for the isolation of mitochondria: 1.2 M sorbitol/20 mM KP_i, pH 7.4 (spheroplasting buffer); 0.6 M mannitol/20 mM HEPES–KOH, pH 7.4–0.1% bovine serum albumin–1 mM phenylmethylsulfonyl fluoride (PMSF) (breaking buffer); 0.6 M mannitol–20 mM HEPES–KOH, pH 7.4 (mitochondrial buffer).

Reagents for immunoprecipitation: Immunoprecipitates are washed with 1% Triton X–100/0.15 M NaCl/5 mM EDTA/50 mM Tris-HCl, pH 8.0 (TNET). *Staphylococcus aureus* cells are grown and fixed as published,[6] except that formaldehyde is replaced by an equal concentration of ultrapure glutaraldehyde (Polysciences, Warrington, Pennsylvania).

Other Reagents: [^{35}S]methionine can be prepared from $^{35}SO_4^{2-}$ at approximately 100 Ci/mmol,[7] or purchased at approximately 1000 Ci/mmol. PMSF and tosyl-L-lysylchloromethyl ketone (TLCK) can be stored as 1 M and 0.1 M stock solutions, respectively, in water-free dimethyl sulfoxide. The methods used for purifying mitochondrial protein antigens and for raising rabbit antisera are described in Daum *et al.*[8]

Procedures

Growth of Yeast and Isolation of Mitochondria

The wild-type *Saccharomyces cerevisiae* strain is grown aerobically on 2% lactate medium at 29° to the early logarithmic phase (110–130 Klett units). The yield is approximately 3 g of cells (wet weight) per liter.

[6] S. W. Kessler, *J. Immunol.* **115**, 1617 (1975).
[7] L. V. Crawford and R. J. Gesteland, *J. Mol. Biol.* **74**, 627 (1973).
[8] G. Daum, P. C. Böhni, and G. Schatz, *J. Biol. Chem.* **257**, 13028 (1982).

Cells are harvested by centrifugation (5 min, 3000 g), washed once with distilled water, and resuspended to 0.5 g of cell wet weight per milliliter in 0.1 M Tris-SO$_4$, pH 9.4–10 mM dithiothreitol. After incubation at 30° for 10 min, the cells are reisolated and washed once with 1.2 M sorbitol and then resuspended in spheroplasting buffer to give 0.15 g of cell wet weight per milliliter. Zymolyase 5000 (5 mg per gram of cell wet weight) is added, and the suspension is incubated at 30° with gentle shaking. Conversion to spheroplasts is checked by lysis of the spheroplasts in water.[9] Complete conversion usually requires 40–60 min.

The spheroplasts are harvested by centrifugation for 5 min at 3000 rpm in a Sorvall GS-3 rotor at room temperature and washed twice with 1.2 M sorbitol. For homogenization, the spheroplasts are resuspended in chilled breaking buffer to a concentration of 0.3 g/ml and homogenized on ice by 10–15 strokes in a tight-fitting Dounce homogenizer. From this point on, all operations are carried out at 0–4°, and centrifugations are done in a Sorvall SS-34 rotor at 2°.

The homogenate is diluted with one volume of breaking buffer and centrifuged for 5 min at 3500 rpm. The supernatant is saved; the pellet is homogenized as before and centrifuged again (3500 rpm, 5 min). Supernatants are combined, and a crude mitochondrial fraction is sedimented (9000 rpm, 10 min). The pellet is carefully resuspended in breaking buffer, and the suspension is centrifuged (3500 rpm, 5 min) to remove residual cell debris. The supernatant is saved and centrifuged (9000 rpm, 10 min) to sediment the mitochondria. The mitochondrial pellet is washed twice by resuspension and centrifugation (9000 rpm, 10 min) in mitochondrial buffer, and resuspended in mitochondrial buffer to a final concentration of 10 mg of protein milliliter. In order to speed up the isolation procedure, the approximate protein content can be measured by diluting 10 μl of the mitochondrial suspension into 1 ml of 0.6% SDS and measuring the absorbance at 280 nm. A mitochondrial suspension containing 10 mg of protein per milliliter will give an absorbance of 0.20.

In Vitro Translation of Yeast mRNA and Filtration of Lysate

Rabbit reticulocyte lysates[10] are programmed with total yeast mRNA (12–18 OD/ml) prepared as described earlier[3] in the presence of 400 μCi of [^{35}S]methionine per milliliter. Translation usually continues for 45 min at 28° and is arrested by chilling the mixture to 2°. The mixture is then centrifuged (40 min, 140,000 g) to sediment polysomes, and aliquots of the supernatants are frozen in liquid N$_2$ and stored at −70°.

---

[9] G. S. P. Groot, L. Kováč, and G. Schatz, Proc. Natl. Acad. Sci. U.S.A. **68**, 308 (1971).
[10] H. R. B. Pelham and R. J. Jackson, Eur. J. Biochem. **67**, 247 (1976).

Before use, the translated lysate is brought to 4° and filtered through a Sephadex G-25 (medium) column, prepared in a 5-ml disposable syringe, and equilibrated with 0.15 M KCl – 20 mM HEPES – KOH, pH 7.4.

To avoid dilution of the lysate, excess equilibration buffer is removed by centrifugation (1500 rpm, 1 min in desk-top centrifuge). The lysate is then applied to the column, and the centrifugation is repeated. This step removes excess [^{35}S]methionine, metal ions, and other small molecules from the translated lysate.

In Vitro Import of Polypeptides into Yeast Mitochondria

To an aliquot (0.1 – 0.13 ml, containing $\sim 3 \times 10^6$ protein-bound counts per minute) of the translated and filtered reticulocyte lysate, ATP, and an ATP-regenerating system (1 mM ATP, 1 mM MgCl$_2$, 5 mM phosphoenolpyruvate, 4 units of pyruvate kinase, 1 mM dithiothreitol) are added. The mixture is adjusted to 0.6 M mannitol and 20 mM HEPES – KOH, pH 7.4, by adding an equal volume of 1.2 M mannitol – 20 mM HEPES – KOH, pH 7.4. Approximately 0.2 mg of isolated mitochondria (20 μl of the final suspension) is added, and the mixture is incubated for 30 min at 27° with gentle shaking.

Two methods for assaying the translocation of radioactive polypeptides across mitochondrial membranes can be used. The first method (processing assay) is applicable for the study of those polypeptides of the inner mitochondrial membrane and matrix space, which are synthesized as larger precursors and are cleaved by a chelator-sensitive protease, localized in the matrix compartment[11] (see also this volume [30]). Because of this location, precursors are processed by intact mitochondria *after* they have been imported. The second method (protection assay) can be used to assay the import of any precursor into any mitochondrial compartment and exploits the fact that the imported polypeptide becomes inaccessible and, hence, resistant to proteases added to the suspending medium.[3]

If the "processing assay" is to be used, it is essential to prevent the processing of precursors by matrix protease that has leaked out of mitochondria during the incubation. This released enzyme can be selectively inhibited by the chelator GTP (Fig. 1). Since GTP cannot penetrate the intact inner mitochondrial membrane, it will not inhibit the protease, which is still properly located in the matrix space (also Fig. 1). Therefore, a high concentration (5 – 8 mM) of GTP is added to the incubation mixture for the "processing assay" of *in vitro* import. Under these conditions a simple

[11] P. C. Böhni, S. Gasser, C. Leaver, and G. Schatz, *in* "The Organization and Expression of the Mitochondrial Genome" (A. M. Kroon and C. Saccone, eds.), p. 423. Elsevier/North-Holland, Amsterdam, 1980.

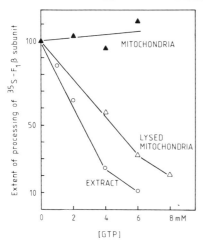

FIG. 1. GTP inhibits processing of the F_1 β-subunit precursor by a matrix fraction and by lysed mitochondria, but not by intact mitochondria. Two hundred micrograms of either intact or hypotonically disrupted mitochondria were incubated for 30 min at 27° with [^{35}S]methionine-labeled precursors as described in the text, with increasing amounts of GTP adjusted to pH 7.0. A hypotonic extract prepared from 200 μg of yeast mitochondria (see this volume [30]) was incubated with radiolabeled precursors under identical conditions, except that 2.5 μg of efrapeptin was added per milliliter to prevent excessive hydrolysis of the added GTP by released F_1-ATPase. After incubation, the intact mitochondria were reisolated and dissociated in 3% SDS. The incubations containing total lysed mitochondria or the membrane-free extract were dissociated directly in SDS. All samples were immunoprecipitated for F_1 β-subunit and analyzed by SDS–polyacrylamide gel electrophoresis, fluorography, and quantitation of the fluorograms. One hundred percent is defined in each case as the radioactivity recovered as the mature form of F_1 β-subunit in the samples containing no GTP.

electrophoretic assay for the conversion from precursor to mature form will measure translocation across the mitochondrial inner membrane. The entire incubation mixture can be dissociated immediately after the incubation at 27°, in 3% SDS–1 mM PMSF for 3 min at 95°.

The "protection assay" is applicable to the import of mitochondrial polypeptides that are not cleaved by the matrix-located protease. The incubation mixture is the same as described prior to the processing assay. After incubation, the mixture is chilled on ice and divided in half: one half receives trypsin (final concentration = 120 μg/ml) and the other half 1 mM TLCK–1 mM PMSF. After 30 min at 0° the trypsin activity is arrested by the addition of 1.2 mg of soybean trypsin inhibitor per milliliter and 1 mM TLCK. The mitochondria are reisolated by centrifugation (10,000 g, 10 min). Pellets and supernatants are dissociated separately in 3% SDS–1 mM PMSF for 3 min at 95°.

Quantitative Immunoprecipitation and Gel Electrophoresis

The dissociated samples from the processing assay are immunoprecipitated for a specific protein in order to check the extent of import and processing. Both the mitochondrial pellets and the supernatants from the protection assay should be subjected to immunoprecipitation, in order to check not only the extent of import, but the efficiency of the trypsin treatment as well. After dissociation in SDS, each supernatant is mixed with 200 μg of SDS-dissociated, unlabeled yeast mitochondrial protein. In this way all samples will contain the same amount of any mitochondrial antigen that would compete with the labeled polypeptides for binding the immunoglobulins. For all immunoprecipitations the samples are diluted 14-fold with TNET, mixed with 50–70 μl of antiserum, shaken for 16 hr at 4°, and then mixed and shaken for 1 hr at room temperature with 250–350 μl of a 10% (w/v) suspension of glutaraldehyde-fixed *Staphylococcus aureus* cells.[6] The *S. aureus*-bound immunoprecipitate is washed five times by centrifugation and resuspension in 1 ml of TNET. Elution of the immunoprecipitate[3] and electrophoresis in an SDS–polyacrylamide slab gel[12] follow published procedures. After electrophoresis, the slab gel is incubated for 30 min in 1 M sodium salicylate, dried at 80° under vacuum on a suitable gel dryer, and exposed to Kodak XS-5 X-ray film at −70°.

The efficiency of import of an immunoprecipitated polypeptide can be determined by scanning the fluorogram with a densitometer or by cutting the bands from the gels and measuring ^{35}S decay directly.[4] To determine the total amount of radioactive precursor present in the reticulocyte lysate, an aliquot of the labeled lysate is dissociated directly in SDS, mixed with an amount of SDS-dissociated mitochondria identical to that which had been used for the import assay, and the mixture is subjected to immunoprecipitation and quantitation exactly as the experimental samples. A typical result from the *in vitro* import "protection assay" is shown in Fig. 2. The samples were immunoprecipitated for cytochrome b_2.

Comments

The translocation of precursors into or across the mitochondrial inner membrane requires an electrochemical potential across that membrane.[4] Such a potential can be generated by including ATP in the import incubation mixture, since hydrolysis of ATP by the mitochondrial ATPase complex will create a proton motive force across the inner membrane. Alternatively, substrate of the respiratory chain (20 mM succinate or 20 mM α-ketoglutarate, adjusted to pH 7.0) can be added. To assay the energy dependence of import, the mitochondria can be pretreated with 2 mM KCN/20 μg of carboxyatractyloside or with 1 μg of valinomycin for 10 min prior to incubation with ATP and the translated, gel-filtered reticulocyte

---

[12] M. Douglas and R. A. Butow, *Proc. Natl. Acad. Sci. U.S.A.* **73**, 1084 (1976).

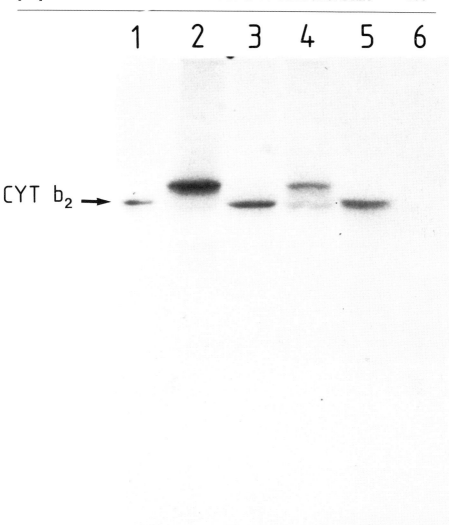

CYT b_2 →

FIG. 2. Cytochrome b_2 is cleaved and becomes protease-resistant during incubation with yeast mitochondria. Isolated yeast mitochondria were incubated with [^{35}S]methionine-labeled *in vitro* translation products; maturation and uptake of the cytochrome b_2 precursor were checked by protease treatment, immunoprecipitation, and fluorography (protection assay). A photograph of the fluorogram is shown: 1, mature cytochrome b_2 immunoprecipitated from ^{35}S-labeled spheroplasts; 2, precursor to cytochrome b_2 immunoprecipitated from reticulocyte lysate; 3, mature form of cytochrome b_2 recovered in mitochondria not treated with protease; 4, precursor and mature form of cytochrome b_2 recovered in the untreated supernatant; 5, cytochrome b_2 recovered in mitochondria after protease treatment; 6, supernatant after protease treatment.

lysate. Under these conditions untreated mitochondria will be able to take up precursors and convert them to their mature forms, whereas the pretreated mitochondria will not.

The efficiency of precursor uptake by isolated mitochondria varies. This may in part reflect the stability of different precursors in the reticulocyte lysate. We have noted that certain precursors are imported and cleaved more efficiently if the translated lysate has not been stored at $-70°$, suggesting that freezing and thawing may alter the conformation of some precursors. The reticulocyte lysate can be gel-filtered and used for the import assay directly after translation, if necessary. Gel filtration of the labeled lysate removes excess radioactivity that is not incorporated into protein, thus eliminating the chance of an artifactual radioactive labeling of mitochondrial protein.[13]

Because of the variation in the efficiency of import, the optimal length of incubation may also vary. Assays for the release of cytochrome b_2, an intermembrane space enzyme,[8] have shown, however, that components in the reticulocyte lysate begin to disrupt the yeast mitochondrial membranes after 30–45 min at 27° (unpublished). The mitochondria become much more sensitive to the subsequent trypsin treatment and release imported and processed polypeptides to the suspending medium. Whereas 45 min is the maximum effective incubation period, a 2–5 min incubation is sufficient for detectable import of certain mitochondrial proteins such as cytochrome b_2[14] and the major 29,000 dalton protein of the outer membrane.[15]

The "protection assay" has been adequate for assessing the *in vitro* import of virtually all polypeptide precursors tested so far. However, it could fail if applied to a precursor that is imported into the outer membrane in such a way that the membrane-bound mature polypeptide remains sensitive to externally added proteases. Such a case has recently been encountered in our laboratory.[15]

[13] M. Suissa, *Anal. Biochem.* **115**, 67 (1981).
[14] G. Daum, S. M. Gasser, and G. Schatz, *J. Biol. Chem.* **257**, 13075 (1982).
[15] S. M. Gasser and G. Schatz, *J. Biol. Chem.* **258**, 3427 (1983).

[33] A Yeast Mitochondrial Chelator-Sensitive Protease That Processes Cytoplasmically Synthesized Protein Precursors: Isolation from Yeast and Assay

By PHYLLIS C. MCADA and MICHAEL G. DOUGLAS

The transport into mitochondria of proteins synthesized on cytoplasmic ribosomes involves posttranslational segregation and processing events that are different from the "vectorial translation" processes that occur for the export of proteins from the cell.[1] Studies from several laboratories have shown that approximately a dozen mitochondrial proteins of nucleocytoplasmic origin are synthesized as discrete higher molecular weight precursors containing from 17 to 50 additional amino acid residues on their amino terminal end.[2] The processing of these transient precursor sequences to form the mature protein has been demonstrated for precursors of F_1-ATPase subunits by *in vivo* as well as *in vitro* methods.[3,4]

More recently, the enzyme that catalyzes this processing of the cytoplasmically synthesized F_1-ATPase precursors has been purified partially from yeast in different studies.[2,5] Unlike the membrane-bound proteases involved in the vectorial translation events in endoplasmic reticulum,[6] the protease responsible for processing precursor proteins imported into mitochondria is soluble[2,5] and is localized within the matrix of mitochondria. The processing of F_1-ATPase subunit precursors by this enzyme is insensitive to the protease inhibitors phenylmethylsulfonyl fluoride (PMSF), tosyllysine chloromethyl ketone (TLCK), tosylphenylalanine chloromethyl ketone (TPCK), and *p*-aminobenzamidine but is inhibited completely by the heavy-metal chelators *o*-phenanthroline or EDTA.[2,5] The chelator-inhibited activity can be restored by readdition of Zn^{2+}, Co^{2+}, or Mn^{2+}, hence this processing activity has been termed the mitochondrial chelator-sensitive protease. The enzyme is most active at pH 7.5. A chelator-sensitive mitochondrial pro-

[1] G. Schatz, *FEBS Lett.* **103**, 201 (1979).

[2] P. Böhni, S. Gasser, C. Leaver, and G. Schatz, *in* "The Organization and Expression of the Mitochondrial Genome" (A. Kroon and C. Saccone, eds.), p. 423. Elsevier/North-Holland, Amsterdam, 1980.

[3] M. Maccechini, Y. Rudin, G. Blobel, and G. Schatz, *Proc. Natl. Acad. Sci. U.S.A.* **76**, 343 (1979).

[4] N. Nelson and G. Schatz, *Proc. Natl. Acad. Sci. U.S.A.* **76**, 4365 (1979).

[5] P. McAda and M. Douglas, *J. Biol. Chem.* **257**, 3177 (1982).

[6] R. Jackson and G. Blobel, *Proc. Natl. Acad. Sci. U.S.A.* **74**, 5598 (1977).

tease has been isolated from rat liver mitochondria[7]; it exhibits almost identical physical properties to the enzyme obtained from yeast mitochondria.[5]

To examine the posttranslational processing of a variety of cytoplasmically synthesized mitochondrial precursors in a soluble system, it is convenient to have available a supply of the processing enzyme. Described here is a convenient procedure for the preparation and assay of large amounts of the chelator-sensitive mitochondrial protease from commercially available cake yeast. The enzyme prepared in this manner is stable and can be stored for up to 12 months in liquid nitrogen with no detectable loss in activity.

Purification of Chelator-Sensitive Mitochondrial Protease

Preparation of Mitochondria from Pressed Yeast

The best yield of intact mitochondria from commercially processed *Saccharomyces cerevisiae* is obtained when the yeast are fresh. It is best to make arrangements with your local supplier to take delivery as soon as fresh yeast become available. All operations described here are performed at $0-4°$. Suspend 1 pound of pressed yeast (Red Star Corporation) into 1 liter of ice-cold $0.6\,M$ mannitol–$20\,mM$ Tris-SO_4, pH 7.4–$1.0\,mM$ PMSF (added fresh). Fill the mixing chamber of the bead beater (Biospec Products) containing 300 g of cooled 0.5 mm glass beads with the yeast suspension. The apparatus is assembled to avoid air in the mixing chamber, then placed in an ice-cold water jacket on top of the homogenizer motor. The suspension is homogenized for a total of 3 minutes with intermittent cooling. Approximately $80-90\%$ of the cells are broken under these conditions. After removal of the cell homogenate, the glass beads are rinsed once with 50 ml of buffer, and the procedure is repeated with fresh suspension. After the entire cell suspension has been homogenized, the homogenate is centrifuged twice at $3500\,g$ for 10 min to remove cell debris. Mitochondria are pelleted from the supernatant at $17,000\,g$ for 20 min. The mitochondrial pellet is resuspended in breaking buffer by hand homogenization in a loose-fitting Dounce homogenizer followed by centrifugation at $17,000\,g$ for 20 min.

Mitochondrial Extraction of the Chelator-Sensitive Mitochondrial Protease

The washed mitochondrial pellet is suspended to 10 mg/ml in 10 mM Tris-HCl, pH 7.4, and 1 mM PMSF (added fresh) by hand homogenization.

[7] S. Miura, M. Mori, Y. Amaya, and M. Tatibana, *Eur. J. Biochem.* **122**, 641 (1982).

The suspension is frozen at $-80°$, thawed, and allowed to sit at 4° for 30 min. Mitochondrial membranes then are pelleted at 75,000 g for 30 min. The supernatant containing 0.7–1.0 mg of extracted mitochondrial protein per milliliter is concentrated approximately 10- to 20-fold at 4° using a Millipore immersible CX-10 filter. The clear concentrate is made 50 mM NaCl and filtered on a Sephadex G-150 column (2 cm × 115 cm), equilibrated with 10 mM Tris-HCl, pH 7.4, 50 mM NaCl at a flow rate of 10 ml/hr. Fractions of 1.5 ml are collected, and every fifth fraction is assayed for processing activity as described below. Alternatively, the gel filtration column may be calibrated with the appropriate molecular weight standards,[5] and fractions corresponding to the molecular weight range 100,000 to 140,000 may be pooled directly for ion-exchange chromatography.

The pooled fractions from the Sephadex G-150 column are pooled and directly applied to a DE-52 ion-exchange column (1.5 cm × 8 cm) equilibrated with 10 mM Tris-HCl, pH 7.4, 50 mM NaCl. After loading, the column is washed with 25 ml of loading buffer followed by elution of the processing activity with a 60 ml of linear salt gradient (0.05 M to 0.2 M NaCl in 10 mM Tris-HCl, pH 7.4). Fractions of 1.0 ml are collected at 4°, and 10–100 μl of every other fraction is assayed for processing activity as described below. The activity elutes from the column at approximately 100 mM NaCl and is stored at 4° during the assay. Fractions containing activity are pooled as soon as possible using either an immersion filter (Millipore CX-10) or an Amicon concentrator with a PM-10 membrane at 10–12 psi. The concentrated activity containing approximately 0.2% of the original mitochondrial protein is quick frozen in small portions (100 μl) in liquid nitrogen and stored in a liquid nitrogen freezer until further use.

The chelator-sensitive processing enzyme purified by this method contains approximately 10 proteins in a molecular weight range of approximately 77,000–24,000, which can be detected on an SDS-polyacrylamide using a sensitive silver-staining technique. A major component of apparent molecular weight 59,000 (Fig. 1, arrow) consistently is observed in different preparations isolated as described above.

Assay of the Chelator-Sensitive Mitochondrial Protease

The substrates for the chelator-sensitive processing enzyme must be synthesized by translation in a messenger RNA-dependent *in vitro* translation system. The nuclease-treated rabbit reticulocyte lysate[8] for the translation of total yeast RNA[3,5] provides the best available cell-free system for a source of labeled substrate. Wheat germ extract contains an endogenous

[8] H. R. B. Pelham and R. J. Jackson, *Eur. J. Biochem.* **67**, 247 (1976).

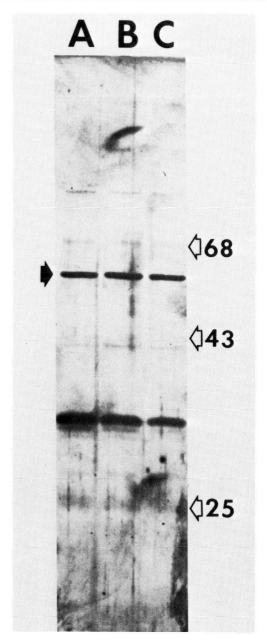

FIG. 1. Sodium dodecyl sulfate (SDS) gel pattern of the yeast mitochondrial chelator-sensitive processing enzyme. Preparations of the chelator-sensitive mitochondrial protease prepared as described were resolved on a 10% SDS gel and stained with a silver stain.[12] The numbers on the right refer to the mobilities of molecular weight standards.

endoprotease[9] which processes mitochondrial precursors and should not be used for these studies. The translation of total yeast RNA in reticulocyte lysates produces a large number of translation products from which the specific products are detected using specific antisera.[2,5] The availability of plasmid DNA containing the structural genes for cytoplasmically synthesized mitochondrial precursors makes possible the hybridization-selection of mRNAs,[10,11] which can be translated into specific substrates for the posttranslational assay without the use of immunoabsorption methods. The detection of precursor and mature forms of the protein in all cases is performed after resolution of the products on an SDS–polyacrylamide gel[12] by fluorography of dried gel slabs containing sodium salicylate.[13]

Yeast RNA Isolation

Yeast cells grown on a nonfermentable carbon source are harvested in mid-log phase ($2-4 \times 10^7$ cells per milliliter). The culture flasks are transferred to an ice-salt bath and rapidly cooled for 15 min prior to centrifugation at $4°$ (3500 g for 10 min). The pellet is washed once with ice-cold water followed by recentrifugation. The pellet then is resuspended to 10 ml per gram wet weight in room temperature 150 mM NaCl, 50 mM Tris-HCl, pH 7.4, 5 mM EDTA, 5% SDS, and rapidly frozen by pipetting the suspension directly into liquid nitrogen. The frozen yeast pellets are either crushed by hand in a liquid nitrogen cooled mortar and pestle or blended for 1 min in a cooled sterile blender. The fine white powder is thawed at $30°$ in the presence of two volumes of phenol–chloroform–isoamyl alcohol (50:50:1) with rapid shaking for 30–60 min.

All centrifugations are performed at room temperature in autoclaved siliconized 150-ml Corex bottles at 5000 g for 5 min. After centrifugation to resolve the two phases, the upper aqueous layer is removed and saved. The lower phenol layer is washed once with shaking for 5 min with a half-volume of breaking buffer followed by centrifugation. The aqueous wash is combined with the aqueous layer from the first spin. The combined mixture is then washed three times with an equal volume of phenol–chloroform–isoamyl alcohol (50:50:1). Ribonucleic acid is precipitated from the washed aqueous phase by addition of 0.11 volume of 2 M sodium acetate, pH 5.2, and 3 volumes of cold 100% ethanol followed by incubation at $-20°$ overnight.

[9] R. Todd, P. McAda, and M. Douglas, *J. Biol. Chem.* **254**, 11134 (1979).
[10] K. O'Malley, P. Pratt, J. Robertson, M. Lilly, and M. Douglas, *J. Biol. Chem.* **257**, 2097 (1982).
[11] K. O'Malley and M. Douglas, this volume [34].
[12] M. Douglas, D. Finkelstein, and R. Butow, this series, Vol. 58G, p. 58.
[13] J. P. Chamberlin, *Anal. Biochem.* **98**, 132 (1979).

The precipitated RNA is collected by centrifugation at 9000 g for 10 min at 4°, drained well, then dried for 1 hr *in vacuo.* The dried pellet is dissolved in 5 – 10 ml of water, and the RNA is reprecipitated by addition of an equal volume of 6 M LiCl at − 20° overnight. The LiCl-precipitable RNA is pelleted at 9000 g for 10 min, dried *in vacuo,* and resuspended in 0.2 M sodium acetate, pH 5.2, followed by ethanol precipitation at − 20°. After two additional ethanol precipitations the RNA is dissolved in 5 mM HEPES-HCl, pH 7.4, and stored at − 80° in small aliquots. An extension of 33.2 A_{260} = 1 mg RNA is used to calculate the concentration of RNA.

In Vitro Translations

Rabbit reticulocyte lysate is either obtained commercially or prepared by the method of Hunt and Jackson[14] from commercially available whole rabbit blood (Pel-Freeze Biologicals, Rogers, Arizona) collected from ace-tylphenylhydrazine-treated rabbits. The lysate is stored in liquid nitrogen in 1-ml aliquots until used. In a standard nuclease digestion, 1 ml of lysate is thawed by hand warming, made to 20 μM hemin, and placed on ice. The lysate is made to 1 mM CaCl$_2$ and incubated at 20° for 10 min in the presence of 4 μg of micrococcal nuclease per milliliter. The reaction is terminated by the addition of 2 mM EGTA and returned to ice. The extent of digestion is determined by the decrease in [^{35}S]methionine incorporation between the control and nucleased lysate in an *in vitro* translation reaction. As a rule, digestion conditions are adjusted to give incorporation values that are approximately 5% of the nonnucleased control. Further, titration of nuclease-treated lysate with each new yeast RNA preparation is required to optimize translation. In general, a 300 – 400 μg/ml final RNA concentration will give optimal incorporation.

In vitro translations are performed in 1-ml batches. Each milliliter of reaction contains 0.5 ml of nuclease-treated lysate, 6 – 12 A_{260} units of total yeast RNA, 1 mM HEPES, pH 7.4, 100 mM KCl, 1 mM MgCl$_2$, 30 μg of creatine kinase, 10 mM creatine phosphate, and 19 unlabeled amino acids at concentrations of 0.05 – 0.25 mM in relative proportion to their content in yeast F_1-ATPase.[15] Unlabeled methionine is replaced with 400 μCi of [^{35}S]methionine (400 – 1065 Ci/mmol) per milliliter of translation lysate. The reaction is carried out for 1 hr at 30°. During the optimization studies, the reactions are carried out in 50 μl total volume. Prior to use in processing reactions, the *in vitro* translation preparation is spun at 166,000 g for 1 hr to

---

[14] T. Hunt and R. C. Jackson, *in* "Modern Trends in Human Leukaemia" (R. Neth *et al.,* eds.), p. 300. Lehmanns Verlag, Munich, 1974.

[15] A. Tzagoloff, *in* "The Enzymes of Biological Membranes" (A. Martonosi, ed.), Vol. 4, p. 103. Plenum, New York, 1976.

obtain a postribosomal translated lysate that can be stored in liquid nitrogen until further use.

Posttranslational Processing Assay

Reagents. In vitro translation of total yeast RNA is performed in a nuclease-treated reticulocyte lysate

Procedure. Fifty microliters of the [^{35}S]methionine-labeled postribosomal lysate (1×10^6 cpm) is incubated in the presence of the mitochondrial extract containing the processing enzyme ($0-50$ μg) or selected column fractions ($10-100$ μl) for 1 hr at 25°. If necessary, volumes can be adjusted with 10 mM Tris-HCl, pH 7.4. After completion of the incubation, the reactions are returned to ice for immunoabsorption. The reaction also can be terminated by the addition of 2 mM o-phenanthroline to the reaction mixture upon completion of incubation.[2,5]

Immunoabsorption of the Reaction Products

Reagents.

  Specific antibody to a nuclear coded mitochondrial protein

  Protein A-Sepharose: 2 mg/25 μl in 10 mM Tris–HCl, pH 7.4, 0.3 M
    NaCl for each absorption. Swollen for 1 hr at 25°.

  Washing solutions

    1. 10 mM Tris-HCl, pH 7.4, 0.3 M NaCl, 1% Triton X-100
    2. 10 mM Tris-HCl, pH 7.4.

  Releasing buffer: 2% SDS, 2 mM EDTA, 2% 2-mercaptoethanol,
    100 mM Tris-HCl, pH 6.8

Procedure. Each processing reaction is made to 1% Triton X-100 and 0.3 M NaCl. Antiserum is added at a ratio of $3-5$ μl per 50 μl of translation mixture and allowed to incubate at 4° for 30 min. The reaction containing the immune complex is then diluted to 1 ml with 10 mM Tris-HCl, pH 7.4, 0.3 M NaCl, and 1% Triton X-100 and 20 μl of the protein A–Sepharose solution is added to each reaction. (One milligram of swollen protein A-Sepharose beads was found to completely absorb the immunoglobulins from 5 μl of rabbit nonimmune serum. A twofold excess is added to ensure complete absorption of the primary complex.) The addition of the Sepharose beads is accomplished by vortexing the solution of protein A-Sepharose beads and quickly pipetting 20 μl with a pipette tip that has been cut off to enlarge the hole. The immunoabsorption mixture is allowed to sit at 4° for 5 min with intermittent mixing. The beads are pelleted in an Eppendorf microfuge for 1 min, and the reaction mixture is removed by gentle suction. The beads are washed a total of three times at room temperature by vortexing the suspension in 1 ml of 10 mM Tris-HCl, pH 7.4, 0.3 M NaCl, and 1% Triton X-100, repelleting in the microfuge, and removing the wash by gentle suction. To remove detergent and salt, the beads are washed an additional three times in 10 mM Tris-HCl, pH 7.4. After removal of the final wash, the immune complex is released into 25 μl of releasing buffer at 95° for 2 min. The released protein is loaded directly onto an SDS-polyacrylamide gel for electrophoresis[12] and subsequent fluorographic detection[13] of the labeled products.

Acknowledgment

The studies were supported by U.S. Public Health Service Grant GM 26713 and the Robert A. Welch Foundation Grant AQ 814.

[34] Selection and Characterization of Nuclear Genes Coding Mitochondrial Proteins: Genetic Complementation of Yeast *pet* Mutants

By KAREN O'MALLEY and MICHAEL G. DOUGLAS

The ability to analyze yeast genes at the molecular level has increased markedly with the development of recombinant DNA technology and yeast transformation methods.[1,2] Previously, yeast genes had been selected by either complementation of *Escherichia coli* mutations,[3] *in situ* immunoassays,[4] or by hybridization with RNA or DNA probes.[5,6] The yeast transformation system together with suitable cloning vehicles[7,8] allows for the direct expression and selection of cloned yeast fragments in yeast. Moreover, this procedure is applicable for any yeast genes for which a mutant allele is available. A collection of defined yeast nuclear mutations exhibiting specific lesions in various mitochondrial enzymatic activities[9,10] has been defined

[1] A. Hinnen, J. Hicks, and G. Fink, *Proc. Natl. Acad. Sci. U.S.A.* **75**, 1929 (1978).

[2] J. D. Biggs, *Nature (London)* **275**, 104 (1978).

[3] L. Clarke, and J. Carbon, *Cell* **9**, 91 (1976).

[4] L. Clarke, R. Hitzeman, and J. Carbon, this series, Vol. 68, p. 436.

[5] T. P. St. John, and R. W. Davis, *Cell* **16**, 443 (1979).

[6] T. D. Petes, J. R. Broach, P. C. Wensink, L. M. Hereford, G. R. Fink, and D. Botstein, *Gene* **4**, 37 (1978).

[7] K. Struhl, D. J. Stinchcomb, S. Scherer, and R. W. Davis, *Proc. Natl. Acad. Sci. U.S.A.* **76**, 1035 (1979).

[8] J. Broach, J. Strathern, and J. Hicks, *Gene* **8**, 121 (1979).

[9] E. Ebner, L. Mennucci, and G. Schatz, *J. Biol. Chem.* **248**, 5360 (1973).

[10] A. Tzagoloff, A. Akai, and R. Needleman, *J. Biol. Chem.* **250**, 8228 (1075).

in earlier studies. Additionally, since the yeast genome is small (9×10^9 daltons),[11] a relatively small number of clones need to be screened to assure the possibility of selecting the gene of interest.

Described here are the procedures used for the preparation of a yeast library encompassing the entire yeast genome and a two-stage screening procedure for the selection of a defined *PET* gene from this library of genomic DNA. In addition, a procedure is described to define and characterize rapidly the mitochondrial protein encoded by the selected *PET* gene.

Principle of the Method

The relative ease in constructing libraries from the DNA of organisms with small genomes, such as *E. coli* and yeast, has been demonstrated by Clarke and Carbon.[3] They calculated that a collection of 4600 independent recombinant clones with an average yeast DNA insert size of 15 kb should contain approximately 99% of the yeast genome. In order to ensure that any given gene is intact on at least a portion of the recombinant fragments, the genomic DNA is randomly sheared either hydrodynamically[3] or by partial restriction endonuclease digestion.[12] The latter method avoids attaching synthetic linkers or tailing both the DNA and vector in order to achieve a random sampling.

The yeast *E. coli* shuttle vector YEp13 was used for the construction of a yeast genomic library. This vector (a gift from J. Broach, Cold Spring Harbor) consists of the entire bacterial plasmid pBR322, the *LEU2* gene of yeast, and the origin of replication from the yeast 2μ circle DNA.[8] In addition to efficiently transforming yeast leu 2^- strains to Leu$^+$ ($10^3 - 10^4$ transformants per microgram of DNA), YEp13 replicates autonomously in multiple copy number in yeast and bacteria as closed circular DNA.[8]

To construct the genomic library, a modification of the procedure of Maniatis *et al.,*[12] is employed in which quasirandom fragments are generated by partial restriction digests. A unique six-base recognition site is chosen in the vector such that its cohesive ends comprise a tetranucleotide recognition site for a related enzyme. Genomic partial digests with the four-base recognition site enzyme will create a subset of cohesive ends that can be directly inserted into the vector by ligation. YEp13 contains a single *Bam*HI site in the tetracycline gene of pBR322. Because *Sau*3A leaves the same four-base overlap as does *Bam*HI, yeast genomic DNA can be partially digested with this enzyme, sized, and inserted directly into the *Bam*HI 1 site of YEp13.

[11] G. Lauer, T. M. Roberts, and L. C. Klotz, *J. Mol. Biol.* **114**, 507 (1977).
[12] T. Maniatis, R. Hardison, E. Lacy, J. Lauer, C. O'Connell, and D. Onon, *Cell* **15**, 687 (1978).

In order to clone by complementation, one needs a mutation in the gene of interest, i.e., an inheritable trait in the parent cell that can be suppressed by the addition of exogenous DNA and quickly scored on plates. To avoid background reversion of selectable markers, the construction of nonreverting recipients is desirable. In the selection of a nuclear gene for a mitochondrial component from a pool of yeast genomic DNA cloned in YEp13, a yeast mutant is constructed from the appropriate *pet* mutant and a strain carrying the *leu 2-3 leu 2-112* mutation. The latter strain was constructed by combining two stable point mutations in the *LEU 2* allele, thereby creating a double mutant at this locus with a reversion frequency greater than 10^{-10}.[1] A haploid strain containing both the double *leu 2* and *pet* mutations is used for the selection.

In theory, one should be able to select for complements of the *leu 2* and *pet* mutations in a single step. However, the simultaneous selection for both phenotypes is inefficient, and problems arise with unstable *pet* alleles that show a high reversion frequency. Therefore, the selection of *leu 2 pet* cotransformants is accomplished using a two-step screening procedure. A flow sheet of this procedure, which has been employed to select the complementing genes of six *pet* mutants showing a wide range of reversion frequencies, is shown in Fig. 1.

Strains

The following yeast strains are used: DC5 *a leu 2-3 leu 2-112 his 3 can 1-11* (J. Hicks, Cold Spring Harbor Laboratories), 6657-4D *a his 3-11 his 3-15 leu 2-112 can 1-11*, GRF18 α *his 3-11 his 3-15 leu 2-3 leu 2-112* (G. Fink, MIT). Nuclear petites in the alpha mating type are as described in earlier studies.[9,10,13] In the present example strain D771-31 α *pet 9 ade 1 ade 2* (D. Deters, University of Texas at Austin) is used. *Escherichia coli* strain RR1 (F⁻ *pro leu thi lacy str^R r⁻ m⁻ endo I⁻*) is used for propagation of all plasmids.

Media

All plates contained 1.5% agar. Tetracycline is used at 25 μg/ml and ampicillin at 50 μg/ml final concentrations. Plates containing antibiotics are used within 10–14 days.
   L Broth: (per liter) 10 g of tryptone (Difco), 5 g of yeast extract (Difco), and 5 g of NaCl. The medium is adjusted to pH 7.5 with NaOH prior to sterilization.
   YPD medium: (per liter) 10 g of yeast extract, 20 g of Bacto-peptone (Difco), and 20 g of dextrose.

[13] L. Kovac, T. M. Lachowicz, and P. P. Slonimski, *Science* **158,** 1564 (1967).

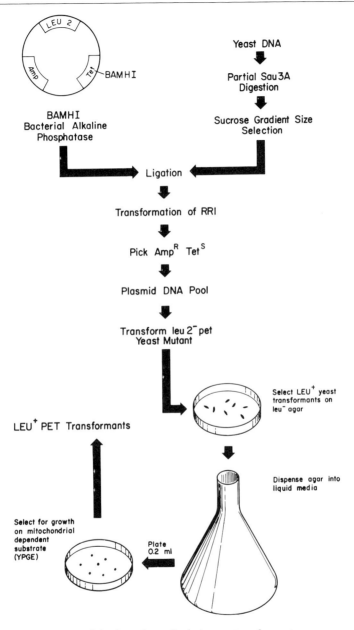

FIG. 1. Selection scheme for *leu⁺ pet* cotransformants.

YPGE medium: (per liter) 10 g of yeast extract, 20 g of Bacto-peptone, and 30 ml of glycerol. After sterilization 30 ml of 100% ethanol are added.

Minimal medium: (per liter) 6.7 g of yeast nitrogen base without amino acids (Difco) and 20 g of dextrose.

Synthetic complete dextrose medium (SCD): (per liter) 20 g of dextrose plus 160 ml of YNB amino acid mixture plus or minus leucine (described below).

Regeneration agar: this is prepared just before use from sterile stock solutions during spheroplast formation (see below), the complete regeneration agar is prepared by mixing 100 ml of 2.5 M sorbitol, 10 ml of 50% dextrose, 2.5 ml of YPD medium, 100 ml of 7.5% agar (microwaved just prior to use), and 40 ml of sterile filtered yeast nitrogen base (YNB) (1.7 g) with 2.5 ml of a stock solution containing adenine, lysine, and histidine (2 mg/ml each). After mixing, the regeneration agar is kept in a water bath at 47–50° until ready for use. This amount of regeneration agar will yield between 20 and 25 plates when poured as a thin layer over minimal medium plates. Alternatively, one could pour the regeneration agar directly into empty petri dishes as a thicker layer.

YNB amino acid mixture: It is sometimes convenient to prepare this mixture for use in the regeneration agar. It is prepared by grinding together 210 g of YNB; 0.6 g of adenine sulfate, uracil, L-tryptophan, L-histidine-HCl, L-arginine-HCl, L-methionine; 0.9 g of L-tyrosine, L-isoleucine, L-lysine-HCl; 1.5 g of L-phenylalanine; and 3.0 g of L-glutamic acid, L-valine, L-threonine, L-serine. L-Leucine (0.9 g) can be added if needed. The mixed powder is stored at room temperature in a brown bottle. For preparation of regeneration agar, 2 g of the YNB amino acid mix are dissolved in 40 ml of water, sterile filtered, and used in place of the YNB above.

Construction of YEp13 Yeast DNA Library

High molecular weight wild-type yeast DNA is prepared from DC5 according to the procedure of Cryer et al.[14] Partial endonuclease digestion conditions are established by digesting 1 μg of yeast DNA with serial dilutions of Sau3A (New England BioLabs). Reactions are incubated for 60 min at 37°, and the extent of digestion is estimated by gel electrophoresis on 0.7% agarose gels with molecular length standards. Based on this information, three large-scale digestions (200 μg each) are performed with 0.5, 1,

[14] D. Cryer, R. Eccleshall, and J. Marmur, Methods Cell Biol. 12, 39 (1975).

and $2\times$ the relative amount of Sau3A needed to generate fragments of 7.5 kb average size. Digestions are pooled and separated according to size on 10 to 40% sucrose gradients.[12] Aliquots of the gradient fractions are analyzed by agarose gel electrophoresis. Fractions containing between 5 and 20 kb DNA were pooled, dialyzed against TE (10 mM Tris, 1 mM EDTA, pH 8.0), ethanol precipitated, and resuspended in TE.

In each ligation, 30 μg of Sau3A digested, yeast genomic DNA are ligated with 10 μg of the yeast vector YEp13 that has been digested with BamHI and then with bacterial alkaline phosphatase.[15] After ligation, the DNA from each ligation is used to transform $E.\ coli$ RR1 to ampicillin resistance. A total of 2.5×10^5 Amp^R colonies are obtained from all the ligations; these are pooled and used to prepare plasmid DNA.[16]

Yeast Transformation Procedure

1. The yeast $leu\ 2\ pet$ mutant is grown overnight to stationary phase in YPD media.
2. One hundred milliliters of YPD are inoculated with 2 ml of the overnight culture, and the cells are grown to early log phase (1 to 2×10^7 cells/ml). A 0.1-ml aliquot of the overnight culture (approximately 5×10^6 cells) is plated on YPGE plates, incubated at 28°, and scored in 4 days for reversion frequency.
3. The cells are centrifuged in Beckman JA-14 sterile bottles at 5000 rpm for 5 min. Unless otherwise indicated, all steps are performed at room temperature.
4. The pellet is resuspended in 10 ml of 1 M sorbitol, transferred to JA-20 screw-cap polycarbonate tubes, and centrifuged for 5 min at 2000 rpm.
5. After the supernatant is decanted, the cells are washed again with 10 ml of 1 M sorbitol and centrifuged as described above.
6. The cells are resuspended in 10 ml of 1 M sorbitol followed by addition of 0.4 mg/ml Zymolyase 6000 (Kirin Breweries, Tokyo, Japan).
7. The cells are incubated at 37° with gentle shaking until approximately 90% of the cells are spheroplasted as judged by microscope analysis (30–60 min with different strains).
8. After spheroplast formation, the cells are again spun down for 3 min at 2000 rpm. The pellet is gently resuspended in 10 ml of 1 M

[15] A. Ullrich, J. Shine, J. Chirgwin, R. Pictet, E. Tischer, W. J. Rutter, and H. M. Goodman, *Science* **196**, 1313 (1977).
[16] L. Elwell, J. de Graff, D. Seibert, and S. Falkow, *Infect. Immun.* **12**, 404 (1975).

sorbitol with a sterile wooden applicator stick before being recentrifuged.

9. The washed pellet is then suspended in 10 ml of 1 M sorbitol – 10 mM Tris, pH 7.5 – 10 mM CaCl$_2$ (STC) and centrifuged as described above.

10. The Ca^{2+}-treated cells are gently resuspended in 0.5 ml of STC, and 5 – 20 μg of the yeast DNA pool in YEp13 in 50 μl are added. The cells are incubated for 15 min at room temperature.[16a]

11. Five milliliters of 10% polyethylene glycol 6000 – 10 mM Tris, pH 7.5 – 10 mM CaCl$_2$ (sterile filtered) are added to each transformation tube, gently mixed and incubated for an additional 10 min.

12. The PEG-treated cells are then spun down and resuspended in 100 μl of 1.2 M sorbitol – 10 mM CaCl$_2$ containing 20 μg of leucine per milliliter plus 50 μl of YPD medium in 1.2 M sorbitol. The transformation mixtures are incubated for 20 min at 30° with gentle shaking, after which 5 ml of STC are added to each tube.

13. After gentle mixing, the transformation mixture is poured directly into 250 ml of complete regeneration agar at 47 – 50°. The agar is immediately poured in a thin layer over minimal medium plates. Plates are incubated at 30° for 3 – 5 days. Leu$^+$ transformants are usually visible in 3 days.

14. After 5 days of growth at 30°, approximately 10^3 LEU$^+$ transformants per microgram of DNA are obtained.

15. Using a sterile spatula, the top agar containing the recombinant colonies is broken into pieces and transferred to a 50 cm^3 syringe. Colonies from 20 plates are mechanically dispersed by passage through the syringe into 1 liter of SCD-leu liquid medium. This mixture is shaken for 1 hr on a rotary shaker at room temperature. Alternatively, the top agar can be transferred into a small sterile Waring blender containing 500 ml of SCD-leu medium and dispersed by blending for approximately 30 sec.

16. Aliquots of 0.2 ml freed of agar particles by brief centrifugation are spread for single colonies on approximately 10 YPGE plates.

17. After incubation for 4 days at 30°, selected glycerol$^+$ transformants are replica plated onto SCD-leu plates to confirm cotransformation of the LEU$^+$ phenotype.

[16a] Transformation of yeast can also be achieved, albeit at lower transformation frequency, by mixing Ca^{2+}-treated spheroplasts with lysed bacterial cells containing plasmid DNA. Generally, a 10-ml overnight culture is pelleted, resuspended in 25% sucrose – 50 mM Tris, pH 7.5 and digested with 1 mg/ml lysozyme. Cells are incubated for 10 min on ice, pelleted, resuspended in 50 μl of STC, and added directly to the transformation mixture.

Small-Scale Preparation of Yeast Plasmid DNA

There are several methods to confirm the location of the functional *PET* gene on the transforming vehicle.[17] The most reliable of these is the demonstration of high-efficiency cotransformation of the *Leu⁺* and *Pet⁺* alleles. Plasmid DNA is first prepared from selected transformants using a modification of a small-scale isolation procedure for bacteriophage DNA.[18]

1. The yeast transformant is grown to stationary phase in 10 ml of selective medium (YPGE).
2. The cells are harvested at 2000 rpm for 5 min and washed once with sterile water.
3. The pellet is resuspended in 0.8 ml of 0.9 M sorbitol, 50 mM potassium phosphate, pH 7.5, 14 mM 2-mercaptoethanol, 1 mg of Zymolyase 5000 per milliliter, then transferred to a 1.5-ml Eppendorf microfuge tube.
4. This cell suspension is incubated at 37° for 1–2 hr, then the spheroplasts are pelleted for 5 min in a microfuge.
5. The pellet is resuspended in 0.4 ml of 0.15 M NaCl, 0.1 M EDTA, pH 8.0.
6. To the suspension is added 10 μl or 10 mg/ml proteinase K and 40 μl of 10% SDS followed by incubation at 37° for 1 hr.
7. Diethyl pyrocarbonate (1 μl) is added to each tube, which is then mixed and incubated an additional 20 min at 60°.
8. Fifty microliters of 5 M potassium acetate are added to each tube, followed by incubation on ice for 60 min.
9. The precipitate is centrifuged at 4° for 10 min, and the supernatant is decanted into a second microfuge tube containing 1 ml of 100% ethanol at room temperature.
10. After incubation at room temperature for 10 min, the yeast DNA is pelleted at room temperature for 10 min.
11. The supernatant is discarded, and the tube is washed once with 1 ml of cold 90% ethanol, then dried well.
12. Approximately 2 μg of DNA prepared by this procedure can be resuspended in 10 mM Tris-HCl, 0.1 mM EDTA, pH 7.4, 5 μg of RNase A per milliliter (heated at 70° for 30 min).

This DNA is then used with the yeast transformation procedure described above to transform the *leu 2 pet* yeast mutant or can be propagated

[17] K. O'Malley, P. Pratt, J. Robertson, M. Lilly, and M. Douglas, *J. Biol. Chem.* **251,** 2097 (1982).
[18] J. Cameron, P. Philippsin, and R. Davis, *Nucleic Acids Res.* **4,** 1429 (1977).

following transformation of *E. coli* RR1 to Amp[R]. In *E. coli,* large amounts of the plasmid can be prepared[19] for further studies.

In Vitro Translation of the Cloned PET Gene Product

The most reliable procedure for confirming the identity of the isolated *PET* gene is the immunological identification of a product translated from RNA that has been selected by the plasmid DNA. Total LiCl-precipitable yeast RNA prepared for log-phase cells grown under derepressed conditions provides a convenient and abundant source of RNA for plasmid-dependent selection of efficiently translated specific mRNAs.[17]

The method utilizes plasmid DNA immobilized on nitrocellulose filters in a hybridization reaction with total yeast RNA. The hybridization-selected mRNAs are eluted from the bound DNA and identified by proteins synthesized from them.[20] *In vitro* translations of the eluted RNAs are performed in nuclease-treated reticulocyte lysates[21] in the presence of [^{35}S]methionine.

Preparation of Plasmid DNA Bound to Nitrocellulose Filters

1. Nitrocellulose filter disks (5 mm in diameter) are prepared by punching out Schleicher & Schuell BA-85, 0.45 μm nitrocellulose filters with a sterile paper hole punch.
2. The filter disks are soaked at room temperature in 0.3 M NaCl, 2.0 mM EDTA, 0.1% SDS, 10 mM Tris-HCl, pH 7.5, 0.1% diethyl pyrocarbonate.
3. Plasmid DNA at a concentration of 1 mg/ml in 0.2 M NaOH – 2 M NaCl is heated at 100° for 90 sec, then immediately transferred to an ice bath for an additional 2 min.
4. The denatured plasmid DNA (2 μg) is then transferred to the presoaked filter disk then baked for 2 hr at 80° in a vacuum oven to fix the DNA to the filter.
5. After baking, the DNA-bound filters are washed by swirling in soaking buffer, then air dried. The filters can be stored indefinitely at room temperature.

Plasmid-Dependent Selection of Translatable RNA

1. Hybridization buffer (2×): 1 M NaCl, 0.004 M EDTA, 0.1 M 1,4-piperazinediethanesulfonic acid (PIPES), pH 7.5, 0.8% SDS, 60%

[19] L. Elwell, J. de Graff, D. Seibert, and S. Falkow, *Infect. Immun.* **12**, 410 (1975).
[20] R. Ricciardi, J. Miller, and B. Roberts, *Proc. Natl. Acad. Sci. U.S.A.* **76**, 4927 (1979).
[21] H. Pelman and R. Jackson, *Eur. J. Biochem.* **67**, 247 (1976).

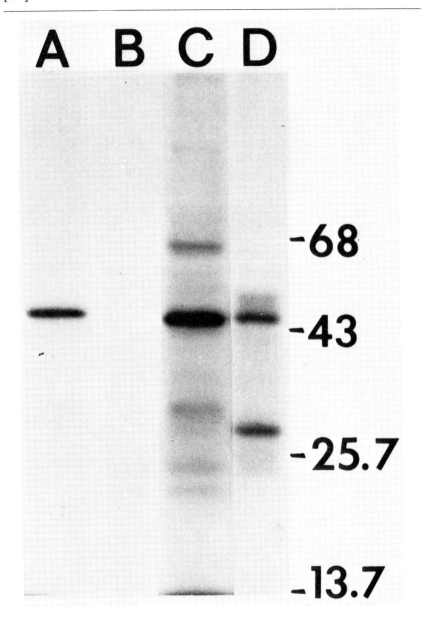

FIG. 2. *In vitro* translation products of plasmid-selected yeast RNA. After hybridization of total yeast RNA with the indicated plasmids, excess RNA was washed away and the hybridized RNA eluted. Resolution and detection of the translation products synthesized from these eluded RNAs are as described.[22] A, YEp13 vector DNA; B, pBR322; C, YEp13 with the F_1-ATPase α subunit gene insert, D, YEp13 with the adenine nucleotide translocator insert.

formamide. The NaCl, EDTA, and PIPES are autoclaved; formamide (deionized) and SDS are added with constant mixing.

2. Beef liver tRNA is phenol extracted, ethanol precipitated, and dissolved at 1 mg/ml in water.

3. All manipulations are performed at room temperature unless otherwise noted. Total yeast LiCl-precipitable RNA (100–200 μg) or 1–2 μg of total yeast poly(A) RNA are suspended in 500 μl of 1× hybridization buffer in an Eppendorf microfuge tube.

4. Plasmid DNA-containing filters that are sliced into several pieces with a sterile razor blade are dropped into the RNA solution followed by a brief vortexing and centrifugation for 10 sec to spin the filter pieces into the solution.

5. The hybridization mixtures are incubated overnight at 47° with mild shaking.

6. The hybridization mixture is aspirated from the filters with a sterile syringe followed by five rapid washes of the filters with 1 ml each of 0.1 M NaCl at 60°.

7. The salt-washed filter is washed an additional five times with 1 ml each of 0.15 M NaCl–0.015 M sodium citrate–0.5% SDS at 60° followed by a final two washes with 1 ml of 2.0 mM EDTA, pH 8.0, at 60°.

8. After traces of EDTA have been removed, 300 μl of 5 mM KCl are added to the tube and the filters are held in boiling water for 90 sec followed by quick transfer to a Dry Ice–methanol bath for 20–30 sec.

9. The frozen RNA-filter mixture is thawed and kept at 4°. The filters are quickly removed from the microfuge tube.

10. The eluted RNA in each tube is precipitated overnight at −20° in the presence of 2 μg of carrier beef liver tRNA by addition to 0.2 M sodium acetate, pH 5.0, and 2.5 volumes of 100% ethanol.

11. The precipitated RNA is pelleted for 15 min at 4° in the microfuge. The supernatant is decanted and dried *in vacuo.*

12. The RNA is resuspended in 10 μl of 5 mM KCl and stored at −20°.

Product Identification of Plasmid-Selected RNA

1. A 30–50 μl nuclease-treated rabbit reticulocyte lysate translation reaction[21] containing 5 μCi of [^{35}S]methionine (400–1065 Ci/mmol) when programmed with 0.5–1 μl of the eluted RNA is sufficient for 3–5 gels.

2. Resolution of labeled products on SDS–polyacrylamide gels followed by autoradiography (Fig. 2) is performed as previously de-

scribed[22] except that the gels are boiled for 10 min in 5% trichloroacetic acid after electrophoresis. This removes major hot acid hydrolyzable bands that interfere with the autoradiographic detection of plasmid-dependent products.

[22] M. Douglas, D. Finkelstein, and R. Butow, this series, Vol. 58, p. 58.

[35] Transformation of Nuclear Respiratory Deficient Mutants of Yeast

By Carol L. Dieckmann and Alexander Tzagoloff

Previously published procedures for the transformation of *Saccharomyces cerevisiae* have been adapted to transform successfully nuclear petite strains (*pet*), with wild-type yeast–*Escherichia coli* hybrid plasmid molecules.[1-3] Selection for transformants carrying a plasmid with the wild-type copy of the mutated gene is made on medium containing the nonfermentable carbon source, glycerol. Described here are methods for (*a*) transformation of *pet* mutants with a plasmid pool, and by selective plating, the isolation of plasmid(s) carrying the wild-type genes complementing the mutations; (*b*) transfer of the selected plasmid into *E. coli* to amplify and purify the wild-type DNA insert for restriction analyses and sequencing; (*c*) transformation of the *pet* mutant with a plasmid preparation from *E. coli* to confirm the presence of the wild-type gene copy in the insert of the isolated plasmid.

Vectors and Host Strains

Vector. The CV13 plasmid (YEp13)[4] has three components: (*a*) pBR322, an *E. coli* plasmid with genes conferring ampicillin and tetracycline resistance and replicative function in *E. coli*, (*b*) an *Eco*RI fragment of the 2 μm yeast plasmid conferring replicative function in *S. cerevisiae;* (*c*) a *Pst*I fragment of nuclear yeast DNA containing the *LEU2* gene, which confers a means of positive selection for *leu2* cells acquiring plasmid during transformation.

[1] J. Beggs, *Nature (London)* 275, 104 (1978).
[2] K. Struhl, D. T. Stinchcomb, S. Scherer, and R. Davis, *Proc. Natl. Acad. Sci. U.S.A.* 76, 1035 (1979).
[3] A. Hinnen, J. B. Hicks, and G. R. Fink, *Proc. Natl. Acad. Sci. U.S.A.* 75, 1929 (1978).
[4] J. R. Broach, J. N. Strathern, and J. B. Hicks, *Gene* 8, 121 (1979).

METHODS IN ENZYMOLOGY, VOL. 97

Vector with Yeast DNA Inserts. A plasmid pool consisting of CV13 vector molecules restricted within the tetracycline resistance gene and ligated with fragments of DNA representing the entire yeast wild-type genome, was constructed by Nasmyth.[5] The fragments of DNA were obtained from *S. cerevisiae* W87 by partial digestion with *Sau*3A. The digest was sized on a sucrose gradient and fractions containing 5–20 kbp fragments were ligated into the *Bam*HI site of the CV13 vector. The ligation mixture was used to transform *E. coli.* Thousands of ampicillin-resistant colonies were pooled, and plasmid DNA was purified.[6] Replica plating on tetracycline shows that 85% of the colonies are sensitive to tetracycline and represent cells transformed with CV13 containing an insert. Some colonies (15%) are resistant to tetracycline and represent cells transformed with CV13 lacking inserts. The purified plasmid preparation has been used as the transforming DNA for isolation of the wild-type genes complementing the nuclear respiratory deficiency mutations.

Host Strain Construction. Although plasmids carrying wild-type alleles of the nuclear petite mutations can be selected directly by plating on glycerol medium, the introduction of *leu2* mutations into strains to be transformed is advantagous for two reasons: (*a*) that it provides a measure of how well any one strain takes up CV13 plasmid DNA in a transformation; (*b*) that it provides a double selection for leu⁺ gycerol⁺ cotransformants, which guards against selection of respiratory competent revertants. A *leu2-3leu2-112* double mutant was constructed specifically to use with plasmids containing *LEU2*. The double mutant does not revert at the *leu2* locus. It is highly recommended that this double mutation be crossed into any nuclear petite yeast strain to be transformed with the CV13 pool. In addition to the double *leu2* mutation, it is useful to introduce at least one additional auxotrophic mutation for easy strain identification of transformed colonies, i.e., leu⁺ glycerol⁺ transformants should still be auxotrophs. Crossing nuclear *pet* strains to the double *leu2* strain produces progeny that are transformed with the CV13 pool at different efficiencies; therefore it is advisable to test a number of spores if the one chosen initially has a low transformation efficiency.

Procedure

All solutions and equipment are sterilized, and sterile technique is used throughout the procedure. The procedure for transforming respiratory deficient mutants with the CV13 pool was modified from that described by Beggs.[1,7,8] The *pet* strain to be transformed is grown on 2% galactose, 2%

[5] K. A. Nasmyth and S. I. Reed, *Proc. Natl. Acad. Sci. U.S.A.* **77**, 2119 (1980).
[6] S. N. Cohen, A. C. Y. Chang, and L. Hsu, *Proc. Natl. Acad. Sci. U.S.A.* **69**, 2110 (1972).
[7] G. Knapp, University of California, San Diego (personal communication).
[8] G. Faye, Université de Paris, Orsay, France (personal communication).

Bactopeptone (Difco), 1% yeast extract (Difco), to a density of 2 to 3×10^7 cells/ml. Using galactose as the carbon source instead of glucose enhances the frequency of transformation. All centrifugations are at 400 g for 5 min at room temperature except as noted, and all washes are with one-half the original volume. The cells are centrifuged, washed once with water, resuspended in one-third the original volume of DTT solution, shaken at 30° for 10 min, centrifuged, and washed twice with 1.2 M sorbitol. The pellets are then resuspended in one-tenth the original volume of glusulase digestion solution and shaken at 30° for 20 min. The resultant spheroplasts are centrifuged, washed twice with 1.2 M sorbitol and a third time with 1.2 M sorbitol–10 mM CaCl$_2$. The spheroplasts are resuspended in one-hundredth the original volume of 1.2 M sorbitol–10 mM CaCl$_2$, and aliquots are transferred to clean tubes for transformation. Eppendorf centrifuge tubes (1.5 ml) are convenient for transforming 0.1-ml spheroplasts (2×10^8).

Solutions

> DTT solution: 1.2 M sorbitol; 0.025 M EDTA, pH 8.0; 7.7 mg/ml dithiothreitol
>
> Glusulase digestion solution: 0.1 M sodium citrate, pH 6.0; 0.01 M EDTA; 1% glusulase (v/v; Endo Laboratories, 1.27×10^5 units/ml)

CV13 pool DNA is added to prepared spheroplasts at a concentration of 200 μg/ml. This concentration of plasmid DNA is 10-fold higher than has been suggested,[1,2,7,8] but has proved to be essential in enhancing the frequency of transformation of the *pet* mutants. Our standard procedure is to add 20 μg in 10 μl of 1.2 M sorbitol to 100 μl of spheroplasts. After 15 min of incubation at room temperature, 0.9 ml of 20% polyethylene glycol $(M_r = 4000)$–10 mM CaCl$_2$–10 mM Tris-HCl, pH 8.0, is added, mixed by inversion, and incubated an additional 5 min. The spheroplasts are then centrifuged at 200 g for 5 min, and the supernatant is removed with a Pasteur pipette. The pellets are resuspended in 0.1 ml of 1.2 M sorbitol–10 mM CaCl$_2$ plus 0.05 ml of 1.2 M sorbitol containing 2% glucose, 2% Bacto-peptone, and 1% yeast extract. The spheroplasts are incubated for 20 min at 30° and then diluted to 0.4 ml with 1.2 M sorbitol (= 1×10^0 dilution).

A dilution series, 1×10^{-2}, 1×10^{-3}, 1×10^{-4}, and 1×10^{-5}, is made in 1.2 M sorbitol. Aliquots of 0.2 ml of each dilution are plated in 7.0 ml overlays as follows:

1. For leu$^+$ glycerol$^+$ transformants, two aliquots of the undiluted spheroplasts are plated on glycerol medium.
2. For leu$^+$ transformants, one aliquot each of the 1×10^{-2} and 1×10^{-3} dilutions are plated on glucose medium.
3. For regenerating spheroplasts, one aliquot of the 1×10^{-4} and

1×10^{-5} dilution are plated on glucose medium supplemented with leucine at 25 μg/ml.

All plating media contain 1.2 M sorbitol, 3% agar, 0.67% yeast nitrogen base without amino acids (Difco), and 25 μg of auxotrophic supplements per milliliter (e.g., methionine, histidine, adenine). Glycerol medium contains 3% glycerol and 0.05% glucose. The low percentage of glucose is essential for the recovery of transformed spheroplasts on glycerol. Glucose medium contains 2% glucose. Previously sterilized 7-ml overlays are melted in boiling water for 15 min and equilibrated at 45° immediately before use. An aliquot of the diluted spheroplasts is added to an overlay, vortexed, and poured onto a 10 cm in diameter petri plate containing 35 ml of the same medium. The plates are incubated at 30°.

Colonies appear in the regeneration overlays (glucose + leucine) after 2–3 days; in the leu$^+$ overlays (glucose only), after 3–5 days; and in the leu$^+$ glycerol$^+$ overlays (glycerol), after 4–10 days. Regeneration of 1–20% of the original number of cells is expected. The frequency of leu$^+$ transformants per regenerate varies from 2×10^{-3} to 1×10^{-4}. The number of leu$^+$ glycerol$^+$ transformants per leu$^+$ transformant expected, based on the composition of the CV13 pool, is 3×10^{-4}.[5]

Purification of leu$^+$ glycerol$^+$ colonies

Colonies appearing in the glycerol overlays are picked and templated on WO medium (2% glucose, 0.67% nitrogen base, 2% agar, 10 μg/ml auxotrophic supplements other than leucine). After 2 days of growth at 30° the plates are replicated on (a) YEPG (3% glycerol, 2% Bacto-peptone, 1% yeast extract, 2% agar); (b) WO medium without auxotrophic supplement (to check for contaminants). Colonies that grow on YEPG but not on WO are confirmed as leu$^+$ glycerol$^+$ and are inoculated into 1 ml of liquid YEPG medium and grown to stationary phase. The culture is diluted and plated for single colonies on YPD medium (2% glucose, 2% Bacto-peptone, 1% yeast extract, 2% agar). After 2 days of incubation at 30°, the plates are replicated on YEPG, WO, and WO plus auxotrophic supplements excluding leucine. This plating is done to purify the transformants and to check whether the wild-type copy of the gene has integrated into chromosomal DNA or is retained in plasmid form. Integration of plasmid DNA into chromosomal DNA appears to be strain (spore) dependent and presumably is enhanced by an element in the nuclear background. In the case of transformants with an integrated wild-type gene, all the colonies plated on YPD are respiratory competent on YEPG. Such strains always grow faster than those with nonintegrated plasmid. In one such strain it was determined genetically that the integration occurred at (or near) the *leu2* locus. Transformants with nonintegrated plasmid segregate into respiratory competent and deficient colonies in the above test. As expected, most respiratory deficient colonies also do not grow on WO medium lacking leucine, signifying loss of both the *PET* and *LEU2* markers on the CV13 plasmid. Occasionally a respiratory deficient colony will grow on WO without leucine, signifying retention of a *LEU2* plasmid. This plasmid may retain the wild-type gene, and the colony may be respiratory deficient owing to a ρ^- deletion in the mitochondrial DNA. The number of these leu$^+$ glycerol$^-$ colonies depends on the frequency of ρ^- production of the nontransformed strain.[9]

Glycerol$^+$ leu$^+$ transformants are purified and maintained on glycerol medium for four reasons: (*a*) to maintain a constant selective pressure for retention of plasmid in the transformed strain; (*b*) to ensure that glycerol$^-$ leu$^+$ cells contaminating glycerol$^+$ leu$^+$ colonies picked from overlays are not propagated; (*c*) to dilute out any *LEU2* plasmids that may have cotransformed with the *PET LEU2* plasmid; (*d*) to select against any plasmids deleting portions of the wild-type gene during growth and manipulation of the strain.

Small-Scale Plasmid Preparations from Yeast

Plasmid DNA from yeast transformants is prepared by a modification of the method of Birnboim and Doly.[10] A stationary phase culture, 10 ml, grown on liquid YEPG, is centrifuged at 400 *g*, washed with 1 ml of H$_2$O, and resuspended in 0.2 ml of zymolyase solution.

Zymolyase solution[5]: 1.0 *M* sorbitol; 0.1 *M* sodium citrate, pH 7.0; 0.06 *M* EDTA; 1 mg/ml zymolyase (Miles Laboratories, 5000 U/g); 0.15 *M* 2-mercaptoethanol

The cells are incubated at 34° for 60 min, or until spheroplast formation is complete. The spheroplasts are transferred to 1.5-ml Eppendorf tubes and washed three times with 1.2 *M* sorbitol. The spheroplasts are resuspended in 0.1 ml of 50 m*M* glucose–25 m*M* Tris-HCl, pH 8.0–10 m*M* EDTA–250 μg of RNase A per milliliter. To this suspension, 0.2 ml of 0.2 *N* NaOH–1% SDS is added immediately. After mixing by repeated inversion, the tubes are incubated on ice for 5 min and 0.15 ml of 3 *M* sodium acetate, pH 4.8, is added. The tubes are mixed by inversion, incubated on ice for 1 hr, and centrifuged for 5 min in an Eppendorf centrifuge. The supernatant is collected, and the DNA is precipitated by the addition of ethanol to a concentration of 70%, freezing at −80°, and centrifuging for 5 min in the

[9] B. Ephrussi, H. Hottinguer, and J. Tavlitzki, *Ann. Inst. Pasteur, Paris* **76**, 419 (1949).
[10] H. Birnboim and J. Doly, *Nucleic Acids Res.* **7**, 1513 (1979).

Eppendorf centrifuge. The ethanol is aspirated, and the pellet is redissolved in 0.3 M sodium acetate–0.1 mM EDTA and reprecipitated with ethanol. This procedure is repeated again, followed by two washes of the DNA pellet with 80% ethanol, drying in a vacuum centrifuge and resuspension of the pellet in 0.1 ml of sterile buffer, 25 mM Tris-HCl, pH 7.5–0.1 mM EDTA.

Transformation of *E. coli*

The *E. coli* strain, RR1 (*pro⁻*, *leu⁻*, *B1⁻*, *lacY*, *galK2*, *xyl5*, *mtl1*, *ara14*, *rpsL20*, *supt44*, *hsdM*, *hsdR*, λ⁻), can be transformed with plasmid preparations described in the preceding section. This strain of *E. coli* has mutations in both the restriction and modification genes, which protect against the destruction of the transforming DNA and eliminates modification of the DNA. The procedure of Petes *et al.*[11] yields 2×10^4 amp^r tet^r transformants/ng of pBR322 plasmid DNA. One-tenth of a small-scale yeast DNA preparation from a glycerol⁺ leu⁺ transformant yields 3×10^1 to 5×10^3 amp^r tet^s transformants by the same procedure. The RR1 strain is also a good producer of plasmid DNA when standard chloramphenicol amplification conditions are imposed,[6] yielding approximately 1 mg of CV13 plasmid DNA per liter of culture.

Confirmation of *pet* Complementation by Purified Plasmid

To complete the identification of the CV13 plasmid isolated in *E. coli* as the carrier of the wild-type gene complementing a yeast nuclear mutation, a small-scale *E. coli* DNA preparation is used to transform the original *pet* mutant.[12] Plasmid DNA is prepared as follows: 5 ml of *E. coli* RR1 cells grown to stationary phase on L broth plus 20 μg of ampicillin per milliliter are centrifuged at 9000 g for 10 min, resuspended in 1.0 ml of 50 mM Tris-HCl, pH 8.0–25% sucrose, and chilled on ice. Lysozyme solution (0.35 ml) is added, and the cells are incubated on ice for 15 min. To this is added 0.5 ml of Triton X-100 solution, mixed by gently rolling the tube, and the preparation is incubated an additional 5 min on ice.

Solutions

Lysozyme solution: 3.5 mg/ml lysozyme (4×10^5 units/mg; Sigma); 18 mM Tris-HCl, pH 8.0; 90 mM EDTA, pH 8.0; 0.7 mg/ml RNase A (80 Kunitz units/mg; Sigma) (RNase A stock is 10 mg/ml in 50 mM sodium acetate, pH 5.0, and has been heated to 80° for 20 min).

---

[11] T. D. Petes, J. R. Broach, P. C. Wensink, L. M. Herefore, G. R. Fink, and D. Botstein, *Gene* **4**, 37 (1978).

[12] R. B. Meagher, R. C. Tait, M. Betlach, and H. W. Boyer, *Cell* **10**, 521 (1977).

Triton X-100 solution: 0.3% Triton X-100; 187 mM EDTA, pH 8.0; 150 mM Tris-HCl, pH 8.0

The lysed cells are centrifuged at 37,000 g for 60 min. The supernatant is diluted with 1 ml of H_2O and extracted with 2.5 ml of water-saturated phenol that is freshly buffered to pH 8.0. The phenol is mixed thoroughly, and the tubes are centrifuged at 20,000 g for 10 min. The supernatant is collected and dialyzed to remove all traces of phenol. The DNA is precipitated by making the solution 0.2 M in NaCl, and then making it 70% in ethanol, freezing at $-80°$, and centrifuging at 20,000 g. The DNA is redissolved and reprecipitated, and the pellet is washed twice with 80% ethanol, dried in a vacuum centrifuge, and resuspended in 0.1 ml of sterile 1.2 M sorbitol. One-tenth of this preparation yields 1×10^2 to 1×10^3 leu$^+$ transformants from 2×10^8 spheroplasts prepared from the original *pet* mutant. A number of leu$^+$ colonies are picked and templated on YEPG medium. If the isolated plasmid does indeed carry the wild-type gene complementing the *pet* mutation, all leu$^+$ colonies should grow on YEPG medium within 2 days.

Acknowledgments

We wish to express our gratitude to Kim Nasmyth for providing us with the CV13 bank of yeast DNA. We wish to thank also Catherine Squires, Gerard Faye, and Gayle Knapp for helpful discussions on techniques of transformation and for providing us with the necessary strains of yeast and *E. coli.*

[36] Analysis of Yeast Mitochondrial Genes

By Carol L. Dieckmann and Alexander Tzagoloff

The mitochondrial genome of *Saccharomyces cerevisiae* is a circular duplex molecule consisting of 70–75 kbp.[1] Two different approaches have been used to sequence yeast mitochondrial genes. The tRNA and ribosomal RNA genes have been sequenced from fragments of mitochondrial DNA (mtDNA) cloned in bacteria.[2-4] Most of the sequences of the protein coding genes have been obtained from the genomes of cytoplasmic petite (ρ^-)

[1] P. Borst and L. A. Grivell, *Cell* **15**, 705 (1978).
[2] D. L. Miller, N. C. Martin, H. D. Pham, and J. E. Donelson, *J. Biol. Chem.* **254**, 11735 (1979).
[3] N. C. Martin, D. Miller, J. Hartley, P. Moynihan, and J. E. Donelson, *Cell* **19**, 339 (1980).
[4] B. Dujon, *Cell* **20**, 185 (1980).

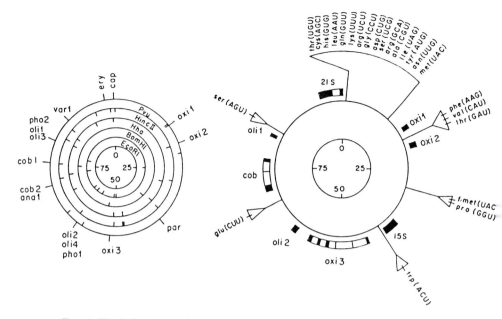

FIG. 1. Physical and genetic maps of *Saccharomyces cerevisiae* D273-10B. *Left:* The restriction sites for five endonucleases are shown on the inner circles. The location of the antibiotic resistance and *mit* loci are indicated on the outer circle. For a more detailed restriction map of the D273-10B genome, see Morimoto and Rabinowitz.[11] *Right:* This map shows the genes that have been sequenced. The dark portions of the 21 S rRNA, *cob* and *oxi*3 loci denote the exon coding regions of the genes. The genes encoded in the various loci are as follows: 21 S (large rRNA), 15 S (small rRNA), *oxi*1 (subunit 2 of cytochrome oxidase), *oxi*2 (subunit 3 of cytochrome oxidase), *oxi*3 (subunit 1 of cytochrome oxidase), *cob* (cytochrome *b*), *oli*1 (subunit 9 of the ATPase), *oli*2 (subunit 6 of the ATPase). The locations of the tRNA genes and their anticodons (3′→5′) are also indicated.

mutants.[5-10] The procedures for the isolation of genetically marked ρ^- mutants, and their use in the analysis of mitochondrial genes will be described here.

Briefly, ρ^- mutants are selected for the retention of known genetic markers in mtDNA. The ρ^- clones are purified and checked for the stability of their genotypes. The physical locations of the ρ^- genomes on the wild-type

[5] F. G. Nobrega and A. Tzagoloff, *J. Biol. Chem.* **255**, 9828 (1980).

[6] S. G. Bonitz, G. Coruzzi, B. E. Thalenfeld, A. Tzagoloff, and G. Macino, *J. Biol. Chem.* **255**, 11927 (1980).

[7] B. E. Thalenfeld and A. Tzagoloff, *J. Biol. Chem.* **255**, 6173 (1980).

[8] G. Macino and A. Tzagoloff, *Cell* **20**, 507 (1980).

[9] G. Coruzzi and A. Tzagoloff, *J. Biol. Chem.* **254**, 9324 (1979).

[10] L. A. Hensgens, L. A. Grivell, P. Borst, and J. L. Bos, *Proc. Natl. Acad. Sci. U.S.A.* **76**, 1663 (1979).

map can usually be established by restriction mapping. Clones with coherent segments of mtDNA free of internal deletions or rearrangements are used for sequence analysis. All our studies of mtDNA were done with the wild-type haploid strain D273-10B (American Type Culture Collection 25657). A partial restriction map of mtDNA in this strain as well as the locations and structures of the genes are presented in Fig. 1 (for a more detailed map, see Morimoto and Rabinowitz[11]).

Isolation of ρ^- Clones

General Properties of ρ^- Clones

Cytoplasmic petite or ρ^- mutants originally described by Ephrussi and co-workers[12] are now known to result from extensive deletions in mtDNA.[13] The deletions can encompass as much as 90% or more of the DNA, and in some cases the entire mitochondrial genome is lost. Mutants lacking all mitochondrial DNA are referred to as ρ^0 clones.

Deletions of mtDNA occur spontaneously, and most wild-type laboratory stocks of *S. cerevisiae* contain a low percentage (0.1 – 1%) of ρ^- mutants. A number of chemical agents induce deletions in mtDNA. The most effective are chemicals that intercalate with DNA, such as acridine and ethidium bromide. Although the deletions occur randomly in any part of the DNA, recent sequence analyses of mtDNA retained in different ρ^- clones indicate that there are preferred sequences at which deletions are initiated. These are commonly direct repeats of six nucleotides or more.[6,14] The nondeleted segment conserves only one of the repeats. Irrespective of the sequence retained, it is always amplified in a manner such that the final content of mtDNA per cell is equivalent to that of wild-type yeast. The amplification results from tandem or palindromic fusion of the retained segment giving rise to dimers, trimers, and higher-order multimers of the repeat unit. In effect, therefore, *S. cerevisiae* acts as a self-cloning vehicle for specific mitochondrial sequences.

In addition to providing a plentiful source of DNA, ρ^- clones can be selected on the basis of their content of selected genetic markers by fairly simple tests. This means that it is possible to enrich for sequences with specific genes or subregions of genes.

[11] R. Morimoto and M. Rabinowitz, *Mol. Gen. Genet.* **170**, 25 (1979).
[12] B. Ephrussi, H. Hottinguer, and J. Tavlitzki, *Ann. Inst. Pasteur, Paris* **76**, 419 (1948).
[13] G. Faye, H. Fukuhara, C. Grandchamp, J. Lazowska, F. Michel, J. Casey, G. S. Getz, J. Locker, M. Rabinowitz, M. Bolotin-Fukuhara, D. Coen, J. Deutsch, B. Dujon, P. Netter, and P. P. Slonimski, *Biochimie* **55**, 779 (1973).
[14] C. Gaillard, F. Strauss, and G. Bernardi, *Nature (London)* **283**, 218 (1980).

While the segments of mtDNA retained in ρ^- clones are generally tandemly repeated and consist of coherent wild-type sequences, there are exceptions. Some ρ^- clones have genomes with palindromic arrangements of the repeat unit. A more serious problem arises when the retained segment may itself have an internal deletion or rearrangement. Such artificial sequences can usually be recognized by comparing the physical maps of independently isolated clones with similar genotypes.

Mutagenesis with Ethidium Bromide

Reagents and Media
Liquid YPD: 1% yeast extract, 2% peptone, 2% glucose
Solid YPD: same as liquid YPD plus 2% agar
Ethidium bromide: 0.5% solution of ethidium bromide, filter sterilized
Sorensen's buffer: 30 ml of 0.1 M Na_2PO_4 plus 70 ml of 0.1 M KH_2PO_4 sterilized by autoclaving

Procedure. A wild-type strain of *S. cerevisiae* is grown to stationary phase in 1 ml of liquid YPD medium. The cells are harvested at 2000 rpm in a clinical centrifuge, washed with 2 ml of Sorensen's buffer, and resuspended in 1 ml of the same buffer. To the cell suspension is added 20 μl of a 0.5% solution of ethidium bromide. After incubation at 30° for 1 hr, the cells are collected by centrifugation and suspended in 1 ml of Sorensen's buffer. The cells are serially diluted and spread on YPD plates to yield 50–100 colonies per plate. This can be done either by a visual adjustment of the initial dilution or by counting the cells in a hemacytometer. The plates are incubated for 2–3 days at 30°. Under the above conditions of mutagenesis with ethidium bromide, 10–30% of the colonies will consist of ρ^- clones easily recognizable by the smaller colony size. For further genetic tests, it is convenient to transfer the small colonies to fresh YPD plates in 50 position grids. The number of ρ^- colonies to be collected depends on the genotype sought. For gross dissection, i.e., clones containing one or two genes on a segment of mtDNA representing 5–10% of the wild-type genome length, it is usually sufficient to analyze 200–400 clones.

First Genetic Screen of ρ^- Clones

Media and Tester Strains
Mit$^-$, syn$^-$ testers: these are haploid strains of *S. cerevisiae* with point mutations in mitochondrial tRNA, rRNA and mRNA coding genes; available upon request
Antibiotic resistant testers: haploid respiratory competent strains of *S. cerevisiae* with point mutations in mtDNA conferring resistance to various antibiotics; available upon request

WO plates: 0.67% Bacto yeast nitrogen base without amino acids, 2% glucose, 2% agar

YEPG plates: 1% yeast extract, 2% peptone, 3% glycerol, 2% agar

Antibiotic media: YEPG supplemented with antibiotics at the concentrations indicated in Table I. The antibiotics should be added after the autoclaved media have cooled to 60–70° just prior to pouring the plates.

Procedure. The primary ρ^- clones obtained by ethidium bromide mutagenesis are checked for the retention of different markers in mtDNA by crosses to appropriate tester strains. The testers include strains with mutations in the protein coding genes (mit⁻ mutants) and in the tRNA and rRNA genes (syn⁻ mutants). Since mtDNA of ρ^- clones can recombine with homologous regions of the full-length genome, the respiratory deficiency of the tester strains will be corrected by clones whose retained segments of mtDNA contain the wild-type alleles corresponding to the mit⁻ or syn⁻ mutations. By crossing the ρ^- clones to the testers with lesions in different mitochondrial genes and scoring the diploid progeny for respiratory competency (growth on glycerol), it is possible to derive a genotype for each clone. The genotype in turn provides an index of the region of mtDNA retained in the clone.[15–18]

If a large number of clones are to be characterized, the crosses are best done by the cross-replication technique.[19] The testers grown in liquid YPD medium are spread as lawns on WO plates. It is sufficient to spread 0.1–0.2 ml of a stationary phase culture per plate. The master plates of the clones (see preceding section) are replicated on YPD to recover the master, on YEPG to verify that the haploid clones do not grow on glycerol, and on the lawn of each mit⁻ or syn⁻ tester. For the crosses, the testers should be of a mating type opposite to that of the original wild-type haploid used for the isolation of the clones. In addition, the testers and ρ^- clones should have different auxotrophic requirements to facilitate prototrophic selection of the diploid progeny from the crosses.

After replication on YPD, YEPG, and the tester lawns, the plates are incubated for 1–2 days at 30°. The YEPG plates are scored, and the WO

[15] P. P. Slonimiski and A. Tzagoloff, *Eur. J. Biochem.* **61**, 27 (1976).

[16] R. J. Schweyen, B. Weiss-Brummer, B. Backhaus, and F. Kaudewitz, *Mol. Gen. Genet.* **159**, 151 (1978).

[17] P. Nagley, K. S. Sriprakash, J. Rytka, K. B. Choo, M. K. Trembath, H. B. Lukins, and A. W. Linnane, *in* "The Genetic Function of Mitochondrial DNA" (C. Saccone and A. M. Kroon, eds.) p. 231. North-Holland Publ., Amsterdam, 1976.

[18] A. Tzagoloff, F. Foury, and A. Akai, *Mol. Gen. Genet.* **149**, 33 (1976).

[19] D. Coen, J. Deutsch, P. Netter, E. Petrochilo, and P. P. Slonimski, *Symp. Soc. Exp. Biol.* **24**, 449 (1970).

plates with the diploid cells are replicated on YEPG. Growth of the diploid cells on glycerol can be scored after 1–2 days of incubation of the YEPG plates at 30°. The ability of a ρ^- clone to confer respiratory competency to a mit⁻ or syn⁻ mutant indicates the presence of the wild-type allele in the mitochondrial genome of the clone. The genotypes of the clones are specified by their marker retention.

In addition to mit⁻ and syn⁻ mutations, which are expressed by a respiratory deficiency, a number of mitochondrial genes (namely, 21 S rRNA, 15 S rRNA, cytochrome b, subunits 6 and 9 of the ATPase complex) have alleles that confer resistance to antibiotics. The retention of this class of markers in ρ^- clones is also easily testable by crosses to a wild-type antibiotic-sensitive tester. In this case, the parental strain used for the derivation of the ρ^- clones must have the antibiotic-resistant allele or alleles. A number of *S. cerevisiae* strains with multiple antibiotic-resistant mutations are currently available. To verify retention of antibiotic resistance markers in the ρ^- genomes, crosses are performed between the clones and a respiratory competent haploid strain whose growth on glycerol is inhibited by the antibiotic. The procedures for the crosses are identical to those described above except that diploid cells are replicated on YEPG medium plus the specific antibiotic to be tested for. The retention in a ρ^- clone of an antibiotic resistance marker is expressed by the ability of the diploid cells issued from a cross of the clone and the tester to grow on glycerol in the presence of the antibiotic. The best-studied antibiotic resistance loci and the genes they represent are listed in Table I.

TABLE I
MITOCHONDRIAL MARKERS AND GENES

Locus	Gene	Phenotype
cob	Cytochrome b	Respiratory deficiency
*oxi*1	Subunit 2 of cytochrome oxidase	Respiratory deficiency
*oxi*2	Subunit 3 of cytochrome oxidase	Respiratory deficiency
*oxi*3	Subunit 1 of cytochrome oxidase	Respiratory deficiency
*pho*1	Subunit 6 of ATPase	Respiratory deficiency
*pho*2	Subunit 9 of ATPase	Respiratory deficiency
*oli*1	Subunit 9 of ATPase	Growth on 5 μg/ml oligomycin
*oli*2	Subunit 6 of ATPase	Growth on 5 μg/ml oligomycin
ery	21 S rRNA	Growth on 2 mg/ml erythromycin
cap	21 S rRNA	Growth on 2 mg/ml chloramphenicol
par	15 S rRNA	Growth on 2 mg/ml paromomycin
ant	Cytochrome b	Growth on 5 μg/ml antimycin

The number of cross-replications that can be done with a single master plate depends on the age of the plate (if the plate is old, the patched cells tend to lift off on the first velvet) and the quality of the velvet. With masters grown for 2 days, it is possible to replicate on six different tester lawns. It should be remembered that each cross-replication requires a fresh velvet.

Purification of the Primary Clones

The primary ρ^- clones may still have a heterogeneous population of mtDNA segments. It is, therefore, essential to repurify the clones after they have been allowed to grow for a sufficient number of generations to give pure segregants.

Primary clones established to have the desired genotypes are inoculated into 10 ml of liquid YPD medium. The inoculum should be small so that at least 20–30 generations ensue before the culture reaches stationary phase. The procedures henceforth are similar to those used for the genetic characterization of the primary clones. The cells are diluted and spread for single colonies on 2 or 3 YPD plates. The plates can be used directly as masters for the cross-replications to the testers.

Quite frequently primary clones with multiple markers give rise to segregants (secondary clones) with different subsets of the original markers. This is especially true when the genotype of the primary clone corresponds to an extensive region of the wild-type mtDNA.

The secondary clones must be further checked for the stability of their genotypes. This is important, since some clones may be unstable and lose part or all of their mtDNA. The stability is determined by growing the purified secondary clone in liquid YPD, spreading for single colonies and testing their genotypes. Clones with 70% stability or more (i.e., 70% of the colonies have identical genotypes) are suitable for physical mapping and DNA sequence analysis.

Fine Dissection of mtDNA

The yeast mitochondrial genes coding for cytochrome b and subunit 1 of cytochrome oxidase contain a number of introns and occupy substantial lengths of mtDNA. There are numerous mutations both in the introns and exons that result in respiratory deficiency.[5,6,20,21] Such mutants provide a convenient set of markers for fine dissection of the genes. The approach used

[20] J. Lazowska, C. Jacq, and P. P. Slonimski, *Cell* **22**, 333 (1980).
[21] D. K. Hanson, M. R. Lamb, H. R. Mahler, and P. S. Perlman, *J. Biol. Chem.* **257**, 3218 (1982).

TABLE II
GENOTYPES OF ρ-CLONES

Clone	Cytochrome b markers[a]									
	M7-40	M8-53	M13-101	M6-200	M8-181	M9-228	M33-119	M21-71	M17-162	M10-152
DS400/A12	+	+	+	+	+	+	+	+	+	+
DS400/N1	–	+	–	–	–	–	–	–	–	+
DS400/M8	–	–	–	–	–	–	+	+	+	+
DS400/M4	–	–	–	+	+	+	+	+	+	+
DS400/N9	–	–	–	–	–	+	+	+	+	+
DS400/N7	–	–	–	–	–	+	–	–	–	–
DS400/N23	–	–	–	–	+	–	–	–	–	–
DS400/N28	–	–	–	–	+	–	–	–	–	–
DS400/N22	–	–	–	+	–	–	–	–	–	–
DS400/N31	–	–	–	+	–	–	–	–	–	–
DS400/N2	+	–	–	–	–	–	–	–	–	–
DS400/M11	+	+	+	–	–	–	–	–	–	–
DS400/N24	–	–	–	–	–	–	–	–	–	–

[a] Plus and minus signs indicate retention and loss of the marker, respectively.

to physically separate different regions of the cytochrome *b* gene in *S. cerevisiae* D273-10B will be briefly described for illustrative purposes.

Ten different mutants each deficient in cytochrome *b* were first used to isolate DS400/A12, a ρ^- clone capable of restoring respiratory competency to all 10 cytochrome *b* mutants (Table II). DS400/A12 was subsequently remutagenized with ethidium bromide, and a second series of clones were isolated with different combinations of cytochrome *b* markers. Some of these were subjected to a third round of mutagenesis to yield clones with still simpler genotypes. The simplest clones obtained by this procedure had unit genome lengths of 0.16 to 0.6 kpb.[5]

A priori the degree to which a particular region of DNA can be resolved by repeated mutageneses is limited only by the distribution of available markers on the DNA. There may be other factors, however, such as the frequency with which deletions are favored by specific sequences, that will determine the occurrence of certain clones.

Physical Characterization of ρ^- Genomes

General Comments

The mitochondrial genomes of ρ^- clones are circular just as are the full length wild-type DNA. Restriction maps of such genomes can be determined by conventional procedures. Segments of several kilobases or more generally have one or more restriction sites that have been mapped on wild-type DNA.[11,22,23] Such sites provide convenient landmarks for alignment of the ρ^- mtDNA segment with the wild-type physical map. For example, the genome of DS400/A12 contains several EcoRI and HincII as well as unique BamHI, BglII, and HindIII sites that correspond to similar sites located between 71 and 78 units of the wild-type map (Fig. 2). The segment of mtDNA retained in this clone, therefore, spans minimally this region of the genome. Clones that fail to have restriction sites matching those of wild-type DNA are more difficult to localize. In this case, alignment is possible if there are other overlapping segments whose origin on the wild-type map are known.

In the following sections some procedures are described for the purification of mitochondrial DNA from ρ^- clones and some comments on their use for physical mapping of mitochondrial markers and sequencing of the

[22] P. Borst, *in* "Biological Chemistry of Organelle Formation" (T. Bücher W. Sebald, and H. Weiss, eds.), p. 27. Springer-Verlag, Berlin and New York, 1980.

[23] L. A. Grivell, A. C. Arnberg, P. H. Boer, P. Borst, J. L. Bos, E. F. J. van Bruggen, G. S. P. Groot, N. B. Hecht, L. A. M. Hensgens, G. J. B. van Ommen, and H. F. Tabak, *ICN-UCLA Symp. Mol. Cell. Biol.* **15**, 130 (1979).

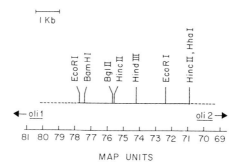

FIG. 2. Location of the EcoRI, BamHI, BglII, HincII, HhaI, and HindIII sites in the DS400/A12 genome. The DS400/A12 mtDNA has been linearized to indicate the correspondence of sites with the wild-type genome of D273-10B (see Fig. 1 of Morimoto and Rabinowitz[11]). The dashed lines represent that portion of the DNA lacking sites previously mapped on wild-type mtDNA. The orientation of the segment with respect to the wild type is provided by the *oli*1 and *oli*2 markers.

genes. The details of restriction mapping and DNA sequencing will not be described, since most of the published procedures are directly applicable to the analysis of mtDNA.

Purification of mtDNA

Reagents

Zymolyase digestion buffer: 1.2 M sorbitol, 0.05 M KPO_4, pH 7.5, 0.14 M 2-mercaptoethanol, 1 mM EDTA, 1 mg of Zymolyase 5000 per milliliter (Miles Laboratory, Inc., Indiana)

Spheroplast lysis buffer: 0.7 M sorbitol, 0.06 M Tris-HCl, pH 7.5, 1 mM EDTA, 0.1% bovine serum albumin

DNase digestion buffer: 0.9 M sorbitol, 0.01 M Tris-HCl, pH 7.5, 10 mM $MgCl_2$, 10 μg of DNase per milliliter equivalent to 20 units of activity per milliliter (Millipore Corporation, Mass.)

Sarkosyl lysis buffer: 2% Sarkosyl, 0.1 M NaCl, 0.1 M Tris-HCl, pH 7.5, 10 mM EDTA

Procedure. The following procedure can be used for the purification of mtDNA from wild-type and ρ^- mutants. The yield of DNA ranges from 30 to 50 μg from a liter of cells at stationary phase.

1. Growth of cells. A New Brunswick 14-liter fermenter cup holding 10 liters of YPD medium is inoculated with 50 ml of a stationary phase culture of yeast. The cells are grown at 30° for 17 hr with good aeration. The cells are collected by centrifugation at 2000 rpm for 10 min in an IEC RCU-5000 centrifuge. They are then suspended in 400 ml of 1.2 M sorbitol and

centrifuged at 7000 g for 10 min. The yield of cells should be 70–80 g wet weight.

2. Preparation of mitochondria. The washed cells are suspended in Zymolyase digestion buffer (10 g of cells per 30 ml of buffer) and incubated at 32° until spheroplast formation is complete. The course of the digestion is followed with a light microscope by mixing a small volume of the cell suspension with a drop of water. At least 90% of the spheroplasts should lyse in order to obtain a good yield of mitochondria. The spheroplast suspension is diluted with an equal volume of cold 1.2 M sorbitol and centrifuged at 7000 g for 10 min. The slightly turbid supernatant is discarded, and the pellet is gently resuspended in 400 ml of 1.2 M sorbitol and centrifuged for 10 min at 7000 g. The spheroplasts are highly aggregated, and it is not necessary to obtain a homogeneous suspension for this washing step.

The washed spheroplasts are suspended in cold spheroplast lysis buffer (50 ml per 10 g of starting cells), homogenized in a Waring blender for 45 sec, and centrifuged at 900 g for 10 min. The turbid supernatant is collected and centrifuged a second time to remove a small amount of remaining cell debris. The mitochondria are recovered from the second supernatant by centrifugation at 33,000 g for 15 min. The mitochondrial pellets are suspended in 100 ml of cold 0.9 M sorbitol, 0.01 M Tris-HCl, pH 7.8, 1 mM EDTA and centrifuged at 33,000 g for 15 min. The washed mitochondria are then suspended in 50 ml of DNase digestion buffer and incubated at 24° for 15 min. The buffer should be prewarmed to room temperature before use. A Potter–Elvehjem homogenizer with a loose-fitting Teflon pestle can be used to suspend the mitochondria in the digestion buffer. The digestion is stopped by the addition of 150 ml of cold 0.9 M sorbitol, 0.01 M Tris-HCl, pH 7.5, and 10 mM EDTA, and the mixture is centrifuged for 10 min at 33,000 g.

3. Extraction of mtDNA. The mitochondrial pellets from step 2 are lysed with Sarkosyl lysis buffer (5 ml/10 g of starting cells) and immediately delivered into an equal volume of water-saturated phenol. The lysis should be done as quickly as possible to minimize degradation of the DNA by the nuclease. The phenol mixture is gently stirred on ice for 10 min and centrifuged at 7000 g for 10 min. The upper aqueous phase is collected and dialyzed overnight against 2 liters of cold 10 mM Tris-HCl, pH 7.5, 5 mM EDTA.

4. Purification of mtDNA. The dialyzate is adjusted to a refractive index of 1.3970–1.3975 with solid CsCl. This requires 5.91 g of CsCl per 5 ml of solution. The solution is dispensed into 10-ml Beckman tubes with screw caps (8 ml/tube), and 10 μl of a 5 mg/ml solution of ethidium bromide and 2 μl of 15% Sarkosyl are added to each tube. The DNA is centrifuged at 36,000 rpm in a Beckman 50 Ti rotor for 36 hr at 24°. Under these condi-

tions mtDNA bands are approximately midway in the gradient and can be visualized with an ultraviolet lamp. The DNA is collected with a syringe. The tip of the needle should be placed just above the DNA band. Ultraviolet-absorbing material below the main DNA band should not be collected, since it represents some residual nuclear DNA. The pooled gradient material is extracted three times with an equal volume of isoamyl alcohol, twice with ether, and dialyzed against cold 10 mM Tris-HCl, pH 7.5, 0.1 mM EDTA overnight. At this point the DNA is ready for restriction analysis and DNA sequencing. The DNA solution can be stored up to several months at 4°. For longer storage it is recommended that it be kept frozen at $^-$80°.

Restriction Analysis

As indicated already, most ρ^- mutants have mitochondrial genomes with tandemly repeated segments of wild-type DNA. Since their genomes are circular, single restriction cuts generate a fragment with a size corresponding to the repeat unit.

Clones with long repeat units (3–10 kbp) usually have restriction sites

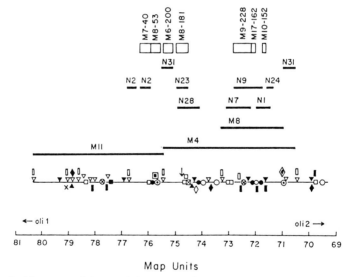

Fig. 3. Alignment of the restriction maps of the low-complexity clones with that of DS400/A12. The restriction map of the DS400/A12 mtDNA segment has been linearized to show the endpoints of the deletions. The physical limits of the smaller genomes retained in the low-complexity clones are indicated by the thick lines. The physical limits of the cytochrome *b* markers are indicated by the boxes. The symbols denote different restriction sites. The orientation of the DS400/A12 segment relative to wild-type mtDNA is provided by the *oli*1 and *oli*2 markers. Note that two of the clones (N2 and N31) have internal deletions in their genomes. Taken from Bonitz *et al.*[6]

that permit the segment to be located and oriented with respect to wild-type DNA. In most instances, particularly if the ρ^- mtDNA segment is to be used for sequencing, it is necessary to isolate less complex clones with repeat units of 1 kbp or less. The mtDNA segments of such clones are useful for constructing detailed restriction maps. For example, genomes of the various ρ^- clones with cytochrome b markers spanned different regions of the parental DS400/A12 clone. Their physical limits could be determined fairly precisely by comparative restriction mapping. Such analyses are helpful in assigning mutations to physical limits on the DNA (Fig. 3) and also in devising a sequencing strategy. In the case of the cytochrome b gene, most of the sequence was derived from the genomes of the less complex clones, although some of the sequences for which relevant clones were not available had to be obtained from preparative restriction fragments of the large DS400/A12 genome.[5]

DNA Sequencing

The protocols described by Maxam and Gilbert for end labeling, strand separation, and chemical derivatization of DNA can be applied directly for sequencing of mitochondrial DNA.[24]

1. Approximately 10 μg of mtDNA are digested with a restriction enzyme or combination of enzymes to give fragments 100–400 bp in length.
2. The mixture of fragments is labeled at the 5′ termini with $[\gamma\text{-}^{32}P]ATP$ in a T4 polynucleotide kinase-catalyzed reaction.
3. The double-stranded, labeled mixture is denatured in 100% formamide and separated into single strands on 6% polyacrylamide gels at high voltage. The single strands are cut out of the gel, eluted, and precipitated.
4. Aliquots of each single-stranded fragment are subjected to each of the four reactions for breaking of the strand at either G's, A's plus G's, C's, and T's plus C's. The reactions products are separated on 20% polyacrylamide for the beginning sequence and on 6–10% polyacrylamide for longer sequences.
5. The entire procedure is repeated with other enzymes until all the restriction sites used for labeling have been crossed.

[24] A. M. Maxam and W. Gilbert, *Proc. Natl. Acad. Sci. U.S.A.* **74**, 560 (1977).

[37] Genetics and Biogenesis of Cytochrome *b*

By Philip S. Perlman and Henry R. Mahler

Respiration-deficit (RD) mutants of Baker's yeast, *Saccharomyces cerevisiae,* were first reported in 1949[1], and since then large numbers of such mutants have been described,[2-4] especially since 1975.[e.g., 5-8] The mitochondrial system for oxidative phosphorylation is comprised of proteins specified by both nuclear and mitochondrial genes, and mutants located in both of these genomes are available. This unicellular organism is well suited for biochemical studies of respiration and oxidative phosphorylation. It is particularly suited for genetic dissections of the component reactions, because RD mutants are viable if cells are maintained on media containing a fermentable carbon and energy source, such as glucose. Mutants affecting individual segments of the system as well as pleiotropic mutants affecting more than one such portion exist, but in only a few instances have they been used to gain insight into the mechanism of electron transport.[e.g., 9-12] In this chapter we survey the types of mutants available for the study of cytochrome *b* (cyt *b*) function and their manner of isolation and manipulation. Both nuclear and mitochondrial mutants are described, but most of our attention is paid to the latter, since methods of mitochondrial genetic analysis are somewhat difficult to retrieve from the primary literature. To a large extent the principles described here can be applied to the dissection of the other parts of the respiratory chain and the mitochondrial ATPase complex with minor modifications.

[1] B. Ephrussi, H. Hottinguer, and A. M. Chimenes, *Ann. Inst. Pasteur Paris* **76,** 351 (1949).

[2] F. Sherman, *Genetics* **48,** 375 (1963).

[3] F. Sherman and P. P. Slonimski, *Biochim. Biophys. Acta* **90,** 1 (1964).

[4] A. Tzagoloff, A. Akai, and R. B. Needleman, *J. Bacteriol.* **122,** 826 (1975).

[5] F. Foury and A. Tzagoloff, *Mol. Gen. Genet.* **149,** 43 (1976).

[6] G. Coruzzi, M. K. Trembath, and A. Tzagoloff, this series, Vol. 56, p. 95.

[7] J. R. Mattoon, J. C. Beck, E. Carvagal, and D. R. Malamud, this series, Vol. 56, p. 117.

[8] B. Dujon, *in* "Molecular Biology of the Yeast Saccharomyces: Life Cycle and Inheritance" (J. Strathern, E. W. Jones, and J. R. Broach, eds.), p. 505. Cold Spring Harbor Lab., Cold Spring Harbor, New York, 1981.

[9] F. Cabral and G. Schatz, *J. Biol. Chem.* **253,** 4396 (1978).

[10] D. Meunier-Lemesle, P. Chevillotte-Brivet, and P. Pajot, *Eur. J. Biochem.* **111,** 151 (1980).

[11] P. Chevillotte-Brivet and D. Meunier-Lemesle, *Eur. J. Biochem.* **111,** 161 (1980).

[12] M. Briquet and A. Goffeau, *Eur. J. Biochem.* **117,** 333 (1981); J. Subik, M. Briquet, and A. Goffeau, *Eur. J. Biochem.* **119,** 613 (1981).

METHODS IN ENZYMOLOGY, VOL. 97

Types of Mutants

Since complex III (the b-c_1 complex, or ubiquinol:cytochrome c reductase) contains at least seven proteins, all but one of which is encoded in the nuclear genome,[13-15] it is possible to obtain mutants deficient in this segment located in either nuclear or mitochondrial DNA. Nuclear respiration-deficient mutants specifically lacking this activity have been available since the early 1960s.[3] Such mutants, given the generic name *pet,* for their Mendelian petite-like phenotype, are readily obtained by screening for mutants that can grow on glucose-containing media but not on media containing nonfermentable carbon and energy sources, such as glycerol or ethanol.[e.g., 16] On media containing glycerol plus a low concentration of glucose (0.1%) all classes of RD mutants yield small colonies. Simple genetic tests, outlined below, distinguish nuclear from cytoplasmic RD mutants.

The best-defined gene affecting complex III is the mitochondrial *cob* gene encoding the cytochrome b apoprotein of apparent $M_r = 30,000$ on polyacrylamide gels; its gene codes for 385 amino acid residues corresponding to a molecular weight of 44,000. The structure of a common form of the *cob* gene found on the mitochondrial genome of many standard laboratory strains is shown in Fig. 1.[17-20] The gene is ~ 7200 bp long and consists of six exons and five introns, three of which contain long open reading frames. The promoter of the gene has not yet been identified, but it is apparently rather distant from the first codon of exon 1, since the mRNA has a long untranslated 5′ leader; it also has a long 3′ untranslated segment.[21] In some strains the structure of the gene is somewhat different, in that the first four exons are fused because of the absence of the first three introns shown in Fig. 1; the remainder of the gene is essentially the same.[17] Other combinations of the five introns have been constructed by genetic means, and so far all forms of the gene are compatible with cyt b synthesis.[22,23]

[13] M. B. Katan, C. Pool, and G. S. P. Groot, *Eur. J. Biochem.* **65**, 95 (1976).

[14] M. B. Katan, N. van Harten-Loosbroek, and G. S. P. Groot, *Eur. J. Biochem.* **70**, 409 (1976).

[15] L. F. H. Lin, L. Clejan, and D. S. Beattie, *Eur. J. Biochem.* **87**, 171 (1978).

[16] A. Tzagoloff, A. Akai, and R. B. Needleman, *J. Biol. Chem.* **250**, 8228 (1975).

[17] F. G. Nobrega and A. Tzagoloff, *J. Biol. Chem.* **255**, 9828 (1980).

[18] J. Lazowska, C. Jacq, and P. P. Slonimski, *Cell* **22**, 333 (1980).

[19] J. Lazowska, C. Jacq, and P. P. Slonimski, *Cell* **27**, 12 (1981).

[20] P. Q. Anziano and P. S. Perlman, unpublished observations.

[21] S. G. Bonitz, G. Homison, B. E. Thalenfeld, A. Tzagoloff, and F. G. Nobrega, *J. Biol. Chem.* **257**, 6268 (1982).

[22] P. S. Perlman, H. R. Mahler, S. Dhawale, D. Hanson, and N. J. Alexander, *in* "The Organization and Expression of the Mitochondrial Genome" (A. M. Kroon and C. Saccone, eds.), p. 161. Elsevier/North-Holland Biomedical Press, Amsterdam, 1980.

[23] C. Jacq, P. Pajot, J. Lazowska, G. Dujardin, M. Claisse, O. Groudinsky, H. de la Salle, C.

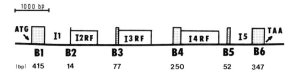

FIG. 1. Schematic diagram of the mitochondrial *cob* gene. The longest form of the mitochondrial *cob* gene consists of six exons (stippled) and five introns. Three of the introns contain a long reading frame in phase with that of the preceding exon, and each intron reading frame ends before the next exon begins.

Four main classes of mutants in the cyt *b* (or *cob*) gene have been described. Most attention has been directed toward the analysis of exon and intron mutants that lack the apoprotein. Intron mutants accumulate partially processed transcripts of the gene that are translated to yield one or more proteins encoded by both exon and intron sequences. Most exon mutants appear to have nonsense or frame-shift mutations that lead to the accumulation of nonfunctional fragments of cyt *b*; depending on the position of the mutation, fragments as small as 10 kd or as large as 29 kd have been detected in mutants all of which lack cyt *b*-associated activities (reviewed by Perlman *et al.*[24]).

Two other classes of exon mutants have been shown to accumulate full-length cyt *b* apoprotein. Most missense mutants exhibit this property and lack spectroscopically detectable cyt *b*,[e.g.,24] but a few have a *b*-like cytochrome with an altered absorption spectrum, and some exhibit partial activity.[10,25] In principle, it should be possible to isolate temperature-sensitive mutant forms of cyt *b* (especially as partial revertants of such chain termination mutants, see below), but so far none has been reported. Several groups have characterized cyt *b* mutants that confer resistance to one or more of the *b*-specific inhibitors antimycin A, diuron, mucidin, and funiculosin (reviewed by Dujon[8]). These mutants are located in more than one exon of the gene, and so it appears that the drug binding sites may be defined by arrays of nonadjacent residues of the protein defined by its conformation in the membrane.

Since the initial discovery of point mutants in the cyt *b* gene in 1975–1976, more than 200 mutants have been reported, and relatively

Grandchamp, M. Labouesse, A. Gargouri, B. Guiard, A. Spyridakis, M. Dreyfus, and P. P. Slonimski, *in* "Mitochondrial Genes" (P. Slonimski, P. Borst, and G. Attardi, eds.), p. 155. Cold Spring Harbor Lab., Cold Spring Harbor, New York, 1982.

[24] P. S. Perlman, N. J. Alexander, D. K. Hanson, and H. R. Mahler, *in* "Gene Structure and Function" (D. Dean, L. Johnson, P. Kimball, and P. Perlman, eds.), p. 211. Ohio State Univ. Press, Columbus, 1980.

[25] A. Haid, R. J. Schweyen, H. Bechmann, F. Kaudewitz, M. Solioz, and G. Schatz, *Eur. J. Biochem.* **94**, 451 (1979).

simple methods for isolating more such mutants have been described. The majority of mutants are nonsense or frame-shift mutants in exons, and the next most frequent class consists of intron mutants. Because most of the investigators studying this gene have been interested primarily in its molecular biology, little attention has been paid to the smaller number of missense and drug-resistant mutants that should be the most valuable tools for studying electron transport processes in complex III (but see references cited in footnotes 10–12). In order to make the use of such mutants more feasible for biochemists, most of this chapter provides a survey of genetic methods that should permit interested investigators to obtain the types of mutants that would be most useful for studies of cyt *b* function. Although this chapter is primarily about cyt *b* itself, some attention is paid to methods for manipulating the nuclear genes encoding the remainder of complex III, since to a large extent it is the interactions of the nuclear and mitochondrial gene products that would be central to the study of complex III function.

Mutant Isolation Procedures

Nuclear Mutants

Pet mutants can be obtained by treating a wild-type strain with a mutagen (e.g., UV light or ethyl methane sulfonate) and screening the surviving viable cells for their ability to grow fermentatively on minimal glucose-containing medium but their inability to grow on rich media containing glycerol or ethanol. This simple test excludes mutants that are defective in processes that are required for both fermentative and oxidative growth. This set of mutants will include both nuclear and cytoplasmic mutants, since no mutagen is absolutely specific for one genome over the other. In addition, mutants defective in other aspects of electron transport or oxidative phosphorylation will be present, so that it is necessary to carry out some simple genetic tests followed by some equally simple biochemical assays in order to obtain mutants affecting only complex III. For simplicity, we refer to all genetically undefined RD mutants as gly$^-$ (unable to grow on glycerol medium) and to their parent strains, revertants, and wild-type recombinants from crosses, as gly$^+$.

Most *pet* mutants are recessive so that a cross with a wild-type (*PET*) tester strain (of opposite mating type) will identify all recessive nuclear mutants regardless of their exact phenotypic defect. Details of how these test crosses can be carried out very quickly and in large numbers can be found in the primary literature.[e.g.,26] Briefly, gly$^-$ colonies are transferred to a glucose

[26] D. Coen, P. Netter, E. Petrochilo, and P. P. Slonimski, *Symp. Soc. Exp. Biol.* **24**, 449 (1970).

plate in the form of a grid, so that a fairly large number of mutants can be tested together on each plate (Fig. 2). Matings are carried out by replica-plating that grid onto a minimal medium glucose plate that has been seeded with a lawn of roughly 10^7 cells of the tester strain. During several days of incubation the mixed cells at each position of the grid mate, and the resulting diploid cells will grow to form areas of heavy growth on the plate. It is useful but not imperative that the parent of the mutant collection and the tester strain be chosen with some care so that each has at least one auxotrophic requirement (e.g., an amino acid, a purine or pyrimidine base) that assures that neither parent can grow on the mating plate; if the auxotrophies of the two parents are different, then the diploids will have no growth factor requirements and they will grow. In practice we often carry out crosses on rich, glucose-containing media and assay the results by replica-plating onto rich glycerol medium; in such cases it is not essential that the parental strains have mutally complementary auxotrophies. It helps to be certain that the

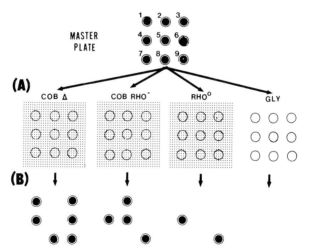

FIG. 2. Test crosses for characterizing respiration-deficient (RD) mutants. In this example nine presumptive RD mutants were placed in a 3×3 matrix (grid) on rich glucose solid medium. After the cells have grown for 24 hr, they are replica-plated onto glycerol medium and glucose media seeded with lawns of the three tester strains shown. After 3 days of incubation, each mating plate in (A) is replica-plated to a glycerol plate that is then incubated for several days. Filled circles in (B) denote growth. These results show that all nine strains are gly−; strains 4 and 9 appear to be recessive nuclear mutants, since they yield gly+ progeny in all three test crosses (especially the *rho-o* cross); strains 2 and 5 appear to be mitochondrial *cob* mutants, since they yield gly+ recombinants in the cross with the petite that retains the wild-type form of the *cob* gene; strains 1, 3, 6, and 8 appear to be mitochondrial mutants in genes other than *cob;* and strain 7 appears to be a petite, since it fails to yield gly+ progeny in all the test crosses. In practice these tests are often done using master plates having 100–200 colonies derived from single cells or in grids larger than those shown here.

two main parental strains selected for use will mate well with each other and that the resulting diploids can yield a high proportion of viable four-spored asci. For this application, the tester strain must be a cytoplasmic petite derivative that lacks the entire mitochondrial genome, known as a *rho-o* strain; thus, two RD strains are mated in each cross, and the progeny are tested for their ability to grow on glycerol medium by replica-plating from the mating plate onto a glycerol plate (Fig. 2). After several days of incubation, each mating is scored for the presence of gly⁺ progeny. If some are present, then the mutant is most likely a recessive nuclear one; if none are present, then the mutant is probably mitochondrial.

Any remaining ambiguity can be resolved by carrying out meiotic analysis of the gly⁺ diploids from each cross or by carrying out several additional test crosses. In practice several test crosses are usually done at the same time. Where no gly⁺ diploids are obtained, the mutant is either a dominant nuclear mutant or else is mitochondrial. If each mutant is mated with the gly⁺ parent of the petite tester strain and the diploids are still gly⁻, then the mutant is most likely a dominant nuclear one.

Meiotic analysis can be carried out only on diploids that are gly⁺ because functional mitochondria are required for meiosis. This process is induced by starving cells for nitrogen on a medium containing acetate, and its products are analyzed by dissecting the tetrads and testing the clones obtained by germinating the ascospores for their ability to grow on glycerol. In the simplest case, where the RD phenotype is due to a single recessive nuclear mutation, the vast majority of tetrads will yield two gly⁺ and two gly⁻ ascospores. Deviation from that result would indicate that the original mutant strain harbors more than one mutation. There is nothing special about the analysis of *pet* mutants in this regard.

Hypothetical dominant nuclear RD mutants cannot be characterized by meiotic analysis, but they can be distinguished from mitochondrial mutants quite easily. When a mitochondrial RD mutant is mated to the *rho-o* tester strain, all the diploid progeny are mutant; however, when mated to the wild-type parent of the *rho-o,* the two types of mitochondrial genomes in the zygotes will segregate mitotically so that some of the progeny will be mutant and others will be wild-type. There is one class of cytoplasmic mutation that can be very difficult to distinguish from a dominant nuclear one: hypersuppressive petites.[27,28] These mutants are phenotypically identical to any other petite, but when mated to a wild-type strain they yield essentially all mutant progeny. Since these two types of mutants are relatively rare they will not be discussed further; they were mentioned only to caution that not all results of test crosses are completely straightforward.

[27] R. Goursot, M. de Zamaroczy, and G. Bernardi, *Curr. Genet.* **1,** 173 (1980).
[28] H. Blanc and B. Dujon, *Proc. Natl. Acad. Sci. U.S.A.* **77,** 3942 (1980).

A rapid and simple spectroscopic method was developed for screening mutants for cytochrome defects[3] and, with patience, is suitable for identifying the subset of *pet* mutants having a specific defect in cyt *b*. One problem with the class of *pet* mutants is that complex III activity can be lost for any one of several reasons, some of which are useful for understanding the biogenesis of cyt *b* whereas others are useful for understanding the composition and function of complex III. For example, mutants in the heme or ubiquinone biosynthetic pathways are of considerable intrinsic interest[7,16,29] and may affect the regulation of expression of the gene encoding components of complex III or its assembly,[30-33] but they are not particularly valuable for studying the reactions carried out by complex III, for which purpose mutants have a partially functional complex III would be more relevant.[cf. 10,11]

What is needed is a rapid method for identifying the subset of nuclear RD mutants that affect complex III function. In particular, one would like to be able to detect those mutants that are in the genes that encode subunits of complex III. That goal has not yet been met, even though very large numbers of potentially useful mutants have been available for years; presumably the lack of progress has been due to the understandable reluctance of investigators who are chiefly biochemists to invest a large amount of effort in screening large numbers of mutants biochemically to find a very small set of interesting strains. The recent development of very powerful recombinant DNA methods appears to have provided a new and more attractive means of identifying these genes. Given a recombinant DNA molecule that contains the wild-type allele of a gene for a protein of complex III, it should be quite easy to screen large numbers of otherwise unselected mutants for those that are complemented by the wild-type gene. This approach avoids the need for biochemical measurements on large numbers of mutants, but forces the experimenter first to obtain clones of the genes in question by relatively brute-force recombinant DNA methods. But since the proteins of complex III are relatively abundant their cloning should be relatively simple, and once cloned it would be possible to obtain large numbers of point mutations in them quite efficiently. Once a single mutant has been identified in this way, it can be used more readily than the recombinant plasmid to identify additional mutants in that gene.

[29] J. M. Haslam and A. M. Astin, this series, Vol. 56, p. 558.
[30] E. Ross and G. Schatz, *J. Biol. Chem.* **251**, 1997 (1976).
[31] G. G. Brown and D. S. Beattie, *Biochemistry* **16**, 4449 (1977).
[32] J. Saltzgaber-Muller and G. Schatz, *J. Biol. Chem.* **253**, 305 (1978).
[33] L. Clejan, D. S. Beattie, E. G. Gollub, K. P. Lin, and D. B. Sprinson, *J. Biol. Chem.* **255**, 1312 (1980).

Mitochondrial Mutants

Several different approaches are used to obtain mitochondrial cyt b deficient mutants. Most begin with the use of $MnCl_2$ as the mutagen, since that agent has been shown to be relatively selective for mtDNA. Details of this mutagenesis have been reported in a series of papers from Putrament's laboratory.[34,35] However, since different strains have been found to respond to this agent somewhat differently, it is important to calibrate the dose response for each strain. We routinely establish an effective dosage by measuring the frequency of induction of chloramphenicol- or erythromycin-resistant mitochondrial mutants, since these mutations can be detected by direct plating on glycerol medium supplemented with an inhibitory concentration of the drug.[35] We have found that a dose that increases the frequency of drug-resistant mutants by 10- to 50-fold will yield an acceptable frequency of RD mutants.

Once the dosage of mutagen has been established, mutants are obtained by treating cells with $MnCl_2$ in rich medium, washing the cells by centrifugation, incubating them for several generations in fresh rich medium, and plating them on glucose-containing medium. Several generations of growth after mutagenesis are needed to permit the segregation of mutant from wild-type mitochondrial genomes; unlike the case where nuclear mutants are induced in a haploid strain with only one copy of each gene, mitochondrial genes are present in many copies per cell, and it is unlikely that the mutagenic treatment would convert all copies of a gene into the same mutant form. Fortunately, cells containing mitochondrial genomes of several genotypes are known to segregate quite rapidly so that cells with exclusively mutant genomes are present after only a few divisions in nonselective medium.[cf. 36] All viable cells in the mutagenized culture should grow on glucose medium and respiration-deficient ones are readily detected by replica-plating each colony onto glycerol medium.

With most parental strains the majority of survivors of the $MnCl_2$ treatment are wild type; most of the RD cells are petite mutants, and so it is necessary to examine thousands of colonies in order to find a few mitochondrial point mutants. By replica-plating onto glycerol medium, all the RD mutants are identified (as gly⁻), and they are picked from the original plates and arranged in grids on glucose plates. Petites are distinguished from point mutants (and nuclear mutants) by carrying out crosses with a *rho-o* tester and a previously isolated mitochondrial mutant. Virtually all nuclear mutants will yield wild-type diploid progeny in both crosses. Petites will yield

[34] A. Putrament, H. Baranowska, and W. Prazmo, *Mol. Gen. Genet.* **126**, 357 (1973).
[35] A. Putrament, H. Baranowska, A. Ejchart, and W. Prazmo, *Methods Cell Biol.* **20**, 25 (1978).
[36] P. S. Perlman, C. W. Birky, Jr., and R. L. Strausberg, this series, Vol 56, p. 139.

no wild-type progeny in the *rho-o* cross and in most cases also in the cross with the mutant. The desired mitochondrial point mutants will yield only mutant progeny in the cross with the petite but will yield wild-type progeny in the cross with the mutant except in the unlikely event that the same mutation present in the test strain was reisolated; thus, all mitochondrial point mutants in the sample tested can be detected by a pair of simple crosses.

It is now necessary to distinguish mutants in the *cob* gene from those in other genes; this can be done by a second set of test crosses using a *rho-o* and another petite that retains the entire *cob* gene but is deleted for the rest of the mitochondrial genome (Fig. 2). All point mutants and most deletion mutants in the *cob* gene will yield some wild-type progeny in the latter cross, but none in the former. If new mutants in a particular region of the gene are desired, then they can be identified and others avoided by simply using a petite deletion mutant that retains only that portion of the gene as the key tester. By now petite testers have been described by several laboratories that divide the *cob* gene into many small segments, and such strains can often be obtained from culture collections.[e.g., 17,18,24,37]

Streamlined Screening Methods

The above steps are not difficult but can be somewhat tedious. Therefore, several groups have devised procedures that streamline the process of mutant screening. For example, Kotylak and Slonimski[38] used the *op1* nuclear mutation for this purpose. *Op1* is a recessive mutation that blocks mitochondrial ADP/ATP transport, but not respiration, and so has a gly⁻ phenotype; derivatives of that strain that are respiration deficient owing to a nuclear or mitochondrial point mutation are viable on glucose medium, but petite derivatives are inviable.[39–41] Therefore, using a *rho⁺ op1* strain as the parent, there will be no petite mutants among the mutants obtained. However, since the parent strain is itself gly⁻, newly induced mutants cannot be detected by the glycerol growth test; instead, the *rho-o* test cross is used to make this distinction. Since the parent strain is incapable of growth on glycerol because of a recessive nuclear mutation, but is otherwise *PET* and *rho⁺*, it yields gly⁺ progeny in that test cross; similarly, newly induced

[37] G. Grosch, C. Schmelzer, and S. Mathews, *Curr. Genet.* **3**, 65 (1981).
[38] Z. Kotylak and P. P. Slonimski, *in* "Mitochondria 1977: Genetics and Biogenesis of Mitochondria" (W. Bandlow, R. J. Schweyen, K. Wolf, and F. Kaudewitz, eds.), p. 83. de Gruyter, Berlin, 1977.
[39] L. Kovac, T. M. Lachowicz, and P. P. Slonimski, *Science* **158**, 1564 (1967).
[40] V. Kovacova, J. Irmlerova, and L. Kovac, *Biochim. Biophys. Acta* **162**, 157 (1968).
[41] J. Kolarov, J. Subik, and E. Racker, *Biochim. Biophys. Acta* **256**, 55 (1982).

nuclear RD mutations will not affect the outcome of that test. However, all strains having a mitochondrial point mutation will yield only mutant progeny in that cross but will yield wild-type progeny in the other test cross noted above. It can be useful to have a collection of mutants derived in this way, because the mutants of *op1* strains tend to have very low reversion frequencies. However, for biochemical purposes, it may be inconvenient to study a set of strains that are both respiration deficient and defective in ADP/ATP transport.

In our group, we have used several different strategies to simplify the process of mutant isolation. First, we have used a parental strain that is auxotrophic for adenine owing to a mutation in the *ade1* or *ade2* gene. Such strains, but not ones with mutations in other *ade* loci, accumulate a red pigment so that colonies on media containing 1 – 2% glucose and adequate but not excessive exogenous adenine are bright red. On that medium the pigment accumulates only if the cells are PET^+ and rho^+; petite mutants have essentially no pigment and yield white colonies, while *pet* and *mit⁻* mutants yield tan colonies, apparently because they have some pigment. We have found that heat-sensitive mutants of this strain usually yield pink colonies even though they can grow on glycerol at the temperature used in this test (usually 30°). Therefore, presumptive respiration-deficient mutants can be identified by visual examination of the initial set of petri dishes obtained by plating the mutagenized culture. In a typical experiment there is about 1 tan colony per 400 colonies examined. These are transferred to grids, and test crosses are carried out on this small subset of the total cells. Media with high concentrations of glucose (e.g., 10%) should be avoided since even petite mutants yield red colonies then.

Even with this visual screening device we have found it useful to improve the efficiency of isolating mutants further. Since most of the colonies are wild-type and only a few presumptive mutants can be obtained per petri dish, hundreds of dishes must be used in the initial steps in order to ensure that a large number of mutants will be obtained. We have found several ways of killing the wild-type cells selectively, so that chiefly RD cells will yield colonies. In this way the number of mutants per dish can be increased greatly. Triphenyltetrazolium chloride (TTC) is reduced by the electron transport chain to form an insoluble formazan that is toxic. Since RD mutants cannot reduce TTC, they are quite resistant to this toxic effect. Therefore, if the mutagenized culture is incubated with TTC before plating, the frequency of useful mutants per plate can be increased.[42] The drug adriamycin is also toxic to wild-type cells (for a reason that is unknown)[43]

[42] H. P. Zassenhaus and P. S. Perlman, *Curr. Genet.* **6**, 179 (1982).
[43] S. C. Hixon, A. Ocak, J. E. Thomas, and J. P. Daugherty, *Antimicrob. Agents Chemother.* **17**, 443 (1980).

and can be used in place of TTC.[44] We have used the colony color character together with the TTC or adriamycin suicide of unwanted wild-type survivors to isolate mutants in many regions of the mitochondrial genome, including the *cob* gene.

All the groups in the field have noted that the vast majority of mutants are in the *oxi3* gene and that only a few percent of otherwise unselected mutants are in the *cob* gene. It is useful, therefore, to carry out a mutant isolation experiment in such a way that little effort is invested in the majority of mutants and that the desired subset of mutants can be identified very early in the screening process. After a successful suicide treatment of mutagenized cells, one can prepare master plates containing several thousand RD colonies each, most of which are petites and screen those nearly confluent plates for the small subset of, say, *cob* mutants by replica-plating those colonies onto a lawn of a petite tester that can restore to a gly⁺ phenotype only the desired mutants.[42] So long as a copy of the master plate is made on another glucose plate, colonies yielding a positive result can be located at least roughly. Positive regions of the master plate are suspended in sterile water and streaked on glucose medium, and tan colonies are selected for retesting. With some perseverance the only limit to the number of mutants that can be obtained is the number of manganese mutable sites, and other mutagens can be used to lessen that limitation.

Drug-resistant mutants are isolated by direct selection for growth on glycerol medium containing sufficient drug to block growth of the sensitive parent strain. Nuclear mutations that can afford resistance exist and they must be distinguished from mitochondrial ones by appropriate crosses. Methods for isolating drug-resistant mutants in the *cob* gene are well documented in the literature and will not be repeated here.[cf. 45-50]

Mapping and Complementation Methods

For most biochemical applications, detailed mapping of nuclear mutants is not needed. However, it is important to avoid carrying out extensive studies of separate isolates of the same mutant and to discern whether different mutations in the same gene have been obtained. For recessive

[44] M. A. Nicholas, Ph.D. Thesis, Ohio State Univ., Columbus, 1983.
[45] G. Burger, B. Lang, W. Bandlow, R. J. Schweyen, B. Backhaus, and F. Kaudewitz, *Biochem. Biophys. Res. Commun.* **72**, 1201 (1976).
[46] D. J. Groot-Obbink, T. W. Spithill, R. J. Maxwell, and A. W. Linnane, *Mol. Gen. Genet.* **151**, 127 (1977).
[47] A. M. Colson and P. P. Slonimski, *Mol. Gen. Genet.* **167**, 287 (1979).
[48] E. Pratje and G. Michaelis, *Mol. Gen. Genet.* **152**, 167 (1977).
[49] G. Michaelis, *Mol. Gen. Genet.* **146**, 133 (1976).
[50] J. Subik, V. Kovacova, and G. Takacsova, *Eur. J. Biochem.* **73**, 275 (1975).

mutants—which should comprise the vast majority of RD mutants—a complementation test usually suffices to show that two mutants are in the same or different genes. To do this, ascospore clones of opposite mating type must be available so that pairwise crosses of mutants can be carried out. Such derivatives of each mutant isolate are routinely obtained from the tetrads dissected in the experiments (see above) aimed at testing whether each mutant phenotype is due to just one mutation. These complementation tests can be quite time consuming in the early stages of building a collection of mutants; however, once an interesting gene has been defined by as little as one mutation, additional mutations in that gene can be obtained from a large collection of candidates very simply. Once that key mutant has been obtained as a derivative of opposite mating type, it can be used to screen all other unscored mutants to identify the subset that fails to complement it.

At this point we encounter a problem that is special for RD mutants. Usually when two nuclear mutants, say two adenine-requiring ones, fail to complement, meiotic analysis is used to determine whether the mutants are identical; if they are, then no ADE recombinants will be obtained among a large number of ascospores. However, since mitochondrial function is required for meiosis and since that requirement cannot be circumvented by changing the sporulation medium, RD diploids will not sporulate.

For mitochondrial mutants, it is relatively easy to obtain an accurate genetic and physical map of a set of mutants. Mapping experiments using three-point crosses, however, are usually not valuable for mitochondrial mutants; instead, a form of deletion mapping is used.[51-53] Such experiments require the availability of a set of petite deletions (of opposite mating type from that of the point mutants), each of which retains a particular portion of the gene. The principle of deletion mapping in this system is illustrated in Fig. 3. There, four partially overlapping petite genomes are shown, each of which yields wild-type recombinant progeny when mated with one or more of the nine point mutants shown. In this way an unambiguous map order is obtained. That map can be oriented with flanking genes in either of two ways: (a) by using additional petites retaining larger portions of the genome so that marker a can be linked with a marker on one side of cob, and marker i with a mutation on the other side; or (b) by aligning the restriction site maps of all four petite genomes with that of the wild-type genome.[cf. 54] It should be noted that this type of deletion mapping is somewhat different from deletion

[51] P. Molloy, A. W. Linnane, and H. B. Lukins, *J. Bacteriol.* **122**, 7 (1975).
[52] R. J. Schweyen, U. Steyrer, F. Kaudewitz, B. Dujon, and P. P. Slonimski, *Mol. Gen. Genet.* **146**, 117 (1976).
[53] N. W. Gillham, "Organelle Heredity." Raven, New York, 1978.
[54] P. Borst, J. P. M. Sanders, and C. Heyting, this series, Vol. 56, p. 182.

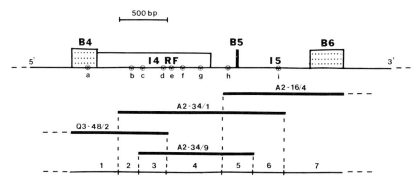

Fig. 3. Petite deletion mapping of *cob* mutants. Each of nine mutants (a through i) was mapped to a short segment of the *cob* gene using four petite mutants. The precise position of each mutation (PZ5, S3, PZ25, M8-219, PZ16, PZ54, A103, G2590, and M6-200 for a to i, respectively) based on direct DNA sequencing is shown by a circled star. The portion of the *cob* gene retained by each petitie mutant, based on restriction-site mapping, is shown by a thick line, and the seven segments of the *cob* gene defined by those four deletion mutations are shown at the bottom of the figure. For the original data, see references cited in footnotes 64, 68, 69, and 78.

mapping in most other systems, because here the deletions are very large and the retained regions are quite small; the presence of gly+ recombinants among the progeny shows that the point mutation is located within the segment of the wild-type genome present in the petite. In many cases simple spot tests are sufficient to make map assignments; for example, if two petites recombine with a given mutant but differ in their ability to recombine with flanking mutations, then the mutant is assigned to the portion of the wild-type genome shared by the two petites. Occasionally petite mutants are found that do not retain a single, continuous segment of the wild-type genome[e.g.,54] ; the use of such multiply deleted strains as testers can yield an ambiguous or contradictory map. Fortunately, such rearranged petites are relatively rare and can be detected easily by restriction-site mapping experiments. With some patience, several groups have been able to divide the *cob* gene into at least 39 physically defined segments, some as small as 100 bp, each of which contains at least one mutation. In the example shown in Fig. 3, the four petites divided the 3′ half of the *cob* gene into seven segments. Similar resolution has been obtained with several other regions of the mitochondrial genome.

Clearly, when two mutants can be distinguished by deletion mapping, they are not identical. When several mutants are assigned to the same segment of the gene—e.g., the overlap between petites A2-16/4 and A2-34/1 in Fig. 3—they may be tested for identity by mating them in pairs and scoring the resulting diploid progeny for gly+ recombinants. Recombination is very efficient in yeast mitochondria, so that this test is a very sensitive one, capable of distinguishing point mutations that are in very close proximity.

To do so it is necessary to prepare a derivative of each mutant having the opposite mating type and suitable auxotrophic markers. In the early studies of mitochondrial point mutants, such derivatives were obtained by mating each mutant with a wild-type strain of opposite mating type followed immediately by meiosis.[55] Since the cytoplasmic genotype of those freshly mated zygotes is mixed, some spores contain mutant and others wild-type mtDNA; some of the ascospore clones retain the original mating type, while others exhibit the desired combination of mutant mitochondria and opposite mating type. Once this nuclear substitution has been effected, pairs of mutants can be mated and their progeny scored for wild-type recombinants.

That approach is quite tedious and has been supplanted since around 1978 by a simpler protocol that takes advantage of the *kar1* mutation.[56,57] Conde and Fink isolated this nuclear mutant and showed that it substantially blocks nuclear fusion in zygotes; it permits mating and the mixing of the cytoplasms from the two parent strains, but buds formed from the zygotes are primarily haploid, retaining, in the vast majority of cases, the entire nuclear genome of one parent or the other.[58,59] The karyogamy deficiency is obtained even if only one of the mating partners has the *kar1* mutation. As outlined in Fig. 4, this mutation permits the substitution of one nuclear genome by another (and hence one mating type and associated auxotrophic markers for another).

For the specific case of obtaining an α derivative of a particular *cob* mutant of mating type a (cf. Fig. 4), the a *cob* strain is mated with a *rho-o* derivative of an α *kar1* strain and α haploids are screened to identify those with the mitochondrial genome that originated in the a parent. Since the α parent of the cross lacks all mitochondrial sequences, all α progeny that are restored to a gly$^+$ phenotype by an appropriate test cross constitute the desired derivative. Since both α *kar1* and a *kar1* strains exist, it is possible to effect nuclear substitutions in both directions. α *kar1* strains with marked mitochondrial genomes can be used in test crosses because the *kar1* deficiency does not block interactions between organelle genomes; also, the block of nuclear fusion is not absolute, so that some diploids result from such crosses. The availability of *kar1* mutants has added a new dimension to

[55] P. Slonimski and A. Tzagoloff, *Eur. J. Biochem.* **61**, 27 (1976).
[56] P. Nagley and A. W. Linnane, *Biochem. Biophys. Res. Commun.* **85**, 585 (1978).
[57] W. E. Lancashire and J. R. Mattoon, *Mol. Gen. Genet.* **170**, 333 (1979).
[58] J. Conde and G. Fink, *Proc. Natl. Acad. Sci. U.S.A.* **73**, 1088 (1976).
[59] S. K. Dutcher, *Mol. Cell. Biol.* **1**, 245 (1981).

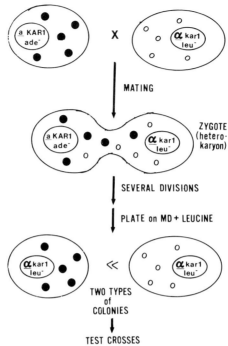

FIG. 4. Nuclear substitution using the *kar1* mutation. Haploid parent strains of nuclear genotype *a KAR ade LEU* and *α kar1 ADE leu* were mated. Filled circles denote mitochondria with *cob⁻* mtDNA, and unfilled circles denote mitochondria having no mtDNA (*rho-o*). Because of the *kar1* mutation, the haploid nuclei fail to fuse in most zygotes, so that the products of mating are nuclear heterokaryons that are also heteroplasmic. The zygotes yield haploid progeny of either nuclear genotype and of either mitochondrial genotype. When cultures containing zygotes, zygotic progeny, and unmated cells are plated on minimal glucose medium (MD) supplemented with leucine, both *cob⁻* and *rho-o* leucine-requiring colonies are obtained; because mating is relatively inefficient and because no effort was made to eliminate unmated *α kar1* parental cells before plating, most colonies obtained at this step are *rho-o*. The desired *α kar1 ADE leu cob⁻* derivatives are identified by a test cross to a *cob⁺* petite strain (of the *a* mating type). Since diploids can grow on the MD + leucine medium, it is necessary to replica-plate the tested cells onto MD medium (lacking leucine) to identify diploids. In this particular example, diploids are gly⁻ and will yield no gly⁺ progeny in the test cross and so cannot be mistaken for the desired strain.

mitochondrial genetic analysis, and several other applications of this useful mutation will be described below.

In general, complementation data are not required for most uses of mitochondrial *cob* mutants. Because a stable heterozygous state cannot be maintained for mitochondrial markers, complementation cannot be obtained in the manner used for nuclear genes. In 1978–1979, several groups reported a complementation assay for mitochondrial RD mutants.[60-62] Mutants of opposite mating type are mated for varying lengths of time, and the mating mixtures are assayed for the rate of oxygen uptake per diploid cell. Since both parents are totally respiration deficient, there is no respiration initially or after very short times of mating; if the mated cells acquire the ability to respire, then that would reflect some mutual interaction between the two mutant organelles, or complementation, provided that recombination can be either excluded or corrected for.

As illustrated in Fig. 5, when two exon mutants in the *cob* gene are mated, no respiration rate is detected for at least 10 hr after mixing. Effective mating is completed by 3–5 hr of incubation, and by 10 hr the mated cells have divided several times. However, when mutants in different genes are mated, respiratory activity is evident within several hours, and a high level is reached by 10 hr. At still longer times of mating (and cell growth) the respiration rate decreases until a lower level is reached. At short times, most mated cells are heteroplasmons; i.e., they contain both mutant genomes in a common cytoplasm but, by later times, few if any cells remain heteroplasmic. Thus, segregation of organelle genomes appears to explain the decline in respiratory rate (per diploid) at later times, and this emphasizes that respiratory enzymes can continue to be made only by cells that have both mutant genomes. In the case where two different mutants in a given gene are mated, the low level of respiration detected only at long times is evidently due to the few wild-type recombinants that result from the cross; most progeny, however, are mutant, and so the respiration rate per diploid is low. Recombinants are also responsible for the low rate observed at long times in crosses between mutants in different complementation groups. Where a high respiratory rate is detected at short times, the two mutants are said to have exhibited complementation; it has been shown that the appearance of a complete respiratory chain depends on mitochondrial protein synthesis, and so it is not merely due to the passive mixing of partially defective membranes. It does not appear to result from recombination, since the ability to maintain a high level of respiratory enzymes segregates at later times; in fact, there is no correlation between the percentage of recombinants obtained and the presence or the absence of a positive result in the complementation test.

When complementation experiments were performed using pairs of nonidentical *cob* mutants, most pairs failed to yield a respiratory rate at

[60] F. Foury and A. Tzagoloff, *J. Biol. Chem.* **253**, 3792 (1978).
[61] P. P. Slonimski, P. Pajot, C. Jacq, M. Foucher, G. Perrodin, A. Kochko, and A. Lamouroux, *in* "Biochemistry and Genetics of Yeasts" (M. Bacila, B. L. Horecker, and A. O. M. Stoppani, eds.), p. 339. Academic Press, New York, 1978.
[62] A. Lamouroux, Thèse de Doctorat de 3ème Cycle, Université Paris XI (1979).

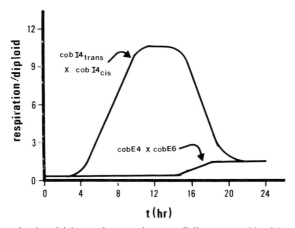

FIG. 5. The mitochondrial complementation test. Cells are mated in rich medium, and at various intervals aliquots of each mating mixture are harvested and washed. The number of diploids per milliliter is measured by plating appropriate dilutions on minimal glucose medium (assuming that the two parent strains are marked by mutually complementing auxotrophies), and the respiration rate per milliliter is measured with a recording oxygen electrode. Typical results for two pairs of nonallelic *cob* mutants are shown. Essential controls (not shown) are comparable matings of each mutant with a *rho-o* derivative of its partner; in those controls no respiration rate is observed even at the longest time of mating. In routine use of this test, each mating is assayed at only one time point, usually 8 or 10 hr, where the maximum distinction between positive and negative results can be obtained.

short times, as expected. Surprisingly, some pairs did complement (cf. Fig. 5), a result that indicated that the *cob* gene consists of several units differing in function.[25,61-64] In fact, the *cob* gene contains four complementation groups; one consists of most mutants in exons plus some intron mutants, and the other three consist of some mutants in introns 2, 3, and 4, respectively. This unexpected result revealed that each of these three introns (but not intron 5) encodes a *trans*-acting factor, which when inactivated by mutation results in a cyt *b* deficiency. These mutants are defective in processing one or more introns from the primary transcript of this mosaic gene,[64-67] and so these introns encode splicing factors that are now widely believed to be proteins encoded largely by intron sequences.[18,23,63,68,69]

[63] B. Weiss-Brummer, G. Rodel, R. J. Schweyen, and F. Kaudewitz, *Cell* **29**, 527 (1982).
[64] M. R. Lamb, P. Q. Anziano, K. R. Glaus, D. Hanson, H. Klapper, P. S. Perlman, and H. R. Mahler, *J. Biol. Chem.* **258**, 1991 (1983).
[65] A. Halbreich, P. Pajot, M. Foucher, C. Grandchamp, and P. P. Slonimski, *Cell* **19**, 321 (1980).
[66] C. Schmelzer, A. Haid, G. Grosch, R. J. Schweyen, and F. Kaudewitz, *J. Biol. Chem.* **256**, 7610 (1981).

All classes of *cob* mutants complement well with mutants in other genes on mtDNA. At one point, it was believed that all intron mutants are *trans*-recessive and that all *cis*-dominant mutations are in exons, but this impression has been corrected by more recent mapping and sequencing studies.[68-70] Introns 2 and 4 of the *cob* gene have been studied in considerable detail; both contain a long reading frame in phase with that of the preceding exon, and in each a number of mutants that block splicing by inactivating the relevant intron-encoded splicing factor have been shown either to result in premature chain termination during translation of the intron reading frame or to alter the primary sequence of this product. Other mutants in the same introns have been shown to block splicing by affecting sequences that appear to be recognized by the (presumed) splicing apparatus. The former group of mutants defines the complementation group for each intron, whereas the latter group fails to complement in crosses with exon mutants, but complements mutants of the other class but located in the same intron. Since many of the *cis*-dominant exon mutants are located rather near the intron–exon boundaries, it is often difficult to determine from mapping studies alone whether a given mutation is located in the intron or the exon. In general, the only way to be confident that a given *cis*-dominant mutant is in an exon is to show that it accumulates the fully processed mRNA or to establish that the altered base pair is in an exon by direct DNA sequencing.

Construction of Double Mutants

Given a collection of point mutants in nuclear and mitochondrial DNA, some of which are found to be useful probes of cytochrome b function, it is likely that characterizing the phenotypes of selected double mutants would be a useful addition to the study. Methods for constructing strains that have two different nuclear mutations, two mitochondrial mutations, or one nuclear and one mitochondrial mutation are relatively straightforward; they will be presented here in some detail because the same methods can be used to dissect initial double mutants and to analyze second-site revertants of interesting mutants.

Strains with two nuclear mutations in different genes are easily obtained by screening meiotic products from doubly heterozygous diploids, for example, those diploids obtained in pairwise crosses that show complemen-

[67] G. J. van Ommen, P. H. Boer, G. S. P. Groot, M. de Haan, E. Roosendaal, L. A. Grivell, A. Haid, and R. J. Schweyen, *Cell* **20,** 173 (1980).

[68] P. Q. Anziano, D. K. Hanson, H. R. Mahler, and P. S. Perlman, *Cell* **30,** 925 (1982).

[69] H. de la Salle, C. Jacq, and P. P. Slonimski, *Cell* **28,** 721 (1982).

[70] P. Netter, C. Jacq, G. Carignani, and P. P. Slonimski, *Cell* **28,** 733 (1982).

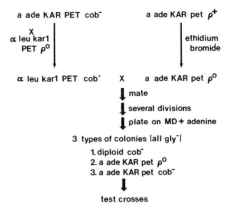

FIG. 6. Construction of a double mutant. Starting with a *pet* and a *cob* mutant derived from the same parent strain (*a KAR ade LEU PET*) an α *kar1 ADE leu* derivative of the *cob* mutant is made as outlined in Fig. 4 and a *rho-o* derivative of the *pet* mutant is made by treating it with ethidium bromide. These two derivatives are then mated and incubated for several hours to permit haploid buds to be formed; cells are then plated on minimal glucose medium (MD) supplemented with adenine to select for *ade* strains (that are also *a KAR LEU pet*). Since most of the resulting colonies are *rho-o*, a test cross with an α *PET cob*⁺ petite strain is used to identify the desired double mutants (*pet cob*⁻). This basic strategy can also be used to construct strains having combinations of mitochondrial mutations as indicated in the test.

tation. Doubly mutant ascospores are those that fail to grow on glycerol and also fail to yield gly⁺ diploids in test crosses with both parents, but yield gly⁺ diploids in a cross with a mutant in another gene.

The *kar1* protocol can be used to combine a given nuclear mutation with a given mitochondrial one. For the likely case where both mutants are derivatives of the same parent strain, the necessary steps are outlined in Fig. 6. A *rho-o* derivative of the *pet* mutant is used as the recipient of mutant mtDNA from a *kar1* derivative of the original mitochondrial mutant. Thus both parental strains must be derivatized before the construction can be carried out. Since both parental strains and the desired double mutant are gly⁻, that recombinant must be identified by screening a number of haploid products of the cross by test crosses. The double mutant will yield wild-type progeny in a cross with a *PET* tester that has a petite genome capable of recombining with the mitochondrial mutant; crosses with a *rho-o* derivative of that tester will yield no gly⁺ progeny.

Double mutants containing two mitochondrial mutations can be obtained in much the same way. Here one of the mutants is obtained as a *kar1 cob*⁻ derivative that is then mated to the other (original) mutant strain.[e.g., 71] Cells exhibiting the nuclear genotype of the recipient strain are then tested

⁷¹ N. J. Alexander, P. S. Perlman, D. K. Hanson, and H. R. Mahler, *Cell* **20**, 199 (1980).

for being double mutants by searching for ones that fail to yield wild-type recombinants in backcrosses to both parental types but that recombine with a petite capable of restoring both mutants to a gly$^+$ phenotype. In some cases all the test crosses can involve petite testers provided that petite strains are available that can discriminate between the two mutations.

Where the double mutant is found to have a phenotype somewhat different from that of either parent, these genetic tests of the success of the construction are sufficient. However, if a presumptive double mutant is phenotypically indistinguishable from one parent (cf. Alexander et al.[71] for several examples), then the best proof is to dissect the double mutant by mating it with a *kar1 rho$^+$* strain and reisolating the two component mutations. Recovering the "extinguished" phenotype in this way is strong evidence that both mutations were present in the constructed strain. Occasionally, double mutants are obtained in the initial mutant isolation protocol[72]; that situation is usually evident from the initial meiotic studies of nuclear mutants or from the mapping studies of mitochondrial ones. For example, a mitochondrial mutant harboring two expressed mutations several hundred (or more) base pairs apart will show close linkage with other single mutants that are not linked to each other or will behave in petite mapping as a deletion, having two mapable end points.

Revertants

Up until this point, we have been discussing the genetic analysis and manipulation of mutants that are totally deficient in complex III function; however, such mutants may not be particularly valuable for biochemical studies of electron transport. For the most obvious types of functional studies, mutants with conditional or leaky respiratory activity are desirable, and these can be obtained as revertants of mutants with severe phenotypic blocks.[cf. 10] Even rare derivatives of mutant strains can be isolated quite readily by plating large numbers of cells on glycerol medium and selecting the few colonies that result. For the case of both missense and chain termination cyt b exon mutants, partially functioning revertants, if analyzed methodically, can provide very useful probes of protein function because they can generate families of partially functional enzymes for biochemical analysis.

Reversion from a gly$^-$ to a gly$^+$ phenotype can result from many different kinds of changes. Where the mutation has been replaced by the original wild-type sequence ("true revertants"), the revertant strain should regain full activity and normal growth properties (at various temperatures) on

[72] F. Foury and A. Tzagoloff, *Mol. Gen. Genet.* **149**, 43 (1976).

glycerol media. Strains that can grow on glycerol only slowly or only at one temperature or another (heat and cold-sensitive revertants) are, by definition, partial revertants and should be very useful. Here it is necessary to learn whether each partial revertant still harbors the original mutant base pair and grows on glycerol because of a second site change or whether the original mutation has been replaced by a new sequence that is more compatible with function. When the revertant is a single or double mutant with alterations in essentially the same place as the original mutant, then it will map there; in crosses with a petite that can restore the original mutant to a full gly$^+$ phenotype, the revertant will be converted from partial to full function by the same cross.

Second-site revertants can fall into several classes. In one, the second site or suppressor mutation is near the original mutation; in another it is physically separable from the original one but in the same gene; in another, it is in another gene on mtDNA; and in the last it is in a gene on nucDNA. So far, in the study of *cob* intron mutants all these classes of suppressor mutations as well as full revertants have been reported,[73-78] and a similar range of reversion events should be found for exon mutants.

For many biochemical purposes it is more important to obtain revertant strains having useful phenotypes than it is to understand the nature of the reversion event itself. However, to understand how two domains of a protein or two different proteins interact it may become necessary to dissect revertants genetically. Such studies are basically similar to the methods already described, but they are a bit more complicated.[73,74] Revertants that result from some change in mtDNA are detected by showing that the gly$^+$ phenotype persists even when the nuclear genome has been replaced by a different one; that is the case when crossing the revertant with a *kar1 rho-o* strain yields haploid progeny having the *kar1* nucleus that can grow on glycerol. If a nuclear suppressor mutation is present, then mating a *kar1* derivative of the original mutant with a *rho-o* derivative of the revertant will regenerate haploid *KAR* cells that grow on glycerol. Given evidence of a nuclear suppressor mutation, it is useful to ask whether that altered nuclear function interferes with the function of wild-type cyt *b*; that is, accomplished by transferring wild-type mtDNA into the *rho-o* derivative of the revertant

[73] G. Coruzzi and A. Tzagoloff, *Genetics* **95**, 891 (1980).
[74] G. Dujardin, P. Pajot, O. Groudinsky, and P. P. Slonimski, *Mol. Gen. Genet.* **179**, 469 (1980).
[75] O. Groudinsky, G. Dujardin, and P. P. Slonimski, *Mol. Gen. Genet.* **184**, 493 (1981).
[76] G. Dujardin, C. Jacq, and P. P. Slonimski, *Nature (London)* **298**, 628 (1982).
[77] T. D. Fox and S. Staempfli, *Proc. Natl. Acad. Sci. U.S.A.* **79**, 1583 (1982).
[78] M. L. Haldi, P. Q. Anziano, and P. S. Perlman, in preparation (1983).

and testing for growth on glycerol. In such, possibly rare, cases, the analysis of revertants can result in very interesting nuclear mutants.

The analysis of second-site revertants in which both changes are in mtDNA resembles the dissection of double mutants.[74,76,77] Again, the suppressor mutation may have a mutant phenotype when the persisting original mutation is removed; this test can be carried out by mating with petites that discriminate a small portion of the *cob* gene containing the mutant site. If the petite restores full growth properties, either the suppressor is located in the same gene segment or else it has no phenotypic defect by itself. These situations can be distinguished by performing crosses with petites that cannot replace the original defect but that, in their aggregate, can replace mutations anywhere else in the *cob* gene (or elsewhere on mtDNA). For example, if a suppressor of a mutation in exon 6 of the *cob* gene is located in exon 1, mating the, say, temperature-sensitive revertant with a petite that can restore any exon 1 mutant will yield some progeny that are phenotypically identical with the original mutant, while a cross with another petite that retains exons 2–5 will yield only ts progeny. Such crosses cannot be scored by replica-plating of the matings; instead, it is necessary to subclone the progeny of the cross and test them by replica plating plus appropriate test crosses. If that cross results in diploid progeny, then test crosses can be done only with difficulty; to avoid that problem we use a petite strain that has the *kar1* mutation and select haploid progeny using selective media so that they can be tested genetically without further manipulation.

Summary

We have reviewed here the genetic methods used for isolating and manipulating nuclear and mitochondrial mutants of bakers' yeast that affect the function and biogenesis of complex III of the mitochondrial respiratory chain. All the methods have been used with success in the past, and it is hoped that this compilation will aid biochemists in using these techniques to study electron transfer.

Acknowledgments

Original studies summarized here were supported by grants from the National Institutes of Health GM19607 and GM21896 to P. S. P. and GM12224 and H. R. M. The preparation of this chapter was supported by GM12224 and by research funds of The Ohio State University.

[38] Synthesis and Intracellular Transport of Mitochondrial Matrix Proteins in Rat Liver: Studies *in Vivo* and *in Vitro*

By GORDON C. SHORE, RICHARD A. RACHUBINSKI, CAROLE ARGAN, RIMA ROZEN, MARCEL POUCHELET, CAROL J. LUSTY, and YVES RAYMOND

We describe here some of the methods we have employed to study uptake of precursor proteins into the matrix fraction of liver mitochondria in rats[1-5] and frogs.[6] We focus on two enzymes, carbamoyl-phosphate synthetase (ammonia) (EC 6.3.4.16) (CPS) and ornithine carbamoyltransferase (EC 2.1.3.3) (OCT); CPS and OCT catalyze the first and second steps, respectively, of the urea cycle.[7,8] The CPS is particularly abundant, accounting for about 30% of matrix protein,[4,9,10] or about 4–5% of total liver protein,[4] whereas OCT represents only about 1% of mitochondrial matrix protein.[11] At least as far as CPS is concerned, the location in liver appears to be restricted to hepatocytes.[12]

Both proteins, CPS (M_r 160,000) and OCT (M_r 36,000), are made as higher molecular weight precursors (pCPS, M_r 165,000[1-4,6,13] and pOCT, M_r 39,000–40,000[14-16]). Work over the past few years[1-4,13-17] appears to have

[1] G. C. Shore, P. Carignan, and Y. Raymond, *J. Biol. Chem.* **254**, 3141 (1979).

[2] Y. Raymond and G. C. Shore, *J. Biol. Chem.* **254**, 9335 (1979).

[3] Y. Raymond and G. C. Shore, *J. Biol. Chem.* **256**, 2087 (1981).

[4] Y. Raymond and G. C. Shore, *Biochim. Biophys. Acta* **656**, 111 (1981).

[5] G. C. Shore, F. Power, M. Bendayan, and P. Carignan, *J. Biol. Chem.* **256**, 8761 (1981).

[6] M. Pouchelet and G. C. Shore, *Biochim. Biophys. Acta* **654**, 67 (1981).

[7] R. L. Metzenberg, L. M. Hall, M. Marshall, and P. P. Cohen, *J. Biol. Chem.* **229**, 1019 (1957).

[8] J. G. Gamble and A. L. Lehninger, *J. Biol. Chem.* **248**, 610 (1973).

[9] S. Clarke, *J. Biol. Chem.* **251**, 950 (1976).

[10] C. J. Lusty, *Eur. J. Biochem.* **85**, 373, (1978).

[11] C. J. Lusty, R. L. Jilka, and E. Nietsch, *J. Biol. Chem.* **254**, 10030 (1979).

[12] M. Bendayan and G. C. Shore, *J. Histochem. Cytochem.* **30**, 139 (1982).

[13] M. Mori, S. Miura, M. Tatibana, and P. P. Cohen, *Proc. Natl. Acad. Sci. U.S.A.* **76**, 5071 (1979).

[14] M. Mori, S. Miura, M. Tatibana, and P. P. Cohen, *Proc. Natl. Acad. Sci. U.S.A.* **77**, 7044 (1980).

[15] M. Mori, T. Morita, F. Ikeda, Y. Amaya, M. Tatibana, and P. P. Cohen, *Proc. Natl. Acad. Sci. U.S.A.* **78**, 6056 (1981).

[16] J. G. Conboy, F. Kalousek, and L. E. Rosenberg, *Proc. Natl. Acad. Sci. U.S.A.* **76**, 5724 (1979).

[17] J. G. Conboy and L. E. Rosenberg, *Proc. Natl. Acad. Sci. U.S.A.* **78**, 3073 (1981).

established at least the outline, if not the details, whereby these nuclear gene products are selectively channeled from their site of synthesis in the cytoplasm to their final destination in the mitochondrial matrix. The precursor polypeptides are synthesized by free polyribosomes, then pass rapidly through the cytosol, cross both mitochondrial membranes in a posttranslational manner, and are finally deposited in the matrix. Endoproteolytic processing takes place at some point either coincident with or immediately following transmembrane uptake of the precursors by mitochondria. A comparison of the kinetics of all of these events for CPS and OCT showed a remarkable similarity between the two proteins.[15]

General Methodology

Choosing Mitochondrial Proteins and Raising Antisera: General Approach

Because denatured proteins in sodium dodecyl sulfate (SDS)–polyacrylamide gels can elicit an immune response when injected into a recipient animal, purification of proteins to homogeneity as a preliminary to raising antibodies can generally be avoided. In the case of CPS, we routinely prepare mitochondrial matrix protein (10–15 mg), fractionate this mixture by DEAE-Sephacel chromatography, and from a single run collect the peak fractions containing enzyme activity.[1] The enriched protein gives a well defined band on preparative SDS–polyacrylamide gels. CPS is excised, and the gel pieces are macerated, mixed with Freund's complete adjuvant, and injected directly into rabbits. Antiserum against rat liver OCT, however, was raised against the purified protein.[11] The quality of antisera against CPS and OCT can be assayed by standard techniques, including their reaction with SDS–polyacrylamide gel profiles of mitochondrial proteins blotted onto nitrocellulose paper. It is important that precise subcellular and submitochondrial localization of the antigens can be determined by a simple but effective immunocytochemical technique[18] based on the ability of exposed antigens on the surface of cut electron microscope sections to react with antibody. Antibody–antigen complexes are then visualized after reaction with protein A-gold. Figure 1 shows the application of this technique[5,12] to a relatively abundant protein, CPS, but this general approach has also been successfully applied to a minor constituent of mitochondria, a protein located in the outer mitochondrial membrane.[5,12]

---

[18] J. Roth, M. Bendayan, and L. Orci, J. Histochem. Cytochem. 26, 1074 (1978).

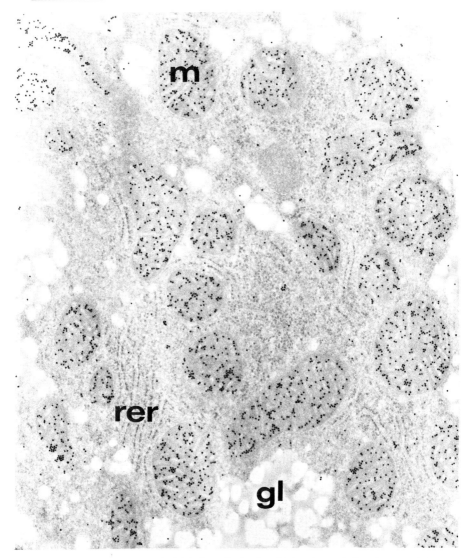

FIG. 1. Immunocytochemical labeling of mitochondrial carbamoyl-phosphate synthetase in rat hepatocytes by the protein A–gold technique. For details, see Shore *et al.*[5] and Bendayan and Shore.[12] The photomicrograph was kindly provided by Dr. Moise Bendayan, Université de Montréal. rer, Rough endoplasmic reticulum; m, mitochondrion; gl, cleared areas where glycogen has been leached from the tissue during the mild fixation procedures employed in this procedure.

RNA and Polyribosome Isolation, Protein Synthesis in Vitro, and Product Analysis

Liver mRNA and polysomes are routinely isolated from adult Sprague-Dawley rats (100–150 g) by standard procedures, with the important exception that isolation media are maintained at an elevated pH (pH 8.5) and at relatively high ionic strength (200 mM Tris acetate and 50 mM KCl).[19] Such conditions are useful for inhibiting ribonuclease activities and usually obviate the necessity for adding exogenous ribonuclease inhibitors. These media are also used when free and membrane-bound polyribosomes are isolated by the rapid procedure of Ramsey and Steele.[20]

For neonatal animals, however, total liver mRNA is extracted by the guanidinium-CsCl method,[21] exactly as described.[2,6] CPS mRNA is large (27 S)[4] and, in our experience, is more efficiently translated *in vitro* in the nuclease-treated[22] reticulocyte lysate system than in standard wheat germ systems. It is imperative, however, to supplement translation reactions with exogenous tRNA (160–200 μg/ml) in order to synthesize full-length pCPS. tRNA is conveniently prepared from wheat germ.[23]

For immunoprecipitation of OCT and CPS translation products from the reticulocyte lysate system, nonspecific entrapment of other polypeptide products can be minimized either by maintaining a high salt concentration (1.4 M NaCl) in the presence of detergent (1% Triton X-100 or 1% Nonident P-40)[1] or by following the SDS procedure for removing nonantigenic contaminants.

Pulse and Pulse–Chase Labeling of Liver Explants[2,3]

Neonatal animals (3–5 days old) are anesthetized with ether, cut along the linea alba to the rib cage, and then cut laterally. The liver and diaphragm are exposed by cutting the peritoneum along the same lines. The xyphoid process is clamped and raised, and the diaphragm is quickly cut to expose the heart fully. An intravenous injection set connected to a 25-gauge needle is used for perfusion. The perfusate consists of PBS+ (0.14 M NaCl, 10 mM sodium phosphate buffer, pH 7.4, 2.7 mM potassium chloride, 0.68 mM CaCl$_2$, and 0.15 mM MgCl$_2$) and is maintained at a pressure head of 3 feet. With the perfusate flowing, the needle is inserted into the left ventricle of the

[19] G. C. Shore and J. R. Tata, *J. Cell Biol.* **72**, 726 (1977).
[20] J. C. Ramsey and W. J. Steele, *Biochemistry* **15**, 1704 (1976).
[21] J. M. Chirgwin, A. E. Przybyla, R. J. MacDonald, and W. J. Rutter, *Biochemistry* **18**, 5294 (1979).
[22] H. R. B. Pelham and R. J. Jackson, *Eur. J. Biochem.* **67**, 247 (1976).
[23] A. Marcus, D. Evron, and D. P. Weeks, this series, Vol. 30, p. 749.

heart while simultaneously cutting the right atrium to allow blood to flow out. The liver should be completely blanched within 30 sec.

Perfused liver lobes are placed in a small volume of PBS$^+$ at room temperature. Pieces are cut from the edges, placed in a Stadie-Riggs microtome, and sliced to give explants approximately 0.5 mm thick. They are transferred to fresh PBS$^+$ until a sufficient number of explants have been accumulated. After a brief blot on tissue paper, the explants are placed directly into sterilized scintillation vials containing 0.7 ml of Eagle's minimal essential medium (α medium minus methionine) supplemented (per milliliter) with 1 μg of unlabeled methionine, 100 units of penicillin, 100 μg of streptomycin, and up to 1 mCi of [^{35}S]methionine. Uncapped vials are agitated in a shaking water bath at 37°. For chase experiments, the incubation medium is replaced with fresh medium containing 3 mg of unlabeled methionine and 20 μg of cycloheximide per milliliter. In both cases (pulse and pulse – chase), incubation media can be retained and clarified by centrifugation in order to analyze secreted proteins (primarily albumin). Several lines of evidence[3] indicate that pCPS is not processed to CPS during tissue homogenization and subsequent fractionation at 4°.

Subcellular Fractionation of Explants[2,3]

A single explant is homogenized at 4° in 0.6 ml of 2 mM HEPES, 70 mM sucrose, 220 mM mannitol, 0.5 mg of BSA per milliliter, 10 mM methionine, and 150 μg of phenylmethylsulfonyl fluoride per milliliter (added before use from a fresh 100$\times$ stock dissolved in ethanol). The pH is adjusted to 7.4 with 1 M KOH just before use. Homogenization is at 4° in a 2-ml Reactiware glass homogenizer (Pierce Chemical Co.) using a Teflon pestle operating at 200 rpm. The homogenate is centrifuged in 2-ml glass tubes at 640 g for 10 min. Mitochondria are recovered from the supernatant by centrifuging at 10,000 g for 10 min, and 150-μl aliquots of the resulting postmitochondrial supernatant are further centrifuged for 4 – 6 min at 100,000 g in a Beckman airfuge to give microsomes (pellet) and cytosol (supernatant). Mitochondrial and microsomal pellets are muddled and suspended in 100 μl of 10 mM Tris-HCl, pH 8.5. Protein content (Coomassie Blue technique[24]) and net incorporation of [^{35}S]methionine into protein (filter disk method[25]) are determined for all fractions. Aliquots containing 15 – 150 μg of protein are subjected to immunoprecipitation with *excess* antiserum against CPS or OCT.

The concentration of CPS is exceedingly high inside mitochondria, so that a certain amount is spuriously released to the cytosol as a result of

[24] M. M. Bradford, *Anal. Biochem.* **72**, 248 (1976).
[25] R. J. Mans and G. D. Novelli, *Arch. Biochem. Biophys.* **94**, 48 (1961).

damage to mitochondria during tissue homogenization.[2,3] It is important, therefore, to ensure that excess antibody is used for immunoprecipitation from all subcellular fractions.

Isolation of Mitochondria for in Vitro Studies

Rat Liver Mitochondria. Mitochondria are isolated from livers of starved (12 hr) animals exactly as described by Greenawalt[26] except that bovine serum albumin (BSA) is not included in the isolation medium. Mitochondria are muddled and suspended in medium containing 10 mM HEPES, pH 7.4, and 0.25 M sucrose. An aliquot (2 mg of protein per milliliter) is mixed with an equal volume (5–50 μl) of reticulocyte lysate containing freshly made translation products under the direction of liver mRNA. Further protein synthesis is arrested by adding 1/100 volumes of cycloheximide (1 mg/ml). This posttranslational mixture is then incubated in capped 1.5-ml plastic Eppendorf tubes at 30°.

Rat Heart Mitochondria.[27,28] A single heart (0.8–1.1 g) is quickly removed from an animal, placed in 10 ml of ice-cold medium A (10 mM HEPES, pH 7.5, 220 mM mannitol, 70 mM sucrose, 1 mM EGTA, 2 mg of BSA per milliliter), and homogenizd for 2 sec with the Brinkmann Polytron at a setting of 6.5. All further manipulations are performed at 2–4°. The homogenate is diluted to 40 ml with medium B (medium A minus BSA) and centrifuged in the Sorvall SS-34 rotor at 1750 rpm for 10 min. The upper three-quarters of the supernatant is removed and further centrifuged at 7000 rpm for 10 min in the same rotor. The resulting crude mitochondrial pellet is uniformly suspended in 500 μl of medium B, to which is added 100 μl of medium B containing twice-recrystallized digitonin (2.5 mg/ml).[28] After incubation for 15 min on ice with periodic swirling, 10 ml of medium B is added and the mixture is centrifuged at 7000 rpm for 10 min. The pellet is resuspended, recentrifuged at 7000 rpm, again resuspended, and centrifuged at 1750 rpm for 10 min. The supernatant is finally centrifuged at 7000 rpm to collect purified mitochondria. Inspection by electron microscopy shows a complete absence of lysosomes and sarcolemmal fragments, but about 10% of mitochondria show some structural damage. Less damage is obtained if the digitonin step is omitted and the mitochondria perform somewhat better during *in vitro* import assays, but the mitochondrial preparation is less pure.

The mitochondria (4–6 mg of protein from one heart) are uniformly suspended in a medium containing 10 mM HEPES, pH 7.4, 0.25 M sucrose,

---

[26] J. W. Greenawalt, this series, Vol. 31, p. 310.
[27] J. W. Palmer, B. Tandler, and C. L. Hoppel, *J. Biol. Chem.* **252**, 8731 (1977).
[28] E. Kun, E. Kirsten, and W. N. Piper, this series, Vol. 55, p. 115.

10 mM sodium succinate, 2.5 mM K$_2$HPO$_4$, pH 7.4, 0.15 mM ADP, and 1 mM dithiothreitol to give a final concentration of 6 mg of protein per milliliter. Aliquots (25–150 μl) are then mixed with an equal volume of reticulocyte lysate containing liver mRNA translation products and 10 μg of cycloheximide per milliliter. Posttranslational incubation is at 30°.

Transport of Precursor Proteins into the Matrix of Mitochondria *in Vivo*

Location of CPS mRNA in Liver. After separation of free and membrane-bound polyribosomes by a modification[5,29] of the rapid procedure of Ramsey and Steele,[20] CPS mRNA is found associated exclusively with the free polyribosome fraction whereas albumin mRNA, a well-known marker coding for a secretory protein, is recovered in the membrane-bound fraction (Fig. 2). In view of the fact that rough endoplasmic reticulum (RER) tends to fragment into a rather heterogeneous population of vesicles and lamellae after tissue homogenization and may even aggregate tightly with mitochon-

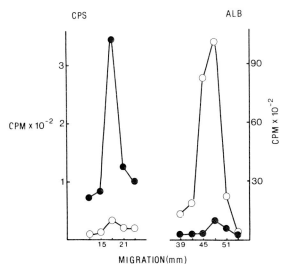

FIG. 2. Distribution of carbamoyl-phosphate synthetase (CPS) and albumin (ALB) poly (A)$^+$-mRNA between free (●——●) and membrane-bound (○——○) polyribosomes in rat liver. Free and bound mRNA were translated in a messenger-dependent rabbit reticulocyte system. Products were subjected to specific immunoprecipitation and electrophoresed in a sodium dodecyl sulfate–polyacrylamide gel, which was then sliced in order to determine radioactivity in the primary translation products of CPS mRNA and albumin mRNA.

[29] R. A. Rachubinski, D. P. S. Verma, and J. J. M. Bergeron, *J. Cell Biol.* **84,** 705 (1980).

dria,[19,30-32] it is important to emphasize that the Ramsey and Steele procedure results in complete recovery of the total population of RER and mitochondria. There is little indication, however, that even a minor fraction of CPS polyribosomes interacts directly with either of these organelles (Fig. 2).

Initial Location of the Primary Translation Product of CPS mRNA in Vivo. The evidence that pCPS is released from free polyribosomes directly to the cytosol simply comes from pulse-labeling of small liver explants followed by classical tissue subfractionation.[2,3] pCPS is recovered from a high-speed supernatant rather than from any familiar particulate fraction; i.e., it does not sediment when 1 ml of homogenate is centrifuged at 500,000 g for 20 min[2] or 150 μl is centrifuged at 100,000 g for 4 min.[3] As in any such study, however, it is difficult to rule out the possibility that a transient or labile interaction exists between pCPS and some structural component of the cytoplasm, e.g., a lipid-rich micelle, a fragile vesicular compartment, cytoskeletal elements, and that these interactions are disrupted or remain undetected after tissue homogenization. If nothing else, however, pulse-labeling studies *in vivo* strongly support the notion that pCPS enters mitochondria by a posttranslational mechanism.

Kinetics of Intracellular Transport and Processing of pCPS. The liver explant system is not suitable for pulse-labeling over periods that are shorter than the turnover time of pCPS, i.e., short enough to label pCPS but not CPS. However, in the one experiment where we pulse-labeled a liver explant for 30 sec with a very high amount of [^{35}S]methionine, radioactivity was observed first in pCPS, and then all the radioactive pCPS was converted to CPS during a subsequent chase (unpublished results). This at least justifies the use of our standard protocol[2,3] for measuring kinetics of intracellular transport and processing of pCPS where longer incubation periods are used to allow greater incorporation of radioisotope. Explants are incubated for 20-30 min with [^{35}S]methionine and are then subjected to chase periods of 1-6 min in the presence of cycloheximide and excess, unlabeled methionine. At the end of the pulse-labeling period (30 min), the ratio between radioactivity in processed CPS vs pCPS is about 20:1 per milligram of homogenate protein (Fig. 3); i.e., the conversion of pCPS to CPS must occur very fast relative to the pulse-labeling time. In an earlier study, measurements of subsequent loss of radioactive pCPS from the cytosol during chase periods showed that 75% of newly synthesized pCPS leaves this compartment by 2 min and that, by 4 min, radioactive pCPS is undetectable in the tissue.[3] Figure 3 presents a balance sheet for whole explant homogenates.

---

[30] G. C. Shore and J. R. Tata, *J. Cell Biol.* **72**, 714, (1977).

[31] G. C. Shore and J. R. Tata, *Biochim. Biophys. Acta* **472**, 197 (1977).

[32] G. C. Shore, *J. Cell Sci.* **38**, 137 (1979).

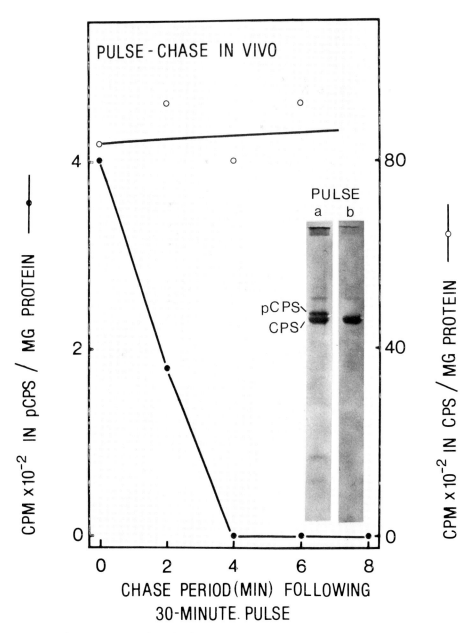

FIG. 3. Pulse-chase studies using liver explants: a balance-sheet analysis of [³⁵S]methionine incorporated into precursor carbamoyl-phosphate synthetase (pCPS) (—●—) and CPS (—○—) and expressed per milligram of homogenate protein. Measurements were made after the homogenate was fractionated to give a postmitochondrial supernatant (inset, lane a) and mitochondria (inset, lane b). The occurrence of processed CPS in the postmitochondrial supernatant arises because of spurious release of CPS from mitochondria during tissue homogenization.[3,4]

The $t_{1/2}$ for transit through the cytosol, uptake into mitochondria, and final processing is ~ 2 min. Endoproteolytic processing of pCPS must occur either coincident with or immediately following its transmembrane uptake by mitochondria because we have been unable to detect even a minor pool of pCPS in association with this organelle.

Posttranslational Uptake and Processing of pCPS and pOCT in Vitro by Homologous and Heterologous Mitochondrial Systems

Liver Mitochondria. When intact liver mitochondria (as judged by electron microscopy) are incubated with reticulocyte lysates containing newly synthesized translation products of rat liver mRNA, subsequent uptake and processing of the precursor to OCT is readily measurable in the presence of cycloheximide (Fig. 4). Nevertheless, it still requires about 60 min to achieve full conversion of pOCT (39 kilodaltons) to OCT (36 kilodaltons). The concentration of mitochondria in these assays (1 mg of protein per milliliter) represents about one-fifth of their "concentration" in the cytoplasm of hepatocytes, based on volumetric criteria.[33]

In addition to processing of pOCT to OCT, however, there is also clear conversion of pOCT to a 37-kilodalton product (Fig. 4). This latter polypeptide appears in lysates even in the absence of mitochondria, although somewhat more so when mitochondria are present (Fig. 4). Nevertheless, it does not further convert to fully processed OCT even after 90 min of posttranslational incubation with mitochondria. In view of the fact that EDTA blocks the appearance of the 37-kilodalton product (unpublished observations), we presume that it arises because of the action of the putative processing enzyme for pOCT, which was observed by Mori and co-workers[14] in extracts of liver mitochondrial matrix fractions. They found that the solubilized EDTA-sensitive enzyme processed pOCT to the 37-kilodalton product but no further. The present results (Fig. 4) can be explained if it is assumed that a similar processing enzyme is present in reticulocyte lysates (Fig. 4, lane a), that the liver mitochondrial processing enzyme is active even during immunoprecipitation in the presence of detergent (Fig. 4, lane b), and perhaps that some processing enzyme leaks from damaged mitochondria during postincubation with reticulocyte lysates and cleaves pOCT to the 37-kilodalton product outside the mitochondria. Loss of the 2000-dalton piece from pOCT apparently renders the precursor incapable of entering mitochondria.

As far as pCPS is concerned, we have not yet been able to identify conditions under which bona fide processing can be achieved using isolated

[33] A. Blouin, R. P. Bolender, and E. R. Weibel, *J. Cell Biol.* **72**, 441 (1977).

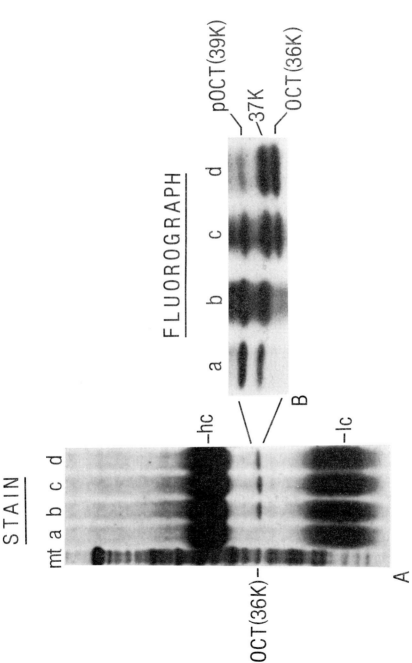

Fig. 4. Posttranslational processing of precursor ornithine carbamoyltransferase (pOCT) by isolated rat liver mitochondria *in vitro*. mt, purified mitochondria; lane a, mitochondria were not added to the reticulocyte lysate system; lanes b, c, and d, mitochondria were added to reticulocyte lysates and postincubation was carried out for 0, 30, and 60 min, respectively. Products were analyzed by specific immunoprecipitation followed by SDS–polyacrylamide gel electrophoresis. The gel was stained and dried (panel A) and fluorographed for 3 days (panel B). OCT, ornithine carbamoyltransferase; pOCT, precursor to OCT; 37K, 37-kilodalton polypeptide; hc, IgG heavy chain; lc, IgG light

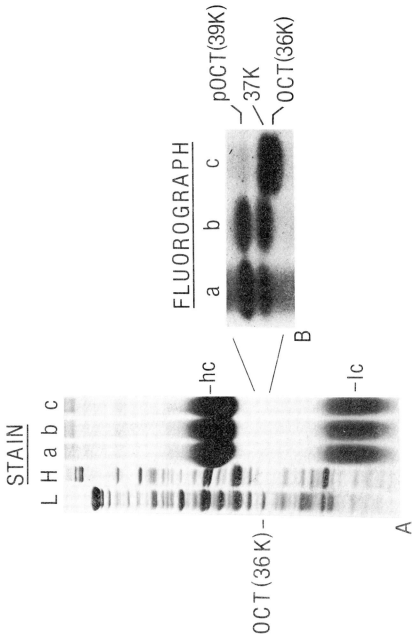

FIG. 5. Posttranslational processing of pOCT by isolated rat heart mitochondria *in vitro*. L, purified mitochondria from rat liver; H, purified mitochondria from rat heart; lane a, mitochondria were not added to the reticulocyte lysate system; lanes b and c, heart mitochondria were postincubated with reticulocyte lysates for 0 and 30 min, respectively. For further details, see Fig. 4.

liver mitochondria *in vitro*. This even includes tests where reticulocyte lysates are incubated with liver mitochondria during translation of pCPS, which suggests that a lack of uptake and processing probably does not arise simply as a result of age-dependent denaturation of the precursor, as was described[34] for M13 procoat protein of *E. coli*.

Heart Mitochondria. The pattern of processing of pOCT by this heterologous system (Fig. 5) is similar to that observed with the homologous liver mitochondrial system described above (Fig. 4). Not surprisingly, those components of mitochondria that function in uptake and processing of pOCT do not appear to be tissue specific. The heart system described here also functions in apparent processing of pCPS, but with exceedingly low efficiency.

[34] J. M. Goodman, C. Watts, and W. Wickner, *Cell* **24**, 437 (1981).

[39] Biosynthesis of Cytochrome *c* and Its Posttranslational Transfer into Mitochondria

By Takashi Morimoto, Shiro Matsuura, and Monique Arpin

Cytochrome *c* is a peripheral membrane protein that is located in the intermembrane space of mitochondria, where it binds to cytochrome oxidase, an integral membrane protein. It is known that cytochrome *c* is coded for by a nuclear gene[1,2] and that the polypeptide is synthesized in cytoplasmic free ribosomes.[3-7] The primary translation product is thus first released into cytoplasm and then transferred posttranslationally into the mitochondria. This protein, therefore, represents an excellent model system for the study of the intracellular mechanisms by which proteins made on cytoplasmic ribosomes are sorted out and segregated to different compartments of subcellular organelles.

In carrying out the posttranslational transfer experiments of newly synthesized cytochrome *c*, several precautions must be taken. These pre-

[1] F. Scherman, J. W. Stewart, E. Margoliash, J. Parker, and W. Campbell, *Proc. Natl. Acad. Sci. U.S.A.* **55**, 1498 (1966).

[2] F. Scherman and J. W. Stewart, *Annu. Rev. Genet.* **5**, 257 (1971).

[3] N. F. Gonzalez-Cadavid, *Subcell. Biochem.* **3**, 275 (1974).

[4] G. Hallermayer, R. Zimmermann, and W. Neupert, *Eur. J. Biochem.* **81**, 533 (1977).

[5] M. A. Harmey, G. Hallermayer, H. Korb, and W. Neupert, *Eur. J. Biochem.* **81**, 533 (1977).

[6] R. Zimmermann, U. Paluch, and W. Neupert, *FEBS Lett.* **108**, 141 (1979).

[7] S. Matsuura, M. Arpin, C. Hannum, E. Margoliash, D. D. Sabatini, and T. Morimoto, *Proc. Natl. Acad. Sci. U.S.A.* **78**, 4368 (1981).

cautions, which are applicable also to the study of the posttranslational transfer of other mitochondrial proteins, involve the following points.

1. It has been found that treatment of animals with 3,3′,5-triiodothyronine (T_3, thyroid hormone), in addition to increasing the levels of cytochrome *c* mRNA, increases the level of cytochrome oxidase mRNAs.

2. Mitochondria should be very fresh and intact to prevent degradation of newly synthesized cytochrome *c* by proteinase released from mitochondria during the posttranslational incubation and to ensure its uptake by mitochondria. This is very critical for mitochondrial proteins such as carbamoyl-phosphate synthetase, which are synthesized in the cytoplasm as large precursors. In this particular example, newly synthesized polypeptides that failed to be transferred into mitochondria were found to be processed in their precursor region.[8]

3. It is necessary to use proteinase inhibitors in order to protect newly synthesized polypeptides from proteolysis. However, the inhibitor effects on the posttranslational transfer of newly synthesized polypeptides should be examined.

I. Methods for the Preparation of Materials

A. Animals

For the preparation of mRNA, albino male Sprague–Dawley rats (approximately 150 g), which received a daily intraperitoneal injection of T_3 (300 μg/150-g rat) for 5 days were used.[9] Injections of T_3 were necessary because the presence of *in vitro*-synthesized cytochrome *c* in translation mixtures programmed with free polysomal mRNA from control rat livers was barely detectable by immunoprecipitation with specific antibodies (less than 0.01% of the hot trichloroacetic acid-insoluble radioactivity was precipitated). However, translation mixtures programmed with mRNA from livers of T_3-treated rats synthesized more than 15-fold higher levels of immunoprecipitatable cytochrome *c* (Fig. 1). Although there were differences between the sodium dodecyl sulfate–polyacrylamide gel electrophoresis (SDS-PAGE) profiles of the control and treated translation products, T_3 treatment did not cause a change of the site of biosynthesis of known cellular proteins.[7] This was demonstrated by comparing the distribution of translatable mRNA coding for preproalbumin in free and bound polysomes. For preparation of mitochondria, unstarved albino Sprague–Dawlay rats (approximately 150 g) were used.

---

[8] M. Arpin, C. Lusty, and T. Morimoto, unpublished observation.
[9] W. W. Winder, K. M. Baldwin, R. L. Terjung, and J. O. Holloszy, *Am. J. Physiol.* **228**, 1341 (1975).

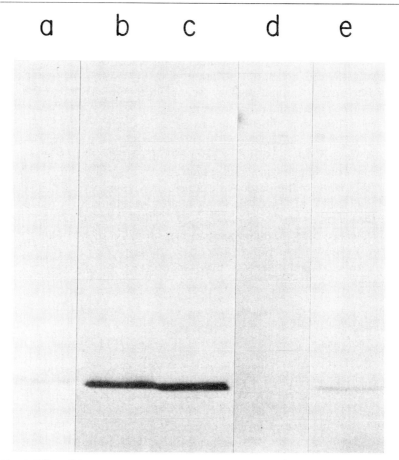

FIG. 1. Effect of T_3 treatment on the level of translatable cytochrome c mRNA in rat liver. Translation was carried out at 28° for 60 min in the reticulocyte lysate (100 μl) with mRNA (0.05 OD_{260} unit) extracted from total polysomes prepared from control (lane a) and T_3-treated rats (lanes b–d) with [^{35}S]methionine as a label. After incubation, translation mixtures were subjected to immunoprecipitation with anticytochrome c, followed by SDS-PAGE as described in the text. In lanes c and d, an acetyl-trapping system[22] was added and translation was carried out with (lane c) and without (lane d) a mixture of the proteinase inhibitors (see text). ^{125}I-labeled horse cytochrome c was used as a marker (lane e).

B. mRNA

Cytoplasmic free ribosomes are preferentially used as the source of cytochrome c mRNA, since it is known that cytochrome c is synthesized in free rather than bound polysomes.[3,7] Free polysomes were prepared from livers of starved rats according to the procedure described elsewhere.[9a,10] For the study of the site of biosynthesis of cytochrome c, free and membrane-bound polysomes were prepared according to the procedure of Ramsey and Steele[10] with the modifications that (a) livers were homogenized in 4 volumes of 0.25 M KCl, 0.25 M sucrose, 50 mM Tris-HCl, pH 7.5, 20 mM MgCl$_2$, 3 mM dithiothreitol (DTT); (b) after treatment with 1% Triton X-100, the resuspended sediment containing nuclei, mitochondria, and microsomes was centrifuged at 8000 rpm for 10 min in a Sorvall HB-4 rotor to remove undissolved material. Occasionally total polysomes were prepared as follows: livers were homogenized as described above, and the homogenate was treated with 1% Triton X-100. Nuclei and mitochondria were then removed by centrifugation at 8000 rpm for 15 min in a Sorvall HB-4 rotor. The supernatant was layered on top of two sucrose layers prepared in centrifuge tubes for a Ti 60 rotor: 5 ml of 2.0 M sucrose, 250 mM KCl, 50 mM Tris-HCl, pH 7.4, 20 mM MgCl$_2$, 3 mM DTT, and 3 ml of a 1:3 (v/v) mixture of the supernatant and 2.0 M sucrose, 0.25 M KCl, 50 mM Tris-HCl, pH 7.4, 20 mM MgCl$_2$, 3 mM DTT.

Since polysomes that carry cytochrome c mRNA are small, in addition to ribosomal pellets, the lower portion of 2.0 M sucrose layer was used for extraction of RNA after the absorbance of 260 nm and 280 nm had been examined. Total RNA was extracted from polysomes using chloroform and isoamyl alcohol (24:1, v/v).[11] Poly(A) containing RNA was isolated by oligo (dT) cellulose chromatography,[12] using a binding buffer containing 0.1% SDS and NaCl instead of KCl.

C. Mitochondria, Mitoplasts, and Other Mitochondrial Membranes

a. Mitochondria. Mitochondria were prepared essentially according to the procedures described by Greenawalt[13,14] with several modifications indicated below. Briefly, livers from control, unstarved rats, were homogenized in cold 220 mM D-mannitol, 70 mM sucrose, 2 mM HEPES, pH 7.4, at 0°, and 1 mM EDTA (medium H) (2 ml per gram of liver) using a Potter–Elvehjem homogenizer with a motor-driven Teflon pestle. The homogenate was mixed with an equal volume of medium H and centrifuged at 660 g for 10 min in a Sorvall SS-34 rotor. The supernatant was then centrifuged at 4400 g for 10 min, whereas the mitochondrial pellet was

---

[9a] S. Gaetani, J. A. Smith, R. A. Feldman, and T. Morimoto, this series, Vol. 96, p. 3.

[10] J. C. Ramsey and N. J. Steele, *Biochemistry* **15**, 1704 (1976).

[11] K. Oda and W. K. Joklik, *J. Mol. Biol.* **27**, 395 (1967).

[12] H. Aviv and P. Leder, *Proc. Natl. Acad. Sci. U.S.A.* **69**, 1408 (1972).

[13] J. W. Greenawalt, this series, Vol. 31, p. 310.

[14] J. W. Greenawalt, this series, Vol. 55, p. 88.

suspended in medium H (4 ml per gram of liver) and centrifuged again at 4400 g for 10 min. This step was repeated and the final mitochondrial pellet was suspended in medium H (2 ml per gram of liver). The protein content was determined by Lowry's procedure.[15]

b. Mitoplast. The mitochondrial suspension was treated with repurified digitonin[16] (1 mg of digitonin per 10 mg of mitochondrial protein) at 0° for 15 min. After treatment, the suspension was diluted 10-fold with cold medium H and centrifuged at 12,000 g for 5 min. The pellet was resuspended in mediun H (an equivalent volume to the diluted suspension) and centrifuged again at 12,000 g for 15 min. This step was repeated twice, and the final mitoplast pellet was suspended in medium H (2 ml per gram of liver).

c. Outer Membranes. Freshly isolated mitochondria were resuspended in two times concentrated medium H. They were then loaded in a French press, and a pressure of 1500 psi was applied. The suspension was extruded at a rate of about 5 ml/min and diluted with an equal volume of two times concentrated medium H. The suspension was centrifuged for 10 min at 12,000 g. The supernatant was removed and saved, and the mitoplast fraction was suspended in medium H and centrifuged again. The supernatants from the first and second centrifugation were combined and centrifuged at 27,000 g for 10 min to remove damaged and broken mitochondria. The resulting supernatant fluid was centrifuged at 144,000 g for 30 min to sediment the outer membrane vesicles. Cytochrome oxidase activity[17] was tested in mitoplast and outer membrane fractions. Less than 10% of total mitochondrial cytochrome oxidase activity was found in the outer membrane fraction.

D. Preparation of Rabbit Anti-Mouse Cytochrome c IgG

Rabbit anti-mouse cytochrome c IgG was prepared according to the procedure described by Urbanski and Margoliash[18] and Jammerson and Margoliash[19] by injecting purified mouse cytochrome c, which has the same amino acid sequence as rat cytochrome c.[20] The antibody thus obtained is a mixture of three different monospecific ones, whose antigenic sites are an alanine residue at position 44, an aspartic acid residue at position 62, and a glycine residue at position 89, respectively. This antibody can form immune

[15] O. H. Lowry, N. J. Rosebrough, A. L. Farr, and R. J. Randall, *J. Biol. Chem.* **193**, 265 (1951).
[16] E. Kun, E. Kirsten, and W. N. Piper, this series, Vol. 55, p. 115.
[17] F. G. Smith and E. Stotz, *J. Biol. Chem.* **179**, 891 (1949).
[18] G. K. Urbanski and E. Margoliash, *J. Immunol.* **118**, 1170 (1977).
[19] R. Jammerson and E. Margoliash, *J. Biol. Chem.* **254**, 12706 (1979).
[20] S. S. Carlson, G. A. Mross, A. C. Wilson, R. T. Mead, L. D. Wolin, S. F. Bowers, N. T. Foley, A. O. Muijser, and E. Margoliash, *Biochemistry* **16**, 1437 (1977).

complexes with both mature and newly formed rat liver apocytochrome c, as well as horse holo- and apocytochrome c and rabbit holo- and apocytochrome c.

E. In Vitro System Used for the Synthesis of Cytochrome c

The rabbit reticulocyte translation system described by Pelham and Jackson[21] was used with the following modifications.

1. The system contained 200 units of trasylol per milliliter (Mobay Chemical Corporation, New York), which was added to protect newly synthesized polypeptides from proteolysis by endogenous proteases (see Section III,B).
2. The system contained a mixture of 1 mM o-phenanthroline, 0.2 mg of pepstatin, and 0.5 mg of L-leucylleucylleucine per milliliter only when a trapping system introduced by Palmiter[22] was used to avoid N-terminal acetylation. In the absence of protease inhibitors, we could not immunoprecipitate newly synthesized cytochrome c, probably because of proteolytic degradation (see Fig. 1).

Each translation mixture contained in 50 μl, 49% lysate, 0.25 mM EGTA, 5 mM NaCl, 0.125 mM DDT, 10 mM HEPES, pH 7.5, 0.018 mM hemin, 36.5 μg of creatine kinase per milliliter, 10 mM creatine phosphate, 100 mM KCl, 0.2 mM spermine, 0.7 mM ATP, 0.14 mM GTP, 1.1–1.2 mM MgCl$_2$, 0.03 mM of each of 19 amino acids, 15 μCi of [^{35}S]methionine (800 Ci/mmol, Amersham Corp., Illinois), 0.05 A_{260} unit of poly(A)$^+$ mRNA, and 200 units trasylol per milliliter (Mobay Chemical Corporation, New York) (see Section III,B).

Translation mixtures were incubated at 28–29°, and 5-μl aliquots were taken after various time intervals to measure total radioactivity incorporation into hot trichloroacetic acid insoluble material. Samples were transferred to Whatman 3 MM paper disks, which were placed in 5% trichloroacetic acid containing 1% casein hydrolyzate and processed according to Mans and Novelli.[23]

The volume of translation mixture should be chosen so as to obtain about 10^4 cpm as immunoprecipitates based on the observation that 0.1–0.3% of total hot trichloroacetic acid-insoluble radioactivity is recovered as immunoprecipitates when free polysomal mRNA from T_3-treated rats is used.

[21] H. R. E. Pelham and R. J. Jackson, Eur. J. Biochem. 67, 247 (1976).
[22] R. D. Palmiter, J. Biol. Chem. 252, 8781 (1977).
[23] R. J. Mans and G. D. Novelli, Arch. Biochem. Biophys. 94, 48 (1961).

F. Immunoprecipitation and Gel Electrophoresis

a. From Total Incubation Mixture. After incubation the translation mixture was made 2% in SDS and heated to 100° for about 2 min. The mixture was then diluted fivefold with a solution containing 190 mM NaCl, 50 mM Tris-HCl, pH 7.4, 6 mM EDTA, 2.5% Triton X-100[24] and centrifuged at 12,800 g for 2 min in an Eppendorf microfuge. The supernatant was removed, mixed with anti-mouse cytochrome c (20 μl of 1.6 mg of IgG was added per milliliter to 100–250 μl of the translation mixture), and kept at 4° overnight. Protein A coupled to Sepharose CL-4B (Pharmacia Fine Chemicals, New Jersey), was added (10 μg of protein A–Sepharose per 20 μl of IgG solution), and the mixture was stirred at room temperature for 2–3 hr. The antigen–antibody–protein A–Sepharose complex was collected by centrifugation at 12,800 g for 2 min in an Eppendorf microfuge and washed at least four times with the same buffered solution by repeated cycles of suspension and centrifugation. Washed immunoprecipitates were dissolved in sample buffer and analyzed by electrophoresis at room temperature in SDS–polyacrylamide gels (10%)[25] using a current of 23 mA for 18 hr. The location of radioactive bands in stained and unstained gels was determined by fluorography.[26]

b. From Mitochondrial and Supernatant Fractions. As soon as post-translational incubation was over, the mixture was centrifuged in an Eppendorf centrifuge at maximum speed for 5 min. The pellet was suspended in 200 μl of 0.25 M sucrose, 10 mM HEPES buffer, pH 7.4, containing chymostatin (0.1 mg/ml), a proteinase inhibitor, and nonradioactive methionine (10^{-3} M), and the suspension was made 2% in SDS. After heating at 100° for 2 min, both mitochondrial and supernatant fractions were processed as described in the preceding sections.

G. Apocytochrome c and CNBr Fragmentation of Apocytochrome c

Heme was removed from horse heart cytochrome c (Sigma) or mitochondrial cytochrome c from other sources, such as rabbit and yeast, according to the method described by Fisher et al.[27] Cytochrome c (80 mg) was dissolved in 2 ml of distilled water, which was mixed with 19.6 ml of Ag_2SO_4 solution, which was prepared as follows: 160 mg of Ag_2SO_4 were dissolved in 18 ml of distilled water by warming up to 50–60°, so that most Ag_2SO_4 would be dissolved, and, while still warm, the Ag_2SO_4 solution was mixed with 1.6 ml of concentrated acetic acid. The mixture was incubated at 40° for 4 hr with occasional shaking, and at the end of the incubation, it

[24] B. M. Goldman and G. Blobel, Proc. Natl. Acad. Sci. U.S.A. 75, 5066 (1978).
[25] J. V. Maizel, Jr., Method Virol. 5, 179 (1971).
[26] R. A. Laskey and A. D. Mills, Eur. J. Biochem. 56, 335 (1975).
[27] W. R. Fisher, H. Taniuchi, and C. B. Anfinsen, J. Biol. Chem. 248, 3188 (1973).

was centrifuged at 10,000 rpm for 10 min in an HB-4 rotor (Sorvall) to remove aggregates. The supernatant was applied to a Sephadex G-25 column (2×90 cm) that was previously equilibrated with 0.1 N acetic acid. Protein was eluted with 0.1 N acetic acid at a void volume (the fraction has a slightly yellow color) and lyophilized.

The lyophilized protein was then dissolved in 2 ml of 0.05 M ammonium acetate, pH 5.0, 6 N guanidine hydrochloride, and 1 M DTT, which was allowed to stand at 20° for 2 hr in the dark. The precipitates of silver mercaptide of DTT were removed by centrifugation at 10,000 rpm for 10 min. The supernatant was applied to a Sephadex G-50 column (3.5×30 cm) that had been equilibrated with 0.05 M ammonium acetate, pH 5.0. Protein eluted near the void volume was pooled and lyophilized.

The cyanogen bromide fragments were prepared according to the procedure described by Gross and Witkop.[28] Lyophilized apocytochrome c prepared from 80 mg of cytochrome c was dissolved in 1 ml of 0.1 N HCl and then mixed with 4 ml of 0.1 N HCl containing 100 mg of CNBr. The mixture was allowed to stand for 30 hr at room temperature with constant stirring in the dark, and thereafter the mixture was lyophilized. The lyophilized protein was dissolved in 1 ml of 10% formic acid and applied to to BioGel P6 column (1×100 cm) that had been previously equilibrated with 10% formic acid, and fragments were eluted with 10% formic acid (5 ml per fraction). The fraction under each peak was collected and lyophilized, and its size was routinely examined by SDS-PAGE containing 6 M urea according to the procedure by Marciani, which was modified from the procedure of Shapiro et al.[29] (see Fig. 2).[30]

II. *In Vitro* Products

A. *Amount of Newly Synthesized Cytochrome c Recovered by Immunoprecipitation*

When a 100-μl translation mixture was programmed with 0.1 OD_{260} unit of mRNA at 28°, [^{35}S]methionine (800 Ci/mmol) was incorporated into hot trichloroacetic acid-insoluble radioactivity linearly for the first 30 min, and the incorporation reached a plateau in 60 min. After 60 min of incubation, an average of 0.3% (2.4×10^4 cpm) of the total hot trichloroacetic acid-insoluble radioactivity was recovered as immunoprecipitate, which represents 0.026 pmol of [^{35}S]methionine, the counting efficiency of the filter disk method being 50%.

[28] E. Gross and B. Witkop, *J. Biol. Chem.* **167**, 1856 (1962).
[29] A. L. Shapiro, E. Viñuela, and J. B. Maizel, *Biochem. Biophys. Res. Commun.* **28**, 815 (1967).
[30] S. Isomura and T. Ikenaka, *J. Biochem. (Tokyo)* **79**, 1182 (1976).

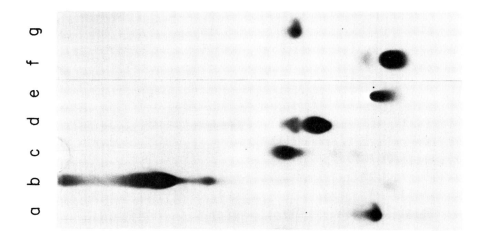

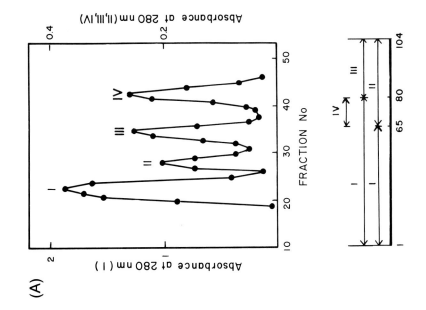

It has been shown[7] that the primary translation product in the presence of trapping agent contains methionine at its N terminus as an extra amino acid. It was also observed that the radioactivity of cytochrome c synthesized *in vitro* in the absence of the trapping agent was two-thirds that of the cytochrome c synthesized in the presence of trapping agent and the former product could not be analyzed by Edman degradation. This indicates that cytochrome c synthesized in the absence of trapping agent has N-acetylated glycine at the N terminus. Since mature cytochrome c contains two internal methionine residues, the amount of newly synthesized cytochrome c is, therefore, calculated as 0.013 pmol based on the specific activity of radioactive methionine.

The amount of newly synthesized cytochrome c was estimated according to the following observations.

1. The amount of ribosomes participating in the protein synthesis programmed with exogenously added mRNA represents about 15% of total ribosomes present in translation mixture (since the 100-μl translation mixture contains 3 OD_{260} units of ribosomes, only 0.45 OD_{260} unit of ribosomes (6.6 pmol) participates in protein synthesis when *Micrococcus* nuclease-treated lysate is used).

2. The rate of elongation in the present translation conditions is estimated to be one amino acid per second base on the time required for the synthesis of rat serum albumin (M_r 68,000) during the linear incorporation phase. As already mentioned, the incorporation is usually linear for the first 30 min and about 5.6×10^6 cpm of hot trichloroacetic acid-insoluble radioactivity is obtained during this period. This accounts for 6.4 pmol of [^{35}S]methionine. Since the average number of methionine residues in proteins is 2% of the total number of amino acid residues,[31] the amount of

---

[31] M. O. Dayhoff, "Atlas of Protein Sequence and Structure," Vol. 5, Suppl. 2. Natl. Biomed. Res. Found., Washington, D.C., 1976.

FIG. 2. Elution profiles of CNBr fragments of horse apocytochrome c (A) and their SDS-PAGE analysis (B). (A) Horse apocytochrome c was subjected to CNBr fragmentation as described in the text. Cytochrome c contains methionyl residues at positions 65 and 80[20] as indicated in the figure. Therefore, incomplete cleavage resulting from the treatment generate five peptide fragments. Separation of the products of gel filtration yielded four fractions (indicated by I–IV) that were identified from their amino acid composition. Fraction I contains segments composed of residues 1–65 (about 90%) and 1–80 (about 10%).   (B) The size and purity of isolated fractions were examined by SDS-PAGE according to the procedure described by Shapiro *et al.*[29] Fractions A, B, and C separated by Sephadex G-25 column chromatography of rat serum albumin which had been treated by CNBr[30] were used as markers, in addition to horse cytochrome c and insulin. a, Albumin fraction C, 5900; b, albumin fraction A, 37,600 as a major component; c, albumin fraction B, 16,200; d, Apocytochrome c fragment I, 7300; e, apocytochrome c fragment II, 5200; f, insulin; 3000; g, horse cytochrome c, 12,500.

endogenous methionine is about 37 times as much as that of input radioactive methionine. The actual amount of newly synthesized cytochrome c is therefore 0.33 pmol/100 μl in 30 min.

B. Partial Amino Acid Sequence Analysis of N-Terminal Region

Cytochrome c was synthesized *in vitro* in a translation mixture containing [^{35}S]methionine or [^{3}H]lysine, to which the acetyl trap introduced by Palmiter[22] to prevent NH$_2$-terminal acetylation was added. In addition, a mixture of proteinase inhibitors (1 mM o-phenanthroline, L-leucylleucyl-leucine (0.5 mg/ml), and pepstatin (0.2 mg/ml)) were added for the reasons described before (see Section I,E). Newly synthesized cytochrome c was purified by immunoprecipitation, followed by preparative SDS-PAGE. A band corresponding to that of mature cytochrome c was extracted from the gel and homogenized in about 2 volumes of distilled water using a Potter glass homogenizer with a Teflon pestle. The homogenate was allowed to stand for several hours or longer in the cold, and then centrifuged at 10,000 rpm for 10 min using a Sorvall HB-4 rotor. The supernatant was transferred into a new tube, and the pellet was rehomogenized in 2 volumes of distilled water. These steps were repeated until no radioactivity was detected in the supernatant. All supernatants were combined and centrifuged at 30,000 rpm for 15 min in a Ti 50 rotor to remove fine acrylamide. The supernatant was mixed with 10 volumes of acetone containing 0.1 N HCl, and the mixture was kept at $-20°$ overnight. The precipitates were collected by centrifugation at 10,000 rpm for 10 min. The pellet was briefly washed with methanol and dried over a vacuum, and the dried sample was dissolved in 50% formic acid. Samples obtained from [^{35}S]methionine-labeled and ^{3}H-labeled immunoprecipitates were combined and applied to the Beckman 890 C sequencer, which was run with the 0.1 M Quadrol program as described by Brauer *et al.*[32] It is recommended to prepare polypeptides that had been labeled separately with two different radioactive amino acids, because it is easier to adjust the radioactivity ratio to the two peptides.

III. Posttranslational Transfer of Newly Synthesized Cytochrome c into Mitochondria

A. Posttranslational Incubation and Subsequent Analysis

Mitochondria were prepared from the liver of unstarved rats as described in Section I, and the amount of mitochondria used was expressed as that of total mitochondrial protein determined as described by Lowry *et al.*[15]

---

[32] A. W. Brauer, M. N. Margoliash, and E. Harber, *Biochemistry* **14**, 3029 (1975).

A 100-μl translation mixture programmed with 0.1 OD_{260} unit of free polysomal mRNA at 28° for 60 min was further incubated with 300 μg of intact mitochondria at 28° for 10–60 min in the presence of cycloheximide (10^{-2} M) to ensure that any uptake of labeled polypeptides into mitochondria during the 1-hr incubation did not involve a cotranslational mechanism and 0.33 M sucrose. After a posttranslational incubation, mitochondria were separated by centrifugation at a maximum speed for 5 min using an Eppendorf centrifuge, and both mitochondria and supernatants were analyzed as described in the preceding section.

B. Comments of Posttranslational Incubation

a. Efficiency of the Posttranslational Transfer.. As described previously, 0.3 pmol of cytochrome c was synthesized in a 100-μl translation mixture, and more than 90% of that was transferred into mitochondria when incubated posttranslationally with 300 μg of mitochondria at 28° for 60 min (Fig. 3). The transfer seemed to be relatively fast, because 80% or more of newly synthesized cytochrome c was already taken up by mitochondria during the first 10 min of incubation (therefore, most transfer experiments were performed by 30 min of incubation).

b. Factors that Affect the Posttranslational Transfer. i. TEMPERATURE. By lowering temperatures for the posttranslational incubation, the amount of newly synthesized cytochrome c taken up by mitochondria was greatly reduced. For example, 50% or less was transferred in the first 10 min when incubated at 4°. Temperatures higher than 30° were not used, because undesirable phenomena, such as proteolysis of newly synthesized peptides and less efficient transfer of newly synthesized cytochrome c were frequently noted.

ii. AMOUNT OF MITOCHONDRIA. The efficiency of posttranslational transfer is strongly dependent upon the amount of newly synthesized cytochrome c in the translation mixture and the number of mitochondria. Usually, we obtained a constant efficiency of the transport when 200–300 μg of mitochondria per 100 μl of posttranslational incubation mixture were used.

iii. INTACTNESS OF MITOCHONDRIA. It is important to start the posttranslational incubation as soon as both mitochondria and translation mixture which had been programmed are ready. The longer the samples were allowed to stand, even in the cold, the less efficient the transfer. For this reason, we usually prepared mitochondria so as to be ready just when the *in vitro* translation was ending. It is also important to avoid disruption of mitochondria that might occur during steps of purification, such as pelleting and subsequent suspending. When intact mitochondria were used, good

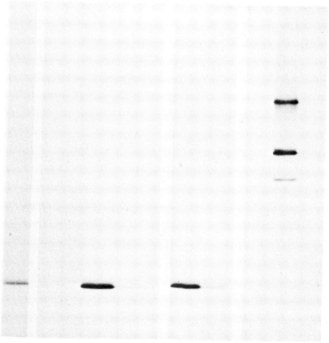

FIG. 3. Posttranslational transfer of *in vitro* synthesized cytochrome *c* into mitochondria. After incubation for translation with [³⁵S]methionine, a mixture programmed with mRNA from free polysomes was divided into three portions (250 μl each), one of which was used directly for immunoprecipitation with anti-rat serum albumin (lane b). The two others (lanes c and d) were incubated at 28° for 60 min in the presence of cycloheximide (10⁻² *M*) with intact mitochondria (750 μg of mitochondria). One of these samples (lane d) was further incubated with trypsin (70 μg/ml) at 4° for 30 min, followed by the addition of soybean trypsin inhibitor (350 μg per milliliter of mixture, 0–4°, 10 min). As a control, a translation mixture that has been programmed with mRNA from bound polysomes was divided into two portions (100 μl each): one of these was subjected to immunoprecipitation with anticytochrome *c* (lane f); the other was further incubated with 300 μg of mitochondria in the presence 10⁻² *M* cycloheximide (lane e). After posttranslational incubation, each mixture was centrifuged and the mitochondrial (m) and supernatants (s) fractions were subjected to immunoprecipitation with anticytochrome *c* (lanes c and d) and antialbumin (lane e), followed by SDS-PAGE as described in the text. ¹²⁵I-labeled cytochrome *c* was used as a marker (lane a).

transfer was observed, even when it was performed in the absence of proteinase inhibitors.

iv. ENERGY REQUIREMENT. Lowering the temperature of posttranslational incubation caused a decrease in the rate of transfer of newly synthesized cytochrome c into mitochondria, which suggests, as in the case of other mitochondrial proteins[33] and chloroplast proteins,[34] that the transfer requires energy. This possibility was examined as follows: A large excess of creatine and creatine kinase were added to a translation mixture that had been programmed with mRNA. The mixture was then reincubated with mitochondria that has been preincubated with sodium azide (10^{-3} M) and 2,4-dinitrophenol (10^{-5} M). Under these conditions a significant reduction was noted in the rate of the posttranslation transfer.

c. *Specificity of the Posttranslational Transfer.* That the posttranslational transfer is very specific was corroborated by the observation that other subcellular membranes, such as microsomes and plasma membranes, failed to take up newly synthesized cytochrome c (Fig. 4) and that the primary translation product of secretory proteins, such as preproalbumin, was not taken up by mitochondria (Fig. 3).

d. *Proteinase Inhibitors.* When this work has initiated, we encountered unspecific (detected as overall decrease of hot trichloroacetic acid-insoluble radioactivity as the incubation continued) and specific proteolysis, particularly when the N-acetyl blocking agent was used.[22] Newly synthesized cytochrome c was degraded, but newly synthesized preproalbumin was not. This degradation could be minimized upon addition of trasylol (for the unspecific proteolysis) and a mixture of proteinase inhibitors (for the specific proteolysis) as described in Section I,E. Proteolysis of newly synthesized cytochrome c was also observed during the posttranslational incubation, which seemed to be very specific to mitochondrial proteins because the proteolysis of newly synthesized preproalbumin was not observed under the same conditions. This proteolysis, however, could be minimized upon addition of chymostatin.

During the course of this study, we noticed that the extent of unspecific proteolysis varied depending upon batches of the reticulocyte lysate preparation. Similarly, the extent of specific proteolysis, in addition to its dependence on the intactness of mitochondria, also varied with the reticulocyte lysate batches. Furthermore, the presence of chymostatin upon solubilization of mitochondria was found to be critical in protecting cytochrome c from proteolysis. Since it is generally not desirable to use proteinase inhibitors during translation and the subsequent incubation with mitochondria for reasons described below, the posttranslational incubation was carried out in the presence of 0.33 M sucrose if the lysate contained undetectable proteolytic activity, and chymostatin was added before mitochondria were solubilized.

In many cases various proteinase inhibitors are required for this kind of work. However, it should be examined whether they affect radioactive

[33] N. Nelson and G. Schatz, *Proc. Natl. Acad. Sci. U.S.A.* **76,** 4365 (1979).
[34] A. Grossman, S. Bartlett, and N. H. Chua, *Nature (London)* **285,** 625 (1980).

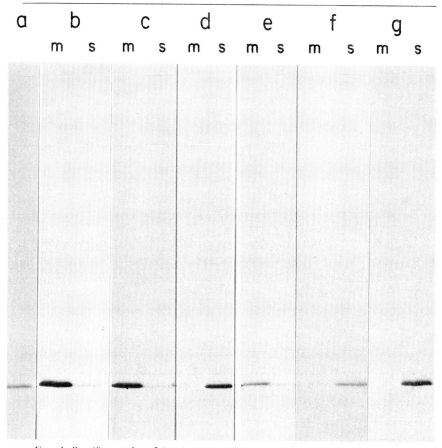

FIG. 4. Specific uptake of *in vitro* synthesized cytochrome *c* by mitochondria. After incubation for protein synthesis (60 min) with [^{35}S]methionine, translation mixtures (250 μl each) programmed with mRNA from free polysomes received cycloheximide (10^{-2} M) and were incubated at 28° for 60 min with mitochondria (750 μg of protein) (lanes b, c, d, and e) or equivalent amounts of rat liver rough microsomal fraction (lane f) or rabbit red blood cell ghosts (lane g). Samples shown in lanes c and d received 80 μg of rabbit holocytochrome c (lane c) or 100 μg of rabbit apocytochrome c (lane d) or 250 μg of polylysine (lane e). After the posttranslational incubation, the mitochondrial or other added membranes were reisolated by centrifugation. Resuspended sediments (m) and supernatants (s) were analyzed as described in the text. A high concentration of antibody was used to recover the *in vitro* product, which in the presence of apo- and holocytochrome *c* (lanes c and d) remained in the supernatant after incubation with mitochondria. ^{125}I-labeled horse cytochrome *c* was run as a marker (lane a).

amino acid incorporation into proteins and posttranslational transfer of newly synthesized polypeptides.

C. Transfer Experiments using ^{125}I-Labeled Horse Apocytochrome c

Holocytochrome *c* (either horse or rabbit) was labeled with ^{125}I by the lactoperoxidase-H_2O_2 method according to the procedure of Osheroff *et al.*[35] This method allows for the preparation of monoiodotyrosyl 74 cytochrome *c*, whose physiological and physicochemical properties are different from those of native cytochrome *c*. Holocytochrome *c* is also labeled with ^{125}I by the chloramine-T method[36] under mild conditions that do not oxidize methionine residues. ^{125}I-labeled apocytochrome *c* was then prepared as described in Section I,G, and was used for the posttranslational transfer experiments using either a reticulocyte lysate translation mixture that does not contain mRNA and radioactive methionine or an energy mixture containing, ATP, GTP, creatine phosphate, and creatine kinase in the same concentration as that used for translation. After incubation, the mixture was analyzed as described in Section I,F.

D. Evidence That Newly Synthesized Cytochrome c is Transferred into Mitochondria

In order to rule out the possibility that newly synthesized cytochrome *c* binds to the cytoplasmic surface of outer mitochondrial membranes through ionic interaction, since it contains a large number of lysine residues, the following three experiments were performed.

a. Effects of Polylysine on the Posttranslational Transfer. Posttranslational incubation was carried out in the presence of an increasing amount of polylysine (up to 1 mg/ml) that had been added before mitochondria (Fig. 4).

b. Resistance of Newly Synthesized Cytochrome c to Trypsin Digestion. After posttranslational incubation, the mixture was chilled down at 4°, and trypsin (70 μg/ml) was added. The mixture was allowed to stand at 4° for 30 min, followed by the addition of soybean trypsin inhibitor (350 μg/ml at 0° for at least 10 min). The mixture was centrifuged to isolate mitochondria from the supernatant. The mitochondrial pellet was suspended in 200 μl of 50 mM Tris-HCl, pH 7.4, buffer containing soybean trypsin inhibitor (350 μg/ml) and chymostatin (0.1 mg/ml).

The suspension (after standing for a few minutes at 4°) was subjected to immunoprecipitation as described before, followed by SDS-PAGE analysis (Fig. 3).

c. Conversion of Apocytochrome c to Holocytochrome c. As will be shown later, holocytochrome *c* cannot be transferred into mitochondria. More direct evidence for the posttranslational transfer can be obtained by showing that newly synthesized cytochrome *c* taken up by mitochondria is holocytochrome. Since the reduced form of holocytochrome *c* is resistant to

[35] N. Osheroff, B. A. Feinberg, and E. Margoliash, *J. Biol. Chem.* **252**, 7743 (1977).
[36] W. H. Hunter, "Atomic Energy Commission Series Radioactive Pharmaceuticals," p. 253. U.S. At. Energy Comm., Oak Ridge, Tennessee, 1966.

trysinization,[37] while the oxidized form as well as the apo form of cytochrome c are sensitive to trypsinization, this property can be used to detect whether the conversion from apo to holo form takes place.

Newly synthesized cytochrome c was isolated by immunoprecipitation from mitochondria that has been posttranslationally incubated, and was dissolved in 200 μl of 50 mM Tris-phosphate, pH 6.7. The cytochrome c was reduced with a few crystals of sodium ascorbate under N_2 atmosphere. A 3-μl trypsin solution (2 mg/ml, 1 mM HCl) was added, and the solution was allowed to stand at room temperature for 20 min, before 6 μl of soybean trypsin inhibitor (5 mg/ml) was added. After 20 min at room temperature, cytochrome c was reimmunoprecipitated with anticytochrome c and subjected to SDS-PAGE. Newly synthesized cytochrome c isolated from a translation mixture by immunoprecipitation was used as control.

E. Structural Feature(s) Required for Recognition by Outer Mitochondrial Membrane

a. Competition with Apo- or Holocytochrome c for the Transfer. A 100-μl translation mixture programmed with free-polysomal mRNA was incubated at 28° for 60 min with 300 μg of mitochondria in the presence of an increasing amount of holocytochrome c or apocytochrome c that was prepared for horse heart holocytochrome c. After incubation, the mixture was analyzed.

b. Interaction between Newly Synthesized Cytochrome c and Mitochondrial Membranes. In order to determine whether other mitochondrial membrane proteins were involved in the recognition of newly synthesized cytochrome c, 300 μg of mitochondria that had been treated with trypsin (10 μg/ml) at 0° for 30 min, followed by incubation with 50 μg of soybean trypsin inhibitor, were then incubated with a 100-μl translation mixture. It was confirmed that this treatment did not cause any changes in the electrophoretic pattern of mitochondrial proteins. It was also observed that the presence of soybean inhibitor did not affect the posttranslational transfer of the newly synthesized cytochrome c. After incubation at 28° for 30 min, the mixture was centrifuged and both mitochondrial and supernatant fractions were subjected to immunoprecipitation, followed by SDS-PAGE. Under these conditions, the transfer was greatly reduced.

To determine whether the interaction is strictly due to the outer mitochondrial membrane, the following experiments were carried out. Two 100-μl translation mixtures that had been programmed with free polysomal mRNA were incubated at 28° for 30 min separately with outer membranes and mitoplasts, which are equivalent in the amount to those contained in

[37] M. Nozaki, H. Mizushima, T. Horio, and K. Okunuki, *J. Biochem.* (*Tokyo*) **45**, 815, (1958).

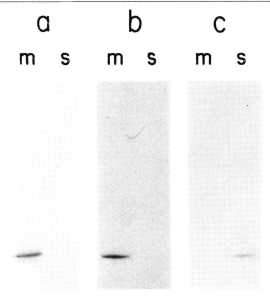

FIG. 5. Inhibition of the mitochondrial uptake of *in vitro* synthesized cytochrome c by a fragment of the protein. Each translation mixture (100 µl) which had been programmed with mRNA (0.05 OD_{260} unit) was reincubated with mitochondria (300 µg) in the presence of cycloheximide (10^{-2} M) without further addition (lane a) or with addition of 1 nmol of fragments I, III, or IV (lane b) and fragment II (lane c). After incubation, the mixture was centrifuged. Resuspended sediments (m) and supernatants (s) were subjected to immunoprecipitation with anticytochrome c. An excess amount of antibody was used for the immunoprecipitation from the supernatant of lane c in which fragment II remained, because it has been observed that the fragment competes with newly synthesized cytochrome c for immunoreaction.

300 µg of mitochondria. Also 10 µg of ^{125}I-labeled holocytochrome were incubated with outer mitochondrial membranes and mitoplasts under the same conditions as described above, namely, in the translation mixture from which [^{35}S]methionine and mRNA were eliminated. After incubation the mixtures were centrifuged at 27,000 g for 10 min (for the isolation of outer membranes) and at 12,000 g for 5 min (for the isolation of mitoplasts). Both sedimented and unsedimented materials were subjected to immunoprecipitation.

It was found that a newly synthesized cytochrome c (apo form) binds to outer mitochondrial membranes, but not to mitoplasts, whereas radioactive holocytochrome c cannot bind to outer membranes, but binds to mitoplasts.

c. Effects of Fragments Generated by CNBr Treatment of Apocytochrome c. Experiments a and b indicate that a putative receptor in the outer mitochondrial membrane recognizes features of the cytochrome c polypeptide that are present in the apocytochrome but not in the fully folded heme containing holoprotein. To test whether a specific segment of the polypeptide plays a role in mitochondrial recognition, fragments produced by CNBr treatment of apocytochrome c were separated as described in the preceding section and used for competition for the transfer of a newly synthesized cytochrome c into mitochondria: A 100-μl translation mixture programmed with free polysomal mRNA was incubated at 28° for 30 min with 300 μg of mitochondria in the presence of an increasing amount of each fragment. After incubation, the mixture was centrifuged to separate mitochondrial and supernatant fractions, both of which were then subjected to immunoprecipitation, followed by SDS-PAGE. It was found that the transfer was almost completely blocked in the presence of 1 – 10 nmol of fragment II, but not in the presence of the other fragments (Fig. 5).

Note. Section V, "Concluding Remarks," appears on p. 621.

[40] Isolation of Mammalian Mitochondrial DNA and RNA and Cloning of the Mitochondrial Genome

By Douglas P. Tapper, Richard A. Van Etten, and David A. Clayton

Analysis of the structure and function of the mammalian mitochondrial genome has been greatly facilitated by improved methods of mitochondrial nucleic acid isolation and the development of specific mitochondrial nucleic acid probes. The ability to clone portions, as well as the entire mouse mitochondrial DNA (mtDNA) genome, has provided an economical and convenient source of DNA substrate that has been utilized to obtain the genomic sequence.[1] These cloned elements have also been employed in analyses of mtDNA replication[2,3] and transcription.[4,5] We describe here our current protocol for the rapid isolation of mtDNA and mitochondrial RNA (mtRNA) from human and mouse tissue culture lines. In addition, we show the ease with which mouse mtDNA can be stably maintained as a cloned species in *Escherichia coli* as compared to human mtDNA.

[1] M. J. Bibb, R. A. Van Etten, C. T. Wright, M. W. Walberg, and D. A. Clayton, *Cell* **26,** 167 (1981).
[2] P. A. Martens and D. A. Clayton, *J. Mol. Biol.* **135,** 327 (1979).
[3] D. A. Clayton, *Cell* **28,** 693 (1982).
[4] J. Battey and D. A. Clayton, *Cell* **14,** 143 (1978).
[5] R. A. Van Etten, M. W. Walberg, and D. A. Clayton, *Cell* **22,** 157 (1980).

METHODS IN ENZYMOLOGY, VOL. 97

Isolation of Mitochondria from Mammalian Cell Cultures

Solutions and Reagents

TD buffer: 133 mM NaCl; 5 mM KCl; 25 mM Tris-HCl, pH 7.5; 0.7 mM Na$_2$HPO$_4$

RSB buffer: 10 mM NaCl; 1.5 mM CaCl$_2$; 10 mM Tris-HCl, pH 7.5

MS buffer: 210 mM mannitol; 70 mM sucrose; 5 mM Tris-HCl, pH 7.5; 5 mM EDTA (added as 100× at pH 8)

STE buffer: 100 mM NaCl; 50 mM Tris-HCl, pH 8.5; 10 mM EDTA

Isolation Procedure. Mammalian spinner culture cells (mouse LA9 or human KB) are routinely concentrated from the culture medium by centrifugation at 1500 rpm (250 g at r_{av}, 9.8 cm) in a Beckman JA-10 rotor. All subsequent steps are performed at 4°. The cells are resuspended and washed in TD buffer and concentrated by centrifugation at 1500 rpm. The cell pellet is then resuspended in RSB to give 10 times the original volume of the packed cells. Sufficient time is given for the cells to swell ($\sim$ 7 – 10 min) prior to disruption with a Dounce homogenizer (pestle A). We routinely find that either insufficient swelling or homogenization leads to low yields of mitochondria. Homogenates should be assayed for cell breakage by phase-contrast microscopy. After disruption, a 2.5× solution of MS buffer is added to give a 1× final concentration of MS. This solution is then gently mixed, and nuclei are removed by three sequential centrifugations at 2500 rpm (1260 g) in a GLC-1 Sorvall centrifuge at 4°. Mitochondria are then isolated by centrifugation at 20,000 g for 15 min. The mitochondria from the equivalent of approximately 10^9 cells are resuspended in 1 ml of STE buffer immediately prior to nucleic acid isolation.

Sucrose Gradient Purification of Mitochondria

Solutions and Reagents

Sucrose-TE: 20% sucrose; 50 mM Tris-HCl, pH 7.5; 10 mM EDTA

Upper sucrose solution: 1.0 M sucrose; 10 mM Tris-HCl, pH 7.5; 5 mM EDTA

Bottom sucrose solution: 1.5 M sucrose; 10 mM Tris-HCl, pH 7.5; 5 mM EDTA

Procedure. The final mitochondrial pellet is resuspended (10^9 cell equivalents per milliliter) in sucrose-TE buffer. Seven milliliters of the resuspended mitochondria are gently layered on a step gradient containing a 15-ml bottom sucrose layer (1.5 M sucrose) and a 15-ml upper sucrose layer (1.0 M sucrose). An SW 27 cellulose nitrate (Beckman) tube is used. The gradient is centrifuged at 22,000 rpm for 20 min at 4°. The mitochondria at the interface of the two sucrose layers can be directly removed with a Pasteur pipette. The mitochondrial solution is diluted with an equal volume of MS solution and mitochondria are concentrated by centrifugation at 20,000 g prior to nucleic acid isolation.

Isolation of Mitochondrial DNA

Solution and Reagents

CsCl solution: 7 M CsCl; 10 mM Tris-HCl, pH 7.6; 1 mM EDTA. This solution is filtered through a 0.45 μm pore size cellulose nitrate filter prior to use.

Ethidium bromide (EthBr):10 mg of EthBr per milliliter in TE

Sodium dodecyl sulfate (SDS): 25% SDS in H_2O

TE buffer: 50 mM Tris-HCl, pH 7.5; 10 mM EDTA

Procedure. Mitochondria are resuspended in STE buffer: two drops of 25% SDS are added per milliliter of solution, then 2.5 ml of 7 M CsCl. EthBr is added to give a final concentration of 500 μg/ml. The refractive index is adjusted to 1.390 (density = 1.598). Gradients are centrifuged at 40,000 rpm in either a type 70 or 70 Ti rotor for at least 36 hr. Closed circular mtDNA can be easily visualized under longwave UV light if cellulose nitrate tubes are used. Replicative intermediates of mtDNA reach equilibrium in a continuum between the closed circular forms in the bottom band and the open circular forms in the upper band. The upper band is readily visible due to significant contamination with nuclear DNA. This contamination can be substantially reduced by prior sucrose-gradient purification of mitochondria. MtDNA is routinely rebanded by simply recentrifuging the DNA from the first gradient in a smaller centrifuge tube (i.e., the initial 70 Ti isolate is recentrifuged in an SW60 Ti tube). After the final centrifugation, the EthBr is removed from the DNA solution with CsCl (or NaCl)-saturated isopropanol and diluted to 1.5 M to 2.0 M CsCl with TE buffer, and the DNA is precipitated with two volumes of ethanol overnight at −20°.

Isolation of Mitochondrial RNA

Solutions and Reagents

Phenol solution: phenol crystals freshly equilibrated with 1 M Tris-HCl, pH 8.5

SDS: 25% SDS in H_2O

Procedure. Mitochondria resuspended after the final pelleting in STE are mixed with an equal volume of phenol solution, containing 1.25% SDS and prewarmed to 70°. The solution is mixed gently at 70° until a uniform single-phase solution is obtained. The solution is allowed to cool to room temperature prior to centrifugation to avoid simultaneous phase transition and phase separation, and a volume of chloroform equivalent to the starting volume of phenol is added. Chloroform is added to ensure quantitative retention of poly(A)$^+$ RNA species in the aqueous phase (J. Battey, unpublished observation). The solution is mixed and centrifuged in a Beckman JA-20 rotor at 15,000 rpm for 20 min at 4°. The aqueous phase is removed and directly added to 70° phenol (no SDS) and mixed until the solution is clear and uniform. As before, chloroform is added after cooling and the solution is centrifuged at 10,000 rpm for 10 min in a Beckman JA-20 rotor at 4°. The extraction is generally repeated one to two more times prior to ethanol precipitation. This procedure results in intact RNA with an A_{260}:A_{280} ratio of 1.8 to 2.1.

Expected yields of RNA vary according to cell type and culture growth conditions. In general, using hot phenol extraction of mitochondria isolated directly from a postnuclear supernatant, from 100 to 300 μg of crude mtRNA is routinely isolated per liter of LA9 spinner culture cells in late logarithmic phase of growth (equivalent of approximately 8×10^8 cells). Discontinuous sucrose gradient purification of mitochondria prior to extraction results in significant enrichment for mitochondrial-specific RNA species (see Fig. 1), but results in reduced yield of 30–100 μg per liter of spinner culture cells. This decreased yield is in close agreement with that observed for isolation of mtDNA using the discontinuous gradient procedure.[6] Yields for human KB cells are consistently 1.5- to 2-fold higher, which may reflect increased cell volume or increased mitochondrial density in these cells as compared to mouse LA9 cells.

The resulting RNA can be analyzed by polyacrylamide gel electrophoresis in 4% gels containing 7 M urea (Fig. 1). In lane A is total mouse cell RNA prepared by the hot phenol method. The predominant bands are 18 S rRNA from cytoplasmic ribosomes, 5.8 S rRNA and 5 S rRNA (tRNAs are not shown). In lane B is LA9 mtRNA isolated without prior sucrose-gradient purification of mitochondria; 18 S rRNA can still be observed in addition to 16 S and 12 S rRNA from mitochondrial ribosomes. In lane C is LA9 mtRNA isolated from mitochondria that have been purified on a sucrose step-gradient prior to hot phenol extraction of the RNA. 18 S rRNA is noticeably absent, with a large enrichment for the 16 S and 12 S rRNA transcripts. In addition, numerous species in both lanes A and B are absent with new species appearing. The sucrose-gradient purification step clearly enriches for mitochondrial-specific transcripts. Nuclear and cytoplasmic contamination still remain, since 5.8 S and 5 S rRNAs are still present. They may be due to the unavoidable circumstance of RNA interacting with or being trapped by the outer mitochondrial membrane.

[6] D. Bogenhagen and D. A. Clayton, *J. Biol. Chem.* **249**, 7991 (1974).

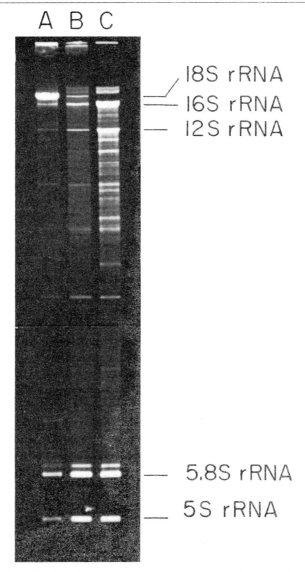

FIG. 1. Polyacrylamide gel electrophoresis of mouse cell RNA. The gel has been stained with EthBr. Lane A: Total mouse cell RNA prepared by the hot phenol method; lane B: mtRNA isolated without sucrose-gradient purification of mitochondria; lane C: mtRNA isolated with sucrose-gradient purification of mitochondria. Prominent rRNA species are indicated.

RNA isolated without a sucrose-gradient purification step (lane B), while not homogeneous mtRNA, is pure enough for many applications, including RNA blotting,[7] direct 5'-end sequence analysis of rRNA,[5] and cDNA cloning. If a homogeneous preparation of an individual mtRNA or RNAs is required, the RNA may be further purified by hybrid selection using a specific cloned region of mtDNA covalently bound to DBM paper or DBM-cellulose.[8] This technique has been used to obtain purified 12 S and 16 S rRNA species for sequence analysis of the 3' termini.[9]

Differential Cloning Capability of Mouse and Human Mitochondrial DNA

The complete mouse mitochondrial genome has been cloned into pACYC177 (pAM 1)[2] and pBR322 (see the table), and the individual HindIII/EcoRI fragments have been cloned into pBR322 (pHR B-G). These clones have been utilized in sequencing the mouse mitochondrial genome and in studies of the replication and transcription of mouse mtDNA. The complete human mitochondrial genome has not been cloned. In studies of D-loop metabolism in human KB cells, a clone spanning the complete D-loop region was desired. Only a partial sequence of the D loop could be cloned, while the remainder of the D loop was refractory to cloning.

Reagents

DNA fragments: *Mbo*I-7 (KB mtDNA 1–740); 2.8 kb-*Taq*I (KB mtDNA 14,956 to 1215)

Vectors: pBR322, M13mp7, M13Goril

Procedure. A 2.8-kb *Taq*I-A fragment spans the D-loop region of human mtDNA (Fig. 2).[10] This fragment was isolated from an agarose gel after digestion of 20 μg of KB cell mtDNA. The ends of this fragment were filled in with the Klenow fragment of DNA polymerase I, and the fragment was then ligated to the M13 cloning vector, mp7, cleaved with *Hinc*II. Insertion of foreign DNA destroys the promoter for transcription of the β-galactosidase gene that is inserted in this vector, thus preventing intracistronic complementation in trans of the *E. coli* strain, JM103, used for transformation. Plaques containing recombinant DNA are then colorless in the colorimetric assay for β-galactosidase activity. Thirty white plaques were picked, and eight of these contained the *Taq*I-A fragment. On further analysis only one orientation was present, that with the H strand (H referring to the heavier of the two strands in alkaline CsCl) in the M13 viral strand. This

[7] J. Battey and D. A. Clayton, *J. Biol. Chem.* **255**, 11599 (1980).
[8] M. L. Goldberg, R. P. Lifton, G. R. Stark, and J. G. Williams, this series, Vol. 68, p. 206.
[9] R. A. Van Etten, J. W. Bird, and D. A. Clayton, *J. Biol. Chem.* **258**, in press.

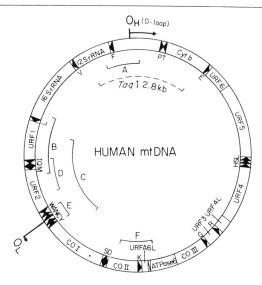

FIG. 2. The genomic map of human mitochondrial DNA (mtDNA). The 2.8-kb *Taq*I fragment is indicated by the dashed line and discussed in the text. The cloned elements are: A, nucleotides 1–740; B, nucleotides 3160–4534; C, nucleotides 3693–6460; D, nucleotides 4121–5274; E, nucleotides 5691–5917; F, nucleotides 7859–8592 (see the table). The nucleotide numbers are those of Anderson *et al.*[10] Filled triangles, tRNA genes, in the polarity shown, are identified by the single-letter amino acid code; CO I, CO II, and CO III, cytochrome *c* oxidase subunits I, II, and III; Cyt *b*, cytochrome *b*; URF, unidentified reading frame; O_H and O_L, the origins of H strand and L strand (light-strand) replication with the direction of replication indicated by the arrow at each origin. For a full description of genomic organization, see Bibb *et al.*[1] and Anderson *et al.*[10]

orientation was very unstable. In an attempt to grow a large quantity of three of the positive isolates, complete deletion of the mitochondrial sequence occurred. In addition, the β-galactosidase coding sequence in the vector was deleted, yielding wild-type virus DNA.

To rule out a problem with the vector, a second attempt was made using the M13 vector, M13Goril. The *Taq*I-A fragment was inserted into a single *Pvu*II site by ligation, and the ligation mixture was digested with *Pvu*II prior to transformation. The DNA was then used to transform *E. coli* HB101 with *E. coli* K37 as the indicator strain. No positive clones were obtained after several attempts.

*Mbo*I fragments 5 and 7 are contained within the *Taq*I-A fragment.[10] Cloning of the *Mbo*I-5 fragment was very inefficient.[11] This fragment repre-

[10] S. Anderson, A. T. Bankier, B. G. Barrell, M. H. L. de Bruijn, A. R. Coulson, J. Drouin, I. C. Eperon, D. P. Nierlich, B. A. Roe, F. Sanger, P. A. Schreier, A. J. H. Smith, R. Staden, and I. G. Young, *Nature (London)* **290**, 457 (1981).
[11] J. Drouin, *J. Mol. Biol.* **140**, 15 (1980).

sents the 3' segment of the human D loop. As a probe for the 5' half of the human D loop, we have cloned the *Mbo*I-7 fragment (710 nucleotides) in both orientations in M13mp7 and have subsequently transferred this fragment to pBR322, where it has been stably maintained. Thus, if specific sequences are present that prevent cloning, they are not present in this fragment (*Mbo*I-7).

Numerous attempts in this laboratory and others have been unsuccessful in cloning the complete human mitochondrial genome. To demonstrate this problem, mtDNA was isolated from a mixture of cells, 95% KB human cells, 5% LA9 mouse cells. The mouse mtDNA was undetectable after restriction enzyme digestion and gel electrophoresis. The mtDNA was linearized with a restriction endonuclease that cleaves human and mouse mtDNA each just once, *Sph*I, and ligated with similarly linearized pBR322. The pBR322 was treated with calf alkaline phosphatase to eliminate self-ligation. A number of positive colonies were picked and analyzed by restriction endonuclease digestion of DNA isolated by the phenol minilysate technique.[12] All isolated contained full-length mtDNA, but surprisingly, all were mouse mtDNA clones. No human mtDNA clones were detectable. If homogeneous *Sph*I-cleaved KB mtDNA alone is used, transformants are obtained, but at an ∼ 50-fold lower frequency. Most such clones contain KB mtDNA, but in all cases the insert (and often the vector as well) has undergone extensive rearrangement and deletion. This demonstrates the existence of a sequence (or sequences) in human mtDNA that prohibits cloning in *E. coli*. If this sequence can be identified (in *Mbo*I-5), it may then be deleted, allowing the remainder of the molecule to be cloned. A current listing of cloned mtDNA elements is presented in the table.

Comments

The isolation procedures outlined here have proved to be rapid and reproducible in isolating intact mitochondrial nucleic acids in high yield. Failure to isolate mitochondrial nucleic acids in expected yields is almost always due to insufficient initial cell breakage in a reasonable time period (∼ 5 min). The cause is usually a pestle that fits the homogenizer too loosely.

With regard to cloning of mammalian mtDNA, the entire mouse mtDNA genome has proved amenable to cloning and has remained stable for years.[1,13] Based on the high degree of conserved gene organization and homology to human mtDNA, there is no obvious explanation for the refractory behavior of human mtDNA to genomic cloning. The extreme

[12] R. D. Klein, E. Selsing, and R. D. Wells, *Plasmid* **3**, 88 (1980).
[13] A. C. Y. Chang, R. A. Lansman, D. A. Clayton, and S. N. Cohen, *Cell* **6**, 231 (1975).

TABLE I
MAMMALIAN MITOCHONDRIAL DNA CLONES

Nucleotides[a]	Restriction sites	Vector
	Human	
1–740	MboI	M13mp7[b]
3,160–4,534	HaeII	pACYC177
3,693–6,460	MboI	pBR322
4,121–5,274	EcoRI	pBR322
5,691–5,917	HincII[c]	M13mp7[b]
7,859–8,592	MboI	pBR322
	Mouse	
1,750–3,807	EcoRI	pACYC177
2,604–2,603[d]	HaeII	pACYC177
2,832–3,807	DdeI/EcoRI[e]	M13mp7[b]
		pBR322
3,807–4,013	EcoRI	pACYC177
4,013–9,136	EcoRI/HindIII	pBR322
4,279–5,884	MboI	M13bla61[f]
4,672–5,125	HaeII/HpaII[c]	M13Goril[g]
9,136–11,089	HindIII	pACYC177
10,753–10,752[d]	SphI	pBR322
11,081–11,969	HindIII	pACYC177
11,969–15,738	HindIII/HpaII[h]	pBR322
11,969–1,750	HindIII/EcoRI	pBR322
15,188–364	BalI/HincII	M13Goril[b]
15,761–10	AluI[c]	pBR322
		M13mp7[f]
16,022–16,127	DdeI[h]	M13mp9[b]

[a] For human nucleotides, see Anderson et al.[10]; for mouse nucleotides, see Bibb et al.[1]
[b] Both orientations.
[c] Blunt-end ligated.
[d] Total genome.
[e] EcoRI linkers added.
[f] H strand in viral (+) strand.
[g] L strand in viral (+) strand.
[h] BamHI linkers added.

variability in the ability to clone various regions of human mtDNA here and elsewhere suggests that specific localized sequences in human mtDNA are responsible for this phenomenon.

Acknowledgments

This work was supported by grants from the National Cancer Institute (CA-12312-12) and American Cancer Society, Inc. (NP-9I). R. A. V. is a Medical Scientist Training Program Trainee of the National Institute of General Medical Sciences (GM-07365).

[41] Analysis of Human Mitochondrial RNA

By GIUSEPPE ATTARDI and JULIO MONTOYA

A large amount of evidence has indicated that, in spite of the considerable degree of conservation of genetic content in mitochondrial DNA (mtDNA), the mitochondrial genetic system exhibits a great diversity of organization in different eukaryotic cells.[1,2] This diversity is most strikingly illustrated by the mitochondrial genomes of yeast and mammalian cells. On one side, there is the loose organization of yeast mtDNA, where the genes are separated by AT-rich stretches and several genes are themselves discontinuous. On the other side, there is the extraordinarily compact and lean gene organization of mammalian mtDNA, with its continuous genes mostly butt-jointed to each other and a nearly complete absence of noncoding stretches.

Until a few years ago, the analysis of the genetic content of the mitochondrial genome in mammalian cells was seriously hampered by the lack of genetic approaches easily applicable to these systems. More recently, however, the developments in DNA, RNA, and protein technologies have opened the way for a detailed structural analysis of these systems. The powerful molecular techniques now available have in fact allowed the dissection of the mammalian, in particular human, mitochondrial genome with a resolution that has approached and often surpassed that of the finest genetic analysis. Two fundamental approaches have been followed in these investigations. First, with the availability of the cloning technology and of rapid sequencing methods, it has been possible to determine the complete sequence of mtDNA from several mammalian species (man,[3] mouse,[4] beef[5]) and the partial sequence of mtDNA from other species. Second, the transcription of mtDNA in HeLa cells has been analyzed in depth, and the structural, metabolic, and mapping properties of the transcription products have been described in detail.[6] Several technological developments have

[1] G. Attardi, *Trends Biochem. Sci.* **6**, 86, 100 (1981).

[2] P. Borst and L. A. Grivell, *Nature (London)* **290**, 443 (1981).

[3] S. Anderson, A. T. Bankier, B. G. Barrell, M. H. L. de Bruijn, A. R. Coulson, J. Drouin, I. C. Eperon, D. P. Nierlich, B. A. Roe, F. Sanger, P. H. Schreier, A. J. H. Smith, R. Staden, and I. G. Young, *Nature (London)* **290**, 457 (1981).

[4] M. J. Bibb, R. A. Van Etten, C. T. Wright, M. W. Walberg, and D. A. Clayton, *Cell* **26**, 157 (1981).

[5] S. Anderson, M. H. L. de Bruijn, A. R. Coulson, I. C. Eperon, F. Sanger, and I. G. Young, *J. Mol. Biol.* **156**, 683 (1982).

[6] G. Attardi, P. Cantatore, A. Chomyn, S. Crews, R. Gelfand, C. Merkel, J. Montoya, and D. Ojala, *in* "Mitochondrial Genes" (P. Slonimski, P. Borst, and G. Attardi, eds.), p. 51. Cold Spring Harbor Lab., Cold Spring Harbor, New York.

METHODS IN ENZYMOLOGY, VOL. 97

played a crucial role in the analysis of HeLa cell mtDNA transcription. Thus, the introduction of a method for obtaining mitochondrial RNA species in pure form (involving degradation of the contaminating extramitochondrial nucleic acids by micrococcal nuclease[7]) has solved what was the major problem in these studies, i.e., the problem arising from the unavoidable contamination of the mammalian mitochondrial fraction by other membrane components, and has opened the way for structural and metabolic studies on mitochondrial RNAs. Furthermore, the availability of a method of HeLa cell growth giving a high yield of cells and the development of *in vivo* labeling procedures producing mitochondrial nucleic acids of high specific activity have alleviated the difficulties associated with the presence of the mitochondrial RNA species in extremely small amounts (estimated to vary between 20 and 0.005 pmol per gram of cell). Finally, very sensitive techniques have been used, and in some cases newly developed, for sequencing, mapping, and studying the metabolism of mitochondrial RNAs. In this chapter, we describe the procedures that have been successfully employed in this laboratory in the study of mtDNA transcription in HeLa cells. These procedures should be useful for investigations concerning mitochondrial RNAs in other mammalian cells and in eukaryotic cells in general.

Experimental Procedures

Cell Growth and Labeling

The S3 clonal strain of HeLa cells is grown in suspension in modified Eagle's phosphate medium supplemented with 5% calf serum (generation time about 22 hr). High yields of cells (up to $\sim 2.0 \times 10^6$ cells/ml) are obtained by an improved method of growth that involves flushing over the medium a mixture of 5% CO_2 and 95% air.[8] The latter method proved to be very useful to obtain the large amounts of cells required for the structural studies on mitochondrial RNA species.

Labeling of Mitochondrial RNA

Until the development of the micrococcal nuclease procedure for the purification of the mitochondrial fraction,[7] the general method for labeling mitochondrial RNA involved the use of inhibitors. Owing to the block of growth induced by the inhibitors, this method is, however, applicable only in relatively short-term experiments. For pulse labeling with [5-³H]uridine,[9] exponentially growing cells are collected by centrifugation and resuspended

[7] S. Crews and G. Attardi, *Cell* **19**, 775 (1980).
[8] G. Attardi and A. Wiseman, unpublished observations.
[9] B. Attardi and G. Attardi, *J. Mol. Biol.* **55**, 231 (1971).

at a concentration of 1 to 3×10^5 cells/ml in warm fresh medium containing 5% dialyzed calf serum. After 10 min, actinomycin D is added at 0.04–0.1 μg/ml to inhibit the synthesis of cytoplasmic rRNA, and 30 min later the cells are exposed for different lengths of time to [5-^{3}H]uridine (1–60 μCi/ml; 25–30 Ci/mmol). After the pulse, the cells are rapidly collected on crushed frozen isotonic salt solution (0.13 M NaCl, 0.005 M KCl, 0.001 M MgCl$_2$, NKM) in a flask immersed in an ice–salt bath. For the labeling of mitochondrial RNA with [^{32}P]orthophosphate,[10] exponentially growing cells, washed three times with warm modified Eagle's medium containing no phosphate, are resuspended in the same medium, supplemented with cold phosphate to 10^{-4} M, at a concentration of 1 to 2×10^6/ml and incubated for 30 min at 37°. Actinomycin D (0.04–0.1 μg/ml) or camptothecin, an inhibitor of nuclear RNA synthesis[11] (20 μg/ml), is then added; 30 min later, carrier-free [^{32}P]orthophosphate (0.25 mCi/ml) is added, and incubation is continued for 3–4 hr.

Labeling of mitochondrial RNA in the absence of inhibitors, until recently, has been used only for the analysis of RNA species identifiable as being encoded in mtDNA on the basis of their hybridization capacity to mtDNA. A more general use of this approach for structural and metabolic studies of mitochondrial RNA has been made possible by the introduction of the micrococcal nuclease procedure,[7] and this has been the method of choice for many of the investigations described in this chapter. Pulse labeling of exponentially growing cells with [5-^{3}H]uridine is carried out as described above, except for the omission of actinomycin D. Long-term labeling with [^{32}P]orthophosphate of exponential cultures is carried out by exposing cell cultures, at an initial concentration of 10^5 cells/ml, to 5 μCi/ml [^{32}P]orthophosphate for 24–48 hr in modified Eagle's medium containing 5×10^{-4} M or 10^{-3} M phosphate, supplemented with 5% dialyzed serum.[12] The specific activity of the RNA under these conditions varies between 10^4 and 10^5 cpm/μg. Uniform labeling of mitochondrial RNA with [5-^{3}H]uridine ($\sim 10^5$ cpm/μg) has been achieved by exposure of exponentially growing HeLa cells for 46 hr to the precursor added in subsequent portions (1.25 μCi/ml at 0 time, 1.25 μCi/ml after 25 hr, 0.87 μCi/ml after 35.5 hr) in modified Eagle's medium with 5% dialyzed calf serum.[13]

Labeling of Mitochondrial DNA

Long-term labeling of HeLa cell mtDNA can be obtained by exposing exponentially growing cells (initial concentration 1.0 to 1.5×10^5 cells/ml) for 48–60 hr to either [2-^{14}C]thymidine (0.03–0.05 μCi/ml; ~ 40–60

[10] F. Amalric, C. Merkel, R. Gelfand, and G. Attardi, *J. Mol. Biol.* **118**, 1 (1978).
[11] H. T. Abelson and S. Penman, *Nature (London), New Biol.* **237**, 144 (1970).
[12] R. Gelfand and G. Attardi, *Mol. Cell. Biol.* **1**, 497 (1981).
[13] Y. Aloni and G. Attardi, *J. Mol. Biol.* **55**, 251 (1971).

μCi/μmol) or [methyl-^{3}H]thymidine (0.37 μCi/ml, $\sim$ 50–80 mCi/μmol, the latter being added four or five times at 12-hr intervals) in modified Eagle's medium supplemented with 5% dialyzed serum. Specific activities of 2700–8000 cpm/μg for [^{14}C]DNA, and of 80,000–100,000 cpm/μg for [^{3}H]DNA have been obtained under such conditions.[14] Long-term labeling of mtDNA with [^{32}P]orthophosphate is carried out by growing cells (initial concentration $\sim$ 10^5 cells/ml) for 60–72 hr in modified Eagle's medium containing 10^{-4} M phosphate, 5% dialyzed calf serum, and 7 μCi of carrier-free [^{32}P]orthophosphate per milliliter. The specific activity of mtDNA thus obtained varies between 1.5 and 2.5 $\times$ 10^5 cpm/μg.[14]

Subcellular Fraction and RNA Extraction

All operations described below are carried out at 2–4°. The labeled cells are washed three times with NKM, and then resuspended in 6 volumes of 0.01 M Tris-HCl, pH 6.7 (25°), 0.01 M KCl, 0.0015 M MgCl$_2$. After 2 min of swelling, the suspension is homogenized with an A. H. Thomas Teflon pestle homogenizer (motor-driven pestle, $\sim$ 1500 rpm), using a sufficient number of strokes to break 60–70% of the cells. After addition of sucrose to 0.25 M, the homogenate is centrifuged at about 1160 g_{av} for 3 min to sediment nuclei, unbroken cells, and large cytoplasmic debris. The supernatant is spun at 8100 g_{av} for 10 min; the pellet thus obtained is resuspended in 0.25 M sucrose in 0.01 M Tris-HCl, pH 6.7, 0.00015 M MgCl$_2$ (one-half of the volume of the homogenate). After a spin at 1100 g_{av} for 2 min to sediment any residual nuclei, the suspension is recentrifuged at 8100 g_{av} for 10 min. In the experiments involving a micrococcal nuclease treatment of the mitochondrial fraction, the final pellet is resuspended in 0.25 M sucrose, 0.01 M Tris-HCl, pH 8.0, 0.001 M CaCl$_2$ (1 ml per 1.5 $\times$ 10^8 cell equivalents), and treated with 100 units ml^{-1} of micrococcal nuclease (Worthington) at 2° for 30 min (whenever a more extensive treatment of the mitochondrial fraction is desirable to remove completely the extramitochondrial rRNA and DNA, the mitochondrial fraction is washed with 0.04 M EDTA and the micrococcal nuclease is used at 250 units ml^{-1} for 15 min at 20°[7]). EGTA is then added to 0.0025 M to inhibit the nuclease activity; after 5 min, the suspension is diluted with 0.25 M sucrose, 0.01 M Tris-HCl, pH 6.7, 0.01 M EDTA (10 ml per 1.5 $\times$ 10^8 cells), and the mitochondria are pelleted by centrifugation at 10,000 rpm in an SS-34 rotor. The pellet is resuspended in 0.01 M Tris-HCl, pH 7.4, 0.15 M NaCl, 0.001 M EDTA (1 ml per 1.5 $\times$ 10^8 cell equivalents), incubated for 2–5 min at room temperature with 100 μg of Pronase (Calbiochem) per milliliter (previously digested

[14] D. Ojala and G. Attardi, *Plasmid* **1**, 78 (1977).

for 2 hr at 37° at a concentration of 2 mg/ml) or 100 μg of proteinase K per milliliter (Merck), lysed with 1% SDS and incubated for an additional 30 min. The nucleic acids are then extracted by shaking twice with an equal volume of a phenol–chloroform–isoamyl alcohol mixture [25:24:1 by volume, using phenol equilibrated with Tris–NaCl–EDTA buffer (0.01 M Tris-HCl, pH 7.4, 0.1 M NaCl, 0.001 M EDTA) supplemented with 0.5% SDS] at room temperature; after ethanol precipitation of the aqueous phase and centrifugation, the pelleted nucleic acids are dissolved in 0.01 M Tris-HCl, pH 7.4, 0.001 M EDTA.

In the experiments not involving micrococcal nuclease treatment of the mitochondrial fraction, this is dissolved in 0.01 M Tris-HCl, pH 7.4, 0.15 M NaCl, 0.001 M EDTA, 1% SDS, and incubated for 30 min at room temperature in the presence of 100 μg of Pronase or proteinase K per milliliter. The nucleic acids are then extracted as described above, ethanol precipitated twice, dissolved in 0.05 M Tris-HCl, pH 6.7, 0.0025 M MgCl$_2$, 0.025 M KCl (1 ml/1.5 × 10^8 cell equivalents), and incubated for 30–60 min at 2° in the presence of 100 μg of DNase I (RNase-free, Worthington) per milliliter. Sodium dodecyl sulfate (1%) and Pronase or proteinase K (100 μg/ml) are then added; after 20 min of incubation at room temperature, the solution is brought to 0.15 M NaCl, and the RNA is reextracted, as described above.

RNA Fractionation and Analysis

In order to separate the poly(A)-containing mitochondrial RNA components from non-poly(A)-containing RNA and from mitochondrial DNA, the total mitochondrial nucleic acids (in 0.001 M Tris-HCl, pH 7.4, 0.001 M EDTA) are denatured by heating at 80° for 5 min and fast cooling, and, after addition of NaCl to 0.12 M, passed through an oligo(dT)-cellulose (T3, Collaborative Research) column; the retained fraction is eluted at room temperature with 0.01 M Tris-HCl, pH 7.4, 0.001 M EDTA, 0.2% SDS, heat-denatured again as described above, and rerun through the oligo(dT)-cellulose column. The final bound fraction (eluted as detailed above) and the nonbound fraction are then collected by ethanol precipitation and fractionated by electrophoresis through 1.4% agarose–0.005 M CH$_3$HgOH slab gels (20 × 14 × 0.2 cm) in 0.05 M boric acid, 0.005 M Na$_2$B$_4$O$_7$ · 10H$_2$O, 0.01 M Na$_2$SO$_4$, 0.0001 M EDTA, pH 8.2.[10] For this purpose, the samples are dissolved in 0.001 M Tris-HCl, pH 7.4, 0.001 M EDTA heated for 5 min at 80°, fast cooled, mixed with 33% Ficoll, 0.2% Bromphenol Blue, and 0.2% xylene cyanole FF, and layered. Electrophoresis is carried out at 120 V (6.0 V/cm) for 5–6 hr, with circulation of the buffer and use of a fan to cool the gel. After the run, for analytical purposes,

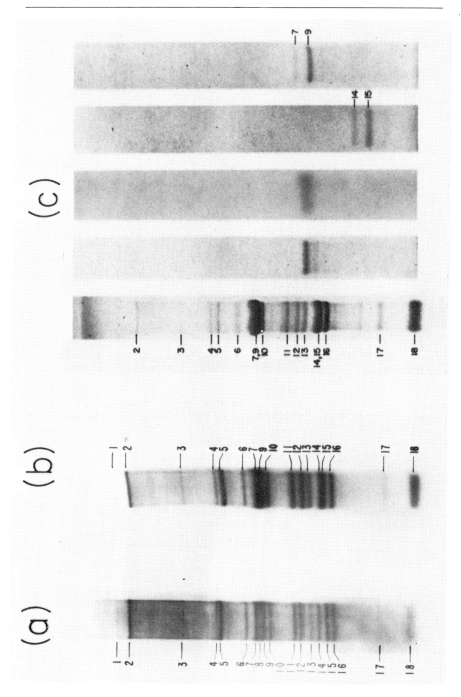

the gel is dried and exposed for autoradiography either at room temperature or at $-70°$ with a DuPont screen intensifier (depending on the amount of radioactivity), using a Kodak X-ray film (XAR-5) (Fig. 1a and 1b). For preparative purposes, the wet gel is stained with ethidium bromide and photographed under UV light and/or exposed for autoradiography at 4° for an appropriate time, and the bands of interest are cut out (Fig. 1c). The RNA is eluted by incubating the minced gel slices in 0.01 M Tris-HCl, pH 7.4, 0.01 M EDTA, 1% SDS (in the presence of 20 μg of yeast tRNA), at 37° for 10–12 hr in a rotary shaker. After ethanol precipitation, each RNA component is either utilized directly or, when it has to be used for sequencing, dissolved in 250 μl of 0.05 M Tris-HCl, pH 6.7, 0.0025 M MgCl$_2$, 0.025 M KCl, incubated with 20 μg of RNase-free DNase I (Boehringer/Mannheim) per milliliter for 30 min at 2°, Pronase–SDS–phenol extracted, and rerun through an agarose–CH$_3$HgOH gel (Fig. 1c). Separation of closely migrating species, like 7 and 9, and 14 and 15, can be obtained by electrophoresis through a 2% agarose–0.005 M CH$_3$HgOH gel at 80 V for 15–25 hr (Fig. 1c).[15]

In experiments where the amount of radioactivity incorporated into the individual RNA components has to be quantitated, in particular in double-labeling experiments, portions of the dried gel corresponding to individual bands identified by autoradiography are cut out and heated to 90° for 20 min in 1.0 ml of 0.01 M Tris-HCl, pH 7.4, 0.001 M EDTA; more than 95% of the radioactivity in the dried gel is eluted by this technique.[12] In some experiments, individual tracks from the dried slab gel or portions of them are sliced sequentially into 1-mm sections on a Mickel gel slicer,[12] and the radioactivity in the individual sections is eluted as described above (Fig. 2).

[15] J. Montoya, D. Ojala, and G. Attardi, *Nature* (*London*) **290**, 465 (1981).

FIG. 1. (a and b) Autoradiograms, after electrophoresis through 1.4% agarose–CH$_3$HgOH slab gels, of the oligo(dT)-bound RNA extracted from the mitochondrial fraction (untreated) of HeLa cells labeled with [^{32}P]orthophosphate in the presence of 20 μg of camptothecin[10] per milliliter (a), or from the micrococcal nuclease-treated mitochondrial fraction of cells labeled with [^{32}P]orthophosphate in the absence of inhibitors of nuclear RNA synthesis[12] (b). (c) Isolation and 5′-end labeling of mitochondrial poly(A)-containing RNA species.[15] Lane 1: electrophoretic pattern, after ethidium bromide staining, of the oligo(dT)-bound RNA extracted from the micrococcal nuclease-treated mitochondrial fraction of HeLa cells (30 g) run through a 1.4% agarose–CH$_3$HgOH slab gel; lanes 2 and 3: RNA species 13 was eluted from several gel tracks similar to that shown in lane 1, treated with RNase-free DNase I, and rerun on a gel under the same conditions (lane 2), then eluted again, labeled at its 5′ end with [γ-^{32}P]ATP and T$_4$ polynucleotide kinase after bacterial alkaline phosphatase treatment and rerun (lane 3). Lanes 4 and 5: rerun of RNA species 14 + 15 (lane 4) and 7 + 9 (lane 5) on 2% agarose-CH$_3$HgOH gels after elution from the gel track shown in lane 1 and treatment with DNase. Adapted from Gelfand and Attardi,[12] and Montoya *et al.*[15]

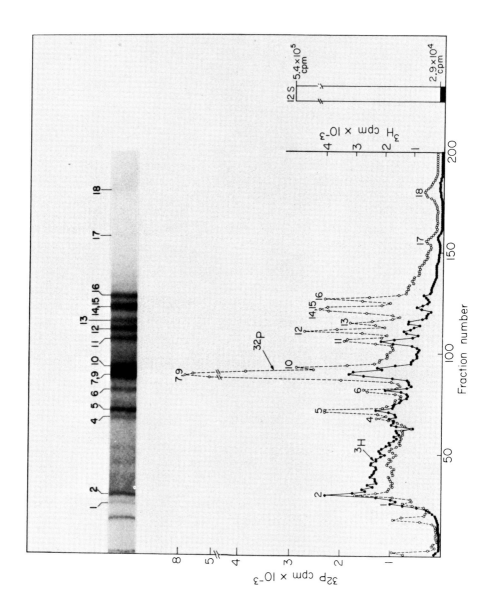

Isolation of Mitochondrial DNA and Separation of Mitochondrial DNA Strands

Preparation of the mitochondrial fraction is carried out as described above, except that the cells are washed three times with NKM and a fourth time with 0.025 M Tris-HCl, pH 7.5, 0.137 M NaCl, 0.005 M KCl, 0.0007 M Na$_2$HPO$_4$, 0.001 M EDTA, and the homogenization buffer and mito-chondrial fraction resuspension buffer contain 0.0001 M EDTA instead of 0.00015 M MgCl$_2$. The final mitochondrial pellet is resuspended in 0.01 M Tris-HCl, pH 7.4, 0.01 M EDTA (0.5–1.0 ml per 1.5 × 10^8 cell equivalents), and lysed with 1.2% SDS. After 10 min at room temperature, one-sixth of the volume of 7 M CsCl is added, and the mixture is incubated on ice for 30 min. The solution is centrifuged at 15,800 g_{av} for 15 min, and 0.741 g of CsCl per milliliter is then added to the supernatant. After incubation on ice for 20 min and centrifugation at 15,800 g_{av} for 20 min, the supernatant is removed (leaving behind the protein "skin") and mixed with one-fiftieth of the volume of an ethidium bromide solution at 10 mg/ml in H$_2$O and with 1.133 g of CsCl per milliliter of ethidium bromide solution. After adjusting (if necessary) the refractive index ($n_D^{25°}$) to 1.39 (ρ = 1.60 g/ml) by addition of either solid CsCl or Tris-EDTA buffer, the solution is centrifuged in a polyallomer tube at 38,000 rpm in a Spinco 65 rotor for 40–48 hr at 20° (or in another fixed-angle or swinging-bucket rotor under equivalent conditions). The closed-circular mtDNA band is collected and rerun in an identical CsCl–ethidium bromide density gradient, this step being repeated if demanded by the purity requirements for mtDNA in the experiments planned. The material in the final closed circular mtDNA band, after removal of the ethidium bromide by shaking twice for a few seconds with isoamyl alcohol, is dialyzed extensively against several changes of 0.01 M Tris-HCl, pH 7.4, 0.001 M EDTA, at 2–4°, to remove the CsCl, and then ethanol-precipitated in the presence of 0.15 M NaCl. After centrif-ugation in an SW 41 Spinco rotor at 30,000 rpm for 30 min, the DNA is dissolved in an appropriate buffer. If further purification of mtDNA from small-molecular-weight RNA and other small molecules is required, as in the case of preparation of ^{32}P-labeled mtDNA to be used for yield measure-ments (see below), the closed-circular mtDNA, after removal of the ethi-dium bromide and dialysis, is passed through a 1 × 50 cm Sephadex G-100

FIG. 2. Analysis by serial gel slicing of the labeling of poly(A)-containing mitochondrial RNA species separated on an agarose–CH$_3$HgOH slab gel.[6] A track containing the RNA from mitochondria of exponentially growing cells labeled for 24 hr with [^{32}P]orthophosphate and for 5 min with [5-^{3}H]uridine was sliced through its entire length, and the RNA in each slice was eluted and counted. For comparison, the band representing 12 S rRNA was cut out from a track in which mitochondrial oligo(dT)-nonbound RNA had been run, and the RNA was eluted and counted. From Attardi et al.[6]

column equilibrated with 0.1 M NaCl, 0.01 M Tris-HCl, pH 7.4, 0.001 M EDTA; the material corresponding to the void volume (recognized by optical density or radioactivity measurements) is pooled and ethanol-precipitated. Alternatively, the mtDNA can be freed of contaminating RNA and other small molecules by sedimentation through a 5 to 20% sucrose gradient in 0.01 M Tris-HCl, pH 8.0, 0.001 M EDTA, 0.5 M NaCl in an SW 41 Spinco rotor for 5 hr at 37,000 rpm at 4°. If required, a brief heating of mtDNA (120 sec at 90° in Tris-EDTA buffer, followed by quenching in ice), prior to the sucrose gradient centrifugation step, will produce the melting from mtDNA of the heavy (H)-strand initiation sequence (7 S DNA[16]) and allow its subsequent centrifugal separation.

For the separation of the complementary strands of HeLa cell mtDNA in an alkaline CsCl gradient,[13] a solution containing closed circular mtDNA in 4.0 ml of 0.05 M K_2HPO_4, 0.05 M glycine, 0.005 M EDTA, 0.01% SDS is brought to a refractive index ($n_D^{25°}$) of about 1.405 ($\rho = 1.76$ g/ml) with solid CsCl (~ 1.43 g/ml) and to pH 12.4 with a solution of 5.8 M CsCl in 2.0 M KOH. The mixture is centrifuged in a polyallomer tube in a Spinco 65 angle rotor at 42,000 rpm for ~ 48 hr at 20°. The fractions corresponding to the peaks of the heavy (H) and light (L) strands are separately pooled and brought to pH 8.0 with 1 M Tris-HCl, pH 6.7. In the case of labeled mtDNA, the specific activities of the H and L strands are determined from OD_{260} measurements (by using a correction factor of 40% for hyperchromicity) and from radioactivity determination on a portion of each sample.

It is known that mammalian mtDNA contains a significant number of alkali-labile sites distributed throughout the genome: these are due to ribonucleotides that are incorporated into DNA, and that are either remnants of RNA primers near the two origins of replication or may result from misincorporation during replication.[17] These alkali-labile sites account for the breakage of the separated strands in the alkaline CsCl gradient. In cases where there is need of recovering relatively intact strands, it is advantageous to reduce the pH of the alkaline CsCl gradient and the time of contact with it of mtDNA to the minimum required for strand separation. This is achieved by preforming a CsCl gradient at pH 11.8 by centrifugal force, and then by introducing the sample into the middle of the gradient and continuing the centrifugation for the minimum time required for banding. In particular, an 8-ml glycine-phosphate buffered alkaline CsCl solution ($\rho = 1.76$ g/ml) is centrifuged in a polyallomer tube at 40,000 rpm for ~ 40 hr at 20° in a fixed-angle 65 Spinco rotor. The run is stopped, and the mtDNA sample in 50–100 μl of CsCl solution ($\rho = 1.76$ g/ml) in 0.01 M Tris-HCl, pH 7.4,

[16] H. Kasamatsu, D. L. Robberson, and J. Vinograd, *Proc. Natl. Acad. Sci. U.S.A.* **68**, 2252 (1971).
[17] A. Brennicke and D. A. Clayton, *J. Biol. Chem.* **256**, 10613 (1981).

0.01 M EDTA is introduced, by the aid of an Eppendorf pipette, into the middle of the gradient. The centrifugation is restarted and continued for ~ 20 hr. The pattern obtained shows two peaks corresponding to the two separated strands near the center of the gradient and a usually smaller peak of undenatured DNA in the upper half of the gradient.

Isolation of Mitochondrial DNA Restriction Fragments and Their Separated Strands

For the fractionation of restriction fragments, depending upon their size and number, either straight (4 or 5%) polyacrylamide gels in Tris–borate–Mg buffer,[18] or 1.5 to 25% polyacrylamide gradient gels in Tris–glycine–EDTA buffer,[19] or straight agarose gels (0.6 to 1.5%) in 0.1 M Tris-HCl, pH 7.4, 0.05 M sodium acetate, 0.0025 M EDTA, are used in our laboratory. After staining with 1 μg of ethidium bromide per milliliter, the patterns are visualized in a UV transilluminator, and the individual bands are cut out. For the recovery of the fragments from the gel, the method most frequently employed in our laboratory involves electroelution of the fragments from the gel slices according to Galibert et al.,[20] followed by ethanol precipitation in the presence of 0.15 M NaCl, and centrifugation for 2 hr at 35,000–40,000 rpm in an SW 50.1 Spinco rotor. Recovery of the fragments by this method is 50–80%. A continuous electroelution method[21] has also been used successfully.[14]

Separation of the strands of the individual fragments (labeled either in vivo with [^{32}P]orthophosphate or in vitro at the 5′ ends with [γ-^{32}P]ATP and T4 polynucleotide kinase or at the 3′ ends with E. coli DNA polymerase I and [α-^{32}P]dNTPs) is carried out by treatment with alkali followed by electrophoresis through polyacrylamide or agarose gels. The optimal conditions of alkali treatment and gel electrophoresis vary for different fragments, depending on the size and probably on the DNA composition and sequence of the fragments, and has to be determined experimentally.[22] After visualization of the separated strands by autoradiography of the wet gel, they are eluted from the minced gel by diffusion (0.001 M Tris-HCl, pH 6.7, 0.001 M EDTA, 0.1% SDS for the agarose gels, or 0.5 M ammonium acetate, 0.1% SDS for the polyacrylamide gels) for 24 hr at 37° in a rotary shaker in the presence of 20 μg of yeast tRNA. In order to determine the strand homology of the separated strands, samples are incubated with an appropriate excess of

[18] T. Maniatis, A. Jeffrey, and H. Van de Sande, Biochemistry 14, 3787 (1975).
[19] F. W. Studier, J. Mol. Biol. 79, 237 (1973).
[20] F. Galibert, J. Sedat, and E. Ziff, J. Mol. Biol. 87, 377 (1974).
[21] A. S. Lee and R. Sinsheimer, Anal. Biochem. 60, 640 (1974).
[22] P. Cantatore and G. Attardi, Nucleic Acids Res. 8, 2605 (1980).

total H or L strands in 0.2 ml of 0.4 M NaCl, 0.01 M Tris-HCl, pH 6.7, for 5 hr at 65°; after a 10-fold dilution of the incubation mixtures with S1 nuclease digestion buffer (0.04 M sodium acetate, pH 4.6, 0.25 M NaCl, 0.001 M ZnSO$_4$) containing 10 μg of denatured HeLa nuclear DNA per milliliter, the samples are digested with 50 units of the *Aspergillus oryzae* nuclease S1 (Sigma) per milliliter for 30 min at 45°, and the amount of trichloroacetic acid-precipitable radioactivity determined.

A convenient approach followed in our laboratory for the quantitation of the fragments or their separated strands consists in mixing an *in vivo* ^{32}P-labeled or ^{3}H-labeled preparation of mtDNA (purified by two cycles of CsCl–ethidium bromide centrifugation, and Sephadex chromatography or sucrose gradient centrifugation, as detailed above) with a 10- or 20-fold excess of unlabeled mtDNA, in determining the specific activity of the mixed preparation by optical density and radioactivity measurements, and in following the recovery of the fragments or their separated strands on the basis of the radioactivity. For the quantitation of the separated strands, one can also label at the 5' ends or 3' ends a known amount of a given fragment, and determine its specific activity.

Structural Analysis of Mitochondrial RNA

Poly(A) or Oligo(A) Tail

All discrete transcripts different from tRNAs derived from the H strand are either polyadenylated (mRNAs or their precursors and a small fraction of the 16 S rRNA[10]) or oligoadenylated (the two rRNA species[23]); the discrete transcripts different from tRNAs derived from the L strand are also polyadenylated. Whereas the polyadenylated RNA species are retained on oligo(dT)-cellulose, the oligoadenylated species are not. A set of RNA components with electrophoretic mobilities in agarose–CH$_3$HgOH gels similar to those of most of the polyadenylated species, and probably related to them, are not retained on oligo(dT)-cellulose[10]; however, it is not known whether or not they possess an oligo(A) tail. The tRNA species do not have any poly(A) or oligo(A) tail.[23] The size of the poly(A) tails of the polyadenylated molecules, isolated as RNase resistant segments [after heating at 95° for 5 min in 0.001 M Tris-HCl, pH 7.4, 0.001 M EDTA, and quick cooling, addition of NaCl to 0.25 M, digestion with 2 μg of RNase A and 100 units of RNase T1 per milliliter (both preheated at 80° for 10 min) for 30 min at 37° in the presence of 25 μg of poly(A) per milliliter, SDS–Pronase–phenol extraction and ethanol precipitation], has been estimated to be 55–60 nucleotides on the basis of their electrophoretic mobility in a 12% polyacryl-

[23] D. T. Dubin, J. Montoya, K. Timko, and G. Attardi, *J. Mol. Biol.* 157, 1 (1982).

amide gel (corresponding to 4 S_E) and their sedimentation behavior in 0.1 M NaCl (s value just below 4 S),[24] or, more directly, from the ratio of adenosine to internal nucleotides after alkaline hydrolysis and analysis of the products by paper electrophoresis.[25] The latter analysis has also indicated that the poly(A) is located at the 3′ end of the RNA molecule. Size determination of the oligo(A) tails has required 3′-end sequencing analysis[23] (see below).

Analysis for Presence of "Cap" Structures in Mitochondrial RNAs

In order to investigate the possible presence of "cap" structures in HeLa cell mitochondrial RNA,[26] samples of poly(A)-containing RNA [selected by two passages through oligo(dT) cellulose] from cells labeled with [^{32}P]orthophosphate in the presence of 20 μg of camptothecin per milliliter were digested either with the nuclease P1 [2–6 μg of RNA in 25 μl of 0.05 M ammonium acetate, 0.001 M EDTA (pH 5.4), 0.5–1.0 μg of nuclease P1 (Yamasa Shoyu, Ltd., Japan),[27] 1 hr at 37°], or with a mixture of RNases [2–6 μg of RNA in 25 μl of 0.05 M ammonium acetate, 0.001 M EDTA (pH 4.5), 0.2 unit of RNase T2 (Sigma), 0.02 μg of RNase U2 (Calbiochem), 0.005 μg of RNase A (Sigma), and 10 units of RNase T1 (Calbiochem), 1 hr at 37°]. The ^{32}P-labeled ribonucleotides and other products of digestion were separated by the two-dimensional electrophoretic technique of Sanger and Brownlee: high voltage electrophoresis (1–2 kV) on Whatman No. 1 paper in pH 3.5 buffer [0.5% pyridine (v/v), 5% acetic acid (v/v), 0.005 M EDTA] was carried out in the first dimension (until the xylene cyanole FF (blue) marker migrated to approximately 10 cm from the origin), and electrophoresis (1 kV) on DEAE paper in the same pH 3.5 buffer in the second dimension (until the xylene cyanole FF marker migrated to about 10 cm from the origin). The ^{32}P-labeled nucleotides and other products of digestion were identified by autoradiography. As a control, samples of ^{32}P-labeled cytoplasmic poly(A)-containing RNA were digested with nuclease P1 or a mixture of RNases, and the digestion products were fractionated as described above [authentic "cap" dinucleotides (P-L Biochemicals, Inc.), identifiable by their blue fluorescence, were included in the P1 digest]. Using the approach described above, "cap" structures were found in the expected amounts in the digests of HeLa cell cytoplasmic RNA, but no such structures were observed in the digests of mitochondrial RNA from the same source, under conditions where their presence in 1 out of 5–10 RNA molecules would have been recognized.[26]

[24] D. Ojala and G. Attardi, *J. Mol. Biol.* **82,** 151 (1974).
[25] M. Hirsch and S. Penman, *J. Mol. Biol.* **80,** 379 (1973).
[26] K. Grohmann, F. Amalric, S. Crews, and G. Attardi, *Nucleic Acids Res.* **5,** 637 (1978).
[27] M. Fujimoto, A. Kuninaka, and H. Yoshino, *Agric. Biol. Chem.* **38,** 1555 (1974).

5'-End Sequencing Analysis of Mitochondrial RNAs[15]

This analysis was carried out by base-specific cleavage of 5'-end-labeled RNA followed by electrophoresis on sequencing gels.[28,29] RNA was extracted from the micrococcal nuclease-treated mitochondrial fraction of 30–100 g of cells, and fractionated by oligo(dT)-cellulose chromatography and agarose-CH_3HgOH slab gel electrophoresis, as detailed above. After ethidium bromide staining and UV visualization, each RNA component was eluted, DNase treated (to eliminate any contaminating DNA fragments), and rerun through an agarose–CH_3HgOH gel; the material in the band observed at the expected position was finally eluted and ethanol precipitated, as described above. One hundred to 500 ng of each RNA species (estimated on the basis of the intensity of the ethidium bromide staining, compared to that of a known amount of 12 S rRNA) were dissolved in 100 μl of 0.01 M Tris-HCl, pH 8.0, and incubated at 65° for 15 min with 100 units of bacterial alkaline phosphatase (BRL) per microgram of RNA. Nitrilotriacetic acid was added to 10 mM and, after 15 min at room temperature, 2 μg of closed-circular pBR322 DNA were added as a carrier, and the nucleic acids were extracted with phenol (equilibrated with Tris–NaCl–EDTA buffer)–chloroform–isoamyl alcohol. The RNA was then ethanol-precipitated in the presence of 0.2 M NaCl and dissolved in 15 μl of polynucleotide kinase buffer (0.05 M Tris-HCl, pH 8.0, 0.01 M $MgCl_2$, 0.005 M dithiothreitol) containing 0.5 nmol of [γ-^{32}P]ATP (8000 Ci/mmol), which had been previously extracted with phenol (equilibrated with Tris–NaCl–EDTA buffer)–chloroform–isoamyl alcohol, freed of phenol residues by shaking twice with ether and lyophilized. After addition of 2 units of polynucleotide kinase (Boehringer-Mannheim), the reaction was carried out at 37° for 15 min and then stopped by addition of 1.0 M ammonium acetate. Ten micrograms of carrier tRNA were added, and the mixture was ethanol-precipitated and run on a 1.4% agarose–0.005 M CH_3HgOH slab gel. The material in the labeled RNA band observed in the autoradiogram at the expected position was eluted and utilized for the 5'-terminal nucleotide analysis and the sequencing reactions. Between 10^4 and 10^5 cpm were usually recovered per RNA species.

5'-Terminal Nucleotide Analysis.[7] A portion of the 5'-end labeled sample with an appropriate amount of radioactivity was dissolved in 15 μl of 0.05 M ammonium acetate, pH 5.4, and digested with 1 μg of nuclease P1, which cleaves phosphodiester linkages in nucleic acids to yield 5'-mononucleotides,[27] for 1 hr at 37°. After addition of 10 nmol each of 5'-AMP, -CMP,

[28] H. Donis-Keller, A. M. Maxam, and W. Gilbert, *Nucleic Acids Res.* **4**, 2527 (1977).
[29] A. Simoncsits, G. G. Brownlee, R. S. Brown, J. R. Rubin, and H. Guilley, *Nature (London)* **269**, 833 (1977).

-GMP, and -UMP, the mixture was applied to a PEI cellulose thin-layer plate and chromatographed in 1.0 M LiCl. The thin-layer plates were examined by autoradiography sensitive enough to detect 1% of the input radioactivity (3 dpm in a 1 mm^2 circle), and then the marker spots were visualized with a shortwave UV lamp and cut out. They were eluted quantitatively with 1 ml of 0.02 M Tris-HCl, pH 7.4, 0.7 M MgCl$_2$, and counted in 10 ml of scintillation fluid.

   RNA Sequencing. Aliquots of the 5′-end-labeled RNA of interest (20 – 40 ng, usually 2 to 8 × 10^3 cpm) were diluted in 20 μl of 0.02 M sodium citrate, pH 5.0, 7 M urea, 0.001 M EDTA, containing 0.25 mg of yeast tRNA per milliliter and 0.025% each of xylene cyanole and Bromphenol Blue, in siliconized Eppendorf tubes. After heating the samples in a water bath at 50° for 5 min, varying amounts of RNase A (CU-specific, Sigma) (0.25 – 1.0 ng), or RNase T1 (G-specific, Sigma) (0.02 – 0.1 unit) or RNase U2 (A-specific, Sankyo) (0.15 – 1.0 unit) were added, and the samples were incubated for 15 min at 50°, then chilled and immediately loaded onto the gel.[28] Digestion with the RNase Phy I (AUG-specific, P-L Biochemicals)[29] was carried out by diluting the 5′-end-labeled RNA and 3 μg of yeast tRNA in 18 μl of 0.01 M sodium acetate buffer, pH 5.0, 0.001 M EDTA, adding 0.02 unit of RNase Phy I, and incubating for 15 min at room temperature; after stopping the reaction by addition of 15 μl of formamide – dye mixture and heating at 100° for 1 min, the sample was loaded onto the gel. Hot formamide degradation of the RNA for the formation of the ladder[29] was carried out by heating 50 – 100 ng of the 5′-end-labeled RNA in 20 μl of formamide (recrystallized) at 100° for 30 min. An alkaline ladder[28] was prepared by heating, at 90°, for 15 min, 50 – 100 ng of the 5′-end-labeled RNA and 5 μg of tRNA in 20 μl of 0.05 M NaHCO$_3$ – Na$_2$CO$_3$ buffer, pH 9.0, 0.001 M EDTA; after mixing with 20 μl of 10 M urea containing 0.05% each of xylene cyanole and Bromphenol Blue, the sample was loaded onto the gel. Electrophoresis was carried out on thin (0.5 mm) 25% polyacrylamide (30 : 1 bisacrylamide)/7 M urea gels in Tris-borate-EDTA buffer[30] at 1000 V for 6 hr for nucleotides 1 – 30, and on 10% polyacrylamide (20 : 1 bisacrylamide) – 7 M urea gels at 1000 V for 3.5 hr for nucleotides 25 – 70.

3′-End Sequencing Analysis of Mitochondrial RNAs[31]

   Poly(A)-Containing RNAs. A 3′-end partial sequence analysis of the HeLa cell mitochondrial poly(A)-containing RNAs for the purpose of aligning the 3′ ends of the encoded portion with the mtDNA sequence was carried out by an adaptation of the "minus sequencing method" using

[30] A. M. Maxam and W. Gilbert, this series, Vol. 65, p. 499.
[31] D. Ojala, J. Montoya, and G. Attardi, *Nature (London)* **290,** 470 (1981).

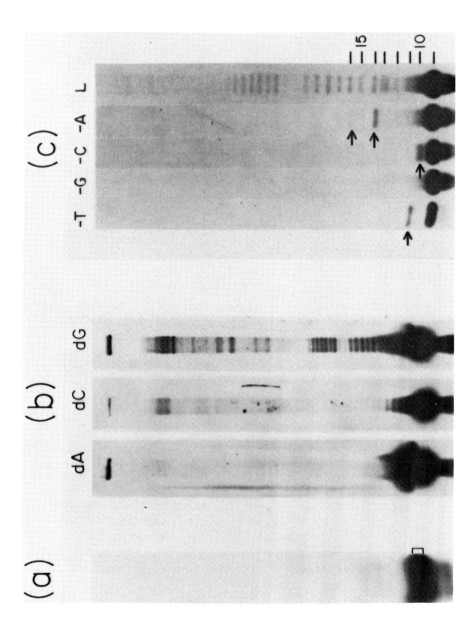

"phased" oligo(dT) primers [p(dT)₈-dA, p(dT)₈-dC, and p(dT)₈-dG] and reverse transcriptase.[31]

5'-END LABELING AND GEL PURIFICATION OF THE OLIGONUCLEOTIDE PRIMERS. Five micrograms ($\sim$ 1.6 nmol) each of p(dT)₈-dA, p(dT)₈-dC, and p(dT)₈-dG, resuspended in 50 μl of 0.01 M Tris, pH 8.0, were dephosphorylated by incubation with bacterial alkaline phosphatase (35 units of BAP per nanomole of primer) for 30 min at 37°. Reaction volumes were adjusted to 100 μl with H₂O, and the mixtures were extracted twice with an equal volume of phenol (equilibrated with Tris–NaCl–EDTA buffer)–chloroform–isoamyl alcohol (25:24:1). The phenol phases were reextracted with 100 μl of H₂O, and the aqueous phases were combined and extracted twice with ethyl ether (to remove remaining phenol). The mixtures were lyophilized to dryness, and each was resuspended in 10 μl of kinase buffer (0.05 M Tris, pH 8.0, 0.01 M MgCl₂, 0.005 M dithiothreitol) containing [γ-³²P]ATP (2 mCi, 0.25 nmol). Two units of polynucleotide kinase were added, and the mixtures were incubated for 30 min at 37°. The reactions were stopped by addition of 10 μl of 10 M urea containing 0.5% each of Bromphenol Blue and xylene cyanole FF dyes. The samples were electrophoresed on a 15% polyacrylamide (20:1 bisacrylamide) gel in Tris–borate–EDTA buffer.[30] The gel was subjected to autoradiography (Fig. 3a), the bands corresponding to the intact primers (nonanucleotides), which contained most of the radioactivity, were excised, and the oligonucleotides were eluted with two washes of 0.5 ml of H₂O. The washes were combined, filtered through glass wool (to eliminate gel particles), lyophilized to dryness,

---

FIG. 3. 3'-End-sequencing analysis of poly(A)-containing RNA 16.[31] (a) Each of the three primers p(dT)₈-dA, p(dT)₈-dC, and p(dT)₈-dG (shown is the p(dT)₈-dC primer) was 5'-end labeled with [γ-³²P]ATP and T4 polynucleotide kinase, after dephosphorylation with bacterial alkaline phosphatase, then run on a 15% polyacrylamide slab gel in Tris-borate-EDTA.[30] After autoradiography, the band corresponding to nonanucleotides, which contained most of the radioactivity, was eluted. (b) In order to determine the nucleotide adjacent to the poly(A) tail in the RNA species 16, this was mixed at 4° with a 100-fold molar excess of each of the three above-mentioned 5'-end-labeled primers and incubated for 3 min at 39° with AMV reverse transcriptase in the presence of the four unlabeled dNTPs. After stopping the reaction by addition of EDTA, the products were ethanol precipitated and run on a thin 10% polyacrylamide–7 M urea sequencing gel. (c) The oligo(dT) primer identified in (b) as capable of priming cDNA synthesis using RNA 16 as a template (p(dT)₈-dG) was incubated with this RNA, dNTPs, and reverse transcriptase in four reaction mixtures, each lacking one of the four dNTPs; in addition, a "ladder" (L) reaction was performed in which all the four dNTPs were present. After stopping the reaction and removing the excess primer by Sephadex G-50 chromatography, the products were analyzed on a 10% polyacrylamide–7 M urea sequencing gel. Arrows indicate the bands interpreted as corresponding to stop positions in the reverse transcriptase reaction. The numbering of the ladder steps indicates the molecular size (number of nucleotides) of the extended primers, starting from the 5'-end nucleotide. Modified from Ojala et al.[31]

and resuspended in 100 μl of H_2O. Final specific activity of the labeled primers was 5 to 10×10^7 cpm/μg.

DETERMINATION OF THE NUCLEOTIDE ADJACENT TO THE POLY(A) TAIL. In order to determine the correct primer to be used with a given RNA under investigation, and, thereby, the identity of the nucleotide adjacent to its poly(A) tail, each primer was tested individually for its ability to prime cDNA synthesis using that RNA as a template. For this purpose, each polyadenylated RNA species ($\sim$ 0.05–0.10 pmol), purified by the micrococcal nuclease procedure and by agarose CH_3HgOH gel electrophoresis, as described above, was mixed (at 4°) with a 100-fold molar excess of each of the three 5'-end-labeled primers and 50 μM of each of the four unlabeled deoxyribonucleoside triphosphates (dNTPs). Buffer conditions were adjusted to 0.05 M Tris-HCl, pH 8.0, 0.05 M KCl, 0.005 M $MgCl_2$, and 0.01 M dithiothreitol. Four units of reverse transcriptase from avian myeloblastosis virus were added, and the mixtures were incubated for 3 min at 39°. cDNA synthesis was terminated by placing the mixtures in ice for $\sim$ 15 sec, and then adding EDTA to 0.01 M and NaCl to 0.3 M, 2 volumes of ethanol and 5 μg of carrier yeast tRNA. (Care was taken to keep the mixtures at temperatures close to 0° to ensure that the shorter complementary DNA chains would not melt from the template and thereby be lost during the ethanol precipitation steps.) The products were collected by centrifugation, washed with cold ethanol (to remove excess labeled primer), recentrifuged, dried, and resuspended in 10 μl of 0.001 M Tris-EDTA; after addition of 10 μl of 10 M urea containing 0.5% each of Bromphenol Blue and xylene cyanole FF dyes, the samples were heated for 2 min at 90°, then layered on a thin (0.5 mm) 10% polyacrylamide–7 M urea sequencing gel in Tris–borate–EDTA buffer.[30] Electrophoresis was carried out for 2.5 hr at 1000 V after a prerun of 1 hr at 800 V. The gels were then subjected to autoradiography at $-70°$ with the aid of an intensifying screen. The correct oligo(dT) primer for each RNA was identified by the occurrence of a prominent band pattern in the autoradiogram, with a highly variable intensity of the individual bands, which was very typical and highly reproducible for each RNA (Fig. 3b).

SEQUENCING REACTIONS. The correct oligo(dT) primer for each RNA was incubated for 5 min, under the conditions described in the preceding section, with the corresponding RNA, dNTPs, and reverse transcriptase in four reaction mixtures, each lacking one of the four dNTPs. Under these conditions, cDNA synthesis was expected to proceed until the transcriptase reached the position in the template for which no complementary dNTP was present in the mixture. Subsequent analysis, on a sequencing gel, of the extended primer then indicated the position in the nucleotide sequence of the particular dNTP missing in the reaction. For each RNA species, a "ladder" reaction was also performed, in which all four dNTPs were present. Transcription was terminated by quick cooling the mixtures in iced water and adding EDTA to 0.01 M. Excess labeled primer was separated from the hybrid by passage of the mixture (after addition of 5 μg of yeast tRNA carrier) through a Sephadex G-50 column (1 $\times$ 25 cm) equilibrated with 0.3 M NaCl, 0.01 M Tris-HCl, pH 7.4, 0.01 M EDTA. As before, this and subsequent steps were carried out at a temperature close to 0° in order to avoid the melting of the extended primers from the template. The material in the void volume was collected, ethanol precipitated, resuspended in 10 μl of 0.01 M Tris-EDTA plus 10 μl of 10 M urea–dye solution, heated for 2 min at 90°, then layered on a 10% polyacrylamide – 7 M urea sequencing gel. Results were visualized by autoradiography (Fig. 3c).

Non-Poly(A)-Containing RNAs. The approach described above for the 3'-end sequencing analysis of the mitochondrial polyadenylated RNA species was not suitable for the nonpolyadenylated species or for those that possess only one or a few As at the 3' end, as is the case for the rRNA components. For these, a different approach involving 3'-end labeling, digestion with RNase A or T1, fingerprinting of the oligonucleotide products and further characterization of these by secondary enzymatic digestion had to be followed. The procedures that have been utilized for the analysis of the 3' ends of the HeLa cell 12 S and 16 S rRNAs[23] are described below.

Samples of 12 S and 16 S rRNA, purified by the micrococcal nuclease – agarose – CH_3HgOH gel electrophoresis procedure, were 3'-end labeled following Peattie's method.[32] Generally, 5 or 10 μg of RNA were incubated in 10 μl containing 6 units of T4 RNA ligase (P-L Biochemicals) and 25–50 μCi of 5'-[^{32}P]pCp (10–20 pmol), for 3 hr at 5°. The samples were then mixed with 15 μl of 83% formamide containing 0.016 M EDTA and 0.1% Bromphenol Blue, held at 65° for 5 min, chilled in ice, and subjected to electrophoresis through a polyacrylamide gradient gel (2.5 to 10% acrylamide) containing 7 M urea. RNA was recovered from the gels as described by Maxam and Gilbert,[30] except that elution was performed for 18 hr at room temperature.

Complete digestion with RNase A (Worthington) or T1 (Calbiochem) was performed using 0.01 μg or 0.8 unit of enzyme, respectively, in 5 μl of 0.01 M Tris-HCl, pH 8.0, containing 10 μg of RNA [mainly added carrier tRNA], for 30 min at 37°. "Standard" fingerprints[33] and two-dimensional "chromatographic" fingerprints[34] were performed as described by Dubin *et al.*[35] (Fig. 4). Secondary analyses of oligonucleotides eluted from finger-

[32] D. A. Peattie, *Proc. Natl. Acad. Sci. U.S.A.* **76**, 1760 (1979).
[33] G. Volckaert, W. Min Jon, and W. Fiers, *Anal. Biochem.* **72**, 443 (1976).
[34] G. Volckaert and W. Fiers, *Anal. Biochem.* **83**, 228 (1977).
[35] D. T. Dubin, K. D. Timko, and R. J. Baer, *Cell* **23**, 271 (1981).

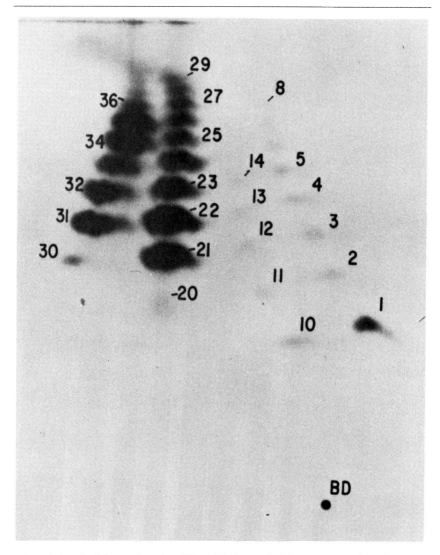

FIG. 4. Standard fingerprint of an RNase T1 digest of 16 S rRNA.[23] The first dimension (cellulose acetate electrophoresis) is from left to right, the second (homochromatography on a PEI plate) is downward. BD (blue dye) indicates the position of xylene cyanole marker. Spots 1 through 8 represent A_1Cp through A_8Cp; 10 through 14, UCp through UA_4Cp; 20 through 29, U_2Cp through U_2A_9Cp; and 30 through 36, U_3Cp through U_3A_6Cp. From Dubin et al.[23]

prints involved digestion of T1 oligonucleotides with RNase A, and of RNase A oligonucleotides with RNase T1, followed by DEAE paper electrophoresis at pH 3.5,[36] or digestion with RNase T2 followed by filter-paper electrophoresis or PEI chromatography.[35]

RNA – DNA Hybridization

A variety of RNA–DNA hybridization experiments have been carried out to measure the extent of transcription of the two strands of HeLa cell mtDNA, to determine the strand homology of the various mitochondrial RNA species and to titrate and map their coding sequences in the mtDNA molecule.

Saturation Experiments with Total Mitochondrial RNA and Separated MtDNA Strands[13,37]

H Strand. The fraction of the H strand that is complementary to mitochondrial RNA was determined by saturation experiments utilizing ^{14}C-labeled H-strand mtDNA bound to nitrocellulose filters and increasing amounts of mitochondrial RNA that had been uniformly labeled with [5-^{3}H]uridine as described in a previous section.[13] In a typical experiment, $0.0125 - 0.025$ μg of H strand was exposed to increasing amounts of RNA in 2.0 ml of $4 \times$ SSC containing 0.01 M Tris-HCl, pH 7.8, for 24 hr at 68°. The filters were washed with $2 \times$ SSC on both sides, incubated with 5 ml of $2 \times$ SSC containing 20 μg of RNase A (preheated at 80° for 10 min) per milliliter for 1 hr at room temperature, then washed with $2 \times$ SSC and dried.

The extent of transcription of the H strand was also determined by analyzing RNA–DNA hybrids formed in solution.[13] For this purpose, H-strand mtDNA $(0.08-0.2$ μg) was mixed with varying amounts of ^{3}H-labeled RNA in $0.5-1.5$ ml of 0.001 M Tris-HCl, pH 8.0, 0.001 M EDTA, then brought to $4 \times$ SSC, 0.1 M Tris-HCl, pH 8.0, and incubated for 5 hr at 70°. After digestion with 200 μg of RNase A and 100 units of T1 RNase per milliliter for 1 hr at room temperature, and chromatography through Sephadex G-100 (0.9 $\times$ 55 cm) equilibrated with $2 \times$ SSC, the material in the fractions corresponding to the peak of radioactivity in the void volume was again treated with RNase A (5 μg/ml) and T1 RNase (5 units/ml) for 45 min at room temperature, SDS–phenol extracted, and, after removing the traces of phenol by shaking the solution with ether and bubbling nitrogen through it, dialyzed for 12 hr against $2 \times$ SSC and then analyzed in a Cs_2SO_4 gradient. The hybrids banded in the Cs_2SO_4 gradient

[36] G. G. Brownlee, "Determination of Sequences in RNA." Am. Elsevier, New York, 1972.
[37] W. I. Murphy, B. Attardi, C. Tu, and G. Attardi, *J. Mol. Biol.* **99,** 809 (1975).

were further examined by electron microscopy for their content of duplex structure. The three approaches described above, i.e., measurement of the H-strand saturation level and determination of density and duplex content of the RNA–DNA hybrids formed at saturation, gave consistent results, indicating that all or almost all of the H-strand sequences are represented in complementary form in HeLa cell mitochondrial RNA.[13]

   L Strand. Exhaustion hybridization experiments in liquid medium had clearly indicated that the L-strand mtDNA is also transcribed in HeLa cells.[38] However, saturation experiments with whole mitochondrial RNA, as described above for H-strand transcripts, could not be performed, owing to the presence in the RNA of an excess of H-strand transcripts that would have competed effectively for hybridization with the L-strand transcripts. Saturation experiments could, on the contrary, be carried out with mitochondrial RNA preparations that had been enriched in L-strand transcripts. This enrichment was achieved by hybridization of heat-denatured total mitochondrial RNA with an excess of L-strand mtDNA, RNase digestion of the H-strand transcripts, Sephadex G-100 chromatography and SDS–phenol–Pronase extraction of the RNA–DNA hybrids, followed by dissociation of the hybrids, digestion of the L-strand mtDNA with DNase I, SDS–Pronase–phenol extraction, and final separation of the partially purified RNA by Sephadex G-100 chromatography.[37] L-strand saturation experiments were carried out in solution with this L-transcript-enriched preparation as described above for the H-strand transcripts. By analyzing the density in a Cs_2SO_4 gradient of the hybrids formed at saturation and the electron microscopic appearance of the hybrids formed with mostly intact L strands, it was clearly shown that mitochondrial RNA from HeLa cells contains sequences that are complementary to the entire, or almost the entire, length of the L strand.[37]

Titration of rRNA and tRNA Genes

   Saturation hybridization experiments using separated strands of ^{14}C-labeled mtDNA bound to nitrocellulose filters and increasing amounts of [5-^{3}H]uridine-labeled ribosomal RNA species (16 S and 12 S rRNA) or tRNA species separated on a sucrose gradient have been carried out to titrate the corresponding genes.[39] In a typical experiment, 0.025 to 0.2 μg of H or L strands (depending upon the specific activity of the RNA) were exposed to increasing amounts of the RNA sample in 2 ml of 4× SSC containing 0.01 M Tris-HCl buffer, pH 7.8, for 24 hr at 68°, and the hybrids were treated with RNase A as described above for the H-strand saturation experiments.

[38] Y. Aloni and G. Attardi, *Proc. Natl. Acad. Sci. U.S.A.* **68**, 1757 (1971).
[39] Y. Aloni and G. Attardi, *J. Mol. Biol.* **55**, 271 (1971).

*Determination of Strand Homology of Mitochondrial
   RNA Species*

For this purpose, an adequate amount of radioactivity of the sample of interest (either total mitochondrial RNA or fractions of it or individual RNA species) was mixed with an amount of H or L strand estimated to contain an appropriate excess of sequences complementary to the RNA species tested, in a volume of 180 μl of 0.001 M Tris-HCl, pH 8.0, 0.001 M EDTA, heat-denatured for 5 min at 95° (to melt any possible RNA–RNA duplex), and then brought to 0.01 M Tris-HCl, pH 8.0, 0.4 M NaCl, 0.01 M EDTA, in a final volume of 200 μl. After 4–5 hr of incubation at 65–66°, the mixtures were treated with 10 μg of RNase A and 10 units of RNase T1 per milliliter for 30 min at room temperature. After addition of KCl to 0.2 M, the mixtures were filtered through nitrocellulose membranes, which were then washed with 100 ml of warm (55–65°) 2× SSC.

Determination of strand homology of [³H]amino acyl (AA)–tRNA complexes required conditions of hybridization apt to preserve the AA–tRNA bond.[40] In particular, the mixture of [³H]AA–tRNA and mtDNA strand was not heat-denatured, and incubation was carried out in a mixture of equal volumes of 8× SSC, adjusted to pH 4.6, and recrystallized formamide (the formamide raises the apparent pH to ~ 5.6) for 4–8 hr at 34–37°. Each mixture was then diluted to 15 ml with 6× SSC (pH 6.0), treated with 5 units of T1 RNase for 30 min at 37°, and passed through a nitrocellulose membrane. The filters were washed with 10 ml of 2× SSC (pH 6.0) at 60° and 10 ml of 2× SSC at room temperature, dried, and counted.

Mapping Experiments

Both DNA and RNA transfer hybridization experiments and S1 nuclease protection experiments have been utilized to construct a detailed transcription map of HeLa cell mtDNA. In this analysis, while the DNA transfer hybridization experiments served to establish the rough location in the DNA physical map of the sequences corresponding to each RNA, the S1 protection experiments localized precisely the RNA coding sequences relative to restriction sites, providing at the same time information on the colinearity of the RNA and DNA sequences. The RNA transfer hybridization experiments corroborated the mapping data obtained by the two other approaches.

DNA Transfer Hybridization Experiments. The mtDNA restriction fragments were resolved on 2% agarose slab gels in 0.01 M Tris-HCl, pH 7.4, 0.05 M sodium acetate, 0.0025 M EDTA and transferred to nitrocellulose

---

[40] D. C. Lynch and G. Attardi, *J. Mol. Biol.* **102**, 125 (1976).

filters.[41] Strips corresponding to individual tracks (containing the digest of 0.5–2.0 μg of mtDNA) were each incubated for 24 hr at 68° with the RNA species under investigation in 2 ml of 6× SSC, 0.5% SDS, in a 20-ml screw-cap tube, which was immersed in a shaking water bath and kept under gentle shaking in the direction of the length of the tube. *In vivo* ^{32}P-labeled RNA species, labeled and isolated as described above, were generally used as probes, with an input of 500–10,000 cpm per hybridization mixture.[42] However, *in vitro* end-labeled RNA species (including "capped" RNA[43]) have been used in some experiments. The RNA sample was denatured by heating in low-salt buffer (0.001 M Tris-HCl, pH 8.0, 0.001 M EDTA) for 5 min at 90° before addition to the hybridization medium. After hybridization, the filter strips were rinsed three times with 10–15 ml of hot (68°) hybridization buffer, then incubated for 2 hr with 10 ml of the same buffer at 68°; the strips were again rinsed twice with buffer at 68°, aligned with an identical strip which had been hybridized to a total mtDNA probe (^{32}P-labeled by nick-translation and denatured), and then autoradiographed.

RNA Transfer Hybridization Experiments. For the experiments of hybridization by the RNA transfer procedure,[42,44] the oligo(dT)-cellulose-bound mitochondrial RNA was fractionated by electrophoresis on 1.6% agarose slab gels (10 × 14 × 0.2 cm) containing 0.005 M CH$_3$HgOH.[42] After electrophoresis, the gels were placed in 200 ml of 0.05 N NaOH containing 0.005 M mercaptoethanol and rocked for 10 min, then treated with 0.2 M phosphate buffer (pH 6.5) containing 0.007 M iodoacetate, and finally washed with 0.025 M phosphate buffer (pH 6.5). Transfer of the RNA to activated cellulose paper (diazobenzyloxymethyl paper, DBM-paper[44]) was carried out as described.[42,44]

Total mtDNA or purified restriction fragments were ^{32}P-labeled *in vitro* by nick-translation using the procedure of Rigby *et al.*[45] or modifications of it. The specific activity obtained was in the range of 5 × 10⁷ to 5 × 10⁸ cpm/μg. Each strip of paper (~ 1 × 20 cm), which had bound to it poly(A)-containing mitochondrial RNA from about 2.5 × 10⁸ cells, was incubated with ^{32}P-labeled DNA (10⁶ to 10⁷ cpm) in 5 ml of hybridization buffer [50% deionized formamide, 0.75 M NaCl, 0.075 M sodium citrate, containing 0.02% (w/v) each of bovine serum albumin, Ficoll (Sigma) and poly(vinylpyrrolidone) (A. H. Thomas)]. Hybridization was carried out by placing the paper strip(s) in a plastic boiling bag (Seal-N-Save, Sears) and rocking gently

[41] E. M. Southern, *J. Mol. Biol.* **98**, 503 (1975).
[42] D. Ojala, C. Merkel, R. Gelfand, and G. Attardi, *Cell* **22**, 393 (1980).
[43] J. Montoya, T. Christianson, D. Levens, M. Rabinowitz, and G. Attardi, *Proc. Natl. Acad. Sci. U.S.A.* **79**, 7195 (1982).
[44] J. C. Alwine, D. J. Kemp, and G. R. Stark, *Proc. Natl. Acad. Sci. U.S.A.* **74**, 5350 (1977).
[45] P. W. J. Rigby, M. Dieckmann, C. Rhodes, and P. Berg, *J. Mol. Biol.* **113**, 237 (1977).

for 36 hr at 42°. The strips were then washed at 42° with six changes of buffer (50% formamide, 0.75 M NaCl, 0.075 M sodium citrate) over a period of 24 hr, blotted with Whatman 3 MM filter paper, and exposed for autoradiography. In general, the strips were first incubated with total ^{32}P-labeled mtDNA to provide a marker pattern, then incubated with individual fragments. In order to reuse the RNA-containing strip, the paper was treated with 99% formamide at 65°: at least six changes of formamide over a period of 24 hr were required to remove the ^{32}P-labeled DNA probe. Autoradiography was used to ensure that all of the ^{32}P-labeled DNA probe had been removed prior to reuse of the strip with another probe.

S1 Nuclease Protection Analysis. Samples of purified RNA species were dried down in 0.5 ml snap-cap vials together with an appropriate amount of the desired double-stranded fragment or of its separated H strands or L strands (10–100 ng). DNA fragments or strands labeled *in vivo,* or labeled *in vitro* at their 5' ends with [γ-^{32}P]ATP and polynucleotide kinase, or at their 3' ends by fill-in with the Klenow fragment of the DNA polymerase I, have been used in different experiments. With double-stranded DNA, hybridization was carried out in 80% formamide in high salt, conditions favoring RNA–DNA association over DNA renaturation.[46] On the contrary, with single-stranded DNA, either the above conditions or hybridization at high temperature was used.

For the hybridizations in high formamide, the dried samples were dissolved in 20 μl of 0.04 M PIPES, pH 7.0, 0.75 M NaCl, 0.001 M EDTA, 80% formamide, heated for 5 min at 68°, and incubated for 8–20 hr at 49° (the above conditions of salt concentration and temperature have been determined to be optimal for the mitochondrial RNA tested; for unknown RNAs, the narrow temperature window, which, at any given salt concentration, allows RNA–DNA association but not DNA renaturation, will have to be experimentally determined). The samples were then diluted with 10 volumes of S1 digestion buffer [0.04 M sodium acetate (pH 4.6), 0.25 M NaCl, 0.001 M ZnSo$_4$], and incubated with 100–500 units of *Aspergillus oryzae* S1 nuclease per milliliter (Sigma) for 30 min at 42–45° in the presence of 20 μg of denatured HeLa nuclear DNA per milliliter. The above conditions of S1 treatment were optimal in the cases tested, as judged from the constancy in size of the protected fragments with increased S1 concentration or time of treatment, and the appearance of heterogeneous, larger size S1-resistant material with less extensive S1 digestion; however, with new RNA species, the optimal conditions will have to be individually worked out. Thus, if the ends of the RNA–DNA hybrid are expected to correspond to an AT-rich region, a lower temperature of S1 treatment (30–37°) may be required.

[46] J. Casey and N. Davidson, *Nucleic Acids Res.* **4**, 1539 (1977).

For the hybridizations at 66° in 0.4 M salt, the dried samples were dissolved in 20 μl of 0.001 M Tris-HCl, pH 8.0, 0.001 M EDTA, heated for 4 min at 90°, fast cooled, brought to 0.4 M NaCl, 0.01 M Tris-HCl, pH 8.0, 0.01 M EDTA, and incubated for 4 hr at 66°; the samples were then diluted with 10 volumes of S1 digestion buffer and treated with this enzyme as described above.

Electrophoretic analysis of the protected DNA segments was carried out either under native conditions in 4 or 5% polyacrylamide slab gels in Tris–borate–Mg^{2+} buffer (TBM),[18,42] or under denaturing conditions in 4% polyacrylamide–7 M urea slab gels in Tris–borate–EDTA buffer (TBE).[18,42] For electrophoresis in TBM buffer, the S1-digested samples were ethanol precipitated in the presence of 5 μg of yeast tRNA, dissolved in buffer, and run through the polyacrylamide slab gel at 10 V/cm for an appropriate time. For electrophoresis under denaturing conditions, the ethanol-precipitated samples were denatured by heating for 5 min at 80° in 0.001 M Tris-HCl (pH 7.4), 0.001 M EDTA, digested with RNase A (10 μg/ml for 10 min at 20° in the presence of 0.1 M NaCl) (if the RNA was labeled), reprecipitated with ethanol in the presence of 5 μg of yeast tRNA, dissolved in 7 M urea, 0.001 M Tris-HCl (pH 7.4), 0.001 M EDTA, heated for 3 min at 80°, fast cooled, and immediately layered on a 4% polyacrylamide–7 M urea slab gel in TBE. Electrophoresis was carried out at 5 V/cm for an appropriate time. After rinsing for 15 min in water to remove the urea, the gels were dried and autoradiographed. HpaII or HpaII–BamHI digests of HeLa cell mtDNA, end-labeled with [α-^{32}P]dNTPs and E. coli DNA polymerase I, were generally used as markers either directly (TBM gels) or after denaturation as described above (urea gels). Higher resolution of the protected DNA fragments was achieved in some experiments by running the denatured samples through 10% polyacrylamide (29:1 bisacrylamide)–7 M urea sequencing gels.[23]

Analysis of Transcription Initiation Sites[22,43]

Two approaches have been followed in this analysis. The first approach involved localizing in mtDNA by S1 protection experiments the 5' ends of the nascent RNA chains isolated from transcription complexes.[22,43] The second approach was based on the assumption that 5'-di- or triphosphates result only from transcriptional initiation; since 5'-diphosphate or 5'-triphosphate terminated molecules are the only substrates for guanylyltransferase ("capping" enzyme),[47] any such molecules present in mitochondrial

[47] S. Venkatesan, A. Gershowitz, and B. Moss, J. Biol. Chem. 255, 903 (1980).

RNA could be labeled *in vitro* with [α-^{32}P]GTP using this enzyme, and the 5′ termini thus labeled could be mapped in the mitochondrial genome by DNA transfer hybridization and S1 protection experiments.[43]

Mapping of Nascent RNA Molecules

Isolation of RNA from mtDNA Transcription Complexes. Mitochondrial DNA transcription complexes were isolated from 5 to 25 g HeLa cells by lysing the mitochondrial fraction with 2% SDS in 0.01 M Tris-HCl, pH 7.0, 0.1 M NaCl, 0.001 M EDTA, and centrifuging the lysate through a 15 to 30% sucrose gradient in 0.01 M Tris-HCl, pH 7.0, 0.1 M NaCl, 0.001 M EDTA, 0.5% SDS (prepared over 7 ml of 64% sucrose in the same buffer) in an SW27 rotor at 27,000 rpm for 15 hr at 20°.[22,48] The fast-sedimenting structures were precipitated with two volumes of ethanol, and the RNA was purified from them by extraction with Pronase-SDS-phenol – chloroform – isoamyl alcohol, DNase I digestion, reextraction, as described in a previous Section (Subcellular Fraction and RNA Extraction) for total mitochondrial RNA; the final material was passed through a Sephadex G-50 column equilibrated with 0.01 M Tris-HCl, pH 6.7, 0.15 M NaCl, 0.001 M EDTA, and collected in the exclusion volume.

Mapping Experiments. S1 protection experiments, utilizing appropriate restriction fragments of mtDNA or separated strands thereof, *in vitro* labeled at their 5′ ends with [γ-^{32}P]ATP and polynucleotide kinase, and increasing amounts of RNA from transcription complexes, were carried out as described above (see Mapping Experiments, S1 Nuclease Protection Analysis).

Mapping of in Vitro "Capped" RNA Molecules[43]

In Vitro "Capping" Reaction. Thirty to 80 μg of poly(A)-containing RNA isolated from micrococcal nuclease-treated mitochondria were dissolved in 3.5 – 8.7 μl of 0.005 M CH$_3$HgOH, and incubated for 5 min at room temperature to minimize secondary structure. The RNA solution was adjusted to 20 – 50 μl containing 0.05 M Tris-HCl, pH 7.5, 0.0025 M MgCl$_2$, 25 – 45 μM [α-^{32}P]GTP (1000 – 3000 Ci/mmol), and 0.001 M dithiothreitol. The reaction was started with the addition of 9 – 12 units of guanylyltransferase (guanine-7-methyltransferase); after 15 min of incubation at 37°, the reaction was terminated by addition of an equal volume of 0.01 M Tris-HCl, pH 7.5, 0.01 M EDTA, 1% SDS, and 5 μg of proteinase K per microliter. After 30 min of incubation at 37°, the mixture was extracted twice with an

[48] Y. Aloni and G. Attardi, *J. Mol. Biol.* **70**, 363 (1972).

equal volume of phenol and chromatographed on a 5-ml Sephadex G-50 column equilibrated with 0.01 M Tris-HCl, pH 7.5, 0.002 M EDTA, and 0.2% SDS. The peak fractions were collected, adjusted to 0.3 M sodium acetate, and precipitated by addition of 3 volumes of ethanol.

Mapping Experiments. A *Hpa*II digest of HeLa cell mtDNA was resolved on a 2% agarose slab gel in Tris-acetate – EDTA buffer and transferred onto a nitrocellulose filter. A strip corresponding to an individual track (containing 2 μg of *Hpa*II digest) was placed in a 20-ml screw-cap tube, washed for 30 min at room temperature with 5 ml of 6 × SSC, incubated for 5 – 8 hr at 68°, under gentle shaking, in 5 ml of hybridization solution [6 × SSC, 1 × Denhardt's solution [0.02% each of bovine serum albumin, Ficoll, and poly(vinylpyrrolidone)], 0.5% SDS, 20 μg of yeast tRNA per milliliter], and finally incubated in 2 ml of the same solution containing the ^{32}P-labeled *in vitro* capped RNA sample (~ 15,000 cpm) for 40 hr at 68°. After hybridization, the filter strips were rinsed as described above and autoradiographed. For a more precise mapping, S1 protection experiments were carried out using the *in vitro* capped RNA and the fragment (*Hpa*II-8) which had given a signal in the DNA transfer hybridization experiments or separated strands of it (unlabeled). For this purpose, samples of the *in vitro* capped RNA were hybridized, under the high formamide conditions, with an excess of the fragment or of each of its separated strands; after S1 digestion, the protected RNA – DNA hybrids were analyzed on a 5% polyacrylamide gel in TBM.

Using the approaches described above, two main initiation sites for H-strand transcription and one for L-strand transcription have been identified near the origin of HeLa cell mtDNA replication: in particular, for the H strand, one at or very near to the 5′ end of the 12 S rRNA gene and the other in the region immediately upstream of the tRNA gene; for the L strand, in the region close to the 5′ end of the 7 S RNA coding sequence.[43]

Metabolism of Mitochondrial RNA

The introduction of the micrococcal nuclease procedure for the isolation in pure form of the mitochondrial RNA species has opened the way for the analysis of the metabolic properties of HeLa cell mitochondrial RNA in the absence of the use of inhibitors.[12] Two main approaches have been followed in this work: (*a*) determination of half-lives and synthesis rate constants of individual RNAs using the data derived from the measurement of the kinetics of labeling of the RNAs and of the changes with time of the precursor pool specific activity; and (*b*) measurement of the decay of label in mitochondrial RNA species after cordycepin block of further mitochondrial RNA synthesis.

*Measurement of the Labeling of Individual RNA Species and of the
Specific Activity of the RNA Precursor Pools[12]*

Exponentially growing HeLa cells were labeled uniformly with
[^{32}P]orthophosphate and then pulse-labeled for varying times with
[5-^{3}H]uridine, as described earlier. The RNA was extracted from the micrococcal nuclease-treated mitochondrial fraction and fractionated by
oligo(dT)-cellulose chromatography and agarose-CH$_3$HgOH slab gel electrophoresis. The individual RNA species were eluted and the ^{3}H:^{32}P ratio
was determined. The ^{32}P radioactivity was utilized to estimate the mass of
the individual RNA species and provided also a normalization factor for
differences in recovery in different experiments. In order to determine the
specific activities of the mitochondrial UTP and CTP pools, the acid-soluble
components were isolated from the mitochondrial fraction of samples of the
labeled cell populations, using the charcoal binding and elution method
described by Humphreys.[49] Separation of the nucleoside triphosphates
(NTPs) was carried out by chromatography on PEI-cellulose thin-layer
plates according to the method of Neuhard *et al.*,[50] procedure 1, or the
method of Cashel *et al.*[51] The radioactivity associated with each NTP spot
(recognizable from the UV absorption of the marker) was analyzed by
eluting the material from the center of the spot and from a ring surrounding
the center: if the chromatography was properly done, the specific activities
of the two samples were substantially identical, arguing against the presence
of closely migrating contaminants.

The RNA and precursor pool labeling data discussed above were used to
calculate the half-lives and synthesis rate constants of the mitochondrial
RNA species by utilizing Eq. (1)[52]

$$dR(t)/dt = k_s S(t) - k_d R(t) \tag{1}$$

where $R(t)$ is the amount of radioactivity in a given RNA species at time t,
$S(t)$ is the specific activity of the precursor pool at time t, k_s is the rate
constant of synthesis, and k_d is the rate constant of decay. It should be noted
that, in the application discussed here, k_s represents the rate of appearance in
mitochondria of the individual RNA species, not the rate of transcription of
their coding sequences. This equation assumes that the rate of synthesis per
cell of the particular RNA species is constant and that this RNA decays with

[49] T. Humphreys, *in* "Molecular Techniques and Approaches in Developmental Biology" (M. Chrispeels, ed.), p. 141. Wiley, New York, 1973.

[50] J. Neuhard, E. Randerath, and K. Randerath, *Anal. Biochem.* **13**, 211 (1965).

[51] M. Cashel, R. A. Lazzarini, and B. Kalbacher, *J. Chromatogr.* **40**, 103 (1969).

[52] G. A. Galau, E. D. Lipson, R. J. Britten, and E. H. Davidson, *Cell* **10**, 415 (1977).

first-order (stochastic) kinetics. The latter assumption has been verified in the case of the mitochondrial RNA species.[12] Given values for $S(t)$ and $R(t)$ for various times, the equation can be solved numerically for the values of k_s and k_d by making use of an appropriate computer program.[52] From k_d, the half-life of the RNA of interest was calculated:

$$t_{1/2} = \ln 2/k_d \qquad (2)$$

In the application discussed here, for calculating the rate constants of decay and the rate constants of synthesis, the UTP and CTP specific activities (^{3}H : ^{32}P) corresponding to each time point, weighted for their proportion in an average H-strand transcript (24.3% U, 31.5% C), were added, and the average NTP ^{3}H : ^{32}P ratio thus obtained was converted to the average ^{3}H per mole of NTP by multiplying it by ^{32}P per mole of NTP (estimated to be equal to ^{32}P per mole of orthophosphate in the medium, multiplied by 3). The incorporation values for a given RNA species (^{3}H : ^{32}P) were expressed on a per-cell basis and normalized to the maximum recovery obtained for that species in the various samples of the same experiment, as follows:

$$\frac{^3H}{^{32}P} \frac{^{32}P \text{ maximum}}{\text{number of cells}} = \text{normalized } ^3H \text{ per cell} \qquad (3)$$

The normalized data of RNA labeling [$R(t)$] and the pool specific activity data [$S(t)$], expressed as indicated above (^{3}H per mole of NTP), were utilized to determine the k_s and k_d values by means of the computer program of Galau *et al.*[52] The ratios k_s/k_d for the individual RNAs provided estimates of their steady-state amounts per cell. The latter could also be estimated directly from the recovery of ^{32}P-labeled RNA species and the specific activity of the orthophosphate in the medium.

Measurement of Decay of Label in Individual RNA Species after Cordycepin Block[12]

For this purpose, exponentially growing cells ($\sim 2 \times 10^5$/ml) were labeled with ^{32}P-orthophosphate (30 μCi/ml) for 2–3 hr in modified Eagle's medium containing 10^{-4} M phosphate with 5% dialyzed calf serum, then cordycepin (3′-deoxyadenosine, Sigma) was added at 50 μg/ml; 500-ml samples were removed at 0 time and at various intervals within the next 60 min, and a constant amount of cells long-term labeled with [5-^{3}H]uridine was added to each sample to correct for differences in recovery. The RNA was extracted from the micrococcal nuclease mitochondrial fraction, fractionated by oligo(dT)-cellulose chromatography and agarose–CH$_3$HgOH slab gel electrophoresis; the individual RNA species were eluted, and the ^{32}P : ^{3}H ratio was determined as described above.

Transcription Products of HeLa Cell mtDNA

Table I shows the list of the discrete transcripts of HeLa cell mtDNA that have been identified so far.[10] It is likely that this list includes all or almost all the main mtDNA-coded RNA species. The amino acid specificity of the tRNAs encoded in mtDNA has been inferred from the mtDNA sequence[3] and, for most of them, directly demonstrated by hybridization experiments involving AA–tRNA complexes.[40] The identification of the majority of the oligo(dT)-bound RNAs as mRNAs is based on their almost perfect correspondence with significant reading frames of human mtDNA,[1,6,42] their structural[10,15,31] and metabolic properties,[12] and their association with polysomes.[10] Table II shows the steady-state amounts of the individual mitochondrial RNAs, their synthesis rates, and their half-lives.[6] The identification of the mtDNA transcripts listed in Tables I and II and the determination of their structural and metabolic properties have been discussed in detail previously.[6,10,12,15,31]

Genetic and Transcription Maps of HeLa Cell mtDNA

Figure 5 shows the genetic and transcription maps of HeLa cell mtDNA.[3,7,23,42,53,54] The positions and identity of the tRNA genes and the positions of the reading frames have been derived from the mtDNA sequence.[3] The specificity of the polypeptides encoded in the reading frames has been determined by correlating the DNA or RNA sequence with protein sequence data or with known yeast gene sequences.[6,55] The mapping positions of the H-strand and L-strand transcripts have been determined by a variety of RNA–DNA hybridization and sequencing methods, as described in previous sections.[15,31,42] There is an almost perfect correspondence between individual reading frames and putative mRNAs. Each mRNA contains only one reading frame, with the exception of RNA 14, which contains a 5'-end proximal unidentified reading frame of 207 nt (URF A6L) overlapping out-of-phase by 46 nt the reading frame for the ATPase 6 polypeptide, and RNA 7, which contains a 5'-end proximal unidentified reading frame of 297 nt (URF 4L), overlapping out of phase by 7 nt the unidentified reading frame URF 4.

The functional and evolutionary implications of the extraordinarily compact gene organization of human, and in general mammalian, mtDNA have been discussed earlier.[1,6] The tight arrangement, with no overlapping or gaps, of the H-strand transcripts and the interspersed distribution of the

[53] D. Ojala and G. Attardi, *J. Mol. Biol.* **138**, 411 (1980).
[54] S. Crews, D. Ojala, J. Posakony, J. Nishiguchi, and G. Attardi, *Nature (London)* **277**, 192 (1979).
[55] A. Chomyn, M. W. Hunkapiller, and G. Attardi, *Nucleic Acids Res.* **9**, 867 (1981).

TABLE I

TRANSCRIPTS OF HeLa CELL MITOCHONDRIAL DNA

RNA species	Molecular length[a] (number of nucleotides)	Functional assignment	Template strand
(a) Oligo(dT)-unbound			
4 S RNA	59–75	Includes 22 tRNAs	14 tRNAs: H; 8 tRNAs L
12 S RNA	954	Structural components of ribosomes	H
16 S RNA	1559		H
(b) Oligo(dT)-bound			
1	~10400		L
2	~7070		L
3	~4155	URF6 mRNAs?	L
4	~2700		H
5	2410	URF 5 mRNA	H
6	1938	Precursor of RNA 9	H
7	1668	(URF4L + URF4)mRNA	H
9	1617	COI mRNA	H
11	1141	Cytochrome b mRNA	H
12	1042	URF2 mRNA	H
13	958	URF1 mRNA	H
14	842	(URF A6L + ATPase 6) mRNA	H
15	784	COIII mRNA	H
16	709	COII mRNA	H
17	346	URF3 mRNA	H
18 (7 S RNA)	~215		L

[a] Of non-poly(A) or non-oligo(A) portion, determined from the length of the coding DNA sequences, except for RNAs 2, 3, and 18 (estimated from S1 protection data) and RNAs 1 and 4 (estimated from electrophoretic mobility).

TABLE II

STEADY-STATE AMOUNT, RATE OF SYNTHESIS, AND HALF-LIFE OF INDIVIDUAL MITOCHONDRIAL RNA SPECIES[a]

RNA species	Steady-state amount (number of molecules/cell)	Rate of synthesis (number of molecules/min per cell)	Half-life (in minutes)	
			Kinetics of incorporation of label	Decay of label after cordycepin block
(a) Oligo(dT)-unbound				
16 S rRNA	ND	ND	215	282
12 S rRNA	34,000	265	208	ND
(b) Oligo(dT)-bound				
2	ND	ND	ND	7
4	44	0.8	ND	39
5	165	3.4	ND	87
6	125	4.4	ND	16
7	960	15.0	47	112
9 COI mRNA	950	10.5	67	116
10 (16 S rRNA)	560	2.1	185	ND
11 Cytochrome b mRNA	570	7.0	53	56
12	720	6.1	73	51
13	650	18.0	ND	ND
14 ATPase 6 mRNA	770	6.8	59⎤	141
15 COIII mRNA	980	9.4	71⎦	
16 COII mRNA	1190	10.0	77	191
17	225	4.5	ND	ND
18 (7 S RNA)	1900	7.3	67	ND

[a] From Attardi et al.[6]
[b] ND, not determined.

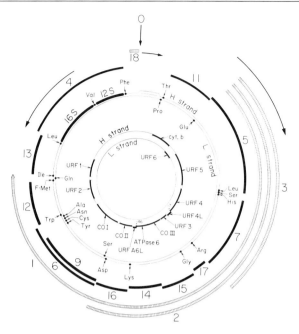

FIG. 5. Genetic and transcription maps of the HeLa cell mitochondrial genome.[3,42] The two outer circles show the positions of the two rRNA genes as derived from mapping and RNA sequencing experiments,[7,23,53] and those of the tRNA genes, as derived from the mtDNA sequence.[3,7] Mapping positions of the polyadenylated H-strand transcripts are indicated by the black bars, those of the L-strand transcripts by the hatched bars.[42] Left and right arrows indicate the direction of H- and L-strand transcription, respectively; the vertical arrow (marked O) and the rightward arrow at the top indicate the location of the origin and the direction of H-strand synthesis.[54] The two inner circles show the positions of the mtDNA reading frames.[3] URF: unidentified reading frame.

tRNA sequences, which separate with nearly absolute regularity the rRNA and protein coding sequences, have suggested a model of H-strand transcription and mitochondrial RNA processing that is discussed elsewhere.[31]

Perspectives

The detailed information concerning the organization of genes and transcripts of the human mitochondrial genome which has emerged from the mtDNA sequence analysis and from the investigation of the structural, mapping and metabolic properties of the mtDNA coded RNAs is expected to provide the framework for future investigations concerning the mechanism of mtDNA transcription and mitochondrial RNA processing and their mode of regulation. While the available data suggest a plausible model of transcription of the H strand and processing of the H-stranded transcripts,[1,6] direct experimental evidence for it is lacking; furthermore, very little is known about the mechanism of transcription of the L strand, the processing of the L-strand transcripts, and their functional role. Further progress will depend to a great extent on the development of suitable *in vitro* transcription systems, either at the level of intact or partially disrupted organelles or at the level of reconstituted systems utilizing mtDNA, the specific mitochondrial RNA polymerase(s), and other mitochondrial components. The identification of the processing enzyme(s), in particular that (those) which putatively process(es) the nascent H-strand transcripts directly into mature products by endonucleolytic cleavages occurring immediately before and after each tRNA sequence,[31] will represent an important step toward understanding the details of the mechanisms and regulation of gene expression in mitochondria.

Acknowledgment

This work was supported by NIH Grant GM-11726.

[42] Isolation of a Hexokinase Binding Protein from the Outer Mitochondrial Membrane

By JOHN E. WILSON, JANICE L. MESSER, and PHILIP L. FELGNER

Reversible interaction of brain hexokinase with the outer mitochondrial membrane is thought to be a significant factor in the regulation of this enzyme.[1] Although the interaction is specific for the outer mitochondrial membrane, it is not specific for the mitochondria from brain, since mitochondria from liver[2,3] as well as other tissues[4,5] have been reported to bind hexokinase. This observation is of great practical significance, since it has permitted the use of rat liver mitochondria, readily available in adequate purity and amount, as a starting point for preparation of a protein that is the outer mitochondrial membrane component responsible for the selective interaction with brain hexokinase. The method described here is that developed by Felgner *et al.*[6] for purification of this protein, designated the "hexokinase binding protein" (HBP).

[1] J. E. Wilson, *Curr. Top. Cell. Regul.* **16**, (1980).
[2] I. A. Rose and J. V. B. Warms, *J. Biol. Chem.* **242**, 1635 (1967).
[3] E. S. Kropp and J. E. Wilson, *Biochem. Biophys. Res. Commun.* **38**, 74 (1970).
[4] B. Font, C. Vial, and D. C. Gautheron, *FEBS Lett.* **56**, 24 (1975).
[5] R. J. Mayer and G. Hubscher, *Biochem. J.* **124**, 491 (1971).
[6] P. L. Felgner, J. L. Messer, and J. E. Wilson, *J. Biol. Chem.* **254**, 4946 (1979).

Assay of Hexokinase Binding

Principle. Membranes or vesicles are incubated with excess brain hexokinase in the presence of Mg^{2+}, which promotes binding to the enzyme.[1,2] After centrifugation, the amount of hexokinase associated with the pelleted membranes or vesicles is determined. A measure of the specificity of the binding is provided by the "G6P:Gal-6-P ratio," which is the ratio of bound enzyme that can be resolubilized by incubation with glucose-6-P to that solubilized with galactose-6-P. Glucose-6-P is very effective at elution of hexokinase from intact mitochondria whereas galactose-6-P is quite ineffective[1,2]; thus, the G6P:Gal-6-P ratio is a measure of the extent to which any observed binding resembles the binding normally seen with intact mitochondria.

Reagents

Brain hexokinase: Rats are decapitated, and the brains are rapidly removed and frozen in liquid nitrogen. They can be stored in liquid nitrogen for several months without deleterious effect. Brains are thawed and homogenized in 0.25 *M* sucrose (10 ml per gram of brain) using a Teflon–glass homogenizer. After centrifugation at 1000 *g* for 10 min, the supernatant is removed and centrifuged at 40,000 *g* for 20 min to provide a crude mitochondrial fraction. This fraction is washed once by rehomogenizing in 0.25 *M* sucrose followed by centrifugation at 40,000 *g* for 20 min. The washed particles are again homogenized in 0.25 *M* sucrose (10 ml per original gram of brain) and made 1 m*M* in glucose-6-P by addition of a 100 m*M* stock solution (disodium salt of glucose-6-P, pH 7). After incubation for 30 min at 25°, the sample is centrifuged at 40,000 *g* for 45 min to yield the solubilized hexokinase. The enzyme can be used directly or stored at −20° for future use; stored enzyme is thawed and recentrifuged at 150,000 *g* for 30 min prior to use in order to remove any aggregated material formed during storage.

Buffer B: 0.25 *M* sucrose, 2 m*M* sodium phosphate, 2 m*M* glucose, 2 m*M* thioglycerol, 0.1 m*M* EDTA, pH 6.6

Procedure. Membranes or vesicles are incubated for 20 min at 25° with excess[7] glucose-6-P-solubilized hexokinase and 3 m*M* $MgCl_2$; the required amount of enzyme can be estimated from preliminary experiments in which the preparation of solubilized enzyme is titrated with increasing amounts of outer mitochondrial membranes, essentially as described by Kropp and Wilson.[3] After incubation, the samples are centrifuged at 150,000 *g* for

---

[7] Verification that the enzyme is in excess is simply done by comparing two samples, one with the estimated excess and another with twice that amount. If the estimated excess was correct, the amount of bound hexokinase in both samples should be the same.

45 min, and the pellets are resuspended in a small volume of buffer B; total bound hexokinase is determined by assay of this suspension. For determination of the G6P:Gal6-P ratio, duplicate aliquots of the suspension are made 1 mM in either glucose-6-P or galactose-6-P, incubated for 10 min at room temperature, then centrifuged for 5 min at full speed in a Beckman Airfuge. Supernatants are assayed for hexokinase, and results are expressed as the ratio of activity solubilized by glucose-6-P to that released with galactose-6-P.

Assay of Hexokinase Activity

Principle. The hexokinase reaction is coupled to NADPH formation with excess glucose-6-P dehydrogenase. NADPH is monitored spectrophotometrically at 340 nm.

Reagents

Assay mix: 3.7 mM glucose, 7.5 mM MgCl$_2$, 11 mM thioglycerol, and 45 mM HEPES (*N*-2-hydroxyethylpiperazine-*N'*-2-ethanesulfonic acid), pH 8.5

NADP, 50 mg/ml, in 0.1 M sodium phosphate, pH 7.0

Glucose-6-P dehydrogenase (from Baker's yeast), 10 units/ml in 0.02% bovine serum albumin–20 mM Tris-HCl, pH 7.5

ATP, 0.22 M, pH 7.0

Procedure. Mix 0.9 ml of assay mix, 0.01 ml of NADP, 0.01 ml of glucose-6-P dehydrogenase, 0.03 ml of ATP, H$_2$O, and enzyme to give a total volume of 1.0 ml in a cuvette and observe the rate of NADPH formation in a spectrophotometer with the sample compartment maintained at 25°. In samples containing no glucose-6-P, the reaction is routinely initiated by addition of hexokinase. However, when the enzyme itself contains glucose-6-P (e.g., when the G6P:Gal-6-P ratio is being determined), the ATP is omitted from the cuvette and oxidation of endogenous glucose-6-P allowed to come to completion (2–3 min) prior to initiation of the hexokinase assay by addition of ATP.

This assay procedure is basically the same as that described previously[8] except that the concentration of NADP used has been increased, and the pH has been increased from 7.5 to 8.5 in order to avoid possible inhibition by traces of Al^{3+} (a frequent contaminant of ATP preparations), which is evident at lower pH values but not at 8.5.[9,10]

[8] A. C. Chou and J. E. Wilson, this series, Vol. 42, p. 20.

[9] F. C. Womack and S. P. Colowick, *Proc. Natl. Acad. Sci. U.S.A.* **76**, 5080 (1979).

[10] L. P. Solheim and H. J. Fromm, *Biochemistry* **19**, 6074 (1980).

Preparation of the Hexokinase Binding Protein

Isolation of Outer Mitochondrial Membranes. The method used represents a slight modification of that described by Sottocasa *et al.*[11] Rats are starved overnight. After decapitation, livers are removed (~ 120 g, from 8 – 12 rats, represents a convenient amount) and immediately placed in ice cold 0.25 M sucrose. After weighing, the livers are minced with scissors, then homogenized in cold 0.25 M sucrose with a Teflon – glass homogenizer, and the volume of the homogenate is adjusted to 10 ml per gram of liver. The homogenate is centrifuged at 600 g for 15 min, and the supernatant is carefully decanted. This is then centrifuged at 6500 g for 20 min to sediment the mitochondria; the supernatant is decanted, and residual supernatant on the side of the tubes is removed with a tissue. The mitochondria are washed twice by resuspension in 0.25 M sucrose (half the original volume, then in one-fourth the original volume) followed by centrifugation at 6500 g for 20 min; after the last wash, the fluffy layer on top of the densely packed brown mitochondrial pellet is removed with a Pasteur pipette. The washed mitochondria are suspended by gently homogenizing in cold 10 mM Tris phosphate, pH 7.5, and the total volume is adjusted to 90 ml (per 120 g original liver). The suspension is kept on ice for 5 min to allow completion of swelling in the hypotonic medium. A solution composed of 30 ml of 1.8 M sucrose, 0.273 ml of 220 mM ATP (pH 7), and 0.6 ml of 100 mM MgSO$_4$ is then added, and the sample is mixed by gently stirring; a strikingly evident increase in turbidity, reflecting shrinking of the inner mitochondrial compartment, should be evident. After a further 5 min at 0°, 20-ml aliquots are sonicated for 20 sec at 3 A and 0° using a Branson Sonifier with microtip; sonication promotes dissociation of the ruptured outer mitochondrial membrane from the intact shrunken inner mitochondrial compartment (mitoplasts). After sonication, the samples are centrifuged at 43,000 g for 20 min, and the supernatant is carefully removed. The supernatant is layered onto freshly prepared discontinuous sucrose density gradients consisting of 5 ml each of 1.32, 1.0, and 0.76 M sucrose; in normal loading, each gradient receives the mitochondrial membranes derived from approximately 20 g of liver. The gradients are centrifuged at 25,000 rpm for 4.5 hr in a Beckman SW 27 rotor. Overlying layers are carefully aspirated off, and the outer mitochondrial membranes that band at the interface between the 1.0 M and 0.76 M sucrose are collected. The suspension is diluted with 3 – 4 volumes of cold water, and the membranes are pelleted by

---

[11] G. L. Sottocasa, G. L. Kuylenstierna, L. Ernster, and A. Bergstrand, this series, Vol. 10, p. 448.

centrifugation at 150,000 g for 45 min. A small dark pellet is sometimes seen to be present under the abundant and clearly distinguishable reddish, somewhat gelatinous layer of outer mitochondrial membranes; this pellet is carefully avoided while removing the membranes, which are then resuspended by gently homogenizing in 0.25 M sucrose. Generally, resuspension using 1 ml of sucrose per 8 g of original liver will provide a membrane suspension containing 3–5 mg of protein per milliliter.

Dissociation of Outer Mitochondrial Membrane and Reconstitution of Hexokinase Binding Vesicles Enriched in the HBP. Purification of the HBP is based on the observation that this protein can be selectively reincorporated into lipid vesicles after dissociation of outer mitochondrial membranes with the detergent octyl-β-D-glucopyranoside (octyl glucoside).

Outer mitochondrial membranes (3–5 mg of protein per milliliter), suspended in 0.25 M sucrose, are diluted with an equal volume of cold buffer A (0.3 M NaCl, 0.05 M sodium phosphate, 1 mM EDTA, pH 7.5); 20% (w/w) aqueous octyl glucoside is added to give a final concentration of 2.6%. Extensive dissociation of the membranes is evident from the dramatic decrease in turbidity after addition of the detergent. The samples are incubated at 25° for 30 min, during which time a slight turbidity develops. The samples are then centrifuged at 150,000 g for 45 min. The pellet, which does not show significant hexokinase binding activity, is discarded. The clear amber supernatant is dialyzed for 12–24 hr at 4° against 0.01 M sodium phosphate, pH 7.0, resulting in the formation of vesicles, referred to as "1×-reconstituted vesicles," which exhibit hexokinase binding ability and which, based on SDS–gel electrophoresis results, are selectively enriched in a single protein component with an estimated subunit molecular weight of 31,000. The 1×-reconstituted vesicles are resuspended by homogenizing in 0.25 M sucrose. Further purification of the HBP can be achieved by repetition of the dissociation-reconstitution process; i.e., the 1×-reconstituted vesicle suspension is diluted with an equal volume of buffer A, followed by incubation at 25°, centrifugation, and dialysis of the supernatant as described above, yielding "2×-reconstituted" vesicles, which are highly enriched in the HBP. Based on staining intensity in SDS–gel patterns, the HBP is estimated to represent approximately 10%, 50%, and >90% of the total protein in outer mitochondrial membranes, 1×-reconstituted vesicles, and 2×-reconstituted vesicles, respectively.

If desired, further purification of the HBP can be achieved by dissociation of the 2×-reconstituted vesicles with buffer A and octyl glucoside, as above, followed by centrifugation through 5 to 20% sucrose gradients containing 2.6% octyl glucoside. Fractions collected from the gradient are analyzed by SDS–gel electrophoresis, and fractions containing the HBP are combined.

Comments

Recovery of Hexokinase Binding Activity and G6P:Gal-6-P Ratios. Substantial decrease in total hexokinase binding activity occurs during purification of the HBP. Taking the binding observed with the original outer mitochondrial membranes as 100%, binding activity recovered in 1×- and 2×-reconstituted vesicles is generally about 50% and 30%, respectively. While this may indicate a considerable lability of this protein, an additional factor may be the accessibility of the protein to added hexokinase. Thus, if the distribution of the HBP in the original outer mitochondrial membranes was asymmetric, with all of it accessible to exogenous hexokinase, while the distribution in 1×-reconstituted vesicles was symmetric (i.e., half of the HBP oriented to the outside of the vesicle while half was oriented toward to the inside and hence inaccessible to added hexokinase), it is evident that approximately a 50% loss of binding ability would be lost even if *no* denaturation of the HBP actually occurred. Since there is no presently available information on the orientation of the HBP in the outer mitochondrial membrane and in reconstituted vesicles, the possible importance of this accessibility factor in evaluating hexokinase binding activity cannot be estimated. It is mentioned here simply to point out that while measures of hexokinase binding ability may be empirically useful, their interpretation may not be simply related to preservation of protein structure, and hence function, per se.

The G6P:Gal6-P ratios seen with intact mitochondria are quite high, generally in the range of 10–20.[12] Substantial decrease in this ratio is seen with isolated outer mitochondrial membranes, with G6P:Gal6-P ratios for 2–4 commonly observed. Further decrease in this ratio occurs as a result of the dissociation–reconstitution process required for purification of the HBP, with ratios of 1.5 commonly found for 2×-reconstituted vesicles. Much of this decrease in G6P:Gal6P ratio seems to be due to a general decrease in the strength of the hexokinase-membrane interaction (i.e., to an increase in the amount of solubilization seen with Gal-6-P), which results from isolation of the outer mitochondrial membranes or preparation of reconstituted vesicles enriched in the HBP. Thus, with intact mitochondria, relatively little (<10%) of the hexokinase is solubilized during incubation with galactose-6-P, whereas with isolated outer mitochondrial membranes or reconstituted vesicles much higher (~30–50%) amounts of the total activity are solubilized during incubation with galactose-6-P (and virtually identical amounts are released during incubation with no added galactose-6-P); this large increase in the denominator clearly can account for the substantial decrease in the G6P:Gal6P ratio. The reason for this general

[12] J. E. Wilson, *Arch. Biochem. Biophys.* **154**, 332 (1973).

weakening in hexokinase binding with increased resolution of the binding site (i.e., progression from intact mitochondria to intact outer mitochondrial membranes to reconstituted vesicles greatly enriched in HBP) is not yet understood. Thus, although there can be little doubt that the HBP represents the membrane component responsible for specific recognition and interaction with hexokinase, it is also likely that additional factors will be found that influence the strength of this interaction.

Nature of the HBP. Incubation of intact outer mitochondrial membranes with proteases such as trypsin or chymotrypsin results in extensive degradation of virtually all proteins of the outer mitochondrial membrane except the HBP; reconstituted vesicles can be prepared from proteolyzed membranes with no effect on recovery of the HBP or hexokinase binding ability. Thus, in the membrane, the HBP is fully protected from exogenous proteases. In contrast, if the outer mitochondrial membrane is first dissociated with octyl glucoside, the HBP is rapidly degraded by added proteases. A reasonable interpretation of these results[6] is that the HBP is an integral protein of the outer mitochondrial membrane, i.e., it is embedded in the intact membrane and hence protected from exogenous protease action. This interpretation is also consistent with the selective reincorporation of the HBP into lipid vesicles, which forms the basis for its purification; such a selective effect can be readily envisaged as being due to preferential incorporation of the HBP into the hydrophobic milieu provided by the reconstituted lipid bilayers, and such interactions are commonly thought to occur with integral membrane proteins.

Although the subunit molecular weight of the HBP is estimated to be 31,000, the actual size of the hexokinase binding entity in the outer mitochondrial membrane or reconstituted vesicles is not known; i.e., it is conceivable that two or more subunits must interact to form an effective hexokinase binding site in the membrane (or vesicles).

Liver mitochondria normally contain little, if any, mitochondrially bound hexokinase.[13] The presence of substantial amounts of the HBP in these organelles would be surprising *if* that were indeed its only function. Hence it seems likely that the HBP might also function in some other role. This now appears to be the case. Two different research groups have recently reported that the HBP is identical to the pore-forming protein of the outer mitochondrial membrane.[14,15] Thus hexokinase apparently binds to the structure through which ATP may exit from the mitochondria, a strategic location indeed for an ATP-requiring enzyme.

[13] J. E. Wilson, and P. L. Felgner, *Mol. Cell. Biochem.* **18,** 39 (1977).
[14] M. Linden, P. Gellerfors, and B. D. Nelson, *FEBS Lett.* **141,** 189 (1982).
[15] C. Fiek, R. Benz, N. Roos, and D. Brdickzka, *Biochim. Biophys. Acta* **688,** 429 (1982).

[43] Protein Synthesis by Isolated Plant Mitochondria[1]

By C. J. Leaver, E. Hack, and B. G. Forde

The demonstration in the early 1960s that both mitochondria and chloroplasts contain their own unique DNA species provided the basis to explain earlier observations that the cytoplasm of higher plants was a source of extranuclear genetic information. The further realization that cytoplasmic genes play an important role in plant development, productivity, and susceptibility to disease,[2] coupled with the desire to define and increase cytoplasmic genetic diversity, has increased interest in the information content and function of plant mitochondrial DNA (mtDNA).

Most of the research of the mitochondrial genetic system has been carried out with yeast (*Saccharomyces* sp.), *Neurospora crassa,* and animal cells. The mtDNA has been shown to contain genes for the ribosomal and transfer RNA components of a mitochondrion-specific protein synthesizing system, whose function is to translate a small number (10–20) of mtDNA-encoded messenger RNAs that specify the same basic set of polypeptides. These polypeptides constitute less than 10% of the total mitochondrial protein, and the best characterized are hydrophobic components of three oligomeric enzyme complexes of the inner mitochondrial membrane: namely, the three largest of the seven subunits of cytochrome c oxidase, the apocytochrome b subunit of the cytochrome bc_1 complex, two subunits of the oligomycin-sensitive ATPase complex, and (in yeast and *Neurospora*) one protein (Var-1) of the small subunit of the mitochondrial ribosome.[3,4]

In view of the observations that the higher plant mitochondrial genome is apparently considerably larger and more complex than the mitochondrial genome of other eukaryotes,[2] it is conceivable that plant mtDNA encodes additional information. One approach to establishing the number and identity of the mitochondrial polypeptides synthesized (and by extrapolation, encoded) by higher plants is by *in vivo* labeling of mitochondrial proteins with radioactive amino acids, in the presence of cycloheximide (a selective inhibitor of cytosolic protein synthesis). The major drawbacks to this *in vivo* approach are the difficulty in labeling the mitochondrial transla-

[1] The research on which this chapter is based has been supported by grants from the Agricultural Research Council and Science Research Council (U.K.) to C. J. Leaver.
[2] C. J. Leaver and M. W. Gray, *Annu. Rev. Plant Physiol.* **33,** 373 (1982).
[3] P. Borst and L. A. Grivell, *Cell* **15,** 705 (1978).
[4] P. Slonimski, G. Attardi, and P. Borst, eds., "Mitochondrial Genes." Cold Spring Harbor Lab., Cold Spring Harbor, New York, 1982.

tion products to a high enough specific activity and the contribution of contaminating bacteria to the observed labeling patterns.

We describe the procedures that we have adopted to isolate[5-7] essentially sterile plant mitochondria that synthesize protein in an energy-dependent, chloramphenicol-sensitive but cycloheximide-insensitive process.[8,9] The spectrum of proteins synthesized by the isolated mitochondria is essentially the same as those labeled *in vivo* in the presence of cycloheximide.

Plant Material

Mitochondria capable of protein synthesis *in vitro* have been isolated from a range of plant tissues including fleshy roots and tubers (artichoke, potato, turnip, sugar beet), etiolated seedling tissue (mung bean hypocotyls, cucumber and castor bean cotyledons, maize, sorghum, and wheat coleoptiles), and green tissue (tobacco, cucumber, petunia leaves, cucumber cotyledons). The major prerequisite is that the mitochondria are uncontaminated with other subcellular structures or bacteria and are morphologically and functionally intact. To this end it is obviously easier to start with nongreen storage organs, such as fleshy roots and tubers or young, dark-grown seedlings.

Growth and Preparation of Material

Storage organs free of obvious fungal or bacterial contamination are selected, washed, and peeled, and surface sterilized by immersion for 5 min in an 1:20 dilution of stock sodium hypochlorite (14% w/v). The tissue is then washed at least four times in sterile, ice-cold distilled water, cut into pieces approximately 1 cm $\times$ 0.5 cm and cooled to 2°.

Seeds are surface sterilized by immersion for 10 min in a 1:15 dilution of stock sodium hypochlorite (14% w/v), followed by washing in at least six changes of sterile distilled water. They are then allowed to germinate on a sterile medium (e.g., cellulose wadding or vermiculite), watered with sterile water, and grown under optimum conditions either in the dark (3–7 days for most seedlings) or the light.

The hypocotyls, coleoptiles, cotyledons, or leaves are harvested, cut into 0.5–1 cm sections, and washed extensively with several changes of sterile distilled water prior to cooling to 2°.

[5] R. Douce, E. L. Christensen, and W. D. Bonner, Jr., *Biochim. Biophys. Acta* **275**, 148 (1972).
[6] G. Laties, this series, Vol. 31, p. 589.
[7] R. Douce, A. L. Moore, and M. Neuburger, *Plant Physiol.* **60**, 625 (1977).
[8] B. G. Forde, R. J. C. Oliver, and C. J. Leaver, *Proc. Natl. Acad. Sci. U.S.A.* **75**, 3841 (1978).
[9] B. G. Forde, R. J. C. Oliver, and C. J. Leaver, *Plant Physiol.* **63**, 67 (1979).

Isolation of Mitochondria

The following operations must be carried out at 2–4°, and all glassware, centrifuge bottles, and homogenizers are sterilized by heating in an oven for 12 hr at 120° or by immersion for 30 min in a 1 : 10 dilution of stock sodium hypochlorite (14% w/v), followed by extensive washing with sterile distilled water. Solutions must be sterilized either by autoclaving or passage through Millipore filters with a pore size of 0.45 μm.

Tissue should be homogenized as soon after harvest as possible in an ice-cold medium containing: 0.4 M mannitol, 1 mM EGTA (ethylene glycol bis(β-aminoethyl ether)-N,N'-tetraacetic acid), 25 mM morpholinopropanesulfonic acid (MOPS), 0.1% (w/v) bovine serum albumin (BSA, Fraction V), and 8 mM cysteine (free base) added just prior to homogenization. The pH of the medium is adjusted to 7.8 with 5 N KOH, and the medium is added in the ratio of 2 ml for each gram fresh weight of nongreen tissue or 4 ml per gram of green-leaf tissue. In addition, 40 mM 2-mercaptoethanol is included in the medium for green tissues. Depending on the tissue, homogenization can be accomplished by one of several methods: (*a*) grinding in a precooled mortar and pestle; (*b*) homogenization in a square-form beaker with either a Moulinex mixer 66 (Alençon, France) for 20–60 sec or a Polytron, Ultra-Turrax tissue blender at 50% full speed for 2 × 10 sec; or (c) homogenization in a Waring blender for 2 × 15 sec at low speed. For nongreen material homogenization is continued until the tissue is just reduced to a creamy consistency, but more active mitochondria are obtained from green leaves by giving a shorter homogenization.

The homogenate is filtered through four layers of sterile muslin into a beaker cooled in ice, the muslin being squeezed to ensure removal of most of the fluid. At this stage, with some tissues the pH may drop, and it is necessary to adjust the filtrate to pH 7.2 to 7.5 by the dropwise addition of 5 N KOH, using a narrow-range pH paper. The filtrate is then passed through two layers of Miracloth (Chicopee Mills, New York) supported flat on a Büchner funnel (7–20 cm in diameter), into cooled 50-, 100-, or 250-ml centrifuge bottles. The bottles are balanced and centrifuged for 5 min at 1000 g (3000 g if starting with green tissue) in the SS-34 (8 × 50 ml) or GSA (6 × 250 ml) angle head, in an RC5-B Sorvall centrifuge at 2°. The supernatant fraction is gently decanted into clean, cold centrifuge bottles, taking care not to transfer any of the pellet, which contains starch, nuclei, cell debris, etc. It is then centrifuged for 15 min at 14,000 g, and the resulting high speed supernatant is discarded. The tan mitochondrial pellet in each bottle is gently resuspended in 5–10 ml of wash medium–0.4 M mannitol, 5 mM MOPS, pH 7.5, 1 mM EGTA, and 0.1% (w/v) BSA, with the aid of a rounded glass rod. The suspended mitochondria are transferred to 50-ml centrifuge tubes, the centrifuge bottles are rinsed twice with 5-ml aliquots of wash medium, and the volume is adjusted to 40 ml. The mitochondria are resuspended further with several strokes of a loose-fitting Teflon or glass pestle and centrifuged at 1000 g (3000 g for green tissue) for 5 min. The supernatant is transferred into clean tubes, and the mitochondria are sedimented at 10,000 g for 15 min. The supernatant is once again discarded; the washed mitochondria are uniformly resuspended in a small volume (1–2 ml per 50–100 g of starting tissue) of wash medium and layered onto gradients for further purification.

Gradient Purification of Mitochondria

While the mitochondria prepared as described above are adequate for many respiratory measurements, they are frequently contaminated with other subcellular fractions and occasionally with bacteria. Further purification, a prerequisite to measuring mitochondrial protein synthesis, is carried out by centrifugation on either sucrose gradients or a silica-sol gradient material, Percoll (Pharmacia Fine Chemicals Ltd., Uppsala, Sweden).

Sucrose gradients are routinely used for the purification of mitochondria from nongreen and etiolated tissue, as they are considerably more effective than Percoll gradients for the removal of contaminating bacteria and microbodies. However, Percoll gradients allow rapid purification, under isosmotic and low viscosity conditions, a particular advantage in preparing mitochondria, which are damaged by the extremes of osmotic shock encountered in sucrose-gradient purification procedures.[10] They also allow improved separation of mitochondria from the chloroplasts and thylakoid fragments encountered in green-leaf mitochondrial preparations.

Sucrose Density Gradients. The washed mitochondria in 1–2 ml of wash medium are layered on top of continuous sucrose gradients (0.6–1.8 M) and centrifuged in a swinging-bucket rotor. The volume of the gradient depends upon the concentration of mitochondria, which in turn depends upon the weight and type of starting tissue. Normally, mitochondria from up to 25 g of tissue can be purified on a 10-ml gradient (< 10 mg of mitochondrial protein) by centrifuging at 50,000 g for 60 min (e.g., MSE 6 × 14 ml rotor or Spinco SW 40 rotor). Mitochondria from up to 75 g of tissue can be purified on a 25-ml gradient (10 to 30 mg of mitochondrial protein) by centrifuging at 40,000 g for 60 min (Spinco SW 25.1 rotor). The gradients are prepared by layering sucrose solutions containing 10 mM Tricine, pH 7.2, 1 mM EGTA, and 0.1% (w/v) BSA into centrifuge tubes in the sequence of concentrations tabulated here.

[10] C. Jackson, J. E. Dench, D. O. Hall, and A. L. Moore, *Plant Physiol.* **64** 150 (1979).

Sucrose concentration (M)	10-ml gradient (ml)	25-ml gradient (ml)
1.8	1.0	2.0
1.45	2.8	6.0
1.2	2.5	6.0
0.9	2.5	6.0
0.6	1.5	5.0

The gradients are allowed to equilibrate overnight at 2° prior to loading of the mitochondrial suspension.

After centrifugation, the mitochondria form a buff-colored band at a density of 1.25–1.35 M sucrose and are recovered by carefully sucking them into a Pasteur pipette having a right-angled bend at the tip. The volume of the mitochondrial suspension recovered is estimated by pipetting into a graduated centrifuge tube standing in ice. The molarity of the suspension is estimated using a refractometer, and the suspension is slowly diluted over a 15-min period by the addition of 0.2 M mannitol, 10 mM Tricine, pH 7.2, 1 mM EGTA to give a final osmotic concentration of approximately 0.6 M. The diluted mitochondria are centrifuged at 10,000 g for 15 min, the supernatant is discarded, and the purified mitochondrial pellet is suspended in a medium containing 0.4 M mannitol, 10 mM Tricine, pH 7.2, 1 mM EGTA at a concentration of 5–20 mg of mitochondrial protein per milliliter. At this stage aliquots can be removed for protein assay; and estimation of the physiological integrity of the mitochondria, by measurement of respiratory control, ADP:0, etc.

Percoll Gradients. A continuous, self-generating gradient[11] is routinely used for purifying mitochondria from a range of tissues. The concentration of Percoll in the starting solution and the time of centrifugation can be varied to optimize a particular separation.

Percoll cannot be autoclaved in the presence of salts and sugars, therefore separate stock solutions of 100% (w/v) Percoll and 0.8 M mannitol, 20 mM Tricine, pH 7.2, 2 mM EGTA, and distilled water are sterilized; a gradient solution containing 35% (w/v) Percoll, 0.4 M mannitol, 10 mM Tricine, pH 7.2, 1 mM EGTA, 0.1% (w/v) BSA is prepared shortly before use. Washed mitochondria (2.5 ml) from up to 100 g of green-leaf tissue are layered over 20 ml of the 35% Percoll solution in a 30-ml centrifuge tube and centrifuged at 27,000 g for 45 min in the SS-34 angle rotor of a Sorvall RC-5B centrifuge. The mitochondria, which form a distinct buff-colored band below the green chloroplast fragments (in preparations from green

[11] Percoll-R, "Methodology and Applications." Pharmacia Fine Chemicals, Uppsala 1, Sweden.

tissue), are collected with care to avoid any green material, using a Pasteur pipette with a bent tip as for the sucrose gradients. Normally, the mitochondria are collected in a volume of 2–4 ml, which is diluted with 10 volumes of 0.4 M mannitol, 10 mM Tricine, pH 7.2, 1 mM EGTA and centrifuged at 10,000 g for 15 min. The mitochondria form a very loose pellet and therefore as much as possible of the supernatant should be removed carefully with a wide-mouthed pipette. The mitochondria are gently resuspended in the remaining supernatant (1–2 ml), transferred to a 1.5-ml conical tube and centrifuged for 5 min at 12,000 g and 2° in a microcentrifuge. The supernatant is removed, and the mitochondrial pellet is resuspended to a concentration of 5–20 mg/ml in 0.4 M mannitol, 10 mM Tricine, pH 7.2, 1 mM EGTA.

When there is significant bacterial contamination, the bacteria are not adequately removed on Percoll gradients and it is necessary further to purify the mitochondria on sucrose gradients.

Protein Synthesis by Isolated Mitochondria

Maximal rates of protein synthesis are dependent upon the isolation of intact, coupled mitochondria free of bacterial contamination. The system described below was optimized for pH, osmotic pressure, temperature, ion and amino acid concentration, adenyl and guanyl nucleotide concentration, oxidizable substrate, and/or an external ATP-generating system. High specific activity [^{35}S]methionine is routinely used as the substrate for mitochondrial protein synthesis. Essentially similar results have been obtained, however, with [^{14}C]leucine or a mixture of ^{14}C-labeled amino acids, provided the components of the unlabeled amino acid mixture are adjusted accordingly. The free amino acid pool in mitochondria isolated from different tissues at successive developmental stages of the same tissue may vary, and this in turn will affect the apparent rates of *in vitro* protein synthesis by the isolated mitochondria.

Incubation Conditions. Purified mitochondria (200–700 μg of protein) are incubated in test tubes (5 × 1 cm) in a final volume of 250 μl. The basic incubation medium contains: mannitol, 250 mM; KCl, 90 mM; MgCl$_2$, 10 mM; Tricine buffer, pH 7.2, 10 mM; potassium phosphate (pH 7.2), 5 mM; EGTA, 1 mM; 19 amino acids except methionine, 25 $\mu$$M$; dithiothreitol, 2 m$M$; GTP, 1 m$M$; L-[^{35}S]methionine (875–1140 Ci/mmol), 5–20 μCi. In addition, for each mitochondrial preparation an incorporation is normally carried out using each of three "energy mixes" with final concentrations: (*a*) 10 mM sodium succinate, 2 mM ADP–ATP generated by respiratory-chainlinked phosphorylations; (*b*) 8 mM creatine phosphate, 25 μg of creatine phosphokinase, 6 mM ATP–ATP generated externally; (*c*)

20 mM sodium acetate — a nonoxidizable substrate that can be utilized by bacteria but not by mitochondria, hence amino acid incorporation in its presence gives an indication of bacterial contamination.

Throughout the incubation the tubes are shaken at approximately 100–150 cycles per minute to ensure adequate aeration and are maintained at 25°.

In practice all the components of the incubation medium except the amino acid mix, dithiothreitol, GTP, and the [^{35}S]methionine are made up as concentrated stocks, autoclaved separately, and stored at −20°. Stock solutions of GTP, dithiothreitol, and the amino acid mix are filter sterilized and stored at −20°. Concentrated solutions (5×) of the energy mixes are prepared from sterile stock solutions just prior to use.

The final incubation medium is prepared by mixing the component stocks together with [^{35}S]methionine in a final volume of 150 μl. To this is added 50 μl of the requisite (5×) energy mix and 50 μl of mitochondria in 0.4 M mannitol, 10 mM Tricine, pH 7.2, 1 mM EGTA.

The time course of the reaction can be followed by removing 5-μl aliquots onto 1.5-cm disks of Whatman 3 MM filter paper. The disks are dried in a stream of air for 30 sec before immersion in ice-cold 10% (w/v) trichloroacetic acid (TCA) for at least 15 min. Unincorporated radioactivity is removed by the procedure of Mans and Novelli,[12] which involves washing in 5% (w/v) TCA at 90° for 15 min; in 5% TCA at room temperature for 5 min (four times); in ethanol:ether (1:1) at 37° for 5 min; and finally in ether at 37° for 10 min. The disks are dried, and the radioactivity incorporated into protein is estimated in a scintillation counter.

Amino acid incorporation is generally linear for at least 90 min with an external ATP-generating system. Depending upon the respiratory integrity of the mitochondria, incorporation is linear for 30–45 min with an oxidizable substrate and ADP, after which it slowly reaches a plateau. Incorporation in the presence of a nonoxidizable substrate should be less than 5% of that obtained with an oxidizable substrate.

After a 60–90-min incubation period, the incorporation of [^{35}S]methionine is stopped by the addition of 1 ml of ice-cold buffer containing 10 mM unlabeled methionine, 0.4 M mannitol, 10 mM Tricine, pH 7.2, and 1 mM EGTA. The mitochondria are pelleted at 12,000 g for 5 min in an Eppendorf microcentrifuge, the supernatant is removed, and the pellet is solubilized in 100–200 μl of sample buffer containing 2% (w/v) sodium dodecyl sulfate (SDS), 10% (w/v) sucrose, 1% (v/v) 2-mercaptoethanol, 60 mM Tris-HCl, pH 6.8, by heating at 90° for 2 min.

SDS–Polyacrylamide Gel Electrophoresis. The solubilized mitochondrial polypeptides (100–200 μg of protein) are fractionated by electrophore-

[12] R. V. Mans and A. D. Novelli, *Arch. Biochem. Biophys.* **94,** 48 (1961).

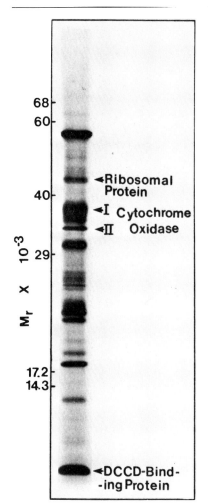

FIG. 1. Polypeptides synthesized by isolated maize mitochondria. Mitochondrial translation products were labeled by incubating for 90 min mitochondria from 5-day-old, dark-grown maize shoots in a medium containing [^{35}S]methionine. The proteins were solubilized in sodium dodecyl sulfate, and electrophoresed in a 15% (w/v) polyacrylamide slab gel; labeled polypeptides were detected by autoradiography of the dried gel. Reprinted from Slonimski et al.,[4] with permission.

sis in either 15% (w/v) or 15 to 20% (w/v) linear gradient, SDS–polyacrylamide slab gels according to the method of Laemmli.[13] The ratio of acrylamide to bisacrylamide should be 30:0.2.

The gels are stained with Coomassie blue, destained, dried onto Whatman 3 MM paper, and exposed for up to 2 weeks to DuPont Cronex-4 X-ray film. The sensitivity of the process can be raised about 10-fold by impregnating the gels with a fluor, such as 2,5-diphenyloxazole (PPO)[14] or sodium salicylate, prior to drying and exposure to X-ray film.[15]

[13] U. K. Laemmli, Nature (London) 227, 680 (1970).
[14] R. A. Laskey and A. D. Mills, Eur. J. Biochem. 56, 335,(1976).
[15] J. P. Chamberlin, Anal. Biochem. 98, 132 (1979).

Figure 1 shows an analysis of the [^{35}S]methionine-labeled polypeptides synthesized by mitochondrial isolated from 5-day-old, dark-grown maize seedlings. Mitochondria from maize and all other higher plants so far examined synthesize an essentially similar spectrum of some 18–20 polypeptides, with apparent molecular weights ranging from 8,000 to 54,000.[2] The two highest molecular weight polypeptides are easily solubilized from broken mitochondria, but the remainder are membrane bound. We have recently shown that the largest translation product with a molecular weight of 58,000 is the α subunit of the mitochondrial F$_1$-ATPase complex. In other organisms this polypeptide is encoded in the nuclear genome and synthesized on cytoplasmic ribosomes. The soluble polypeptide with an apparent molecular weight of 44,000 copurifies with mitochondrial ribosomes and is probably equivalent to the Var-1 protein described in yeast mitochondria.

Two further mitochondrial translation products have been tentatively identified as subunits I and II of cytochrome oxidase by immunoprecipitation with monospecific antibodies prepared against purified yeast cytochrome c oxidase subunits. The smallest plant mitochondrial translation product has an estimated molecular weight of about 8000, is soluble in a chloroform–methanol mixture or in butanol, and specifically binds [^{14}C]DCCD. These characteristics suggest that it is equivalent to the proteolipid subunit of the mitochondrial proton-translocating ATPase (subunit 9).

The remaining plant mitochondrial translation products remain to be identified. It does appear, however, that plant mitochondria synthesize, and by extrapolation probably code for, additional polypeptides not synthesized by mitochondria from animals and fungi.

Section III

Chloroplasts

[44] Synthesis and Assembly of Thylakoid Membrane Proteins in Isolated Pea Chloroplasts

By RICHARD S. WILLIAMS and JOHN BENNETT

A major aim of current research on the biogenesis of the photosynthetic membrane of the chloroplast is to determine the sites of synthesis and the mechanisms of assembly of thylakoid polypeptides. Like other chloroplast proteins, thylakoid proteins are under dual genetic control.[1,2] Probably a majority of chloroplast proteins are encoded by nuclear DNA and synthesized by cytoplasmic ribosomes prior to transport into the chloroplast. However, a substantial minority are encoded by chloroplast DNA and synthesized by chloroplast ribosomes. This chapter describes some of the methods for studying the synthesis and assembly of those thylakoid proteins that originate within the chloroplast. Methods for studying the synthesis, transport, and assembly of thylakoid polypeptides of cytoplasmic origin are described in this volume.[3]

Methods

Plant Growth. The isolation of intact chloroplasts showing high rates of protein synthesis *in vitro* is favored by the use of young, starch-free tissue. In the case of pea plants (*Pisum sativum* L. var. Feltham First), excellent results are obtained by sowing seeds in moist potting fiber, allowing seedlings to develop under a 12-hr day length at day/night temperature of 22/19°, and harvesting the leaves 9–11 days after sowing. Illumination is provided by warm white fluorescent tubes that emit photosynthetically active radiation (400–700 nm) with a quantum flux density of 50 μmol/m² sec⁻¹. Peas grow well under these conditions and accumulate little starch, but at a quantum flux density in excess of 100 μmol/m² sec⁻¹, starch grains accumulate and are liable to pierce the chloroplast envelope during centrifugation and greatly reduce the yield of intact organelles. With other species of plants, it is undoubtedly worth conducting a preliminary search for growth conditions that give maximal rates of protein synthesis *in vitro*. This may involve finding growth conditions that minimize not only starch levels, but also levels of polyphenols and mucilage. Intact chloroplasts capable of

---

[1] N.-H. Chua and G. W. Schmidt, *J. Cell Biol.* **81**, 461 (1979).
[2] R. J. Ellis, *Annu. Rev. Plant Physiol.* **32**, 111 (1981).
[3] J. E. Mullet and N.-H. Chua, this volume [45].

METHODS IN ENZYMOLOGY, VOL. 97

high rates of protein synthesis *in vitro* have been isolated from a number of species of plants,[2] including spinach,[4] maize,[5] and the alga *Euglena*.[6]

Solutions. The following solutions are required for the isolation of intact pea chloroplasts, their incubation in the presence of L-[^{35}S]methionine to label translation products, and the analysis of their translation products.

Sucrose isolation medium (autoclaved without D-isoascorbate): 25 mM
HEPES buffer, pH 7.6; 350 mM sucrose; 2 mM EDTA (plus 2 mM
D-isoascorbate, added just prior to use; should not be autoclaved)

KCl resuspension medium (autoclaved): 66 mM Tricine–KOH buffer,
pH 8.3; 200 mM KCl; 6.6 mM MgCl$_2$

Sorbitol resuspension medium (autoclaved): 50 mM HEPES–KOH
buffer, pH 8.0; 330 mM sorbitol

ATP, 50 mM, in water

L-[^{35}S]Methionine (Amersham International, Amersham, Bucks, England), at least 600 Ci/mmol

100% (w/v) trichloroacetic acid (TCA)

5% TCA, containing 0.5% L-methionine

10× Tris-glycine buffer, pH 8.0: 62 mM Tris; 480 mM glycine

Sorbitol–EDTA buffer: 5 mM EDTA, pH 7.8; 50 mM sorbitol

80% (v/v) aqueous acetone: 80 ml of acetone, water to 100 ml

10% Sodium dodecyl sulfate (SDS): 10 g SDS, water to 100 ml

5% Triton X-100 in Tris-glycine buffer: 5 g of Triton X-100, 10 ml of
10× Tris–glycine buffer, water to 100 ml

1 M sucrose: 342 g of sucrose, water to 1000 ml

75% glycerol: 75 ml glycerol, water to 100 ml

Chloroplast Isolation and Incubation. This procedure is based on that described by Blair and Ellis.[7] Muslin is cut into squares (20 cm × 20 cm), and two sets of eight layers of muslin are washed with distilled water and then with cold sucrose isolation medium to remove detergents and other contaminants. Pea leaves (20 g) are harvested, placed in a frozen Perspex container (5 cm × 5 cm × 20 cm) and covered with a semifrozen slurry of sucrose isolation medium (100 ml). The probe of a Polytron homogenizer (Kinematica GmbH, Lucerne, Switzerland) is lowered into the container and the leaves are homogenized by two or three 4-sec bursts with the Polytron set at position 7. The homogenate is squeezed through one set of eight layers of muslin and then poured through a second set of eight layers. The muslin retains unhomogenized leaves, debris, and most whole cells.

[4] W. Bottomley, D. Spencer, and P. R. Whitfeld, *Arch. Biochem. Biophys.* **164**, 106 (1974).
[5] A. E. Grabanier, D. M. Coen, A. Rich, and L. Bogorad, *J. Cell Biol.* **78**, 734 (1978).
[6] A. C. Vasconcelos, *Plant Physiol.* **58**, 719 (1976).
[7] G. E. Blair and R. J. Ellis, *Biochim. Biophys. Acta* **319**, 223 (1973).

The filtrate is immediately centrifuged at 5° and 2800 g for 1 min in a swing-out rotor of a centrifuge equipped with a rapid breaking system. After centrifugation the supernatant is decanted, the tubes are wiped to remove froth from the inner surfaces, and the pellets are resuspended in buffered, isotonic solution. Two osmotica are commonly used for this purpose: KCl and sorbitol.[7-9] It is desirable to resuspend the pellets initially in a minimum volume (1–2 ml) of resuspension medium. This may be achieved either by a gentle swirling motion or by the use of a clean, soft paint brush or a small ball of cotton wool attached to a glass rod.

Intactness of the chloroplast preparation may be quickly assessed by examination of a drop of the preparation by phase contrast microscopy. Chloroplasts that have retained their double envelope membranes (class A and class B[10]) are phase-bright, whereas broken, envelope-free chloroplasts are phase-dark. Alternatively, the degree of intactness may be measured using the ferricyanide reduction assay in the presence and in the absence of uncoupler[11,12]; ferricyanide cannot penetrate the chloroplast envelope and can therefore act as a Hill oxidant only in the case of broken chloroplasts. Intact chloroplasts usually account for 40–80% of isolated chloroplasts, the percentage of intact chloroplasts being higher for chloroplasts resuspended in sorbitol medium compared with KCl medium.[13] A small sample of the chloroplast suspension is also removed for chlorophyll determination[12]; the chloroplasts are then diluted to about 500 μg of chlorophyll per milliliter in the same resuspension medium.

The water bath used for incubation of the chloroplasts with radioisotope is made of glass and is illuminated from below by an unfiltered tungsten lamp (Photolita, 500 W, Philips, Eindhoven, The Netherlands). The lamp is placed at such a distance from the bath that the quantum flux density of the photosynthetically active radiation is 500 μmol/m^2 sec^{-1}. The water bath is maintained at 20°, and a fan circulates air underneath the bath to prevent overheating of the glass bottom by the lamp.

Samples of the chloroplast suspension are pipetted into incubation vessels. For small volumes (about 200 μl) microcentrifuge tubes are suitable, whereas for larger volumes (about 2 ml) flat-bottomed flasks are advisable. With a chlorophyll concentration of 500 μg/ml, it is important to illuminate the sample approximately uniformly, and this can be achieved only with

[8] R. J. Ellis, *Biochim. Biophys. Acta* **463**, 185 (1977).
[9] J.-J. Morgenthaler and L. R. Mendiola-Morgenthaler, *Arch. Biochem. Biophys.* **172**, 52 (1976).
[10] S. G. Reeves and D. O. Hall, this series, Vol. 69, p. 85.
[11] U. Heber and K. A. Santarius, *Z. Naturforsch., B: Anorg. Chem., Org. Chem; Biochem., Biophys., Biol.* **25B**, 718 (1970).
[12] D. A. Walker, this series, Vol. 69, p. 94.
[13] J. F. Allen and J. Bennett, *FEBS Lett.* **123**, 67 (1981).

thin layers of chloroplast suspension. For most experiments it is adequate simply to add radioisotope, ATP, inhibitors and any other additions to the incubation vessel prior to the addition of the chloroplast suspension. The final concentration of L-[35S]methionine is 500 μCi/ml. When the incubation is to be performed in darkness, the tubes and flasks are wrapped in foil.

Assaying Incorporation of Radioisotope into Chloroplast Proteins. For both the optimization of protein synthesis by isolated intact chloroplasts *in vitro* and the study of the translation process itself, it is useful to have a quick way of following the progress of the incubation. A convenient procedure is to remove 5-μl samples from the incubation, pipette them on to a grid of numbered squares marked on a filter paper and then to immerse the filter paper in hot (90–100°) 5% TCA containing nonradioactive methionine (0.5%). Hot TCA hydrolyzes methionyl-tRNA without hydrolyzing protein. After 5 min, the filter paper is washed once in 5% TCA–0.5% methionine, twice with ethanol, and once with ether. This procedure removes all pigment and TCA from the filter. The squares are then cut from the paper and counted in a liquid scintillation spectrometer. By this method, the effects of numerous parameters on total chloroplast protein synthesis can be studied.

Analysis of Translation Products. Translation products include stromal proteins,[4-9] thylakoid proteins,[5,6,8,14] and proteins of the chloroplast envelope.[9,15] Thylakoid proteins may be separated from stromal and envelope proteins by bursting chloroplasts by hypotonic shock and then sedimenting and washing the thylakoid membranes. The suspension of intact, 35S-labeled chloroplasts is diluted with 5–10 volumes of water. As water is drawn into the chloroplasts in response to the difference in osmotic pressure between the medium and the stroma, they swell and burst. After 5 min on ice, the suspension is centrifuged at 10,000 g for 2 min, the supernatant is retained, and the thylakoid pellet is washed twice by resuspension in 1 × Tris-glycine buffer to 100 μg of chlorophyll per milliliter and recentrifuged at 10,000 g. The washed thylakoid pellet is resuspended in Tris-glycine buffer to 250 μg of chlorophyll per milliliter, and SDS is added to 1%. At the same time the retained supernatant, containing mainly soluble proteins, is made 5% in TCA, the precipitated proteins are collected by centrifugation at 10,000 g for 1 min, and the pellet is drained completely of supernatant and resuspended in the same volume of Tris-glycine buffer as that used for the resuspension of the washed thylakoids. Then SDS is added to the suspension to 1%. Duplicate 5-μl samples are withdrawn from the thylakoid protein and soluble protein preparations and assayed for 35S-labeled protein as described above. To the remainder of each protein preparation, glycerol is added to

[14] A. R. J. Eaglesham and R. J. Ellis, *Biochim. Biophys. Acta* **335**, 396 (1974).
[15] K. W. Joy and R. J. Ellis, *Biochim. Biophys. Acta* **378**, 143 (1975).

7.5% to facilitate loading of 50-μl samples for polyacrylamide slab gel electrophoresis (see below).

Assaying Assembly. This procedure exploits the ability of the nonionic detergent Triton X-100 to release multisubunit complexes and chlorophyll–protein complexes from the thylakoid membrane. It is based on the procedures of Arntzen and co-workers[16,17] and involves separation of the released complexes by sucrose density gradient centrifugation. Isolated chloroplasts (1 mg of chlorophyll) are resuspended in 2 ml of KCl medium supplemented with 2 mM ATP and incubated in the light with 1 mCi of L-[^{35}S]methionine for 30–60 min. At the end of the incubation the chloroplast suspension is diluted with 8 ml of cold KCl medium and centrifuged at 2800 g for 2 min. The pellet is resuspended in 15 ml of sorbitol–EDTA medium to lyse the chloroplasts and unstack grana, and centrifuged at 10,000 g for 5 min. The resulting pellet is resuspended in 10 ml of Tris-glycine buffer and centrifuged at 15,000 g for 10 min. The pellet of washed thylakoids is dispersed in about 1 ml of Tris-glycine buffer, and a 10-μl sample is withdrawn and added to 1 ml of 80% aqueous acetone for chlorophyll determination.[12] The thylakoids are then diluted to 500 μg of chlorophyll per milliliter with Tris-glycine buffer, and Triton X-100 in the same buffer is added to 0.5%. The solubilized membrane is stirred at room temperature for 30 min and centrifuged at 30,000 g for 30 min to give a small, soft, pale-green pellet, which is discarded, and a dark-green supernatant. The supernatant (about 2 ml) is loaded above a 0.1 to 0.7 M sucrose density gradient containing 0.02% Triton X-100 in Tris-glycine buffer. The total volume of the gradient is about 15 ml. The gradient is centrifuged at 100,000 g_{av} for 16 hr at 5° in a swing-out rotor. After centrifugation the gradient is collected from the top with a descending gradient fractionator (Densiflow IIC, Buchler Instruments, Fort Lee, New Jersey). Twenty-two fractions (0.7 ml) are collected, and the green pellet is resuspended in 0.7 ml of Tris-glycine buffer. The fractions and the resuspended pellet are made 1% in SDS, while the pellet and gradient fractions 1–3 (numbering from the top of the gradient) are also made 7.5% in glycerol. Samples are removed from each fraction for electrophoresis (50 μl), chlorophyll (100 μl), and ^{35}S-labeled protein (5 μl).

Electrophoresis. The electrophoretic procedure is basically that of Laemmli[18] for polyacrylamide slab gel electrophoresis. The separating gel is a 15 × 15 × 0.15 cm vertical slab consisting of a 10 to 30% exponential gradient of acrylamide. The acrylamide:bisacrylamide ratios at the top and bottom of the gradient are 37 and 200, respectively. The stacking gel consists

[16] J. J. Burke, C. L. Ditto, and C. J. Arntzen, *Arch. Biochem. Biophys.* **187**, 252 (1978).
[17] J. E. Mullet, J. J. Burke, and C. J. Arntzen, *Plant Physiol.* **65**, 814 (1980).
[18] U. K. Laemmli, *Nature (London)* **227**, 680 (1970).

of 5% polyacrylamide, with an acrylamide:bisacrylamide ratio of 20. The buffer in the separating gel consists of 378 mM Tris-HCl, pH 8.8, containing 0.1% SDS; the buffer in the stacking gel consists of 120 mM Tris-HCl, pH 6.8, containing 0.1% SDS; and the buffer in the electrode chambers is 25 mM Tris, 200 mM glycine, containing 0.1% SDS.

Protein samples (50 μl) are subjected to electrophoresis at 20° and 15 mA for 16 hr. After electrophoresis the gel is stained with Coomassie Brilliant Blue R250,[18] dried under reduced pressure at 80° for 2 hr, and autoradiographed using Kodak X-Omat S film.

For a combination of high resolution and preservation of chlorophyll–protein complexes, the Laemmli procedure[18] may be replaced by those of Delepelaire and Chua[19] and Machold and Meister.[20]

Optimum Conditions for Protein Synthesis by Isolated Intact Pea Chloroplasts

Energy Source. The incorporation of L-[^{35}S]methionine into protein by isolated intact pea chloroplasts is dependent on an energy source such as light or exogenous ATP (Table I). When chloroplasts are resuspended in KCl medium at 500 μg of chlorophyll per milliliter, incorporation over a 30-min period is stimulated 9-fold by light, 5-fold by 2 mM ATP in the dark, and 13-fold by light and ATP together. The effects of light and ATP appear to be additive. The uncoupler carbonyl cyanide m-chlorophenylhydrazone (CCCP), which is a potent inhibitor of the endogenous generation of ATP by photophosphorylation, blocks light-driven protein synthesis without inhibiting ATP-driven protein synthesis.

TABLE I

ENERGY SOURCE FOR PROTEIN SYNTHESIS BY ISOLATED INTACT PEA CHLOROPLASTS

Incubation conditions	Incorporation of L-[^{35}S]methionine into protein (%)
Light	100[a]
Light + 5 μM CCCP[b]	28
Light + 2 mM ATP	160
Dark	12
Dark + 2 mM ATP	68
Dark + 2 mM ATP + 5 μM CCCP	70

[a] Typically, over 10^6 cpm of ^{35}S are incorporated per milligram of chlorophyll in 30 min.
[b] CCCP, carbonyl cyanide m-chlorophenylhydrazone.

[19] P. Delepelaire and N.-H. Chua, *Proc. Natl. Acad. Sci. U.S.A.* **76,** 111 (1979).
[20] O. Machold and A. Meister, *Photobiochem. Photobiophys.* **1,** 213 (1980).

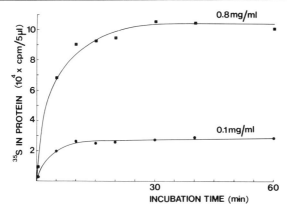

FIG. 1. Time course of incorporation of L-[35S]methionine into protein by isolated intact pea chloroplasts. Chloroplasts were resuspended in KCl medium containing 2 mM ATP to a chlorophyll concentration of either 0.1 or 0.8 mg/ml and were illuminated in the presence of radioactive methionine. At intervals, 5-μl samples were withdrawn for measurement of 35S in protein.

Time Course. When intact pea chloroplasts are resuspended in KCl medium and illuminated in the presence of 2 mM ATP, the time course of protein labeling is curvilinear (Fig. 1). Net incorporation ceases after about 30 min at chlorophyll concentrations in the range 0.1–1.0 mg/ml. Although the reason for the decline in protein synthesis with time is not clear, it is known that isolated intact chloroplasts not only elongate polypeptide chains that were initiated *in vivo* (the minimum requirement for protein labeling), but also terminate correctly (see below) and reinitiate.[21]

Optimal Chlorophyll Concentration. The quantity of translation products accumulated during the first 30 min of the incubation is linearly related to chlorophyll concentration up to 0.3 mg/ml (Fig. 2). In the range 0.3–0.8 mg/ml, the absolute rate of synthesis increases, but the synthetic rate per unit chlorophyll declines. Above 0.8 mg/ml, absolute incorporation is reduced, presumably as a result of self-shading of the chloroplasts at the light intensity employed.

Effects of Resuspension Medium and ATP on the Labeling of Soluble and Thylakoid Polypeptides. Isolated pea chloroplasts incorporate L-[35S]methionine in the light into both soluble and thylakoid polypeptides (Table II). In the absence of exogenous ATP, the KCl medium and the sorbitol medium are equally suitable for protein synthesis, but in the presence of 2 mM ATP the KCl medium is superior, especially for the labeling of thylakoid proteins. ATP stimulates labeling of thylakoid proteins by 30% in KCl medium but

21 P. E. Highfield and R. J. Ellis, *Biochim. Biophys. Acta* **447**, 20 (1976).

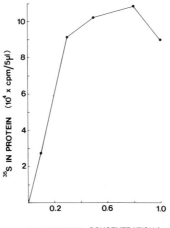

FIG. 2. Chloroplast protein synthesis as a function of chlorophyll concentration. Chloroplasts were resuspended in KCl medium containing 2 mM ATP to a chlorophyll concentration in the range 0.1–1.0 mg/ml and illuminated in the presence of radioactive methionine. After 30 min, 5-μl samples were withdrawn for measurement of ^{35}S in protein.

inhibits by 60% in sorbitol medium. The reasons for these effects of ATP are not yet clear. Analysis of the translation products by SDS–polyacrylamide gel electrophoresis and autoradiography shows that the stimulatory and inhibitory effects of ATP in the different media are due to general effects on protein synthesis as a whole rather than to the preferential stimulation or inhibition of the synthesis of individual polypeptides.

TABLE II

INCORPORATION OF L-[^{35}S]METHIONINE INTO SOLUBLE AND THYLAKOID PROTEINS BY ISOLATED INTACT CHLOROPLASTS. EFFECTS OF ATP AND INCUBATION MEDIUM

		Incorporation of ^{35}S into protein[a]	
Osmoticum	ATP	Soluble proteins	Thylakoid proteins
KCl	—	1.83	2.25
	2 mM	3.42	2.93
Sorbitol	—	2.07	1.62
	2 mM	1.60	0.61

[a] 10^6 cpm incorporated per milligram of chlorophyll in 30 min.

Analysis of Translation Products and the Assay of Assembly

Electrophoresis. Sodium dodecyl sulfate – polyacrylamide gel electrophoresis of soluble and thylakoid polypeptides, followed by autoradiography of the stained and dried gel, reveals that isolated intact chloroplasts synthesize a number of discrete translation products (Fig. 3). The major soluble translation product is the 54,000 M_r large subunit of the CO_2-fixing enzyme, ribulosebisphosphate carboxylase/oxygenase (EC 4.1.1.39).[7] The fact that isolated chloroplasts synthesize the full-length large subunit (Fig. 3, c,f) indicates that they are capable of both elongation and correct termination. The 14,000 M_r small subunit of this enzyme is not labeled in isolated chloroplasts because it is synthesized in the cytoplasm and transported into the chloroplast by a posttranslational mechanism.[1,2]

The major thylakoid-bound translation product is a 32,000 – 33,500 M_r polypeptide (Fig. 3, d,e). This polypeptide is known variously as peak D[14], photogene 32 protein,[22] the 32,000 M_r thylakoid protein,[23] and one of the herbicide-binding proteins of photosystem II.[24] Its biological role is believed to be to bind or shield a quinone on the reducing side of photosystem II. There is evidence that the 32,000 M_r polypeptide is subject to turnover *in vivo.*[2,23] It is for this reason that such a major chloroplast translation product (Fig. 3, d,e) does not correspond to a major staining band (Fig. 3, a,b).

Among other membrane-bound translation products are the α and β subunits of chloroplast coupling factor (CF_1), which is part of the chloroplast ATP synthase.[25] These subunits have molecular weights of 59,000 and 56,000 respectively, and form a conspicuous doublet on SDS – polyacrylamide gels, both when visualized by staining (Fig. 3, a,b) and when labeled in intact chloroplasts (Fig. 3, d,e).

Rigorous Identification of Translation Products. Most of the minor soluble and thylakoid-bound translation products of isolated chloroplasts are as yet unidentified. In those cases where identification has been achieved, one or more of the following techniques have been employed: (*a*) immunoprecipitation; (*b*) partial proteolytic digestion; (*c*) purification of the protein in question. Thylakoid polypeptides that have been shown to be synthesized in isolated chloroplasts include five of the subunits of the ATP

[22] L. Bogorad, S. O. Jolly, G. Kidd, G. Link, and L. McIntosh, *in* "Genome Organization and Expression in Plants" (C. J. Leaver, ed.), p. 291. Plenum, New York, 1980.

[23] A. K. Mattoo, U. Pick, H. Hoffman-Falk, and M. Edelman, *Proc. Natl. Acad. Sci. U.S.A.* **78**, 1572 (1981).

[24] K. Pfister, K. E. Steinback, G. Gardner, and C. J. Arntzen, *Proc. Natl. Acad. Sci. U.S.A.* **78**, 981 (1981).

[25] L. R. Mendiola-Morgenthaler, J. J. Morgenthaler, and C. A. Price, *FEBS Lett.* **62**, 96 (1976).

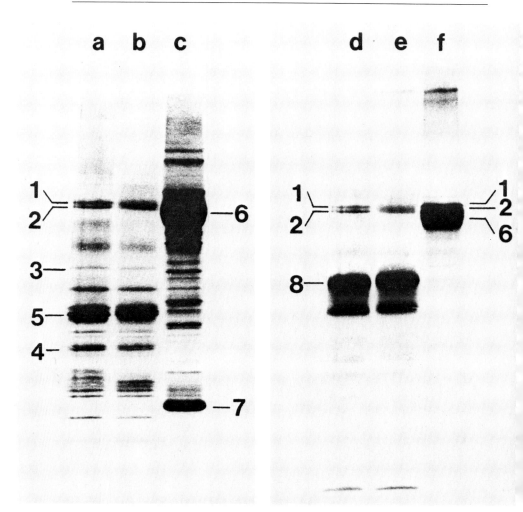

FIG. 3. Electrophoretic analysis of the polypeptides synthesized in isolated intact pea chloroplasts. Chloroplasts were resuspended in KCl medium containing 2 mM ATP to a chlorophyll concentration of 0.5 mg/ml. After illumination in the presence of L-[^{35}S]methionine for 60 min, chloroplasts were osmotically ruptured and separated into a thylakoid fraction (tracks a, b, d, e) and a soluble fraction (tracks c, f). Some of thylakoid preparation was heated to 70° (tracks b, e). Samples were subjected to SDS–polyacrylamide gel electrophoresis. Tracks a–c: stained gel; tracks d–f: autoradiogram. The following chloroplast polypeptides are indicated: 1–4, the α, β, γ, and δ subunits of the coupling factor (CF$_1$); 5, the light-harvesting chlorophyll a/b protein; 6 and 7, the large and small subunits of ribulosebisphosphate carboxylase; 8, the 32,000–33,500 M_r herbicide-binding protein of photosystem II.

synthase (α, β, ϵ, I, and III),[25-27] cytochrome f[28], cytochrome b_{559},[29] and the 68,000 M_r apoprotein of the 130,000 M_r chlorophyll a-binding protein of photosystem I,[29-31] and the apoproteins of two minor chlorophyll a complexes thought to be part of photosystem II.[32]

The immunological approach is especially useful in those cases where the translation product differs from the mature protein in electrophoretic mobility (as in the case of a precursor) or where the translation product represents only a very minor percentage of total labeled protein.

Unlabeled Thylakoid Proteins. It is clear from Fig. 3 that many thylakoid proteins do not become labeled when isolated chloroplasts are incubated with L-[35S]methionine. The absence of label from any given thylakoid polypeptide could be due to any one of the following reasons: (*a*) that the protein does not contain methionine; (*b*) that the protein is synthesized in the cytoplasm rather than in the chloroplast; (*c*) that the protein is synthesized in the chloroplast, but as a result of faulty assembly *in vitro* it is recovered in the soluble phase of the chloroplast rather than in the thylakoid fraction; (*d*) that synthesis of the protein is inhibited under the incubation conditions employed *in vitro.*

The presence of methionine in the protein can be checked by labeling *in vivo* with L-[35S]methionine. If the protein is not labeled *in vivo* with this amino acid, then a much more common amino acid, such as leucine, could be used as tracer. If a tritium-labeled amino acid is chosen, autoradiography must be replaced by fluorography.[33]

The site of synthesis of the protein may be investigated in two general ways. First, the use *in vivo* of selective inhibitors of cytoplasmic and chloroplast protein synthesis can often provide good evidence for or against the involvement of chloroplast ribosomes in the synthesis of any given polypeptide.[34] Second, if an antibody is available, a strong indication of the site of synthesis of a protein can be obtained by *in vitro* translation of cytoplasmic mRNA and chloroplast mRNA, followed by immunoprecipitation to detect the protein in question or a precursor form.[1-3,35] A reasonably representative preparation of cytoplasmic mRNA may be obtained by

[26] N. Nelson, H. Nelson, and G. Schatz, *Proc. Natl. Acad. Sci. U.S.A.* **77**, 1361 (1980).
[27] A. Doherty and J. C. Gray, *Eur. J. Biochem.* **98**, 87 (1979).
[28] A. Doherty and J. C. Gray, *Eur. J. Biochem.* **108**, 131 (1980).
[29] R. E. Zielinski and C. A. Price, *J. Cell Biol.* **85**, 435 (1980).
[30] B. R. Green, *Biochim. Biophys. Acta* **609**, 107 (1980).
[31] W. Hachtel, *Z. Pflanzenphysiol.* **107**, 383 (1982).
[32] B. R. Green, *Eur. J. Biochem.* **128**, 543 (1982).
[33] R. A. Laskey and A. D. Mills, *Eur. J. Biochem.* **56**, 335 (1975).
[34] R. J. Ellis, *in* "Methods in Chloroplast Molecular Biology" (M. Edelman, R. B. Hallick, and N.-H. Chua, eds.), p. 559. Elsevier/North-Holland, Amsterdam.
[35] M. R. Hartley, A. M. Wheeler, and R. J. Ellis, *J. Mol. Biol.* **91**, 67 (1975).

isolation of the polyadenylated mRNA fraction of the cell, while chloroplast mRNA may be obtained by the extraction of nucleic acids from isolated intact chloroplasts of high purity. Among the thylakoid proteins that are synthesized outside the chloroplast are the 26,000 M_r light-harvesting chlorophyll a/b protein[1-3] and the γ and δ subunits of CF_1[36]; these three polypeptides are not labeled in isolated intact chloroplasts (Fig. 3).

The KCl medium is an excellent incubation medium for protein synthesis by isolated pea chloroplasts, but it is a very poor medium for CO_2-dependent O_2 evolution.[13] This implies that there are important chloroplast reactions that are inhibited in KCl medium. If such reactions are involved in the synthesis of a particular thylakoid protein or in its attachment to the photosynthetic membrane, then the sorbitol medium (or the mannitol medium used in the case of *Acetabularia* chloroplasts[30]) may be preferable to the KCl medium, despite the generally lower rate of labeling of thylakoid proteins in sorbitol medium when supplemented with ATP (Table II).

Assembly of Translation Products into Multisubunit Complexes. Most of the proteins involved in photosynthetic electron transport and photophosphorylation can be isolated in the form of multicomponent complexes after solubilization of the thylakoids with detergent.[37,38] These complexes (including the ATP synthase, photosystem I, photosystem II, plastoquinol:plastocyanin oxidoreductase) are believed to correspond closely with functional groupings of thylakoid proteins *in vivo*. There is considerable current interest in discovering how these complexes are formed. Most if not all of the complexes are composed of two classes of polypeptide: those synthesized on chloroplast ribosomes and those synthesized on cytoplasmic ribosomes and then transported into the chloroplast. Such a situation could pose problems for the study of the assembly process in isolated chloroplasts, since assembly of each complex would require not only the synthesis of the subunits of chloroplast origin, but also the presence of pools of subunits of cytoplasmic origin. However, studies on CF_1 suggest that both of these conditions may be satisfied in the case of this particular complex.[9,26,36]

The study of the assembly of chloroplast translation products into multicomponent thylakoid complexes involves four steps: (*a*) *in vitro* labeling of isolated chloroplasts; (*b*) solubilization of the thylakoid membrane with detergents; (*c*) purification of the complex in question; and (*d*) analysis of the complex for the presence of radioactivity in the subunits that originate within the chloroplast. Procedures for steps (*a*) and (*d*) have already been

[36] R. Nechushtal, N. Nelson, A. K. Mattoo, and M. Edelman, *FEBS Lett.* **125**, 115 (1981).
[37] C. J. Arntzen, *Curr. Top. Bioenerg.* **8**, 111 (1978).
[38] E. Hurt and G. Hauska, *Eur. J. Biochem.* **117**, 591 (1981).

discussed. In this section we consider possible approaches to steps (*b*) and (*c*).

The best approach is to study the assembly of each complex individually, employing the solubilization and purification procedures most appropriate to that complex. There are several published procedures for the isolation of each complex. These should be compared for their applicability to the species and age of tissue under consideration. However, for those interested in a general survey of thylakoid protein assembly, a useful initial approach is to eschew selective solubilization procedures and lengthy purification protocols and to employ instead the procedure described in this chapter. It involves the complete solubilization of unstacked thylakoids with the nonionic detergent Triton X-100, followed by the partial separation of multicomponent complexes by sucrose density gradient centrifugation. Its main virtue is that it enables the extent of assembly of a given translation product to be assessed. This will be illustrated by reference to the α and β subunits of CF_1.

Isolated intact chloroplasts are labeled with L-[35S]methionine and then lysed by hypotonic shock. Thylakoid membranes are washed to unstack grana lamellae and solubilized with Triton X-100 at a detergent:chlorophyll mass ratio of 10:1. The solubilized membrane components are fractionated by overnight centrifugation through a $0.1-0.7\ M$ sucrose density gradient.[16,17] Figures 4 and 5 show the results of two separate experiments. In

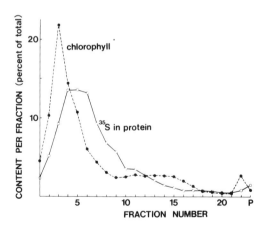

FIG. 4. Sedimentation of chlorophyll and 35S-labeled proteins after solubilization of thylakoids in Triton X-100. Pea chloroplasts were suspended in KCl medium containing 2 m*M* ATP to a chlorophyll concentration of 0.5 mg/ml and incubated for 60 min with L-[35S]methionine in the light. Thylakoids were isolated, unstacked, and solubilized with Triton X-100. The solubilized material was subjected to sucrose density gradient centrifugation for 16 hr at 100,000 g_{av}. The distributions of chlorophyll and 35S-labeled protein were determined. P, pellet.

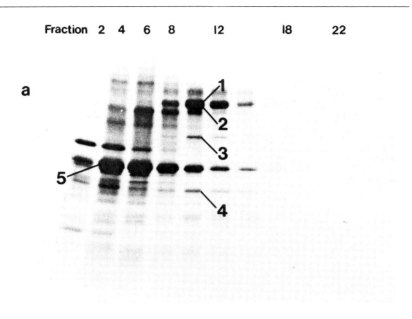

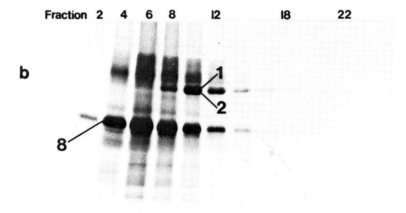

FIG. 5. Assembly of chloroplast translation products into multisubunit complexes of the thylakoid membrane. In an experiment similar to that described in the legend to Fig. 4, sucrose density gradient fractions were analyzed for protein composition by SDS–polyacrylamide gel electrophoresis. a, Stained gel; b, autoradiogram. The numbering of the polypeptides is described in the legend to Fig. 3.

the first experiment, the distribution of chlorophyll in the gradient is compared with that of ^{35}S-labeled protein. In the second, the polypeptide composition of the gradient fractions is analyzed by SDS–polyacrylamide gel electrophoresis.

Most of the polypeptides released from the thylakoids by Triton X-100 sediment faster than their subunit molecular weights would indicate. This is due to their existence as multisubunit complexes under these conditions. This can be seen clearly for the α, β, γ, and δ subunits of CF_1 (Fig. 5a). They sediment, together with the ϵ subunit, as a 325,000 M_r complex and are found mainly in fractions 8–12 of the gradient. The newly synthesized α and β subunits of CF_1 are also found in these same fractions (Fig. 5b), indicating that all the labeled α and β subunits that become associated with the thylakoid membrane in isolated intact chloroplasts do so as complete CF_1 complexes. This implies the existence within chloroplasts of pools of the cytoplasmically synthesized γ and δ subunits.[36]

The 32,000–33,500 M_r photosystem II polypeptide that is heavily labeled in isolated intact chloroplasts (Fig. 3, d,e) is found to sediment mainly to fractions 4–8 (Fig. 5b), implying a molecular weight of about 150,000. This sedimentation behavior is consistent with the idea that it is largely, if not entirely, associated with photosystem II, which is known to sediment to this position on these gradients.[16] However, proof that this translation product is assembled *in vitro* in isolated chloroplasts can be provided only by the demonstration that the ^{35}S-labeled protein is present in purified photosystem II particles. Procedures for the purification of photosystem II particles from higher plants have been published[39,40] and may prove to be suitable for the further analysis of this question.

It has been reported that the 130,000 M_r chlorophyll a–protein complex of photosystem I is assembled *in vitro*. This result, which has been obtained for spinach and *Vicia faba* chloroplasts resuspended in sorbitol medium[29,31] and for *Acetabularia* chloroplasts resuspended in mannitol medium,[30] indicates that the 68,000 M_r apoprotein of this complex is first synthesized *in vitro* and then assembled with chlorophyll a to form the complex. Concerning photosystem I, we should like to point out that the sedimentation behavior of this complex is rather variable, especially when isolated from young tissue. It has been reported that photosystem I particles sediment to the bottom of the centrifuge tube during sucrose density gradient centrifugation under conditions very similar to those described here.[17] In our experience, this is true only for mature leaves; the photosystem I particles released by Triton X-100 from the thylakoids of young leaves sediment about as fast as CF_1.

[39] J. E. Mullet and C. J. Arntzen, *Biochim. Biophys. Acta* **635**, 236 (1981).
[40] D. A. Berthold, G. T. Babcock, and C. F. Yocum, *FEBS Lett.* **134**, 231 (1981).

Conclusion

Isolated intact chloroplasts synthesize many thylakoid proteins. Some of these have been identified and shown to be assembled into the correct multicomponent complexes. Others remain to be identified and studied in this way. Further progress will depend on a wider application of immunochemical analysis; antibodies will need to be raised against a greater number of thylakoid polypeptides to facilitate identification and analysis.

Acknowledgments

We thank Professor R. J. Ellis for useful discussions during the writing of this chapter. Financial support from the Science and Engineering Research Council is gratefully acknowledged.

[45] *In Vitro* Reconstitution of Synthesis, Uptake, and Assembly of Cytoplasmically Synthesized Chloroplast Proteins

By JOHN E. MULLET and NAM-HAI CHUA

The biogenesis of the chloroplast is dependent on the posttranslational uptake of polypeptides that are coded by nuclear genes and translated on cytoplasmic ribosomes. The process by which nuclear coded polypeptides are assembled into the chloroplast has been reconstituted *in vitro*[1,2] and involves three sequential events: (*a*) translation of cytoplasmic RNA; (*b*) uptake of translation products into the chloroplasts; and (*c*) sorting and assembly of polypeptides to sites in the envelope, stroma, and thylakoid membrane. The methodology that has been utilized to study these processes is described here. Similar procedures have also been described elsewhere.[3]

Isolation and Translation of mRNA

Isolation of mRNA. Poly(A) mRNA is prepared as described by Cashmore[4] or, if nuclease degradation of mRNA is suspected, by the guanidine

[1] N.-H. Chua and G. Schmidt, *J. Cell Biol.* **81**, 461 (1979).
[2] P. E. Highfield and R. J. Ellis, *Nature (London)* **271**, 420 (1978).
[3] S. Bartlett, A. R. Grossman, and N.-H. Chua, *in* "Methods of Chloroplast Molecular Biology" (M. Edelman, R. Hallick, and N.-H. Chua, eds.), p. 1081–1091. Elsevier/North-Holland, Amsterdam, 1981.
[4] A. R. Cashmore, *in* "Methods of Chloroplast Molecular Biology" (M. Edelman, R. Hallick, and N.-H. Chua, eds.), p. 387–392. Elsevier/North Holland, Amsterdam, 1981.

thiocyanate extraction procedure of Chirgwin *et al.*[5] The general methodology of mRNA isolation, including isolation of selected mRNAs, is described by P. M. Lizardi, this series, Vol. 96 [2].

Translation of mRNA. Prepare translation mixes as described in Table I, and incubate for 1.5 hr at 27°. After incubation, transfer the translation mix to an ultracentrifuge tube and spin at 105,500 g_{av} (40,000 rpm in a Beckman 50 Ti rotor) for 1 hr to pellet ribosomes. Pipette a 2- to 5-μl aliquot of the translation mix onto a filter paper circle for an estimation of total incorporation into protein. Save about 10–50 μl of the postribosomal supernatant, depending on the incorporation rate, for SDS-gel electrophoresis. Using pea poly(A) RNA (1 A_{260}/ml), we obtain incorporation rates of 250,000 cpm ^{35}S per microliter if 150 μCi of [^{35}S]methionine is used per milliliter in the translation mix.

Uptake of Translation Products by Isolated Chloroplasts

Plant Materials. Seedlings (peas, barley, maize) are used when they are 7–14 days old. Spinach is purchased at a local market. Pea and barley plants are cut off at ground level. Pea shoots are homogenized whole. Barley leaves are cut into small segments with a razor blade prior to homogenization. Maize leaf sheaths are removed, and the blades are prepared as for barley. Spinach leaves are broken into approximately 1 inch pieces.

Preparation of Intact Chloroplasts. The method for isolation of intact chloroplasts reported here is adapted from that of Morgenthaler *et al.*[6] The use of Percoll gradients for isolation of intact chloroplasts has several advantages over conventional methods employing sucrose gradients or high KCl concentrations including low viscosity and low osmolality, nontoxicity, and inability to permeate membranes. We refer the reader to publications by Morgenthaler *et al.*[6] and Schmitt and Herrmann[7] as well as the Pharmacia bulletin for background reading on the rationale of the method.

Place plant materials in a 2-liter beaker and cover with a minimum of 1000 ml of ice-cold GR buffer (Table II). Grind leaves with a Polytron homogenizer (PT 35K probe) set at 6–8. Addition of a small amount of bovine serum albumin (BSA) (about 0.5 g/2 liters) helps overcome the tendency of grass leaves to float on top of the buffer. Filter brei through four layers of Miracloth into 500-ml centrifuge bottles and spin to 4229 g_{av} (5000 rpm in a precooled Sorvall GS-3 rotor) and stop immediately. Pour off

[5] J. M. Chirgwin, A. E. Przybyla, R. J. MacDonald, and W. J. Rutter, *Biochemistry* **18**, 5294 (1979).
[6] J. J. Morgenthaler, M. P. F. Marsden, and C. A. Price, *Arch. Biochem. Biophys.* **168**, 289 (1975).
[7] J. M. Schmitt and R. G. Herrmann, *Methods Cell Biol.* **15**, 177 (1978).

TABLE I

COMPOSITION OF *in Vitro* PROTEIN SYNTHESIS SYSTEM[a]

Reagent	Amount per milliliter (μl)	Final concentration
Group A		
HEPES-KOH, 1.0 mM, pH 8.0	Enough to bring to 1.0 ml	
ATP, 100.0 mM	5.0	0.5 mM
GTP, 12.0 mM	1.3	16.0 μM
Creatine phosphate, 840.0 mM	10.0	8.4 mM
Amino acid mix, 1.2 mM for each amino acid except methionine	20.0	24.0 μM each
>600 Ci/mM [³⁵S]methionine[b]	15.75	0.3–1.5 μM
5 mCi/0.5 ml	—	150–750 μCi
Group B		
Creatine phosphate kinase, 4 mg/ml in 10% glycerol	10.1	40 μg/ml
mRNA		1 A_{260}/ml
Wheat germ[c]	400.0	
Compensating buffer	100.0	
HEPES-KOH, pH 8.0, 1.0 M	50.0	20.0 mM
Potassium acetate, 2.0 M	135.7	110.0 mM[b]
Dithiothreitol, 1.0 M	5.0	2.0 mM
Magnesium acetate, 0.5 M	4.0	0.8 mM[b]
Spermine, 31.0 mM, neutralized	20.0	0.25
H₂O	34.0	—

[a] Make up group A and adjust pH to 7.6 with 1.0 mM KOH. Less than 5 μl of KOH is required to adjust the pH of 1.0 ml of translation mix. Check using narrow-range pH paper. Then add group B reagents. Stocks of mRNA generally are adjusted to 40 A_{260} per milliliter. However, if this is not possible, the volume of the translation mix can be kept constant by changing the amount of H₂O used. [³⁵S]Cysteine can also be used, but we get considerably less incorporation than with [³⁵S]methionine using this isotope in the wheat germ system.

[b] Add immediately before use.

[c] R. Roman, J. D. Brooker, S. N. Seal, and A. Marcus, *Nature (London)* **260**, 359 (1976).

TABLE II
COMPOSITION OF GR BUFFER[6]

Stock solution	Per liter	Final concentration
Sodium-EDTA, 0.2 M, pH 7.0	10 ml	2.0 mM
MgCl$_2$, 1.0 M	1 ml	1.0 mM
MnCl$_2$, 1.0 M	1 ml	1.0 mM
HEPES-KOH, 1.0 M, pH 7.5	50 ml	50 mM
Solid sorbitol	60.13 g	0.33 M
Sodium ascorbate[a]	0.99 g	5.0 mM

[a] Add immediately before use.

supernatant. Gently resuspend the green pellet in a few milliliters of cold GR buffer, and load onto precooled 10 to 80% linear Percoll gradients or 40 to 80% Percoll step gradients (Table III). The green pellet obtained from about 1000–1500 ml of homogenate can be pooled and loaded onto a single gradient. Spin at 14,600 g_{max} (9500 rpm in HB-4 rotor) for 6 min. Two green bands are generated. The upper band contains predominantly broken chloroplasts, and the lower band contains intact chloroplasts.

Discard the broken chloroplasts. Transfer the intact chloroplasts to a clean Corex tube, and add 5–10 volumes of cold GR buffer to dilute out the Percoll. Spin to 4300 g_{av} (6000 rpm in SS-34 rotor) and stop immediately (with brake off). Discard the supernatant, removing the last bit by aspira-

TABLE III
PROCEDURE FOR MAKING PERCOLL GRADIENTS[a]

PBF-Percoll	Amount	Final concentration
Polyethylene glycol 4000	0.36 g	3%
Borine serum albumin	0.12 g	1%
Ficoll	0.12 g	1%
Percoll to	12 ml	

	10% Percoll	80% Percoll
PBF-Percoll	1.31 ml	10.5 ml
EDTA, 0.2 M, pH 7.0	0.131 ml	0.131 ml
MgCl$_2$, 1.0 M	13.1 μl	13.1 μl
HEPES-KOH, 1.0 M, pH 7.5	0.66 ml	0.66 ml
Solid sorbitol	0.79 g	0.76 g
Sodium ascorbate	12.87 mg	12.87 mg
Glutathione	2.25 mg	2.25 mg
Water to	13.13 ml	13.13 ml

[a] Make gradient using 12.5 ml of 10% Percoll and 11.5 ml of 80% Percoll.

tion. Resuspend the chloroplast pellet in a small amount (0.2 – 1.0 ml) of 0.33 M sorbitol – 50 mM HEPES – KOH, pH 8.0, with gentle agitation. Measure chlorophyll content[8] (5 – 10 μl of chloroplasts diluted to 1 ml in 80% acetone), and adjust the chlorophyll concentration of intact chloroplasts to 4 mg/ml with additional sorbitol – HEPES. This preparation can be held on ice for several hours without noticeable deterioration of the chloroplasts' ability to import proteins.

We routinely obtain intact chloroplasts equivalent to 4 mg or more of chlorophyll from a single gradient. Yields from the dicot leaves are about 50 – 80 μg of chlorophyll in intact chloroplasts per gram fresh weight of starting material. Yields from barley and corn are much lower, about 10 μg/g and less than 5 μg/g, respectively. The low yields from monocot leaves probably are due to the prolonged homogenization necessary to macerate these fibrous leaves. Yields from corn are especially variable and seem to be somewhat proportional to the size of the leaf segments used as starting material; the smaller the segment, the better the yields. All our chloroplast preparations from mature leaves contain about 1.5 to 2.0 × 10^9 chloroplasts per milligram of chlorophyll.

Incubation for Uptake. Chloroplasts are combined with the translation mix, and the volume is adjusted so that the final chlorophyll concentraton is 660 μg/ml. A typical 300-μl incubation mix contains 50 μl of chloroplasts at 4 mg of chlorophyll per milliliter, 100 μl of translation mix, 15 μl of 1.0 M HEPES – KOH, pH 8.0, 40 μl 2.0 M sorbitol, 10 μl of 250 mM methionine, and 85 μl of water. Chloroplasts should be added last.

Uptake tubes are covered with Parafilm and incubated in the light for 1 hr at room temperature. We use 15-ml Corex tubes for 300 μl of uptake mix and place them on a rotary shaker at an angle to increase the surface area exposed to the light. Shaker speed should be just fast enough to keep the chloroplasts suspended.

Post-uptake Treatment. After incubation, dilute each uptake mix with 5 ml of cold 0.33 M sorbitol – 50 mM HEPES – KOH, pH 8.0. Spin to 4340 g_{av} at 4° and stop immediately (no brake). Pour off supernatant and resuspend the chloroplast pellet in 0.5 ml of 0.33 M sorbitol – 50 mM HEPES – KOH, pH 8.0. Add 30 μl of the trypsin plus chymotrypsin stock (4 mg of each per milliliter), and incubate on ice for 30 min. After incubation, dilute the suspension with 5 ml of 0.33 M sorbitol – 50 mM HEPES – KOH, pH 8.0, containing the protease inhibitors 1.0 mM phenylmethylsulfonyl fluoride, 1.0 mM benzamidine, and 5 mM ε-aminocaproic acid. Spin to 4340 g_{av}, and stop immediately. The supernatant is again discarded. If reisolation of intact chloroplasts is necessary, resuspend the pellet in 2 ml of 0.33 M sorbitol – 50 mM HEPES – KOH, pH 8.0, plus inhibitors and carefully

---

[8] D. I. Arnon, *Plant Physiol.* **24**, 1 (1949).

TABLE IV
40% PERCOLL GRADIENTS FOR CHLOROPLAST REISOLATION[a]

Reagent	Amount	Final concentration
Percoll	4.0 ml	40%
HEPES-KOH, 1.0 M, pH 8.0	0.5 ml	50 mM
Sorbitol, 2.0 M	1.65 ml	0.33 M
BSA	0.04 g	4 mg/ml
Ficoll	0.04 g	0.4%
Water to:	10 ml	

[a] Use 2.5 ml of 40% Percoll for each uptake mixture.

underlay with 2.5 ml of 40% Percoll (Table IV).[9] Spin at 4340 g (6000 rpm in a Sorvall SS-34 rotor) for 3 min. Intact chloroplasts sediment to the bottom of the tube, while broken ones remain at the sorbitol–HEPES/Percoll interface.

Sorting and Assembly of Uptake Products

Analysis of Uptake Products. Translation products imported by chloroplasts are sorted to the envelope membrane, stroma, and the thylakoid membrane. A technique to isolate chloroplast envelope from intact chloroplasts has been described.[10] To separate stromal proteins from thylakoid membranes, intact chloroplasts are lysed by resuspension in 20 mM HEPES–KOH, pH 8.0, 5 mM MgCl$_2$. The thylakoid membranes are pelleted by centrifugation at 5000 g_{av} for 5 min. The chloroplast stromal proteins are removed and prepared for analysis on SDS–polyacrylamide gels. Thylakoid membranes may be resuspended in 0.1 M Na$_2$CO$_3$–0.1 M DTT and 2% SDS for gel analysis or resolved into submembrane protein complexes (see below).

Assembly of Thylakoid Membrane Polypeptides. The chloroplast thylakoid membrane contains four major protein complexes: ATPase, Cyt b_6/f complex, photosystem I and photosystem II. Each of these complexes consists of polypeptides synthesized within the chloroplast as well as polypeptides synthesized on cytoplasmic ribosomes.[11] Isolation of a thylakoid membrane complex following the uptake of ^{35}S-labeled translation products provides a method to test for assembly of the newly imported polypeptides and to determine the site of synthesis of polypeptides in a given protein

[9] W. R. Mills and K. W. Joy, *Planta* 148, 75 (1980).

[10] R. Douce, R. B. Holtz, and A. A. Benson, *J. Biol. Chem.* 248, 7215 (1973).

[11] J. T. O. Kirk and R. A. E. Tillney-Basset, "The Plastids." Elsevier/North-Holland Biomedical Press, New York, 1978.

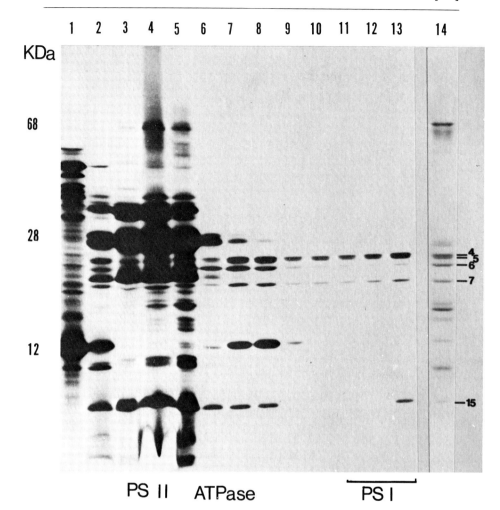

FIG. 1. Fluorogram of sodium dodecyl sulfate–polyacrylamide–8 M urea gel (12–18%) showing polypeptides posttranslationally imported into the chloroplast. Stromal proteins are shown in lane 1 and thylakoid membrane proteins in lane 2. Photosystem (PS) I complexes were prepared from the thylakoid membranes by Triton X-100 solubilization followed by separation on sucrose gradients (fluorographed in lanes 3–13). The stained profile of the polypeptides in the PS I complexes is shown in lane 14. The corresponding fluorogram (lane 13) shows that 5 subunits of PS I are posttranslationally imported into the chloroplast and assembled into PS I complexes *in vitro*. The relative positions of the PS II complex, ATPase, and the PS I complex in the gradients are indicated beneath the figure. Mass markers (kilodaltons, kDa) are at the left of the figure.

complex. This approach is described in detail for the photosystem I complex. The isolation method used was developed by Mullet et al.[12]

Intact chloroplasts (1 mg of chlorophyll) are reisolated after uptake of ^{35}S-labeled translation products. The choroplasts are lysed in 2 ml of cold 20 mM HEPES–KOH, pH 8.0, 5 mM MgCl$_2$ and centrifuged at 7500 g_{av} for 5 min. The supernatant is removed, and the pelleted thylakoid membranes are resuspended in 5 ml of cold 5 mM sodium EDTA, pH 7.5. Centrifuge the thylakoid membranes at 12,000 g_{av} for 10 min, and resuspend the chlorophyll-containing pellet in 1 mM HEPES–KOH, pH 8.0, to a chlorophyll concentration of 0.8 mg/ml. Add Triton X-100 to a final concentration of 0.76% (see Mullet et al.[12] for titration of Triton X-100 vs. chlorophyll for different plants). Stir the solution slowly for 30 min at 4°. Then centrifuge at 42,000 g_{av} for 30 min. Load the supernatant on a 0.1 to 1.0 M linear sucrose gradient containing 1 mM HEPES–KOH, pH 8.0, which is layered over a 2 M sucrose cushion. A Beckman SW 56 rotor with 4-ml capacity tubes containing a 0.25 ml 2 M sucrose cushion, 0.5–0.75 ml sample, and a 3 ml 0.1–1.0 M linear sucrose gradient is convenient for small samples. Centrifugation at 44,000 g_{av} for 10–15 hr causes sedimentation of the PS I complex to the 2 M sucrose cushion and distribution of the other thylakoid membrane complexes in the gradient (Fig. 1). Fractionate the gradients into 0.25-ml samples. Adjust each sample to 10% trichloroacetic acid, and wash the precipitate once with acetone. Analyze samples by SDS–polyacrylamide gel electrophoresis followed by fluorography. This analysis allows visualization of the proteins and radioactivity distributed in the gradient (Fig. 1).

The results in Fig. 1 demonstrate assembly of uptake products into the PS I complex and identify the cytosolic site of synthesis of five subunits of the PS I complex.[13] A similar approach could be used to analyze the assembly of the chloroplast ATPase (isolation method of Pick and Racker[14]) and the cytochrome b_6/f complex (isolation method of Hurt and Hauska[15]).

Acknowledgments

We thank Drs. Sue Bartlett, Arthur Grossman, and Gregory Schmidt for their collaboration in the development and optimization of the uptake procedure described in this chapter. This work was supported in part by N.I.H. Grant GM-25114. John E. Mullet is the recipient of an N.I.H. National Research Service Postdoctoral Award 5F32 GM08114.

[12] J. E. Mullet, J. J. Burke, and C. J. Arntzen, *Plant Physiol.* **65**, 814 (1980).
[13] J. E. Mullet, A. R. Grossman, and N.-H. Chua, *Cold Spring Harbor Symp. Quant. Biol.* **46**, 879 (1982).
[14] U. Pick and E. Racker, *J. Biol. Chem.* **254**, 2793 (1979).
[15] E. Hurt and G. Hauska, *Eur. J. Biochem.* **117**, 591 (1981).

[46] Structure and Synthesis of Chloroplast ATPase

By NATHAN NELSON

The proton–ATPase complex from chloroplast is composed of eight different polypeptides[1]; five of them, which constitute the catalytic sector (CF_1) of the enzyme, were designated as α, β, γ, δ, and ϵ, and their molecular weights were determined to be 59,000, 56,000, 37,000, 17,500, and 13,000, respectively. The membrane sector of the enzyme (CF_0) is composed of three different polypeptides that were designed as subunits I, II, and III in the order of decreasing molecular weights of 15,000, 12,500, and 8,000, respectively.[1] A specific function for each individual subunit of CF_0–CF_1 in energy transduction was proposed. The function of the β subunit in the catalytic activity of the enzyme, the δ subunit in the binding of CF_1 to CF_0, and the function of subunit III as a proton channel seems to have survived the test of time. A comprehensive description of the purification procedures of CF_1 and the chloroplast proteolipid (subunit III of CF_0) have been described in previous volumes.[2,3] In this chapter the purification of CF_0–CF_1 and its biogenesis will be described.

Purification of the Chloroplast Proton ATPase (CF_0–CF_1)

The isolation of the proton–ATPase complex from chloroplast is based upon liberation of the complex from the membrane by detergent treatment.[4] Purification procedures for CF_0–CF_1 have been described.[5,6]

Reagents
Sodium cholate, 20%, pH 7.5
Octyl-β-D-glucopyranoside (octyl glucoside), 10%
Triton X-100, 20%
Tricine-NaOH, 0.5 M, pH 8
Tris-succinate, 0.3 M, pH 6.5
4% Soybean asolectin vesicles prepared by sonication in 50 mM Tricine–NaOH (pH 8)
ATP, 0.2 M; adjusted to pH 7 with NaOH
EDTA, 0.2 M; adjusted to pH 7 with NaOH

[1] N. Nelson, *Curr. Top. Bioenerg.* **11**, 1 (1981).
[2] S. Lien and E. Racker, this series, Vol. 23 [49].
[3] N. Nelson, this series, Vol. 69 [27].
[4] C. Carmeli and E. Racker, *J. Biol. Chem.* **248**, 8281 (1973).
[5] U. Pick and E. Racker, *J. Biol. Chem.* **254**, 2793 (1979).
[6] N. Nelson, H. Nelson, and G. Schatz, *Proc. Natl. Acad. Sci. U.S.A.* **77**, 1361 (1980).

METHODS IN ENZYMOLOGY, VOL. 97

Procedure. The proton–ATPase complex should be isolated from purified chloroplast membranes that are essentially free of the enzyme ribulose bisphosphate carboxylase. In biogenesis studies the starting material is often intact chloroplasts that should be lysed, and the membranes should be washed thoroughly prior to the isolation of CF_0–CF_1. Chloroplast suspension containing down to 1 mg of chlorophyll are centrifuged at 10,000 g for 10 min. The pellet is homogenized by a glass–Teflon homogenizer in 2 ml of water and centrifuged at 10,000 g for 10 min. The pellet is washed with 5 ml of 10 mM Tricine (pH 8) then with 10 mM Tricine (pH 8) containing 0.15 M NaCl once again with 10 mM Tricine (pH 8), and finally it is homogenized in 1 ml of solution containing 50 mM Tricine, pH 8, and 250 mM sucrose. The chlorophyll concentration is about 1 mg/ml; however, in order to get the preparation of CF_0–CF_1 active in the ATP–P_i exchange reaction, the chlorophyll concentration in the Tricine–sucrose solution should be higher than 2 mg/ml. Octyl glucoside and sodium cholate are added at 0° under gentle stirring to final concentrations of 1% and 0.5%, respectively. After incubation for 20 min, the suspension is centrifuged at 200,000 g for 1 hr. The pellet is used for purification of the photosystem reaction center.[7,8] To the supernatant, saturated (at room temperature) ammonium sulfate is added to 37% saturation. After 20 min at 0°, the suspension is centrifuged at 10,000 g for 10 min. The pellet is discarded, and the supernatant is brought to 48% saturation with saturated ammonium sulfate solution. After 20 min at 0°, the suspension is centrifuged as before and all the supernatant is carefully removed by suction. To the pellet is added 0.5 ml of a solution containing 30 mM Tris–succinate (pH 6.5). 0.2% Triton X-100, 0.1 mM ATP, 0.5 mM EDTA, and 0.1% soybean asolectin (add a 4% sonicated vesicles). The sample is applied on to a linear 7 to 30% sucrose gradient containing all the components listed above. After centrifugation for 15 to 18 hr in a Spinco SW 41 rotor at 35,000 rpm, or for shorter times in high-speed rotors, the bottom of each tube is punctured and eleven 1-ml fractions are collected. Figure 1 depicts SDS–polyacrylamide gel containing fractions 4 to 11. Usually the pure H^+–ATPase complex appeared in gradient fractions 5 and 6. Residual ribulose bisphosphate carboxylase is apparent in fraction 4.

Remarks. The purified proton-ATPase complex can be stored at $-70°$ for several months with a little loss of its specific activity. In order to get preparation of CF_0–CF_1 active in ATP–P_i exchange, the chloroplast membranes should be treated with 100 mM dithiothreitol for 30 min prior to the purification procedure.

[7] C. Bengis and N. Nelson, *J. Biol. Chem.* **252**, 4564 (1977).
[8] R. Nechushtai, N. Nelson, A. K. Mattoo, and M. Edelman, *FEBS Lett.* **125**, 115 (1981).

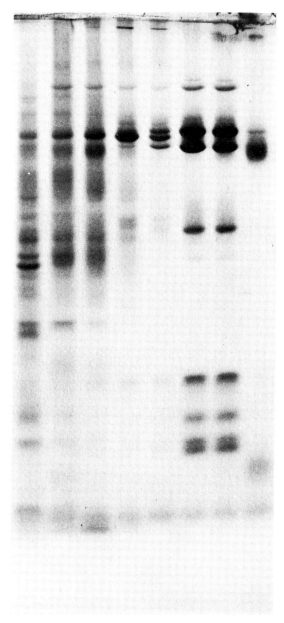

FIG. 1. Sodium dodecyl sulfate gel of fractions from the sucrose gradient of the last step in the purification of spinach CF_0–CF_1 complex. Fractions 4–11 were dissociated in the presence of 2% SDS and 2% 2-mercaptoethanol and electrophoresed in a gradient gel of 10 to 15% acrylamide.[15] The gel was stained by Coomassie Brilliant Blue and destained by a solution containing 20% methanol and 7.5% acetic acid.

Assay of the Proton – ATPase Complex

The ATPase activity of the complex can be assayed as previously described for CF_1.[3] The $ATP - P_i$ exchange and ADP-phosphorylation activities of the enzyme are assayed after reconstitution of the enzyme into phospholipid vesicles.

Reagents

Reaction mixture for $ATP - P_i$ exchange: 100 mM Tricine (pH 8), 10 mM ATP, 10 mM $MgCl_2$, 6 mM sodium phosphate (pH 8), and 30 μCi of $^{32}P_i$ per milliliter. Store frozen; can be frozen and thawed several times.

Reaction mixture for ADP phosphorylation: 100 mM Tricine (pH 8), 100 mM NaCl, 6 mM ADP, 10 mM $MgCl_2$, 4 mM sodium phosphate (pH 8), 5 μCi of $^{32}P_i$ per milliliter, 10 mM ascorbate, 100 μM N-methylphenazonium methosulfate, 40 mM glucose, and 2 IEU of hexokinose. Store frozen in the dark; can be frozen and thawed twice. For ADP phosphorylation by bacteriorhodopsin vesicles, N-methylphenazonium methosulfate and ascorbate are omitted.

4% Soybean asolectin vesicles (prepared by sonication in 50 mM Tricine – NaOH, pH 8)

Trichloroacetic acid, 40%

Bacteriorhodopsin vesicles: Prepare by sonication of 3 mg of bacteriorhodopsin with 40 mg of asolectin in 1 ml of 50 mM Tricine (pH 8).

Photosystem I reaction center vesicles: Prepare by sonication of photosystem I reaction center containing 0.2 mg of chlorophyll with 30 mg of asolectin in 1 ml of 20 mM Tricine (pH 8), 100 mM KCl by sonicator with microtip for 10 min.[9]

ATP-P_i Exchange. Twenty-five microliters of $CF_0 - CF_1$ complex (prepared from DTT-treated chloroplasts) containing about 0.3 mg of protein per milliliter are added to 25 μl of preformed asolectin vesicles. After addition of 100 μl of water, the mixture is incubated for 5 min at room temperature. Addition of 100 mM DTT during this stage might double the activity. Optionally, the mixture can be frozen in liquid nitrogen and thawed slowly at room temperature. The mixture is then diluted with additional 0.35 ml of water and 0.5 ml of the $ATP - P_i$ exchange-reaction mixture. It is then incubated for 10 min at 37°, and the reaction is terminated by the addition of 100 μl of 40% trichloroacetic acid. After centrifugation, 0.5 ml of the supernatant is assayed for ^{32}P incorporation as previously described.[3]

ADP Phosphorylation.[10] Twenty-five microliters of photosystem I reac-

[9] G. Orlich and G. Hauska, *Eur. J. Biochem.* **111**, 525 (1980).
[10] G. Hauska, D. Samoray, G. Orlich, and N. Nelson, *Eur. J. Biochem.* **111**, 535 (1980).

tion center vesicles or bacteriorhodopsin vesicles are incubated at room temperature for 5 min with 25 μl of CF_0-CF_1 complex and 100 μl of water. Then 0.35 ml of water and 0.5 ml of the reaction mixture for ADP phosphorylation are added. The mixture is illuminated for 10 min by white light for bacteriorhodopsin or red light for photosystem I reaction center. The reaction is terminated by 100 μl of 40% trichloroacetic acid, and the $^{32}P_i$ incorporation is assayed as previously described.[3]

Electrotransfer of Polypeptides from Slab Gels to Nitrocellulose Paper and Immunodecoration by ^{125}I-Labeled Protein A

This technique is very useful for identification of a given protein in SDS gels. Specific proteins can be detected, and their quantity in crude cell extracts can be estimated. The procedure is very convenient, and it requires very simple instrumentation.[11,12] It is based upon the specific transfer of proteins by electrophoresis from the gel to a nitrocellulose paper in such a way that a replica of the original gel pattern is obtained. After blocking all the additional binding sites on the paper, specific proteins are detected by the sequential incubation with specific antibodies and ^{125}I-labeled protein A or any other substance with a marked specific affinity, such as ^{125}I-labeled lectins or secondary antibodies.

Reagents

10 × Electrotransfer buffer containing 1.6 M glycine, 0.25 M Tris, and 0.2% SDS

Methanol

Albumin buffer containing 25 mM Tris-HCl (pH 7.6), 140 mM NaCl, and 1% bovine serum albumin (Sigma fraction V)

Washing buffer containing 25 mM Tris-HCl (pH 7.6) and 140 mM NaCl

Triton-washing buffer containing 25 mM Tris-HCl (pH 7.6), 140 mM NaCl, and 1% Triton X-100

Slab gels containing SDS and exponential gradients of 10 to 15% acrylamide

Methyl green, 0.1% in sample buffer

Nitrocellulose paper: 0.45 μm pore size (Schleicher & Schuell No. 401180)

Filter paper (Whatman No. 3)

^{125}I-Labeled protein A containing about 0.2 mg of protein per milliliter and about 100 μCi of ^{125}I per milliliter

[11] H. Towbin, T. Staehelin, and J. Gordon, *Proc. Natl. Acad. Sci. U.S.A.* **76**, 4350 (1979).
[12] R. Rott and N. Nelson, *J. Biol. Chem.* **256**, 9224 (1981).

Procedure. Before the application of the protein samples, the lanes of the gel are marked by 5 μl of 0.1% methyl green. The green stain is introduced into the gel by electrophoresis for 10 min, and the protein samples are supplied. After electrophoresis overnight, when the green bands are about 4 cm from the bottom of the gel, additional aliquots of 5 μl of methyl green are applied into the wells containing the protein samples. The stain is electrophoresed until it passes the stacking gel and penetrates into the resolving gel. The gels are removed from the glass plates, and the position of the methyl green bands are marked by cutting off small parts of the edges from the sides of the gels. One of the gels is stained by Coomassie Brilliant Blue, and the protein band from the second gel is electrotransferred into the nitrocellulose paper.

Two liters of running buffer are prepared by mixing of 200 ml of 10 × electrotransfer buffer with 1400 ml of water and 400 ml of methanol. The gel is soaked in the running buffer and placed on a soaked Whatman filter paper. It is covered with soaked nitrocellulose paper and additional filter paper, then tied by rubber bands in a sandwich of two porous plexiglass supports and scouring pads of Scotch Brite. The sandwich is placed in running buffer between two platinum electrodes (made of 0.2 mm platinum wire containing 8 strips and 17 × 15 cm porous Perspex plates), in a 20 × 16 × 8 cm Perspex container placed in ice. The nitrocellulose sheet should face the anode. The protein bands are electrotransferred at 40 V for 2 hr. After finishing the transfer, the assembly is dismantled and part of the nitrocellulose paper could be stained with 0.1% Amido Black in 20% methanol and 7.5% acetic acid solution. The other part is soaked in 100 ml of the albumin buffer solution for 1 hr with gentle shaking. Then, 10–100 μl of antibody are added to the entire nitrocellulose sheet. Alternatively, the sheet might be cut into strips according to the methyl green marks, and different antibodies might be tested. After incubation at room temperature overnight with gentle shaking, the nitrocellulose is washed five times with 50 ml of the albumin buffer (for strips, 10 ml portions are enough). Then, the nitrocellulose paper is placed in 50 ml of the washing buffer, and iodinated protein A containing about 5 μCi of ^{125}I is added. After incubation for 1 hr under gentle shaking, the paper is washed with 50 ml of washing buffer, then with 50 ml of the Triton-washing buffer, and three additional times with the washing buffer. The nitrocellulose paper is then dried and exposed overnight to X-ray film.

Preparation of Subunit Specific Antibodies

Some of the biogenesis studies are heavily dependent on subunit specific antibodies. Potentially, monoclonal antibodies are the ultimate technique of choice; however, for the time being, technical difficulties in immunoprecipitation prevent their general use. We have always insisted on using antigens that were electroeluted from sliced strips of SDS gels.[13] This procedure eliminates contamination of the given antigen with peptides having different molecular weights. Failure to follow this procedure was probably the source for the erroneous observation of the "polyprotein" as a precursor for the cytoplasmically made subunits of cytochrome oxidase.[14-16]

Reagents

Chloroplast membranes containing 0.5 mg of chlorophyll per milliliter, dissociated in a dissociation buffer containing 2% SDS and 2% 2-mercaptoethanol

Purified CF_1 or CF_0-CF_1 complex containing about 0.5 mg of protein per milliliter dissociated as above

Slab gels containing SDS and gradients of acrylamide at concentrations of 10–15% and 12–17%

Buffer for partial cleavage containing 0.1 M Tris-HCl (pH 6.8), 5 mM EDTA, 0.1% Bromphenol Blue, 10% glycerol

Staphylococcus protease (V-8), 250 $\mu g/ml$ in H_2O

Reagents for electrotransfer and immunodecoration

Procedure. The subunits are isolated from purified protein complexes. The purified protein complex containing about 5 mg of protein is dissociated in 2% SDS and 2% 2-mercaptoethanol at a protein concentration of 0.5–1 mg/ml and electrophoresed on 10 slab gels.[17] The gels are fixed, stained, and destained, and the specific protein bands are sliced out of the gel. After a brief lyophilization to remove the methanol and acetic acid, the polypeptides are electroeluted into dialysis tubes. The samples are assayed on SDS gels and injected into rabbits or mice as previously described.[13,18]

The specificity of the antibodies is assayed by the electrotransfer immunodecoration technique as previously described.[11,12] The above-described procedure is fairly safe for preventing contaminations of antibodies against protein having molecular weights different from that of the given antigen. However, very small amounts of a highly immunogenic polypeptide can escape this rigorous purification procedure and give rise to a contaminating antibody. It is of great importance to detect the presence of such a contamination and by doing so to prevent wrong conclusions in the biogenesis studies. For example, the large subunit of ribulose bisphosphate

[13] N. Nelson, D. W. Deters, H. Nelson, and E. Racker, *J. Biol. Chem.* **248**, 2049 (1973).
[14] R. O. Poyton and E. McKemmie, *J. Biol. Chem.* **254**, 6763 (1979).
[15] R. O. Poyton and E. McKemmie, *J. Biol. Chem.* **254**, 6772 (1979).
[16] A. S. Lewin, I. Gregor, T. L. Mason, N. Nelson, and G. Schatz, *Proc. Natl. Acad. Sci. U.S.A.* **77**, 3998 (1980).
[17] M. G. Douglas and R. A. Butow, *Proc. Natl. Acad. Sci. U.S.A.* **73**, 1083 (1976).
[18] J. E. Lamb, H. Riezman, W. M. Becker, and C. J. Leaver, *Plant Physiol.* **62**, 754 (1978).

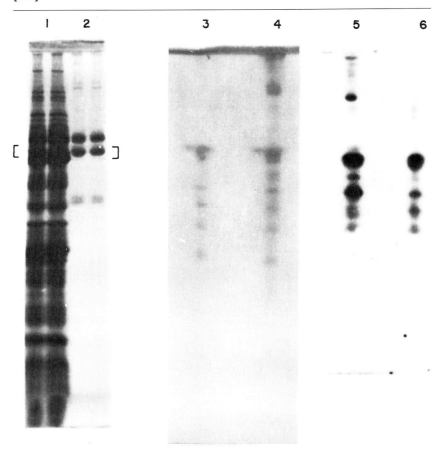

FIG. 2. Examination of the specificity of antibody against β subunit of CF₁ by immunodecoration of one-dimensional proteolytic fingerprinting. Chloroplast membranes containing about 100 μg of protein and CF₁ containing about 10 μg of protein were dissociated in the presence of 2% SDS and 2% 2-mercaptoethanol. Four lanes of each preparation were run on a gradient gel of 10 to 15% acrylamide containing SDS. The gel was fixed, stained, destained, and dried on a filter paper (lane 1 depicts the stained membrane preparation and lane 2 the purified CF₁). The β subunits of CF₁ were sliced out as well as pieces·from the same position of the membrane lanes as marked in lanes 1 and 2. The gel slices then underwent one-dimensional proteolytic fingerprinting and immunodecoration as described in the text. Lane 3: stained gel of the β subunit slice that was treated with 1 μg of V-8 protease and electrophoresed on a gradient gel of 12 to 17% acrylamide. Lane 4: stained gel of a membrane slice treated as in lane 3. Lane 5: autoradiograph of experiment similar to lane 3 except that the protein bands were electrotransferred into nitrocellulose paper. After saturation with 1% bovine serum albumin, the paper was incubated with 10 μl of antibody against β subunit of CF₁ in 20 ml of albumin buffer. After proper washing, the paper was incubated with 125 I-labeled protein A for 1 hr. The paper was then washed from excess protein A and the nitrocellulose paper was exposed to X-ray film. Lane 6: as for lane 5, except that the gel slice was as in lane 4 (a slice from the membrane preparation).

carboxylase and at least one more major polypeptide of the chloroplast membrane (see Fig. 1) coincide with the β subunit of CF_1 on SDS gels. A simple procedure to detect such a contamination is described below. It is based on one-dimensional proteolytic fingerprinting[19] and immunodecoration with the tested antibody and iodinated protein A.

The membrane preparation and the CF_1 preparation (50 μl of each) are electrophoresed on 10 to 15% gradient SDS-gel. After staining and destaining, the slab gel is dried on a Whatman No. 3 paper (see Fig. 2). The β subunit of CF_1 is cut by a razor blade, and a similar piece of the membrane fraction from the same position in the gel is also removed. The dry gel slices are briefly soaked in water, the paper is separated from the gel, and the slices of the gel are then introduced into two similar 12 to 17% gradient gels with wide wells. One hundred microliters of the buffer for partial cleavage with or without 1 μg of V-8 protease are applied on the gel slices. After incubation for 20 min, the gels are electrophoresed overnight. One of the gels is fixed, stained, and destained. The proteins from the second gel are electrotransferred to cellulosenitrate paper as previously described.[12] The paper is then allowed to interact with antibody against the β subunit of CF_1 and subsequently decorated with ^{125}I-labeled protein A.

Figure 2 depicts such an experiment. It shows that the pattern of the immunodecoration is similar for the purified β subunit of CF_1 and part of the crude membranes, and it matches the stained pattern of the cleaved β subunit. Contaminated antibody would give additional radioactive bands in the membrane preparation.

Assay of Synthesis of Protein Complexes and Individual Subunits by Isolated Chloroplasts

Intact chloroplasts are isolated by published procedures from spinach leaves or any other plant source.[20] The chloroplasts are labeled with [35S]methionine, and the protein complexes are isolated and assayed for incorporation of [35S]methionine. Alternatively, the reaction is quenched by the addition of SDS, and individual polypeptides are assayed for 35S incorporation by immunoprecipitation.[6]

Reagents

Salts–buffer solution containing 25 mM Tris-HCl (pH 7.4), 200 mM KCl, 1.3 M sorbitol, and 2 mM MgCl$_2$

GTP, 20 mM

[19] D. W. Cleveland, S. G. Fischer, M. W. Kirschner, and U. K. Laemmli, *J. Biol. Chem.* **252**, 1102 (1977).
[20] D. A. Walker, this series, Vol. 69 [9].

ATP, 0.2 M

20 common amino acids except methionine, each 1 mM

[^{35}S]Methionine, 100–300 Ci/mmol

Protease inhibitors, dissolved in dimethyl sulfoxide at 0.1 M each of phenylmethylsulfonyl fluoride, N-tosyl-L-phenylalanylchloromethane, and N-tosyl-L-lysylchloromethane

Triton buffer: 25 mM Tris-HCl (pH 7.6), 140 mM NaCl, and 1% Triton X-100

Intact chloroplasts prepared as previously described at final chlorophyll concentration of 2 mg/ml

The reaction mixture contained in a final volume of 1 ml: 0.2 ml of the salts–buffer solution, 20 μl of 20 mM GTP, 20 μl of 0.2 M ATP, 25 μl of the 20 amino acids solution, and 0.1–0.5 mCi of [^{35}S]methionine.

Procedure. The reaction is started by addition of 0.2 ml of a chloroplast suspension containing 2 mg of chlorophyll per milliliter. After incubation at 30° for 40 min, 10 μl of protease inhibitors are added. After an additional 5 min of incubation, the reaction is terminated by centrifugation at 10,000 g for 10 min at 0°. The pellet is either lysed by water for the isolation of the protein complexes, like CF_0–CF_1 or the photosystem I reaction center,[6,8] or it is extracted by acetone for immunoprecipitation.[6] In the first procedure only the assembled subunits are assayed; and in the second one, free subunits and precursors can be followed. It is recommended to include in the experiment samples that are labeled in the presence of chloramphenicol or cycloheximide. In this case, reciprocal labeling of polypeptides should be observed, and in good isolated chloroplasts chloramphenicol should eliminate completely the labeling by [^{35}S]methionine. This control is necessary to eliminate possible erroneous observations due to artificial chemical labeling of proteins by a contamination present in the [^{35}S]methionine preparations.[21]

Immunoprecipitation of Individual Polypeptides. The immunoprecipitation is performed with the aid of *Staphylococcus aureus* cells that were fixed with glutaraldehyde.[22] The labeled chloroplasts are washed twice with five volumes of 80% acetone. The two extractions by 80% acetone are necessary because residual chlorophyll prevents the interaction between the immune complexes and the *Staphylococcus* protein A in the subsequent immunoprecipitation. The pellet obtained from the last acetone wash is dried by a stream of nitrogen and dissolved in 1 ml of a solution containing 25 mM Tris-HCl (pH 7.5), 5% SDS, and 1 mM protease inhibitors. After incubation for 1 hr at room temperature, 5 ml of Triton-buffer solution is added, it is then centrifuged at 100,000 g for 30 min. To the supernatant, an additional

[21] M. Swissa, *Anal. Biochem.* **115,** 67 (1981).
[22] S. W. Kessler, *J. Immunol.* **115,** 1617 (1975).

20 ml of Triton–buffer and about 0.1 ml of the desired antibody are added, and the solution is incubated overnight at 4°. Then 0.5 ml of a 10% (w/v) suspension of fixed *S. aureus* cells are added, and the mixture is shaken at room temperature for 1 hr. The *S. aureus* cells are centrifuged down at 5000 *g* for 10 min and washed four times with 5 ml of the Triton–buffer solution. The antigen is extracted for 10 min at room temperature with 0.25 ml of a solution containing 2% SDS, 0.1 *M* Tris-HCl (pH 6.8), 2 m*M* EDTA, 0.03 mg of dithiothreitol per milliliter, and 10% (w/v) glycerol. The cells are removed by centrifugation, and the supernatant is analyzed by SDS–polyacrylamide gel electrophoresis followed by autoradiography or fluorography.[6]

Assembly of polypeptides into protein complexes is followed by the isolation of the pure complexes from the labeled chloroplasts (for the isolation of CF_0-CF_1 complex see the section on purification above, and for the isolation of photosystem I reaction centers, see Bengis and Nelson[7] and Nechushtai *et al.*[8]). In the interpretation of such experiments, especially after labeling of isolated chloroplasts, attention should be given to possible exchange between newly synthesized subunits and preassembled complexes.

In Vivo Labeling of Protein Complexes in Chlamydomonas reinhardi Cells

Inorganic [^{35}S]sulfate is widely used for labeling experiments in *Escherichia coli* and yeast cells. Since it is the least expensive radiochemical, we worked out conditions for effective labeling of *Chlamydomonas* cells.[23] *Chlamydomonas reinhardi* (strain Y-1) is grown as described by Ohad *et al.*[24] About 6 liters of cell suspension containing 10–20 μg of chlorophyll per milliliter are harvested by centrifugation at 1500 *g* for 3 min and washed twice with growth medium in which the $MgSO_4$ was replaced by an equal weight of $MgCl_2$. The cells are suspended in 2 liters of the same medium and incubated overnight at room temperature under fluorescent light (5000 ergs cm^{-2} sec^{-1}). Usually the suspension is then divided into three equal parts, control and protein synthesis inhibitors (cycloheximide at final concentration of 100 μg/ml or chloramphenicol at final concentration of 2 m*M*). To each flask 0.5 mCi of [^{35}S]sulfate is added, and after incubation for 30 min under fluorescent light, over 80% of the ^{35}S is incorporated in the control experiments. The cells are harvested by centrifugation and broken in a French pressure cell; chloroplast membranes are isolated, and the protein complexes are purified.

[23] R. Nechushtai and N. Nelson, *J. Biol. Chem.* **256,** 11624 (1981).
[24] I. Ohad, P. Siekevitz, and G. E. Palade, *J. Cell Biol.* **35,** 521 (1967).

For the preparation of the membranes, cells are washed twice with about 200 ml of solution containing 0.4 M sucrose, 0.01 M NaCl solution of protease inhibitors (each 1 mM), and 0.01 M Tricine NaOH (pH 8). The cells are suspended in 30 ml of the same solution and are broken twice at about 4° in a French pressure cell at 4000 psi. The suspension is centrifuged at 1500 g for 3 min, and NaCl is added to the supernatant, from a stock solution of 4 M, to give a final concentration of 0.25 M. The suspension is then centrifuged at 20,000 g for 15 min, and the pellet is homogenized in about 40 ml of 10 mM Tricine-NaOH (pH 8). After centrifugation at 60,000 g for 20 min, the pellet is homogenized in a solution containing 10 mM Tricine (pH 8) and 0.15 M NaCl. After similar centrifugation, the pellet is further washed in 10 mM Tricine (pH 8) and suspended in a solution containing 250 mM sucrose and 50 mM Tricine (pH 8) at a chlorophyll concentration of 1.5–2 mg/ml. Octyl glucoside and sodium cholate are added to give final concentrations of 1% and 0.5%, respectively. After 20 min of incubation at 0° the suspension is centrifuged at 200,000 g

FIG. 3. Sodium dodecyl sulfate gel of purified photosystem I reaction from *Chlamydomonas reinhardi*. Photosystem I reaction center containing about 20 µg of protein was electrophoresed on exponential 10 to 15% acrylamide gel as described in Fig. 1.

for 60 min. The supernatant is used for the preparation of the proton–ATPase complex as described above.

The pellet is homogenized in a solution containing 25 mM Tris-HCl (pH 8) and 2% Triton X-100 to give a final chlorophyll concentration of 0.5 mg/ml. It is then centrifuged at 10,000 g for 10 min, and the supernatant is kept at $-20°$ until used. A sample of about 10 ml is thawed and centrifuged

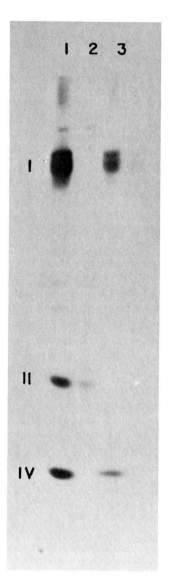

FIG. 4. Effect of chloramphenicol and cycloheximide on the labeling of subunits of photosystem I reaction center. *Chlamydomonas* cells were labeled with [^{35}S]sulfate as described in the text. One flask was served as a control; the second flask was labeled in the presence of 2 mM chloramphenicol, and the third flask was labeled in the presence of 100 μg of cycloheximide per milliliter. From each one of the three flasks, photosystem I reaction center was prepared. After electrophoresis of samples containing about 20 μg of protein on exponential 10 to 15% acrylamide gel, it was exposed for autoradiography. Lane 1, control; lane 2, labeling in the presence of chloramphenicol; lane 3, labeling in the presence of cycloheximide.

for 10 min at 10,000 g, and the supernatant is applied to a DEAE-cellulose column (1 $\times$ 12 cm) equilibrated with a solution containing 20 mM Tris-HCl (pH 8) and 2% Triton X-100. The reaction center is eluted with a linear NaCl gradient of 0 to 300 mM (20 ml in each chamber) in a solution containing 20 mM Tris-HCl (pH 8) and Triton X-100. The tubes of the first green peak (containing more than 0.2 mg of chlorophyll per milliliter) are pulled, and fractions of 1 ml are applied to linear sucrose gradients of 5% to 30% sucrose in a solution containing 20 mM Tris-HCl (pH 8) and 0.2% Triton X-100. The gradients are centrifuged in a SW 41 rotor at 35,000 rpm for 15 hr at 2°. The lower green band is collected and used promptly or stored frozen at $-70°$. When desired, it could be concentrated on a small DEAE-cellulose column.

Figure 3 depicts a stained SDS gel of the purified *Chlamydomonas* photosystem I reaction center. It contains four different subunits that are designated as subunits I, II, III, and IV with molecular weights of about 70,000, 19,000, 10,000, and 8,000, respectively.

Figure 4 illustrates typical autoradiographs of purified photosystem I reaction centers that were prepared from [^{35}S]sulfate labeled cells in the absence or the presence of protein synthesis inhibitors. It can be concluded from such an experiment that subunit III probably does not contain any cysteine or methionine, that subunits I and IV are synthesized on chloroplast ribosomes, and that subunit II is a product of cytoplasmic ribosomes. Similarly, the site of synthesis of subunits of the CF_0-CF_1 complex could be followed, and, by the use of specific antibodies, the assembly of the protein complexes can be studied. These simple experiments, together with the studies on the molecular biology of the chloroplast and the nucleus DNA should in the near future shed some light on the biogenesis of protein complexes in the chloroplast membranes.[25]

---

[25] P. Westhoff, N. Nelson, H. Bunemann, and R. G. Herrmann, *Curr. Genet.* **4**, 109 (1981).

[47] Cloning and Physical Mapping of Maize Plastid Genes

By Lawrence Bogorad, Earl J. Gubbins, Enno Krebbers, Ignacio M. Larrinua, Bernard J. Mulligan, Karen M. T. Muskavitch, Elizabeth A. Orr, Steven R. Rodermel, Rudi Schantz, Andre A. Steinmetz, Guido De Vos, and Yukun K. Ye

The circular DNA molecules of chloroplasts, and presumably of other plastids, range in size from about 129 to 215 kilobase pairs (kbp).[1] The methods described here have been used for physically mapping genes for ribosomal RNAs (rRNAs), transfer RNAs (tRNAs), and proteins on the *Zea mays* plastid chromosome and have been applied to other chloroplast chromosomes; they seem applicable to analyses of modest sized DNA molecules generally. Some of the experimental procedures mentioned here are modifications of widely used methods; we have described our applications of a few of them in detail where it seems appropriate.

The initial step in mapping plastid genes physically is to establish a restriction endonuclease recognition site map of the plastid DNA molecule. In the original work[2] this was done entirely with DNA isolated from plastids. With currently available technology the approach of choice would include the following steps: (*a*) analyzing the sizes of the fragments produced by digestion of chloroplast DNA isolated from plastids with a few restriction endonucleases, i.e., determination of agarose gel electrophoresis patterns of the DNA digested with a few restriction endonculeases; (*b*) cloning of partial digests of the plastid DNA using one of these restriction enzymes into vehicles such as Charon phages[3] capable of carrying relatively large fragments; (*c*) mapping restriction endonuclease recognition sites on the plastid DNA sequences inserted into these phages; (*d*) ordering the cloned fragments from overlaps of chloroplast sequences, and thus generating a map of restriction sites for the entire plastid DNA molecule; and (*e*) as discussed in more detail below, the "purification" feature of molecular cloning in plasmids of products of terminal restriction endonuclease digestion can be taken advantage of to resolve ambiguities.

The physical mapping of genes on maize plastid chromosomes has been achieved following two general patterns, depending upon the gene products.

[1] J. R. Bedbrook and R. Kolodner, *Annu. Rev. Plant Physiol.* **30**, 593 (1979).
[2] J. R. Bedbrook and L. Bogorad, *Proc. Natl. Acad. Sci. U.S.A.* **73**, 4309 (1976).
[3] N. Sternberg, D. Tiemeier, and L. Enquist, *Gene* **1**, 255 (1977).

First, genes whose final products are stable RNAs, such as rRNAs and tRNAs, have been located by molecular hybridization of rRNAs or tRNAs, prepared from plastids or plants, against plastid DNA fragments immobilized on nitrocellulose by the method of Southern[4,5] or by electron microscopy.[5] Genes for proteins have been mapped by two methods. The gene for the large subunit of ribulosebisphosphate carboxylase in maize, for example, was mapped by identifying the cloned fragment of plastid DNA that directed the *in vitro* transcription-translation of an [^{35}S]methionine-labeled product that was identified as the large subunit of ribulosebisphosphate carboxylase by both size and immunoprecipitation[6] and then by limited proteolysis[6] and tryptic digestion.[7] Another approach has been followed to identify genes that are differentially transcribed under varying developmental conditions. This method was first adopted with photogene 32,[8] a gene coding for a 34,500 dalton polypeptide, which is processed *in vivo* to a 32,000 dalton photosynthetic membrane protein[9] that binds the herbicide analog azidoatrazine in susceptible plants.[10] RNAs are isolated from plants at different developmental stages (or from purified plastids of these plants). The RNA is then used to direct translation *in vitro* as a way of identifying polypeptides produced under one condition or the other. In addition, the RNA is labeled *in vitro* (because of the level of *in vivo* labeling is not high enough for some of the subsequent analyses) and is then used as a hybridization probe against restriction fragments of plastid DNA. An improved method for identifying DNA fragments that are transcribed differently at different developmental stages or under different environmental conditions is presented here as a "prehybridization method."

The direct identification of genes of the maize chloroplast chromosome by *in vitro* linked transcription – translation of DNA sequences is feasible because of the relatively small size of the plastid genomes. Identification of differentially expressed genes using RNA, as in the prehybridization method, should be of wider applicability as long as it is possible to obtain relatively large amounts of RNAs.

[4] E. M. Southern, *J. Mol. Biol.* **98**, 503 (1975).

[5] J. R. Bedbrook, R. Kolodner, and L. Bogorad, *Cell* **11**, 739 (1977).

[6] D. M. Coen, J. R. Bedbrook, and L. Bogorad, *Proc. Natl. Acad. Sci. U.S.A.* **74**, 5487 (1977).

[7] J. R. Bedbrook, D. M. Coen, A. R. Beaton, L. Bogorad, and A. Rich, *J. Biol. Chem.* **254**, 905 (1979).

[8] J. R. Bedbrook, G. Link, D. M. Coen, L. Bogorad, and A. Rich, *Proc. Natl. Acad. Sci. U.S.A.* **75**, 3060 (1978).

[9] A. E. Grebanier, D. M. Coen, A. Rich, and L. Bogorad, *J. Cell Biol.* **78**, 734 (1978).

[10] K. E. Steinbeck, L. McIntosh, L. Bogorad, and C. J. Arntzen, *Proc. Natl. Acad. Sci. U.S.A.* **78**, 7463 (1981).

Growing Maize

The *Zea mays* strain FR 9^{Cms} X FR 37 (Illinois Foundation Seed Co., Champaign, Illinois) has been used in these experiments, in part because of its ability to produce large leaves and grow well in the dark for up to 12 days at 30°. Seeds are ordinarily shipped in a fungicide. If not, they should be treated with one, such as Captan 50-WP (Stauffer Chemicals, Westport, Connecticut) before storage at 4°.

Seeds are prepared for planting by soaking them in tap water for 6 – 16 hr to promote synchronous germination. In order to inhibit fungal contamination during growth, the fungicide is not rinsed off completely. Maize is grown in vermiculite (Terra Lite Vermiculite Grade 2, W. R. Grace Co., Cambridge, Massachusetts 02140) in Rubbermaid dishpans (28 × 32 × 14 cm). Dishpans are filled approximately halfway with vermiculite, water is added, and the vermiculite is allowed to soak for 6 – 16 hr. The vermiculite is drained in a standard greenhouse strainer and placed in the dishpans to a depth of 5 cm. A dense monolayer of soaked seeds is spread on the packed vermiculite, and another 1 cm of wet vermiculite is placed on top of the seeds; 400 ml of seeds are soaked per tray to be planted.

Maize to be used for the isolation of chloroplast DNA or thylakoid membranes can be grown on a greenhouse bench or in a growth chamber. At a temperature of about 27° and 16-hr day, seedlings will be ready for harvest in 6 – 8 days. Maize to be used for the isolation of chloroplast DNA is moved to a darkroom 24 hr before harvesting to reduce the amount of starch; the starch grains disrupt chloroplasts during processing. One tray of corn grown in the greenhouse normally yields approximately 100 g of leaves. Plants to be used for preparation of chloroplast RNA or etioplast membranes are grown at 30° in a darkroom and are ready for harvest in approximately 7 days. Corn to be used to prepare RNA from greening seedlings is moved from the darkroom and greened under fluorescent lights for the desired number of hours. Corn grown in the darkroom, greened or not, yields about 30 g of tissue per tray. The darkroom should be lighttight and equipped with lights covered with green filters. All plants should receive some water if they dry out or, in any case, within 2 days prior to harvesting—this makes grinding smoother.

Isolation of Plastid DNA

This procedure is for the isolation of chloroplast DNA from 1 kg of maize seedling leaves (modified from Kolodner and Tewari[11,12] and Ko-

[11] R. Kolodner and K. K. Tewari, *J. Biol. Chem.* **247**, 6355 (1972).
[12] R. Kolodner and K. K. Tewari, *Biochim. Biophys. Acta* **402**, 372 (1975).

lodner et al.[13]). Unless otherwise noted, all manipulations are to be performed at $0-4°$.

Materials

1. Buffer A: 0.3 M mannitol, 50 mM Tris-HCl, 3 mM disodium EDTA, 1 mM 2-mercaptoethanol, 0.1% bovine serum albumin (BSA), pH 8.0
2. DNase solution: 5 mg/ml DNase (Sigma DN-100), 1 $\times$ SSC (0.15 M NaCl, 0.015 M sodium citrate), 5 mM MgCl$_2$. Make fresh just before use.
3. Buffer B: 0.3 M sucrose, 50 mM Tris-HCl, 20 mM disodium EDTA, pH 8.0
4. Sucrose pad solution 1: 0.5 M sucrose, 50 mM Tris-HCl, 20 mM disodium EDTA, pH 8.0
5. Resuspension buffer 1: 50 mM Tris-HCl, 20 mM disodium EDTA, pH 8.0
6. Resuspension buffer 2: 5 mM Tris-HCl, 0.25 mM disodium EDTA, pH 8.0
7. Sucrose gradient solutions A: 5% (or 30%) sucrose, 1 M NaCl, 10 mM disodium EDTA, pH 8.0

Procedures

Maize leaves (1 kg) are cut into $1-2$-cm fragments and homogenized in 7 or 8 lots in a total of 2.5 liters of ice-cold buffer A (No. 1) with two 5-sec bursts in a Waring blender at low and high speed. The homogenates are filtered through one, then four, layers of Miracloth (Calbiochem), and the filtrate is centrifuged at 1500 g for 20 min at 4°. This centrifugation is conveniently performed in a Sorvall GS-3 rotor at 3000 rpm using 500-ml bottles with sealing caps. Pellets are gently resuspended in a small volume (approximately 15 ml) of buffer A (No. 1) with a small camel hair brush, the tubes are rinsed, and the chloroplasts are pooled into a single Sorvall GSA tube; the final volume is brought up to 200 ml with buffer A. The tube is centrifuged at 1500 g (3000 rpm) for 20 min at 4° in a Sorvall GSA rotor, the supernatant is aspirated off, and the pellet is gently resuspended in 60 ml of buffer A. The suspension is put into a chilled tissue homogenizer (Wheaton, 60-ml capacity type "A"), and the plastids are dispersed by three gentle passes. The homogenate is transferred to a flask on ice. The GSA tube is rinsed (40 ml of buffer A); after passage through the homogenizer, this suspension is added to the flask.

To remove contaminating nuclear and mitochondrial DNA from the crude chloroplast preparation, 2 ml of DNase solution (No. 2) is added to

[13] R. Kolodner, K. K. Tewari, and R. C. Warner, *Biochim. Biophys. Acta* **477**, 144 (1976).

the flask along with $MgCl_2$ to a final concentration of 10 mM, and the suspension is incubated for 1 hr on ice. Following DNase treatment, two methods have been utilized to isolate the chloroplast DNA.

Method 1. To remove the DNase, the plastids are layered onto 100 ml of sucrose pad solution (No. 4) in a GSA tube and centrifuged at 6000 g (6000 rpm) for 20 min at 4°. The supernatant is aspirated off, and the plastids are washed twice by gentle resuspension in 200 ml of buffer B (No. 3) and centrifugation at 6000 g for 20 min. The final pellet is suspended in 20 ml of resuspension buffer 1 (No. 5) and transferred to a 250-ml flask. Sodium Sarkosyl-NL 97 (ICN Pharmaceuticals) and Pronase (Calbiochem grade B) are then added to the flask as follows: (*a*) the GSA tube is rinsed with 6 ml of 10% sodium Sarkosyl in resuspension buffer 1 (No. 5), and 5 ml of this rinse is added to the flask to give a final Sarkosyl concentration of 2%; (*b*) 1.5 mg of Pronase in 0.3 ml of distilled water is added to the flask to give a final Pronase concentration of about 60 μg/ml. The Pronase solution must be self-digested for 2 hr at 37° to eliminate contaminating nucleases. The plastid suspension is incubated for 1 hr at 37° in a gently shaking water bath. After this incubation, the lysed chloroplast suspension is gently extracted three times with equal volumes of phenol saturated with 0.1 M Tris, pH 8.0. The organic and aqueous phases are separated by centrifugation for 10 min at 8000 g (7000 rpm in a Sorvall SS-34 rotor). The nucleic acids are precipitated from the final aqueous phase by the addition of 2.5 volumes of cold 100% ethanol (at $-80°$ for at least 30 min, or $-20°$ overnight) and collected by centrifugation at 12,000 g (10,000 rpm in a Sorvall SS-34 rotor) for 10 min at 4°. The pellet is washed with cold 70% ethanol, briefly desiccated, and resuspended in 2.0 ml of resuspension buffer 1 (No. 5). The material is then loaded onto four 12.8-ml 5–30% sucrose gradients (No. 7) and centrifuged at 21,000 rpm in a Beckman SW 40 rotor for 15 hr at 4°. All but the bottom 4 cm of each gradient is removed, and the tubes are then carefully filled with resuspension buffer 1 (No. 5) and recentrifuged at 35,000 rpm for 16 hr at 4°. After this centrifugation, the supernatants are poured off rapidly, and the pellets are drained of excess sucrose by inverting the tubes; any remaining sucrose is then wiped from the walls of the tubes. The pellets from all four tubes are resuspended gently in a total of 0.5 ml of resuspension buffer 2 (No. 6), and the absorbance at 260 nm is determined (A_{260} of a 1 mg/ml solution is 20). One kilogram of maize leaves typically yields 100–200 μg of chloroplast DNA by this procedure.

Method 2. After incubation with DNase, as described above, the plastid suspension is brought to 20 mM EDTA, pH 8.0, layered onto 200 ml of sucrose pad solution (No. 4) in a GS-3 tube, and centrifuged at 6000 g for 20 min at 4°. The supernatant is aspirated off, and the pellet is resuspended gently in 20 ml of resuspension buffer 1 (No. 5). The suspension is transferred to a 250-ml flask, after which the GS-3 tube is rinsed with 6 ml of 10% sodium Sarkosyl (w/v); 5 ml of this rinse is added to the flask, which is then incubated for 30 min at 65°. After incubation, the lysed chloroplast suspension is transferred to two SS-34 tubes and extracted three times with equal volumes of phenol saturated with 0.1 M Tris, pH 8.0. The nucleic acids are precipitated from the aqueous phase with ethanol and collected, as described above. The resulting pellets are drained of excess ethanol, gently resuspended in 10 ml of 2 M LiCl, and incubated on ice for several hours. The tubes are centrifuged at 12,000 g for 10 min at 4°, and the supernatants are dialyzed overnight against two changes of 5 mM Tris, 0.25 mM EDTA, pH 8.0, at 4°. (The pellets consist mainly of RNA species larger than approximately 4 S and the supernatants of DNA and of RNA species smaller than approximately 4 S.) The nucleic acids are precipitated from the dialyzate by the addition of 2.5 volumes of cold 100% ethanol, and the nucleic acid fraction is collected by centrifugation at 12,000 g for 10 min at 4°. The pellets are reextracted with 2 M LiCl, the supernatants are collected and dialyzed, and the DNA is precipitated from the dialyzate, as described above. The resulting pellets are resuspended gently in a total of 2 ml of resuspension buffer 2 (No. 6), and the absorbance at 260 nm is determined. One kilogram of maize leaves isolated by this procedure typically yields 200–300 μg of chloroplast DNA. Although this method requires far less time than Method 1 and yields DNA of high quality, these chloroplast DNA preparations contain RNA species smaller than approximatley 4 S. However, the latter do not interfere with most of the procedures to be described later (e.g., endonuclease digestion) except for obscuring DNA fragments in the same size class on ethidium bromide-stained gels. It is possible to remove the RNA with RNase treatment if necessary.

The most important qualities of a DNA preparation are (a) the amount of nuclear and mitochondrial DNA contamination; and (b) the degree of shearing during the isolation procedure. Both of these parameters can be assayed by performing restriction endonuclease digestion and agarose gel electrophoresis analysis of the chloroplast DNA using an enzyme that gives a known pattern of bands on an ethidium bromide-stained gel. With maize, 1 μg of chloroplast DNA is conveniently digested with 8–10 units of the appropriate restriction endonuclease for 1–2 hr at 37°, and the digest is run on a 1% agarose gel. After ethidium bromide staining, contaminating DNA and a high degree of breakage are visible as smearing or discrete unexpected bands in the gel lane. (See Mapping Restriction Endonuclease Cleavage Sites, below.)

Also of concern in chloroplast DNA preparations is the presence of material that absorbs at 280 nm that may interfere with many of the procedures to be described later (e.g., endonuclease digestion, nick translation). An $A_{260}:A_{280}$ ratio of 2.0 is optimal; a lower ratio indicates the presence of contaminating material that can often be removed by repeating the phenol extraction procedures described above. Other contaminants, such as polyphenols (which absorb at 280 nm) and polysaccharides (which absorb at 230 nm) often can be removed by repeating the sucrose gradient procedure outlined above.

Isolation of RNAs

Deciding which of the techniques presented in this section one should use for the isolation of large RNAs depends both on what the RNA is to be used for and on the amount of plant material available. Isolation of RNA from purified plastids (plastid RNA) yields the purest preparation of plastid-specific RNA but, unfortunately, does not preserve all the RNA as intact transcripts. Although not suitable for RNA-blot ("Northern") analysis,[14] plastid RNA should be used for *in vitro* translation experiments because the purity of the RNA is very important and the yield of intact transcripts is sufficient for translation into detectable radioactive products. RNA isolated from plants that have been grown in darkness and then exposed to light for 16 hr gives the best results in *in vitro* translation experiments. This is presumably because mRNA constitutes a larger percentage of the RNA population in greening maize plastids than in the mature chloroplasts of light-grown plants. Isolation of plastid RNA is normally also the method of choice when preparing RNA for use as a hybridization probe. Either of the total leaf RNA isolation procedures presented below will yield intact transcripts suitable for RNA-blot analysis if sufficient care is taken to exclude RNase contamination. Total RNA isolation procedures are also recommended whenever less than 50 g of leaf tissue are available.

Materials

8. Grinding buffer: 0.5 M sucrose, 0.1 M Tris-HCl, pH 8.0, 10 mM $MgCl_2$, 0.1% (w/v) BSA, 40 mM 2-mercaptoethanol
9. Sucrose gradient solutions B: 30 or 65% sucrose in 50 mM Tris-HCl, pH 8.0, 10 mM $MgCl_2$, 0.1% (w/v) BSA, 40 mM 2-mercaptoethanol
10. Resuspension buffer 3: 50 mM Tris-HCl, pH 8.0, 10 mM $MgCl_2$, 40 mM 2-mercaptoethanol
11. TNS lysis buffer: 50 mM Tris-HCl, pH 7.6, 2.5 mM $MgCl_2$, 1% (w/v) NaCl, 2% (w/v) triisopropylnaphthalenesulfonic acid (TNS) (Eastman Kodak; sodium salt). Store at room temperature, as the TNS will precipitate at $0-4°$.

[14] J. C. Alwine, D. J. Kemp, and G. R. Stark, *Proc. Natl. Acad. Sci. U.S.A.* **74**, 5350 (1977).

12. DNase buffer: 50 mM Tris-HCl, pH 7.8, 5 mM MgCl₂
13. Sucrose pad solution 2: 7.5% (w/v) sucrose in 10 mM Tris-HCl, pH 8.0, 0.5 M NaCl
14. Sodium dodecyl sulfate (SDS) lysis buffer: 0.2 M sodium borate, pH 9.0, 1% (w/v) SDS, 30 mM ethylene glycol bis(β-aminoethyl ether)-*N,N'*-tetraacetic acid (EGTA) (Eastman Kodak)

To reduce the chance of RNase contamination, all solutions and materials used in the preparation of RNAs, except those used for grinding, should be sterile.

Isolation of Plastid RNA

Grow maize as described above. Harvest leaf tissue, grind it in a blender, filter the homogenate, and pellet the plastids as described for the isolation of plastid DNA with the following modifications. First, use a grinding buffer (No. 8) that contains sucrose. Second, if isolating etioplasts, homogenize 120-g batches of leaf tissue in 250 ml of grinding buffer and carry out these procedures under green light. Grinding and all subsequent procedures are carried out at 0–4°. After pelleting, use a small camel hair brush carefully to resuspend the plastids in no more than the volume of the grinding buffer (No. 8) needed for loading on the sucrose gradients (see below).

Linear 30 to 65% sucrose gradients (No. 9) can be prepared up to 24 hr before use. The Beckman SW 25.2 rotor holds three 50-ml gradients, while the SW 27 holds six 32-ml gradients. Because the viscosity of the sucrose solution makes gradient preparation a long process, the gradients are usually poured at room temperature the day before harvesting the maize and then cooled to 4° before use. The resuspended crude plastids from about 60–90 g of leaf tissue are layered onto each gradient; 10 ml are layered on each SW 25.2 gradient and 8 ml on each SW 27 gradient. Centrifugation is for 30–45 min at 22,000 rpm at 4°. When preparing etioplasts do not exceed 30 min, or they will band too far down in the tube. Chloroplasts will give two bands. The upper, about halfway down the gradient, consists of broken chloroplasts; the lower contains intact chloroplasts. Etioplasts band about three quarters of the way down the tube. Collect the band of intact plastids from the top of the tube by first aspirating off the solution above it and then pulling off the plastid band from its bottom using a syringe with a bend (L-shaped) 18-gauge needle.

Dilute the plastid removed from the sucrose gradient with an equal volume of resuspension buffer 3 (No. 10). Pellet the plastids by centrifugation at 2000–2500 *g* for 15 min using more force for etioplasts than for chloroplasts. Then resuspend the plastids in a minimal volume of resuspension buffer. If desired, the plastids can be stored at −80° or in liquid nitrogen.

To lyse the plastids and extract the RNA, add four volumes of TNS lysis buffer (No. 11) and then immediately an equal volume of phenol that has been saturated with 0.1 M Tris-HCl, pH 8.0. Centrifuge to separate the phases, and remove the aqueous layer to a clean tube. Repeat the phenol extraction twice and pool the aqueous layers. Then extract the aqueous solution three to six times with an equal volume of diethyl ether, or twice with chloroform. Transfer the aqueous phase to a clean tube.

Depending on the volume and personal preference, add either 0.2 volume of 2 M sodium or ammonium acetate and 0.5 volume of isopropanol, or 0.05 volume of 4 M sodium or ammonium acetate and 2.5 volumes of ethanol. After cooling to $-80°$ for at least 30 min or to $-20°$ for at least 2 hr, pellet the precipitated RNA by centrifugation at 12,000 g for 10 min. Rinse the pellet with 70% ethanol, centrifuge, and then dry in a vacuum desiccator.

The DNA contaminating the RNA preparation can be removed in either of two ways. The first method utilizes digestion by DNase. Resuspend the precipitated RNA in 1 ml of DNase buffer (No. 12). Add 10 μg of RNase-free DNase (prepared by the method of Maxwell et al.[15]) and incubate at 5° for 30 min. Extract twice with an equal volume of buffer-saturated phenol and then twice with either diethyl ether or chloroform. Following addition of 0.05 volume of 4 M sodium or ammonium acetate and 2.5 volumes of ethanol, cool to $-80°$ for at least 30 min. Pellet the RNA, rinse with 70% ethanol, and dry in a vacuum desiccator.

The second method utilizes the differential solubility of nucleic acids in LiCl. Resuspend the pellet thoroughly in 1 – 2 ml of 2 M LiCl. Repellet the undissolved RNA, leaving the tRNA and DNA in the supernatant (see more detailed description under Isolation of Plastid DNA).

Resuspend the RNA pellet, now freed of other contaminating nucleic acids, in distilled water, reprecipitate with ethanol, dissolve in distilled water, and estimate the concentration spectrophotometrically. A solution containing 40 μg of RNA per milliliter will have an absorbance of 1.0 at 260 nm. RNA that is free of protein contamination will have an A_{260}/A_{280} of approximately 2.0

Isolation of Total Leaf RNA

Grow the maize as described above. Harvest the leaf tissue with scissors, cutting it into 1 – 2-cm pieces. Keep the cut tissue cold. In a VirTis tissue homogenizer vessel, combine the leaf tissue and 1 – 2 ml each of TNA lysis buffer (No. 11) and buffer-saturated phenol per gram of tissue. Grind with the homogenizer until a uniform suspension is produced (about 30 sec at a medium speed setting). All subsequent steps are to be carried out at $0-4°$

---

[15] I. H. Maxwell, F. Maxwell, and W. E. Hahn, *Nucleic Acids Res.* **4**, 241 (1977).

with sterile buffers and vessels, unless otherwise noted. If the volume is greater than about 20 ml, filter through 8 layers of cheesecloth before centrifugation at 12,000 g for 10 min to separate the phases. With smaller volumes, do not filter, since too much liquid is lost during filtration and the solid material is easily removed as a pellet during the centrifugation. Transfer the upper, aqueous phase to a clean tube and repeat the extraction with an equal volume of phenol once or twice until there is only a small amount of material at the interface. Extract the pooled aqueous layers once or twice with an equal volume of chloroform and then precipitate the RNA as described above for the isolation of plastid RNA.

To remove the contaminating DNA and polyphenols use either of the following procedures.

1. Resuspend the pellet of precipitated RNA in 1.5 ml of 10 mM Tris-HCl, pH 8.0, 0.5 M NaCl. Centrifuge to remove any undissolved material. Layer the resuspension on top of 3.0 ml of sucrose pad solution (No. 13) in a Beckman SW 56 rotor tube. Centrifuge for 36 hr at 50,000 rpm in the SW 56 rotor to pellet the RNA through the sucrose pad. Discard the supernatant fluid and resuspend the pellet in 1.0 ml of DNase buffer (No. 12). Proceed with the DNase treatment, extractions, and precipitation as described above for the isolation of plastid RNA.

2. Thoroughly resuspend the pellet of precipitated RNA in approximately 10 ml of 2 M LiCl per 30 g of leaf tissue. This may require several hours, or overnight in some cases. Thorough resuspension is essential. Repellet the undissolved RNA at 12,000 g for 10 min. Decant the supernatant, which contains the polyphenols, tRNA, and DNA, leaving a white RNA pellet. If the pellet is still brown, repeat the extraction with 2 M LiCl.

Dissolve the RNA pellet in distilled water and determine the RNA concentration spectrophotometrically as described above for the isolation of plastid RNA.

In many respects, the following variant for isolation of total leaf RNA, in which the leaf tissue is first broken in liquid nitrogen rather than phenol, is very convenient. To do this, harvest the leaves as described for grinding in phenol, but place the tissue in a mortar that is precooled with, and contains, liquid nitrogen. This procedure has been used with up to about 60 g of leaf tissue. It is equally useful with one or a few grams. Grind the tissue with the mortar and pestle, replenishing the liquid nitrogen if necessary, until a fine powder is formed. Transfer the powder to an unchilled beaker, and then quickly add 1 ml of boiling SDS lysis buffer (No. 14) per gram of tissue and solid dithiothreitol (DTT, Sigma) to 6 mg/ml. The buffer should not freeze if the powder is allowed to warm to about $-20°$ in the unchilled beaker. Immediately filter the mixture through Miracloth into a flask containing an equal volume of phenol saturated with 0.1 M Tris-HCl, pH 8.0. After centrifugation to separate the phases, remove the aqueous phase to a clean tube and proceed in the same manner as described above for grinding in phenol.

Comments

In the isolation of plastids for preparation of RNA, the continuous 30 to 65% sucrose gradients can be replaced by discontinuous step gradients. Chloroplasts band in a gradient at 1.4 M or 48% sucrose. Etioplasts will band on top of the 65% layer in a 10, 39, 47, 56, 65% sucrose step gradient. An equal volume of each layer is used.

After sucrose gradient sedimentation and pelleting of the banded plastids, resuspension in buffer (No. 10) is not essential to the procedure. Plastids can be resuspended directly in TNS lysis buffer (No. 11) when phenol extraction will follow immediately, or plastid pellets can be stored at $-80°$ without resuspension. Regardless of the details of the procedure used, phenol should be added to the resuspended plastids as soon as possible to minimize the action of any contaminating RNases. For this reason, rapid thawing of frozen plastids is recommended. This is easily accomplished by shaking in a 37° water bath until just thawed or adding hot TNS lysis buffer, or both, to the frozen pellet.

RNA isolations are complicated by the formation of polyphenolic compounds when whole-leaf tissue is disrupted in phenol. The formation of polyphenols can be minimized by keeping the solutions on ice, protecting them from light, and minimizing the time that the tissue is in contact with phenol. However, polyphenol formation cannot usually be avoided entirely. Since these compounds can inhibit nucleases and kinases, interfere with transfer to nitrocellulose and activated paper, and make absorbance estimates of nucleic acid concentration impossible, it is essential to minimize them. Both of the techniques described above work well. It has not been determined whether residual Li$^+$ in RNA prepared using LiCl fractionation is inhibitory in *in vitro* translation reactions.

These isolation procedures normally yield 1–2 μg of plastid RNA per gram of leaf tissue and 50–100 μg of total RNA per gram of tissue.

The quality, both in terms of purity and integrity, of the isolated RNA can be checked by gel electrophoresis either in polyacrylamide in the presence of 7 M urea[16] or in agarose after glyoxalation of the RNA.[17] The 23 S (2.9 kbp) and 16 S (1.5 kbp) rRNAs of the plastids should be visible, as well as a fragment of 1.1 kbp, which is a specific breakage product of the 23 S

[16] T. Maniatis and A. Efstratiadis, this series, Vol. 65, p. 299.
[17] G. G. Carmichael and G. K. McMaster, this series, Vol. 65, p. 380.

rRNA. Even the purest preparations are contaminated by RNAs of 3.3 and 1.9 kbp. These are presumably the rRNAs of either the mitochondria, which have a density close to that of plastids, or of cytoplasmic ribosomes that were trapped on the plastids during the isolation.

Cloning Plastid DNA Fragments

Cloning Plastid DNA in Bacterial Plasmids

In vitro construction of recombinant plasmids was first described by Cohen *et al.*[18] The procedure whereby a fragment of chloroplast DNA is inserted into a bacterial plasmid consists of the following steps: (*a*) digesting the chloroplast DNA with a restriction endonuclease and, if a specific fragment and no others are desired, isolating or enriching for the fragment to be cloned; (*b*) digesting the vehicle to be used with the restriction endonuclease and treating it with alkaline phosphatase; (*c*) ligating the two DNAs; (*d*) transforming competent bacteria with the chimeric plasmid; and (*e*) identifying the cloned plastid DNA fragment.[19]

Materials

15. Luria broth (LB): Bacto-tryptone (Difco), 10 g/liter; yeast extract (Difco), 5 g/liter; NaCl, 5 g/liter; pH 7.2
16. Luria broth plates (LB plates) 1.5%: Add 15 g of Bactoagar (Difco) per liter
17. LB antibiotic plates: Add ampicillin, 30 mg/liter; tetracycline, 20 mg/liter; or 20 mg/l kanamycin, 20 mg/liter
18. 5× Ligation buffer: 330 mM Tris-HCl, pH 7.6, 33 mM MgCl$_2$, 50 mM DTT
19. STET: 8% sucrose, 5% Triton X-100, 50 mM EDTA, 50 mM Tris-HCl, pH 8

Vehicles

We have used plasmids pMB9,[20] pML21,[21] RSF1030,[22] and pBR322.[20,23] The last of these has proved most generally useful.

[18] S. N. Cohen, A. C. Y. Chang, H. W. Boyer, and R. B. Helling, *Proc. Natl. Acad. Sci. U.S.A.* **70**, 3240 (1973).
[19] F. Bolivar and K. Backman, this series, Vol. 68, p. 245.
[20] F. Bolivar, M. Betlach, H. L. Heynecker, H. L. Heynecker, J. Shine, R. Rodriguez, and H. W. Boyer, *Gene* **2**, 95 (1977).
[21] M. A. Lovett and D. R. Helinski, *J. Bacteriol.* **127**, 982 (1972).
[22] S. E. Conrad, M. Wold, and J. L. Campbell, *Proc. Natl. Acad. Sci. U.S.A.* **76**, 736 (1979).
[23] J. G. Sutcliffe, *Nucleic Acids Res.* **5**, 2721 (1978).

Plastid DNA

From 1 to 10 μg of plastid DNA are used in a cloning experiment. If ligating a single fragment, 1 or 2 μg should be sufficient. In a shotgun experiment, in which the entire chloroplast DNA is digested and the resulting mixture of fragments is used in cloning, it is advisable to use 5 μg or more depending upon the number of different fragments produced by the restriction endonuclease. Calculate the amount of DNA to use by assuming you will need equimolar amounts of vehicle DNA and plastid DNA fragment(s).[24]

After incubation of plastid or vehicle DNA with the restriction endonuclease, digests are phenol extracted to remove the endonuclease, extracted with chloroform, and precipitated with ethanol.

Alkaline Phosphatase Treatment of the Vehicle

In order to prevent recircularization, the endonuclease-cleaved vector is treated with bacterial or calf intestinal alkaline phosphatase.[25] When using calf intestine alkaline phosphatase, the suspension in ammonium sulfate (Boehringer-Mannheim) is spun down; 1 μl of the supernatant is removed and diluted 100-fold in 0.1 M Tris, pH 9. From this dilution, 20 μl are used per microgram of digested vehicle DNA. The DNA plus phosphatase is incubated at 37° for 60 min; 0.25 M EDTA is added to bring the mixture to a final concentration of 20 mM EDTA, and then the solution is extracted once with phenol saturated with 0.1 M Tris-HCl, pH 8.0, and twice with chloroform. Sodium acetate is added to a final concentration of 1 M, and the DNA is precipitated by the addition of 2.5 volumes of cold 100% ethanol and transferred to −20° for 15–30 min. After centrifugation, the pellet is dried and resuspended in a small volume of sterile H_2O (see ligation calculation below).

The Ligation Reaction

The final volume of the ligation mixture is determined by the amount of DNA to be used. The concentration of DNA should be in the range of 5 to 30 μg/ml.

The vehicle and chloroplast DNAs are combined in ligation buffer (No. 18), and the mixture is placed at 65° for 10 min to dissociate sticky ends. It is cooled on ice; then nuclease-free BSA (Bethesda Research Laboratories) is added to a final concentration of 50 μg/ml, and ATP (Calbiochem) is added to a final concentration of 1 mM. Finally, 15 units of T4 ligase (Bethesda

[24] A. Dugaiczk, H. W. Boyer, and H. M. Goodman, *J. Mol. Biol.* **96,** 171 (1975).
[25] A. Ullrich, J. Shine, J. Chirgwin, R. Pictet, E. Tischer, W. J. Rutter, and H. M. Goodman, *Science* **196,** 1313 (1977).

Research Laboratories) are added, and the mixture is incubated at 10° overnight.[26]

Yeast tRNA (Sigma type III) is added to a final concentration of 20 μg/ml, and the DNA is precipitated with ethanol. The pellet is dried and resuspended in 30 mM CaCl$_2$ to a DNA concentration of 10–100 μg/ml.

Transformation

The preparation and transformation of competent cells is based on the method described by Cohen *et al.*[27] A flask containing 50 ml of LB (No. 15) is inoculated with 1/100 volume of an overnight culture of the host strain (this laboratory has used mostly *E. coli* HB 101). Cells are grown at 37° in a shaking water bath to mid-log phase, which takes approximately 2–2.5 hr.

The bacteria are pelleted at 3000 rpm (Sorvall SS-34 rotor) for 10 min (4°) and gently resuspended in 25 ml of chilled 30 mM CaCl$_2$. The tubes are kept on ice for 20–30 min. Again the bacteria are pelleted and resuspended in 1 ml of chilled 30 mM CaCl$_2$. (Modification of Mandel and Higa.[28])

Competent bacteria (200 μl) are added to 100 μl of the CaCl$_2$ solution containing 5 to 10 μg of ligated DNA and incubated on ice for 1–2 hr. The bacteria are then subjected to a heat pulse at 42° for 2 min.

The cells are diluted 10-fold in LB medium (No. 15) and grown for 1 hr at 37°. The culture is plated on selective medium (No. 17), and plates are incubated overnight.

Subcloning

If in a shotgun cloning two or more fragments ligate into a single plasmid, a very simple subcloning procedure can be followed to yield single insert clones. About 2 μg of the plasmid DNA is digested with the appropriate enzyme and the enzyme is then inactivated (see above.) After chilling on ice, ligation buffer, BSA, ligase, and ATP are added to the DNA. The mixture is incubated at 10° overnight. Transformation is as described above.

Screening

Colonies of transformed bacteria can be screened for the presence of cloned plastid DNA fragments by replica plating[22] or directly. The rapid-boil method of Holmes and Quigley[29] is used to prepare small amounts of plasmid DNA for analysis. Twelve or more colonies are streaked on an appropriate antibiotic plate and grown overnight. An inoculating loopful of

[26] V. Sgaramella and H. G. Khorana, *J. Mol. Biol.* **72,** 427 (1972).
[27] S. N. Cohen, A. C. Y. Chang, and L. Hsu, *Proc. Natl. Acad. Sci. U.S.A.* **69,** 2110 (1972).
[28] M. Mandel and A. Higa, *J. Mol. Biol.* **53,** 159 (1970).
[29] D. S. Holmes and M. Quigley, *Anal. Biochem.* **114,** 193 (1981).

cells is collected, carefully avoiding picking up any agar, an inhibitor of restriction endonucleases, and is resuspended in 50 μl of STET (No. 19) in a microfuge tube; 4 μl of a fresh 10 mg/ml lysozyme solution are quickly added, and the tubes are shut as tightly as possible and placed in a boiling water bath for 40 sec. The samples are then spun in a microfuge for 10 min at room temperature, and the supernatant is transferred to a new microfuge tube. Samples are precipitated with an equal volume of isopropanol, chilled for 5 min at −80°, spun for 10 min, washed with 70% ethanol, and vacuum desiccated. The entire sample is digested in a final volume of 25 μl in the appropriate buffer using 10–15 units of restriction enzyme per digest for 2 hr. After the addition of loading dye, the samples are loaded on an agarose gel of appropriate percentage for electrophoresis.

Cloning Plastid DNA in Charon 4A

The development and use of *in vitro* phage packaging systems for cloning of large DNA fragments have been described in detail.[3,30–32] We have used a scaled-down procedure including features of several methods.

Preparation of Charon 4A DNA

Charon 4A phage was grown and purified as described in Sternberg *et al.*[3] Preparation of phage DNA and isolation of end fragments (arms) for cloning was carried out exactly as has been described.[33]

Materials

20. *Escherichia coli* strains: NS 428:N205(A*am11 b2 red3 cITs857 Sam7*) and NS 433:N205(*Eam4 b2 red3 cITs857 Sam7*)
21. LB plates, 1%: as for LB medium (No. 15) with the addition of 10 g of Bactoagar (Difco) per liter
22. Phage dilution buffer: 10 mM Tris-HCl, pH 7.5, 10 mM MgSO$_4$
23. Packaging buffer: 40 mM Tris-HCl, pH 8, 10 mM MgSO$_4$, 10 mM spermidine · 3HCl, 10 mM putrescine · 2HCl, 0.1% 2-mercaptoethanol, 7% DMSO
24. Sucrose gradient solutions C: 10 or 40% sucrose in 20 mM Tris-HCl, pH 8.0, 10 mM EDTA, 1.0 M NaCl

[30] B. Hohn, this series, Vol. 68, p. 299.
[31] L. Enquist and N. Sternberg, this series, Vol. 68, p. 281.
[32] R. W. Davis, D. Botstein, and J. R. Roth, "Advanced Bacterial Genetics," p. 130. Cold Spring Harbor Lab., Cold Spring Harbor, New York, 1980.
[33] T. Maniatis, R. C. Hardison, E. Lacy, J. Lauer, C. O'Connell, D. Quon, G. K. Sim, and A. Efstratiadis, *Cell* 15, 687 (1978).

Preparation of Packaging Extracts

Bacterial strains used for the preparation of packaging extracts are stored at $-20°$ as glycerol stocks derived from single tested colonies.[31]

Overnight cultures of the two *E. coli* strains (No. 20) are grown at 30–32°. Two prewarmed flasks of LB (50 ml each) are inoculated separately with 100 μl of each of the overnight cultures, and growth is continued at 32° to an absorbance of 0.3–0.35 at 600 nm. This takes 3–4 hr.

For prophage induction, both flasks are transferred to a 45° shaking water bath for 20 min and then to a 37–39° incubator for a further 3 hr with vigorous shaking. Cultures are checked for induction of prophage by addition of a drop of chloroform to 1 ml of cells, shaking briefly, and maintaining at room temperature for 5 min. Induced cultures are lysed as indicated by reduced turbidity.[31] The cultures are then mixed and stored overnight at 4° (optional).

The induced cells are sedimented at 3000 g for 10 min, gently resuspended in 20 ml of cold dilution buffer (No. 22), resedimented, and carefully resuspended in 250 μl of packaging buffer (No. 23) brought to 2 mM ATP (Calbiochem). Twenty-microliter aliquots of the suspension are pipetted (using a "sawn-off" Eppendorf pipette tip) into 1.5-ml microfuge tubes, immediately frozen in liquid nitrogen, and then transferred to $-80°$ for long-term storage. Enough material for about 10 packaging reactions is obtained by this procedure.

The efficiency of the extracts for packaging exogenous DNA is tested using purified Charon 4A DNA.[30] In our experience, even low packaging efficiencies of 10^5 to 10^6 plaque-forming units per microgram of Charon 4A DNA are more than adequate for producing libraries of plastid DNA.

The likely causes of failure of packaging have been discussed.[31] In our experience, poor induction of lysogens is the most common stumbling block. The induction conditions given above are suitable only for small volumes of culture.

Preparation of 20 kbp EcoRI Partials of Plastid DNA

Approximately 200 μg of plastid DNA (see section on DNA isolation) are digested with 300 units of *Eco*RI (Bethesda Research Laboratories) for approximately 30 min in a total reaction mixture of 300 μl. The reaction is stopped by the addition of 20 μl of 250 mM EDTA, pH 8, and stored at 4° until further use. The sample is centrifuged for 16 hr at 30,000 rpm in an SW 40 rotor through a 10 to 40% sucrose gradient (No. 24).[33] The gradients are fractionated into approximately 500-μl aliquots and precipitated, after the addition of 50 μg of *E. coli* tRNA (Boehringer-Mannheim) to each fraction, with two volumes of ethanol, and resuspended in 1–5 μl of H_2O.

An aliquot of each fraction is run on a 0.8% agarose gel in TAE (No. 25, below) using SalI-digested chloroplast DNA from maize as a molecular weight marker. Those fractions containing fragments approximately 20 kbp in length are pooled and dephosphorylated with bacterial alkaline phosphatase (Bethesda Research Laboratories) using the recommended procedures. The samples are then extracted with phenol, precipitated with ethanol, washed with 70% ethanol, and resuspended in 1–5 μl of double-distilled water.

Ligation

Two micrograms of Charon 4A arms prepared as described[33] are resuspended in 7 μl of 100 mM Tris-HCl, pH 8.0, 10 mM MgCl$_2$ and heated to 42° for 1 hr to reanneal the arms. To this is added: 1.0 μl of dephosphorylated 20 kbp partials of plastid DNA (1 μg/μl), 1.1 μl of 4 mM ATP, 1.1 μl of 0.1 M DTT, 16 units T4 DNA ligase (Bethesda Research Laboratories), H$_2$O to 5 μl.

This mixture is incubated at 9° for 18 hr. One microliter is run on an 0.8% agarose gel using SalI-digested maize chloroplast DNA as size standard to test for the efficiency of the ligation.

Packaging of Recombinant Phages and Screening of Recombinant Library

The packaging reaction is done exactly as described by Davis *et al.*[34] except that the reaction is carried out at 25°. The entire packaging extract is plated on five petri dishes, 152 mm in diameter, using *E. coli* CSH 18 as host. Individual phage plaques are picked as described by Hohn[30] and incubated with 50–100 μl of a saturated culture of CSH 18 grown in LB (No. 15) with 10 mM MgSO$_4$ added for 20 min at 37° with shaking, to allow for phage adsorption. The culture is then diluted to 10 ml with the same growth medium and incubated at 37° with shaking until lysis is visible. The rest of the library can be lifted from the plates by overlaying with phage dilution buffer (No. 22) for 24 hr at 4° and then taking the overlay solution. Both the individual phages and the library are stored at 4° in the presence of chloroform.

The screening of the individual phages is done as described by Davis *et al.*[34] with the following modifications: 80 μl of the phage lysate is diluted to 400 μl with H$_2$O in a 1.5-ml microfuge tube. No diethyloxydiformate is added. The SDS, Tris-HCl, pH 8.0, and EDTA are combined into one 10× buffer. The final DNA pellet is resuspended in 40 μl of H$_2$O and digested with 10 units of *Eco*RI for 3 hr. The samples are run on a 1% agarose gel in

[34] R. W. Davis, D. Botstein, and J. R. Roth, "Advanced Bacterial Genetics," p. 109. Cold Spring Harbor Lab., Cold Spring Harbor, New York, 1980.

TBE (No. 26) using *Eco*RI-digested maize chloroplast DNA as reference standards. Large quantities of purified phage DNA may be obtained using the method of Sternberg *et al.*[3]

Mapping Restriction Endonuclease Cleavage Sites

The strategy utilized for mapping the cleavage sites of a particular restriction endonuclease on a genome the size of the 139 kbp plastid DNA depends to some degree on the frequency with which the enzyme cleaves the DNA. The cleavage sites of infrequently cutting enzymes can be mapped by simply using cross-hybridization analysis, but for enzymes that cleave the genome frequently to yield many relatively small fragments, it is more productive to work from libraries of cloned DNA fragments.

Materials

25. Tris-acetate buffer (TAE)[35]: 40 mM Tris base, 20 mM acetic acid, 2 mM Na$_2$ EDTA, pH 8.1
26. Tris-borate buffer (TBE)[35]: 89 mM Tris base, 89 mM boric acid, 2.5 mM Na$_2$ EDTA, pH 8.3

Procedures

When mapping the cleavage sites of restriction endonucleases, one must first establish the number, size, and stoichiometry of the DNA fragments produced by at least two enzymes. The identification can be confused by the production of comigrating fragments, a complication that is more likely to occur if the enzyme cleaves the DNA frequently. Densitometric scanning of photographic negatives of ethidium bromide-stained gels (or fluorometric scanning of the gels themselves) will reveal bands containing multiple fragments. The problem of multiple fragments in a single band can sometimes be resolved by varying both the agarose concentration and/or the gel buffer system used in gel electrophoresis in order to obtain the maximum resolution. A Tris-acetate buffer (No. 25) gives better resolution of fragments larger than 3 kbp, while Tris-borate (No. 26) is superior for smaller fragments.

There are at least three different techniques that can be used for cross-hybridization analysis to detect overlaps between fragments produced by different enzymes. In the first, overlaps between a fragment produced by one restriction endonuclease and the fragments generated by a second endonuclease are detected by isolating and radiolabeling fragments produced by

[35] R. W. Davis, D. Botstein, and J. R. Roth, "Advanced Bacterial Genetics," p. 148. Cold Spring Harbor Lab., Cold Spring Harbor, New York, 1980.

one restriction endonuclease and hybridizing them to a Southern blot[4] of all of the fragments produced by a second endonuclease. The DNA is digested and run on a gel, and the individual bands are cut out of the gel. The fragment in each band is eluted from the gel (we find electroelution[36] to be the most reliable technique), a radioactive probe is produced by either nick-translation or enzymatic synthesis of cRNA (see other sections of this paper), and then used for hybridization.[37] Spermine precipitation[38] of the DNA after elution is an effective way to remove the gel residue, which frequently inhibits the enzymes used for labeling. Other techniques that might be useful for the hybridization analysis are a "cross-Southern" technique,[39] and labeling of genomic DNA before restriction and electrophoresis to yield radiolabeled fragments that can be used without further processing, for hybridization.[40]

If the enzymes used produce relatively few fragments that migrate uniquely, it may be possible to deduce the map for the cleavage sites from the fragment sizes, stoichiometries, and overlaps. It should be noted that such an analysis, which focuses on relatively large fragments, can overlook smaller fragments produced by the endonucleases used.

If, however, the restriction endonucleases of interest produce so many fragments that mapping by fragment overlaps is impractical, cloned DNA fragment libraries are very helpful if not essential. It is best to have two types of libraries: first, a recombinant lambda phage or cosmid library into which have been cloned large DNA fragments produced by partial digestion of the genome (lambda phage Charon 4A has been used in this laboratory for this purpose); second, libraries of individual DNA fragments produced by the enzyme used for the partial digestion and at least one other endonuclease cloned in a plasmid vector.

An initial map of cleavage sites for the enzyme used to make the lambda phage library can be obtained by isolating DNA from individual recombinant phages, cutting with the restriction enzyme, and analyzing the resulting fragments by electrophoresis using techniques described above for screening of the recombinant library. By determining which phages contain some, but not all, fragments in common, a group of phages that comprise an overlapping set can be selected for further study. The overlaps will serve to order the phage inserts and so provide an initial map. It should be stressed, however, that this map is preliminary, since multiple fragments of similar size, which

[36] U. Wienand, Z. Schwarz, and G. Feix, *FEBS Lett.* **98**, 319 (1978).
[37] R. W. Davis, D. Botstein, and J. R. Roth, "Advanced Bacterial Genetics," p. 174. Cold Spring Harbor Lab., Cold Spring Harbor, New York, 1980.
[38] B. C. Hoopes and W. R. McClure, *Nucleic Acids Res.* **9**, 5493 (1981).
[39] S. Sato, C. A. Hutchinson, Jr., and J. I. Harris, *Proc. Natl. Acad. Sci. U.S.A.* **74**, 542 (1977).
[40] J. D. Palmer, *Nucleic Acids Res.* **10**, 1593 (1982).

are commonly produced by frequently cutting restriction endonucleases, will cause confusion.

The production of a definitive map for the cleavage sites of the enzymes of interest requires that the location of sites for each enzyme be accurately determined in the plastid DNA sequences of each phage in the overlapping set. Analysis of the fragments produced by simultaneous cleavage of the phage DNA with more than one enzyme is useful, but sometimes the production of many fragments makes this difficult. It is at this point that the individually cloned fragments become essential. One can easily and accurately locate the cleavage sites for the enzymes of interest in the individual DNA fragments using the recombinant plasmid library. More important, the recombinant DNA plasmids, when radiolabeled, become a source of probe for hybridization analyses. This is an important point, because a restriction endonuclease may produce multiple DNA fragments of the same size and the genomes of higher plant plastids usually have repeated sequences. Such hybridization analysis is also the only way to confirm the order, identity, and overlap of the DNA fragments produced by the enzymes of interest, and thus the overlaps among the selected group of recombinant phages. By combining the maps for the set of overlapping recombinant phages, the genomic map is derived.

The basic strategy outlined above is applicable to genomes other than those of plastids. For larger genomes, up to about 300 kbp, the random approach described here for identifying a set of phages carrying overlapping DNA fragments should be more time efficient than directed "walking."[41] A mixture of random and directed approaches may be necessary to complete the set of overlapping phages in the most direct manner. Genomes larger than 300 kbp will be most efficiently mapped by a strategy that is directed rather than random. When "walking" along a genome, it is important to be mindful of the problems caused by repeated sequences.[41]

Use of Computers for Map Generation

A computer program for generating the maps of the restriction endonuclease sites has been prepared. The information for the map is stored as a linear array of characters on a disk file. Sites are represented by any desired character string, much as in an actual sequence but at a reduced scale. The separation between the sites is indicated by the total character count. Comments can be included as part of the data in portions of the array that lack sites. Any desired data can be included, each different type with its own specific character string. A different character string is used for each restriction enzyme site.

[41] T. H. Eickbush, C. W. Jones, and F. C. Kafatos, *in* "Developmental Biology Using Purified Genes" (D. D. Brown, ed.), p. 135. Academic Press, New York, 1981.

In practice, the program is given a list of each of the strings to be located. Each group of sites is then located, and a list of them, including the exact location in the map and the separations between them, is printed. The program then draws a circular map using an x-y plotter. On the plotter, sites are indicated by radial lines on the circle. Several special-purpose characters can be used to indicate whether the radial should be drawn inwardly or outwardly directed, which can be used to indicate uncertainties or the

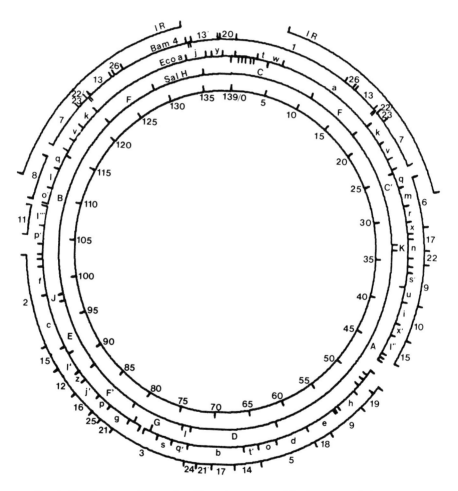

FIG. 1. Circular map of the maize chloroplast genome. The innermost circle represents a coordinate system in kilobase pairs. The next three circles are maps of the restriction endonuclease sites for *Sal*I, *Eco*RI, and *Bam*HI. Inward slashes indicate uncertainty in the order of the adjoining fragments. Outward slashes indicate that the placement of the adjoining fragments is known. The outermost circle indicates the extent of the inverted repeats (IR).

direction of transcription of genes. Other characters can define unknown regions which are not drawn. Successive circular maps are drawn concentrically, with a reference circle containing 5 kbp markings innermost.

The program is written in FORTRAN IV. Copies will be provided on request.

A map of recognition sites for *Eco*RI, *Sal*I, and *Bam*HI on the maize plastid genome produced by this program is shown in Fig. 1.

Nick Translation

Radiolabeling of DNA to a specific activity of at least 5×10^7 cpm/μg may be achieved using the nick translation reaction of *E. coli* DNA polymerase I. Suitable reaction conditions have been well described.[42-45]

The reaction is terminated by addition of an equal volume of 20 mM EDTA, 0.2% SDS, 2 mg of sonicated calf thymus DNA per milliliter, followed by 0.1 volume of a saturated solution (4–5 mg/ml) of Blue Dextran 2000 (Pharmacia). To remove unincorporated ^{32}P-labeled nucleotides, the mixture is passed through a 2–3-cm column of autoclaved Sephadex G-50 or BioGel P-60 in a Pasteur pipette. Labeled DNA elutes at the void volume along with the Blue Dextran.

Prior to use as a probe in hybridizations, the labeled DNA is denatured by addition of 0.1 volume of 1 M NaOH. After 10 min at room temperature, 0.2 volume of 1 M Tris-HCl, pH 7.5, is added, and the probe is immediately mixed with the hybridization solution. Alternatively, the probe may be denatured by incubation for 10 min in a boiling water bath.

Copy RNA

In certain situations, such as when a DNA template has been removed from an agarose gel, it is advantageous to create a radioactive copy RNA (cRNA) probe rather than nick-translated DNA, since RNA polymerase appears to be less sensitive than DNA polymerase to agarose. The procedure is somewhat shorter than that for nick translation.

Materials

27. 5× Polymerase buffer: 0.25 M KCl, 0.25 M Tris-HCl, pH 8.0, 0.5 mM DTT, 50% glycerol

[42] P. Rigby, C. Rhodes, M. Dieckmann, and P. Berg, *J. Mol. Biol.* **113**, 237 (1977).
[43] T. Maniatis, A. Jeffrey, and D. G. Kleid, *Proc. Natl. Acad. Sci. U.S.A.* **72**, 1184 (1975).
[44] M. F. Thomashow, R. Nutter, A. L. Montoya, M. P. Gordon, and E. W. Nester, *Cell* **19**, 729 (1980).
[45] R. W. Davis, D. Botstein, and J. R. Roth, "Advanced Bacterial Genetics," p. 168. Cold Spring Harbor Lab., Cold Spring Harbor, New York, 1980.

28. [α-^{32}P]UTP, 300–400 Ci/mmol, concentrated aqueous solution (Amersham)
29. 20 mM stocks each of unlabeled CTP, GTP, and ATP (Calbiochem)

Procedure

Before each reaction a fresh reaction cocktail of 2 mM each CTP, GTP, and ATP in 0.1 M MgCl$_2$ is made up from stocks. The reaction volume is determined by the concentration of the DNA, but a typical volume is 50 $μ$l, as follows: 10 $μ$l of DNA (0.5–2 $μ$g), 10 $μ$l of 5× polymerase buffer (No. 27), 5 $μ$l of reaction cocktail, 5 $μ$l (50 $μ$Ci) of concentrated aqueous [α-^{32}P]UTP (No. 28), 1 $μ$l of RNA polymerase (*E. coli* enzyme, New England Nuclear), 19 $μ$l of H$_2$O.

Incubate at 37° for 1 hr. If higher specific activities are required, [α-^{32}P]CTP or [α-^{32}P]ATP can be added (and the cold CTP or ATP deleted from the 10× magnesium – nucleotide mix).

For homologous hybridizations, where stringent washing conditions can be used, it may not be necessary to run a column to remove unincorporated [α-^{32}P]UTP. Instead the reaction is brought to 200 $μ$l with 0.3 M sodium acetate after incubation, 10 $μ$g of carrier tRNA, and 750 $μ$l of 100% ethanol are added, and the sample is precipitated as usual. Otherwise a column can be run as described under nick translation.

Physical Mapping of Maize Plastid Genes

Mapping Genes for Plastid rRNAs

Materials

30. Lysis buffer: 100 mM Tris-HCl, pH 8.0, 25 mM MgCl$_2$, 100 mM KCl, 1% Triton X-100
31. Resuspension buffer 4: 100 mM Tris-HCl, pH 8.0, 5 mM MgCl$_2$, 100 mM KCl
32. Sucrose pad solution 3: 1.0 M sucrose, 5 mM MgCl$_2$, 100 mM KCl
33. Elution-purification buffers: 50 mM ammonium acetate, 350 mM ammonium acetate, 2.0 M ammonium acetate

Preparation of Plastid Ribosomes and Ribosomal RNAs

Chloroplasts were isolated as described for the isolation of plastid DNA through the 1500 g GSA centrifugation. The chloroplasts can be stored at −20° as a pellet until further use.

The chloroplasts are broken in lysis buffer (No. 30) at a chloroplast/buffer concentration of 1 g/ml. Debris is removed by centrifugation for 20 min in a Sorvall SS-34 rotor at 15,000 rpm. The cleared supernatant is spun in a Beckman Ti 60 rotor for 3 hr at 39,000 rpm. The green pellet is taken up in resuspension buffer (No. 31) containing 0.1% Triton X-100 and clarified in an SS-34 at 10,000 rpm for 10 min. The supernatant is then layered over 20 ml of the sucrose pad solution (No. 32) and spun for at least 12 hr in a Ti 60 rotor at 39,000 rpm. The ribosome pellet is taken up in 2 ml of resuspension buffer (No. 31), clarified, and used as the source of rRNA.

The rRNAs can be prepared from the ribosomes by extraction with phenol saturated with 0.1 M Tris-HCl, pH 8.0, and precipitation with two volumes of ethanol for 2 hr at $-20°$. The 4.5 S and 5 S RNAs can be separated from the higher molecular weight RNAs by extracting the dried RNA pellet with 3 M potassium acetate, pH 6.0. The lower molecular weight RNAs dissolve, but the higher molecular weight RNAs do not. After two ethanol precipitations, the 4.5 S and the 5 S RNAs can be separated from each other on 12% polyacrylamide gels,[46] while contaminating large molecular RNAs will not enter the gel. The RNAs can then be electroeluted from the acrylamide gel.[36] For radiolabeling (see kinasing RNA), the RNAs must be cleaned of the contaminating agarose or acrylamide as follows: DE-52 (Whatman) is prepared as per manufacturer's instructions and stored at 4° in 50 mM ammonium acetate and 0.1% sodium azide with a 1:2 (v/v) ratio of settled DE-52 per storage solution. Before use, shake thoroughly and remove an appropriate volume (usually 10 μl of solution to 1 μg of the expected yield of RNA). Pellet the DE-52, remove the excess buffer, add the RNA in 50 mM ammonium acetate, and shake for 1 min. Pellet the DE-52 and wash, once with 50 mM ammonium acetate and once with 350 mM ammonium acetate. Elute the RNA from the DE-52 twice with 10 μl of 2 M ammonium acetate per 10 μg of expected RNA. Precipitate with 2 volumes of cold ethanol, wash with 70% ethanol, and resuspend in an appropriate volume of 10 mM Tris-HCl, 1 mM disodium EDTA, pH 8.0.[47]

The 16 S and 23 S RNA can be separated from one another on 15 to 30% sucrose density gradients[48] in a Beckman SW 27 rotor for 16 hr at 25,000 rpm, followed by fractionation of the gradient and ethanol precipitation of the appropriate fractions.

The mapping of the rRNA genes on the chloroplast genome by electron microscopy and by the Southern technique after kinasing the rRNAs has been described.[5]

Kinasing RNA

Because of the large pool sizes of ortho- and polyphosphate in maize, it has not been possible to obtain RNAs sufficiently radioactive for Southern

[46] J. J. Dunn and F. W. Studier, *Proc. Natl. Acad. Sci. U.S.A.* **70**, 1559 (1973).
[47] Z. Schwarz, personal communication.
[48] E. H. McConkey, this series, Vol. 12, p. 620.

hybridization[4] by labeling *in vivo*. Isolated RNAs can be labeled *in vitro*[2,8,49] for hybridization against plastid DNA fragments. The specific activity of kinased RNA depends on the number of ends labeled. Therefore, the RNA is first hydrolyzed to fragments of 80–100 bases. It follows that the RNA used need not be of very high quality. The duration of the hydrolysis reaction must to some extent be determined empirically, as it depends on the initial condition of the RNA.

Materials

34. 10× Kinase buffer: 700 mM Tris-HCl, pH 7.6, 100 mM MgCl$_2$, 50 mM DTT

Procedure

For hydrolysis, 10 μg of RNA are brought up in 9 μl of H$_2$O, 1 μl of 0.5 M glycine-NaOH pH 9.5 is added, and the mixture is incubated at 90° for 3–5 min. The reaction is stopped by putting the tube on ice, 2 μl are removed for the kinasing reaction, and the remainder is stored for future use as an ethanol precipitate. The kinasing reaction is done as follows: Combine 2.0 μl of RNA solution, 4.0 μl of 10× kinase buffer (No. 34); 1.0 μl of 5 units/μl T4 polynucleotide Kinase (Bethesda Research Laboratories), 0.2 μl of 50 μg/μl BSA, 5–10 μl (50 μCi) of [γ-^{32}P]ATP (2000–3000 Ci/mmol, New England Nuclear), H$_2$O to 40 μl final volume. Incubate for 1 hr at 37°. At the conclusion of the reaction, the labeled RNA is isolated using a column as described under nick translation.

Mapping Genes for Plastid tRNAs

Two methods have been used: (*a*) hybridization of individually purified radioactive tRNAs to restriction fragments of plastid DNA on nitrocellulose filters[50] by the method of Southern[4]; (*b*) catching (i.e., hybridization) of radioactive tRNAs from an unresolved mixture by a cloned plastid DNA fragment immobilized on a nitrocellulose filter, discharge of the radioactive tRNA into an unresolved mixture of nonradioactive tRNAs, two-dimensional polyacrylamide gel electrophoresis of the mixture[51] containing the discharged radioactive tRNA, staining of the gel with ethidium bromide,

[49] N. Maizels, *Cell* **9**, 431 (1976).
[50] A. Steinmetz, M. Mubumbila, M. Keller, G. Burkard, J. H. Weil, A. J. Driesel, E. J. Crouse, K. Gordon, H. J. Bohnert, and R. G. Hermann, *in* "Transfer RNA: Biological Aspects" (D. Soll, J. N. Abelson, and P. R. Schimmel, eds), p. 281. Cold Spring Harbor Lab., Cold Spring Harbor, New York, 1980.
[51] A. Fradin, H. Gruhl, and H. Feldman, *FEBS Lett.* **50**, 185 (1975).

and autoradiography.[52] The tRNA or tRNAs hybridizing to the cloned DNA fragment are identified by their location on the gel (autoradiogram) relative to the unlabeled marker tRNAs and a two-dimensional gel electrophoretic map of the plastid tRNAs of the species being studied.

Mapping Plastid Genes for Proteins

Two approaches for mapping plastid genes are described. The first is based upon analysis of the polypeptide(s) produced by an *in vitro* transcription–translation system directed by a known plastid DNA fragment. The second determines the developmental class of the genes through a comparison of the *in vitro* translation products of RNAs obtained from plastids at different developmental states, e.g., etioplasts of dark-grown seedlings vs mature or maturing chloroplasts, and a comparison of the hybridization patterns of these RNAs to Southern blots[4] of restriction endonuclease-digested plastid DNA.

A logical progression of steps for the first approach follows.

1. Use total plastid RNA to gain preliminary evidence, e.g., immunochemically, that there are plastid transcripts for the gene being sought and to determine what type of translation system, e.g., an *E. coli* preparation, wheat germ preparation or rabbit reticulocyte lysate, is best for translating the specific message. Different plastid transcripts in a population are favored for translation by each of these translation systems.[53]

2. Using the translation system that has been selected, pick a transcription system (e.g., *E. coli,* wheat germ, or other DNA-dependent RNA polymerase preparations) to link to the translation system using undigested plastid DNA to direct the reaction. (Obviously, coupled transcription-translation systems from a single species, e.g., *E. coli,* are the easiest to try first.)

3. Test digested plastid DNA to determine that the restriction endonuclease that may be used for cloning does not disrupt the synthesis of the polypeptide of interest. The approach to the gene of interest can be made most directly by testing large cloned fragments, such as those cloned in Charon phages, as templates before proceeding to test with smaller cloned fragments.

Translation of RNA

Materials and Procedure for Preparation of Reticulocyte Lysate

35. Saline: 140 mM NaCl, 1.5 mM magnesium acetate, 5.0 mM KCl; for collecting blood, add 0.001% heparin.

---

[52] Z. Schwarz, S. O. Jolly, A. A. Steinmetz, and L. Bogorad, *Proc. Natl. Acad. Sci. U.S.A.* **78,** 3423 (1981).

[53] D. M. Coen, Ph.D. Thesis, Massachusetts Institute of Technology, Cambridge, 1978.

36. Acetylphenylhydrazine (Sigma Chemical Co.), 1.2% solution in saline neutralized to pH 7.5 with 1 M HEPES, pH 7.0. Keep in a dark bottle at room temperature since this compound is both light-sensitive and near saturation at this concentration.

37. Lysing buffer: 40 μg of hemin (Eastman Kodak) per milliliter in H_2O. A 4 mg/ml hemin stock is made in the following manner: 20 mg of hemin is dissolved in 0.4 ml of 0.2 M KOH and made up to 1 ml with H_2O. The solution is brought to pH 7.8 with 0.1 ml of 1 M Tris-HCl, pH 7.8, and then diluted to 5 ml with ethylene glycol. Vortex after each addition.

38. Micrococcal nuclease (P-L Biochemicals) is stored in 50 mM glycine, 5 mM CaCl$_2$, pH 9.2, at 150,000 units/ml. Storage at lower concentrations causes rapid inactivation.

This procedure is essentially that of Villa-Komaroff[54] with the following modifications.[53,55] One day of rest was inserted between the acetylphenylhydrazine regime and the bleeding of the rabbits. The rabbits were bled from the ear for 2–3 days depending on the health of the animals. On the final day, the rabbits were exsanguinated by a heart puncture. The cells were lysed with hemin, No. 37, the known activator of the system, at 40 μg/μl.

Endogenous mRNAs are removed by digestion with micrococcal nuclease in the presence of hemin. To treat 1 ml of rabbit reticulocyte lysate, add 10 μl of the 4 mg/ml hemin stock and 20 μl of 50 mM CaCl$_2$. Mix and equilibrate in a 20° water bath for 5 min. Add 0.5 μl of micrococcal nuclease, mix, and incubate at 20° for 15 min. Add 20 μl of 100 mM EGTA (Eastman Kodak) and 4 μl of 40 mg/ml creatine phosphokinase (40 mg/ml, Sigma type I in 50% glycerol). Mix and store in liquid nitrogen. The preparation is stable for about 12 months.

See section on isolation of RNAs for preparation of RNA.

Translation Conditions

39. 10× Translation cocktail: 1 ml of 1.0 M HEPES-KOH, pH 7.5, 100 mg of creatine phosphate (Sigma) in 1 ml of H_2O, 5.1 mg of spermidine trihydrochloride (Sigma) in 1 ml of H_2O, and 1 ml of H_2O. This gives 4 ml of 10× translation cocktail.

40. 2.5× Translation mix: 25 μl of 10× translation cocktail (No. 39), 25 μl of 1.0 M potassium acetate, 3.75 μl of 50 mM magnesium acetate, 2.50 μl of 0.1 M DTT, 250 μCi of [^{35}S]methionine (Amersham), H_2O to 100 μl, enough for 10 reactions

[54] L. Villa-Komaroff, M. McDowell, D. Baltimore, and H. F. Lodish, this series, Vol. 30, p. 709.
[55] H. R. B. Pelham and R. L. Jackson, *Eur. J. Biochem.* **67**, 147 (1976).

Translation reaction mixture: 10 μl of 2.5× translation mix (No. 40), 10 μl of rabbit reticulocyte lysate, 10–20 μg of total chloroplast RNA H$_2$O to 25 μl. Incubation time is 60 min at 37°.

Summary of translation reaction mixture components, in final concentrations: 25 mM HEPES-KOH, pH 7.5, 0.75 mM magnesium acetate, 100 mM potassium acetate, 7.64 mM creatine phosphate, 1 mM DTT, 875 μM spermidine trihydrochloride, 1 μCi/μl [³⁵S]methionine (Amersham), 10 μl of rabbit reticulocyte lysate. The total volume is 25 μl.

Linked Transcription – Translation of DNA

DNA, which may be restricted before use, is extracted with 0.1 M Tris-HCl, pH 8.0, saturated phenol. The aqueous phase is adjusted to 1 M ammonium acetate and precipitated with ethanol. The sample is washed with 70% ethanol, resuspended in sterile water, brought to 250 mM ammonium acetate, and precipitated and washed as before. The sample is vacuum desiccated and resuspended at a high DNA concentration (usually 1 μg/μl) in sterile water. The samples are stored at −20°.

Transcription Step[56]

41. 10× Transcription buffer: 200 mM HEPES-KOH, pH 7.8, 100 mM magnesium acetate, 2 M potassium acetate
42. 10× NTP stock: 5 mM ATP, 5 mM GTP, 5 mM CTP, 5 mM UTP. All nucleotide triphosphates are from Calbiochem.

Store the Nos. 41 and 42 at −80°.

Transcription reaction mixture: 1 μl of 10× transcription buffer (No. 41), 1 μl of 10× NTP stock (No. 42), 1 μl of RNA polymerase stock (1 unit/μl) (New England Nuclear *E. coli* RNA polymerase), 2–3 μg of DNA, H$_2$O to 10 μl. Incubation time is 30 min at 37°.

Translation Step

43. 10× translation mix: 50 μl of 10× translation cocktail (No. 39), 30 μl of 1.0 M potassium acetate, 6.25 μl of 0.1 M trisodium EDTA, H$_2$O to 100 μl, enough for 10 reactions

Translation reaction mixture: 10 μl of transcription reaction mixture, 10 μl of 10× translation mix (no. 43), 20 μl of rabbit reticulocyte lysate, micrococcal nuclease treated, 50 μCi of [³⁵S]methionine, H$_2$O to 50 μl. Incubation time is 60 min at 37°.

Summary of reaction mixture components, in final concentration, for linked transcription – translation: 28 mM HEPES-KOH, pH 7.6,

⁵⁶ B. E. Roberts, M. Gorecki, R. M. Mulligan, K. J. Danna, and A. Rich, *Proc. Natl. Acad. Sci. U.S.A.* **72**, 1922 (1975).

2 mM magnesium acetate, 100 mM potassium acetate, 8 mM creatine phosphate, 1 mM DTT, 500 μM spermidine trihydrochloride, 1 μCi of [^{35}S]methionine per microliter, 20 μl of rabbit reticulocyte lysate, 1.25 mM trisodium EDTA, 0.10 mM each ATP, GTP, CTP, UTP, 1 unit of $E.$ $coli$ RNA polymerase. Total volume is 50 μl.

Notice that sodium EDTA has been added to the linked transcription-translation reaction mixture to lower the effective concentration of Mg^{2+} to 0.75 mM.

Comments

No Ca^{2+} can be introduced, since the endogenous mRNA is degraded with calcium-dependent micrococcal nuclease that is subsequently inactivated by binding the calcium with EGTA. Cell-free in $vitro$ translation is sensitive to trace amounts of ethanol, detergents, or Na^+. All reactions and mixtures are done in or stored in siliconized glass tubes or in plastic microfuge tubes.

Both simple translation and linked transcription – translation reactions are performed at the same concentrations of potassium and magnesium. The concentration of these ions in the reticulocyte lysate is not measured; added Mg^{2+} is 0.75 mM, and the added K^+ is 100 mM. The optimal concentrations of these ions should be checked experimentally with each lysate made and for different RNA templates. No amino acids are added to the system, since the rabbit reticulocyte lysate is not amino-dependent.

Analyses of Polypeptides Synthesized in Vitro

Supernatants of plastid RNA-directed translation reactions or of plastid DNA fragment-directed linked transcription – translation reactions may be applied to polyacrylamide gels[6] after ribosomes have been removed by centrifugation. Newly synthesized polypeptides may be immunoprecipitated and/or compared with their suspected in $vivo$ counterparts by limited proteolysis.[6] Immunoprecipitation can be accomplished conveniently as described by Tobin.[57]

Identification of Plastid Genes by Developmental Class

An example of a differential RNA hybridization method for identifying DNA sequences of developmentally regulated genes illustrates a general approach that should be widely applicable when reasonable amounts of RNA of even poor quality can be obtained.

A comparison of the hybridization of in $vitro$ radiolabeled plastid RNA from etiolated and greening maize seedlings to Southern filters[4] containing

[57] E. M. Tobin, $Plant$ $Physiol.$ **67,** 1078 (1981).

restriction fragments of plastid DNA has been used to identify a fragment that codes for a major nonribosomal chloroplast RNA species that either appears or increases in abundance during light-induced plastid development.[8] One limitation in this procedure is that hybridization to the same DNA fragment by rRNA transcripts, transcripts from constitutively expressed plastid genes, or from long-lived etioplast transcripts can obscure the hybridization of transcripts from other light-induced genes (genes whose transcription is restricted to, or greatly increased in, greening plastids).

As a result of these limitations, a technique of competitive hybridization has been utilized to identify regions of the chloroplast genome that contain light-induced genes. In this procedure (modified from Koller and Bogorad[58]), a Southern filter containing restriction fragments of chloroplast DNA is prehybridized with unlabeled etioplast RNA and subsequently hybridized with labeled plastid RNA from greening seedlings. As a control for the amount of hybridization masked by the prehybridization of unlabeled etioplast RNA, another Southern filter is directly hybridized with labeled plastid RNA from greening seedlings. As a control for the "efficiency" of prehybridization with the unlabeled etioplast RNA, a third Southern filter is prehybridized with unlabeled etioplast RNA and then hybridized with an aliquot of radioactively labeled etioplast RNA. Titration experiments demonstrate that as the concentration of unlabeled etioplast RNA is increased in the prehybridization mixture, the restriction fragments containing rRNA genes become less and less visible on autoradiographs after hybridization of labeled RNA from the light-induced plastids; in parallel, hybridization to other fragments becomes more visible. Because rRNA constitutes the bulk of the RNA pool in etioplasts and greening plastids, it can be assumed that at concentrations of unlabeled RNA sufficiently high to mask the hybridization of labeled rRNA, any hybridization seen is due to the presence of transcripts from light-induced genes.

Competitive hybridization has been used to identify regions of the chloroplast genome containing genes that are expressed only in etioplasts, as well as genes that are expressed in etioplasts and greening plastids. It has also been useful for identifying DNA fragments containing other kinds of differentially transcribed genes.

Materials

44. Hybridization solution: 50% formamide; $3 \times SSC$ ($1 \times SSC$ is $0.15\ M$ NaCl, $0.015\ M$ sodium citrate, pH 7.0); $1 \times$ Denhart's solution: 0.02% BSA (w/v), 0.02% poly(vinylpyrrolidone) (w/v) in $3 \times SSC$, 0.02% Ficoll (w/v); 0.1 mg/ml sonicated calf thymus DNA, 0.1% SDS

[58] K. P. Koller and L. Bogorad, in preparation.

Procedure

Three identical Southern filters[4] (1×20 cm) containing fragments from a restriction digest of chloroplast DNA (1 μg) are presoaked at 37° for at least 2 hr in approximately 1 ml of hybridization solution (No. 44) in individual 8×9-inch "Seal-N-Save Boilable Cooking Pouches" (Sears, Roebuck and Co.). The presoak solution is squeezed out of each bag, fresh hybridization solution is added (approximately 1 ml/pouch), and at least 10 μg of unlabeled plastid RNA from 7-day-old dark-grown seedlings are pipetted into each of two bags. The filters are incubated at 37° for 16–24 hr, after which 5×10^5 to 10^7 cpm of kinased plastid RNA (see RNA Kinasing) from seedlings greened for a desired number of hours is added to one of the bags containing unlabeled RNA and also to the bag lacking unlabeled RNA (the nonprehybridized control). The same number of counts per minute of kinased etioplast RNA is added to the remaining pouch containing unlabeled etioplast RNA (the prehybridized control). After hybridization at 37° for 16–24 hr, filters are removed from the bags and vigorously washed in a shaking water bath, once for 45 min at 65° in $2 \times$ SSC, 0.3% SDS, and twice for 45 min each at 65° in $1 \times$ SSC, 0.1% SDS. The filters are then air dried, mounted on a glass plate, and exposed to X-ray film.

Acknowledgments

We are grateful for the contributions of all the past members of the group. We thank Diane Headen for excellent assistance in the preparation of this manuscript. Portions of the above work have been funded by the Maria Moors Cabot Foundation of Harvard University and by grants from the National Institute of General Medical Sciences, The National Science Foundation, and the Competitive Research Grants Office of the United States Department of Agriculture.

[48] Role of Thylakoid Polypeptide Phosphorylation and Turnover in the Assembly and Function of Photosystem II

By M. Wettern, J. C. Owens, and I. Ohad

I. Introduction

The greening of the *y-1* mutant of *Chlamydomonas reinhardi* can be used as an experimental system for the study of chloroplast membrane biogenesis.[1] Various aspects of this process can be demonstrated including

---

[1] I. Ohad, *in* "Membrane Biogenesis" (A. Tzagoloff ed.), pp. 279–350. Plenum, New York, 1975.

light regulation,[2] chlorophyll and protein synthesis, and subcellular origin,[3,4] transport and assembly of various thylakoid polypeptides,[2,5] as well as time sequence of the various events leading to the formation of a functional membrane.[1-6]

The methodology to be used in order to generate *in vivo* membranes containing chlorophyll–protein complexes serving as energy-collecting antennae but lacking photochemical activity and the subsequent synthesis and assembly of the reaction centers has been described in a previous volume of this series.[7] In the present chapter posttranslation modification of supramolecular complexes affecting their assembly and function are described. These modifications include light-dependent protein phosphorylation and turnover of specific thylakoid polypeptides.

Recently, a membrane-bound protein kinase activity has been demonstrated in thylakoids of higher plants[8,9,10] and algae.[11,12] The activation of this kinase *in vitro* is light dependent and controlled by the redox state of the plastoquinone pool.[10] This kinase phosphorylates the polypeptides forming the chlorophyll *a-b* light-harvesting complex (LHC) of photosystem II (PS II), polypeptides of 44–47 kilodaltons forming the PS II reaction center, and the 32–34 kilodalton polypeptide(s) participating in the formation of the PS II specific herbicide-binding site and possibly the Q–B complex.[12,13] A membrane-bound, permanently active phosphatase is responsible for the dephosphorylation of these polypeptides.[14] Based on measurements of fluorescence emission spectra and kinetics[9] and ultrastructural analysis,[10,15] it has been proposed that in the dephosphorylated condition LHC remains mostly associated with and transfers energy to reaction center II. Upon phosphorylation of PS II polypeptides, LHC dissociates from the reaction center II and might become associated with the photosystem I complex and transfers the absorbed energy to this complex. Thus assembly and dissocia-

[2] J. M. Gershoni and I. Ohad, *J. Cell Biol.* **86**, 124 (1980).
[3] S. Bar-Nun, R. Schantz, and I. Ohad, *Biochim. Biophys. Acta* **459**, 451 (1977).
[4] S. Bar-Nun and I. Ohad, *Plant Physiol.* **59**, 161 (1977).
[5] J. M. Gershoni, S. Shochat, S. Malkin, and I. Ohad, *Plant Physiol.* **70**, 637 (1982).
[6] G. C. Owens and I. Ohad, *Biochim. Biophys. Acta* **722**, 234 (1983).
[7] S. Bar-Nun and I. Ohad, this series, Vol. 69, p. 363.
[8] R. Alfonso, N. Nelson and E. Racker, *Plant Physiol.* **65**, 730 (1979).
[9] J. Bennet, K. E. Steinback, and C. J. Arntzen, *Proc. Natl. Acad. Sci. U.S.A.* **77**, 5253 (1980).
[10] P. Haworth, D. J. Kyle, P. Horton, and C. J. Arntzen, *Photochem. Photobiol.* **36**, 743 (1982).
[11] R. Beliveau and G. Bellmare, *Biochim. Biophys. Res. Commun.* **88**, 757 (1979).
[12] G. C. Owens and I. Ohad, *J. Cell. Biol.* **93**, 712 (1982).
[13] S. Shochat, G. C. Owens, P. Hubert, and I. Ohad, *Biochim. Biophys. Acta,* **681**, 21 (1982).
[14] J. Bennet, *Eur. J. Biochem.* **104**, 85 (1980).
[15] D. J. Kyle, L. A. Staehelin, and C. J. Arntzen, *Archiv. Biochem. Biophys.* **222**, 527 (1983).

tion as well as energy distribution between these complexes seem to be regulated by the phosphorylation–dephosphorylation of PS II polypeptides.[10] Furthermore, the function of the 32–34 kilodalton herbicide-binding polypeptide seems also to be affected by its phosphorylation and assembly with reaction center II polypeptides.[13] It has also been demonstrated that this polypeptide, which is coded for by the chloroplast genome and translated by the chloroplast ribosomes,[16-18] turns over significantly faster than all other thylakoid polypeptides.[18,19] The breakdown and synthesis of this polypeptide is light regulated. Membranes depleted of the 32–34 kilodalton polypeptides can be formed by incubation of cells in the light in the presence of chloramphenicol. The polypeptides can be synthesized and assembled into the depleted membrane following removal of the inhibitor and incubation of the cells in the light.

Experiments described to demonstrate the phosphorylation and turnover of the 32–34 kilodalton polypeptide in mature and developing membranes are described below.

II. Materials and Methods

A. Cultivation of Cells

Chlamydomonas reinhardi y-1 cells are grown on a mineral medium containing acetate as a carbon source.[20] Both continuous cultures or batch-type cultures can be used. In order to maintain the pH of the cultures between pH 7.4 and 7.8, 5% CO_2 in air can be bubbled through the cultures at a rate of 5–10 ml/min using a sterile filter to prevent bacterial contamination.[20] Cells will grow reasonably well (10–12 hr/div) in batch cultures without CO_2 bubbling until the pH will rise to about 8.5–9.2. Cells obtained from such cultures should be washed once in fresh growth medium, resuspended in growth medium at a final concentration of 10^7 cell/ml, and incubated in the light (or dark, for dark-grown cells) for 1–2 hr at 25° with stirring before use.

[16] K. E. Steinback, L. McIntosh, L. Bogorad, and C. J. Arntzen, *Proc. Natl. Acad. Sci. U.S.A.* **78**, 7463 (1981).
[17] A. Reisfeld and M. Edelman, *in* "Genome Organization and Expression in Plants" (C. J. Leaver, ed.), pp. 353–362. Plenum, New York, 1980.
[18] A. K. Matoo, U. Pick, H. Hoffman-Falk, and M. Edelman, *Proc. Natl. Acad. Sci. U.S.A.* **78**, 1572 (1981).
[19] Owens, G. C., M. Wettern, N. Lavintman, D. Ish-Shalom, G. Schuster, D. Kirilowsky, and I. Ohad, *FEBS Symp.* **65**, 223 (1982).
[20] I. Ohad, P. Siekevitz, and G. E. Palade, *J. Cell Biol.* **35**, 521, 553 (1967).

B. Greening of Dark-Grown Cells and Preparation of Thylakoid Membranes

Cells grown in the dark for 5–6 generations (5 days) are collected by centrifugation (3000 $g \times 2$ min), washed once in fresh growth medium, resuspended in growth medium at a final concentration of $1-2 \times 10^7$ cell/ml, and exposed to white fluorescent light for 6–8 hr.[20] Formation of membranes devoid of active photochemical reaction centers of photosystem II and I (RC II, RC I) can be obtained by adding a chloroplast translation inhibitor, chloramphenicol (D-*threo* form), at a final concentration of 200 μg/ml.[1,5] Resumption of photochemical activity can be obtained by washing away the inhibitor and further incubating the cells in the dark or in the light with addition of cycloheximide (1 μg/ml). In both condition synthesis of chlorophyll and addition of thylakoid polypeptides of cytoplasmic origin is prevented.[4,5,7]

To obtain thylakoid membranes, cells are washed and resuspended in the appropriate breaking medium at a concentration of $\geqslant 5.0 \times 10^8$/ml and passed through a French pressure cell,[20] or sonicated using an immersion sonicator tip at 0–5°. In the second case, preliminary trials should be carried out to define the minimal sonication conditions that will cause about 95% cell breakage. It is advisable to use repeated short bursts of sonication (5 sec) with interposed cooling periods (15–20 sec) to prevent excessive heating. As one example, sonication 4–5 times for 5 sec each with a Branson sonicator using the microtip at an instrumental setting of 5 or 6 is sufficient. Photosynthetic activities of both photosystems are preserved in these conditions.

Purified membranes can be obtained by the procedure described by de Petrocellis *et al.*[21] Membranes of sufficient purity to be used for thylakoid phosphorylation or turnover assays can be obtained by a modification of this procedure in which only one round of centrifugation is used but an additional layer of 1.75 M sucrose is added to the discontinous gradient of 2.0, 1.5, and 1.0 M sucrose.[12] When quick isolation of membranes is necessary, as in the case of measuring kinetics of phosphorylation *in vivo*, the homogenate can be overlayered on a 2-step sucrose gradient (1.0 and 2.0 M sucrose) prepared in Tricine (50 ml) containing 50 mM NaCl, pH 7.4, in 2 ml conical plastic tubes of a minicentrifuge (Beckman or Eppendorf type) and spun at top speed for 10–15 min in the cold. In this case the green layer at the 2.0 M interphase is collected. Membranes prepared in this way are contaminated to various degrees with other cellular membranes showing as lightly stained bands in the MW range above 65,000–68,000. The yield of membranes obtained by this procedure can be increased if 2 mM Mg^{2+} is added to the homogenization medium and the sucrose gradient.

---

[21] B. de Petrocellis, P. Sieckewitz, and G. E. Palade, *J. Cell Biol.* **44**, 618 (1970).

C. Preparation of Grana and Stroma Lamellae

Subfractionation of thylakoid membranes into grana and stroma lamellae is accomplished by digitonin treatment.[22] Purified thylakoids prepared as described above[21] are suspended in phosphate buffer (1.5 mM, pH 7.4) containing $MgCl_2$ (10 mM), benzamidine (2 mM), and $NaMoO_4$ (20 mM) at a concentration of 500 μg chlorophyll per milliliter. Digitonin is added to a final concentration of 0.5% (w/v), and the membranes are incubated in the dark at 4° with continuous stirring for 30 min. Chloroplast fractions are separated by differential centrifugation, 1000 g for 10 min, 10,000 g for 30 min, 100,000 g for 30 min, and 144,000 g for 60 min. The pellets obtained at the second and last step are collected (10K fraction or grana and 144K fraction or stroma).

D. Fractionation of Thylakoid Membrane Polypeptides and Chlorophyll Protein Complexes

Chloroplast thylakoid membrane proteins are fractionated by LDS–PAGE according to the methodology of Chua.[23] In long-term labeling experiments in which membrane phospholipids are significantly labeled membranes are precipitated with TCA (10%) delipidated with acetone (90%) and diethyl ether prior to electrophoresis. Gels are stained for protein with Coomassie Brilliant Blue R, destained, dried, and exposed to X-ray film. When the total radioactivity loaded per sample is less than about 10,000 dpm, an enhancement screen might be used to reduce exposure time. Fluorography is recommended when using ^{14}C or ^{35}S radioisotopes.

E. Determination of Specific Radioactivity, Chlorophyll, and Protein Concentrations

To determine the amount of isotope incorporated into total cellular or membrane proteins, aliquots of labeled material are precipitated onto disks of Whatman 3MM cellulose paper with cold chloroacetic acid (10%). The disks are washed once with cold (10%, w/v), then once with hot TCA (90° for 20 min). Lipids and pigments are extracted with acetone and the disks are finally washed in diethyl ether and dried prior to counting in scintillation fluid. Total chlorophyll is quantitated by the method of Arnon,[24] and the concentrations of chlorophyll a and b separately are measured as described

[22] J. M. Anderson and N. K. Boardman, *Biochim. Biophys. Acta,* **112,** 403 (1966).
[23] N. H. Chua, this series, Vol. 60, p. 434.
[24] D. I. Arnon, *Plant Physiol.* **24,** 1 (1979).

by Mackinney.[25] Protein concentration is determined according to Lowry *et al.*[26] using bovine serum albumin as a standard.

F. Herbicide Binding

Tritiated DCMU or atrazine of specific radioactivity of at least 25 μCi/μmol can be used in the binding assay. Isolated chloroplast membranes (50 – 100 μg chlorophyll) are incubated in the presence of the ligand for 3 min at 25° in Tris-HCl buffer (50 mM, pH 8.0) with addition of KCl (30 mM). The membranes are sedimented (10,000 g for 10 min) and aliquots of the supernatant are removed and counted in Triton X-100 enriched scintillation fluid. The data are analyzed using double reciprocal plots of free vs bound herbicide, and, from the straight lines, fitted by linear regression, binding constants and the number of binding sites can be obtained.[13]

G. *In Vivo* Pulse Labeling with [^{32}P]Orthophosphate or [1-^{14}C]Acetate

Cells are harvested, washed once, and resuspended in Tris-HCl buffered growth medium containing only 10 μM potassium dihydrogen phosphate to a final concentration of 2×10^7 cells per milliliter (nondividing conditions[20]). Aliquots of 25 ml are preincubated for 15 min at 25° with continuous agitation prior to the addition of the radioisotope. Either [^{32}P]orthophosphate or [1-^{14}C]acetate are then added to give final specific activities of 0.5 Ci/mmol and 0.01 Ci/mmol, respectively, and the incubation is continued for a further 30 min. Illumination, when required, is provided by cool fluorescent lamps (1.6×10^4 ergs cm^{-2} s^{-1}). In pulse-chase experiments the labeled cells are pelleted, washed twice, and resuspended in medium containing 10 mM potassium dihydrogen phosphate and 10 mM sodium acetate. Thylakoid membranes are obtained by sonication in an ice bath as described above (see Section II,B). All manipulations are carried out at 4° and sterilized glassware should be used for the incubations.

H. *In Vitro* Labeling with [γ-^{32}P]ATP

Membranes for the *in vitro* assay of protein kinase are prepared by the modified procedures of de Petrocellis *et al.*[21] Only membranes at the 1.5 M : 1.75 M interface are collected, washed, and pelleted (see Section II,B).

---

[25] G. MacKinney, *J. Biol. Chem.* **140**, 315 (1971).
[26] O. H. Lowry, N. J. Rosebrough, A. L. Farr, and R. J. Randall, *J. Biol. Chem.* **193**, 265 (1951).

Labeling of thylakoid membranes is carried out at 25° with illumination provided by a quartz halogen fiber light (Dolan Jenner Industries, Inc., Woburn, Massachusetts, 2.5×10^6 ergs cm^{-2} s^{-1}). The reaction mixture contains Tris-HCl (50 mM, pH 8.0), MgCl$_2$ (10 mM), ATP (0.05 mM), and [γ-^{32}P]ATP to give a final specific radioactivity of 0.15 Ci/mmol. The reaction is terminated by addition of an equal volume of a solution containing SDS (5%), 2-mercaptoethanol (70 mM), EDTA (50 mM), KH$_2$PO$_4$ (25 mM), and ATP (25 mM), and aliquots are chromatographed on Whatman DE-81 paper with ammonium formate (200 mM, pH 5.5) as the developer. For electrophoretic separation of polypeptides the reaction is terminated by rapid centrifugation in an Eppendorf microfuge followed by resuspension in electrophoresis sample mixture. For labeling times longer than 5 min, benzamidine (2 mM) and sodium molybdate (20 mM) are added to the reaction mixture to prevent possible proteolysis and dephosphorylation. Additional details are given in Ref. 12.

I. Dephosphorylation *in Vitro*

Dephosphorylation of isolated thylakoid membranes phosphorylated *in vivo* or *in vitro* is performed at 25° in the dark in Tris-HCl (50 mM, pH 8.0), MgCl$_2$ (10 mM), ZnCl$_2$ (0.1 mM), and benzamidine (2 mM). The reaction is stopped as described above (see Section II,H).

J. Identification of the Phosphorylated Polypeptides

The phosphorylated bands can be divided into three groups according to molecular weight:

1. Two and possibly three polypeptides in the molecular size range of 32 to 35 kilodaltons corresponding to polypeptides D1, 9/10 (the individual polypeptides are designated by the numbers proposed by Chua[23])
2. Four polypeptides in the molecular size range of 22 to 28 kilodaltons probably corresponding to 11, 12, 15, and 17
3. At least one polypeptide in the molecular weight size of 12 to 20 kilodaltons, possibly polypeptide 24.

Recent results have shown that a polypeptide of 32–34 kilodaltons possibly equivalent to polypeptide D1 in *Chlamydomonas* constitutes the binding site for atrazine in higher plants[16] and DCMU in cyanobacteria.[27] In addition polypeptides 10, 14, and 15 have been shown to belong to the

[27] C. Astier and F. Joset-Espardelier, *FEBS Lett.* **129**, 77 (1981).

antenna of photosystem II.[28] It is well established that polypeptides 11 and 17 are components participating in the formation of the LHC[3,29], and their assignment as phosphoproteins was confirmed first by electrophoresing the isolated chlorophyll–protein complex II following heating, and second by chloroform:methanol extraction.[23] Phosphorylation of polypeptide 12 was revealed after chloroform:methanol extraction of polypeptides 11, 14, 15, 16, and 17. Polypeptide 12 appears to be a component of photosystem II, as evidenced by its absence in the PS II-deficient *Chlamydomonas* F34 mutant[30] and in membranes of they *y-1* mutant lacking photosystem II due to the inhibition of chloroplast translation by chloramphenicol during greening.[4]

To ascertain whether these ^{32}P-labeled membrane polypeptides are indeed phosphoproteins, gel slices containing labeled polypeptides can be treated with TCA (10%, 10 min, 90°), NaOH (1 M, 10 min, 90°), and hydroxylamine succinate (1 M, pH 7.0, 30 min, 37°).[31] The radioactivity is only removed from the slices treated with NaOH, as expected for esterified phosphate. Label is also completely removed following treatment with *E. coli* alkaline phosphatase (15 units per milligram membrane protein) and bovine pancreatic trypsin (0.5 mg per milligram membrane protein).[12] Phosphorylated membranes can be hydrolyzed in HCl (6 M) *in vacuo* for 3 hr[32] and the hydrolyzate chromatographed on Whatman 3MM paper with butanol:acetic acid:water (60:15:25 v/v/v) as the developer. Autoradiography of the chromatogram reveals the presence of phosphothreonine and a smaller amount of phosphoserine as determined by the use of appropriate markers. Phosphoserine is known to be less resistant to acid treatment than phosphothreonine.[32]

In order to establish whether the turnover of the phosphate esterified to the thylakoid membrane proteins is independent of protein turnover, the same population of light-grown y-1 cells can be pulse labeled in the light with either [1-^{14}C]acetate or [^{32}P]orthophosphate and subsequently chased in the light for 1 and 4 hr. Membrane polypeptides are separated as in Section II,G and the slab gels containing the labeled polypeptides autoradiographed as in Section II,D. The results show that turnover of the esterified phosphate is not coupled to protein turnover per se. The half-life of the esterified phosphate appeared to be less than 60 min in the light, and when the chase of [^{32}P]orthophosphate pulse-labeled cells is carried out in the dark the turnover appeared to be even faster.[12]

[28] P. Delepelaire and N. H. Chua, *J. Biol. Chem.* **256**, 9300 (1982).
[29] F. Blomberg and N. H. Chua, *J. Biol. Chem.* **254**, 215 (1979).
[30] N. H. Chua and P. Bennoun, *Proc. Natl. Acad. Sci. U.S.A.* **72**, 2175 (1978).
[31] J. Avruch, G. R. Leone, and D. B. Martin, *J. Biol. Chem.* **251**, 1505 (1976).
[32] B. B. Bylund and T. S. Huang, *Anal. Biochem.* **73**, 477 (1976).

K. *In Vivo* Pulse Labeling and Chase with ^{35}S

For labeling of the fast turning over 32–35 kilodalton polypeptide, cells are suspended in fresh growth medium (2.0×10^7 cells/ml) in which the sulfate concentration has been reduced to 0.01 μmol/ml. After a short (20 min) preincubation at 25° in the light or dark (according to the following incorporation conditions) $^{35}SO_4^{2-}$ is added to give a final specific radioactivity of 5–10 μCi/μmol. Incubation is continued with stirring for the desired period of time (15 min to 2 hr) and incorporation stopped by addition of nonradioactive sulfate at a final concentration of 10 μmol/ml followed by centrifugation and washing of the cells in normal growth medium twice. Cells are then processed for membrane preparation as above (Section II,B) or further incubated in the same medium (chase) for various lengths of time (7–24 hr). Detection of ^{35}S-labeled polypeptides can be carried out following electrophoretic separation and fluorography as above (Sections II,B and II,D).

III. Modulation of Thylakoid Polypeptide Phosphorylation Pattern in Mature and Developing Chloroplasts

A. Phosphorylation in Mature Chloroplasts

Pulse labeling of cells in white light with ^{32}P or to phosphate results in the labeling of all the above-mentioned PS II polypeptides. Similar results are obtained if incorporation of ^{32}P is carried out in the dark or in red or far-red light (650 nm or 710 nm, respectively).[12] The label is quickly chased in all incubation conditions if the cells are transferred to nonradioactive medium (half-life 30 min).[12] As opposed to these results, labeling *in vitro* with ^{32}P and ATP is light dependent and inhibited by photosynthetic electron flow inhibitors such as DCMU or DBMIB.[12] These results can be explained if one considers that the plastoquinone pool *in vivo* is partially reduced even in the dark[33] and thus the thylakoid protein kinase remains in an active state in all illumination conditions. However, this is not the case in isolated chloroplast membranes in which the plastoquinone pool reduction depends solely on electron flow via PS II.

Fractionation of ^{32}P-labeled thylakoids into grana and stroma fraction shows a difference in the phosphorylation pattern. First, the amount of LHC polypeptides is significantly reduced in the stroma fraction and the ratio of phosphorylation of the lower mass polypeptide (22, 26 kilodaltons) is higher relative to the 28 kilodalton polypeptide.[16] Possibly the LHC present in the grana is mostly found in the oligomeric form[34] tightly bound within the PS II

[33] P. Bennoun, *Proc. Natl. Acad. Sci. U.S.A.* **79**, 4352 (1982).
[34] J. H. Argiroudi-Akoyonoglou and G. Akoyonoglou, *FEBS Lett.* **104**, 78 (1979).

complex and thus less available to the kinase. In addition, most of the reduction of the plastoquinone pool occurs in the grana where most of the PS II complex is segregated. This will reduce the amount of active kinase (see Section I). The kinase seems to be associated with the PS II complex as indicated by its partition with isolated PS II enriched particles[35] obtained by detergent fractionation of membranes.[36]

B. Changes in the Polypeptide Phosphorylation Pattern during Synthesis and Assembly of Chloroplast Membranes

Residual chloroplast membranes present in dark-grown cells[20] are not phosphylated *in vivo.*[6] During the early phase of greening where mostly reaction centers are formed[1] the 44–47 kilodalton polypeptides are preferentially phosphorylated. In the later phase of the greening when both LHC and the chlorophyll–protein complexes forming the PS I antennae are accumulated the LHC polypeptides become phosphorylated and the general phosphorylation pattern acquires the characteristics of the mature membrane (Fig. 1). This stage corresponds to the formation of completely assembled PS II and PS I units and their segregation into grana and stroma lamellae.[16]

The 32–34 kilodalton polypeptides are phosphorylated at all stages of the greening process. Measurements of the DCMU binding indicates that the affinity of the binding site for this ligand is practically constant throughout the development process while the amount of binding sites on a chlorophyll basis increases, indicating continuous synthesis and assembly of binding sites.[19] Dephosphorylation of the membranes *in vitro* causes a reduction in the affinity of ligand binding and an increase in the nonspecific binding.[13] The presence of an active reaction center II is also required for the formation of the high-affinity herbicide-binding site.[13]

These results indicate that the assembly of a PS II complex exhibiting photosynthetic electron flow and susceptible to DCMU inhibition requires the presence of the 44–47 kilodalton polypeptides and the phosphorylated 32–34 kilodalton polypeptides. This complex appears to be formed almost from onset of the growing process prior to the accumulation of LHC. The synthesis and assembly of the LHC into a complete PS II unit and the modulation of this assembly process by phosphorylation of the LHC polypeptides occur mostly toward the end of the greening process.

The LHC of membranes formed in the presence of chloramphenicol which lack the polypeptides of RC II and the 32–34 kilodalton polypeptides are not phosphorylated.[6] Addition of the reaction centers without the addition of the 32–34 kilodalton polypeptides using the procedure de-

[35] J. C. Owens, Ph.D. Thesis, Hebrew University, Jerusalem, Israel, 1982.
[36] B. A. Diner and F. A. Wollman, *Eur. J. Biochem.* **110,** 521 (1980).

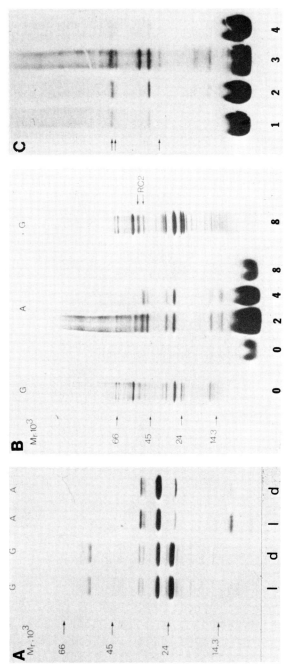

FIG. 1. Phosphorylation of thylakoid polypeptides *in vivo* in normal and developing chloroplasts. (A) Light-grown cells pulse labeled in the light (L) or dark (D). (B) Dark-grown cells greened in the light for various times as indicated (hr), then pulse labeled for 30 min in the light. (C) Dark-grown cells exposed to the light with addition of chloramphenicol (200 μg/ml) for 6 hr, then washed free of the inhibitor and pulse labeled in the light (1) or after further incubation for 8 hr in the dark (2), light (3), or light with addition of cycloheximide (2 μg/ml).[4] RC II, polypeptides 44–47 kilodaltons of photosystem II reaction center; LHC, the lower molecular weight polypeptides of the light-harvesting complex; G, Coomassie Brilliant Blue stained gels; A, autoradiograph.

scribed in Section II,B or addition of all these polypeptides without the formation of chlorophyll–protein interconnecting antenna serving as a link between LHC and RC II[5] does not result in the assembly of a normal PS II unit and in the phosphorylation of its polypeptides.[6]

C. Turnover of the 32–34 Kilodalton Polypeptides and Assembly of the PS II Unit

Pulse labeling of nondividing cells with $^{35}SO_4^{2-}$ in the light for 10–20 min results in the preferential labeling of the 32–34 kilodalton polypeptide bands and only a slight labeling of the LHC and other membrane polypeptides. As the labeling duration is increased, this difference diminishes. The labeling of the 32–34 kilodalton polypeptide reaches a plateau after about 2 hr. The radioactivity of the 32–34 kilodalton polypeptides is preferentially reduced following a chase period of 12–24 hr in nondividing cells incubated in the light (Fig. 2). The half-life time of this polypeptide(s) can be estimated as 4–7 hr. Practically no loss of radioactivity is noticed from the other thylakoid polypeptides under these conditions. The half-life time of thyla-

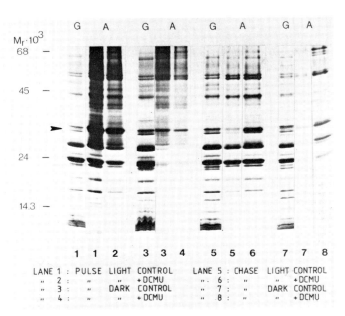

FIG. 2. Turnover of the 32–34 kilodalton polypeptide(s) in *Chlamydomonas reinhardi* y-1 cells. Light-grown cells were pulse labeled in the light or dark for 30 min with $^{35}SO_4^{2-}$; DCMU was added to a final concentration of 5×10^{-6} *M*; cells labeled in the light (as in lane 1) were then "chased" for 12 hr as indicated; membranes were prepared and fluorography of the LDS–PAGE resolved polypeptides was carried out as described in Section II.

koid polypeptides in dividing cells is similar to the duration of one generation (8–12 hr in the light; 20–22 hr in the dark). Measurements of the initial rate of radioactivity loss from the 32–34 kilodalton polypeptide labeled in short pulses (10–15 min) show no loss of radioactivity during the first 1.5–2.5 hr. This phenomenon can be explained if one considers that the insertion of the newly synthesized polypeptides might occur at the ribosomal binding site on the stroma lamellae.[37] The time required to reach the site of degradation might be correlated with transfer of the newly synthesized polypeptide to the grana and its assembly and phosphorylation within active PS II units. One should note that, unlike higher plants in which the herbicide-binding polypeptide is inserted as a precursor of the 34.5 kilodalton form and then processed to the mature form (32.5 kilodaltons)[16,18], no such phenomenon is observed in *Chlamydomonas* even when pulse labeling is of short duration (5–10 min). The turnover of this polypeptide in nondividing cells in completely light dependent and inhibited by DCMU, indicating that its proteolysis is induced by or follows changes occurring during active electron flow via PS II. It has been reported that the 32–34 kilodalton polypeptide is susceptible to trypsin digestion which results in the formation of several intermediates until a fragment of about 16 kilodaltons is obtained.[16] Such a fragment does not appear to accumulate in the membranes following turnover of the 32–34 kilodalton polypeptide *in vivo.* Degradation of the 32–34 kilodalton polypeptide in the light can be demonstrated also in isolated thylakoids *in vitro,* indicating that the protease involved might be membrane bound.

IV. Conclusions

The experimental procedures and summary of results obtained as described here demonstrate the adequacy of this experimental system for the investigation of the process of assembly and modulation of photosystem II organization in *Chlamydomonas* thylakoids.

The conclusions presented regarding (1) the mechanism of protein kinase activation, (2) the role of LHC phosphorylation and the reversible association–dissociation with the PS II reaction center, and (3) the role of 32–34 kilodalton polypeptide phosphorylation and its association with RC II to form an active PS II unit containing a high-affinity binding site for herbicides such as DCMU are rather tentative. Particularly, the light activation of the 32–34 kilodalton polypeptide turnover, including its synthesis, assembly, and degradation, is important for those interested in the genetic engineering of herbicide-resistant crop plants. The study and understanding

[37] M. A. Michaels, personal communication.

of these phenomena are still in infancy and we believe that the experimental system described above can be extremely useful in pursuing this line of research.

Acknowledgments

This work was supported partially by a grant from the government of Niedersachsen, Federal Republic of Germany, awarded to G. Galling, Institute of Botany, Braunschweig University, Braunschweig, and I. Ohad. We wish to thank Dr. Galling for his advice and encouragement in performing the work on the turnover of the 32–36 kilodalton polypeptide.

Section IV

Summary of Membrane Proteins

[49] Membrane Proteins: A Summary of Known Structural Information

By DAVID R. NELSON and NEAL C. ROBINSON

Our knowledge concerning the structure of intrinsic membrane proteins has greatly advanced during the past decade due to the purification and characterization of hundreds of these proteins. Unfortunately, much of the information about the three-dimensional structure, size, shape, molecular weight, amino acid sequence and amino acid composition of these proteins is scattered throughout the literature making comparisons difficult. In order to facilitate the study of this class of proteins, we have compiled a list of the intrinsic membrane proteins that have been purified, together with literature references that will direct the reader to either review articles or literature reports containing this type of structural information.

We have divided the proteins that are listed here into two categories based upon the extent to which they have been characterized. The proteins that have been well characterized, e.g., those proteins for which a reasonably accurate structural model has been determined, are listed in Table I. The remainder of the proteins, e.g., those that have only been purified or for which limited structural information is available, are listed in Table II. In addition, a few extrinsic membrane proteins have also been included in these tables if the amount of data that had been collected about them warranted their mention, e.g., clathrin and the extrinsic proteins of the erythrocyte membrane.

Molecular weight values for each protein are listed in Tables I and II. Nearly all of the included values were obtained by one of the following three methods of molecular weight determination: (*a*) sedimentation equilibrium centrifugation with correction for detergent bound to the protein; (*b*) calculation from either the known amino acid or nucleic acid sequence; or (*c*) polyacrylamide gel electrophoresis in sodium dodecyl sulfate. Molecular weight values that were obtained by other methods, e.g., gel filtration or sucrose density gradient centrifugation, were generally not included, since such determinations are dependent upon the detergent that is used, the amount of bound detergent, and the shape of the protein itself. We have, however, made reference to that type of information if other more accurate measurements had not been reported. The reader should be aware that such data are more useful if used to evaluate the size and shape of a solubilized protein rather than to evaluate the protein molecular weight. The excellent articles by Tanford *et al.* (1) and Reynolds and Tanford (2) should be

METHODS IN ENZYMOLOGY, VOL. 97

consulted for the proper methods to be used for determining accurate protein molecular weights.

Although a great deal of physical information is available for many of the proteins listed in these tables, we decided not to list the known values for the sedimentation, diffusion, and frictional coefficients, radii of gyration, Stokes' radii, hydrodynamic volumes, or values for bound detergent. We imposed this limitation since these values are so dependent upon the detergent that has been used in their evaluation. We felt that it would have been more confusing than useful to list multiple conflicting results. However, literature references to this information are given so that the reader can accurately assess the methods used when the values were collected.

We may have neglected to include references to some structural studies that have been reported in the literature and some intrinsic membrane proteins that have been purified; we regret any oversights.

TABLE I
WELL-CHARACTERIZED MEMBRANE PROTEINS[a]

Protein	$\dfrac{M_r^{\,b}}{10^{-3}}$	M_r refs.	Puri-fication	Hydrodynamic properties; detergent[c] binding	AA comp.	AA seq.[d]	3-D structure; EM[e]	Other structural studies[f]	Reviews
Bacterial outer membrane proteins									
1. Outer membrane protein F of *E. coli* (Omp F) Exists as a trimer of identical subunits	37.205s	3–6	3,4	3,4	3	6(7)	3–5,8	3,9	10
2. Outer membrane lipoprotein of *E. coli* Monomeric protein	7.174s	11	—	—	—	11	—	—	11,12
3. Outer membrane proteins C, D, and F of *Salmonella typhimurium* Exist as trimers of identical subunits	C: 39.8 D: 38.0 F: 39.3	13	13	14	13	—	—	—	—
Brush border hydrolases									
1. Aminopeptidases A and N (porcine intestine and kidney; rabbit intestine) Exists as a dimer of identical subunits: Molecular weight species dependent	A: 120 N: 120–160	15–21	15–21	—	15,16 18,19	22i	17	—	23
2. Sucrase isomaltase (porcine, rabbit, rat intestinal) Contains 2 different subunits derived from a single precursor	140–150	24	24	25	—	26i	27	27–29	—
Erythrocyte membrane proteins									
1. Glycophorins Usually exist as dimers in the membrane	29.0h	30,32,33	30–32	33,34	—	35,36	—	37–39	40,41

(Continued)

TABLE I (*Continued*)

Protein	$\dfrac{M_r}{10^{-3}}$ [b]	M_r refs.	Purification	Hydrodynamic properties; detergent[c] binding	AA comp.	AA seq.[d]	3-D structure; EM[e]	Other structural studies[f]	Reviews
Histocompatibility antigens									
1. H-2K[b]	45.5/7s 13.0	42	42	—	—	42,43	44	—	43,44
Contains a membrane-bound heavy chain and β_2-microglobulin									
Immunoglobulins: membrane bound									
All three classes are divalent (2 membrane-bound heavy chains and 2 light chains). IgD may also exist in a monomeric form. M_r of light chains is 22,500.	—	—	—	—	—	—	—	—	45,46
1. IgD heavy chain	48.5	45	—	—	—	(45)	—	—	—
2. IgG heavy chain	70	46	—	—	—	(46)	—	—	—
3. IgM heavy chain	67	48	—	—	—	(47)i	—	48	—
Microsomal proteins									
1. Cytochrome b_5 (microsomal membrane) Monomeric protein	16.2h 16.7	49,50	49,50	51,52	—	53–59	60,61	51,62–65	66,67,98
2. NADH-cytochrome b_5 reductase Monomeric protein	43	68,69	68,69	—	68	70i	—	71,72	66
Mitochondrial inner membrane electron transport complexes									
1. Succinate: ubiquinone oxidoreductase (also known as Complex II) Contains 4–6 polypeptide subunits, depending upon the preparation.	—	See Col. 10	See Col. 10	77,78	76	—	—	—	79–82
a. Succinate dehydrogenase (2 subunits)	70,27	See Col. 10	—	—	—	—	—	—	79–82
b. Cytochrome b_{560}	15,13.5	73	73	—	73	—	—	—	—

Component	Mol. wt. ($\times10^{-3}$)								
c. Q binding protein	15	75	75	—	75	—	—	—	—
d. Proteins C_{II-3} and C_{II-4}	13.5,7.0	74	74	—	—	—	—	—	—
2. Ubiquinone:cytochrome c oxidoreductase (also known as Complex III or the cytochrome bc_1 complex) Contains 8 or 9 different polypeptide subunits: 2 Rieske core proteins, cytochrome b protein(s), cytochrome c_1, Rieske iron sulfur protein, and 3 smaller protein subunits.	440–550h	77,83	See Col. 10	77,83	110,111	—	84–88	115	89–92
a. Cytochrome b protein(s)	45s	94,95	112	93	110–113	(95,96)	—	—	97–99
	62h	93	93–94	—	93	—	—	—	—
	30	93	—	—	—	—	—	—	—
b. Cytochrome c_1 protein	29–30.6	100–105	100–105	105	100–105	106	—	—	—
	27.874s	106,114	114	—	—	—	—	—	—
	24.5	107,108	107,108	—	—	—	—	—	—
c. Rieske iron-sulfur protein (also known as oxidation factor)		77,116–118	See Col. 10	—	—	—	—	—	109
3. Cytochrome c oxidase (mitochondrial inner membrane) In nondenaturing detergents the complex exists as a dimer of the heme aa_3 monomeric unit, which contains 7 nonidentical polypeptide subunits; the molecular weights of the individual subunits is species dependent.	320–385h	—	See Col. 10	77,116–118	—	119–122b	123,124	See Col. 10	125–127
Receptors									
1. Acetylcholine receptor (isolated from the electric organs of *Torpedo californica* and *Electrophorus electricus*) Contains 4 different polypeptide subunits with the stoichiometry: $\alpha_2\beta\gamma\delta$	250h	135,136, 152	See Col. 10 and 128–130	131–136	137	(138),(139)i, 152	136,140–144	145–147	148–153

(Continued)

TABLE I (*Continued*)

Protein	$\frac{M_r{}^b}{10^{-3}}$	M_r refs.	Purification	Hydrodynamic properties; detergent[c] binding	AA comp.	AA seq.[d]	3-D structure; EM[e]	Other structural studies[f]	Reviews
Transport proteins									
1. ADP/ATP carrier (isolated from mitochondrial inner membrane) Exists as a dimer of identical subunits	30 63h	154 155	154	155	154	156i	—	—	157,158
2. H⁺ translocating ATPase (isolated from bacterial plasma, chloroplast, and mitochondrial inner membranes) The complex usually contains 7–10 different polypeptide subunits, although the mammalian ATPase contains 12–16 subunits. The hydrophilic, enzymic F_1 portion contains 5 subunits with the stoichiometry: $\alpha_3\beta_3\gamma\delta\epsilon$ The F_0 membrane-embedded portion contains the other 2–5 subunits (7–11 subunits in the mammalian enzyme). The M_r of the F_0 subunits are dependent upon the source, and the stoichiometries are not agreed on.	— F, 325–380, 382s	See Col. 10	See Col. 10	159–161	See Cols. 7,10	(96),162–164 (165–172)	173-177	See Col. 10	178–187, 162, 223 179,223
3. Ca²⁺,Mg²⁺-ATPase (sarcoplasmic reticulum) Usually exists as a monomer of a single subunit in detergent, but there is evidence for an oligomeric structure in the membrane.	102	189,191–193	188,190	191–195	189,190	196–199i	190,200	201–206	207,208 179,223

4. Na+,K+-ATPase (renal medulla, shark rectal gland, and *Electricus*) The purified ATPase contains 2 different subunits and has a stoichiometry of either $(\alpha\beta)_2$ or $\alpha_2\beta_4$ in detergent solution; a proteolipid is also associated with it in the membrane.	265–380h	216,217	210–212, 214,215	34,216–218	209,212, 214,218	—	219,220	—	179,215 221–224
5. Bacteriorhodopsin (*Halobacterium* purple membrane) Exists as a trimer of identical subunits	25.1	235	See Col. 10	225	—	226,227 237	228–233	234	179,223 235–238
Virus proteins									
1. fl coat protein (also known as fd or M13 coat protein) Exists as a dimer of identical subunits in phospholipid bilayers or in detergent solution	5.240s	239,240	239,240	239,240	—	240–242	243–244	239,240, 244–247	—
2. Pfl coat protein	5.0	248	248	—	—	248,249	248,249	—	—
3. Influenza virus hemagglutinin Exists as a trimer of identical subunits	74.880s	250	250	—	—	248,249 (251) 252 (253,254)	250	—	250,255
4. Semliki forest virus glycoproteins E₁, E₂, and E₃ These proteins are derived from a common precursor; the viral spike contains one copy of each protein	E₁ 51 E₂ 52 E₃ 11	256	257–259	258	257,259	256	—	259	—
Visual proteins									
1. Rhodopsin and opsin Exists as a monomer in detergent solution	38.5	264	260,261, 263,264	34,260–264	—	—	263,265	266–268	269

(Continued)

TABLE I (*Continued*)

Proteins	$\dfrac{M_r}{10^{-3}}$ [b]	M_r refs.	Puri-fication	Hydrodynamic properties; detergent[c] binding	AA comp.	AA seq.[d]	3-D structure; EM[e]	Other structural studies[f]	Reviews
Miscellaneous proteins									
1. Clathrin	180	270,272	See Col. 10	270	—	—	271,272	270,273	272,274
Associates into pentamers and hexamers that form basket-like lattices composed of 84–1000 monomers									
2. Crambin (a plant seed protein)	4.720s	275	286	—	—	275,276	275,276	277	—
Monomeric in detergent solution									
3. Gap junction protein	28.0	278	278	—	—	278i	278–281	278	—
Exists as a hexamer of identical subunits									
4. Melittin (bee venom)	2.840s	282	282,283	284	—	282,284, 285,287	282,285, 287–288	283,284, 289,290	—
Exists as a tetramer in aqueous solution; probably exists as a monomer in the phospholipid bilayer									

[a] The numbers in columns 3–10 refer to entries in the reference list. Abbreviations: Cols. 2 and 3 — M_r, molecular weight. Cols. 6 and 7 — AA comp., amino acid composition; AA seq., amino acid sequence. Col. 8 — 3-D, three-dimensional; EM, electron micrographs.

[b] Molecular weight values that are followed by "s" were calculated from the known amino acid or nucleic acid sequence; those followed by "h" were determined by hydrodynamic methods (usually sedimentation equilibrium centrifugation correcting for the bound detergent). Values that are not followed by a letter are based on polyacrylamide gel electrophoresis in sodium dodecyl sulfate. The references for these values are given in the adjacent column (Col. 3).

[c] In this column (Col. 5) are references to articles describing hydrodynamic measurements. The following parameters are evaluated in these articles: sedimentation, diffusion, and frictional coefficients; radii of gyration; Stokes' radii as well as evaluation of the protein molecular weight by a wide variety of techniques, e.g., sedimentation equilibrium, sucrose density centrifugation, sedimentation velocity, light scattering, and gel filtration. Some of the values reported in these references are more reliable than others, since some of the measurements were not corrected for bound detergent or for the nonspherical shape of the proteins. Detergent and phospholipid binding measurements have also been included in this column.

[d] References not in parentheses refer to reported amino acid sequences; those within parentheses refer to reported nucleic acid sequences. An "i" following a reference number indicates that the sequence is incomplete.

[e] The three-dimensional structures referred to in these reports were obtained by X-ray diffraction, electron diffraction, computer-enhanced image reconstruction of electron micrographs, and neutron diffraction. References are made also to published electron micrographs of sufficient quality to show individual protein molecules.

[f] The references listed below describe biochemical and physical studies that are not listed in the other columns, e.g., nuclear magnetic resonance, circular dichroism and neutron diffraction studies, denaturation studies of domain structure, topology investigations using vectorial labeling and limited proteolysis, peptide mapping, and some isoelectric points.

TABLE II
Purified Membrane Proteins[a]

Protein	M_r	M_r refs	Purification	Hydrodynamic properties; detergent binding	AA comp.	AA seq.	3-D structure and EM	Other structural studies	Reviews
Adenylate cyclase and related proteins									
1. Adenylate cyclase	220,000h	291–294	295	291–293	—	—	—	294	296
2. (G/F) Adenylate cyclase regulatory protein	130,000	294,297,298	297	298	—	—	—	—	—
3. Adrenergic receptors (listed under Receptors)									
4. Guanylate cyclase	182,000	299	299	—	—	—	—	—	—
Bacterial membrane proteins									
1. Outer membrane proteins (Omp) of E. coli									10
a. Omp A	35,159s	301	—	—	—	301(302)	—	—	—
b. Omp C	—	—	303	—	—	—	303	9	—
c. Omp F (refer to Table I)									
d. Lam B	47,400s	304,305	305,306	—	—	304	—	—	—
e. Nmp AB	40,000	307	307	—	—	—	—	—	—
f. Nmp C and Lc protein (protein 2)	—	—	308	—	—	—	—	—	—
g. Colicin E receptor; also receptor for phage BF23 and vitamin B₁₂	60,000	309	309	—	—	—	—	—	—
h. Colicin K receptor; also receptor for phage T6 (Tsx protein)	26,000	310	310	—	310	—	—	—	—
i. Colicin M receptor; also receptor for ferrichrome and phages φ80, T1, T5	85,000	311	311	—	—	—	—	—	—
j. Lipoprotein (refer to Table I)									

2. Outer membrane proteins of other bacteria

	MW							
a. Omp of *Chlamydia trachomatis*	39,500	312	312	—	—	—	—	—
b. Omp of *Chromatium vinosum*	42,000	313	313	313	313i	313	313	—
c. Omp I and IIa of *Neisseria gonorrhoeae*	36,000	314	314	—	—	—	—	—
d. Omp of *Neisseria meningitidis*	40,000	315	315	—	—	—	—	—
e. Omp of *Proteus mirabilis*	39,000	316	316	—	—	—	—	—
f. Omp of *Pseudomonas aeruginosa*		317	317	317	—	317	317	—
g. Omp C, D, and F of *Salmonella typhimurium* (refer to Table I)	Multiple	—	—	—	—	—	—	—

3. Plasma membrane proteins

  a. Electron transport and oxidative phosphorylation complexes

	MW							
i. H^+-translocating ATPases (refer to Table I)	See Col. 10	—	—	—	—	—	—	—
ii. Cytochrome *c* oxidase (2 subunits)	110,000	—	—	—	—	—	—	318,319
iii. Formate dehydrogenase	Multiple	320	320	320	—	—	—	—
iv. Fumarate reductase (2 subunits)	Multiple	321	321	321	—	—	—	—
v. Lactate dehydrogenase	74,000	322	322	322	—	—	—	—
vi. Nitrate reductase (2 subunits)	—	—	—	—	—	—	—	323

  b. Penicillin binding proteins

	MW							
i. Four penicillin binding proteins of *Bacillus megaterium*	—	—	324	—	—	—	—	—
ii. Penicillin binding proteins of *B. stearothermophilus* and *B. subtilis*	Multiple	325–327	325–327	328i	—	—	—	—
iii. Two D-alanine carboxypeptidases I of *E. coli*	Multiple	329	329	—	—	—	—	—

(Continued)

TABLE II (Continued)

Protein	M_r	M_r refs	Purification	Hydrodynamic properties; detergent binding	AA comp.	AA seq.	3-D structure and EM	Other structural studies	Reviews
c. Phosphotransferase system									
i. Constitutive enzymes II (*E. coli*)	36,000	330	330	—	—	—	—	—	—
ii. Lactose-specific enzyme II (*Staphylococcus aureus*)	36,000	331	331	—	—	—	—	—	—
iii. Mannitol-specific enzyme II (*E. coli*)	60,000	332	332	—	—	—	—	332	—
d. Pilins (Isolated from *Caulobacter crescentus, E. coli, Klebsiella pneumoniae, Moraxella nonliquifasciens, Neisseria gonorrhoeae, N. meningitidis,* and *Pseudomonas aeruginosa*)	Multiple ~20,000	333–337	333–337	—	333–337	333–338i	—	300,335,336	—
e. Other plasma membrane proteins									
i. ATPases (listed under Transport proteins)									
ii. Bacteriorhodopsin (refer to Table I)									
iii. C-55 isoprenoid alcohol phosphokinase	17,000	339–341	339,340	—	340,342	—	—	342	341
iv. Chemotactic receptor for aspartate (*Salmonella typhimurium*)	60,000	343	343	—	—	—	—	—	—
v. Ferrochelatase	53,000	344	344	—	—	—	—	—	—
vi. Lactose permease	46,504s	345	—	—	—	(345)	—	—	—
vii. Spiralin of *Spiroplasma citri*	26,000	346	346	—	—	—	—	—	—

582

	MW							
4. Additional cell envelope proteins (outside of the outer membrane)								
a. Cell envelope protein of *Aeromonas salmonicida*	54,000	347	347	—	—	347	—	—
Cytochrome P₄₅₀ and related proteins								
1. Cytochrome P-448								
a. Rat liver (induced by 3-methylcholanthrene)	53,000	348,349	348,349	—	—	—	—	—
b. Rat liver (induced by polychlorinated biphenyl)	52,000	349	349	—	—	—	349	364–366
2. Cytochrome P-450	—	—	—	—	—	—	—	—
a. Rabbit Liver								
i. Cytochrome P-450 LM1 (constitutive)	47,000	350	350	—	—	—	—	—
ii. Cytochrome P-450 LM2 (induced by phenobarbital)	50,000	350–352	350–354	—	—	—	353	—
iii. Cytochrome P-450 LM3 (induced by triacetyloleadomycin)	52,000	355	355	—	—	—	—	—
iv. Cytochrome P-450 LM3b (constitutive)	52,000	356	356	—	—	—	—	—
v. Cytochrome P-450 LM4 (induced by β-naphtho-flavone or 3-methylcholan-threne)	54,000	350,357	350,354,357	—	—	—	357	—
vi. Cytochrome P-450 LM7	60,000	350	350	—	—	—	—	—
b. Rat Liver								
i. Cytochrome P-450 (induced by phenobarbital)	51,500	348,349	348,349	—	—	—	—	—
ii. Cytochrome P-450 (induced by β-naphtho-flavone)	53,400	358	358	358	358	—	—	—
iii. Cytochrome P-450 (induced by pregnenolone 16α-carbonitrile)	51,000	359	359	—	359	—	359	—

(Continued)

TABLE II (*Continued*)

Protein	M_r	M_r refs	Purification	Hydrodynamic properties; detergent binding	AA comp.	AA seq.	3-D structure and EM	Other structural studies	Reviews
iv. Cytochrome *P*-450 RLM3 (constitutive)	50,000	360	360	—	360	360i	—	360	—
v. Cytochrome *P*-450 RLM5 (constitutive)	51,000	360	360	—	360	360i	—	360	—
c. Mitochondrial									
i. Cytochrome *P*-450 scc from bovine corpus luteum	48,000	361	—	—	361	—	—	—	—
ii. Cytochrome *P*-450 scc from bovine adrenal cortex	51,000	362	362	362	—	—	—	—	—
iii. Cytochrome *P*-450 11-β from bovine adrenal cortex	46,000	362	362	362	—	—	—	—	—
iv. Cytochrome *P*-450 D1α from bovine kidney (also known as 25-hydroxyvitamin D₃1α-hydroxylase)	49,000	363	363	—	—	—	—	—	—
3. NADPH-Cytochrome *P*-450 Reductase									
i. From rabbit liver microsomes	74,000	367,368	367,368	—	367,368	—	—	—	364,375
ii. From rat liver microsomes	79,000	369,370	369,370	370	—	—	—	367,368	—
iii. Membrane-binding peptide from rat liver microsomes	6,400	371	371	—	371	—	—	—	—
4. Related proteins									
i. Epoxide hydrase (rat liver)	53,000	372	372	—	—	—	—	—	—
ii. Epoxide hydrolase (human liver)	50,000	373	373	—	373	373i	—	373	—

584

	MW								
iii. Mixed function amine oxidase (pig liver)	60,000	—	—	—	—	—	—	—	—
Erythrocyte membrane proteins									47
Integral membrane proteins		374	374	374					
1. Glycophorins (refer to Table I)	88,000	376,377	376,378	376,379	34	—	—	380–384	385.386
2. Anion transporter (band 3)	144,000	387–390	387–389	—	—	—	—	391	391
3. Ca^{2+}, Mg^{2+}-ATPase	55,000	392–395	392–393	392	—	—	392	396,397	398,399
4. Glucose transporter (band 4.5)	195,000	400	400	—	—	—	—	—	—
5. Complement receptor		—	—	—	—	—	—	—	—
Peripheral membrane proteins									
1. Cytoskeleton proteins									
a. Spectrin (bands 1 and 2)	240,000, 220,000	405–408	405–408	405–409	379,407,410	—	401 / 407,411	402,412–417	402–404 / 418
b. Ankyrin (band 2.1)	210,000	379,419,420	379,419–421	—	379,420	—	—	391	—
c. Band 4.1	82,000	419,420	419,420	—	420	—	—	419,420	—
d. Actin (band 5)	43,000	422,423	422,423	—	—	—	422	—	—
e. Torin	20,000	424,425	424,425	424	—	—	424	425	428
f. Cylindrin	747,000	424–427	424–427	424,426	—	—	424,426,427	425,427	428
g. Acetylcholinesterase	80,000	429–431	429,430,432	429,432,433	430,432	—	—	429–431, 433,434	—
h. Protein kinases	Multiple	435–437	435–438	—	—	—	—	—	—
i. Phospholipase A_2	18,500	439	439	—	—	—	—	—	—
j. Protease	50,000	440	440	—	440	—	—	—	—
Histocompatibility antigens									
1. Human (HLA)									
a. HLA	—	—	—	—	—	(442)	—	—	441
b. HLA A, B, and C	—	—	443	—	—	443i	—	—	441
c. HLA A2 and B7	—	—	444–446	—	—	444i;445i, 447i	—	446	—
d. HLA DR	29,000, 34,000	448	448	448	—	—	—	—	—
e. HLA-DW2.2; DR2.2	—	—	449	449	449	449i	—	—	—
2. Mouse (H2)									
a. H-2	—	—	—	—	—	(450,451)i	—	44,441	44,441

(Continued)

TABLE II (*Continued*)

Protein	M_r	M_r refs	Purification	Hydrodynamic properties; detergent binding	AA comp.	AA seq.	3-D structure and EM	Other structural studies	Reviews
b. H-2D[b]	—	—	—	—	—	447i	—	—	—
c. H-2D[d]	—	—	—	—	—	452i	—	—	—
d. H-2K[b] (refer to Table I)	—	—	—	—	—	—	—	—	—
e. H-2K[d]	—	—	—	—	—	447i	—	—	—
f. H-2K[k]	47,000, 13,000	453	453	—	—	—	—	454	—
g. Ia antigens H-2[b] H-2[d] H-2[k] and H-2[s]	—	—	—	—	—	454i	—	454	—
Hydrolases *Proteases*									
1. Aminooligopeptidase from rat intestinal brush border	70,000	455	455	455	—	—	—		—
2. Aminopeptidases A and N (refer to Table I)	—	—	—	—	—	—	—		
3. Arylamidase (human placenta and kidney)	193,000,211,000	456	456	—	—	—	—	—	—
4. Dipeptidyl peptidase IV (porcine kidney brush border and rat liver plasma membranes)	120,000h	457,458	457,458	458	457,458	457i	457	457	—
5. Enterokinase (porcine and bovine)	Not agreed	459–461	459–461	459	459	—	—		—
6. Membrane peptidases (rat intestinal brush border)	140,000	467	467	—	467	—	—	467	—
7. Membrane protease of rat liver mitochondria	23,500	462	462	—	—	—	—		—
8. Membrane protease of the erythrocyte membrane (listed under Erythrocyte membrane proteins)									

586

9. Metalloendopeptidase (mouse, rat, and rabbit kidney; bovine pituitaries)	Multiple	464–466,662	463–466	—	—	—	—	465	—
10. Signal peptidase (yeast and *E. coli*)	59,000,39,000	469, 468	469, 468	—	—	—	—	—	—
Phospholipases									
1. Phospholipase A$_1$ (*E. coli* and *Mycobacterium phlei*)	29,000	470,471	470,471	—	—	—	—	—	—
2. Phospholipase A$_2$ (rat ascites hepatoma 10BA cells (also refer to section on red blood cell membrane proteins)	—	—	472	—	—	—	—	—	—
Additional hydrolases									
1. Acetylcholinesterase (also listed under Erythrocyte membrane proteins)	1,050,000	655–660	655,659	—	655–658	—	656–658, 660	661	655,658
2. Alkaline phosphatase	64,000	473,474	473–475	—	—	—	—	476	—
3. Glucosylceramidase (mouse intestinal brush border)	130,000	477	477	—	—	—	—	—	—
4. Lactase/phlorizin hydrolase (human and monkey brush border)	160,000	478	478,479	—	—	—	478	—	—
5. Maltase/glucoamylase (rat intestine)	—	—	631	—	631	—	—	—	—
6. Maltase (porcine intestinal brush border)	10,000,8.300	18	18	—	18	—	—	—	—
7. α-Mannosidase (rabbit liver microsomes)	—	—	480	—	—	—	—	—	—
8. 5′-Nucleotidase (rat heart, rat liver, and *Vibrio costicola*)	74,000	481,482	481–485	—	—	—	—	486	—
9. Nucleotide pyrophosphatase (rat liver plasma membrane)	137,000	458,487	458,487	458	—	—	—	—	—
10. Sucrase/isomaltase (refer to Table I)									
Immunoglobulins–membrane bound (refer to Table I)									

(Continued)

TABLE II (Continued)

Protein	M_r	M_r refs	Purification	Hydrodynamic properties; detergent binding	AA comp.	AA seq.	3-D structure and EM	Other structural studies	Reviews
Kinases									
1. Epidermal growth factor receptor–kinase complex	—	—	—	—	—	—	—	—	488
2. Membrane-bound casein kinase (guinea pig mammary gland)	100,000h	489	489	—	—	—	—	—	—
3. Erythrocyte membrane kinases (listed with Erythrocyte membrane proteins)									
4. Bovine heart mitochondrial kinases	Multiple	654	654	—	—	—	—	—	—
Microsomal electron transport proteins									
1. Cytochrome b_5 (refer to Table I)									
2. NADH-Cytochrome b_5 reductase (refer to Table I)									
3. NADPH-cytochrome c reductase (listed under Cytochrome P-450 and related proteins as NADPH-cytochrome P-450 reductase)									
Microsomal lipid synthetic enzymes									
1. Acyl-CoA synthetase-long chain fatty acid type (rat liver and Candida lipolytica)	Multiple	490–494	490–494	491,492	491,493	—	—	491	—
2. Arachidonyl-CoA : CDPmono-acylglycerol acyltransferase (rat liver)	35,000	495	495	—	—	—	—	—	—
3. CDPdiacylglycerol : inositol transferase (rat liver)	60,000	496	496	—	—	—	—	—	—

588

No. Enzyme	MW										
4. CTPphosphatidic acid cytidyltransferase (*S. cerevisiae*)	45,000, 19,000	497	497	—	—	—	—	—	—	—	—
5. 1,2-Diacylglycerol:CDP choline cholinephosphotransferase	—	—	498	—	—	—	—	—	—	—	—
6. 1,2-Diacylglycerol:CDP ethanolamine-phosphotransferase	—	—	498	—	—	—	—	—	—	—	—
7. Diacylglycerol acyltransferase (rat liver)	—	—	499	—	—	—	—	—	—	—	—
8. Hydroxymethylglutaryl-CoA reductase	51,000	515–517	515–518	—	—	—	—	—	—	—	515,516
9. Phosphatidylethanolamine methyltransferase	—	—	500	—	—	—	—	—	—	—	—
10. Phosphatidylglycerophosphate synthetase	—	—	501	—	—	—	—	—	—	—	—
11. Squalene synthetase	54,500	502	502	—	—	—	—	502	—	—	—
12. Steryl-CoA desaturase (rat liver)	53,000	503	503	—	—	—	—	—	—	—	**504**
Microsomal glycosyltransferases											
1. N-Acetylgalactosaminyltransferase (porcine submaxillary glands)	100,000	505	505	—	—	—	—	505	—	—	—
2. α(1 → 3)-Mannoside β(1 → 2)-N-acetylglucosaminyltransferase (rabbit liver)	—	506	506	—	—	—	—	—	—	—	—
3. α-D-Mannose: β-1,2-N-acetylglucosaminyltransferase (porcine submaxillary glands)	—	—	507	—	—	—	—	508	—	—	—
4. β-Galactoside α(1 → 2)-fucosyltransferase (porcine submaxillary glands)	—	508	508	—	—	—	—	508	—	—	—
5. Galactosyltransferase (ovine mammary glands)	—	509	509	—	—	—	—	—	—	—	—
6. UDPgal:N-acetylgalactosaminide mucin: β-1,3-galactosyltransferase (porcine submaxillary glands)	84,000	510	510	—	—	—	—	—	—	—	—

(Continued)

TABLE II (*Continued*)

Protein	M_r	M_r refs	Purification	Hydrodynamic properties: detergent binding	AA comp.	AA seq.	3-D structure and EM	Other structural studies	Reviews
7. UDP-glucuronosyltransferase (rat liver and guinea pig liver)	59,000	511,512	511	—	511	—	—	511	—
8. α-N-Acetylgalactosaminide α(2 → 6)-sialyltransferase (pig submaxillary gland)	—	513	513	—	—	—	—	—	—
9. β-Galactoside α(2 → 3)-sialyltransferase	50,000	514	514	514	—	—	—	—	—
Mitochondrial inner membrane proteins	—	—							519–521
1. NADH:ubiquinone oxidoreductase (Complex I) ~26 nonidentical polypeptides	Multiple	523–525	523–525	526	525	—	527	—	80–82, 522
a. NADH dehydrogenase (3 nonidentical subunits)	I 51,000 / II 24,000 / III 9,000	525	—	—	525	—	—	—	81,82
2. Succinate:ubiquinone oxidoreductase (Complex II) (refer to Table I)									
3. Succinate:cytochrome c oxidoreductase (the subunits of this complex are listed in Table I under Succinate:ubiquinone oxidoreductase and Ubiquinone:cytochrome c oxidoreductase)	Multiple	528,529	528	—	—	—	—	—	530
4. Ubiquinone:cytochrome c oxidoreductase (Complex III) (refer to Table I)									
5. Cytochrome c oxidase (refer to Table I)									

	MW							
6. H⁺ translocating ATPase (oligomycin-sensitive ATPase) (refer to Table I)								
7. β-Hydroxybutyrate dehydrogenase	31,500	531	531	—	531	—	532,533	—
8. Choline dehydrogenase	—	534,535	534	—	—	—	—	536
9. α-Glycerophosphate dehydrogenase	—	—	537	—	—	—	—	536
10. Electron transfer factor dehydrogenase	65,000	538	538	—	—	—	—	—
11. Nicotinamide nucleotide transhydrogenase	97–120,000	539–541	539–542	—	—	—	—	542–543
12. Calciphorin	3,000	544	544	—	—	—	—	—
13. Carnitine acyltransferase	67,000	545	545,546	545	545	—	—	—
14. Ferrochelatase	42,000	547	547	—	547	—	—	—
15. ADP/ATP carrier (refer to Table I)								
16. Glutamate carrier	—	—	548,549	—	—	—	—	—
17. Phosphate translocase	34,000	550	550	—	—	—	—	—
18. Tricarboxylate carrier	—	—	551	—	—	—	—	—
19. Uncoupling protein from brown adipose tissue (a dimer in Triton X-100)	32,000	552	552	552	—	—	—	553
Mitochondrial outer membrane proteins								
1. Monoamine oxidase (amine oxidase)	59,000–62,000	554,555	554–557	—	—	—	558	—
2. Porin	31,000–32,000	559,560	559,560	560	559	—	—	—
3. Voltage-dependent anion selective protein (VDAC protein)	~ 30,000	561	561	—	—	—	—	—
Myelin proteolipids								
1. Lipophilin (also called N2 and P7) (brain)	23,000–27,000	562,563	562–564	—	562–564	562–563i	565	562,566, 651
2. PO Protein (peripheral nerve myelin)	28,000	567–569	567–569	—	567,569	568,569i	—	—

591

(Continued)

TABLE II *(Continued)*

Protein	M_r	M_r refs	Purification	Hydrodynamic properties; detergent binding	AA comp.	AA seq.	3-D structure and EM	Other structural studies	Reviews
Photosynthetic membrane proteins									
Bacterial photosynthetic complexes									
1. Light-harvesting antennae complexes (contains $B_{800-850}$) Isolated from *Rhodopseudomonas capsulata, R. palustris, R. sphaeroides, R. viridis,* and *Rhodospirillum rubrum.* Contains 2 or 3 different polypeptides	8,000–9,000 9,000–10,000 12,000–14,000	575–584	575–584	581	577–578, 580–583	585–586	—	578	570–574
2. Reaction center complexes (contains B_{870}) Isolated from *Rhodopseudomonas capsulata, R. sphaeroides, R. viridis,* and *Rhodospirillum rubrum.* Contains 3 different polypeptides: L, M, and H	L 20,000–21,000 M 23,000–24,000 H 27,000–28,000	584,587–593	584,587–593	—	593	—	605	594	—
Algal and plant photosynthetic complexes									
1. H-translocating ATPase (refer to Table I: listed with bacterial plasma membrane and mitochondrial H⁺ translocating ATPases)	—	—	—	—	—	—	—	—	—
2. Photosystem I (The P_{700}-Chlorophyll a Reaction Center). Contains 6 different polypeptides	—	600–606	600–606	—	—	—	—	—	574,595–599
a. The P_{700} protein	64,000–70,000	600–601,605	—	—	—	—	—	607,630	608–609

592

	Molecular weight								
3. Photosystem II. Contains 2–5 different polypeptides. Contains 2–5 nonidentical subunits	—	610–611	610–611	—	—	—	—	—	510
4. Light-harvesting antennae proteins (chlorophyll a/b proteins). Several different chlorophyll a/b protein complexes have been isolated	—	604,606, 612–615	604,606, 612–615	—	—	—	616	—	—
5. Cytochrome protein complexes	—	—	—	—	—	—	—	—	—
a. Cytochrome b_6–cytochrome f electron transport complex. Contains 5 nonidentical polypeptides	—	618	618–619	—	—	—	—	—	617
i. Cytochrome b_6 (contains 3 nonidentical polypeptides)	32,000–36,000, 26,600[h]	620	620	620	—	—	—	—	—
ii. Cytochrome f (single polypeptide)	45,900[h]	621–629,621	621–629	621,627	622,626–629	628[i]	—	—	—
b. Cytochrome b_{559} (single polypeptide)	—	632	632	632	—	—	—	—	—
6. Phosphate translocator (single polypeptide)	29,000	633	633	—	—	—	—	—	—
Platelet membrane proteins									
1. Glycoprotein Ib and glycocalicin (derived from Ib)	160,000,150,000	634,653	634	—	—	—	—	634	—
2. Glycoprotein IIb and IIIa (a two-polypeptide complex)	140,000,115,000	635,636	635,636	—	636	—	—	635,636	—
3. Glycoprotein V (a peripheral protein)	82,000	637	637	—	—	—	—	637	—
Receptors and binding proteins									
1. Complement receptors	Multiple	400,638–640	638–640	—	—	—	—	—	—
2. Hepatic lectin receptors (bovine, chicken, and rat liver)	Multiple; chicken = 26,259[s]	641–645	641–645	—	—	641	—	—	646

593

(Continued)

TABLE II (Continued)

Protein	M_r	M_r refs	Purification	Hydrodynamic properties; detergent binding	AA comp.	AA seq.	3-D structure and EM	Other structural studies	Reviews
3. Hormone and growth factor receptors									
a. α₁-Adrenergic receptor	—	—	647	—	—	—	—	—	—
b. β-Adrenergic receptors (Calf lung, frog, and turkey erythrocytes and S49 lymphoma cells)	Multiple	291,294, 648,649	648–650	291	—	—	—	649	—
c. Chemotactic peptide receptor of neutrophils	Multiple	664	664	—	664	—	—	—	—
d. Epidermal growth factor receptors (A-431 carcinoma cells and human placenta)	170,000	652,663	652,663	—	—	—	—	—	488
e. Follitropin receptor (calf testis)	146,000	665	665	—	—	—	—	—	—
f. FSH receptor (calf testis)	134,000	666	666	—	—	—	—	—	—
g. Gonadotropin and human chorionic gonadotropin receptors	90,000	667	667,668	—	—	—	—	—	—
h. Growth hormone receptor	75,000	669	669	—	—	—	—	669	—
i. Insulin receptor	Multiple	670–677	670,672,677	670–672	—	—	—	670,672, 674,676	678,679
j. Luteinizing hormone receptor	65,000	680	680	—	—	—	—	—	—
k. Prolactin receptor	99,800h	681,682	—	682	—	—	—	—	—
l. Somatomedin receptor	140,000	683	683	—	—	—	—	—	—
4. Fc receptors									
a. Fc receptors (specific for IgE)	45,000–50,000	685	685	—	—	—	—	—	685
b. Fc receptors (specific for IgG)	46,000	686–689	686–689	—	688	—	—	688	—

594

	Molecular weight								
5. Neurotransmitter receptors									
a. Acetylcholine receptor (refer to Table I)									
b. γ-Aminobutyric acid receptor	55,000	691	691	690,691	692	—	—	—	—
c. Glutamate receptor	13,800	693	693	693	693	693	—	—	—
d. Benzodiazepine receptor	62,000	691	691	691	—	—	—	—	—
e. Dopamine receptor	—	—	—	694	—	—	—	—	695
f. Opiate receptor	Multiple	696	696	696	—	—	—	—	—
6. Prostaglandin receptors									
a. Prostaglandin E receptor	105,000h	697	697	—	697	—	—	—	—
b. Prostaglandin F$_{2a}$ receptor	107,000h	698	698	—	698	—	—	—	—
7. Additional receptors									
a. Folate receptor	38,500	699	699	699	—	699	—	—	—
b. Intrinsic factor-cobalamin receptor	222,000	701–705	701–705	701–705	704	701	—	—	700,701
c. Low density lipoprotein (LDL) receptor	164,000	706	706	706	—	—	—	—	706
d. Platelet von Willebrand factor receptor (see Glycoprotein 1b of Platelet membrane proteins)									
e. Saxitoxin receptor	316,000	707	707	707	—	—	—	—	—
f. Transcobalamin II receptor (human placenta)	37,000	708	708	708	708	708	—	—	—
g. Transferrin receptor	93,000	709–711	709–711	709,711	—	—	—	—	—
Transport proteins									
1. ATPases									
a. Erythrocyte Ca^{2+}, Mg^{2+}-ATPase (listed under Erythrocyte membrane proteins)									
b. Plasma membrane H$^+$ translocating ATPase									
i. From *Neurospora crassa*	104,000	712	712	712	—	—	—	—	—
ii. From *Schizosaccharomyces pombe*	105,000	713,714	713	713	714	—	—	—	—

(Continued)

TABLE II (*Continued*)

Protein	M_r	M_r refs	Purification	Hydrodynamic properties; detergent binding	AA comp.	AA seq.	3-D structure and EM	Other structural studies	Reviews
c. Renal medulla Na⁺, K⁺-ATPase (refer to Table I)									
d. Sarcoplasmic reticulum Ca²⁺, Mg²⁺-ATPase and other SR proteins									
1. Ca²⁺, Mg²⁺-ATPase (refer to Table I)									
2. Calsequestrin	44,000	715	715	—	715	—	—	—	—
3. 53,000 M_r glycoprotein	53,000	716	716	—	716	—	—	716	—
4. Phospholamban (tetramer)	5,500	717–719	717–718	—	717,718	—	—	—	—
5. Proteolipid	6,000	720	720	—	720	—	—	—	—
2. Other transport proteins									
a. Bacteriorhodopsin (refer to Table I)									
b. Bilitranslocase (2 subunits)	35,500, 37,000	721	721	—	—	—	—	—	—
c. Endoplasmic reticulum protein translocator	Multiple	722	722	—	—	—	—	—	—
d. Glucose transporter (erythrocyte) (listed under Erythrocyte membrane proteins)									
e. γ-Glutamyltransferase	Multiple	724–728	724–726,728	726	724,725	—	—	723,729	—
Virus proteins									
Filamentous bacterial virus coat proteins									
1. f1, fd, and M13 coat proteins (all identical) (refer to Table I)									
2. Ifl coat protein	—	—	—	—	—	730	—	—	—
3. IKe coat protein	—	—	—	—	—	730	—	—	—
4. Pf1 coat protein (refer to Table I)									

Protein									
5. Xf coat protein	4.343	731	—	—	—	731	—	—	—
6. ZJ-2 coat protein (1 amino acid difference from f1 coat protein)	—	—	—	—	—	730	—	—	—
Animal virus envelope proteins									
1. Avian myeloblastosis virus major envelope glycoprotein	80,000	732.733	732.733	733	732.733	—	—	732	—
2. Influenza virus hemagglutinin (refer to Table I)	—	—	—	—	—	(734.735)	—	—	—
3. Influenza virus neuraminidase	—	—	736	—	—	—	—	737	—
4. Newcastle disease virus glycoproteins	—	—	738	—	—	—	—	—	—
5. Rauscher murine leukemia virus, glycoprotein gp70	—	—	—	—	—	—	—	—	—
6. Rous sarcoma virus glycoprotein complex (gpB5. gp35)	85,000. 35,000	739	—	—	—	—	—	—	—
7. Semliki Forest virus glycoproteins E_1, E_2, and E_3 (Refer to Table I)	—	742	—	—	—	—	741	—	—
8. Sendai virus glycoproteins F and HN	F 48–54.000	—	741	—	—	—	—	740.741.743	—
9. Sindbis virus glycoproteins E_1 and E_2	—	—	—	—	—	(744)746i	—	745.746	—
10. Vesicular stomatitis virus (VSV) G protein	—	—	749	—	—	(747)	—	747.748.750	—
Miscellaneous membrane proteins									
1. Clathrin (refer to Table I)	1,600,000	751.752	751.752	751.753–755	—	—	751.752	—	—
2. Complement membrane attack complex (C6-9 complex)	—	—	—	—	—	—	—	—	756
3. Crambin (refer to Table I)	—	—	—	757	—	—	—	—	—
4. Cytochrome b_{561} (chromomembrin B from chromaffin granule membrane)	20,500	757.758	757.758	—	—	—	—	758	—
5. Dopamine β-hydroxylase	—	—	759	—	—	—	—	—	—
6. Gap junction protein (refer to Table I)	—	—	—	—	—	—	—	—	—

(*Continued*)

597

TABLE II (Continued)

Protein	M_r	M_r refs	Purification	Hydrodynamic properties; detergent binding	AA comp.	AA seq.	3-D structure and EM	Other structural studies	Reviews
7. Hexose-6-phosphate dehydrogenase	108,000	760	760	760	—	—	—	760	—
8. Intrinsic membrane calcium-binding protein (IMCal) (human and porcine)	20,500	761,762	761,762	—	—	—	—	—	—
9. D-Lactate dehydrogenase	75,000	763–765	763–765	763,765	—	—	—	—	—
10. Ly 2/3 antigen	30,000, 35,000	710	710	—	—	—	—	—	—
11. Melittin (refer to Table I)									
12. Milk fat globule membrane glycoprotein	70,000	766,767	766,767	—	766,767	—	—	—	—
13. Murine cell surface glycoprotein	110,000	768	768	—	—	—	—	768	—
14. Rhodopsin and opsin (refer to Table I)									
15. Rat liver plasma membrane glycoprotein	110,000	769	769	—	—	—	—	769	—
16. Thy-1 glycoprotein (rat brain)	17,500h	770,771	770,771	—	770	771	—	—	—
17. Thyroid plasma membrane glycoproteins	Multiple	772	772	—	—	—	—	—	—

[a] The types of information in each column as well as the symbols that are used are given in the footnotes to Table I. The label "multiple" in the molecular weight columns indicates that the protein complex contains several nonidentical subunits.

References

1. C. Tanford, Y. Nozaki, J. A. Reynolds, and S. Makino, *Biochemistry* **13**, 2369 (1974).
2. J. A. Reynolds and C. Tanford, *Proc. Natl. Acad. Sci. U.S.A.* **73**, 4467 (1976).
3. J. P. Rosenbusch, *J. Biol. Chem.* **249**, 8019 (1974).
4. J. P. Rosenbusch, R. M. Garavito, D. L. Dorset, and A. Engel, *Protides Biol. Fluids* **29**, 171 (1982).
5. A. C. Steven, B. T. Heggeler, R. Müller, J. Kistler, and J. P. Rosenbusch, *J. Cell Biol.* **72**, 292 (1977).
6. R. Chen, C. Krämer, W. Schmidmayr, and U. Henning, *Proc. Natl. Acad. Sci. U.S.A.* **76**, 5014 (1979).
7. N. Mutoh, K. Inokuchi, and S. Mizushima, *FEBS Lett.* **137**, 171 (1982).
8. R. M. Garavito and J. P. Rosenbusch, *J. Cell Biol.* **86**, 327 (1980).
9. I. Crowlesmith, K. Gamon, and U. Henning, *Eur. J. Biochem.* **113**, 375 (1981).
10. M. J. Osborn and H. C. P. Wu, *Annu. Rev. Microbiol.* **34**, 369 (1980).
11. V. Braun, *Biochim. Biophys. Acta* **415**, 335 (1975).
12. M. Inouye, *Biomembranes* **10**, 141 (1979).
13. H. Tokunaga, M. Tokunaga, and T. Nakae, *Eur. J. Biochem.* **95**, 433 (1979).
14. M. Tokunaga, H. Tokunaga, Y. Okajima, and T. Nakae, *Eur. J. Biochem.* **95**, 441 (1979).
15. A. Benajiba and S. Maroux, *Eur. J. Biochem.* **107**, 381 (1980).
16. A. Benajiba and S. Maroux, *Biochem. J.* **197**, 573 (1981).
17. E. M. Danielsen, O. Norén, H. Sjöström, J. Ingram, and A. J. Kenny, *Biochem. J.* **189**, 591 (1980).
18. S. Maroux and D. Louvard, *Biochim. Biophys. Acta* **419**, 189 (1976).
19. H. Feracci and S. Maroux, *Biochim. Biophys. Acta* **599**, 448 (1980).
20. J. B. McClellan, Jr. and C. W. Garner, *Biochim. Biophys. Acta* **613**, 160 (1980).
21. H. Sjöström and D. Norén, *Eur. J. Biochem.* **122**, 245 (1982).
22. H. Feracci, S. Maroux, J. Bonicel, and P. Desnuelle, *Biochim. Biophys. Acta* **684**, 133 (1982).
23. P. Desnuelle, *Eur. J. Biochem.* **101**, 1 (1979).
24. H. Sjöström, D. Norén, L. Christiansen, H. Wacker, and G. Semenza, *J. Biol. Chem.* **255**, 11332 (1980).
25. M. Speiss, H. Hauser, J. P. Rosenbusch, and G. Semenza, *J. Biol. Chem.* **256**, 8977 (1981).
26. H.-P. Hauri, H. Wacker, E. E. Rickli, B. Bigler-Meier, A. Quaroni, and G. Semenza, *J. Biol. Chem.* **257**, 4522 (1982).
27. Y. Nishi, R. Tamura, and Y. Takesue, *J. Ultrastruct. Res.* **73**, 331 (1980).
28. A. Herscovics, A. Quaroni, B. Bugge, and K. Kirsch, *Biochem. J.* **197**, 511 (1981).
29. M. Spiess, J. Brunner, and G. Semenza, *J. Biol. Chem.* **257**, 2370 (1982).
30. V. T. Marchesi and E. P. Andrews, *Science* **174**, 1247 (1971).
31. H. Furthmayr, *J. Supramol. Struct.* **9**, 79 (1978).
32. A. S. B. Edge and P. Weber, *Arch. Biochem. Biophys.* **209**, 697 (1981).
33. S. P. Grefrath and J. A. Reynolds, *Proc. Natl. Acad. Sci. U.S.A.* **71**, 3913 (1974).
34. S. Clarke, *J. Biol. Chem.* **250**, 5459 (1975).
35. M. Tomita, H. Furthmayr, and V. T. Marchesi, *Biochemistry* **17**, 4756 (1978).
36. K. Honma, M. Tomita, and A. Hamada, *J. Biochem. (Tokyo)* **88**, 1679 (1980).
37. T. H. Schulte and V. T. Marchesi, *Biochemistry* **18**, 275 (1979).
38. J. A. Cramer, V. T. Marchesi, and I. M. Armitage, *Biochim. Biophys. Acta* **595**, 235 (1980).
39. T. Irimura, T. Tsuji, S. Tagami, K. Yamamoto, and T. Osawa, *Biochemistry* **20**, 560 (1981).

40. V. T. Marchesi, *Semin. Hematol.* **16**, 3 (1979).
41. D. J. Anstee, *Semin. Hematol.* **18**, 13 (1981).
42. H. Uehara, J. E. Coligan, and S. G. Nathenson, *Biochemistry* **20**, 5940 (1981).
43. J. E. Coligan, T. J. Kindt, H. Uehara, J. Martinko, and S. G. Nathenson, *Nature* (*London*) **291**, 35 (1981).
44. S. G. Nathenson, H. Uehara, B. M. Ewenstein, T. J. Kindt, and J. E. Coligan, *Annu. Rev. Biochem.* **50**, 1025 (1981).
45. H.-L. Cheng, F. R. Blattner, L. Fitzmaurice, J. F. Mushinski, and P. W. Tucker, *Nature* (*London*) **296**, 410 (1982).
46. J. Rogers, E. Choi, L. Souza, C. Carter, C. Word, M. Kuehl, D. Eisenberg, and R. Wall, *Cell* **26**, 19 (1981).
47. M. Kehry, S. Ewald, R. Douglas, C. Sibley, W. Raschke, D. Fambrough, and L. Hood, *Cell* **21**, 393 (1980).
48. F. W. Alt, A. L. M. Bothwell, M. Knapp, E. Siden, E. Mather, M. Koshland, and D. Baltimore, *Cell* **20**, 293 (1980).
49. A. Ito and R. Sato, *J. Biol. Chem.* **243**, 4922 (1968).
50. L. Spatz and P. Strittmatter, *Proc. Natl. Acad. Sci. U.S.A.* **68**, 1042 (1971).
51. N. C. Robinson and C. Tanford, *Biochemistry* **14**, 369 (1975).
52. L. Visser, N. C. Robinson, and C. Tanford, *Biochemistry* **14**, 1194 (1975).
53. A. Tsugita, M. Kobayashi, S. Tani, S. Kyo, M. A. Rashid, Y. Yoshida, T. Kajihara, and B. Hagihara, *Proc. Natl. Acad. Sci. U.S.A.* **67**, 442 (1970).
54. F. G. Nóbrega and J. Ozols, *J. Biol. Chem.* **246**, 1706 (1971).
55. J. Ozols and C. Gerard, *Proc. Natl. Acad. Sci. U.S.A.* **74**, 3725 (1977).
56. J. Ozols and C. Gerard, *J. Biol. Chem.* **252**, 8549 (1977).
57. P. J. Fleming, H. A. Dailey, D. Corcoran, and P. Strittmatter, *J. Biol. Chem.* **253**, 5369 (1978).
58. K. Kondo, S. Tajima, R. Sato, and K. Narita, *J. Biochem.* (*Tokyo*) **86**, 1119 (1979).
59. Y. Takagaki, G. E. Gerber, K. Nihei, and H. G. Khorana, *J. Biol. Chem.* **255**, 1536 (1980).
60. F. S. Mathews, P. Argos, and M. Levine, *Cold Spring Harbor Symp. Quant. Biol.* **36**, 387 (1971).
61. P. Argos and F. S. Mathews, *J. Biol. Chem.* **250**, 747 (1975).
62. S. Tajima, K.-I. Enomoto, and R. Sato, *Arch. Biochem. Biophys.* **172**, 90 (1976).
63. R. Bisson, C. Montecucco, and R. A. Capaldi, *FEBS Lett.* **106**, 317 (1979).
64. R. M. Keller and K. Wüthrich, *Biochim. Biophys. Acta* **621**, 204 (1980).
65. S. Tajima and R. Sato. *J. Biochem.* (*Tokyo*) **87**, 123 (1980).
66. F. S. Mathews and E. W. Czerwinski, *in* "The Enzymes of Biological Membranes" (A. Martonosi ed.), Vol. 4, p. 143. Plenum, New York, 1976.
67. R. P. Hughey and N. P. Curthoys, *in* "Multifunctional Proteins" (H. Bisswanger and E. Shmincke-Ott, eds.), p. 235. Wiley, New York, 1980.
68. L. Spatz and P. Strittmatter, *J. Biol. Chem.* **248**, 793 (1973).
69. K. Mihara and R. Sato, *J. Biochem.* (*Tokyo*) **78**, 1057 (1975).
70. J. W. Crabb, G. E. Tarr, K. T. Yasunobu, T. Iyanagi, and M. J. Coon, *Biochem. Biophys. Res. Commun.* **95**, 1650 (1980).
71. S. Tajima, K. Mihara, and R. Sato, *Arch. Biochem. Biophys.* **198**, 137 (1979).
72. K. Mihara, R. Sato, R. Sakakibara, and H. Wada, *Biochemistry* **17**, 2829 (1978).
73. Y. Hatefi and Y. M. Galante, *J. Biol. Chem.* **255**, 5530 (1980).
74. B. A. C. Ackrell, M. B. Ball, and E. B. Kearney, *J. Biol. Chem.* **255**, 2761 (1980).
75. C.-A. Yu and L. Yu, *Biochemistry* **19**, 3579 (1980).
76. R. A. Capaldi, J. Sweetland, and A. Merli, *Biochemistry* **16**, 5707 (1977).
77. H. Weiss and H. J. Kolb, *Eur. J. Biochem.* **99**, 139 (1979).

78. G. Dooijewaard, E. C. Slater, P. J. Van Dijk, and G. J. M. De Bruin, *Biochim. Biophys. Acta* **503**, 405 (1978).
79. B. A. C. Ackrell, E. B. Kearney, and T. P. Singer, this series, Vol. 53, p. 466.
80. Y. Hatefi and D. L. Stiggall, *in* "The Enzymes" (P. D. Boyer, ed.), 3rd ed., Vol. 13, p. 175. Academic Press, New York, 1976.
81. T. Ohnishi, *in* "Mitochondria and Microsomes" (C. P. Lee, G. Schatz, and G. Dallner, eds.), p. 191. Addison-Wesley, Reading, Massachusetts, 1981.
82. T. P. Singer, R. R. Ramsay, and C. Paech, *in* "Mitochondria and Microsomes" (C. P. Lee, G. Schatz, and G. Dallner, eds.), p. 155. Addison-Wesley, Reading, Massachusetts, 1981.
83. G. von Jagow, H. Schägger, P. Riccio, M. Klingenberg, and H. J. Kolb, *Biochim. Biophys. Acta* **462**, 549 (1977).
84. H. Weiss, P. Wingfield, and K. Leonard, *in* "Membrane Bioenergetics" (C. P. Lee, G. Schatz, and L. Ernster, eds.), p. 119. Addison-Wesley, Reading, Massachusetts, 1979.
85. P. Wingfield, T. Arad, K. Leonard, and H. Weiss, *Nature (London)* **280**, 696 (1979).
86. S. Hovmöller, K. Leonard, and H. Weiss, *FEBS Lett.* **123**, 118 (1981).
87. K. Leonard, P. Wingfield, T. Arad, and H. Weiss, *J. Mol. Biol.* **149**, 259 (1981).
88. J. Lepault, H. Weiss, J.-C. Homo, and K. Leonard, *J. Mol. Biol.* **149**, 275 (1981).
89. J. S. Rieske, *Biochim. Biophys. Acta* **456**, 195 (1976).
90. T. E. King, *in* "Chemiosmotic Proton Circuits in Biological Membranes" (V. P. Skulachev and P. C. Hinkle, eds.), p. 147. Addison-Wesley, Reading, Massachusetts, 1981.
91. B. D. Nelson, *in* "Mitochondria and Microsomes" (C. P. Lee, G. Schatz, and G. Dallner, eds.), p. 217. Addison-Wesley, Reading, Massachusetts, 1981.
92. M. Wikström, K. Krab, and M. Saraste, *Annu. Rev. Biochem.* **50**, 623 (1981).
93. G. von Jagow, H. Schägger, W. D. Engel, W. Machleidt, and I. Machleidt, *FEBS Lett.* **91**, 121 (1978).
94. S. Yoshida, Z. P. Zhang, and T. E. King, *Biochem. Int.* **4**, 1 (1982).
95. F. G. Nóbrega and A. Tzagoloff, *J. Biol. Chem.* **255**, 9828 (1980).
96. S. Anderson, A. T. Bankier, B. G. Barrell, M. H. L. de Bruijn, A. R. Coulson, J. Drouin, I. C. Eperon, D. P. Nierlich, B. A. Roe, F. Sanger, P. H. Schreier, A. J. H. Smith, R. Staden, and I. G. Young, *Nature (London)* **290**, 457 (1981).
97. H. Weiss, *Biochim. Biophys. Acta* **456**, 291 (1976).
98. G. von Jagow and W. Sebald, *Annu. Rev. Biochem.* **49**, 281 (1980).
99. E. C. Slater, *in* "Chemiosmotic Proton Circuits in Biological Membranes" (V. P. Skulachev and P. C. Hinkle, eds.), p. 69. Addison-Wesley, Reading, Massachusetts, 1981.
100. C.-A. Yu, L. Yu, and T. E. King, *J. Biol. Chem.* **247**, 1012 (1972).
101. B. L. Trumpower and A. Katki, *Biochemistry* **14**, 3635 (1975).
102. E. Ross and G. Schatz, *J. Biol. Chem.* **251**, 1991 (1976).
103. N. C. Robinson and L. Talbert, *Biochem. Biophys. Res. Commun.* **95**, 90 (1980).
104. B. W. König, L. T. M. Schilder, M. J. Tervoort, and B. F. Van Gelder, *Biochim. Biophys. Acta* **621**, 283 (1980).
105. Y. Li, K. Leonard and H. Weiss, *Eur. J. Biochem.* **116**, 199 (1981).
106. S. Wakabayashi, H. Matsubara, C. H. Kim, K. Kawai, and T. E. King. *Biochem. Biophys. Res. Commun.* **97**, 1548 (1980).
107. J. S. Rieske, D. H. MacLennan, and R. Coleman, *Biochem. Biophys. Res. Commun.* **15**, 338 (1964).
108. B. L. Trumpower and C. A. Edwards, *J. Biol. Chem.* **254**, 8697 (1979).
109. B. L. Trumpower, *Biochim. Biophys. Acta* **639**, 129 (1981).
110. R. L. Bell and R. A. Capaldi, *Biochemistry* **15**, 996 (1976).
111. L. Yu, C.-A. Yu, and T. E. King, *Biochim. Biophys. Acta* **495**, 232 (1977).
112. C.-A. Yu, L. Yu, and T. E. King, *Biochem. Biophys. Res. Commun.* **66**, 1194 (1975).
113. M. B. Katan, L. Pool, and G. S. P. Groot, *Eur. J. Biochem.* **65**, 95 (1976).

114. T. E. King, this series, Vol. 53, p. 181.

115. C.-A. Yu and L. Yu, *Biochemistry* **19**, 3579 (1980).

116. S. H. Phan and H. R. Mahler, *J. Biol. Chem.* **251**, 257 (1976).

117. N. C. Robinson and R. A. Capaldi, *Biochemistry* **16**, 375 (1977).

118. M. Saraste, T. Penttilä, and M. Wikström, *Eur. J. Biochem.* **115**, 261 (1981).

119. G. Buse and G. J. Steffens, *Hoppe-Seyler's Z. Physiol. Chem.* **359**, 1005 (1978).

119a. G. J. Steffens and G. Buse, *Hoppe-Seyler's Z. Physiol. Chem.* **360**, 613 (1979).

119b. R. Sacher, G. J. Steffens, and G. Buse, *Hoppe-Seyler's Z. Physiol. Chem.* **360**, 1385 (1979).

119c. G. C. M. Steffens, G. J. Steffens, and G. Buse, *Hoppe-Seyler's Z. Physiol. Chem.* **360**, 1641 (1979).

120. M. Tanaka, M. Haniu, K. T. Yasunobu, C.-A. Yu, L. Yu, Y.-H. Wei, and T. E. King, *J. Biol. Chem.* **254**, 3879 (1979).

121. M. Tanaka, K. T. Yasunobu, Y.-H. Wei, and T. E. King. *J. Biol. Chem.* **256**, 4832 (1981).

122. B. E. Thalenfeld and A. Tzagoloff, *J. Biol. Chem.* **255**, 6173 (1980).

122a. G. Coruzzi and A. Tzagoloff, *J. Biol. Chem.* **254**, 9324 (1979).

122b. S. G. Bonitz, G. Coruzzi, B. E. Thalenfeld, A. Tzagoloff, and G. Macino, *J. Biol. Chem.* **255**, 11927 (1980).

123. R. Henderson, R. A. Capaldi, and J. S. Leigh, *J. Mol. Biol.* **112**, 631 (1977).

124. S. D. Fuller, R. A. Capaldi, and R. Henderson, *J. Mol. Biol.* **134**, 305 (1979).

125. B. G. Malmström, *Biochim. Biophys. Acta* **549**, 281 (1979).

126. R. A. Capaldi, L. Prochaska, and R. Bisson, *Adv. Exp. Med. Biol.* **132**, 197 (1980).

127. M. Wikström, K. Krab, and M. Saraste, *Annu. Rev. Biochem.* **50**, 623 (1981).

128. A. Karlin, C. L. Weill, M. G. McNamee, and R. Valderrama, *Cold Spring Harbor Symp. Quant. Biol.* **40**, 203 (1975).

129. M. A. Raftery, R. L. Vandlen, K. L. Reed, and T. Lee, *Cold Spring Harbor Symp. Quant. Biol.* **40**, 193 (1975).

130. A. Sobel, M. Weber, and J.-P. Changeux, *Eur. J. Biochem.* **80**, 215 (1977).

131. J. C. Meunier, R. W. Olsen, and J.-P. Changeux, *FEBS Lett.* **24**, 63 (1972).

132. M. Martinez-Carrion, V. Šator, and M. A. Raftery, *Biochem. Biophys. Res. Commun.* **65**, 129 (1975).

133. S. J. Edelstein, W. B. Beyer, A. T. Eldefrawi, and M. E. Eldefrawi, *J. Biol. Chem.* **250**, 6101 (1975).

134. R. E. Gibson, R. D. O'Brien, S. J. Edelstein, and W. R. Thompson, *Biochemistry* **15**, 2377 (1976).

135. J. A. Reynolds and A. Karlin, *Biochemistry* **17**, 2035 (1978).

136. D. S. Wise, A. Karlin, and B. P. Schoenborn, *Biophys. J.* **28**, 473 (1979).

137. R. L. Vandlen, W. C.-S. Wu, J. C. Eisenach, and M. A. Raftery, *Biochemistry* **18**, 1845 (1979).

138. M. Noda, H. Takahashi, T. Tanabe, M. Toyosato, Y. Furutani, T. Hirose, M. Asai, S. Inayama, T. Miyata, and S. Numa, *Nature (London)* **299**, 793 (1982).

139. K. Sumikawa, M. Houghton, J. C. Smith, L. Bell, B. M. Richards, and E. A. Barnard, *Nucleic Acids Res.* **10**, 5809 (1982).

140. M. J. Ross, M. W. Klymkowsky, D. A. Agard, and R. M. Stroud, *J. Mol. Biol.* **116**, 635 (1977).

141. E. Holtzman, D. Wise, J. Wall, and A. Karlin, *Proc. Natl. Acad. Sci. U.S.A.* **79**, 310 (1982).

142. J. Kistler, R. M. Stroud, M. W. Klymkowsky, R. A. Lalancette, and R. H. Fairclough, *Biophys. J.* **37**, 371 (1982).

143. D. S. Wise, B. P. Schoenborn, and A. Karlin, *J. Biol. Chem.* **256**, 4124 (1981).

144. D. S. Wise, J. Wall, and A. Karlin, *J. Biol. Chem.* **256**, 12624 (1981).

145. F. J. Barrantes, *J. Mol. Biol.* **124**, 1 (1978).
146. M. W. Klymkowsky and R. M. Stroud, *J. Mol. Biol.* **128**, 319 (1979).
147. R. Tarrab-Hazdai and V. Goldfarb, *Eur. J. Biochem.* **121**, 545 (1982).
148. T. Heidmann and J.-P. Changeux, *Annu. Rev. Biochem.* **47**, 317 (1978).
149. F. J. Barrantes, *Annu. Rev. Biophys. Bioeng.* **8**, 287 (1979).
150. D. M. Fambrough, *Physiol. Rev.* **59**, 165 (1979).
151. A. Karlin, *in* "The Cell Surface and Neuronal Function" (C. W. Cotman, G. Poste, and G. L. Nicolsen, eds.), p. 191. Elsevier/North-Holland Biomedical Press, New York, 1980.
152. M. A. Raftery, M. W. Hunkapiller, C. D. Strader, and L. E. Hood, *Science* **208**, 1454 (1980).
153. B. M. Conti-Tronconi and M. A. Raftery, *Annu. Rev. Biochem.* **51**, 491 (1982).
154. M. Klingenberg, P. Riccio, and H. Aquila, *Biochim. Biophys. Acta* **503**, 193 (1978).
155. H. Hackenberg and M. Klingenberg, *Biochemistry* **19**, 548 (1980).
156. W. Babel, E. Wachter, H. Aquila, and M. Klingenberg, *Biochim. Biophys. Acta* **670**, 176 (1981).
157. M. Klingenberg, *Trends Biochem. Sci.* **4**, 249 (1979).
158. M. Klingenberg, *in* "Mitochondria and Microsomes" (C. P. Lee, G. Schatz, and G. Dallner, eds.), p. 293. Addison-Wesley, Reading, Massachusetts, 1981.
159. A. Tzagoloff and P. Meagher, *J. Biol. Chem.* **246**, 732B (1971).
160. H. H. Paradies and U. D. Schmidt, *J. Biol. Chem.* **254**, 5257 (1979).
161. M. Satre and G. Zaccaï, *Hoppe-Seyler's Z. Physiol. Chem.* **361**, 1471 (1980).
162. W. Sebald and J. Hoppe, *Curr. Top. Bioenerg.* **12**, 1 (1981).
163. B. Frangione, E. Rosenwasser, H. S. Penefsky, and M. E. Pullman, *Proc. Natl. Acad. Sci. U.S.A.* **78**, 7403, (1981).
164. J. E. Walker, A. D. Auffret, A. Carne, A. Gurnett, P. Hanisch, D. Hill, and M. Saraste, *Eur. J. Biochem.* **123**, 253 (1982).
165. N. J. Gay and J. E. Walker, *Nucleic Acids Res.* **9**, 2187 (1981).
166. N. J. Gay and J. E. Walker, *Nucleic Acids Res.* **9**, 3919 (1981).
167. M. Saraste, N. J. Gay, A. Eberle, M. J. Runswick, and J. E. Walker, *Nucleic Acids Res.* **9**, 5287 (1981).
168. K. Mabuchi, H. Kanazawa, T. Kayano, and M. Futai, *Biochem. Biophys. Res. Commun.* **102**, 172 (1981).
169. H. Kanazawa, K. Mabuchi, T. Kayano, F. Tamura, and M. Futai, *Biochem. Biophys. Res. Commun.* **100**, 219 (1981).
170. H. Kanazawa, T. Kayano, K. Mabuchi, and M. Futai, *Biochem. Biophys. Res. Commun.* **103**, 604 (1981).
171. H. Kanazawa, K. Mabuchi, T. Kayano, T. Noumi, T. Sekiya, and M. Futai, *Biochem. Biophys. Res. Commun.* **103**, 613 (1981).
172. H. Kanazawa, T. Kayano, T. Kiyasu, and M. Futai, *Biochem. Biophys. Res. Commun.* **105**, 1257 (1982).
173. V. Spitzberg and R. Haworth, *Biochim. Biophys. Acta* **492**, 237 (1977).
174. L. M. Amzel and P. L. Pederson, this series, Vol. 55, p. 333.
175. Y. Kagawa, this series, Vol. 55, p. 372.
176. J. W. Soper, G. L. Decker, and P. L. Pederson, *J. Biol. Chem.* **254**, 11170 (1979).
177. H. H. Paradies, G. Mertens, R. Schmid, E. Schneider, and K. Altendorf, *Biophys. J.* **37**, 195 (1982).
178. N. Nelson, *Curr. Top. Bioenerg.* **11**, 1 (1981).
179. Y. Kagawa, *Biochim. Biophys. Acta* **505**, 45 (1978).
180. B. A. Baird and G. Hammes, *Biochim. Biophys. Acta* **549**, 31 (1979).
181. H. S. Penefsky, this series, Vol. 55, p. 297.
182. R. H. Fillingame, *Curr. Top. Bioenerg.* **11**, 35 (1981).

183. Y. Kagawa, S. Ohta, M. Yoshida, and N. Sone, *Ann. N.Y. Acad. Sci.* **358**, 103 (1980).
184. P. L. Pederson, L. M. Amzel, N. Cintrón, J. W. Soper, J. Hullihen, and J. Wehrle, *Adv. Exp. Med. Biol.* **132**, 317 (1980).
185. L. M. Amzel, *J. Bioenerg. Biomembr.* **13**, 109 (1981).
186. E. Racker, *in* "Mitochondria and Microsomes" (C. P. Lee, G. Schatz, and G. Dallner, eds.), p. 337. Addison-Wesley, Reading, Massachusetts, 1981.
187. J. Houštěk, J. Kopecký, P. Svoboda, and Z. Drahota, *J. Bioenerg. Biomembr.* **14**, 1 (1982).
188. D. H. MacLennan, *J. Biol. Chem.* **245**, 4508 (1970).
189. D. H. MacLennan, P. Seeman, G. H. Iles, and C. C. Yip, *J. Biol. Chem.* **246**, 2702 (1971).
190. G. Meissner, G. E. Conner, and S. Fleischer, *Biochim. Biophys. Acta* **298**, 246 (1973).
191. M. le Maire, K. E. Jørgensen, H. Røigaard-Peterson, and J. V. Møller, *Biochemistry* **15**, 5805 (1976).
192. M. le Maire, J. V. Møller, and C. Tanford, *Biochemistry* **15**, 2336 (1976).
192a. L. J. Rizzolo, M. le Maire, J. A. Reynolds, and C. Tanford, *Biochemistry* **15**, 3433 (1976).
193 W. L. Dean and C. Tanford, *Biochemistry* **17**, 1683 (1978).
194. M. le Maire, J. V. Møller, and A. Tardieu, *J. Mol. Biol.* **150**, 273 (1981).
195. A. J. Murphy, M. Pepitone, and S. Highsmith, *J. Biol. Chem.* **257**, 3551 (1982).
196. G. Allen, *Biochem. J.* **187**, 545 (1980).
197. G. Allen, *Biochem. J.* **187**, 565 (1980).
198. G. Allen, R. C. Bottomley, and B. J. Trinnaman, *Biochem. J.* **187**, 577 (1980).
199. G. Allen, B. J. Trinnaman, and N. M. Green, *Biochem. J.* **187**, 591 (1980).
200. A. Saito, C.-T. Wang, and S. Fleischer, *J. Cell Biol.* **79**, 601 (1978).
201. L. J. Rizzolo and C. Tanford, *Biochemistry* **17**, 4044 (1978).
202. L. J. Rizzolo and C. Tanford, *Biochemistry* **17**, 4049 (1978).
203. P. Fellman, J. Andersen, P. F. Devaux, M. le Maire, and A. Bienvenue, *Biochem. Biophys. Res. Commun.* **95**, 289 (1980).
204. R. A. F. Reithmeier and D. H. MacLennan, *J. Biol. Chem.* **256**, 5957 (1981).
205. N. Ikemoto, A. Miyao, and Y. Kurobe, *J. Biol. Chem.* **256**, 10809 (1981).
206. J. P. Andersen, P. Fellman, J. V. Møller, and P. F. Devaux, *Biochemistry* **20**, 4928 (1981).
207. N. Ikemoto, *Annu. Rev. Physiol.* **44**, 297 (1982).
208. D. H. MacLennan and P. C. Holland, *Annu. Rev. Biophys. Bioeng.* **4**, 377 (1975).
209. J. Kyte, *J. Biol. Chem.* **247**, 7642 (1972).
210. P. L. Jørgensen, *Biochim. Biophys. Acta* **356**, 53 (1974).
211. J. F. Dixon and L. E. Hokin, *Arch. Biochem. Biophys.* **163**, 749 (1974).
212. J. R. Perrone, J. F. Hackney, J. F. Dixon, and L. E. Hokin, *J. Biol. Chem.* **250**, 4178 (1975).
213. W. S. Craig and J. Kyte, *J. Biol. Chem.* **255**, 6262 (1980).
214. J. H. Collins, B. Forbush III, L. K. Lane, E. Ling, A. Schwartz, and A. Zot, *Biochim. Biophys. Acta* **686**, 7 (1982).
215. L. C. Cantley, *Curr. Top. Bioenerg.* **11**, 201 (1981).
216. D. F. Hastings and J. A. Reynolds, *Biochemistry* **18**, 817 (1979).
217. M. Esmann, C. Christiansen, K.-A. Karlsson, G. C. Hansson, and J. C. Skou, *Biochim. Biophys. Acta* **603**, 1 (1980).
218. J. W. Freytag and J. A. Reynolds, *Biochemistry* **20**, 7211 (1981).
219. L. E. Hokin, J. L. Dahl, J. D. Deupree, J. F. Dixon, J. F. Hackney, and J. F. Perdue, *J. Biol. Chem.* **248**, 2593 (1973).
220. E. Skriver, A. B. Maunsbach, and P. L. Jørgensen, *FEBS Lett.* **131**, 219 (1981).
221. J. C. Skou and J. G. Nørby, eds., "Na, K -ATPase: Structure and Kinetics." Academic Press, New York, 1979.

222. P. L. Jørgensen, *Physiol. Rev.* **60**, 864 (1980).

223. A. S. Hobbs and R. W. Albers, *Annu. Rev. Biophys. Bioeng.* **9**, 259 (1980).

224. P. L. Jørgensen, *Biochim. Biophys. Acta.* **694**, 27 (1982).

225. J. A. Reynolds and W. Stoekenius, *Proc. Natl. Acad. Sci. U.S.A.* **74**, 2803 (1977).

226. Y. A. Ovchinnikov, N. G. Abdulaev, M. Y. Feigina, A. V. Kiselev, and N. A. Lobanov, *FEBS Lett.* **84**, 1 (1977).

227. H. G. Khorana, G. E. Gerber, W. C. Herlihy, C. P. Gray, R. J. Anderegg, K. Nihei, and K. Biemann, *Proc. Natl. Acad. Sci. U.S.A.* **76**, 5046 (1979).

228. R. Henderson and P. N. T. Unwin, *Nature (London)* **257**, 28 (1975).

229. R. Henderson and D. Shotton, *J. Mol. Biol.* **139**, 99 (1980).

230. H. Michel and D. Oesterhelt, *Proc. Natl. Acad. Sci. U.S.A.* **77**, 1283 (1980).

231. D. M. Engelman and G. Zaccai, *Proc. Natl. Acad. Sci. U.S.A.* **77**, 5894 (1980).

232. D. M. Engelman, R. Henderson, A. D. McLachlan, and B. A. Wallace, *Proc. Natl. Acad. Sci. U.S.A.* **77**, 2023 (1980).

233. D. A. Agard and R. M. Stroud, *Biophys. J.* **37**, 589 (1982).

234. B. Becher and T. G. Ebrey, *Biochem. Biophys. Res. Commun.* **69**, 1 (1976).

235. R. Henderson, *Annu. Rev. Biophys. Bioeng.* **6**, 87 (1977).

236. Y. A. Ovchinnikov, *in* "Chemiosmotic Proton Circuits in Biological Membranes" (V. P. Skulachev and P. C. Hinkle, eds.), p. 311. Addison-Wesley, Reading, Massachusetts, 1981.

237. Y. A. Ovchinnikov, N. G. Abdulaev, and N. N. Modyanov, *Annu. Rev. Biophys. Bioeng.* **11**, 445 (1982).

238. W. Stoeckenius and R. A. Bogomolni, *Annu. Rev. Biochem.* **51**, 587 (1982).

239. S. Makino, J. L. Woolford, Jr., C. Tanford, and R. E. Webster, *J. Biol. Chem.* **250**, 4327 (1975).

240. Y. Nozaki, J. A. Reynolds, and C. Tanford, *Biochemistry* **17**, 1239 (1978).

241. V. F. Asbeck, K. Beyreuther, H. Köhler, G. von Wettstein, and G. Braunitzer, *Hoppe-Seyler's Z. Physiol. Chem.* **350**, 1047 (1969).

242. Y. Nakashima and W. Konigsberg, *J. Mol. Biol.* **88**, 598 (1974).

243. D. W. Banner, C. Nave, and D. A. Marvin, *Nature (London)* **289**, 814 (1981).

244. R. A. Grant, T.-C. Lin, W. Konigsberg, and R. E. Webster, *J. Biol. Chem.* **256**, 539 (1981).

245. Y. Nozaki, B. K. Chamberlain, R. E. Webster, and C. Tanford, *Nature (London)* **259**, 335 (1976).

246. I. Ohkawa and R. E. Webster, *J. Biol. Chem.* **256**, 9951 (1981).

247. H. D. Dettman, J. H. Weiner, and B. D. Sykes, *Biophys. J.* **37**, 243 (1982).

248. Y. Nakashima, R. L. Wiseman, W. Konigsberg, and D. A. Marvin, *Nature (London)* **253**, 68 (1975).

249. L. Makowski, D. L. D. Caspar, and D. A. Marvin, *J. Mol. Biol.* **140**, 149 (1980).

250. I. A. Wilson, J. J. Skehel, and D. C. Wiley, *Nature (London)* **289**, 366 (1981).

251. A. L. Hiti, A. R. Davis, and D. P. Nayak, *Virology* **111**, 113 (1981).

252. C. W. Ward and T. A. Dopheide, *Biochem. J.* **193**, 953 (1981).

253. R. Fang, W. Min Jou, D. Huylebroeck, R. Devos, and W. Fiers, *Cell* **25**, 315 (1981).

254. G. Winter, S. Fields, and G. G. Brownlee, *Nature (London)* **292**, 72 (1981).

255. G. Laver and G. Air, eds., "Structure and Variation in Influenza Virus." Elsevier/North-Holland, New York, 1980.

256. H. Garoff, A.-M. Frischauf, K. Simons, H. Lerach, and H. Delius, *Nature (London)* **288**, 236 (1980).

257. H. Garoff and K. Simons, *Virology* **61**, 493 (1974).

258. K. Simons, A. Helenius, and H. Garoff, *J. Mol. Biol.* **80**, 119 (1973).

259. G. Utermann and K. Simons, *J. Mol. Biol.* **85**, 569 (1974).

260. R. Hubbard, *J. Gen. Physiol.* **37**, 381 (1954).
261. H. B. Osborne, C. Sardet, and A. Helenius, *Eur. J. Biochem.* **44**, 383 (1974).
262. M. S. Lewis, L. C. Krieg, and W. D. Kirk, *Exp. Eye Res.* **18**, 29 (1974).
263. C. Sardet, A. Tardieu, and V. Luzzati, *J. Mol. Biol.* **105**, 383 (1976).
264. D. R. McCaslin and C. Tanford, *Biochemistry* **20**, 5212 (1981).
265. J. M. Corless, D. R. McCaslin, and B. L. Scott, *Proc. Natl. Acad. Sci. U.S.A.* **79**, 1116 (1982).
266. C. W. Wu and L. Stryer, *Proc. Natl. Acad. Sci. U.S.A.* **69**, 1104 (1972).
267. H. B. Osborne and E. Nabedryk-Viala, *Eur. J. Biochem.* **89**, 81 (1978).
268. M. Yeager, B. Schoenborn, D. Engelman, P. Moore, and L. Stryer, *J. Mol. Biol.* **137**, 315 (1980).
269. R. R. Birge, *Annu. Rev. Biophys. Bioeng.* **10**, 315 (1981).
270. H. T. Pretorius, P. K. Nandi, R. E. Lippoldt, M. L. Johnson, J. H. Keen, I. Pastan, and H. Edelhoch, *Biochemistry* **20**, 2777 (1981).
271. E. Ungewickell and D. Branton, *Nature (London)* **289**, 420 (1981).
272. B. M. F. Pearse and M. S. Bretscher, *Annu. Rev. Biochem.* **50**, 85 (1981).
273. P. K. Nandi, H. T. Pretorius, R. E. Lippoldt, M. L. Johnson, and H. Edelhoch, *Biochemistry* **19**, 5917 (1980).
274. J. L. Goldstein, R. G. W. Anderson, and M. S. Brown, *Nature (London)* **279**, 679 (1979).
275. M. M. Teeter, J. A. Mazer, and J. J. L'Italien, *Biochemistry* **20**, 5437 (1981).
276. W. A. Hendrickson and M. M. Teeter, *Nature (London)* **290**, 107 (1981).
277. A. De Marco, J. T. J. Lecomte, and M. Llinás, *Eur. J. Biochem.* **119**, 483 (1981).
278. B. J. Nicholson, M. W. Hunkapiller, L. B. Grim, L. E. Hood, and J.-P. Revel, *Proc. Natl. Acad. Sci. U.S.A.* **78**, 7594 (1981).
279. M. V. L. Bennet and D. A. Goodenough, *Neurosci. Res. Program Bull.* **16**, 379 (1978).
280. P. N. T. Unwin and G. Zampighi, *Nature (London)* **283**, 545 (1980).
281. S. S. Sikerwar, J. P. Tewari, and S. K. Malhotra, *Eur. J. Cell Biol.* **24**, 211 (1981).
282. D. Anderson, T. C. Terwilliger, W. Wickner, and D. Eisenberg, *J. Biol. Chem.* **255**, 2578 (1980).
283. E. Knöppel, D. Eisenberg, and W. Wickner, *Biochemistry* **18**, 4177 (1979).
284. J. Lauterwein, C. Bösch, L. R. Brown, and K. Wüthrich, *Biochim. Biophys. Acta* **556**, 244 (1979).
285. D. Eisenberg, T. C. Terwilliger, and F. Tsui, *Biophys. J.* **32**, 252 (1980).
286. C. H. Van Etten, H. C. Nielsen, and J. E. Peters, *Phytochemistry* **4**, 467 (1965).
287. L. R. Brown, W. Braun, A. Kumar, and K. Wüthrich, *Biophys. J.* **37**, 319 (1982).
288. T. C. Terwilliger, L. Weissman, and D. Eisenberg, *Biophys. J.* **37**, 353 (1982).
289. J. Lauterwein, L. R. Brown, and K. Wüthrich, *Biochim. Biophys. Acta* **622**, 219 (1980).
290. L. R. Brown, J. Lauterwein, and K. Wüthrich, *Biochim. Biophys. Acta* **622**, 231 (1980).
291. T. Haga, K. Haga, and A. G. Gilman, *J. Biol. Chem.* **252**, 5776 (1977).
292. E. J. Neer, *J. Biol. Chem.* **253**, 1498 (1978).
293. D. Stengel and J. Hanoune, *Eur. J. Biochem.* **102**, 21 (1979).
294. T. B. Nielson, P. M. Lad, M. S. Preston, E. Kempner, W. Schlegel, and M. Rodbell, *Proc. Natl. Acad. Sci. U.S.A.* **78**, 722 (1981).
295. C. Homcy, S. Wrenn, and E. Haber, *Proc. Natl. Acad. Sci. U.S.A.* **75**, 59 (1978).
296. E. M. Ross and A. G. Gilman, *Annu. Rev. Biochem.* **49**, 533 (1980).
297. P. C. Sternweis, J. K. Northup, E. Hanski, L. S. Schleifer, M. D. Smigel, and A. G. Gilman, *Adv. Cyclic Nucleotide Res.* **14**, 23 (1981).
298. A. C. Howlett and A. G. Gilman, *J. Biol. Chem.* **255**, 2861 (1980).
299. D. L. Garbers, *J. Biol. Chem.* **251**, 4071 (1976).
300. D. Moore, B. A. Sowa, and K. Ippen-Ihler, *J. Bacteriol.* **146**, 251 (1981).

301. R. Chen, W. Schmidmayr, C. Krämer, U. Chen-Schmeisser, and U. Henning, *Proc. Natl. Acad. Sci. U.S.A.* **77**, 4592 (1980).
302. N. R. Movva, K. Nakamura, and M. Inouye, *J. Mol. Biol.* **143**, 317 (1980).
303. H. Yamada and S. Mizushima, *Eur. J. Biochem.* **103**, 209 (1980).
304. J. M. Clément and M. Hofnung, *Cell* **27**, 507 (1981).
305. T. Nakae, *Biochem. Biophys. Res. Commun.* **88**, 774 (1979).
306. M. Luckey and H. Nikaido, *Proc. Natl. Acad. Sci. U.S.A.* **77**, 167 (1980).
307. T.-J. Chai and J. Foulds, *J. Bacteriol.* **139**, 418 (1979).
308. D. R. Lee, C. A. Schnaitman, and A. P. Pugsley, *J. Bacteriol.* **138**, 861 (1979).
309. S. F. Sabet and C. A. Schnaitman, *J. Biol. Chem.* **248**, 1797 (1973).
310. P. A. Manning and P. Reeves, *Mol. Gen. Genet.* **158**, 279 (1978).
311. V. Braun and H. Wolff, *FEBS Lett.* **34**, 77 (1973).
312. H. D. Caldwell, J. Kromhout and J. Schachter, *Infect. Immun.* **31**, 1161 (1981).
313. B. C. Lane and R. E. Hurlbert, *J. Bacteriol.* **143**, 349 (1980).
314. L. T. James and J. E. Heckles, *J. Immunol. Methods* **42**, 223 (1981).
315. T. B. Helting, G. Guthöhrlein, F. Blackkolb, and H.-J. Ronneberger, *Acta Pathol. Microbiol. Scand. Sect. C* **89**, 69 (1981).
316. K. Nixdorff, H. Fitzer, J. Gmeiner, and H. H. Martin, *Eur. J. Biochem.* **81**, 63 (1977).
317. T. Mizuno and M. Kageyama, *J. Biochem. (Tokyo)* **86**, 979 (1979).
318. B. Ludwig, *Biochim. Biophys. Acta* **594**, 177 (1980).
319. N. Sone, *in* "Chemiosmotic Proton Circuits in Biological Membranes" (V. P. Skulachev and C. P. Hinkle, eds.), p. 197. Addison-Wesley, Reading, Massachusetts, 1981.
320. A. Kröger, E. Winkler, A. Innerhofer, H. Hackenberg, and H. Schägger, *Eur. J. Biochem.* **94**, 465 (1979).
321. G. Unden, H. Hackenberg, and A. Kröger, *Biochim. Biophys. Acta* **591**, 275 (1980).
322. G. Kaczorowski, L. D. Kohn, and H. R. Kaback, this series, Vol. 53, p. 519.
323. P. B. Garland, *in* "Chemiosmotic Proton Circuits in Biological Membranes" (V. P. Skulachev and P. C. Hinkle, eds.), p. 211. Addison-Wesley, Reading, Massachusetts, 1981.
324. H. A. Chase, *J. Gen. Microbiol.* **117**, 211 (1980).
325. J. N. Umbreit and J. L. Strominger, *J. Biol. Chem.* **248**, 6759 (1973).
326. D. J. Waxman, D. M. Lindgren, and J. L. Strominger, *J. Bacteriol.* **148**, 950 (1981).
327. D. J. Waxman and J. L. Strominger, *J. Biol. Chem.* **256**, 2059 (1981).
328. D. J. Waxman and J. L. Strominger, *J. Biol. Chem.* **256**, 2067 (1981).
329. T. Tamura, Y. Imae, and J. L. Strominger, *J. Biol. Chem.* **251**, 414 (1976).
330. W. Kundig and S. Roseman, *J. Biol. Chem.* **246**, 1407 (1971).
331. T. Korte and W. Hengstenberg, *Eur. J. Biochem.* **23**, 295 (1971).
332. G. R. Jacobson, C. A. Lee, and M. H. Saier, Jr., *J. Biol. Chem.* **254**, 249 (1979).
333. L. O. Frøholm and K. Sletten, *FEBS Lett.* **73**, 29 (1977).
334. M. A. Hermodson, K. C. S. Chen, and T. M. Buchanan, *Biochemistry* **17**, 442 (1978).
335. W. Paranchych, P. A. Sastry, L. S. Frost, M. Carpenter, G. D. Armstrong, and T. H. Watts, *Can. J. Microbiol.* **25**, 1175 (1979).
336. J. Smit, M. Hermodson, and N. Agabian, *J. Biol. Chem.* **256**, 3092 (1981).
337. R. C. Fader, L. K. Duffy, C. P. Davis, and A. Kurosky, *J. Biol. Chem.* **257**, 3301 (1982).
338. W. Paranchych, L. S. Frost, and M. Carpenter, *J. Bacteriol.* **134**, 1179 (1978).
339. H. Sandermann, Jr. and J. L. Strominger, *J. Biol. Chem.* **247**, 5123 (1972).
340. H. Sandermann, Jr. and J. L. Strominger, *Proc. Natl. Acad. Sci. U.S.A.* **68**, 2441 (1971).
341. R. B. Gennis and J. L. Strominger, *in* "The Enzymes of Biological Membranes" (A. Martonosi, ed.), Vol. 2, p. 327. Plenum, New York, 1976.
342. R. B. Gennis and J. L. Strominger, *J. Biol. Chem.* **251**, 1264 (1976).

343. E. A. Wang and D. E. Koshland, Jr., *Proc. Natl. Acad. Sci. U.S.A.* **77,** 7157 (1980).
344. H. A. Dailey, Jr., *J. Bacteriol.* **132,** 302 (1977).
345. D. E. Büchel, B. Gronenborn, and B. Müller-Hill, *Nature* (*London*) **283,** 541 (1980).
346. H. Wróblewski, K.-E. Johansson, and S. Hjérten, *Biochim. Biophys. Acta* **465,** 275 (1977).
347. D. Evenberg and B. Lugtenberg, *Biochim. Biophys. Acta* **684,** 249 (1982).
348. D. Ryan, A. Y. H. Lu, and W. Levin, this series, Vol. 52, p. 117.
349. J. A. Goldstein, P. Linko, M. I. Luster, and D. W. Sundheimer, *J. Biol. Chem.* **257,** 2702 (1982).
350. D. A. Haugen, T. A. van der Hoeven, and M. J. Coon, *J. Biol. Chem.* **250,** 3567 (1975).
351. T. A. van der Hoeven, D. A. Haugen, and M. J. Coon, *Biochem. Biophys. Res. Commun.* **60,** 569 (1974).
352. Y. Imai and R. Sato, *Biochem. Biophys. Res. Commun.* **60,** 8 (1974).
353. J. S. French, F. P. Guengerich, and M. J. Coon, *J. Biol. Chem.* **255,** 4112 (1980).
354. M. J. Coon, T. A. van der Hoeven, S. B. Dahl, and D. A. Haugen, this series, Vol. 52, p. 109.
355. C. Bonfils, I. Dalet-Beluche, and P. Maurel, *Biochem. Biophys. Res. Commun.* **104,** 1011 (1982).
356. D. R. Koop and M. J. Coon, *Biochem. Biophys. Res. Commun.* **91,** 1075 (1979).
357. C. Hashimoto and Y. Imai, *Biochem. Biophys. Res. Commun.* **68,** 821 (1976).
358. T. Saito and H. W. Strobel, *J. Biol. Chem.* **256,** 984 (1981).
359. N. A. Elshourbagy and P. S. Guzelian, *J. Biol. Chem.* **255,** 1279 (1980).
360. K.-C. Cheng and J. B. Schenkman, *J. Biol. Chem.* **257,** 2378 (1982).
361. K. Kashiwagi, R. E. Carraway, and H. A. Salhanick, *Biochem. Biophys. Res. Commun.* **105,** 110 (1982).
362. M. Katagiri, S. Takemori, E. Itagaki, and K. Suhara, this series, Vol. 52, p. 124.
363. A. Hiwatashi, Y. Nishii, and Y. Ichikawa, *Biochem. Biophys. Res. Commun.* **105,** 320 (1982).
364. A. Y. H. Lu and S. B. West, *Pharmacol. Ther. Part A* **2,** 337 (1978).
365. M. J. Coon and R. E. White, *Met. Ions Biol.* **2,** 73 (1980).
366. R. E. White and M. J. Coon, *Annu. Rev. Biochem.* **49,** 315 (1980).
367. S. D. Black, J. S. French, C. H. Williams, Jr., and M. J. Coon, *Biochem. Biophys. Res. Commun.* **91,** 1528 (1979).
368. J. S. French and M. J. Coon, *Arch. Biochem. Biophys.* **195,** 565 (1979).
369. J. L. Vermilion and M. J. Coon, *Biochem. Biophys. Res. Commun.* **60,** 1315 (1974).
370. H. W. Strobel and J. D. Dignam, this series, Vol. 52, p. 89.
371. J. R. Gum and H. W. Strobel, *J. Biol. Chem.* **256,** 7478 (1981).
372. A. Y. H. Lu and W. Levin, this series, Vol. 52, p. 193.
373. G. C. Dubois, E. Appella, D. E. Ryan, D. M. Jerina, and W. Levin, *J. Biol. Chem.* **257,** 2708 (1982).
374. D. M. Ziegler and L. L. Poulsen, this series, Vol. 52, p. 142.
375. J. W. DePierre and G. Dallner, *Biochim. Biophys. Acta* **415,** 411 (1975).
376. J. Yu and T. L. Steck, *J. Biol. Chem.* **250,** 9170 (1975).
377. J. Yu and T. L. Steck, *J. Biol. Chem.* **250,** 9176 (1975).
378. M. F. Lukacovic, M. B. Feinstein, R. I. Sha'afi, and S. Perrie, *Biochemistry* **20,** 3145 (1981).
379. V. Bennet and P. J. Stenbuck, *J. Biol. Chem.* **255,** 2540 (1980).
380. R. A. F. Reithmeier, *J. Biol. Chem.* **254,** 3054 (1979).
381. M. J. A. Tanner, D. G. Williams, and R. E. Jenkins, *Ann. N.Y. Acad. Sci.* **341,** 455 (1980).
382. K. Drickamer, *Ann. N.Y. Acad. Sci.* **341,** 419 (1980).
383. S. Markowitz and V. T. Marchesi, *J. Biol. Chem.* **256,** 6463 (1981).

384. T. Tsuji, T. Irimura, and T. Osawa, *J. Biol. Chem.* **256,** 10497 (1981).

385. T. L. Steck, *J. Supramol. Struct.* **8,** 311 (1978).

386. P. A. Knauf, *Curr. Top. Membr. Transp.* **12,** 249 (1979).

387. V. Niggli, J. T. Penniston, and E. Carafoli, *J. Biol. Chem.* **254,** 9955 (1979).

388. K. Gietzen, M. Tejčka, and H. U. Wolf, *Biochem. J.* **189,** 81 (1980).

389. K. Gietzen, R. Konrad, M. Tejcka, S. Fleischer, and H. U. Wolf, *Acta Biol. Med. Ger.* **40,** 443 (1981).

390. T. R. Hinds and T. J. Andreasen, *J. Biol. Chem.* **256,** 7877 (1981).

391. J. Yu and S. R. Goodman, *Proc. Natl. Acad. Sci. U.S.A.* **76,** 2340 (1979).

392. D. C. Sogin and P. C. Hinkle, *J. Supramol. Struct.* **8,** 447 (1978).

393. S. A. Baldwin, J. M. Baldwin, F. R. Gorga, and G. E. Lienhard, *Biochim. Biophys. Acta* **552,** 183 (1979).

394. D. C. Sogin and P. C. Hinkle, *Proc. Natl. Acad. Sci. U.S.A.* **77,** 5725 (1980).

395. J. Cuppoletti and C. Y. Jung, *J. Biol. Chem.* **256,** 1305 (1981).

396. F. R. Gorga, S. A. Baldwin, and G. E. Lienhard, *Biochem. Biophys. Res. Commun.* **91,** 955 (1979).

397. G. V. Marinetti and K. Cattieu, *Biochim. Biophys. Acta* **685,** 109 (1982).

398. S. A. Baldwin and G. E. Lienhard, *Trends Biochem. Sci.* **6,** 208 (1981).

399. M. N. Jones and J. K. Nickson, *Biochim. Biophys. Acta* **650,** 1 (1981).

400. N. J. Dobson, J. D. Lambris, and G. D. Ross, *J. Immunol.* **126,** 693 (1981).

401. C. M. Cohen, J. M. Tyler, and D. Branton, *Cell* **21,** 875 (1980).

402. S. E. Lux, *Nature (London)* **281,** 426 (1979).

403. S. E. Lux, *Semin. Hematol.* **16,** 21 (1979).

404. D. Branton, C. M. Cohen, and J. M. Tyler, *Cell* **24,** 24 (1981).

405. W. B. Gratzer and G. H. Beaven, *Eur. J. Biochem.* **58,** 403 (1975).

406. N. M. Schechter, M. Sharp, J. A. Reynolds, and C. Tanford, *Biochemistry* **15,** 1897 (1976).

407. D. M. Shotton, B. E. Burke, and D. Branton, *J. Mol. Biol.* **131,** 303 (1979).

408. J. C. Dunbar and G. B. Ralston, *Biochim. Biophys. Acta* **667,** 177 (1981).

409. G. B. Ralston, *Biochim. Biophys. Acta* **455,** 163 (1976).

410. J. M. Anderson, *J. Biol. Chem.* **254,** 939 (1979).

411. J. S. Morrow and V. T. Marchesi, *J. Cell Biol.* **88,** 463 (1981).

412. A. Elgsaeter, *Biochim. Biophys. Acta* **536,** 235 (1978).

413. A. Mikkelsen and A. Elgsaeter, *Biochim. Biophys. Acta* **536,** 245 (1978).

414. H. W. Harris and S. E. Lux, *J. Biol. Chem.* **255,** 11512 (1980).

415. J. S. Morrow, D. W. Speicher, W. J. Knowles, C. J. Hsu, and V. T. Marchesi, *Proc. Natl. Acad. Sci. U.S.A.* **77,** 6592 (1980).

416. A. Mikkelsen and A. Elgsaeter, *Biochim. Biophys. Acta* **668,** 74 (1981).

417. B. T. Stokke and A. Elgsaeter, *Biochim. Biophys. Acta* **640,** 640 (1981).

418. V. T. Marchesi, *J. Membr. Biol.* **51,** 101 (1979).

419. J. M. Tyler, W. R. Hargreaves, and D. Branton, *Proc. Natl. Acad. Sci. U.S.A.* **76,** 5192 (1979).

420. J. M. Tyler, B. N. Reinhardt, and D. Branton, *J. Biol. Chem.* **255,** 7034 (1980).

421. W. R. Hargreaves, K. N. Giedd, A. Verkleij, and D. Branton, *J. Biol. Chem.* **255,** 11965 (1980).

422. M. P. Sheetz, R. G. Painter, and S. J. Singer, *Biochemistry* **15,** 4486 (1976).

423. T. Ohnishi, *Br. J. Haematol.* **35,** 453 (1977).

424. J. R. Harris and I. Naeem, *Biochim. Biophys. Acta* **537,** 495 (1978).

425. J. R. Harris and I. Naeem, *Biochim. Biophys. Acta* **670,** 285 (1981).

426. M. D. White and G. B. Ralston, *Biochim. Biophys. Acta* **554,** 469 (1979).

427. H. L. Malech and V. T. Marchesi, *Biochim. Biophys. Acta* **670,** 385 (1981).

428. J. R. Harris, *Nouv. Rev. Fr. Hematol.* **22,** 411 (1980).
429. P. Ott, B. Jenny, and U. Brodbeck, *Eur. J. Biochem.* **57,** 469 (1975).
430. E. Niday, C. S. Wang, and P. Alaupovic, *Biochim. Biophys. Acta* **469,** 180 (1977).
431. G. Beauregard, M. Potier, and B. D. Roufogalis, *Biochem. Biophys. Res. Commun.* **96,** 1290 (1980).
432. P. Ott and U. Brodbeck, *Eur. J. Biochem.* **88,** 119 (1978).
433. G. Beauregard and B. D. Roufogalis, *Biochem. J.* **179,** 109 (1979).
434. H. Grossmann, K.-P. Ruess, and M. Liefländer, *Experientia* **35,** 1545 (1979).
435. P. Boivin, M. Garbarz, and C. Galand, *Int. J. Biochem.* **12,** 445 (1980).
436. M. Tao, R. Conway, and S. Cheta, *J. Biol. Chem.* **255,** 2563 (1980).
437. K. Suzuki, T. Terao, and T. Osawa, *J. Biochem. (Tokyo)* **89,** 1 (1981).
438. B. F. Grant, T. B. Breithaupt, and E. B. Cunningham, *J. Biol. Chem.* **254,** 5726 (1979).
439. P. Zahler and R. Kramer, this series, Vol. 71, p. 690.
440. A. M. Golovtchenko-Matsumoto, I. Matsumoto, and T. Osawa, *Eur. J. Biochem.* **121,** 463 (1982).
441. H. L. Ploegh, H. T. Orr, and J. L. Strominger, *Cell* **24,** 287 (1981).
442. M. Malissen, B. Malissen, and B. R. Jordan, *Proc. Natl. Acad. Sci. U.S.A.* **79,** 893 (1982).
443. L. Trägårdh, L. Rask, K. Wiman, J. Fohlman, and P. A. Peterson, *Proc. Natl. Acad. Sci. U.S.A.* **77,** 1129 (1980).
444. R. J. Robb, C. Terhorst, and J. L. Strominger, *J. Biol. Chem.* **253,** 5319 (1978).
445. H. T. Orr, J. A. López de Castro, D. Lancet, and J. L. Strominger, *Biochemistry* **18,** 5711 (1979).
446. J. S. Pober, B. C. Guild, J. L. Strominger, and W. R. Veatch, *Biochemistry* **20,** 5625 (1981).
447. E. S. Kimball, S. G. Nathenson, and J. E. Coligan, *Biochemistry* **20,** 3301 (1981).
448. J. F. Kaufman and J. L. Strominger, *Proc. Natl. Acad. Sci. U.S.A.* **76,** 6304 (1979).
449. H. Kratzin, C.-Y. Yang, H. Gotz, E. Pauly, S. Kolbel, G. Egert, F. P. Thinnes, P. Wernet, P. Altevogt, and N. Hilschmann, *Hoppe-Seyler's Z. Physiol. Chem.* **362,** 1665 (1981).
450. M. Steinmetz, J. G. Frelinger, D. Fisher, T. Hunkapiller, D. Pereira, S. M. Weissman, H. Uehara, S. Nathenson, and L. Hood, *Cell* **24,** 125 (1981).
451. F. Brégégère, J. P. Abastado, S. Kvist, L. Rask, J. L. Lalanne, H. Garoff, B. Cami, K. Wiman, D. Larhammar, P. A. Peterson, G. Gachelin, P. Kourilsky, and B. Dobberstein, *Nature (London)* **292,** 78 (1981).
452. R. Nairn, S. G. Nathenson, and J. E. Coligan, *Eur. J. Immunol.* **10,** 495 (1980).
453. S. H. Herrmann and M. F. Mescher, *J. Biol. Chem.* **254,** 8713 (1979).
454. J. M. Cecka, M. McMillan, D. B. Murphy, H. O. McDevitt, and L. Hood, *Eur. J. Immunol.* **9,** 955 (1979).
455. M. Maze and G. M. Gray, *Biochemistry* **19,** 2351 (1980).
456. K. Hiwada, T. Ito, M. Yokoyama, and T. Kokubu, *Eur. J. Biochem.* **104,** 155 (1980).
457. R. D. C. MacNair and A. J. Kenny, *Biochem. J.* **179,** 379 (1979).
458. J. Elovson, *J. Biol. Chem.* **255,** 5807 (1980).
459. J. Baratti, S. Maroux, D. Louvard, and P. Desnuelle, *Biochim. Biophys. Acta* **315,** 147 (1973).
460. L. E. Anderson, K. A. Walsh, and H. Neurath, *Biochemistry* **16,** 3354 (1977).
461. J. J. Liepnieks and A. Light, *J. Biol. Chem.* **254,** 1677 (1979).
462. M. Jusic, S. Seifert, E. Weiss, R. Haas, and P. C. Heinrich, *Arch. Biochem. Biophys.* **177,** 355 (1976).
463. M. A. Kerr and A. J. Kenny, *Biochem. J.* **137,** 477 (1974).
464. P. T. Varandani and L. A. Shroyer, *Arch. Biochem. Biophys.* **181,** 82 (1977).
465. R. J. Beynon, J. D. Shannon, and J. S. Bond, *Biochem. J.* **199,** 591 (1981).

466. M. Orlowski and S. Wilk, *Biochemistry* **20,** 4942 (1981).
467. Y. S. Kim and E. J. Brophy, *J. Biol. Chem.* **251,** 3199 (1976).
468. C. Zwizinski and W. Wickner, *J. Biol. Chem.* **255,** 7973 (1980).
469. P. C. McAda and M. G. Douglas, *J. Biol. Chem.* **257,** 3177 (1982).
470. C. J. Scandella and A. Kornberg, *Biochemistry* **10,** 4447 (1971).
471. M. Nishijima, Y. Akamatsu, and S. Nojima, *J. Biol. Chem.* **249,** 5658 (1974).
472. Y. Natori, M. Nishijima, S. Nojima, and H. Satoh, *J. Biochem.* (*Tokyo*) **87,** 959 (1980).
473. G. W. Cyboron and R. E. Wuthier, *J. Biol. Chem.* **256,** 7262 (1981).
474. A. Colbeau and S. Maroux, *Biochim. Biophys. Acta* **511,** 39 (1978).
475. E. Mössner, M. Boll, and G. Pfleiderer, *Hoppe-Seyler's Z. Physiol. Chem.* **361,** 543 (1980).
476. M. G. Low and D. B. Zilversmit, *Biochemistry* **19,** 3913 (1980).
477. T. Kobayashi and K. Suzuki, *J. Biol. Chem.* **256,** 7768 (1981).
478. H. Skovbjerg, H. Sjöström, and O. Norén, *Eur. J. Biochem.* **114,** 653 (1981).
479. S. Ramaswamy and A. N. Radhakrishnan, *Biochim. Biophys. Acta* **403,** 446 (1975).
480. W. T. Forsee and J. S. Schutzbach, *J. Biol. Chem.* **256,** 6577 (1981).
481. C. Bengis-Garber and D. J. Kushner, *J. Bacteriol.* **146,** 24 (1981).
482. Y. Naito and J. M. Lowenstein, *Biochemistry* **20,** 5188 (1981).
483. C. C. Widnell, this series, Vol. 32, p. 368.
484. M. Slavik, N. Kartner, and J. R. Riordan, *Biochem. Biophys. Res. Commun.* **75,** 342 (1977).
485. J. P. Luzio, E. M. Bailyes, A. C. Newby, and K. Siddle, *Protides Biol. Fluids* **29,** 33 (1982).
486. V. Panagia, C. E. Heyliger, P. C. Choy, and N. S. Dhalla, *Biochim. Biophys. Acta* **640,** 802 (1981).
487. E. Bischoff, T.-A. Tran-Thi, and K. F. A. Decker, *Eur. J. Biochem.* **51,** 353 (1975).
488. T. Maciag, *Trends Biochem. Sci.* **7,** 197 (1982).
489. J. C. Pascall, A. P. Boulton, and R. K. Craig, *Eur. J. Biochem.* **119,** 91 (1981).
490. J. Bar-Tana, G. Rose, and B. Shapiro, *Biochem. J.* **122,** 353 (1971).
491. J. Bar-Tana and G. Rose, *Biochem. J.* **131,** 443 (1973).
492. E. Maes and J. Bar-Tana, *Biochim. Biophys. Acta* **480,** 527 (1977).
493. T. Tanaka, K. Hosaka, M. Hoshimaru, and S. Numa, *Eur. J. Biochem.* **98,** 165 (1979).
494. K. Hosaka, M. Mishina, T. Tanaka, T. Kamiryo, and S. Numa, *Eur. J. Biochem.* **93,** 197 (1979).
495. W. Thompson and G. MacDonald, *J. Biol. Chem.* **254,** 3311 (1979).
496. T. Takenawa and K. Egawa, *J. Biol. Chem.* **252,** 5419 (1977).
497. G. Belendiuk, D. Mangnall, B. Tung, J. Westley, and G. S. Getz, *J. Biol. Chem.* **253,** 4555 (1978).
498. H. Kanoh and K. Ohno, *Eur. J. Biochem.* **66,** 201 (1976).
499. M. A. Polokoff and R. M. Bell, *Biochim. Biophys. Acta* **618,** 129 (1980).
500. W. J. Schneider and D. E. Vance, *J. Biol. Chem.* **254,** 3886 (1979).
501. W. C. McMurray and E. C. Jarvis, *Can. J. Biochem.* **56,** 414 (1978).
502. W. S. Agnew and G. Popják, *J. Biol. Chem.* **253,** 4574 (1978).
503. P. Strittmatter and H. G. Enoch, this series, Vol. 52, p. 188.
504. T. A. Beyer, J. E. Sadler, J. I. Rearick, J. C. Paulson, and R. L. Hill, *Adv. Enzymol.* **52,** 23 (1981).
505. M. Schwyzer and R. L. Hill, *J. Biol. Chem.* **252,** 2338 (1977).
506. C. L. Oppenheimer and R. L. Hill, *J. Biol. Chem.* **256,** 799 (1981).
507. J. Mendicino, E. V. Chandrasekaran, K. R. Anumula, and M. Davila, *Biochemistry* **20,** 967 (1981).
508. T. A. Beyer, J. E. Sadler, and R. L. Hill, *J. Biol. Chem.* **255,** 5364 (1980).

509. C. A. Smith and K. Brew, *J. Biol. Chem.* **252**, 7294 ((1977).
510. J. Mendicino, S. Sivakami, M. Davila, and E. V. Chandrasekaran, *J. Biol. Chem.* **257**, 3987 (1982).
511. J. P. Gorski and C. B. Kasper, *J. Biol. Chem.* **252**, 1336 (1977).
512. O. M. P. Singh, A. B. Graham, and G. C. Wood, *Eur. J. Biochem.* **116**, 311 (1981).
513. J. E. Sadler, J. I. Rearick, and R. L. Hill, *J. Biol. Chem.* **254**, 5934 (1979).
514. J. E. Sadler, J. I. Rearick, J. C. Paulson, and R. L. Hill, *J. Biol. Chem.* **254**, 4434 (1979).
515. Z. H. Beg, J. A. Stonik, and H. B. Brewer, Jr., *FEBS Lett.* **80**, 123 (1977).
516. D. A. Kleinsek and J. W. Porter, *J. Biol. Chem.* **254**, 7591 (1979).
517. Z. H. Beg, J. A. Stonik, and H. B. Brewer, Jr., *J. Biol. Chem.* **255**, 8541 (1980).
518. D. A. Kleinsek, S. Ranganathan, and J. W. Porter, *Proc. Natl. Acad. Sci. U.S.A.* **74**, 1431 (1977).
519. J. W. DePierre and L. Ernster, *Annu. Rev. Biochem.* **46**, 201 (1977).
520. S. Fleischer and L. Packer, eds., this series, Vol. 53.
521. Y. Hatefi, Y. M. Galante, D. L. Stigall, and C. I. Ragan, this series, Vol. 56, p. 577.
522. C. I. Ragan, S. Smith, F. G. P. Earley, and V. M. Poore, *in* "Chemiosmotic Proton Circuits in Biological Membranes" (V. P. Skulachev and P. C. Hinkle, eds.), p. 59. Addison-Wesley, Reading, Massachusetts, 1981.
523. C. I. Ragan, *Biochim. Biophys. Acta* **456**, 249 (1976).
524. C. Heron, S. Smith, and C. I. Ragan, *Biochim. J.* **181**, 435 (1979).
525. Y. M. Galante and Y. Hatefi, *Arch. Biochem. Biophys.* **192**, 559 (1979).
526. G. Dooijewaard, G. J. M. De Bruin, P. J. Van Dijk, and E. C. Slater, *Biochim. Biophys. Acta* **501**, 458 (1978).
527. E. J. Boekema, J. F. L. Van Breemen, W. Keegstra, E. F. J. Van Bruggon, and S. P. J. Albracht, *Biochim. Biophys. Acta* **679**, 7 (1982).
528. C.-A. Yu, L. Yu, and T. E. King, *J. Biol. Chem.* **247**, 1012 (1972).
529. C.-A. Yu, L. Yu, and T. E. King, *J. Biol. Chem.* **249**, 4905 (1974).
530. T. E. King, T. Ohnishi, C.-A. Yu, and L. Yu, *in* "Structure and Function of Energy Transducing Membranes" (K. van Dam and B. F. van Gelder, eds.), p. 49. Elsevier/North-Holland Biomedical Press, New York, 1977.
531. H.-G. Bock and S. Fleischer, *J. Biol. Chem.* **250**, 5774 (1975).
532. P. Gazzotti, H.-G. Bock, and S. Fleischer, *J. Biol. Chem.* **250**, 5782 (1975).
533. D. C. Phelps and Y. Hatefi, *Biochemistry* **20**, 459 (1981).
534. H. Tsuge, Y. Nakano, H. Onishi, Y. Futamura, and K. Ohashi, *Biochim. Biophys. Acta* **614**, 274 (1980).
535. H. Tsuge, Y. Futamura, Y. Nakano, and K. Ohashi, *Biochem. Int.* **1**, 519 (1980).
536. T. P. Singer and D. E. Edmondson, this series, Vol. 53, p. 397.
537. M. E. Wernette, R. S. Ochs, and H. A. Lardy, *J. Biol. Chem.* **256**, 12767 (1981).
538. F. J. Ruzicka and H. Beinert, *J. Biol. Chem.* **252**, 8440 (1977).
539. B. Höjeberg and J. Rydström, *Biochem. Biophys. Res. Commun.* **78**, 1183 (1977).
540. W. M. Anderson and R. R. Fisher, *Arch. Biochem. Biophys.* **187**, 180 (1978).
541. B. Höjeberg and J. Rydström, this series, Vol. 55, p. 275.
542. J. Rydström, *Biochim. Biophys. Acta* **463**, 155 (1977).
543. J. Rydström, *in* "Mitochondria and Microsomes" (C. P. Lee, G. Schatz, and G. Dallner, eds.), p. 317. Addison-Wesley, Reading, Massachusetts, 1981.
544. A. E. Shamoo and J. M. Brenza, *Ann. N.Y. Acad. Sci.* **358**, 73 (1980).
545. P. R. H. Clarke and L. L. Bieber, *J. Biol. Chem.* **256**, 9861 (1981).
546. B. Kopec and I. B. Fritz, *Can. J. Biochem.* **49**, 941 (1971).
547. S. Taketani and R. Tokunaga, *J. Biol. Chem.* **256**, 12748 (1981).
548. E. K. Michaelis, *Biochem. Biophys. Res. Commun.* **65**, 1004 (1975).

549. D. C. Gautheron and J. H. Julliard, this series, Vol. 56, p. 419.
550. H. Wohlrab, *J. Biol. Chem.* **255,** 8170 (1980).
551. F. Palmieri, G. Genchi, I. Stipani, and E. Quagliariello, *in* "Structure and Function of Energy Transducing Membranes" (K. van Dam and B. F. van Gelder, eds.), p. 251. Elsevier/North-Holland, Amsterdam, 1977.
552. C. S. Lin, H. Hackenberg, and E. M. Klingenberg, *FEBS Lett.* **113,** 304 (1980).
553. M. Klingenberg, H. Hackenberg, R. Krämer, C. S. Lin, and H. Aquila, *Ann. N.Y. Acad. Sci.* **358,** 83 (1980).
554. J. I. Salach, Jr., this series, Vol. 53, p. 495.
555. R. M. Denney, R. R. Fritz, N. T. Patel, and C. W. Abell, *Science* **215,** 1400 (1982).
556. R. G. Dennick and R. J. Mayer, *Biochem. J.* **161,** 167 (1977).
557. M. A. Stadt, P. A. Banks, and R. D. Kobes, *Arch. Biochem. Biophys.* **214,** 223 (1982).
558. R. M. Cawthon and X. O. Breakefield, *Nature* (*London*) **281,** 692 (1979).
559. H. Freitag, W. Neupert, and R. Benz, *Eur. J. Biochem.* **123,** 629 (1982).
560. N. Roos, R. Benz, and D. Brdiczka, *Biochim. Biophys. Acta* **686,** 204 (1982).
561. M. Colombini, *Ann. N.Y. Acad. Sci.* **341,** 552 (1980).
562. J. M. Boggs and M. A. Moscarello, *Biochim. Biophys. Acta* **515,** 1 (1978).
563. J. L. Nussbaum, J. F. Rouayrenc, and P. Mandel, *Biochem. Biophys. Res. Commun.* **57,** 1240 (1974).
564. J. Gagnon, P. R. Finch, D. D. Wood, and M. A. Moscarello, *Biochemistry* **10,** 4756 (1971).
565. D. D. Wood, J. M. Boggs, and M. A. Moscarello, *Neurochem. Res.* **5,** 745 (1980).
566. M. J. Schlesinger, *Annu. Rev. Biochem.* **50,** 193 (1981).
567. K. Kitamura, M. Suzuki, and K. Uyemura, *Biochim. Biophys. Acta* **455,** 806 (1976).
568. K. Kitamura, A. Suzuki, M. Suzuki, and K. Uyemura, *FEBS Lett.* **100,** 67 (1979).
569. A. Ishaque, M. W. Roomi, I. Szymanska, S. Kowalski, and E. H. Eylar, *Can. J. Biochem.* **58,** 913 (1980).
570. W. W. Parson and R. J. Cogdell, *Biochim. Biophys. Acta* **416,** 105 (1975).
571. G. Drews, *Curr. Top. Bioenerg.* **8,** 161 (1978).
572. P. A. Loach, this series, Vol. 69, p. 155.
573. J. P. Thornber, T. L. Trosper, and C. E. Strouse, *in* "The Photosynthetic Bacteria" (R. K. Clayton and W. R. Sistrom, eds.), p. 133. Plenum, New York, 1978.
574. R. K. Clayton, "Photosynthesis: Physical Mechanisms and Chemical Patterns." Cambridge Univ. Press, London and New York, 1980.
575. N. N. Firsow and G. Drews, *Arch. Microbiol.* **115,** 299 (1977).
576. S. J. Tonn, G. E. Gogel, and P. A. Loach, *Biochemistry* **16,** 877 (1977).
577. R. Feick and G. Drews, *Biochim. Biophys. Acta* **501,** 499 (1978).
578. K. Sauer and L. A. Austin, *Biochemistry* **17,** 2011 (1978).
579. R. J. Cogdell and J. P. Thornber, *in* "Chlorophyll Organization and Energy Transfer in Photosynthesis" (G. Wolstenholme and D. W. Fitzsimons, eds.), p. 61. Excerpta Medica, New York, 1979.
580. R. J. Cogdell, J. G. Lindsay, G. P. Reid, and G. D. Webster, *Biochim. Biophys. Acta* **591,** 312 (1980).
581. L. K. Cohen and S. Kaplan, *J. Biol. Chem.* **256,** 5909 (1981).
582. L. K. Cohen and S. Kaplan, *J. Biol. Chem.* **256,** 5901 (1981).
583. J. A. Shiozawa, P. A. Cuendet, G. Drews, and H. Zuber, *Eur. J. Biochem.* **111,** 455 (1980).
584. R. M. Broglie, C. N. Hunter, P. Delepelaire, R. A. Niederman, N.-H. Chua, and R. K. Clayton, *Proc. Natl. Acad. Sci. U.S.A.* **77,** 87 (1980).
585. R. A. Brunisholz, P. A. Cuendet, R. Theiler, and H. Zuber, *FEBS Lett.* **129,** 150 (1981).

586. G. E. Gogel, P. S. Parkes, R. A. Brunisholz, H. Zuber, and P. A. Loach, *Biophys. J.* **37**, 109a (1982).
587. H. Nöel, M. Van der Rest, and G. Gingras, *Biochim. Biophys. Acta* **275**, 219 (1972).
588. R. K. Clayton and R. Haselkorn, *J. Mol. Biol.* **68**, 97 (1972).
589. M. Y. Okamura, L. A. Steiner, and G. Feher, *Biochemistry* **13**, 1394 (1974).
590. K. F. Nieth, G. Drews, and R. Feick, *Arch. Microbiol.* **105**, 43 (1975).
591. G. Feher and M. Y. Okamura, in "The Photosynthetic Bacteria" (R. K. Clayton and W. R. Sistrom, eds.), p. 349. Plenum, New York, 1978.
592. C. Vadeboncoeur, M. Mamet-Bratley, and G. Gingras, *Biochemistry* **18**, 4308 (1979).
593. C. Vadeboncoeur, H. Nöel, L. Poirier, Y. Cloutier, and G. Gingras, *Biochemistry* **18**, 4301 (1979).
594. T. D. Marinetti, M. Y. Okamura, and G. Feher, *Biochemistry* **18**, 3126 (1979).
595. J. M. Anderson, *Biochim. Biophys. Acta* **416**, 191 (1975).
596. J. P. Thornber, *Annu. Rev. Plant Physiol.* **26**, 127 (1975).
597. J. P. Thornber, and R. S. Alberte, in "The Enzymes of Biological Membranes" (A. Martonosi, ed.), Vol. 3, p. 163. Plenum, New York, 1976.
598. D. W. Krogmann, in "The Enzymes of Biological Membranes" (A. Martonosi, ed.), Vol. 3, p. 143. Plenum, New York, 1976.
599. N. Nelson and G. Hauska, in "Membrane Bioenergetics" (C. P. Lee, G. Schatz, and L. Ernster, eds.), p. 189. Addison-Wesley, Reading, Massachusetts, 1979.
600. N.-H. Chua, K. Matlin, and P. Bennoun, *J. Cell Biol.* **67**, 361 (1975).
601. C. Bengis and N. Nelson, *J. Biol. Chem.* **252**, 4564 (1977).
602. J. P. Markwell, S. Reinman, and J. P. Thornber, *Arch. Biochem. Biophys.* **190**, 136 (1978).
603. J. M. Anderson and J. Barrett, in "Chlorophyll Organization and Energy Transfer in Photosynthesis" (G. Wolstenholme and D. W. Fitzsimons, eds.), p. 81. Excerpta Medica, New York, 1979.
604. J. P. Markwell, J. P. Thornber, and R. T. Boggs, *Proc. Natl. Acad. Sci. U.S.A.* **76**, 1233 (1979).
605. K. R. Miller, *Nature (London)* **300**, 53 (1982).
606. J. P. Markwell, H. Y. Nakatani, J. Barber, and J. P. Thornber, *FEBS Lett.* **122**, 149 (1980).
607. G. Hauska, D. Samoray, G. Orlich, and N. Nelson, *Eur. J. Biochem.* **111**, 535 (1980).
608. N. Nelson and B.-E. Notsani, in "Bioenergetics of Membranes" (L. Packer, G. C. Papageorgiou, and A. Trebst, eds.), p. 233. Elsevier/North-Holland Biomedical Press, New York, 1977.
609. R. Malkin, *Annu. Rev. Plant Physiol.* **33**, 455 (1982).
610. K. Satoh, *Biochim. Biophys. Acta* **546**, 84 (1979).
611. B. A. Diner and F.-A. Wollman, *Eur. J. Biochem.* **110**, 521 (1980).
612. J. J. Burke, C. L. Ditto, and C. J. Arntzen, *Arch. Biochem. Biophys.* **187**, 252 (1978).
613. J. M. Anderson, *Biochim. Biophys. Acta* **591**, 113 (1980).
614. R. S. Alberte, A. L. Friedman, D. L. Gustafson, M. S. Rudnick, and H. Lyman, *Biochim. Biophys. Acta* **635**, 304 (1981).
615. P. Delepelaire and N.-H. Chua, *J. Biol. Chem.* **256**, 9300 (1981).
616. J. Li and C. Hollingshead, *Biophys. J.* **37**, 363 (1982).
617. W. A. Cramer and J. Whitmarsh, *Annu. Rev. Plant Physiol.* **28**, 133 (1977).
618. E. Hurt and G. Hauska, *Eur. J. Biochem.* **117**, 591 (1981).
619. N. Nelson and J. Neumann, *J. Biol. Chem.* **247**, 1817 (1972).
620. A. L. Stuart and A. R. Wasserman, *Biochim. Biophys. Acta* **376**, 561 (1975).
621. J. Singh and A. R. Wasserman, *J. Biol. Chem.* **246**, 3532 (1971).
622. N. Nelson and E. Racker, *J. Biol. Chem.* **247**, 3848 (1972).

623. E. Matsuzaki, Y. Kamimura, T. Yamasaki, and E. Yakushiji, *Plant Cell Physiol.* **16**, 237 (1975).
624. M. Takahashi and K. Asada, *Plant Cell Physiol.* **16**, 191 (1975).
625. P. M. Wood, *Eur. J. Biochem.* **72**, 605 (1977).
626. K. Tanaka, M. Takahashi, and K. Asada, *J. Biol. Chem.* **253**, 7397 (1978).
627. J. C. Gray, *Eur. J. Biochem.* **82**, 133 (1978).
628. K. K. Ho and D. W. Krogmann, *J. Biol. Chem.* **255**, 3855 (1980).
629. H. Böhme, S. Brütsch, G. Weithmann, and P. Böger, *Biochim. Biophys. Acta* **590**, 248 (1980).
630. G. Orlich and G. Hauska, *Eur. J. Biochem.* **111**, 525 (1980).
631. L. M. Y. Lee, A. K. Salvatore, P. R. Flanagan, and G. G. Forstner, *Biochem. J.* **187**, 437 (1980).
632. H. S. Garewal and A. R. Wasserman, *Biochemistry* **13**, 4072 (1974).
633. U. I. Flügge and H. W. Heldt, *Biochim. Biophys. Acta* **638**, 296 (1981).
634. K. J. Clemetson, H. Y. Naim, and E. F. Lüscher, *Proc. Natl. Acad. Sci. U.S.A.* **78**, 2712 (1981).
635. L. L. K. Leung, T. Kinoshita, and R. L. Nachman, *J. Biol. Chem.* **256**, 1994 (1981).
636. R. P. McEver, J. U. Baenziger, and P. W. Majerus, *Blood* **59**, 80 (1982).
637. M. C. Berndt and D. R. Phillips, *J. Biol. Chem.* **256**, 59 (1981).
638. R. J. Schneider, A. Kulczycki, Jr., S. K. Law, and J. P. Atkinson, *Nature (London)* **290**, 789 (1981).
639. J. D. Lambris, N. J. Dobson, and G. D. Ross, *Proc. Natl. Acad. Sci. U.S.A.* **78**, 1828 (1981).
640. J. Gerdes and H. Stein, *Protides Biol. Fluids* **29**, 435 (1982).
641. K. Drickamer, *J. Biol. Chem.* **256**, 5827 (1981).
642. G. G. Sahagian, J. Distler, and G. W. Jourdian, *Proc. Natl. Acad. Sci. U.S.A.* **78**, 4289 (1981).
643. M. A. Lehrman and R. L. Hill, *Fed. Proc. Fed. Am. Soc. Exp. Biol.* **40**, 1820, (1981).
644. Y. Mizuno, Y. Kozutsumi, T. Kawasaki, and I. Yamashina, *J. Biol. Chem.* **256**, 4247 (1981).
645. T. Kawasaki and G. Ashwell, *J. Biol. Chem.* **252**, 6536 (1977).
646. G. Ashwell and J. Harford, *Annu. Rev. Biochem.* **51**, 531 (1982).
647. R. M. Graham, H.-J. Hess, and C. J. Homcy, *Proc. Natl. Acad. Sci. U.S.A.* **79**, 2186 (1982).
648. J. C. Venter, C. M. Fraser, A. I. Soiefer, D. R. Jeffrey, W. L. Strauss, R. R. Charlton, and R. Greguski, *Adv. Cyclic Nucleotide Res.* **14**, 135 (1981).
649. R. G. L. Shorr, R. J. Lefkowitz, and M. G. Caron, *J. Biol. Chem.* **256**, 5820 (1981).
650. G. Vauquelin, P. Geynet, J. Hanoune, and A. D. Strosberg, *Eur. J. Biochem.* **98**, 543 (1979).
651. J. Folch-Pi and P. J. Stoffyn, *Ann. N.Y. Acad. Sci.* **195**, 86 (1972).
652. S. Cohen, H. Ushiro, C. Stoscheck, and M. Chinkers, *J. Biol. Chem.* **257**, 1523 (1982).
653. K. J. Clemetson, H. Y. Naim, and E. F. Lüscher, *Protides Biol. Fluids* **29**, 359 (1982).
654. Y. Kitagawa and E. Racker, *J. Biol. Chem.* **257**, 4547 (1982).
655. T. L. Rosenberry, *Adv. Enzymol.* **43**, 103 (1975).
656. T. L. Rosenberry and J. M. Richardson, *Biochemistry* **16**, 3550 (1977).
657. L. Anglister and I. Silman, *J. Mol. Biol.* **125**, 293 (1978).
658. I. Silman and L. Anglister, *Monogr. Neural Sci.* **7**, 55 (1980).
659. Z. Rakonczay, J. Mallol, H. Schenk, G. Vincendon, and J.-P. Zanetta, *Biochim. Biophys. Acta* **657**, 243 (1981).
660. C. Mays and T. L. Rosenberry, *Biochemistry* **20**, 2810 (1981).
661. S. Bon and J. Massoulié, *Proc. Natl. Acad. Sci. U.S.A.* **77**, 4464 (1980).

662. M. A. Kerr and A. J. Kenny, *Biochem. J.* **137**, 489 (1974).
663. R. A. Hock, E. Nexø, and M. D. Hollenberg, *J. Biol. Chem.* **255**, 10737 (1980).
664. E. J. Goetzel, D. W. Foster, and D. W. Goldman, *Biochemistry* **20**, 5717 (1981).
665. H. Abou-Issa and L. E. Reichert, Jr., *Biochim. Biophys. Acta* **631**, 97 (1980).
666. B. I. Bluestein and J. L. Vaitukaitis, *Biol. Reprod.* **24**, 661 (1981).
667. M. L. Dufau, D. W. Ryan, A. J. Baukal, and K. J. Catt, *J. Biol. Chem.* **250**, 4822 (1975).
668. F. S. Khan, P. Rathnam, and B. B. Saxena, *Biochem. J.* **197**, 7 (1981).
669. M. J. Waters and H. G. Friesen, *J. Biol. Chem.* **254**, 6815 (1979).
670. S. Jacobs, Y. Shechter, K. Bissell, and P. Cuatrecasas, *Biochem. Biophys. Res. Commun.* **77**, 981 (1977).
671. M. D. Baron, M. H. Wisher, P. M. Thamm, D. J. Saunders, D. Brandenburg, and P. H. Sönksen, *Biochemistry* **20**, 4156 (1981).
672. T. W. Siegel, S. Ganguly, S. Jacobs, O. M. Rosen, and C. S. Rubin, *J. Biol. Chem.* **256**, 9266 (1981).
673. U. Lang, C. R. Kahn, and L. C. Harrison, *Biochemistry* **19**, 64 (1980).
674. J. Massague, P. F. Pilch, and M. P. Czech, *J. Biol. Chem.* **256**, 3182 (1981).
675. C. C. Yip, M. L. Moule, and C. W. T. Yeung, *Biochemistry* **21**, 2940 (1982).
676. R. J. Pollet, E. S. Kempner, M. L. Standaert, and B. A. Haase, *J. Biol. Chem.* **257**, 894 (1982).
677. H. E. Meyer, H.-J. Bubenezer, L. Herbertz, L. Kuehn, and H. Reinauer, *Hoppe-Seyler's Z. Physiol. Chem.* **362**, 1621 (1981).
678. S. Jacobs and P. Cuatrecasas, *Endocr. Rev.* **2**, 251 (1981).
679. M. P. Czech, J. Massague, and P. F. Pitch, *Trends Biochem. Sci.* **6**, 222 (1981).
680. K. Metsikkö and H. Rajaniemi, *Biochem. Biophys. Res. Commun.* **95**, 1730 (1980).
681. R. P. C. Shiu and H. G. Friesen, *J. Biol. Chem.* **249**, 7902 (1974).
682. R. C. Jaffe, *Biochemistry* **21**, 2936 (1982).
683. B. Bhaumick, R. M. Bala, and M. D. Hollenberg, *Proc. Natl. Acad. Sci. U.S.A.* **78**, 4279 (1981).
684. A. Kulczycki, Jr., *J. Reticuloendothel. Soc.* **28**, 29s (1980).
685. A. Kulczycki, Jr. and C. W. Parker, *J. Biol. Chem.* **254**, 3187 (1979).
686. I. S. Mellman and J. C. Unkeless, *J. Exp. Med.* **152**, 1048 (1980).
687. A. Kulczycki, Jr., V. Krause, C. C. Killion, and J. P. Atkinson, *J. Immunol.* **124**, 2772 (1980).
688. B. J. Takacs, *Mol. Immunol.* **17**, 1293 (1980).
689. R. Matre, G. Kleppe, and O. Tönder, *Acta Pathol. Scand., Sect. C* **89**, 209 (1981).
690. A. Arce, D. M. Beltramo, S. C. Kivatinitz, and R. Caputto, *FEBS Lett.* **127**, 149 (1981).
691. M. Gavish and S. H. Snyder, *Proc. Natl. Acad. Sci. U.S.A.* **78**, 1939 (1981).
692. F. A. Stephenson, A. E. Watkins, and R. W. Olsen, *Eur. J. Biochem.* **123**, 291 (1982).
693. E. K. Michaelis, M. L. Michaelis, H. H. Chang, R. D. Grubbs, and D. R. Kuonen, *Mol. Cell. Biochem.* **38**, 163 (1981).
694. Y. C. Clement-Cormier, *Adv. Biochem. Psychopharmacol.* **21**, 159 (1980).
695. Y. C. Clement-Cormier and P. E. Kendrick, *Biochem. Pharmacol.* **30**, 2197 (1981).
696. J. M. Bidlack, L. G. Abood, P. Osei-Gyimah, and S. Archer, *Proc. Natl. Acad. Sci. U.S.A.* **78**, 636 (1981).
697. M. Smigel and S. Fleischer, *J. Biol. Chem.* **252**, 3689 (1977).
698. U. Kyldén and S. Hammarström, *Eur. J. Biochem.* **109**, 489 (1980).
699. A. C. Antony, C. Utley, K. C. Van Horne, and J. F. Kolhouse, *J. Biol. Chem.* **256**, 9684 (1981).
700. I. Kouvonen and R. Gräsbeck, *Protides Biol. Fluids* **29**, 385 (1982).
701. B. Seetharam, D. H. Alpers, and R. H. Allen, *J. Biol. Chem.* **256**, 3785 (1981).
702. I. Kouvonen and R. Gräsbeck, *J. Biol. Chem.* **256**, 154 (1981).

703. I. Kouvonen and R. Gräsbeck, *Biochem. Biophys. Res. Commun.* **86**, 358 (1979).
704. G. Marcoullis and R. Gräsbeck, *Biochim. Biophys. Acta* **499**, 309 (1977).
705. B. Seetharam, S. S. Bagur, and D. H. Alpers. *J. Biol. Chem.* **257**, 183 (1982).
706. W. J. Schneider, U. Beisiegel, J. L. Goldstein, and M. S. Brown, *J. Biol. Chem.* **257**, 2664 (1982).
707. R. P. Hartshorne and W. A. Catterall, *Proc. Natl. Acad. Sci. U.S.A.* **78**, 4620 (1981).
708. P. A. Seligman and R. H. Allen, *J. Biol. Chem.* **253**, 1766 (1978).
709. P. A. Seligman, R. B. Schleicher, and R. H. Allen, *J. Biol. Chem.* **254**, 9943 (1979).
710. J. W. Goding and A. W. Harris, *Proc. Natl. Acad. Sci. U.S.A.* **78**, 4530 (1981).
711. B. Ecarot-Charrier, V. L. Grey, A. Wilczynska, and H. M. Schulman, *Can. J. Biochem.* **58**, 418 (1980).
712. B. J. Bowman, F. Blasco, and C. W. Slayman, *J. Biol. Chem.* **256**, 12343 (1981).
713. J.-P. Dufour and A. Goffeau, *J. Biol. Chem.* **253**, 7026 (1978).
714. J.-P. Dufour and A. Goffeau, *Eur. J. Biochem.* **105**, 145 (1980).
715. D. H. Maclennan and P. T. S. Wong, *Proc. Natl. Acad. Sci. U.S.A.* **68**, 1231 (1971).
716. K. P. Campbell and D. H. Maclennan, *J. Biol. Chem.* **256**, 4626 (1981).
717. J. M. Bidlack, I. S. Ambudkar, and A. E. Shamoo, *J. Biol. Chem.* **257**, 4501 (1982).
718. C. J. Le Peuch, D. A. M. Le Peuch, and J. G. Demaille, *Biochemistry* **19**, 3368 (1980).
719. M. A. Kirchberger and T. Antonetz, *Biochem. Biophys. Res. Commun.* **105**, 152 (1982).
720. D. H. Maclennan, C. C. Yip, G. H. Iles, and P. Seeman, *Cold Spring Harbor Symp. Quant. Biol.* **37**, 469 (1972).
721. G. Lunazzi, C. Tiribelli, B. Gazzin, and G. Sottocasa, *Biochim. Biophys. Acta* **685**, 117 (1982).
722. P. Walter and G. Blobel, *Proc. Natl. Acad. Sci. U.S.A.* **77**, 7112 (1980).
723. R. P. Hughey, P. J. Coyle, and N. P. Curthoys, *J. Biol. Chem.* **254**, 1124 (1979).
724. S. Takahashi, R. S. Zukin, and H. M. Steinman, *Arch. Biochem. Biophys.* **207**, 87 (1981).
725. S. S. Tate and A. Meister, *J. Biol. Chem.* **250**, 4619 (1975).
726. R. P. Hughey and N. P. Curthoys, *J. Biol. Chem.* **251**, 7863 (1976).
727. A. Tsuji, Y. Matsuda, and N. Katunuma, *J. Biochem.* (*Tokyo*) **87**, 1567 (1980).
728. J. L. Ding, G. D. Smith, and T. J. Peters, *Biochim. Biophys. Acta* **657**, 334 (1981).
729. T. Frielle and N. P. Curthoys, *Biophys. J.* **37**, 193 (1982).
730. Y. Nakashima, B. Frangione, R. L. Wiseman, and W. H. Konigsberg, *J. Biol. Chem.* **256**, 5792 (1981).
731. B. Frangione, Y. Nakashima, W. Konigsberg, and R. L. Wiseman, *FEBS Lett.* **96**, 381 (1978).
732. D. J. Marciani and J. D. Papamatheakis, *J. Biol. Chem.* **255**, 1677 (1980).
733. R. W. Green and J. H. Shaper, *Biochim. Biophys. Acta.* **668**, 439 (1981).
734. A. L. Hiti and D. P. Nayak, *J. Virol.* **41**, 730 (1982).
735. S. Fields, G. Winter, and G. G. Brownlee, *Nature* (*London*) **290**, 213 (1981).
736. A. Scheid and P. W. Choppin, *J. Virol.* **11**, 263 (1973).
737. T. Kohama, W. Garten, and H.-D. Klenk, *Virology* **111**, 364 (1981).
738. V. S. Kalyanaraman, M. G. Sarngadharan, and R. C. Gallo, *J. Virol.* **28**, 686 (1978).
739. G. Pauli, W. Rhode, and E. Harms, *Arch. Virol.* **58**, 61 (1978).
740. M. Ozawa and A. Asano, *J. Biol. Chem.* **256**, 5954 (1981).
741. M.-C. Hsu, A. Scheid, and P. W. Choppin, *J. Biol. Chem.* **256**, 3557 (1981).
742. A. Scheid and P. W. Choppin, *Virology* **80**, 54 (1977).
743. D. S. Lyles, *Proc. Natl. Acad. Sci. U.S.A.* **76**, 5621 (1979).
744. C. M. Rice and J. H. Strauss, *Proc. Natl. Acad. Sci. U.S.A.* **78**, 2062 (1981).
745. M. F. G. Schmidt and M. J. Schlesinger, *J. Biol. Chem.* **255**, 3334 (1980).
746. C. M. Rice, J. R. Bell, M. W. Hunkapiller, E. G. Strauss, and J. H. Strauss, *J. Mol. Biol.* **154**, 355 (1982).

747. J. K. Rose and C. J. Gallione, *J. Virol.* **39**, 519 (1981).
748. C. L. Reading, E. E. Penhoet, and C. E. Ballou, *J. Biol. Chem.* **253**, 5600 (1978).
749. R. Gibson, S. Kornfeld, and S. Schlesinger, *J. Biol. Chem.* **256**, 456 (1981).
750. M. F. G. Schmidt and M. J. Schlesinger, *Cell* **17**, 813 (1979).
751. S. Bhakdi and J. Tranum-Jensen, *Proc. Natl. Acad. Sci. U.S.A.* **78**, 1818 (1981).
752. E. R. Podack and H. J. Müller-Eberhard, *J. Biol. Chem.* **256**, 3145 (1981).
753. E. R. Podack and H. J. Müller-Eberhard, *J. Immunol.* **121**, 1025 (1978).
754. C. F. Ware, R. A. Wetsel, and W. P. Kolb, *Mol. Immunol.* **18**, 521 (1981).
755. E. R. Podack, G. Biesecker, and H. J. Müller-Eberhard, *Proc. Natl. Acad. Sci. U.S.A.* **76**, 897 (1979).
756. M. M. Mayer, *Ann. N.Y. Acad. Sci.* **358**, 43 (1980).
757. T. Flatmark and M. Grønberg, *Biochem. Biophys. Res. Commun.* **99**, 292 (1981).
758. D. K. Apps, J. G. Pryde, and J. H. Phillips, *Neuroscience* **5**, 2279 (1980).
759. P. Fleming and A. Saxena, *Biophys. J.* **37**, 99 (1982).
760. Y. Hino and S. Minakami, *J. Biol. Chem.* **257**, 2563 (1982).
761. D. Schachter and S. Kowarski, *Fed. Proc. Fed. Am. Soc. Exp. Biol.* **41**, 84 (1982).
762. S. Kowarski and D. Schachter, *J. Biol. Chem.* **255**, 10834 (1980).
763. L. D. Kohn and H. R. Kaback, *J. Biol. Chem.* **248**, 7012 (1973).
764. E. A. Pratt, L. W.-M. Fung, J. A. Flowers, and C. Ho, *Biochemistry* **18**, 312 (1979).
765. M. Futai, *Biochemistry* **12**, 2468 (1973).
766. A. Imam, D. J. R. Laurence, and A. M. Neville, *Biochem. J.* **193**, 47 (1981).
767. L. D. Snow, D. G. Colton, and K. L. Carraway, *Arch. Biochem. Biophys.* **179**, 690 (1977).
768. E. N. Hughes and J. T. August, *J. Biol. Chem.* **257**, 3970 (1982).
769. W. Kreisel, B. A. Volk, R. Büchsel, and W. Reutter, *Proc. Natl. Acad. Sci. U.S.A.* **77**, 1828 (1980).
770. A. F. Williams, A. N. Barclay, M. Letarte-Muirhead, and R. J. Morris, *Cold Spring Harbor Symp. Quant. Biol.* **41**, 51 (1977).
771. D. G. Campbell, J. Gagnon, K. B. M. Reid, and A. F. Williams, *Biochem. J.* **195**, 15 (1981).
772. Y. Okada and R. G. Spiro, *J. Biol. Chem.* **255**, 8865 (1980).

Addendum

Addendum to Article [39]

By TAKASHI MORIMOTO, SHIRO MATSUURA, and MONIQUE ARPIN

V. Concluding Remarks

We described the techniques for the study of the mechanisms by which newly synthesized cytochrome c is transferred posttranslationally into rat liver mitochondria, using a reticulocyte lysate system programmed with mRNA from liver free polysomes of T_3-treated rats. These techniques, the principles behind them, and their precautions can be used for the study of the other mitochondrial proteins which are synthesized on cytoplasmic ribosomes and then transferred posttranslationally into mitochondria.

As with the posttranslational transfer of newly synthesized cytochrome c, we found that (a) the primary translation product was different in amino acid sequence from the mature protein in that it contained an NH_2 terminal methionine. The methionine residue appeared to be cleaved before the cytochrome c was taken up by mitochondria. (b) The *in vitro* product was incorporated specifically into purified rat liver mitochondria and became inaccessible to exogenously added trypsin. (c) The incorporation of the *in vitro* product into mitochondria was not inhibited upon addition of polylysine. The incorporation was, however, completely abolished when mitochondria had been previously treated with trypsin (under very mild conditions) just before use, which suggests the presence of a specific interaction between the *in vitro* product and the outer mitochondrial membrane protein. (d) Apocytochrome c prepared from horse cytochrome c and some other animal cytochrome c, but not holocytochrome c, could compete with the *in vitro* product for its transfer into mitochondria. (e) The *in vitro* product was converted to holocytochrome c after being taken up by mitochondria. (f) The *in vitro* product could bind to the isolated outer mitochondrial membrane, but not to the mitoplast that lacks the outer mitochondrial membrane. On the contrary, holocytochrome c could bind to the mitoplast, but not to the isolated outer mitochondrial membrane. These results suggest that the newly synthesized cytochrome c is transferred to its destination through four steps: (1) interaction between the newly formed product and the outer mitochondrial membrane (recognition process); (2) transfer of the new product across the outer mitochondrial membrane into the intermembrane space (vectorial transfer process); (3) conversion of the new product (apo-form) to the holo-form (modification process); and (4) binding to the inner mitochondrial membrane, site of its function (disposition process). In addition, the primary structural features of apocytochrome c may serve as an addressing signal for mitochondria and the failure of the

METHODS IN ENZYMOLOGY, VOL. 97

holocytochrome c, which contains the same domain, to compete with the *in vitro* product for its transfer is due to the inaccessibility of the domain by being masked after the acquisition of Heme, that is, the apo-form is essential to maintain the polypeptide in a conformation capable of interacting with the outer mitochondrial membrane protein. This interpretation of the role of the apo-form in the posttranslational transfer can be applied for the role of the N-terminal transient segment of the other mitochondrial proteins which are cytoplasmically synthesized as a large molecular weight precursor and cleaved after being taken up by mitochondria.

We have recently identified a peptide segment which was generated by CNBr treatment of horse apocytochrome c (defined as fragment II, see Fig. 2) contains a domain which would serve as an addressing signal.[7,38] The fragment strongly competes with the *in vitro* product for its transfer into mitochondria by interfering in Step 1. Furthermore, fragment II can inhibit the posttranslational transfer of some other mitochondrial proteins, including subunits IV and V of cytochrome oxidase,[38] which are known to be synthesized cytoplasmically as a large molecular weight precursor,[39,40] suggesting that several mitochondrial proteins share a common site of entry to mitochondria and therefore are likely to use similar or closely related structural signals directing their recognition by the outer mitochondrial membrane.

As with the role of the structural features present in other fragments, our recent observation suggests that the features present within fragment I (see Fig. 2), which has no effect on the recognition process, may play an important role in Steps 2 (vectorial transfer process) and 3 (modification process), and that cytochrome c synthetase, which is present in the intermembrane space and is involved in binding of Heme to apocytochrome c, may also play an important role in Step 2.

Synthesis of membrane proteins and their transfer from sites of synthesis to sites of function are essential for the correct implementation of a genetic program in cells. However, intracellular mechanisms by which proteins are segregated and sorted out during this transfer remain obscure. This is true for those proteins synthesized in the cytoplasm, but destined for compartments within mitochondria and chloroplasts. Cytochrome c is very useful for studying various aspects of this sorting out mechanism for the following reasons: (a) The protein contains a primary addressing signal that insures its transfer from cytoplasmic free ribosomes to the mitochondria. (b) Several

[38] M. Arpin, S. Matsuura, E. Margoliash, D. D. Sabatini, and T. Morimoto, *Eur. J. Cell Biol.* **22**, 152 (1980).

[39] K. Mihara and G. Blobel, *Proc. Natl. Acad. Sci. U.S.A.* **77**, 4160 (1980).

[40] A. S. Lewin, I. Gregor, T. L. Mason, N. Nelson, and G. Schatz, *Proc. Natl. Acad. Sci. U.S.A.* **77**, 3998 (1980).

steps involved in the posttranslational transfer can be differentiated from one another experimentally by using mitochondrial subfractions, chemically modified apocytochrome c and the fragments generated from apocytochrome c, or by using deuterhemin, a specific inhibitor for cytochrome c synthetase,[41] or by a combination of the above. (c) The properties and primary structure of c-type cytochromes from various sources have been extensively studied, which will allow for further study and insight into the sorting out mechanisms at a molecular level, focusing on the essential features within fragments I and II for their respective function in the posttranslational transfer steps.

[41] B. Hennig and W. Neupert, *Eur. J. Biochem.* **121**, 203 (1981).

Author Index

Numbers in parentheses are footnote reference numbers and indicate that an author's work is referred to although his name is not cited in the text.

Subject Index

A

pulse-labeling, with [^{35}S]methionine, 114–116

S-30 extract, preparation, 192–194

S-30 system, mRNA-directed, in analysis of penicillinase processing, 157

secA gene, 9, 10

secB gene, 9, 10

shuttle vector YEp13, 345, 355

strains

    genotypes, 198

    for study of lactose permease, 161, 162

transformants, with halobacterial DNA insert, colony hybridization with oligonucleotide probe, 233, 234, 237, 238

transformation

    with halobacterial genome, 231, 232, 237

    with yeast plasmid, 360

unc operon

    cloning, 198

    DNA sequence, 196–203

    promoter, 210–213

    restriction map, 177, 193

    subunits, sequence analysis, 195–218

Eukaryote, *see also* specific eukaryote

protein export, 3–4

F

Fc receptor, 594

Ferrochelatase, 582, 591

Fibroin, cotranslational processing, electrophoretic analysis, 79

Folate, receptor, 595

Follicle-stimulating hormone, receptor, 594

Follitropin, receptor, 594

Formate dehydrogenase, bacterial, characterization, 581

Fumarate reductase, bacterial, characterization, 581

G

β-Galactosidase, 197

altered enzymatic properties, in gene fusions, 5, 6, 9

    selection procedure based on, 28, 29

envelope protein fusion, lethality, 16, 17

in gene fusion techniques, suitability, 12, 13

hybrid protein, formation, 5

lacZ gene fusion, *in vitro*, 20, 21

β-Galactoside α-fucosyltransferase, 589

β-Galactoside α-sialyltransferase, 590

Galactosyltransferase, 589

Gene

deletion mutation, isolation, *in vitro*, 30–32

fusion

    cloning, 20, 21

    construction, 12–22

    growth inhibitory effects, 5

        selection procedure based on, 25–28

    lacZ

        in vitro, 20–22

        in vivo, 14–20

        target genes, 14

    lethal to cell, 16, 17

    Mud phages used for, 14–16

        construction of λ derivatives, 17, 18

    ompF-lacZ, genetic verification, 22, 23

    phoA-lacZ, 21

        isolation, 30

    selection procedures based on, 5–7, 25–30

    in signal sequence analysis, 5, 6

    strains, characterization, 22–24

    study of protein localization in *E. coli,* 11

point mutation, isolation, *in vitro,* 32, 33

Gene-polypeptide relationships, determination, 188, 189

Globomycin

for isolation of *E. coli* lipoprotein mutants, 124–126

structure, 126

M

Maize
  growth, for isolation of plastid genes, 526
  mitochondria, isolated, protein synthesis, 483, 484
  plastid genes, 524–555
Maltase, characterization, 587
Maltase/glucoamylase, 587
Maltodextrin, transport, role of LamB protein, 109, 110
Maltose, transport, role of LamB protein, 109, 110
Maltose-binding protein, cotranslational processing, electrophoretic analysis, 78, 79, 84
α-D-Mannose: β-1,2-N-acetylglucosaminyltransferase, 589
α-Mannosidase, characterization, 587
α-Mannoside β-N-acetylglucosaminyltransferase, 589
Melittin, characterization, 578
Membrane
  bacterial
    assembly
      leader peptidase in, 44–46
      in intact cells, pulse-labeling studies, 57–61
    protein insertion into, 138–146
      membrane potential and, 146–153
    structure, 124
  cytoplasmic, comparison to rough endoplasmic reticulum, 3, 4
  erythrocyte, proteins, 573
  inner, see Inner membrane
  M13 procoat assembly into, 130–138
  outer, see Outer membrane
  pore, 286, 287
    exclusion limit, evaluation, 91–94
    penetration rate, evaluation, 94–98
  protein, see Protein
  purple, biogenesis, 218–226
  thylakoid, C. reinhardi
    chlorophyll protein complex, fractionation, 558
    chlorophyll quantitation, 558, 559
    dephosphorylation in vitro, 560
    herbicide binding, assay, 559

in vitro labeling, 559, 560
in vivo pulse labeling, 559
32–34 kd polypeptides, 555, 556, 563
  labeling, 562
  turnover, 565, 566
44–47 kd polypeptides, 555, 563, 564
polypeptides
  fractionation, 558
  in vivo pulse labeling and chase with ³⁵S, 562
  phosphorylated, identification, 560, 561
  phosphorylation pattern, modulation with chloroplast development, 562–566
  preparation, 557
  protein quantitation, 559
  specific radioactivity, determination, 558
  subfractionation, 558
vesicle, preparation, for protein insertion study, 145
Metalloendopeptidase, characterization, 587
Milk fat globule, membrane glycoprotein, 598
Mitochondria
  apocytochrome c
    binding, 274
    immunoprecipitation, 268–271
    import, in vitro, 272–274
  G6P: Gal-6-P ratio, and hexokinase binding activity, 474, 475
  holocytochrome c, immunoprecipitation, 268–271
  human, RNA, analysis, 435–469
  inner membrane, electron transport complexes, characterization, 574, 575
  mammalian
    DNA, clones, 431–434
    DNA and RNA, isolation, 428–431
    genome, analysis, 435, 436
    isolation, from cell cultures, 427, 428
    matrix proteins, synthesis, 396–408
    uptake of cytochrome c, 418–426

in vitro capped, mapping, 461, 462
mammalian, isolation, 428–431
mapping experiments, 457–460
metabolism, 462–464
nascent, mapping, 461
oligo(A) tail, structure, 446, 447
poly(A) tail, structure, 446, 447
S1 nuclease protection analysis,
459, 460
species
half-lives, 467
steady-state amount, 467
strand homology
determination, 457
synthesis, rates, 467
transcription initiation sites,
analysis, 460–462
transfer hybridization
experiments, 458, 459
non-poly(A)-containing, RNase
digestion, 453–455
plastid, 525
in vitro translation products, at
various developmental stages,
549, 552–554
isolation, 530–532, 534, 535
kinasing, 547, 548
ribosomal, mapping genes for, 546,
547
transfer, mapping genes for, 548,
549
translation, on rabbit reticulocyte
lysate, 549–551
poly(A)-containing, 3'-end sequencing,
449–453
rat liver, isolation, 399
ribosomal
gene titration experiments, 456
yeast, antibiotic resistance loci,
366
transfer, gene titration experiments,
456
Ribosome, plastid, preparation, 546, 547
Ribulosebisphosphate carboxylase, maize,
large subunit, gene mapping, 525
Ribulosebisphosphate
carboxylase/oxygenase, subunits,
synthesis, by chloroplasts, 495, 496
Rieske iron-sulfur protein,
characterization, 575

Rough endoplasmic reticulum,
comparison to bacterial cytoplasmic
membrane, 3, 4

S

Saccharomyces cerevisiae, see Yeast
Salmonella typhimurium
LamB protein variant, 112
outer membrane, proteins C, D, and
F, characterization, 573
porins, 86–89, 293
protein synthesis, 3
Sarcoplasmic reticulum
proteolipid, 596
transport proteins, characterization,
596
Saxitoxin, receptor, 595
Shigella, LamB protein, 110–112
Signal peptidase, characterization, 587
Signal sequence, 40, 41, 77, 79, 83
genetic analysis, 4–7
mutagenesis, *in vitro,* 6
mutant
lamB, isolation, 25–28
malE, isolation, 28, 29
unlinked mutations restoring
secretion in, 10
mutations, gene fusions for selection
of, 5, 6
prokaryotic vs. eukaryotic, 3
Somatomedin, receptor, 594
Spectrin, characterization, 585
Spiralin, *S. citri,* characterization, 582
Squalene synthetase, 589
Staphylococcus aureus
penicillinase, 154, 155
phosphotransferase system, protein,
characterization, 582
Steryl-CoA desaturase, 589
Succinate: cytochrome *c* oxidoreductase,
590
Succinate dehydrogenase, mitochondrial
inner membrane, characterization,
574
Succinate: ubiquinone oxidoreductase,
mitochondrial inner membrane,
characterization, 574
Sucrase isomaltase, characterization, 573